Laser-based Mid-infrared Sources and Applications

Laser-based Mid-infrared Sources and Applications

Konstantin L. Vodopyanov

University of Central Florida
Orlando, FL, USA

WILEY

SCIENCE WISE
PUBLISHING

This edition first published 2020

Registered Office
John Wiley & Sons, Inc., 111 River Street, Hoboken, NJ 07030, USA

Editorial Office
111 River Street, Hoboken, NJ 07030, USA

For details of our global editorial offices, customer services, and more information about Wiley products visit us at www.wiley.com.

Wiley also publishes its books in a variety of electronic formats and by print-on-demand. Some content that appears in standard print versions of this book may not be available in other formats.

Library of Congress Cataloging-in-Publication data applied for

ISBN: 9781118301814

Cover Image: © merrymoonmary/Getty Images
Cover Design: Wiley

Set in 10/12pt Warnock by SPi Global, Pondicherry, India

FSC www.fsc.org
MIX
Paper from responsible sources
FSC® C013604

In memory of Feliciana Ignatievna Vergunas

Table of Contents

About the Author *xi*
Preface *xiii*

1 Mid-IR Spectral Range *1*
1.1 Definition of the Mid-IR *1*
1.2 The World's Second Laser *3*
1.3 Internal Vibrations of Molecules *4*
References *5*

2 Solid-state Crystalline Mid-IR Lasers *7*
2.1 Rare-Earth-based Tm^{3+}, Ho^{3+}, and Er^{3+} Lasers *7*
2.1.1 Tm^{3+} Lasers *7*
2.1.2 Ho^{3+} Lasers *10*
2.1.3 Er^{3+} Lasers *13*
2.2 Transition Metal Cr^{2+} and Fe^{2+} Lasers *18*
2.2.1 Spectroscopic Properties of Cr^{2+} and Fe^{2+} *18*
2.2.2 Lasers Based on Chalcogenide Crystals Doped with Cr^{2+} *21*
2.2.2.1 Broadly Tunable Cr^{2+} Lasers *21*
2.2.2.2 High-power Continuous-wave Cr^{2+} Lasers *23*
2.2.2.3 High-power Cr^{2+} CW Laser Systems Operating at 2.94 μm *23*
2.2.2.4 Gain-switched High-power Cr^{2+} Lasers *24*
2.2.2.5 Microchip Cr^{2+} Lasers *25*
2.2.2.6 Waveguide and Thin-disk Cr:ZnSe Lasers *26*
2.2.2.7 Mode-locked Cr:ZnS/Cr:ZnSe Lasers *27*
2.2.3 Lasers Based on Chalcogenide Crystals Doped with Fe^{2+} *30*
2.2.3.1 Free-running Pulsed Fe:ZnSe/ZnS Lasers *30*
2.2.3.2 Gain-switched Regime of Fe^{2+} Lasers at Room Temperature *32*
2.2.3.3 Continuous-wave Fe^{2+} Lasers *33*
2.2.3.4 Tunable Fe^{2+} Lasers at Room Temperature *35*
2.2.3.5 Ultrafast Amplifier in the 3.8–4.8 μm Range *35*
2.3 Summary *35*
References *36*

3 Fiber Mid-IR Lasers *43*
3.1 Introduction *43*
3.2 Continuous-wave Mid-IR Fiber Lasers *44*
3.2.1 Tm-based Fiber Lasers *44*
3.2.2 Ho-based Fiber Lasers *47*
3.2.3 Er-based Fiber Lasers *49*
3.2.4 Dy-based Fiber Lasers *52*
3.2.5 Raman Fiber Lasers *52*
3.3 Q-switched Mid-IR Fiber Lasers *54*
3.4 Mode-locked Mid-IR Fiber Lasers *56*
3.5 Summary *60*
References *61*

4 Semiconductor Lasers *65*
4.1 Heterojunction Mid-IR Lasers *65*
4.1.1 GaSb-based Diode Lasers *66*
4.1.2 Distributed Feedback GaSb-based Lasers *70*
4.2 Quantum Cascade Lasers *73*
4.2.1 High Power and High Efficiency QCLs *76*
4.2.2 Single-mode Distributed Feedback (DFB) QCLs *79*
4.2.3 Broadly Tunable QCLs with an External Cavity *82*
4.2.4 Short-wavelength (<4 μm) QCLs *85*
4.2.5 QCLs at Long (16–21 μm) Wavelengths *86*
4.3 Interband Cascade Lasers *87*
4.4 Optically Pumped Semiconductor Disk Lasers (OPSDLs) *94*
4.4.1 (AlGaIn)(AsSb)-based OPSDL at $\lambda \approx 2.3\,\mu m$ *95*
4.4.2 PbS-based OPSDL at $\lambda = 2.6–3\,\mu m$ *96*
4.4.3 PbSe-based OPSDL at $\lambda = 4.2–4.8\,\mu m$ *96*
4.4.4 PbTe-based OPSDL at $\lambda = 4.7–5.6\,\mu m$ *98*
4.5 Summary *100*
References *100*

5 Mid-IR by Nonlinear Optical Frequency Conversion *109*
5.1 Two Approaches to Frequency Downconversion Using Second-order Nonlinearity *109*
5.1.1 Difference Frequency Generation *111*
5.1.2 Optical Parametric Oscillators (OPOs) *112*
5.1.3 Brief Review of $\chi^{(2)}$ Nonlinear Crystals for Mid-IR *115*
5.1.3.1 Periodically Poled Oxides *116*
5.1.3.2 Birefringent Crystals *116*
5.1.3.3 Emerging QPM Nonlinear Optical Materials *119*
5.2 Continuous-wave (CW) Regime *121*
5.2.1 DFG of CW Radiation *121*
5.2.2 CW OPOs *123*

5.3 Pulsed Regime *130*
5.3.1 Pulsed DFG *130*
5.3.2 Pulsed OPOs *133*
5.3.2.1 Broadly Tunable Pulsed OPOs *133*
5.3.2.2 Narrow-linewidth Pulsed OPOs *143*
5.3.2.3 High Average Power OPOs *147*
5.3.2.4 High Pulse Energy OPOs *150*
5.3.2.5 Waveguide OPOs *152*
5.4 Regime of Ultrashort (ps and fs) Pulses *153*
5.4.1 Ultrafast DFG *153*
5.4.2 Intra-pulse DFG (Optical Rectification) *157*
5.4.3 Ultrafast OPOs *161*
5.4.3.1 Picosecond Mode *161*
5.4.3.2 Femtosecond Mode *163*
5.4.4 Ultrafast OPGs *165*
5.4.5 Ultrafast OPAs *167*
5.5 Raman Frequency Converters *168*
5.5.1 Crystalline Raman Converters *169*
5.5.2 Fiber Raman Converters *169*
5.5.3 Silicon Raman Converters *170*
5.5.4 Diamond Raman Converters *171*
5.5.5 Other Raman Converters *172*
5.6 Summary *174*
References *174*

6 Supercontinuum and Frequency Comb Sources *189*
6.1 Supercontinuum Sources *189*
6.1.1 SC from Lead-silicate Glass Fibers *191*
6.1.2 SC from Tellurite Glass Fibers *192*
6.1.3 SC from ZBLAN Fibers *194*
6.1.4 SC from Chalcogenide Glass Fibers *196*
6.1.5 SC from Waveguides *203*
6.1.6 SC from Bulk Crystals *207*
6.1.7 Other SC Sources *212*
6.2 Frequency Comb Sources *213*
6.2.1 Direct Comb Sources from Mode-locked Lasers *214*
6.2.2 Combs Produced by Spectral Broadening in NL Fibers and Waveguides *215*
6.2.3 Combs Produced by Difference Frequency Generation *217*
6.2.4 OPO-based Combs *220*
6.2.5 Combs Based on Optical Subharmonic Generation *226*
6.2.6 Microresonator-based Kerr Combs *229*
6.2.7 Combs from Quantum Cascade Lasers *234*

6.2.8 Combs from Interband Cascade Lasers *235*
6.3 Summary *235*
References *236*

7 Mid-IR Applications *247*
7.1 Spectroscopic Sensing and Imaging *247*
7.1.1 QCLs for Spectroscopy and Trace-gas Analysis *248*
7.1.2 Spectroscopy with ICLs *252*
7.1.3 Spectroscopy with DFG and OPO Sources *252*
7.1.4 Broadband Spectroscopy with Frequency Combs *253*
7.1.5 Hyperspectral Imaging *255*
7.2 Medical Applications *258*
7.2.1 Laser Tissue Interactions *258*
7.2.1.1 Holmium and Thulium Surgical Lasers *258*
7.2.1.2 Er:YAG Lasers (λ = 2.9 μm) *259*
7.2.1.3 Importance of the Spectral Band of 6–7 μm *260*
7.2.2 Medical Breath Analysis *261*
7.2.2.1 Ethane (C_2H_6) *262*
7.2.2.2 NO *262*
7.2.2.3 NH_3 *263*
7.2.2.4 CO *263*
7.2.2.5 OCS *263*
7.2.2.6 Optical Frequency Comb Spectroscopy for Breath Analysis *264*
7.3 Nano-IR Imaging and Chemical Mapping *265*
7.4 Plasmonics in the Mid-IR *267*
7.5 Infrared Countermeasures *269*
7.6 Extreme Nonlinear Optics and Attosecond Science *270*
7.7 Other Applications *273*
7.7.1 Laser Wake-field Accelerators *273*
7.7.2 Laser Acceleration in Dielectric Structures *274*
7.7.3 Free-space Communications *274*
7.7.4 Organic Material Processing *275*
References *276*

Index *287*

About the Author

Konstantin L. Vodopyanov is an Endowed Chair and Professor of Optics and Physics at CREOL, the College of Optics and Photonics at the University of Central Florida. He is a Fellow of The Optical Society of America (OSA), the International Optical Engineering Society (SPIE), the American Physical Society (APS), and the UK Institute of Physics (IOP). K.L. Vodopyanov is a world expert in mid-IR lasers, laser–matter interactions, nonlinear optics, and laser spectroscopy. He has pioneered several new laser sources, optical parametric devices, and methods for mid-IR and THz wave generation, including generation of ultrabroadband frequency combs; he is the author of more than 400 technical publications on the subject and a co-editor of the book *Solid-State Mid-Infrared Laser Sources* (Springer, 2003). His research interests include the use of nonlinear effects for mid-IR and THz generation, spectroscopic applications of frequency combs, nano-IR spectroscopy, and biomedical applications of lasers.

Preface

The field of mid-infrared (mid-IR) photonics is rapidly expanding driven by a growing number of applications in fundamental science, technology, defense, medicine, biology, environmental monitoring, among others. The last 25 years have seen remarkable advances in the field of mid-IR lasers starting with the quantum cascade laser pioneered by Lucent Technologies in 1994 and followed by a growing number of diverse innovative approaches to coherent light generation in this spectral range. Because the last comprehensive book on the subject, *Solid-State Mid-Infrared Laser Sources,* which I coedited with I. Sorokina, was published by Springer in 2003, it was both my and the publisher's understanding that this material needed to be significantly updated to include the impressive number of new techniques and applications.

The main goal of this book is to introduce the reader to the state-of-the-art technologies used to generate coherent mid-IR light, and to discuss their most important applications. The book assembles an array of methods developed by several scientific communities, which include solid-state physics, semiconductor physics, materials science, crystal growth, nonlinear optics, and nanofabrication, in their search to create an efficient and inexpensive solid-state mid-IR laser source.

The book loosely defines mid-IR range as 2–20 μm. It examines a variety of state-of-the-art approaches from diverse areas of photonics: solid-state lasers based on rare-earth and transition metals; fiber lasers; semiconductor lasers including intra- and intersubband cascade lasers; nonlinear-optical frequency converters including difference frequency generators, optical parametric oscillators and amplifiers, and Raman converters. It also discusses several emerging technologies such as "white light" and frequency combs generation in microresonators, waveguides, and microstructured fibers. In the final chapter, the book provides an overview of the most significant applications of mid-IR, such as chemical sensing and imaging including nano-imaging, medical and defense applications, plasmonics, extreme nonlinear optics, attosecond science, and particle acceleration. Such mature fields as free-electron lasers, CO_2 and CO gas lasers, synchrotron radiation, and cryogenic lead-salt semiconductor lasers are outside of the scope of this book, since the reader can find published material on these subjects.

The book is based on the short courses that I taught at major laser conferences, including the Conference on Lasers and Electro-Optics (CLEO) and SPIE Photonics West. Each chapter begins with a self-contained description of the underlying principle for a given method, and gradually brings the reader to the discussion of the latest achievements. I made every effort to make the narrative comprehensible to a broader community. However, it is assumed that the reader is familiar with basic concepts of laser physics, such as population inversion, Q-switching, and mode-locking, as well as of nonlinear optics, such as frequency mixing and nonlinear refraction.

The book should be useful to students, academics, researchers, and engineers, and to those who would like to learn about state-of-the-art and major trends in the development of mid-IR laser sources, and their current and upcoming applications.

I would like to thank Dr. Sergey Vasilyev, Prof. Sergey Mirov, Prof. Ken Schepler, Prof. Stuart Jackson, Prof. Gregory Belenky, Prof. Leon Shterengas, Dr. Jerry Meyer, Dr. Igor Vurhaftman, Prof. Arkadiy Lyakh, and Prof. Jerome Faist for reading the book chapters and making valuable suggestions. Finally, my wife Mila has earned my endless gratitude for her optimism, continuous support, helpful edits of the text, and for her bearing my spending long late-night and weekend hours on writing this book.

January 2020

Konstantin L. Vodopyanov
Orlando, Florida

1

Mid-IR Spectral Range

1.1 Definition of the Mid-IR

Infrared radiation was unknown before the year 1800 when Friedrich Wilhelm Herschel – a German-born musician, who moved to England to work as a music band conductor, but later became obsessed with astronomy and eventually landed the position of the King's Astronomer – discovered infrared radiation. He made this finding while exploring sunlight, dispersed into its colors by a glass prism, with the aid of a liquid thermometer with a blackened bulb to absorb radiation (a prototype of a modern microbolometer). His experimentation led to the conclusion that there must be an invisible form of light beyond the visible spectrum [1].

Further experiments showed that this invisible radiation is electromagnetic radiation with a lower frequency than the red in the visible spectrum. Modern science further divides the infrared spectral region into near-infrared, mid-infrared, and far-infrared.

According to the Encyclopedia Britannica, the "middle infrared" (mid-infrared or mid-IR) region of the electromagnetic spectrum covers, in wavelength, the portion between 2.5 and 50 μm (6–120 THz in frequency or 200–4000 cm^{-1} in wavenumbers).[1] (The wavenumber is the inverse of the vacuum wavelength, λ, expressed in cm^{-1}; it is also equal to the optical frequency divided by the speed of light, ν/c.)

However, the definitions of the "mid-IR" vary substantially in the technical literature, depending on a field-specific community. For example, the detector-based community subdivides the IR into four spectral bands, based on transmission windows of the atmosphere,[2] as can be seen in Figure 1.1: shortwave infrared (SWIR), 1–3 μm; mid-wave infrared (MWIR), 3–5 μm;

1 https://www.britannica.com/science/infrared-radiation
2 https://en.wikipedia.org/wiki/Infrared_vision

Laser-based Mid-infrared Sources and Applications, First Edition. Konstantin L. Vodopyanov.

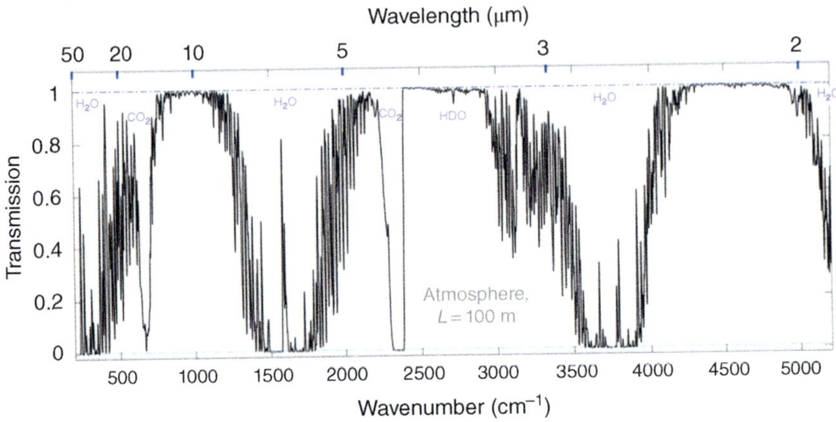

Figure 1.1 Transmission spectrum of a 100-m path in the "standard" atmosphere (spectral resolution $4\,\mathrm{cm}^{-1}$). The plot uses data from the HITRAN database (US model, mean latitude, summer, and zero elevation) [2]. The labels indicate the molecules that are responsible for transmission dips in the corresponding spectral regions.

longwave infrared (LWIR), 8–14 μm; and very-long wave infrared (VLWIR), 14–30 μm.

Also, it is not uncommon in the current literature to refer to mid-IR as "multi-terahertz range," especially when the authors generate few-cycle mid-IR transients combined with electro-optic methods of their detection, which is typical for terahertz science.

This book loosely defines the mid-IR range as 2–20 μm. This definition allows, on the short-wavelength side, to encompass a few categories of solid-state and fiber lasers, as well as certain types of microresonator-, nonlinear fiber-, and waveguide-based sources. On the long-wavelength side, 20 μm is a suitable practical limit set by the atmospheric transparency.

Heat energy is often transferred in the form of infrared radiation, which is given off from an object as a result of atomic and molecular motion. The mid-IR region overlaps with the spectral range of heat (blackbody) radiation at temperatures close to room temperature. Based on the Planck's law, the peak of the infrared radiation (in terms of power per unit wavelength) emitted by a human body at 310 K is at $\lambda \approx 9.35\,\mu\mathrm{m}$. Overall, our body emits 52 mW of mid-IR radiation per square centimeter; that radiation can be easily detected by a thermal microbolometer-based camera, as shown in Figure 1.2.

However, this book is about coherent laser sources, and the difference between the diffuse light of a heated body and a monochromatic laser-like beam is that the latter has a well-defined frequency and phase.

Figure 1.2 This is what the author's laboratory looks like in the mid-IR at 8–12 μm.

1.2 The World's Second Laser

Interestingly, the world's second laser – after the Maiman's ruby laser – was a mid-IR solid-state laser based on trivalent uranium-doped calcium fluoride (U^{3+}:CaF_2) [3]. It was operating at a wavelength $\lambda = 2.49\,\mu m$ and was developed by Peter Sorokin and Mirek Stevenson at the IBM research labs, in the same year as the Maiman's ruby laser, 1960.

The laser was pumped by a pulsed flashlamp and was cooled by liquid helium. The energy-level diagram of U^{3+}:CaF_2 is shown in Figure 1.3. Broadband pumping in the visible part of the spectrum causes transitions to excited U^{3+} bands. These pumping transitions are followed by rapid, nonradiative transitions to the two metastable upper laser levels. The thick arrow shows the 2.49-μm laser transition observed by the authors. The laser oscillation takes place in a transition from a metastable state to a level that is approximately $515\,cm^{-1}$ above the ground state. At liquid helium temperatures, this state is depopulated by at least a factor of 10^{10} relative to the ground state. Hence, this was the first demonstration of a four-level solid-state laser.

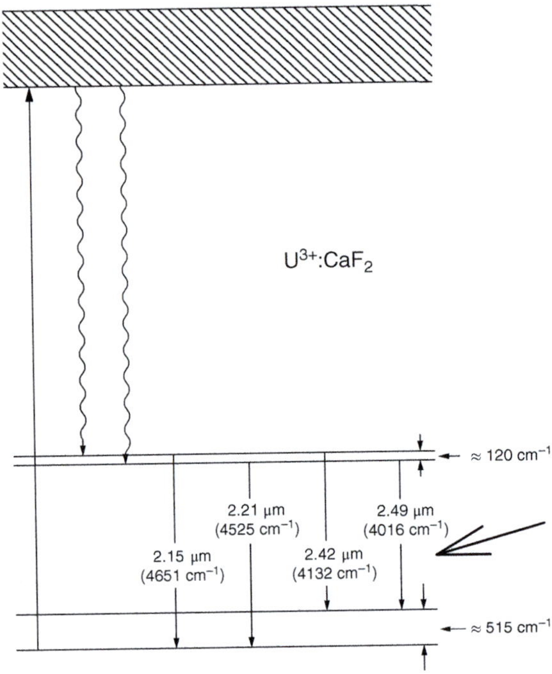

Figure 1.3 Energy-level diagram of trivalent uranium in calcium fluoride [3]. Broadband pumping light applied in the blue and green visible spectrum causes transitions to excited bands. These pumping transitions are followed by rapid, nonradiative transitions to the two metastable upper laser levels. The thick arrow shows the 2.49 μm laser transition observed by the authors. *Source:* reproduced from figure 1 of [3], with permission of APS.

Furthermore, the 1960 work by Sorokin and Stevenson coined a few significant keywords used in this book: *trivalent metal cation, and the 4-level system.*

1.3 Internal Vibrations of Molecules

Molecules typically have a characteristic absorption spectrum in the mid-IR, which is often used for identifying their structure. This results from the fact that mid-IR frequencies coincide with the strongest main-tone vibrational (strictly speaking, rotational–vibrational) frequencies of most of the molecules. In the gas phase, molecules possess dozens of distinct, sharp, and strong absorption features (with the exception of symmetric diatomic molecules like nitrogen, N_2, whose vibrations are not infrared active). This makes the mid-IR range especially important for chemical sensing, molecular

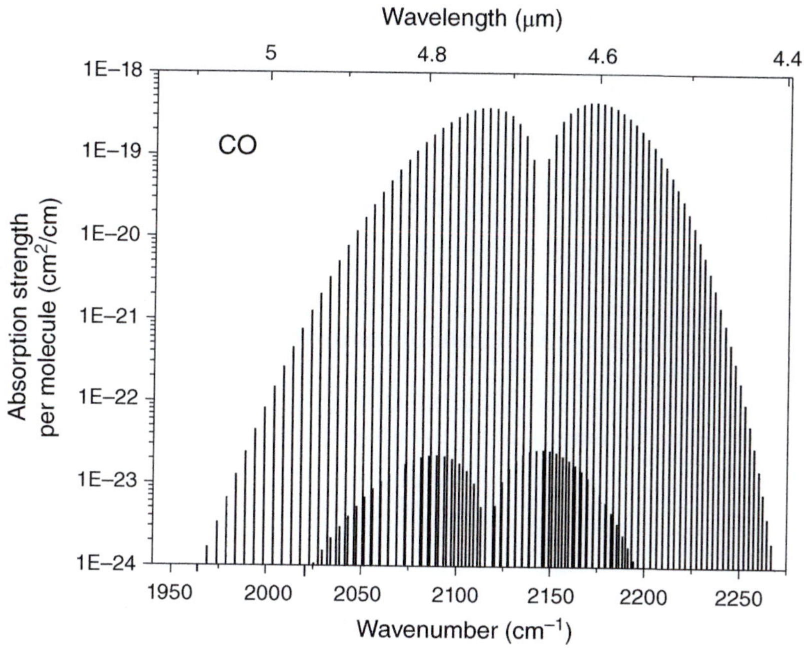

Figure 1.4 Rotational–vibrational absorption band of the CO molecule at 4.5–5 μm. The plot uses data from the HITRAN database [2].

spectroscopy, and molecular fingerprinting. Figure 1.4 shows a rotational–vibrational band of the CO molecule, located between 4.5 and 5 μm in wavelength. The band is represented by a regular sequence of sharp and extremely strong absorption peaks. For example, a 1-mm path of pure CO gas at 1 atm pressure would absorb >99.5% of the incoming mid-IR light, given that the light is tuned to one of the resonances. In theory, these resonances can even serve as a reference for high-precision molecular clocks for time and frequency metrology.

Characteristic vibrational transitions in the mid-IR are also present in the solid and liquid phases of matter, and also in 2D materials with exotic properties, such as graphene [4].

References

1 Herschel, W. (1800). Experiments on the refrangibility of the invisible rays of the sun. Philos. Trans. R. Soc. Lond. 90: 284.

2 Gordon, I.E., Rothman, L.S., Hill, C., Kochanov, R.V., Tan, Y., Bernath, P.F., Birk, M., Boudon, V., Campargue, A., Chance, K.V., Drouin, B.J., Flaud, J.-M.,

Gamache, R.R., Hodges, J.T., Jacquemart, D., Perevalov, V.I., Perrin, A., Shine, K.P., Smith, M.-A.H., Tennyson, J., Toon, G.C., Tran, H., Tyuterev, V.G., Barbe, A., Császár, A.G., Devi, V.M., Furtenbacher, T., Harrison, J.J., Hartmann, J.-M., Jolly, A., Johnson, T.J., Karman, T., Kleiner, I., Kyuberis, A.A., Loos, J., Lyulin, O.M., Massie, S.T., Mikhailenko, S.N., Moazzen-Ahmadi, N., Müller, H.S.P., Naumenko, O.V., Nikitin, A.V., Polyansky, O.L., Rey, M., Rotger, M., Sharpe, S.W., Sung, K., Starikova, E., Tashkun, S.A., VanderAuwera, J., Wagner, G., Wilzewski, J., Wcisło, P., Yu, S., and Zak, E.J. (2017). The HITRAN 2016 molecular spectroscopic database. J. Quant. Spectrosc. Radiat. Transf. 203: 3.

3 Sorokin, P.P. and Stevenson, M.J. (1960). Stimulated infrared emission from trivalent uranium. Phys. Rev. Lett. 5: 557.

4 Mak, K.F., Ju, L., Wang, F., and Heinz, T.F. (2012). Optical spectroscopy of graphene: from the far infrared to the ultraviolet. Solid State Commun. 152: 1341.

2

Solid-state Crystalline Mid-IR Lasers

Crystalline mid-IR lasers are *direct* sources of coherent light in the sense that they require a minimal number of energy conversion steps. When combined with laser diode pumping, these lasers are efficient, simple, and compact. The gain medium of a crystalline laser is a host crystal doped with active ions. These active ions (also referred to as impurity ions) doped into a crystalline matrix acquire, due to energy-level splitting, characteristic set of energy levels, not present in free ions. For rare earths, the primary cause of energy-level splitting is the interaction of electron spins of the dopant ion with the orbital angular momentum of electrons (spin–orbit interactions), while in transition metal ions, it is mostly due to the interaction of the optically active electron with the crystalline electric field of the host (the Stark effect).

The most common active media for mid-IR crystalline lasers are based on triply ionized rare-earth thulium (Tm), holmium (Ho), and erbium (Er) ions in yttrium aluminum garnet ($Y_3Al_5O_{12}$ or YAG), yttrium lithium fluoride ($LiYF_4$ or YLF), yttrium-scandium-gallium garnet ($Y_3Sc_2Ga_3O_{12}$ or YSGG), or other crystalline hosts. Alternatively, transition-metal-doped (Cr^{2+}, Fe^{2+}) II–VI zinc chalcogenide crystals (ZnSe, ZnS) or other chalcogenides (CdSe, CdS, ZnTe, and CdMnTe) can serve as active media for mid-IR lasers with an extremely broad gain bandwidth.

2.1 Rare-Earth-based Tm^{3+}, Ho^{3+}, and Er^{3+} Lasers

2.1.1 Tm^{3+} Lasers

The energy-level diagram for the trivalent thulium ion is shown in Figure 2.1a. Tm-doped crystalline lasers can provide tunable operation in two spectral regions: 1.8–2.2 μm using the $^3F_4$–$^3H_6$ transition and 2.2–2.4 μm using the $^3H_4$–$^3H_5$ transition.

Laser-based Mid-infrared Sources and Applications, First Edition. Konstantin L. Vodopyanov.

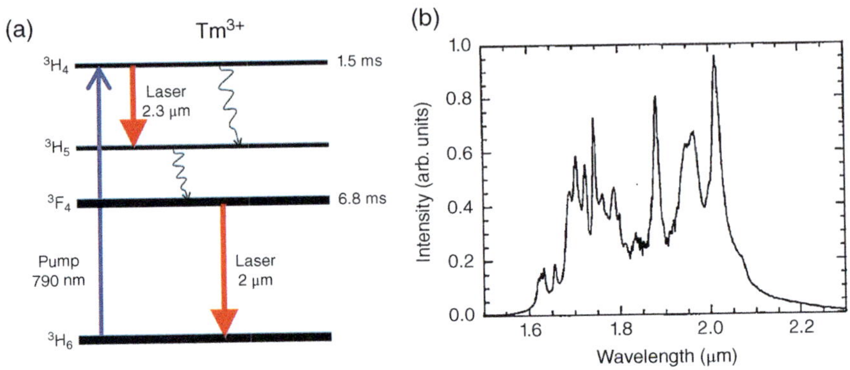

Figure 2.1 (a) Energy-level diagram for the trivalent thulium. Wavy arrows indicate nonradiative phonon-assisted decay. Laser upper-state lifetimes are also indicated. (b) Spectrum of fluorescence from the $^3F_4$ level of Tm^{3+} in YAG. *Source:* reproduced from figure 1 of [1], with permission of OSA, The Optical Society.

At first glance, a laser that uses $^3F_4$–$^3H_6$ transition appears as a three-level system. In such a system the lower laser level is the ground state. Nevertheless, it should be mentioned that both $^3F_4$ and $^3H_6$ levels consist of manifolds of energy levels split due to the Stark effect in the electric field of the crystalline lattice, so that the lower laser level is not necessarily the lowest energy state. Due to the fast energy relaxation between these sublevels, the system becomes virtually a four-level system. An additional effect is that the presence of partially overlapping manifolds of Stark levels within the upper and lower laser state broadens the bandwidth of fluorescence, resulting in broad emission linewidths. Furthermore, thulium lasers (e.g. Tm:YAG and Tm:YSGG) are characterized by large phonon broadening. The phonons (vibrations of the ions of the host crystal) modulate the crystal field at the site of the "lasing" dopants (thulium ions in this case), which in turn broadens the energy levels [1]. Both of these effects allow laser tunability over several hundreds of nanometers. The room-temperature (RT) fluorescence spectrum from the $^3F_4$ state to the $^3H_6$ ground state of Tm^{3+} in YAG is shown in Figure 2.1b.

YAG crystal is one of the most commonly used host materials for thulium because of its unique thermal–mechanical and optical properties. Typically, Tm^{3+}-doped solid-state lasers are pumped ($^3H_6 \rightarrow {}^3H_4$) by commercially available high-power AlGaAs diode bars at ~800 nm. There is a "two-for-one" cross-relaxation process that can lead to pumping quantum efficiencies approaching a factor of two. (The pumping quantum efficiency indicates how many laser photons are emitted per one absorbed pump photon.) The essence of this effect (Figure 2.2) is that because of the fortuitous proximity (resonance) of the energy spacing $^3H_4$–$^3F_4$ and $^3F_4$–$^3H_6$, the $^3F_4$ upper laser level is populated through the cross-relaxation process $^3H_4 + {}^3H_6 \rightarrow {}^3F_4 + {}^3F_4$ [1]. The effectiveness of this

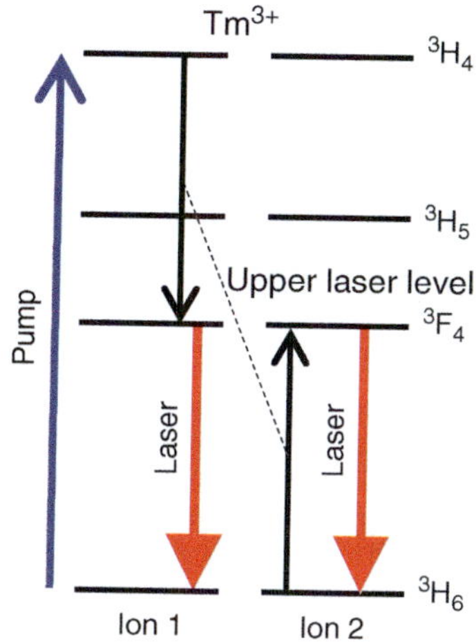

Figure 2.2 Resonant pumping diagram for the 2-μm Tm^{3+} laser. The "two-for-one" cross-relaxation process leads to the efficient transfer of all absorbed pump energy to the excitation of Tm ions to the $^3F_4$ level.

two-body cross-relaxation process increases with Tm^{3+} doping concentrations; it becomes significant, typically at >3 at.% concentration. For example, it was shown that at Tm^{3+} concentration of 12% for Tm:YAG and Tm:YSGG, the cross-relaxation totally dominates the decay of the $^3H_4$ state. As a result, a slope efficiency of as high as 59% has been demonstrated in Tm:YAG, considerably larger than the 39% maximum expected from the quantum defect alone [1]. (The quantum defect is defined as the ratio of the energy of the lasing photon to that of the pump.)

With a 785-nm laser pumping, broadly tunable continuous wave (CW) laser emission over the ranges 1.87–2.16 μm in Tm:YAG and 1.85–2.14 μm in Tm:YSGG has been reported [1]. Similarly, tunable Tm laser emission using a different host crystal – Tm:YALO (Tm:$YAlO_3$) – was observed over the range 1.93–2.0 μm [2].

The $^3F_4$–$^3H_6$ transition in Tm is especially attractive for high-power applications because of the ability to use highly efficient AlGaAs diodes or diode stacks operating around 800 nm as a pump source. A compact diode-pumped Tm:YAG laser capable of generating 115 W of CW power at 2.01 μm has been demonstrated at the 805-nm pumping power of 360 W [3]. Another high-power CW Tm:YAG laser used a linear laser cavity with three laser rods, each side-pumped by arrays of laser diode bars with central wavelength of 785 nm, arranged in fivefold symmetry around each laser crystal (Figure 2.3). The laser was water-cooled at 8 °C and yielded a maximum output CW power of 267 W at 2.07 μm, with the total laser-diode pump power of 1.3 kW. The corresponding optical-to-optical conversion efficiency was 20.7%, with slope efficiency of 29.8% [4].

The Q-switching performance of Tm:YAG near 2 μm is facilitated by a long fluorescence lifetime of 11 ms, which leads to a high energy storage capability [5].

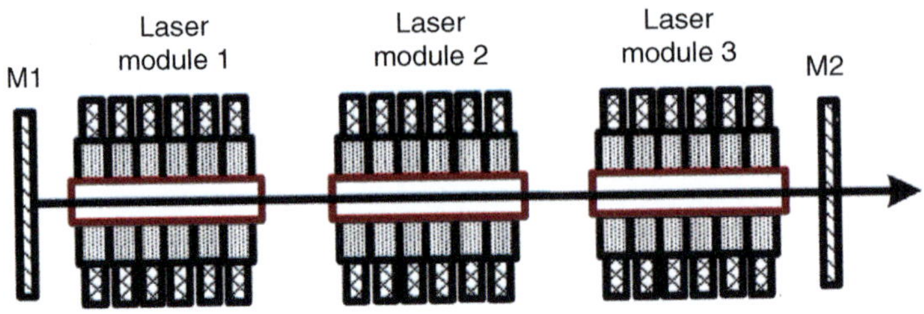

Figure 2.3 High-power (267 W) Tm:YAG laser system at 2.07 μm based on a linear laser cavity containing three laser rods. The laser cavity was formed by two plane mirrors: M1 with reflectivity $R > 99.5\%$, and the outcoupler mirror M2 with transmission $T = 5\%$ around 2 μm. Each rod (Tm concentration of 3.5 at.%, 4 mm in diameter, and 69 mm in length) was side-pumped by an array of laser diode bars at 785 nm. Undoped YAG end caps were bonded to end faces of the rods to reduce thermal effects as well as reabsorption losses in the unpumped regions. *Source:* reproduced from figure 1 of [4], with permission of OSA, The Optical Society.

For example, Q-switched pulses at 2.016 μm with 20.4 mJ energy and 69 ns duration were demonstrated at 500 Hz pulse repetition rate; a Tm:YAG ceramic slab laser was end-pumped by a diode laser (the absorbed pump power 53 W) [6]. In general, because of the low gain due to low stimulated emission cross section of Tm ions, Q-switched Tm lasers operate by necessity at high intracavity energy density (fluence), close to the material damage threshold [5]. The reason is that in order to achieve the laser threshold in this low-gain medium, one needs high population inversion; this in turn results in a high amount of energy stored in the medium that is eventually released as an energetic Q-switched laser pulse that may cause material damage.

2.1.2 Ho^{3+} Lasers

Figure 2.4a represents the energy-level diagram for trivalent holmium. With Ho^{3+} as a laser ion, one can obtain oscillation using the $^5I_7$–$^5I_8$ transition in the 1.95–2.15 μm range (the corresponding emission spectrum for this transition is shown in Figure 2.4b), as well as $^5I_6$–$^5I_7$ transition at 2.85–3.05 μm, $^5I_5$–$^5I_6$ transition at 3.94 μm, and $^5S_2$–$^5F_5$ transition near 3.2 μm. Holmium lasers are more favorable (as compared to thulium lasers) to operation in the Q-switched mode due to high stimulated emission cross section (high gain); also, they have long (~10 ms, similar to thulium) fluorescence lifetime for the upper laser energy manifold $^5I_7$.

Since there are no convenient schemes for direct diode pumping of Ho^{3+} lasers (at least with available high-efficiency AlGaAs or InGaAs laser diodes), holmium laser crystals are often codoped (sensitized) with other ions. For example, crystals doped with a combination of Tm^{3+} and Ho^{3+} ions have proven to be efficient sources for diode-pumped laser action in the 2-μm region using

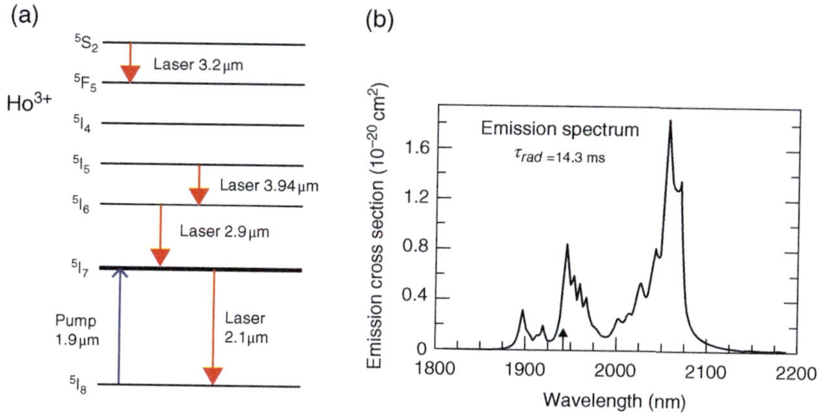

Figure 2.4 (a) Energy-level diagram for the trivalent holmium. Red arrows indicate laser transitions. (b) Emission spectrum of Ho^{3+} (the $^5I_7$–$^5I_8$ transition) in $LiYF_4$ host. *Source:* reproduced from figure 3c of [7], with permission of IEEE.

the following mechanism: thulium ions provide absorption of the AlGaAs diode-laser pump around 780–790 nm and permit efficient population of the $^3F_4$ state of Tm^{3+}. This excitation is transferred to the $^5I_7$ energy level of Ho^{3+} through the dipole–dipole interaction between nearby Tm^{3+} and Ho^{3+} ions, and laser action takes place on the Ho^{3+} $^5I_7$–$^5I_8$ transition at 2.1 μm [8].

Although Tm and Ho lasers operate in the similar wavelength range near 2 μm, there are at least two reasons why Ho lasers are preferable, especially in the Q-switched regime: Ho^{3+} has higher stimulated-emission cross section ($\sigma = 1.2 \times 10^{-20}$ cm^2 at $\lambda = 2.09$ μm in Ho:YAG) compared with that in Tm^{3+} ($\sigma = 1.5 \times 10^{-21}$ cm^2 at $\lambda = 2.01$ μm in Tm:YAG) [7, 9]. This enables Q-switching at high repetition rates and makes compact low-threshold devices feasible. Besides, Ho lasers operate at slightly longer wavelengths that make them more suitable for coherent Doppler light detection and ranging (LIDAR) applications because of much smaller absorption in the atmosphere at $\lambda > 2$ μm (see Figure 1.1).

Ho lasers can be directly pumped by Tm-doped lasers (including both solid-state and fiber-based lasers). For example, while the absorption spectrum of Tm:YLF falls well within the emission spectrum of commercially available laser diodes emitting in the 792–793 nm range, its emission spectrum at ~1.91 μm aligns well with the absorption spectrum of Ho:YAG [9]. An efficient diode-pumped cascaded Tm–Ho system (both CW and Q-switched) was reported in [10]. Using a high-brightness 1.91-μm output from a diode-pumped Tm:YLF laser, the authors resonantly populated the $^5I_7$ manifold in Ho:YAG. An end-pumped geometry was adopted for both the Tm (1.91 μm) and Ho (2.09 μm) lasers (Figure 2.5). The Tm laser used two laser crystals to enable the use of four

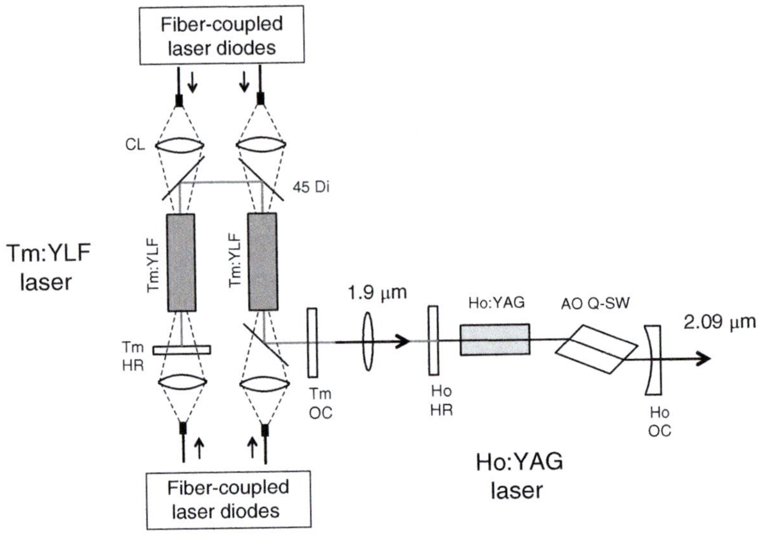

Figure 2.5 Ho:YAG laser ($\lambda = 2.09\,\mu m$) resonantly pumped by a Tm:YLF laser ($\lambda = 1.91\,\mu m$) [10]. HR, high reflector; 45Di, 45° dichroic mirror; CL, coupling lens; AO Q-SW, acousto-optic Q-switch (not used in the CW experiment); OC, output coupler mirror.

fiber-coupled laser diodes as the pump and to distribute the thermal load. With this design, 36 W of Tm laser output was achieved with the optical-to-optical conversion efficiency (diode-to-Tm laser) of 32%. The secondary holmium laser was pumped by the Tm:YLF laser and used a 5-mm-diameter, 20-mm-long Ho:YAG crystal. With 36 W of Tm laser pump, 19 W of continuous-wave output at 2.09 µm was demonstrated [10]. This corresponds to Tm:YLF to Ho:YAG optical-to-optical efficiency of 56% and gross laser-diode-to-Ho:YAG conversion efficiency of 18%.

The Q-switched mode of Ho:YAG in the above-described experiment was achieved by inserting an acousto-optic modulator in the Ho laser cavity. In the repetitively Q-switched (9–50 kHz) configuration of Ho:YAG pumped by the CW Tm:YLF laser, the average power reached 16 W with the corresponding Tm:YLF to Ho:YAG optical-to-optical efficiency of 50% (overall diode-to-Ho:YAG conversion efficiency of 15%). At a Q-switch frequency of 15 kHz, the beam propagation factor was $M^2 < 1.5$, the pulse duration 35 ns, and the energy per pulse 1.1 mJ [10].

Another laser material, Ho-doped $YLiF_4$ (Ho:YLF) is an excellent gain medium for obtaining high-energy Q-switched 2-µm pulses. Ho:YLF has a weak thermal lensing and is naturally birefringent, which allows one to maintain polarization. A setup that delivered single-frequency pulses with up to 330-mJ pulse energy (pulse width 350 ns, $\lambda = 2.064\,\mu m$, repetition rate 50 Hz) was reported [11]. A double-pass Ho:YLF slab amplifier was end-pumped with a diode-pumped Tm:YLF slab laser at 1.89 µm (a wavelength that matched one of the strong absorption peaks of Ho:YLF), and seeded with 50-mJ single-frequency pulses. The seed laser in this case was a single-frequency ring-cavity Ho laser system, pumped with an 80-W 1940-nm Tm:fiber laser.

There have been a number of reports on fiber-bulk hybrid lasers combining Q-switched Ho:YAG and Ho:YLF laser system with CW Tm-doped fiber laser (1908 and 1940 nm, respectively) as an optical pump. This approach enables the efficient (>50%) conversion of a low-cost and reliable high-power fiber laser's output into high-energy nanosecond pulses. The recent achievement on the fiber-bulk hybrids including 1-kHz, nanosecond Ho:YAG and Ho:YLF systems with 52 and 35 mJ pulse energy, respectively, are summarized in [12]. The Ho:YLF hybrids with pulse energy in excess of 100 mJ were also reported [13].

Codoping holmium with ytterbium as sensitizer offers the possibility of using well-developed InGaAs (970 nm) diode lasers for pumping Ho lasers. In [14], the researchers used Yb (at 20 at.%) and Ho (1 at.%) codoped YSGG as the active material to achieve laser action in holmium near 3 μm. YSGG was chosen as the host material because of the lower (as compared to YAG) phonon energies and thus longer lifetime of the upper laser level ($^5I_6$) of the 3-μm transition. First, the Yb^{3+} absorbs a pump photon at 970 nm; this excitation is then transferred to the $^5I_6$ upper laser level of the Ho^{3+}. Experiments were performed under quasi-CW excitation. A high-power pulsed laser diode array (500 W peak power) was used at a pulse length of 0.8 ms and repetition rate 15 Hz. A maximum pulse energy of 10.5 mJ was obtained at the laser wavelength 2.844 μm (peak power 13 W). In the CW regime, the output power was a few milliwatts; the reason for the low power is a typical self-terminating behavior of the laser transition because of the long-lived lower laser level $^5I_7$ [14]. (Ideally, the lifetime of the lower laser level should be shorter than that of the upper level to avoid a "bottleneck" effect.)

More information on Ho lasers doped with multiple (Cr^{3+}, Er^{3+}, and Tm^{3+}) dopants, cascaded lasing at 2.1 and 2.9 μm, and on flashlamp-pumped Ho lasers can be found in [15]. Flashlamp-pumped 2-μm holmium lasers at low repetition rates are widely used in medicine. For example, commercially available Ho:YAG lasers with free-running pulse energies exceeding 3 J are commonly exploited for renal stone removal [16]. The 2-μm radiation has numerous other applications. Strong absorption at this wavelength by human tissue is attractive for laser surgery (see Chapter 7); low atmospheric absorption makes this system useful for range-finding, remote sensing, and wind-shear detection onboard an aircraft with coherent laser radar. Also, 2-μm lasers are widely used for pumping optical parametric oscillators to reach longer mid-IR wavelengths (see Chapter 5).

2.1.3 Er^{3+} Lasers

Considerable interest in the lasers operating in the 2.7–3 μm range, based on trivalent erbium (Er:YAG, Er:YSGG, and others), has been observed in the last 30 years. The reason is that these lasers are extremely appealing to specialists from different branches of medicine (dentistry, laser therapy, surgery, and microsurgery), due to the fact that the laser wavelength near 2.94 μm corresponds to the strongest absorption peak of water (absorption strength $>10^4\,cm^{-1}$), and water is present in abundance in soft and hard biological tissues. Furthermore, 3-μm lasers are suitable pump sources for mid-IR nonlinear optical frequency downconversion (see Chapter 5).

A great majority of literature before 1994 is dedicated to flashlamp-pumped Er:YAG (2.94 μm) and Er:YSGG (2.8 μm) lasers operating in the free-running [15, 17], Q-switched (see [18] and references therein), and actively mode-locked [19] regimes. Q-switched operation with pulse energies of 10–100 mJ in the fundamental TEM_{00} mode, and mode-locked operation with pulse energies up to 4 mJ in 100-ps pulse duration are accessible with these lasers. In the free-running pulsed mode, flashlamp-pumped erbium lasers can deliver several joules of pulse energy and are commercially available from a variety of vendors.

The Er^{3+} energy-level diagram relevant to the 3-μm laser transition is shown in Figure 2.6a. The laser transition is between the $^4I_{11/2}$ (upper) and the $^4I_{13/2}$ (lower) states (not to be confused with more common 1.5-μm Er laser transition, which is between $^4I_{13/2}$ and $^4I_{15/2}$ energy levels). For the 3-μm transition, the upper laser state $^4I_{11/2}$ can be directly pumped at 970 nm, or at 800 nm (in the latter case the upper laser level is populated via a fast nonradiative transition from $^4I_{9/2}$ to $^4I_{11/2}$). The corresponding emission spectrum for the 3-μm transition is shown in Figure 2.6b. The lifetime of the lower laser level $^4I_{13/2}$ (6.4 ms in Er:YAG) is much longer than the lifetime of the upper laser level $^4I_{11/2}$ (0.1 ms in Er:YAG), which would lead, under normal circumstances, to self-termination of the laser transition. However, at high Er^{3+} concentrations

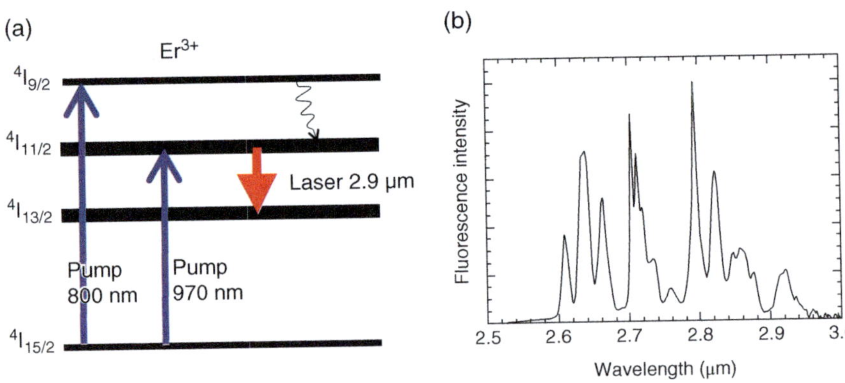

Figure 2.6 (a) Energy-level diagram for the trivalent erbium. (b) Emission spectrum of Er^{3+} in GSGG ($Gd_3Sc_2Ga_3O_{12}$) host at room temperature. *Source:* reproduced from figure 2 of [20], with permission of OSA, The Optical Society.

and under intense pumping, cross-relaxation between neighboring Er^{3+} ions depletes the $^4I_{13/2}$ level of one ion while simultaneously providing a pumping path to the upper laser level via the $^4I_{9/2}$ energy level of a neighboring atom in the $^4I_{13/2} + {}^4I_{13/2} \rightarrow {}^4I_{9/2} + {}^4I_{15/2}$ resonant upconversion process [20–22]. This effect, illustrated in Figure 2.7, prevents self-termination of the lasing process, while at the same time providing two photons at the laser wavelength for every absorbed pump photon. There is a certain similarity in the "two-for-one" pumping, between the 3-μm erbium and the 2-μm thulium laser system described above.

Efficient CW operation of the Er^{3+}:GSGG (gadolinium scandium gallium garnet, $Gd_3Sc_2Ga_3O_{12}$) $^4I_{11/2}$–$^4I_{13/2}$ laser transition at 2.795 μm was demonstrated in [20] with 130-mW output power. The $^4I_{11/2}$ upper laser state was directly pumped by a Ti:Sapphire laser at 970 nm. The slope efficiency for the 2.795-μm laser transition was 36%, at the laser-to-pump photon ratio of only 34.7%, indicating a greater than unity quantum efficiency. This excess efficiency resulted from the recycling of population through the "two-for-one" cross-relaxation process mentioned before.

The authors in [23] reported on CW operation of monolithic (3 mm long and 3 mm in diameter) lasers based on three different host crystals: Er:YAG, Er:GGG (gadolinium gallium garnet, $Gd_3Ga_5O_{12}$), and Er:YSGG (yttrium scandium gallium garnet $Y_3Sc_2Ga_3O_{12}$) at $\lambda \sim 3$ μm, using pumping with an InGaAs laser diode at 970 nm. High doping concentrations of erbium were used: 30% for Er:YSGG and Er:GGG, and 33% for Er:YAG. Figure 2.8 shows the input–output curves for the three laser materials. In terms of maximum power and efficiency, the best performance was demonstrated in Er:YSGG, where the output power of 0.5 W and slope efficiency of 31% were achieved. In addition, by using an Er:YAG monolithic laser with the cavity length of 1 mm, a single-frequency output was demonstrated with an output power of 70 mW [23].

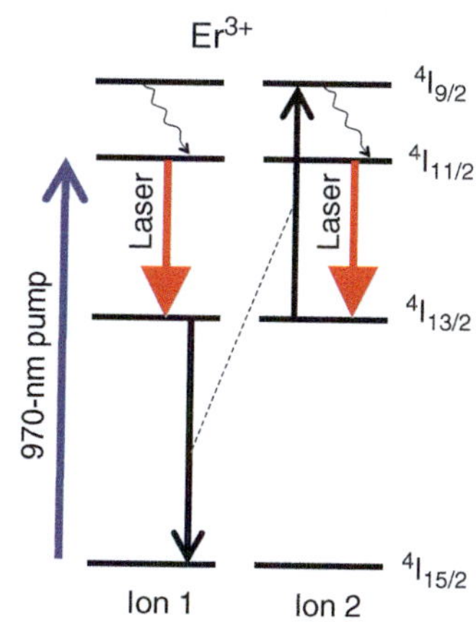

Figure 2.7 The "two-for-one" cross-relaxation process in a 3-μm erbium laser. The upconversion process recycles population from the $^4I_{13/2}$ lower laser state into the upper laser state.

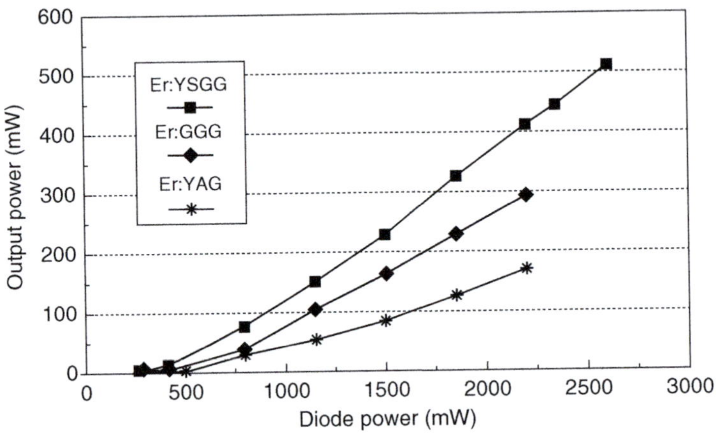

Figure 2.8 Input–output curves for three different Er-doped monolithic lasers, Er:YSGG, Er:GGG, and Er:YAG, pumped by two 1-W diode lasers. *Source:* reproduced from figure 2 of [23], with permission of OSA, The Optical Society.

Subsequently, CW output powers exceeding 1 W were achieved by several authors. Jensen et al. demonstrated 1.1 W at 2.8 μm from a diode-pumped Er^{3+}:YLF [24] and Chen et al. produced 1.15 W from a diode-pumped Er^{3+}:YAG laser system [25].

Sheaumann Laser, Inc. made a diode-pumped 2.94-μm Er:YAG laser producing 1.5 W continuous output power in a TEM_{00} beam [26]. Now a compact diode-pumped 2.94-μm Er:YAG laser with 1 W continuous output power in a TEM_{00} beam with the beam quality factor $M^2 = 1.12$ is commercially available from this company (Figure 2.9).[1] The laser is hermetically sealed, has free-space or fiber-coupled output, and has wall-plug efficiency of 10%. A side-pumped laser using 50-at.% Er:YAG crystal, based on the slab concept, which employs multiple total internal reflections (TIR), was developed by Ziolek et al. [27]. The cavity consists of a plane-mirror resonator of 40-mm total length and uses three (out of five) TIR regions within the laser crystal as pump facets to efficiently couple the laser-diode pump energy into the resonant laser mode (Figure 2.10). When the laser was pumped by 400-μs laser diode pulses at 100 Hz rate (duty cycle 4%), the average output power reached 3.2 W.

In the medical application of tissue removal with 3-μm lasers, the main mechanism is fast ablation due to extremely high optical absorption. The extent of thermal damage to surrounding tissue can be reduced if the laser pulse duration is shortened. Therefore, the Q-switching mode in a diode-pumped configuration is the preferred one for medical use. A Q-switched

1 www.sheaumann.com

Figure 2.9 A commercial fiber-coupled diode-laser-pumped 2.94-μm Er:YAG laser from Sheaumann Laser, Inc., operating at 1 W continuous wave output power. *Source:* reproduced with permission of Sheaumann Laser, Inc.

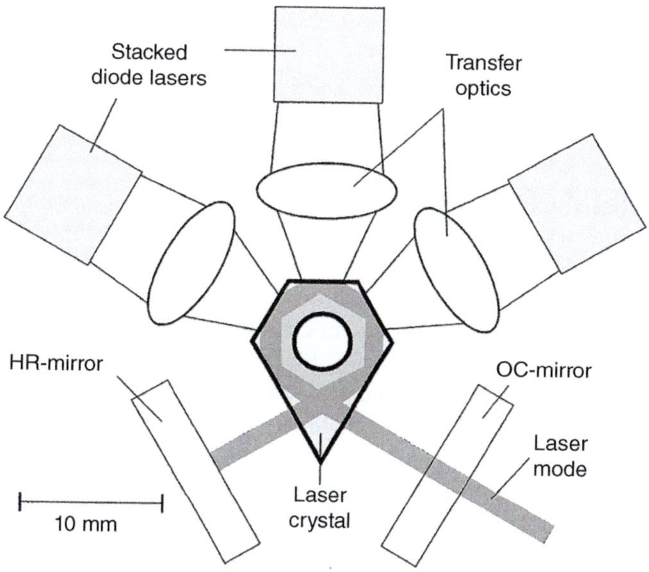

Figure 2.10 Setup of the total internal reflection 3-μm Er:YAG laser being pumped at three facets using quasi-CW stacked-array diode lasers emitting near 962 nm. The laser crystal is mounted between two water-cooled copper heat sinks (out of the figure plane) to remove the deposited heat. HR, high reflection mirror; OC, output coupler. *Source:* reproduced from figure 1 of [27], with permission of OSA, The Optical Society.

diode-pumped Er:YSGG laser (λ = 2.797 μm) that yielded 0.5-mJ pulses with 77-ns pulse duration was demonstrated in [28]. The authors used a $LiNbO_3$ electro-optic crystal for the cavity loss control and side pumping in a single-bounce geometry at a repetition rate of 140 Hz and pump pulse duration of 500 μs.

The main results for the 3-μm Er^{3+} lasers are listed in Table 2.1.

Table 2.1 Summary of the 3-μm Er^{3+} lasers.

Laser crystal	Laser operating mode	Output parameters	Ref.
Er:YSGG	CW diode pumped	Output power 0.5 W	[23]
Er:YAG	CW diode pumped	Output power 1.5 W	[26]
Er:YAG	Long pulses, diode pumped	Average power 3.2 W @ 4% duty cycle (400 μs, 100 Hz)	[27]
Er:YAG	Free-running, flashlamp pumped	Pulse energy 27 J (single shots), 700 μs	[29]
Er:YSGG	Q-switched, diode pumped	Pulse energy 0.5 mJ, 77 ns, 140 Hz	[28]
Er:YAG	Q-switched, flashlamp pumped	Pulse energy 85 mJ, 130 ns, 2 Hz	[18]
Er, Cr:YSGG	Mode-locked, flashlamp pumped	Pulse energy 4 mJ, 100 ps, 2 Hz	[19]

2.2 Transition Metal Cr^{2+} and Fe^{2+} Lasers

A new class of laser materials, based on tetrahedrally coordinated transition metals chromium (Cr^{2+}) or iron (Fe^{2+}) that are doped into II–VI crystals (ZnS, ZnSe), was pioneered in [30, 31]. This represents a new category of mid-IR lasers that are widely tunable and operate at RT. Electronic transitions in the doubly ionized Cr and Fe atoms are strongly coupled to phonons in the lattice of a host crystal, which leads to significant homogeneous broadening, typically as much as 30% of the center laser frequency. Chalcogenide hosts such as ZnSe and ZnS are particularly well suited for laser action, since they support this "vibronic" broadening [15, 31]. At the same time, undesirable nonradiative multi-phonon relaxation is sufficiently suppressed – even at RT for Cr^{2+} – due to their low phonon energies. Optically pumped RT lasers based on binary (e.g. ZnS, ZnSe, CdSe, CdS, and ZnTe) and ternary (e.g. CdMnTe, CdZnTe, and ZnSSe) chalcogenide crystals doped with transition metal Cr^{2+} or Fe^{2+} provide access to the spectral region, which spans, without gaps, from below 2 to about 6 μm, with continuous tunability in excess of 1000 nm for a given laser medium.

2.2.1 Spectroscopic Properties of Cr^{2+} and Fe^{2+}

Transition-metal-doped II–VI lasers can be regarded as mid-IR analogs of the familiar near-IR Ti:Sapphire ($Ti:Al_2O_3$) laser, in terms of their spectroscopic and laser characteristics. The tetrahedral crystal field of the II–VI host crystal splits the ^5D ground state of Cr^{2+} and Fe^{2+} into a triplet $^5T_2$ and a doublet ^5E. The triplet $^5T_2$ is the ground state of the *chromium* ions, while the doublet ^5E is the ground state of the *iron* ions. Comparatively small energy splitting between these levels in the tetrahedral coordination corresponds, in terms of photon energy, to the mid-IR spectral range. The main similarities between transition metal-doped II–VI lasers and the Ti:Sapphire laser are described in [15, 30, 31] and include:

- High-emission cross sections ($\sigma \sim 10^{-18}\,\text{cm}^2$) due to spin-allowed laser transitions.
- Comparatively short (μs-range) upper-level lifetimes.
- Transitions to higher-lying levels are spin forbidden, which precludes excited-state absorption.
- Strong interaction of the electronic states with lattice vibrations (phonons). This vibrational–electronic (vibronic) interaction leads to a strong homogeneous broadening and to a large gain bandwidth.
- A strong wavelength shift (Franck–Condon shift) exists between absorption and emission, due to the fact that the energy minima in the configurational coordinate space (lattice displacement) for the lower and upper states do not coincide. This allows the four-level laser action.

These similarities allow both Cr^{2+} and Fe^{2+} laser media to be used in broadly tunable as well as in ultrafast lasers.

The absorption and emission spectra for the Cr^{2+} ion are shown in Figure 2.11. Broad absorption bands spanning over the 1.5–2.2 μm spectral range allow the use of a variety of pumping sources [32]. From the same figure it follows that Cr^{2+}-doped II–VI crystals have extremely broad emission bands, which are beneficial for a broad and continuous mid-IR tunability, as well as for the ultrashort pulse generation. The inset to Figure 2.11 plots the temperature dependence of the luminescence lifetime of the $^5E \leftrightarrow {}^5T_2$ transition of chromium ion. There is no luminescence quenching in ZnSe crystal at RT and the luminescence quantum yield is close to unity in these crystals. It is one of the reasons why Cr:ZnSe and similar crystals are currently used as gain media in efficient RT lasers operating over the 2–3 μm spectral range. It is remarkable that the peak absorption and emission cross sections of Cr^{2+} for all II–VI hosts ($\sigma \sim 10^{-18}\,\text{cm}^2$) are about two orders of magnitude greater than in rare-earth ions.

The absorption and emission spectra of Fe^{2+}-doped chalcogenide crystals are shifted to longer wavelengths, as compared to Cr^{2+}. As an example, the spectra of Fe^{2+}:ZnSe and Fe^{2+}:CdMnTe are depicted in Figure 2.12. The maximum absorption is at 3.1 μm (Fe:ZnSe) and 3.6 μm (Fe:CdMnTe); thus, available pump sources for Fe^{2+} are not as diverse as for Cr^{2+} ions. The temperature dependences of the excited state lifetime of Fe^{2+} ions in ZnSe and CdMnTe crystals are shown in the inset. The decreasing luminescence lifetime at $T > 120\,\text{K}$ is due to thermally activated nonradiative decay (e.g. multiphonon quenching). Nonradiative processes result in the upper state lifetime as short as several hundreds of nanoseconds and low luminescence quantum yield at RT ($\eta \sim 1\%$) [32].

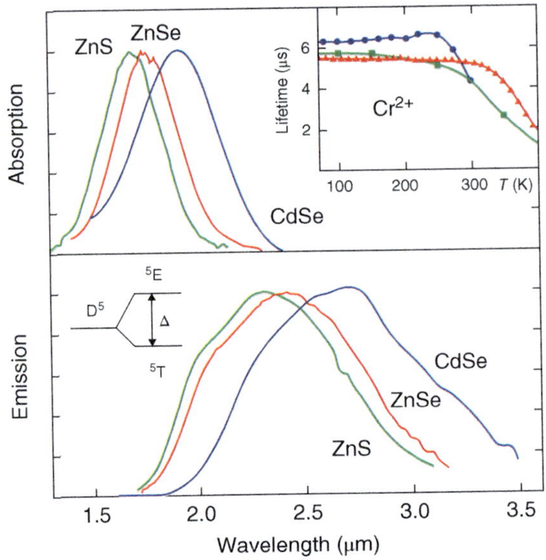

Figure 2.11 (Top) Normalized room temperature absorption and (bottom) emission-gain spectra of Cr^{2+} ions in ZnS, ZnSe, and CdSe crystals. The inset shows the temperature dependences of Cr^{2+} excited state lifetimes in ZnS (circles), ZnSe (triangles), and CdSe (squares) crystals. *Source:* reproduced from figure 1 of [32], with permission of Wiley.

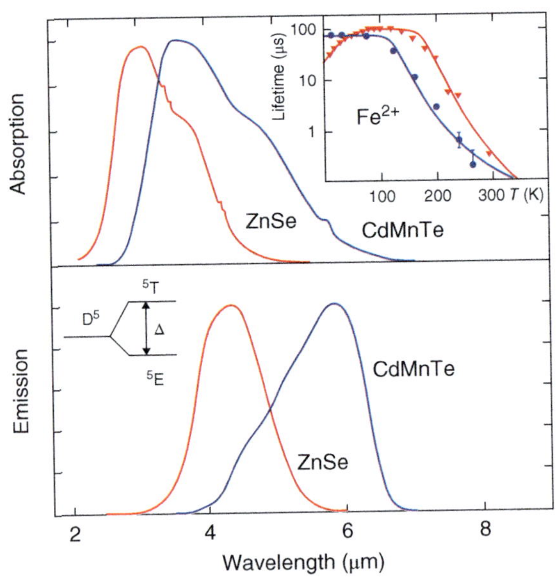

Figure 2.12 Normalized room temperature absorption (top) and emission-gain (bottom) spectra of Fe^{2+} ions in ZnSe and CdMnTe crystals. The inset shows the temperature dependences of Fe^{2+} upper-state lifetimes in ZnSe (triangles) and CdMnTe (circles) crystals. *Source:* reproduced from figure 2 of [32], with permission of Wiley.

The most recent measurements indicate that in Fe:ZnSe the lifetime of the upper lasing level decreases from 60 μs at $T = 80$ K to 370 ns at RT [33].

2.2.2 Lasers Based on Chalcogenide Crystals Doped with Cr^{2+}

Currently Cr^{2+}-doped zinc chalcogenide lasers are widely used as compact RT coherent sources emitting mid-IR radiation between 1.9 and 3.6 μm in the CW, gain-switched, mode-locked, and single-frequency regimes (see in-depth reviews [12, 34–37]). These lasers possess a favorable combination of thermal, optical, and spectroscopic properties and can be conveniently pumped by erbium and thulium fiber lasers, or directly by laser diodes.

2.2.2.1 Broadly Tunable Cr^{2+} Lasers

In the pioneering work by Page et al. [31], the authors used a pulsed Co^{2+}:MgF_2 laser ($\lambda = 1.85$–1.9 μm) to pump a Cr^{2+}:ZnSe laser. Using a diffraction grating inside a Cr:ZnSe cavity, they achieved a tuning range of 2.15–2.8 μm with a linewidth of ~1 nm. With no tuning elements in the cavity, both the Cr:ZnS and Cr:ZnSe lasers operated at the emission peak of 2.35 μm with a bandwidth of 40 nm. The first continuous-wave RT operation of a broadly tunable Cr^{2+}:ZnSe laser was demonstrated in [38]. The pump source was a CW diode-pumped Tm:YALO laser operating at 1.94 μm with an output power of 1 W. The Cr:ZnSe laser delivered the output power of 250 mW at 2.4 μm, with a slope efficiency as high as 63% (related to the absorbed optical power), with the absolute efficiency of 42% (again, related to the absorbed optical power). By adding an intracavity tuning prism made of ZnSe (Figure 2.13), the Cr:ZnSe laser was tuned continuously from 2.14 to 2.76 μm.

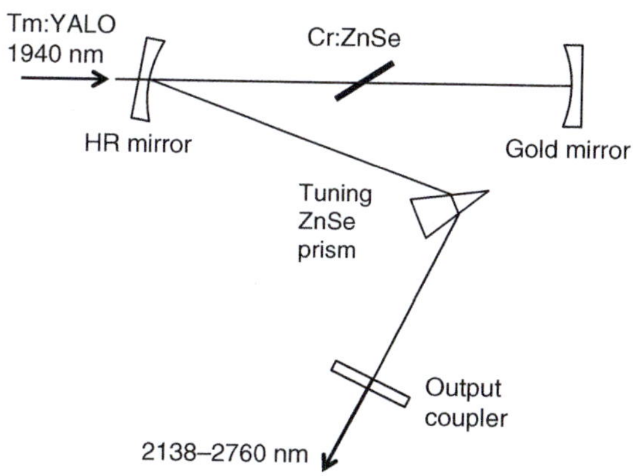

Figure 2.13 Simplified schematic of the tunable Cr:ZnSe laser of [38]. The laser was pumped by a 1-W Tm:YALO laser at 1940 nm. HR, high reflector. *Source:* reproduced with permission of OSA, The Optical Society.

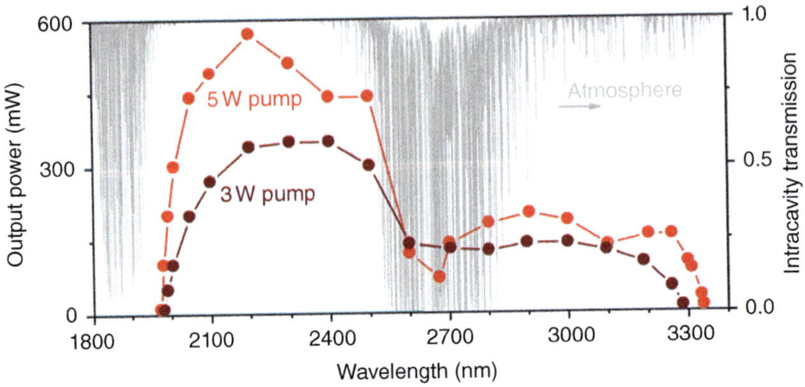

Figure 2.14 Tunability range of a CW Cr:ZnSe laser. Transmission of the atmosphere is shown in the background. *Source:* reproduced from figure 2 of [39], with permission of OSA, The Optical Society.

With the improved quality of the Cr-doped laser crystals, progress in pump lasers, and in fabrication of broadband mirrors, the tuning range of Cr^{2+} lasers has been dramatically extended. With the 1607-nm pump wavelength from a commercial 5-W Er-fiber laser, and with the intracavity prism, the laser tuning achieved was 1.973–3.349 μm in Cr:ZnSe (Figure 2.14) and 1.962–3.195 μm in Cr:ZnS, with a single set of optics, with up to 600 mW of the output power, and < 10 GHz (<0.2 nm) linewidth [39]. Most recently, a Cr:ZnSe laser system with a continuous tuning of 2.4–3.0 μm and with up to 60-W power was reported [12].

The emission spectrum of a Cr:CdSe-based laser is shifted to longer wavelengths (see Figure 2.11). Using 300-μs pump pulses from a free-running 1.94-μm Tm:YAP laser, a continuous 2.26–3.61 μm tunability of Cr:CdSe laser was achieved using a resonator containing a prism [40]. Without spectral selection, the Cr^{2+}:CdSe laser emitted up to 17 mJ at 2.65 μm with the *quantum* slope efficiency of 63% related to the absorbed pump energy.

Single-frequency performance in the Cr:ZnSe tunable laser was obtained via employing an intracavity diffraction grating in combination with two etalons. Gross tuning in this experiment was obtained by tuning the grating angle, while fine-tuning was achieved by adjusting the cavity length [41]. Fast-tunable single-frequency Cr:ZnSe laser with 150 mW output power was reported in [42]. The laser cavity used a diffraction grating at a grazing incident angle of 75° and a flat tuning mirror, mounted on a piezo-driven fast "shaker" mirror. Rapid (4500 nm/s) single-frequency tunability over 120 nm around $\lambda = 2.5\,\mu m$ was demonstrated with a single-longitudinal-mode linewidth of 120 MHz.

Cr^{2+} lasers can be directly pumped by laser diodes [31], which offers simplicity and compactness. By using (AlGaIn)(AsSb)-based laser diodes operating at

1.9 or 2.0 μm with an output power of ~1 W, efficient CW operation of a Cr^{2+}:ZnSe laser was demonstrated [43]. With the 1.9-μm pump, the output power at $\lambda = 2.5\,\mu m$ was 105 mW and the slope efficiency of 35% with respect to the absorbed pump power. When a Cr^{2+}:CdMnTe laser ($\lambda = 2.6\,\mu m$) was tested under the same conditions, the output power was only 6 mW. This was accounted for by the poor thermal conductivity and strong thermal lensing in CdMnTe. According to the authors of [43], this host material might be restricted to only pulsed or low-power CW operation.

2.2.2.2 High-power Continuous-wave Cr^{2+} Lasers

Cr^{2+}-doped lasers currently serve as efficient (>40%) converters of the outputs of high-power commercial erbium or thulium fiber lasers to the 2–3 μm spectral region, thanks to the high-quality Cr:ZnSe/ZnS structures that were developed recently (see, for example, reviews [32, 44]). A high-power master oscillator–power amplifier (MOPA) CW laser system based on Cr^{2+}:ZnSe was demonstrated by Berry et al. [45]. With a 1.9-μm Tm-fiber laser as a pump, the output power of 14 W was extracted at $\lambda \approx 2.44\,\mu m$. Using a Cr^{2+}:ZnS laser system, Moskalev et al. extracted 10 W of CW output near $\lambda = 2.38\,\mu m$ using a 1560-nm Er-fiber laser as a pump source [46]. The optical efficiency with respect to the incident pump power was 40% with no noticeable output power roll-off at high pump powers. Also, broadband high-power wavelength tunability spanning 1.94–2.78 μm was achieved in the same work, using a diffraction grating in the Littrow configuration. The output power exceeded 5 W within the 2.12–2.59 μm wavelength range, and was >2 W over the whole range of 1.97–2.76 μm [46]. When pumped by 50-W Tm fiber laser at 1908 nm, a Cr^{2+}:ZnS laser delivered more than 20 W of power near its gain maximum at 2.4 μm [44].

One of the most notable practical applications of 2.4 μm lasers is processing of polymer materials (e.g. polypropylene, polycarbonate, and others) that have strong absorption features in this wavelength range [44]. A record high-power CW operation of a Cr:ZnSe laser near the gain maximum of 2.5 μm was reported using a spinning-ring gain medium, which allowed to eliminate thermal lensing effects and obtain 140 W of the output power with the optical-to-optical conversion efficiency of 62% with respect to the 1.91-μm Tm-fiber pump [47].

A CW operation of a Cr^{2+}:CdSe laser was reported by Akimov et al. [48]. When pumped by a 1.908-μm thulium fiber laser, the laser generated 1.07 W at $\lambda = 2.623\,\mu m$ with the quantum slope efficiency of 60% with respect to the absorbed power.

2.2.2.3 High-power Cr^{2+} CW Laser Systems Operating at 2.94 μm

There is a great demand in the biomedical community for lasers operating near 2.94 μm. This wavelength, corresponding to a very strong absorption peak in human tissue (due to high water content), is highly suitable for skin treatment, surgery, and microsurgery. Despite the fact that 2.94 μm is at the wing of the Cr^{2+}:ZnSe gain curve (25% of its maximum), an efficient multi-watt Cr^{2+}:ZnSe system operating at 2.94 μm was reported, thanks to the homogeneous emission line broadening in Cr^{2+} [44]. The laser operation at 2.94 μm was forced by using dielectric mirrors that suppressed lasing at shorter wavelengths. Using a 15-W 1.91-μm Tm-fiber pump laser, the laser produced 5 W of CW output at 2.94 μm. In the MOPA design, this power was boosted to 7.5 W. Furthermore, using a spinning-ring laser gain medium, Moskalev et al. demonstrated as much as 32 W of CW output at $\lambda = 2.94$ μm with an optical-to-optical conversion efficiency of 29% measured with respect to the 1.91-μm pump wavelength [47].

2.2.2.4 Gain-switched High-power Cr^{2+} Lasers

Due to very short energy storage time of Cr-doped chalcogenides (typically 2–10 μs), Q-switched Cr lasers need to be pumped with *pulsed* (duration < 1 μs) sources – in order to maintain high optical conversion efficiency. Carrig et al. [41] reported on a *gain-switched* (instead of Q-switched) regime of a Cr:ZnSe laser, where the laser was pumped by a high-repetition-frequency (40 kHz) Q-switched Tm:YALO laser with ~400 ns pulse duration at 1.94-μm wavelength. With the average Q-switched Tm-laser power of 45 W, the average power of the gain-switched Cr:ZnSe laser was as high as 18.5 W (pulse energy 0.46 mJ) at a $\lambda \approx 2.5$ μm.

Higher pulse energies were obtained in the gain-switched regime at lower repetition rates. With a Q-switched Ho:YAG pump laser operating at 2.09 μm, repetition rate 100 Hz, pulse energy up to 15 mJ, pulse duration 15–50 ns, and pulse energies of up to 3 mJ were demonstrated around $\lambda = 2.43$ μm (bandwidth 0.5 nm) [44]. Wavelength tuning over 2.35–2.96 μm in the gain-switched mode was achieved in the same work using a laser cavity containing a diffraction grating in the Littrow configuration [44]. Under pumping at 2.095 μm (repetition rate 3 Hz, 13 mJ, 90 ns), 3.1 mJ of output energy was demonstrated from a Cr:ZnSe laser at ≈2.47 μm with 52% slope efficiency [49]. Further increase in the Cr:ZnSe laser pulse energy – up to 10 mJ at 2.35 μm – was accomplished by using a Raman shifted Q-switched Nd:YAG laser pump (Stokes line from H_2 gas at 1.906 μm) with a repetition rate of 10 Hz and pulse duration of 7 ns [49]. Finally, pulse energy as high as 52.2 mJ at $\lambda = 2.43$ μm with the optical conversion efficiency of 44.4% was reported in a Cr:ZnSe oscillator–amplifier configuration using a three-stage amplifier at a repetition rate of 10 Hz. The system was driven by a diode-pumped Q-switched Tm:YAG laser (130 mJ, 160 ns) operating at 2.01 μm. In addition, the Cr:ZnSe system was tunable within 2.12–2.67-μm range by tuning the Cr:ZnSe oscillator with an acousto-optic filter [50].

2.2.2.5 Microchip Cr^{2+} Lasers

A compact CW Cr:ZnSe laser consisting of only two elements: a flat input mirror and an uncoated rectangular-shape plane-parallel 9-mm-long Cr:ZnSe crystal was reported [32, 42]. The laser crystal was placed in close proximity to the input dielectric mirror (coated for high transmission at the pump and high reflection at the lasing wavelength) with a small (10–100 μm) adjustable air gap such that the input mirror and the input facet of the gain element formed a low-finesse Fabry–Pérot interferometer. By adjusting this gap, it was possible to minimize the intracavity losses. The output facet of the gain element served as an outcoupler with 18% Fresnel reflection. The laser was pumped by a linearly polarized 9-W Er-fiber laser at 1560 nm. The input–output characteristics and the spectrum of this compact laser are shown in Figure 2.15. The laser operated around 2.36 μm (20-nm linewidth) and delivered up to 3 W of the output power at 7.5-W pump, with 41% optical efficiency.

Microchip Cr^{2+} laser operation was also achieved in the gain-switched mode. The laser consisted of just one element – a plane-parallel 5.5-mm-long ZnSe crystal with uncoated faces, such that Fresnel reflections served as a feedback, and was pumped at 1.56 μm using a D_2 (gas-phase deuterium) Raman-shifted output of an Nd:YAG laser operating at 10 Hz with pulse duration of 5 ns and 1.5-mm beam diameter. The output pulse energy of 1 mJ was obtained at $\lambda \approx 2.28$ μm (20-nm linewidth) with the pump pulse energy of 24 mJ [51].

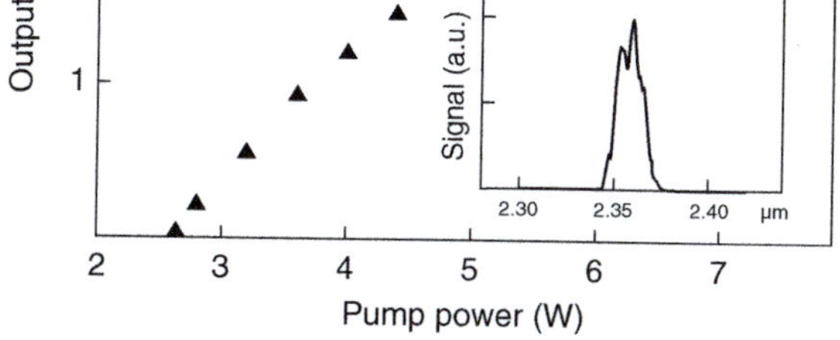

Figure 2.15 Output–input dependence of a CW microchip Cr:ZnSe laser pumped by an Er-fiber laser at 1560 nm. The inset shows the output spectrum. *Source:* reproduced from [32], with permission of Wiley.

2.2.2.6 Waveguide and Thin-disk Cr:ZnSe Lasers

Waveguide structures represent a solution for creating low-threshold lasers. Furthermore, the effects of thermal lensing that typically limit the maximum average power in such lasers may be dramatically reduced due to a guided-wave geometry. A Cr:ZnSe waveguide laser operating at 2.5 μm has been demonstrated, where inscription with an ultrafast laser was used to fabricate a buried channel waveguide (so-called depressed-index cladding waveguide) in a polycrystalline Cr:ZnSe sample [52, 53]. Channel waveguides were created via nonlinear absorption that resulted in localized structural modifications and, as a result, change in the refractive index. By using tightly focused 750-fs pulses from an Yb-fiber MOPA system at $\lambda = 1047$ nm, with μJ-level pulse energies at 100 kHz repetition rate, multiple elements of reduced refractive index were inscribed, forming a circular cross section as shown in Figure 2.16. The waveguides were typically 6-mm long, with a diameter of 40, 60, 80, 120, 160, and 200 μm. When pumped by a Tm-fiber laser at 1928 nm, the maximum power output of 285 mW was achieved with 1.1 W of pump power, with slope efficiency as high as 45% [53]. With an elevated pump power of 9.3 W, the maximum 2.5-μm output power was 1.7 W [54].

A thin-disk operation of a Cr^{2+} laser is a promising method for high average power scaling. The authors of [55] showed that the face-cooled disk laser can effectively reduce thermal effects in Cr lasers and demonstrated 4.2 W average power from a Cr^{2+}:ZnSe laser operating in the gain-switched regime. The laser disk was 1 mm thick and was pumped with 9-W laser power at $\lambda = 2.05$ μm in a multipass pump geometry.

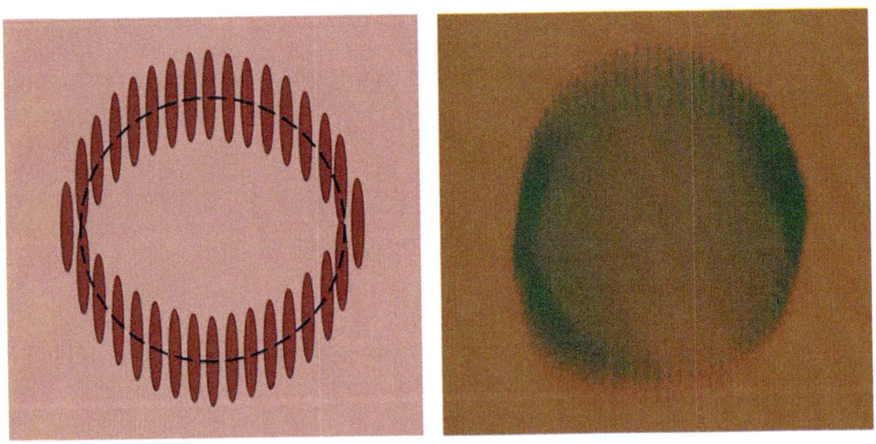

Figure 2.16 Cross section of a depressed-index cladding waveguide in Cr:ZnSe. Design (left) and a microscope image of a 120-μm diameter sample (right). The cladding is created by inscribing multiple elongated (in the normal to the page direction) elements with reduced refractive index by a subpicosecond 1.047-μm laser. The waveguide laser beam exits normal to the page. *Source:* reproduced from figure 1 of [54], with permission of OSA, The Optical Society.

2.2.2.7 Mode-locked Cr:ZnS/Cr:ZnSe Lasers

Because of their extra-large (over 1300 nm) gain bandwidth, Cr:ZnS and Cr:ZnSe lasers are especially suitable for generating ultrashort mid-IR pulses, with pulse durations down to two optical cycles (~15 fs). In their very first demonstration of a mode-locked Cr^{2+}:ZnSe laser ($\lambda \approx 2.5\,\mu m$), with the active (acousto-optic) mode-locking, Carrig et al. produced 4.4-ps transform-limited pulses with 82 mW of average power [56]. This was followed by demonstration of passive mode-locking using semiconductor saturable absorber mirrors (SESAMs) [57–60], as well as Kerr-lens mode-locking (KLM) [61–63].

Using Cr^{2+}:ZnS as an active medium, 70-fs pulses were demonstrated from a KLM oscillator (center wavelength 2.39 μm) with the output power of 1 W and 20% optical conversion efficiency with respect to an Er-fiber pump [63, 64]. Subsequently, a KLM Cr^{2+}:ZnS laser (center wavelength 2.3 μm) with the average power of 2 W and pulse duration of 67 fs was reported [44]. Cr-doped ZnS and ZnSe active media have very similar spectroscopic and laser parameters. However, according to [44], ZnS host is better suited for high-power output because of its higher thermal conductivity, higher thermal shock parameter, and lower thermal lensing.

Cizmeciyan et al. used graphene as a fast saturable absorber inside a laser cavity to attain passive mode-locking in a Cr:ZnSe laser centered at 2.5 μm. The laser generated a stable train of 176-fs bandwidth-limited pulses at a pulse repetition rate of 78 MHz with 185 mW of the average power [65, 66]. Using graphene that was deposited directly onto the high reflector end mirror (Figure 2.17) of a Cr^{2+}:ZnS laser, pulses as short as 41 fs in duration (~5 optical cycles) were reported with the laser spectrum centered at 2.4 μm (spectral bandwidth 190 nm). The average output power was 250 mW at 108 MHz repetition rate [67].

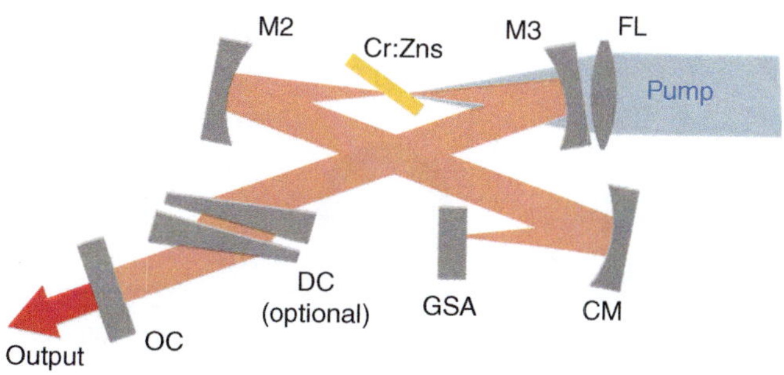

Figure 2.17 X-folded resonator of the graphene mode-locked Cr:ZnS laser producing 41-fs pulses at 2.4 μm. The pump beam of the Er-fiber laser at 1.61 μm was directed onto the laser crystal using a focusing lens FL. M2 and M3, highly reflective concave mirrors; GSA, graphene-based saturable absorber mirror; DC, YAG wedge pair for dispersion compensation; OC, output coupler; CM, chirped high-reflectivity concave mirror for the group-delay dispersion compensation. *Source:* reproduced from figure 1 of [67], with permission of OSA, The Optical Society.

Thanks to the simplicity of KLM and the availability of the off-the-shelf erbium and thulium fiber lasers for pumping (optical-to-optical conversion efficiency can be as high as 37%), there has been speedy progress, over the past few years, in the development of ultrafast Cr:ZnS and Cr:ZnSe lasers [12, 68–74]. Also, demonstration of KLM lasers based on *polycrystalline* Cr:ZnS and Cr:ZnSe [68] has led to considerable improvements in terms of average power, pulse energy, and pulse duration. The pulse width of current Cr:ZnSe and Cr:ZnS oscillators is predominantly limited by the dielectric coatings that provide dispersion control.

By control of the second- and third-order dispersion in the cavity of a polycrystalline Cr^{2+}:ZnS laser, Vasilyev et al. obtained sub-30-fs pulses at 2.4 μm with the emission spectrum spanning over 950 nm [69]. Figure 2.18 illustrates the effects of the peak power inside a KLM laser cavity on its spectral and temporal characteristics. With an output coupler with low (R_{OC} = 40%) reflectivity, a smooth spectrum with 16-THz FWHM bandwidth (dashed curve in Figure 2.18a) was obtained with 21-nJ output pulse energy and 40-fs pulse duration. The corresponding intracavity peak power was 1.5 MW. With R_{OC} = 90%, the output pulse energy decreased to 6.3 nJ, but an increased intracavity peak power (2.5 MW) resulted in strong broadening of the spectrum to 31 THz (solid curve in Figure 2.18a) and shorter pulse duration of 26 fs.

Overall, KLM oscillators based on Cr^{2+}-doped gain media can operate at repetition rates from below 80 MHz to above 1 GHz (see Table 2.2).

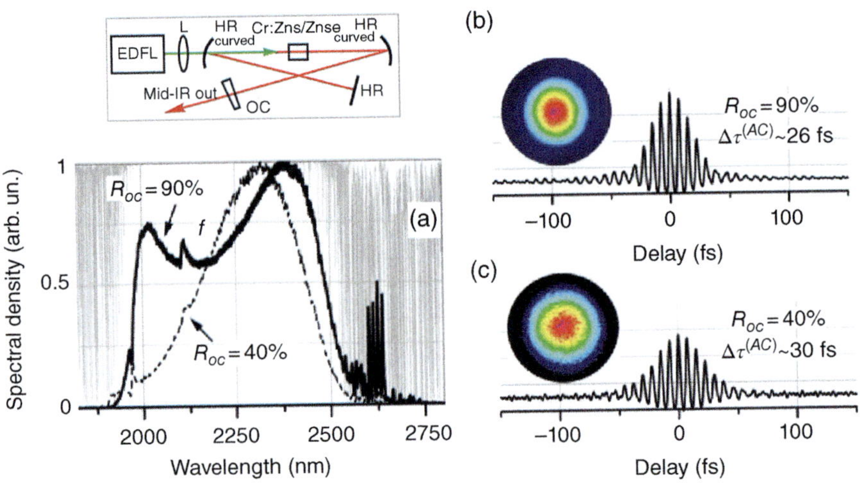

Figure 2.18 (a) Two spectra for a Cr:ZnS KLM oscillator with 84-MHz repetition rate – for different output couplers with reflectivity (R_{OC}) of 90% and 40% that provide different peak power levels inside the resonator (2.5 and 1.5 MW, respectively) and different pulse durations. Shown in gray is transmission of 1-m standard air. (b, c) Second-order autocorrelation traces corresponding to different R_{OC}. The insets show the laser schematic and measured beam profiles. *Source:* reproduced from figure 4 of [74], with permission of OSA, The Optical Society.

Table 2.2 Summary of lasers based on Cr^{2+}-doped chalcogenide crystals.

Gain medium	Laser operating mode	Output parameters	Ref.
Cr:ZnSe	CW	Output power 12 W @ 2.425 μm	[75]
Cr:ZnSe	CW, MOPA	Output power 14 W @ 2.45 μm	[45]
Cr:ZnSe	CW, spinning-ring gain medium	Output power 140 W @ 2.5 μm; 32 W @ 2.94 μm	[47]
Cr:ZnSe	CW, tunable	Range 1.97–3.35 nm	[40]
Cr:ZnSe	CW, single frequency	Linewidth <1 MHz at >1 W; range 1.9–3.0 μm	http://www.ipgphotonics.com/Mid_ir_lasers.htm
Cr:ZnSe	Gain-switched	Average power 18.5 W @ 10 kHz	[41]
Cr:ZnSe	Gain-switched	Pulse energy 10 mJ, 5 ns, 10 Hz	[49]
Cr:ZnSe	Gain-switched, MOPA	Pulse energy 52 mJ, 160 ns, 10 Hz	[50]
Cr:ZnSe	Graphene mode-locked	176 fs, 185 mW, 78 MHz, 2.5 μm	[67]
Cr:ZnSe	Cr:ZnS/Cr:ZnSe MOPA, spinning-ring amplifier	60 fs, 27.5 W, 81 MHz, 2.4 μm	[76]
Cr:ZnSe	MOPA, chirped-pulse amplification	Pulse energy 1 mJ, 184 fs, 1 kHz, 2.5 μm, peak power 5 GW	[77]
Cr:ZnS	CW	Output power 10 W @ 2.38 μm	[45]
Cr:ZnS	CW, MOPA	Output power 30 W @ 2.4 μm	[44]
Cr:ZnS	CW, tunable	Tuning range 1.962–3.195 μm	[39]
Cr:ZnS	Graphene mode-locked	41 fs, 250 mW, 108 MHz, 2.4 μm	[67]
Cr:ZnS	KLM (Kerr-lens mode-locked)	40 fs, 1.7 W, 79 MHz, 2.38 μm	[72]
Cr:ZnS	KLM	67 fs, 2 W (10 W pump), 95 MHz, 2.3 μm	[44]
Cr:ZnS	KLM	29 fs, 400 mW, 100 MHz, 2.4 μm	[69]
Cr:ZnS	KLM	125 fs, 800 mW, 1.06 GHz, 2.35 μm	[74]
Cr:ZnS	KLM	50 fs, 120 mW, 1.2 GHz, 2.4 μm	[71]

(*Continued*)

Table 2.2 (Continued)

Gain medium	Laser operating mode	Output parameters	Ref.
Cr:ZnS	KLM, MOPA	27 fs, 7 W, 79 MHz, 2.38 μm	[72]
Cr:CdS	Free-running, pulsed, tunable	Tuning range 2.18–3.32 μm, single pulses	[78]
Cr:CdSe	Free-running, pulsed, tunable	Tuning range 2.26–3.61 μm, single pulses	[40]

A simple and robust MOPA arrangement permitted access to even shorter pulses with broader instantaneous spectra. A favorable interplay between nonlinear interactions and dispersion inside the amplifier gain element allowed simultaneous amplification, spectral broadening, and compression of femtosecond pulses (all in one pass through the amplifier). For instance, when 40-fs pulses of a 1.7-W, 80-MHz master oscillator with central wavelength 2.4 μm were amplified to the average power of 7.1 W, they were simultaneously compressed to 27 fs, which corresponds to the peak power of 3 MW [72]. Using the same strategy, pulses with 22-fs duration and with the spectral span as broad as 1.78–2.97 μm were generated [12].

Furthermore, using spinning-ring amplifiers, the average power of such femtosecond sources can be boosted to tens of watts [76].

Finally, Slobodchikov et al. demonstrated energy scaling of femtosecond Cr^{2+}:ZnSe laser pulses to a millijoule level [79]. The seed pulses were provided by a mode-locked laser oscillator that was pumped by a CW Tm-fiber (1940 nm) laser. By using chirped-pulse amplification (CPA) in a regenerative amplifier, a technology similar to the one used with Ti:sapphire lasers, the researchers generated energetic Cr:ZnSe laser pulses centered at 2.48-μm at 1-kHz pulse repetition rate. After pulse compression, they obtained 300-fs pulses with 0.3 mJ energy, corresponding to 1 GW peak power. Key to the system were (i) a chirped volume Bragg grating (CVBG) element to stretch the oscillator pulse before amplification and (ii) a Q-switched Ho:YLF laser (2050 nm) providing 11-mJ pulses for pumping the regenerative amplifier [79]. Most recently, a 1-kHz Cr:ZnSe-based femtosecond laser system with 1-mJ pulse energy, 184-fs pulse duration, and a peak power of 5 GW at $\lambda \approx 2.5$ μm has been reported [77].

The main results for Cr^{2+} II–IV lasers are presented in Table 2.2.

2.2.3 Lasers Based on Chalcogenide Crystals Doped with Fe^{2+}

2.2.3.1 Free-running Pulsed Fe:ZnSe/ZnS Lasers

The first laser action with the Fe^{2+} ion doped into a chalcogenide crystal was demonstrated by Adams et al. in Fe:ZnSe at cryogenic temperatures (15–180 K) using, as pump, a 2.7-μm Er:YAG laser operating in a free-running pulsed mode with 48-μs pulse duration [80]. The output wavelength of the Fe:ZnSe laser was tuned with temperature from 3.98 μm at 15 K to 4.54 μm at 180 K. The maximum energy per pulse of 12 μJ was produced at T = 130 K and the maximum laser slope efficiency of 8.2% was observed at 150 K [80]. Later, the lasing characteristics of the Fe^{2+}:ZnSe laser were studied in a broader (85–255 K) temperature range using, as pump, a free-running flashlamp-pumped Er:YAG laser at λ = 2.94 μm with 750 mJ output energy and a 200-μs pulse duration [81, 82]. For the maximum pump energy of 733 mJ (absorbed pump energy 470 mJ), the authors obtained the output energy of 187 mJ at λ = 4.1 μm at T = 85 K. That corresponds to the slope efficiency of 43% and quantum efficiency of 59% (with respect to the absorbed pump energy). The former value dramatically dropped (to 9%) when the temperature was elevated to 255 K.

Most recent results show that Fe:ZnSe lasers operating in the long-pulse free-running regime with pulse duration of hundreds of microseconds can generate joule-level pulses in the 3.5–5 μm wavelength range. Fedorov et al. [83] reported a pulsed Fe:ZnSe laser operating at T = 77–250 K. The laser gain element, mounted in a liquid nitrogen cryostat, was pumped by a free-running (250-μs pulse duration) flashlamp-pumped Er:YAG (2.94 μm) or Er:YSGG (2.78 μm) laser at a repetition rate of 11 Hz. At T = 77 K, the authors demonstrated the output energy of 0.5 J at 4.16 μm, with 32% optical conversion efficiency. Only a small decline (<30%) in output energy was observed when the operating temperature was raised to T = 200 K. The laser wavelength was tuned with temperature from 4.16 μm (at 77 K) to 4.65 μm (at 250 K). With an intracavity CaF_2 prism, the tuning range was extended to 3.9–5.1 μm with the linewidth below 3 nm [83]. Frolov et al. further scaled the output energy of the Fe:ZnSe laser to achieve 2.1-J pulses at T = 85 K under irradiation by 750-μs-long pump pulses with 8-J energy from an Er:YAG laser [84]. The same team also developed a thermoelectrically cooled (T = 220 K) Fe:ZnSe laser with a free-running Er:YAG laser pump with an energy of up to 27 J per pulse and pulse duration of 700 μs. The laser delivered output pulses at 4.3 μm with the energy of 7.5 J and the optical conversion efficiency of 30% in single-shot (1 pulse/minute) operation. By using an intracavity prism, a continuous 3.75–4.82-μm tuning was obtained with up to 3.1-J output pulse energy (at 12 J of pump) [29]. (See Table 2.3 for the summary of results.)

A high-average-power free-running pulsed Fe:ZnSe laser was reported by Mirov et al. [91]. As a pump, the authors used four free-running 2.94-μm Er:YAG lasers with the maximum repetition rate of 100 Hz. The laser gain was provided by two antireflection coated Fe^{2+}-doped ZnSe elements in the form of 2-mm-thick rectangular slabs mounted in the liquid nitrogen cryostat at 77 K. The best results were obtained in a flat–flat laser cavity with a 70% output coupler. At a 100-Hz repetition rate and 1.2 J of pumping energy, the maximum output average power reached 35 W with a center wavelength of 4.15 μm and a pulse duration of 150 μs. By adding an intracavity YAG prism (Figure 2.19), it became possible to tune the laser from 3.88 to 4.17 μm with the maximum output power reaching 23 W.

Table 2.3 Summary of lasers based on Fe^{2+}-doped chalcogenide crystals.

Gain medium	Laser operating mode	Output parameters	Ref.
Fe:ZnSe	CW, pump 2.6–2.9 μm, T = 140 K	Output power 1.6 W @ 4.1 μm	[44]
Fe:ZnSe	CW, tunable, pump 2.6–3.1 μm, T = 78–170 K	Tuning range 3.74–4.95 μm	[83]
Fe:ZnSe	CW, pump 2.94 μm, T = 77 K	Output power 9.2 W @ 4.15 μm	[85]
Fe:ZnSe	CW, tunable, pump 2.6–2.9 μm, T = 140 K	Tuning range 3.65–4.8 μm	[44]
Fe:ZnSe	Gain-switched, high pulse energy, pump 2.6–3.1 μm, T = 292 K	Pulse energy 192 mJ (50 ns) @ 4.6 μm	[86]
Fe:ZnSe	Gain-switched, high pulse energy, 20 Hz, pump 2.6–3.1 μm, T = 292 K	Pulse energy 1.6 J (250 ns) @ 4.6 μm, average power 20 W	[87]
Fe:ZnSe	Gain-switched, tunable, pump 2.94 μm, T = 292 K	Tuning range 3.95–5.05 μm	[81]
Fe:ZnS	Gain-switched, tunable, pump 2.94 μm, T = 292 K	Tuning range 3.49–4.65 μm	[88]
Fe:ZnTe	Gain-switched, tunable, pump 2.94 μm, T = 292 K	Tuning range 4.35–5.45 nm	[89]
Fe:CdSe	Gain-switched, tunable, pump 2.94 μm, T = 292 K	Tuning range 4.6–5.9 μm	[90]
Fe:ZnSe	Free-running pulsed, pump 2.94 μm, T = 85 K	Pulse energy 2.1 J @ 4.1 μm	[84]
Fe:ZnSe	Free-running pulsed, pump 2.94 μm, 1 pulse/min, T = 220 K	Pulse energy 7.5 J @ 4.3 μm	[29]
Fe:ZnSe	Free-running pulsed, tunable, T = 220 K	Tuning range 3.75–4.82 μm	[29]
Fe:ZnSe	Pulse-periodic, 100 Hz, 150 μs, pump 2.94 μm, T = 77 K	Average power 35 W at 4.1 μm	[91]
Fe:ZnSe	Pulse-periodic, tunable, 100 Hz, 150 μs, pump 2.94 μm, T = 77 K	Tuning range 3.88–4.17 μm	[91]

2.2.3.2 Gain-switched Regime of Fe^{2+} Lasers at Room Temperature

Despite a very short upper-state lifetime caused by thermally induced nonradiative decay in Fe^{2+}: II–VI at RT – e.g. 370 ns in Fe:ZnSe – those crystals can be used for RT lasing in the gain-switched regime. Moreover, due to a big emission cross section, the laser effect can be observed at RT without any coatings, just due to a feedback provided by Fresnel reflections ($R \sim 17\%$) from the laser

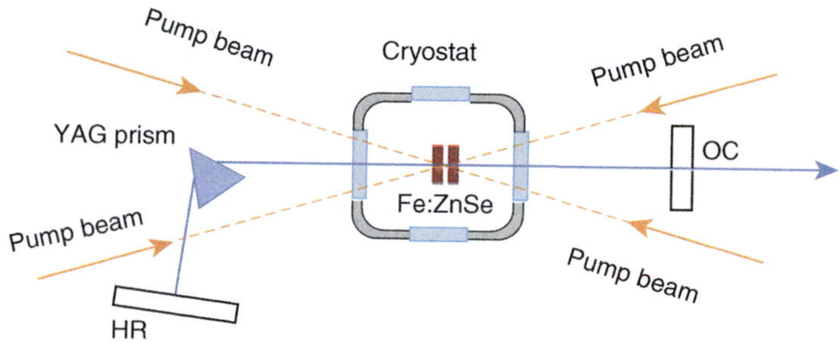

Figure 2.19 Schematic of the tunable high-average-power pulsed Fe:ZnSe laser. HR, high reflector; OC, output coupler with 70% reflectivity. *Source:* reproduced from figure 3 of [91], with permission of OSA, The Optical Society.

crystal surfaces. Gain-switched Fe^{2+}:ZnSe laser oscillation at RT was achieved using nanosecond pump pulses from a 2.92-μm Raman-shifted (the second Stokes in D_2 [deuterium gas]) Nd:YAG laser [92] or from a passively Q-switched 2.94-μm Er:YAG laser [81]. In the latter case, with the pump pulse energy 10 mJ and pump pulse duration of 60 ns, the laser oscillation at 4.4 μm was obtained with the output energy of 0.37 mJ and slope efficiency of 13%. Under stronger laser excitation using a Q-switched 2.8-μm Er, Cr:YSGG laser (pump energy 33 mJ, pulse duration 20 ns, repetition rate 6.7 Hz), the maximum output energy of a Fe^{2+}:ZnSe laser reached 4.7 mJ at 4.3 μm (T = 236 K) and 3.6 mJ at 4.37 μm (300 K). With an increase of temperature from T = 236 K to 300 K, the laser threshold (~8 mJ) was practically unchanged and the slope efficiency decreased only slightly, from 19 to 16% [93]. Both Fe^{2+}:ZnS (λ = 3.98 μm) and Fe^{2+}:ZnSe (λ = 4.36 μm) lasers also operated in the high repetition rate (~1 kHz) gain-switched regime at RT with 0.6-mJ pulse energy [44]. Both lasers were pumped by a gain-switched Cr^{2+}:ZnSe laser (pumped in turn by a Ho:YAG laser) tuned to 2.7–2.9 μm – close to the Fe^{2+} absorption peak.

With pumping by an electric-discharge HF laser (λ = 2.6–3.1 μm, pulse duration 125 ns) in the transverse geometry, the output energy of a gain-switched Fe:ZnSe laser operating at RT at λ = 4.6 μm was scaled to 30.6 mJ [94]. When the longitudinal pumping geometry was used with pulse duration of 50 ns, the output energy of up to 192 mJ per pulse was reached, with the slope efficiency of 34% [86]. Subsequently, Velikanov et al. achieved further energy scaling of a RT Fe:ZnSe laser using 250-ns HF laser excitation pulses. At the repetition rate of 20 Hz, the average laser power reached 20 W (1 J per pulse) with 40% efficiency with respect to the absorbed pump laser power [87].

2.2.3.3 Continuous-wave Fe^{2+} Lasers

The first continuous-wave lasing in the Fe^{2+}:ZnSe laser medium was reported at liquid nitrogen temperature (T = 77 K) [95]. The pump was a 0.6-W CW Cr^{2+}:CdSe laser tuned to a wavelength of 2.97 μm. The output power of the

Fe^{2+}:ZnSe laser centered at 4.1 μm was 160 mW, corresponding to 56% slope efficiency with respect to the absorbed pump power. Taking into account the quantum defect (the ratio of photon energies for the laser and for the pump), this corresponds to a quantum slope efficiency of as high as 76%. CW laser oscillation in Fe:ZnSe at T = 77 K was also demonstrated using a 2.94-μm Er:YAG pump. The laser crystal was pumped from the two ends by two 1.5-W Er:YAG lasers from Sheaumann Laser, Inc. The output power achieved was 840 mW with the center wavelength at 4.14 μm and the linewidth 80 nm. The beam quality was measured to be $M^2 \leq 1.2$ and the maximum slope efficiency was 47% [96].

Very often, highly efficient Cr:ZnSe lasers serve as a pump for Fe^{2+}-based lasers. With 5.5 W of CW pump power provided by a Cr:ZnSe laser tunable over 2.6–2.9 μm, Mirov et al. demonstrated 1.6 W of a CW output power from a 4.1-μm Fe:ZnSe laser operating at T = 140 K [44]. With a diffraction grating in the autocollimation regime inside the laser cavity, the authors demonstrated a tuning range of 3.65–4.8 μm (Figure 2.20) with hundreds of milliwatts of CW power [44]. Vasilyev et al. demonstrated a high-power CW Fe:ZnSe laser (T = 77 K) that operated at 4.15-μm center wavelength and was pumped with a 25-W CW Cr:ZnSe laser tuned to λ = 2.94 μm. The laser output power reached 9.2 W with the slope efficiency of 41% [85].

CW operation of a Fe^{2+} laser at RT has not been achieved so far because of the strong drop in the upper-level lifetime. Nevertheless, a Fe:ZnSe laser can operate in the CW mode at $T \approx 220$ K; this can be provided by a two-stage thermoelectric cooler [90].

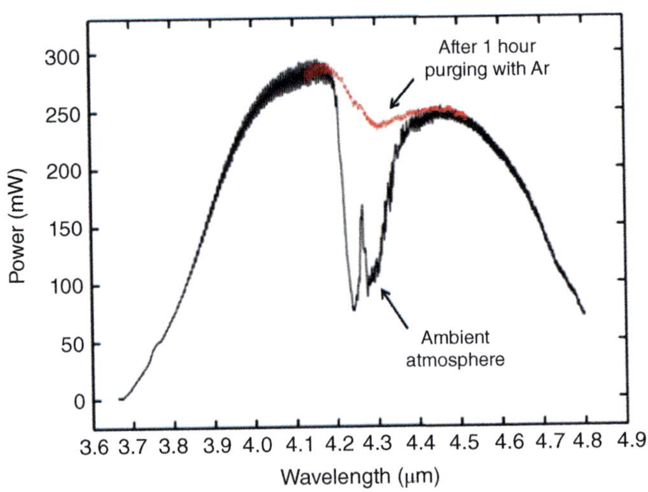

Figure 2.20 Tunable output of a CW Fe:ZnSe laser operating at T = 140 K. Carbon dioxide absorption feature can be seen due to absorption of the ambient atmosphere before purging. *Source:* reproduced from figure 27 of [44], with permission of IEEE.

2.2.3.4 Tunable Fe^{2+} Lasers at Room Temperature

Several groups reported broadband tunable RT operation of Fe^{2+}-based lasers in the gain-switched pulsed mode, using a variety of chalcogenide hosts (ZnSe, ZnS, ZnTe, and CdSe). Typically, nanosecond pulses from a Q-switched Er:YAG laser (λ = 2.94 μm, pulse duration 40–60 ns, pulse energy 10–30 mJ) were used as a pump source, and the output spectrum of the laser was tuned using an intracavity prism made of a mid-IR transparent material (e.g. CaF_2). A continuous tuning range from 3.95 to 5.05 μm (50-nm linewidth) was achieved in a gain-switched Fe:ZnSe laser [81], tuning of 3.49–4.65 μm was obtained in a gain-switched Fe:ZnS laser [88], 4.35–5.45 μm tuning – in a gain-switched Fe:ZnTe system [89], and the most longwave tuning of 4.6–5.9 μm was attained from a gain-switched Fe:CdSe laser [91].

2.2.3.5 Ultrafast Amplifier in the 3.8–4.8 μm Range

Finally, efficient amplification of tunable (3.8–4.8 μm) mid-IR femtosecond pulses from an $AgGaS_2$-based optical parametric source to a gigawatt level in a multi-pass Fe^{2+}:ZnSe amplifier has been demonstrated [97]. An Fe:ZnSe crystal was optically pumped by nanosecond pulses from a Q-switched Cr, Yb, Ho:YSGG laser at 2.85 μm. With a total amplifier gain of $G = 2000$ and an input pulse energy of 40 nJ, the output pulse energy reached 80 μJ. At a pulse duration of 200 fs, this corresponds to 0.4-GW peak power.

The main results for the Fe^{2+} II–IV lasers are presented in Table 2.3.

2.3 Summary

Mid-IR crystalline lasers based on rare-earth Tm^{3+}, Ho^{3+}, and Er^{3+} dopants operate both in the pulsed and CW mode at RT and deliver >100 W (Tm^{3+}), >35 W (Ho^{3+}), and >3 W (Er^{3+}) of average power. (One can immediately notice that there is a strong decline of the achievable average power as one goes to longer wavelengths.) Thulium and erbium lasers can be conveniently pumped by laser diodes, while holmium lasers are typically pumped by thulium lasers (including fiber-based thulium lasers), with a very small quantum defect and high (>50%) conversion efficiency. Both Tm^{3+} and Er^{3+} lasers take advantage of the "two-for-one" cross-relaxation mechanism. There has been a remarkable progress in the development of transition-metal lasers based on Cr^{2+} and Fe^{2+} ions that, most typically, use ZnS or ZnSe as a host. The advantages of Cr^{2+} lasers include RT performance with >50% optical conversion efficiency, very broadband tuning, and convenient pumping by erbium or thulium fiber lasers. Fe^{2+}-based lasers complement Cr^{2+} lasers in terms of longer wavelength tunability. They are typically pumped at 2.5–3.3 μm, e.g. by Cr:ZnS/ZnSe or Er:YAG lasers, and can cover the whole 3.5–6 μm spectral range. At RT, they operate only in the pulsed regime, while at temperatures below ~150 K, they can run CW. One important feature of Fe:ZnSe (and its family) is an excellent energy storage capability. For example, at $T < 150\,K$, the upper-state lifetime is on the order of 100 μs, which enables cost-efficient mid-IR amplifiers to achieve high-energy ultrashort pulses at 4–5 μm, e.g. for high-field physics and harmonic generation (see Chapter 7).

Finally, for the useful information on other (less advanced) transition-metal mid-IR lasers (Dy^{2+}, Co^{2+}, and Ni^{2+}), the reader can refer to [15].

References

1 Stoneman, R.C. and Esterowitz, L. (1990). Efficient, broadly tunable, laser-pumped Tm:YAG and Tm:YSGG cw lasers. *Opt. Lett.* 15: 486.

2 Stoneman, R.C. and Esterowitz, L. (1995). Efficient 1.94-μm Tm:YALO laser. *IEEE J. Sel. Topics Quantum Electron.* 1: 78.

3 Honea, E.C., Beach, R.J., Sutton, S.B., Speth, J.A., Mitchell, S.C., Skidmore, J.A., Emanuel, M.A., and Payne, S.A. (1997). 115-W Tm:YAG diode-pumped solid-state laser. *IEEE J. Quantum Electron.* 33: 1592.

4 Wang, C.L., Niu, Y.X., Du, S.F., Zhang, C., Wang, Z.C., Li, F.Q., Xu, J.L., Bo, Y., Peng, Q.J., Cui, D.F., Zhang, J.Y., and Xu, Z.Y. (2013). High-power diode-side-pumped rod Tm:YAG laser at 2.07 μm. *Appl. Opt.* 52: 7494.

5 Suni, P.J.M. and Henderson, S.W. (1991). 1-mJ/pulse Tm:YAG laser pumped by a 3-W diode laser. *Opt. Lett.* 16: 817.

6 Zhang, S., Wang, M., Xu, L., Wang, Y., Tang, Y., Cheng, X., Chen, W., Xu, J., Jiang, B., and Pan, Y. (2011). Efficient Q-switched Tm:YAG ceramic slab laser. *Opt. Express* 19: 727.

7 Payne, S.A., Chase, L.L., Smith, L.K., Kway, W.L., and Krupke, W.F. (1992). Infrared cross-section measurements for crystals doped with Er^{3+}, Tm^{3+}, and Ho^{3+}. *IEEE J. Quantum Electron.* 28: 2619.

8 Fan, T.Y., Huber, G., Byer, R.L., and Mitzscherlich, P. (1988). Spectroscopy and diode laser-pumped operation of Tm, Ho:YAG. *IEEE J. Quantum Electron.* 24: 924.

9 Stoneman, R.C. and Esterowitz, L. (1992). Intracavity-pumped 2.09-μm Ho:YAG laser. *Opt. Lett.* 17: 736.

10 Budni, P.A., Lemons, M.L., Mosto, J.R., and Chicklis, E.P. (2000). High-power/high-brightness diode-pumped 1.9-μm thulium and resonantly pumped 2.1-μm holmium lasers. *IEEE J. Sel. Topics Quantum Electron.* 6: 629.

11 Strauss, H.J., Preussler, D., Esser, M.J.D., Koen, W., Jacobs, C., Collett, O.J.P., and Bollig, C. (2013). 330 mJ single-frequency Ho:YLF slab amplifier. *Opt. Lett.* 38: 1022.

12 Mirov, S.B., Moskalev, I.S., Vasilyev, S., Smolski, V., Fedorov, V.V., Martyshkin, D., Peppers, J., Mirov, M., Dergachev, A., and Gapontsev, V. (2018). Frontiers of mid-IR lasers based on transition metal doped chalcogenides. *IEEE J. Sel. Topics Quantum Electron.* 24: 1601829.

13 Dergachev, A., Armstrong, D., Smith, A., Drake, T.E., and Dubois, M. (2008). High-power, high-energy ZGP OPA pumped by a 2.05-μm Ho:YLF MOPA system. *Proc. SPIE* 6875: 687507.

14 Diening, A. and Kück, S. (2000). Spectroscopy and diode-pumped laser oscillation of Yb^{3+}, Ho^{3+}-doped yttrium scandium gallium garnet. *J. Appl. Phys.* 87: 4063.

15 Sorokina, I.T. (2003). Crystalline mid-infrared lasers. In: *Solid-State Mid-Infrared Laser Sources* (eds. I.T. Sorokina and K.L. Vodopyanov), Topics Appl. Phys., Vol. 89. Berlin: Springer.

16 Jou, Y.C., Shen, C.H., Cheng, M.C., Lin, C.T., and Chen, P.C. (2007). High-power holmium: yttrium-aluminum-garnet laser for percutaneous treatment of large renal stones. *Urology* 69: 22.

17 Visuri, S.R., Gilbert, J.L., Wright, D.D., Wigdor, H.A., and Walsh, J.T. (1996). Shear strength of composite bonded to Er:YAG laser-prepared dentin. *J. Dental Res.* 75: 599.

18 Vodopyanov, K.L., Shori, R., and Stafsudd, O.M. (1998). Generation of Q-switched Er:YAG laser pulses using evanescent wave absorption in ethanol. *Appl. Phys. Lett.* 72: 2211.

19 Vodopyanov, K.L. (1993). Parametric generation of tunable infrared radiation in $ZnGeP_2$ and GaSe pumped at 3 μm. *J. Opt. Soc. Am. B* 10: 1723.

20 Stoneman, R.C. and Esterowitz, L. (1992). Efficient resonantly pumped 2.8-μm Er^{3+}:GSGG laser. *Opt. Lett.* 17: 816.

21 Zhekov, V.I., Lobachev, V.A., Murina, T.M., and Prokhorov, A.M. (1983). Efficient cross-relaxation laser emitting at $\lambda = 2.94\,\mu m$. *Sov. J. Quantum Electron.* 13: 1235.

22 Bagdasarov, Kh.S., Zhekov, V.I., Lobachev, V.A., Murina, T.M., and Prokhorov, A.M. (1983). Steady-state emission from a $Y_3Al_5O_{12}:Er^{3+}$ laser ($\lambda = 2.94\,\mu m$, $T = 300\,K$). *Sov. J. Quantum Electron.* 13: 262.

23 Dinerman, B.J. and Moulton, P.F. (1994). 3-μm CW laser operations in erbium-doped YSGG, GGG, and YAG. *Opt. Lett.* 19: 1143.

24 Jensen, T., Diening, A., Huber, G., and Chai, B.H.T. (1996). Investigation of diode-pumped 2.8-μm $Er:LiYF_4$ lasers with various doping levels. *Opt. Lett.* 21: 585.

25 Chen, D.W., Fincher, C.L., Rose, T.S., Vernon, F.L., and Fields, R.A. (1999). Diode-pumped 1-W continuous-wave Er:YAG 3-μm laser. *Opt. Lett.* 24: 385.

26 Sousa, J.G., Welford, D., and Foster, J. (2010). Efficient 1.5 W CW and 9 mJ quasi-CW TEM_{00} mode operation of a compact diode-laser-pumped 2.94-μm Er:YAG laser. *Proc. SPIE* 7578: 75781E-1.

27 Ziolek, C., Ernst, H., Will, G.F., Lubatschowski, H., Welling, H., and Ertmer, W. (2001). High-repetition-rate, high-average-power, diode-pumped 2.94-μm Er:YAG laser. *Opt. Lett.* 26: 599.

28 Arbabzadah, E.A., Phillips, C.C., and Damzen, M.J. (2013). Free-running and Q-switched operation of a diode pumped Er:YSGG laser at the 3 μm transition. *Appl. Phys. B* 111: 333.

29 Frolov, M.P., Korostelin, Yu.V., Kozlovsky, V.I., Podmar'kov, Yu.P., and Skasyrsky, Ya.K. (2018). High-energy thermoelectrically cooled Fe:ZnSe laser tunable over 3.75–4.82 μm. *Opt. Lett.* 43: 623.
30 DeLoach, L.D., Page, R.H., Wilke, G.D., Payne, S.A., and Krupke, W.F. (1996). Transition metal-doped zinc chalcogenides: spectroscopy and laser demonstration of a new class of gain media. *IEEE J. Quantum Electron.* 32: 885.
31 Page, R.H., Schaffers, K.I., DeLoach, L.D., Wilke, G.D., Patel, F.D., Tassano, J.B., Payne, S.A., Krupke, W.F., Chen, K.T., and Burger, A. (1997). Cr^{2+}-doped zinc chalcogenides as efficient, widely tunable mid-infrared lasers. *IEEE J. Quantum Electron.* 33: 609.
32 Mirov, S., Fedorov, V., Moskalev, I., Martyshkin, D., and Kim, C. (2010). Progress in Cr^{2+} and Fe^{2+} doped mid-IR laser materials. *Laser Photon. Rev.* 4: 21.
33 Myoung, N., Fedorov, V.V., Mirov, S.B., and Wenger, L.E. (2012). Temperature and concentration quenching of mid-IR photoluminescence in iron doped ZnSe and ZnS laser crystals. *J. Luminesc.* 132: 600.
34 Sorokina, I.T. (2004). Cr^{2+}-doped II-VI materials for lasers and nonlinear optics. *Opt. Mater.* 26: 395.
35 Mirov, S.B., Fedorov, V.V., Moskalev, I.S., and Martyshkin, D.V. (2007). Recent progress in transition metal doped II-VI mid-IR lasers. *IEEE J. Sel. Topics Quantum Electron.* 13: 810.
36 Sorokin, E., Naumov, S., and Sorokina, I.T. (2005). Ultrabroadband infrared solid-state lasers. *IEEE J. Sel. Topics Quantum Electron.* 11: 690.
37 Mirov, S.B., Fedorov, V.V., Martyshkin, D.V., Moskalev, I.S., Mirov, M.S., and Gapontsev, V.P. (2011). Progress in mid-IR Cr^{2+} and Fe^{2+} doped II-VI materials and lasers. *Opt. Mater. Express* 1: 898.
38 Wagner, G.J., Carrig, T.J., Page, R.H., Schaffers, K.I., Ndap, J., Ma, X., and Burger, A. (1999). Continuous-wave broadly tunable Cr^{2+}:ZnSe laser. *Opt. Lett.* 24: 19.
39 Sorokin, E., Sorokina, I.T., Mirov, M.S., Fedorov, V.V., Moskalev, I.S., and Mirov, S.B. (2010). Ultrabroad continuous-wave tuning of ceramic Cr:ZnSe and Cr:ZnS lasers, Technical Digest. In: *Advanced Solid State Photonics.* Washington, DC: Optical Society of America, Paper AMC2.
40 Akimov, V.A., Kozlovsky, V.I., Korostelin, Yu.V., Landman, A.I., Podmar'kov, Yu.P., Skasyrsky, Ya.K., and Frolov, M.P. (2008). Efficient pulsed Cr^{2+}:CdSe laser continuously tunable in the spectral range from 2.26 to 3.61 μm. *Quantum Electron.* 38: 205.
41 Carrig, T.J., Wagner, G.J., Alford, W.J., and Zakel, A. (2004). Chromium-doped chalcogenide lasers. *Proc. SPIE* 5460: 74–82.
42 Moskalev, I.S., Fedorov, V.V., and Mirov, S.B. (2008). Tunable, single-frequency, and multi-watt continuous-wave Cr^{2+}:ZnSe lasers. *Opt. Express* 16: 4145.
43 Mond, M., Albrecht, D., Heumann, E., Huber, G., Kück, S., Levchenko, V.I., Yakimovich, V.N., Shcherbitsky, V.G., Kisel, V.E., Kuleshov, N.V., Rattunde, M., Schmitz, J., Kiefer, R., and Wagner, J. (2002). 1.9-μm and 2.0-μm laser diode pumping of Cr^{2+}:ZnSe and Cr^{2+}:CdMnTe. *Opt. Lett.* 27: 1034.

44 Mirov, S., Fedorov, V., Martyshkin, D., Moskalev, I., Mirov, M., and Vasilyev, S. (2015). Progress in mid-IR lasers based on Cr and Fe doped II-VI chalcogenides. *IEEE J. Sel. Topics Quantum Electron.* 21: 1601719.
45 Berry, P.A. and Schepler, K.L. (2010). High-power, widely-tunable Cr^{2+}:ZnSe master oscillator power amplifier systems. *Opt. Express* 18: 15062.
46 Moskalev, I.S., Fedorov, V.V., and Mirov, S.B. (2009). 10-W, pure continuous-wave, polycrystalline Cr^{2+}:ZnS laser. *Opt. Express* 17: 2048.
47 Moskalev, I., Mirov, S., Mirov, M., Vasilyev, S., Smolski, V., Zakrevskiy, A., and Gapontsev, V. (2016). 140 W Cr:ZnSe laser system. *Opt. Express* 24: 21090.
48 Akimov, V.A., Kozlovsky, V.I., Korostelin, Yu.V., Landman, A.I., Podmar'kov, Yu.P., Skasyrsky, Ya.K., and Frolov, M.P. (2007). Efficient cw lasing in a Cr^{2+}:CdSe crystal. *Quantum Electron.* 37: 991.
49 Fedorov, V.V., Moskalev, I.S., Mirov, M.S., Mirov, S.B., Wagner, T.J., Bohn, M.J., Berry, P.A., and Schepler, K.L. (2011). Energy scaling of nanosecond gain-switched Cr^{2+}:ZnSe lasers. *Proc. SPIE* 7912: 79121E.
50 Saito, N., Yumoto, M., Tomida, T., Takagi, U., and Wada, S. (2012). All-solid-state rapidly tunable coherent 6–10 μm light source for lidar environmental sensing. *Proc. SPIE* 8526: 852605.
51 Mirov, S.B., Fedorov, V.V., Graham, K., Moskalev, I.S., Sorokina, I.T., Sorokin, E., Gapontsev, V., Gapontsev, D., Badikov, V.V., and Panyutin, V. (2003). Diode and fibre pumped Cr^{2+}:ZnS mid-infrared external cavity and microchip lasers. *IEEE Optoelectron.* 150: 340.
52 Macdonald, J.R., Beecher, S.J., Berry, P.A., Schepler, K.L., and Kar, A.K. (2013). Compact mid-infrared Cr:ZnSe channel waveguide laser. *Appl. Phys. Lett.* 102: 161110.
53 Macdonald, J.R., Beecher, S.J., Berry, P.A., Brown, G., Schepler, K.L., and Kar, A.K. (2013). Efficient mid-infrared Cr:ZnSe channel waveguide laser operating at 2486 nm. *Opt. Lett.* 38: 2194.
54 Berry, P.A., Macdonald, J.R., Beecher, S.J., McDaniel, S.A., Schepler, K.L., and Kar, A.K. (2013). Fabrication and power scaling of a 1.7 W Cr:ZnSe waveguide laser. *Opt. Mater. Express* 3: 1250.
55 McKay, J.B., Roh, W.B., and Schepler, K.L. (2002). 4.2 W Cr^{2+}:ZnSe face cooled disk laser. In: *Conference on Lasers and Electro-Optics*. Washington, DC: Optical Society of America, Paper CMY3.
56 Carrig, T.J., Wagner, G.J., Sennaroglu, A., Jeong, J.Y., and Pollock, C.R. (2000). Mode-locked Cr^{2+}:ZnSe laser. *Opt. Lett.* 25: 168.
57 Pollock, C., Brilliant, N., Gwin, D., Carrig, T.J., Alford, W.J., Heroux, J.B., Wang, W.I., Vurgaftman, I., and Meyer, J.R. (2005). Mode locked and Q-switched Cr:ZnSe laser using a semiconductor saturable absorbing mirror (SESAM), Technical Digest. In: *Advanced Solid-State Photonics*. Washington, DC: Optical Society of America, Paper TuA6.
58 Sorokina, I.T., Sorokin, E., and Carrig, T. (2006). Femtosecond pulse generation from a SESAM mode-locked Cr:ZnSe laser, Technical Digest. In: *Conference on Lasers and Electro-Optics*. Washington, DC: Optical Society of America, Paper CMQ2.

59 Sorokin, E., Tolstik, N., Schaffers, K.I., and Sorokina, I.T. (2012). Femtosecond SESAM-modelocked Cr:ZnS laser. *Opt. Express* 20: 28947.

60 Bernhardt, B., Sorokin, E., Jacquet, P., Thon, R., Becker, T., Sorokina, I.T., Picqué, N., and Hänsch, T.W. (2010). Mid-infrared dual-comb spectroscopy with 2.4 μm Cr^{2+}:ZnSe femtosecond lasers. *Appl. Phys. B* 100: 3.

61 Cizmeciyan, M.N., Cankaya, H., Kurt, A., and Sennaroglu, A. (2009). Kerr-lens mode-locked femtosecond Cr^{2+}:ZnSe laser at 2420 nm. *Opt. Lett.* 34: 3056.

62 Cizmeciyan, M.N., Cankaya, H., Kurt, A., and Sennaroglu, A. (2012). Operation of femtosecond Kerr-lens mode-locked Cr:ZnSe lasers with different dispersion compensation methods. *Appl. Phys. B* 106: 887.

63 Tolstik, N., Sorokin, E., and Sorokina, I.T. (2013). Kerr-lens mode-locked Cr:ZnS laser. *Opt. Lett.* 38: 299.

64 Sorokin, E., Tolstik, N., and Sorokina, I.T. (2013). 1 W femtosecond mid-IR Cr:ZnS laser. *Proc. SPIE* 8599: 859916.

65 Cizmeciyan, M.N., Kim, J.W., Bae, S., Hong, B.H., Rotermund, F., and Sennaroglu, A. (2013). Graphene mode-locked femtosecond Cr:ZnSe laser at 2500 nm. *Opt. Lett.* 38: 341.

66 Cizmeciyan, M.N., Kim, J.W., Bae, S., Hong, B.H., Rotermund, F., and Sennaroglu, A. (2013). Graphene mode-locked Cr:ZnSe laser, Technical Digest. In: *Advanced Solid-State Lasers*. Washington, DC: Optical Society of America, Paper MW1C.4.

67 Tolstik, N., Sorokin, E., and Sorokina, I.T. (2014). Graphene mode-locked Cr:ZnS laser with 41 fs pulse duration. *Opt. Express* 22: 5564.

68 Vasilyev, S., Mirov, M., and Gapontsev, V. (2014). Kerr-lens mode-locked femtosecond polycrystalline Cr^{2+}:ZnS and Cr^{2+}:ZnSe lasers. *Opt. Express* 22: 5118.

69 Vasilyev, S., Moskalev, I., Mirov, M., Mirov, S., and Gapontsev, V. (2015). Three optical cycle mid-IR Kerr-lens mode-locked polycrystalline Cr^{2+}:ZnS laser. *Opt. Lett.* 40: 5054.

70 Vasilyev, S., Mirov, M., and Gapontsev, V. (2015). Mid-IR Kerr-lens mode-locked polycrystalline Cr^{2+}:ZnS laser with 0.5 MW peak power, Technical Digest (Online). In: *Advanced Solid State Lasers*. Washington, DC: Optical Society of America, Paper AW4A.3.

71 Vasilyev, S., Moskalev, I., Mirov, M., Smolski, V., Mirov, S., and Gapontsev, V. (2016). Kerr-lens mode-locked middle IR polycrystalline Cr:ZnS laser with a repetition rate 1.2 GHz, Technical Digest (Online). In: *Lasers Congress ASSL, LSC, LAC*. Washington, DC: Optical Society of America, Paper AW1A.2.

72 Vasilyev, S., Moskalev, I., Mirov, M., Mirov, S., and Gapontsev, V. (2016). Multi-watt mid-IR femtosecond polycrystalline Cr^{2+}:ZnS and Cr^{2+}:ZnSe laser amplifiers with the spectrum spanning 2.0–2.6 μm. *Opt. Express* 24: 1616.

73 Vasilyev, S., Moskalev, I., Mirov, M., Mirov, S., and Gapontsev, V. (2016). 7 W few-cycle mid-infrared laser source at 79 MHz repetition rate, Technical Digest (Online). In: *High-Brightness Sources and Light-Driven Interactions*. Washington, DC: Optical Society of America, Paper MT2C.4.

74 Vasilyev, S., Moskalev, I., Mirov, M., Smolski, V., Mirov, S., and Gapontsev, V. (2017). Ultrafast middle-IR lasers and amplifiers based on polycrystalline Cr:ZnS and Cr:ZnSe. *Opt. Mater. Express* 7: 2636.

75 Moskalev, I.S., Fedorov, V.V., Mirov, S.B., Berry, P.A., and Schepler, K.L. (2009). 12-W CW polycrystalline Cr^{2+}:ZnSe laser pumped by Tm-fiber laser. In: *Advanced Solid-State Photonics*. Denver, CO: OSA, Paper WB30.

76 Vasilyev, S., Moskalev, I., Smolski, V., Peppers, J., Mirov, M., Mirov, S., and Gapontsev, V. (2018). 27 W middle-IR femtosecond laser system at 2.4 μm, Technical Digest. In: *Advanced Solid State Lasers*. Washington, DC: Optical Society of America, Paper AW3A.1.

77 Slobodchikov, E., Chieffo, L.R., and Wall, K.F. (2016). High peak power ultrafast Cr:ZnSe oscillator and power amplifier. *Proc. SPIE* 9726: 972603.

78 Akimov, V.A., Frolov, M.P., Korostelin, Yu.V., Kozlovsky, V.I., Landman, A.I., Podmar'kov, Yu.P., Skasyrsky, Ya.K., and Voronov, A.A. (2009). Pulsed broadly tunable room-temperature Cr^{2+}:CdS laser. *Appl. Phys. B* 97: 793.

79 Slobodchikov, E. and Moulton, P. (2011). 1-GW-peak-power, Cr:ZnSe laser, Technical Digest. In: *Conference on Lasers and Electro-Optics*. Washington, DC: Optical Society of America, Paper PDPA10.

80 Adams, J.J., Bibeau, C., Page, R.H., Krol, D.M., Furu, L.H., and Payne, S.A. (1999). 4.0–4.5 μm lasing of Fe:ZnSe below 180 K, a new mid-infrared laser material. *Opt. Lett.* 24: 1720.

81 Fedorov, V.V., Mirov, S.B., Gallian, A., Badikov, D.V., Frolov, M.P., Korostelin, Yu.V., Kozlovsky, V.I., Landman, A.I., Podmar'kov, Yu.P., Akimov, V.A., and Voronov, A.A. (2006). 3.77–5.05-μm tunable solid-state lasers based on Fe^{2+}-doped ZnSe crystals operating at low and room temperatures. *IEEE J. Quantum Electron.* 42: 907.

82 Voronov, A.A., Kozlovskii, V.I., Korostelin, Yu.V., Landman, A.I., Podmar'kov, Yu.P., and Frolov, M.P. (2005). Laser parameters of a Fe:ZnSe laser crystal in the 85-255 K temperature range. *Quantum Electron.* 35: 809.

83 Fedorov, V., Martyshkin, D., Mirov, M., Moskalev, I., Vasilyev, S., Peppers, J., Gapontsev, V., and Mirov, S. (2013). Fe-doped binary and ternary II-VI mid-infrared laser materials, Technical Digest. In: *Advanced Solid-State Lasers*. Washington, DC: Optical Society of America, Paper AW1A8.

84 Frolov, M.P., Korostelin, Yu.V., Kozlovsky, V.I., Mislavskii, V.V., Podmar'kov, Yu.P., Savinova, S.A., and Skasyrsky, Ya.K. (2013). Study of a 2-J pulsed Fe:ZnSe 4-μm laser. *Laser Phys. Lett.* 10: 125001.

85 Vasilyev, S., Moskalev, I., Mirov, M., Smolsky, V., Martyshkin, D., Fedorov, V., Mirov, S., and Gapontsev, V. (2017). Progress in Cr and Fe doped ZnS/Se mid-IR CW and femtosecond lasers. *Proc. SPIE* 10193: 101 930 U.

86 Firsov, K.N., Gavrishchuk, E.M., Kazantsev, S.Yu., Kononov, I.G., and Rodin, S.A. (2014). Increasing the radiation energy of ZnSe:Fe^{2+} laser at room temperature. *Laser Phys. Lett.* 11: 085001.

87 Velikanov, S.D., Gavrishchuk, E.M., Zaretsky, N.A., Zakhryapa, A.V., Ikonnikov, V.B., Kazantsev, S.Yu., Kononov, I.G., Maneshkin, A.A.,

Mashkovskii, D.A., Saltykov, E.V., Firsov, K.N., Chuvatkin, R.S., and Yutkin, I.M. (2017). Repetitively pulsed Fe:ZnSe laser with an average output power of 20 W at room temperature of the polycrystalline active element. *Quantum Electron.* 47: 303.

88 Kozlovsky, V.I., Korostelin, Yu.V., Landman, A.I., Mislavskii, V.V., Podmar'kov, Yu.P., Skasyrsky, Ya.K., and Frolov, M.P. (2011). Pulsed Fe^{2+}:ZnS laser continuously tunable in the wavelength range of 3.49–4.65 μm. *Quantum Electron.* 41: 1.

89 Frolov, M.P., Korostelin, Yu.V., Kozlovsky, V.I., Mislavsky, V.V., Podmar'kov, Yu.P., Skasyrsky, Ya.K., and Voronov, A.A. (2011). Laser radiation tunable within the range of 4.35–5.45 μm in a ZnTe crystal doped with Fe^{2+} ions. *J. Russ. Laser Res.* 32: 528.

90 Kozlovsky, V.I., Akimov, V.A., Frolov, M.P., Korostelin, Yu.V., Landman, A.I., Martovitsky, V.P., Mislavskii, V.V., Podmar'kov, Yu.P., Skasyrsky, Ya.K., and Voronov, A.A. (2010). Room-temperature tunable midinfrared lasers on transition-metal doped II-VI compound crystals grown from vapor phase. *Phys. Status Solidi B* 247: 1553.

91 Mirov, S., Fedorov, V., Martyshkin, D., Moskalev, I., Mirov, M., and Vasilyev, S. (2015). High average power Fe:ZnSe and Cr:ZnSe mid-IR solid state lasers, Technical Digest. In: *Advanced Solid State Lasers.* Washington, DC: Optical Society of America, Paper AW4A.1.

92 Kernal, J., Fedorov, V.V., Gallian, A., Mirov, S.B., and Badikov, V.V. (2005). 3.9–4.8 μm gain-switched lasing of Fe:ZnSe at room temperature. *Opt. Express* 13: 10608.

93 Myoung, N., Martyshkin, D.V., Fedorov, V.V., and Mirov, S.B. (2011). Energy scaling of 4.3 μm room temperature Fe:ZnSe laser. *Opt. Lett.* 36: 94.

94 Velikanov, S.D., Danilov, V.P., Zakharov, N.G., Il'ichev, N.N., Kazantsev, S.Yu., Kalinushkin, V.P., Kononov, I.G., Nasibov, A.S., Studenikin, M.I., Pashinin, P.P., Firsov, K.N., Shapkin, P.V., and Shchurov, V.V. (2014). Fe^{2+}:ZnSe laser pumped by a nonchain electric-discharge HF laser at room temperature. *Quantum Electron.* 44: 141.

95 Voronov, A.A., Kozlovsky, V.I., Korostelin, Yu.V., Landman, A.I., Podmar'kov, Yu.P., Skasyrsky, Ya.K., and Frolov, M.P. (2008). A continuous-wave Fe^{2+}:ZnSe laser. *Quantum Electron.* 38: 1113.

96 Evans, J.W., Berry, P.A., and Schepler, K.L. (2012). 840 mW continuous-wave Fe:ZnSe laser operating at 4140 nm. *Opt. Lett.* 37: 5021.

97 Potemkin, F.V., Migal, E.A., Pushkin, A.V., Sirotkin, A.A., Kozlovsky, V.I., Korostelin, Yu.V., Podmar'kov, Yu.P., Firsov, V.V., Frolov, M.P., and Gordienko, V.M. (2016). Mid-IR (4–5 μm) femtosecond multipass amplification of optical parametric seed pulse up to gigawatt level in Fe^{2+}:ZnSe with optical pumping by a solid state 3 μm laser. *Laser Phys. Lett.* 13: 125403.

3

Fiber Mid-IR Lasers

3.1 Introduction

Fiber lasers can produce, in a compact device, outputs with remarkable beam quality and high output power and can operate with long life, high power efficiency, and low cost of ownership. In many cases, fiber lasers are replacing solid-state lasers in research, defense, and industry. Extending the emission wavelength into the mid-infrared is one of the avenues of fiber laser development, which will clearly benefit numerous existing and future applications, e.g. in the field of medicine, remote sensing, welding of polymers, and for pumping longer-wavelength mid-IR or terahertz optical parametric oscillators.

Fiber lasers are typically pumped by laser diodes with free-space or fiber-coupled output (Figure 3.1) and consist of a special optical fiber whose core has been infused with active ions that create the optical gain. For mid-IR fiber lasers, the dopants in the gain medium are rare-earth ions such as thulium (Tm^{3+}), erbium (Er^{3+}), holmium (Ho^{3+}), or dysprosium (Dy^{3+}) [1].

Although the gain media for fiber lasers are similar to those of solid-state bulk lasers, the waveguiding effect and the small effective mode area usually lead to substantially different properties. For example, fiber lasers often operate with much higher laser gain and higher resonator losses. Also, the spatial beam quality in fiber lasers is typically close to diffraction-limited. Similar to solid-state crystalline lasers, the lasing wavelength is determined by the energy levels of the rare-earth cation doped into the core of the fiber.

Figure 3.2 shows the laser transitions responsible for infrared emission of rare-earth-cation-doped fiber lasers, while the emission cross sections of rare-earth laser transitions used in mid-IR fiber lasers as a function of emission wavelength are plotted in Figure 3.3.

Laser-based Mid-infrared Sources and Applications, First Edition. Konstantin L. Vodopyanov.

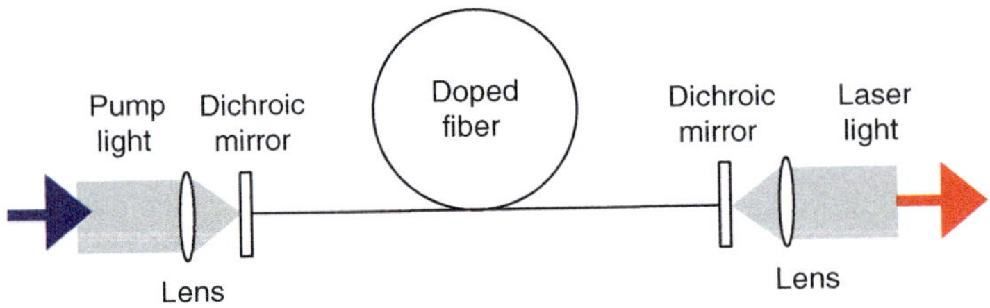

Figure 3.1 Setup of a simple fiber laser. Pump light is launched from the left-hand side through a dichroic mirror into the core of the doped fiber. The generated laser light is extracted on the right-hand side.

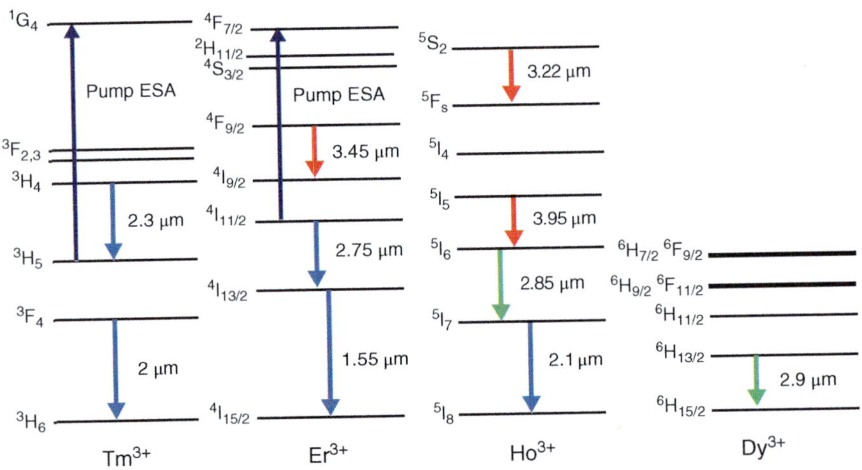

Figure 3.2 Laser transitions of rare-earth cations that produce mid-IR emission wavelengths. ESA, excited state absorption. *Source:* reproduced from figure 3 of [1], with permission of Springer Nature.

3.2 Continuous-wave Mid-IR Fiber Lasers

3.2.1 Tm-based Fiber Lasers

Fiber lasers employing the Tm^{3+} cation [2–5] with emission at around 2 μm are the most powerful, efficient, and developed of the mid-IR fiber lasers. These devices are pumped with robust diode lasers emitting at ~790 nm, and their output can be tuned from 1.86 to 2.09 μm [6]. With high concentrations of Tm^{3+} dopant ions, exceeding 2.5 weight%, cross-relaxation between neighboring Tm^{3+} cations can nearly double the slope efficiency [7]. Similar to the solid-state Tm-based lasers reviewed in Chapter 2, the cross-relaxation between

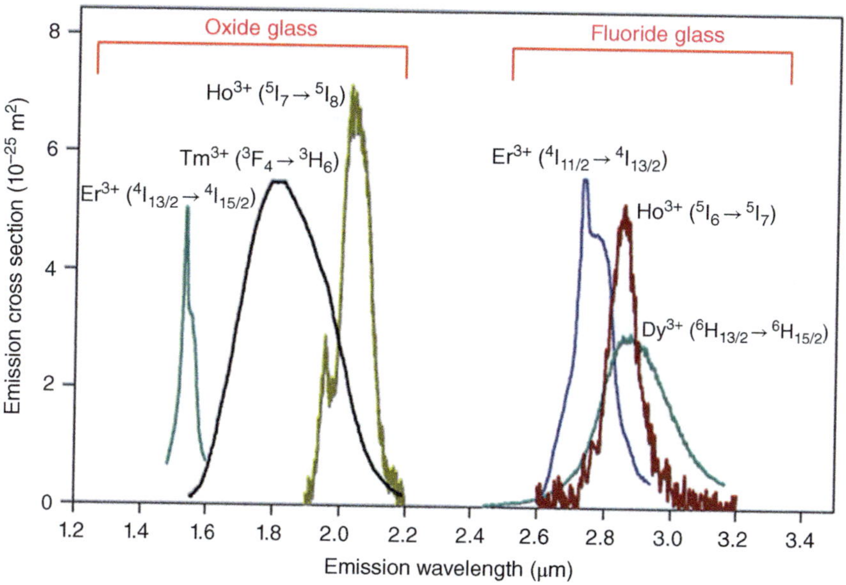

Figure 3.3 The emission cross section of common rare-earth laser transitions used in fiber lasers emitting in the mid-IR, as a function of emission wavelength. The energy levels that participate in each of the laser transitions are shown as labels. The oxide glass here refers to aluminosilicate and the fluoride glass refers to ZBLAN (ZrF_4–BaF_2–LaF_3–AlF_3–NaF composition). *Source:* reproduced from figure 4 of [1], with permission of Springer Nature.

Tm^{3+} cations creates two excited Tm^{3+} cations in the upper laser level for every pump photon excitation (Figure 3.4). This "two-for-one" excitation is resonant in silicate glass and provides an efficient way to generate 2 μm light using a commercial laser diode pump.

Kilowatt-class thulium-doped silica fiber laser with the central wavelength of 2.045 μm was reported in [5, 8]. The layout of such a laser is depicted in Figure 3.5 and is based on a 50-W seed laser at 2.045 μm and a two-stage power amplifier configuration. High-power operation of the Tm-fiber laser has been made possible due to (i) development of high-brightness, fiber-coupled 790-nm pump lasers with nearly 40% electrical-to-optical conversion, (ii) fiber-based pump-coupling optics that provide a highly efficient combining of light from several multimode sources into one fiber, and (iii) "two-for-one" photon recycling due to the cross-relaxation process. The power levels achieved in this work are: >500 W at 62% slope efficiency with respect to diode pump optical power with 6 pump lasers and 1.05 kW at 53% slope efficiency with 12 pump lasers (Figure 3.6). With the 20-μm core diameter of the fiber amplifier, the laser showed single-spatial-mode operation. These results represent the highest continuous wave (CW) power level ever generated in this wavelength range [8].

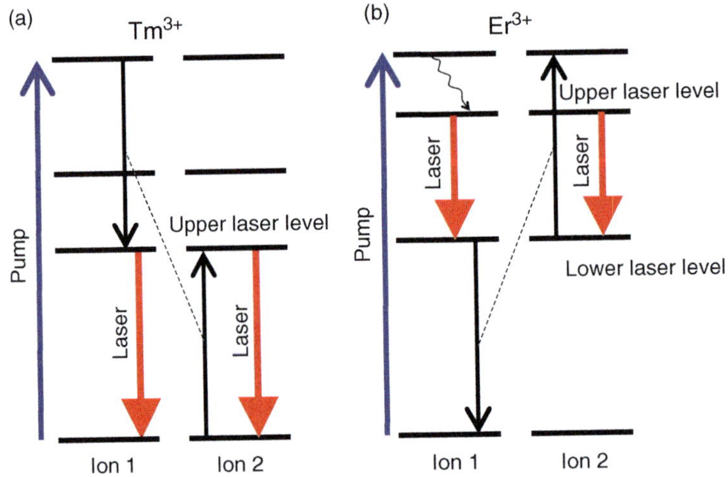

Figure 3.4 Energy-transfer processes between neighboring cations relevant to the operation of 2-μm Tm^{3+} and 3-μm Er^{3+} fiber lasers. (a) Cross-relaxation between Tm^{3+} cations creates two excited Tm^{3+} cations in the upper laser level for every pump photon excitation. (b) Energy transfer between excited Er^{3+} cations forces the de-excitation of one Er^{3+} cation and the further excitation of the other Er^{3+} cation taking part in the process.

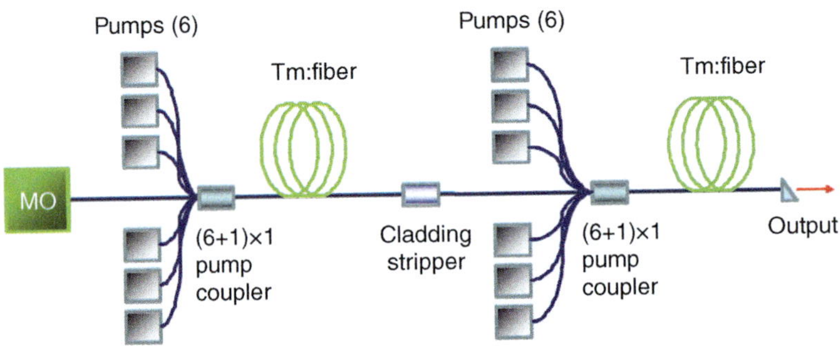

Figure 3.5 Two-stage power amplifier of the 1.05-kW Tm-doped silica fiber laser operating at $\lambda = 2.045\,\mu m$. MO, master oscillator. *Source:* reproduced from [8], with permission of Q-peak, Inc.

The Tm-fiber laser technology is now commercialized. For example, single-spatial-mode Tm-fiber lasers with the wavelength selectable anywhere between 1900 and 2000 nm and with output power in the range of 1–200 W are now standard commercial products offered by IPG Photonics.[1]

1 http://www.ipgphotonics.com.

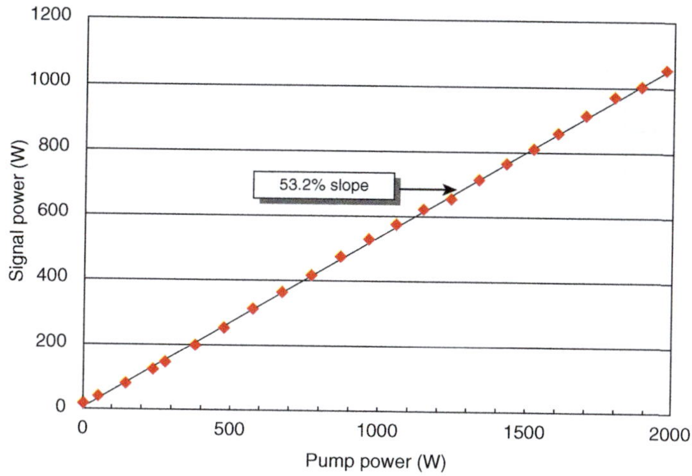

Figure 3.6 Output power versus diode pump optical power of the 1.05-kW single-spatial-mode Tm-doped silica fiber laser at 2.045 μm. *Source:* reproduced from [8], with permission of Q-peak, Inc.

3.2.2 Ho-based Fiber Lasers

The $^5I_7 \rightarrow {}^5I_8$ transition of Ho^{3+} has a peak emission wavelength of 2.1 μm, which overlaps with the important atmospheric transmission window. This Ho transition can be pumped by 1150-nm laser diodes, but it is also well suited for resonant pumping by the output of Tm^{3+}-doped silicate glass fiber lasers, where the quantum defect can be as small as 7% [9].

Alternatively, a very practical solution is a codoped Tm^{3+}, Ho^{3+}-fiber laser, where thulium provides efficient excitation using standard 790-nm laser diodes, and where the resonant energy transfer process efficiently populates the Ho^{3+} ion. With such a fiber, and using a bidirectional 793-nm diode pumping, 83 W of CW output power has recently been achieved at 2.105 μm, with up to 42% slope efficiency [10].

The long-wavelength $^5I_6 \rightarrow {}^5I_7$ ($\lambda \approx 2.9$ μm) laser transition of Ho^{3+} is self-terminating, because the lifetime of the upper laser level is shorter than that of the lower laser level. A simplified energy-level diagram for the cascade laser is shown in Figure 3.7. As a solution to this problem, a *cascaded* lasing of both the 3 μm ($^5I_6 \rightarrow {}^5I_7$) and 2.1 μm ($^5I_7 \rightarrow {}^5I_8$) transitions of Ho^{3+} has recently been demonstrated [11]. The laser used Ho^{3+}-doped ZBLAN fiber (ZBLAN is the most common fluoride glass used in mid-IR fiber lasers with the following composition: ZrF_4–BaF_2–LaF_3–AlF_3–NaF) and two commercial 1150-nm diode lasers (total launched pump power of 7.6 W) to pump directly the upper laser level $^5I_6$. Cascading in this case helps to quickly unload the long-living $^5I_7$ energy level, which serves as a lower laser level for the 3-μm transition. The laser

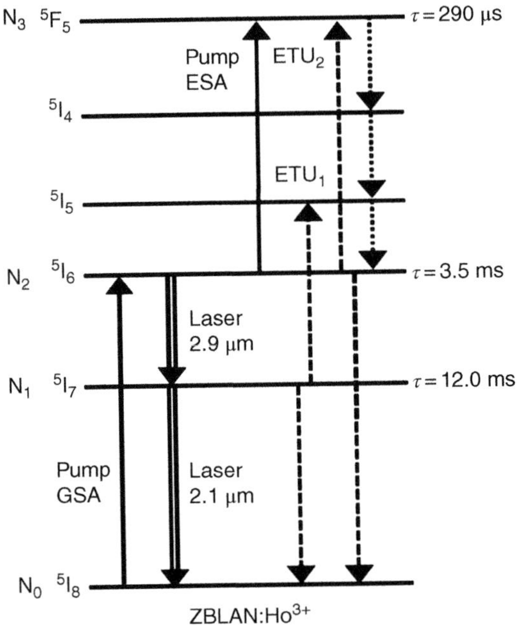

Figure 3.7 Energy-level diagram for a Ho^{3+}-doped cascaded ZBLAN fiber laser showing the pump, laser, and energy-transfer upconversion processes. GSA, ground state absorption; ESA, excited state absorption. *Source:* reproduced from figure 1 of [11], with permission of OSA, The Optical Society.

produced 0.77 W at 3.002 μm and 0.24 W at 2.1 μm, making this system the first watt-level fiber laser operating at ~3 μm in the mid-IR.

De-excitation of the lower laser level can also be achieved by codoping of Ho^{3+}-fibers with praseodymium (Pr^{3+}) ions. An efficient high-power operation of a 2.94-μm Ho^{3+}, Pr^{3+}-doped fluoride glass fiber laser was demonstrated in [12]. The core of the laser was made of a Ho^{3+}, Pr^{3+}-doped ZBLAN and the diode pumping was performed through fiber cladding. The laser produced a maximum output power of 2.5 W at 32% slope efficiency (that is 82% of the Stokes limit) with respect to the diode laser pump at 1150 nm. It is noteworthy that the emission wavelength of 2.94 μm measured at maximum pump power is particularly well suited for medical applications because it overlaps with the fundamental OH absorption peak of water in human tissue.

Finally, the laser action at the long-wavelength $^5I_5 \rightarrow {}^5I_6$ transition ($\lambda = 3.9$ μm) with a cascade laser scheme in a Ho^{3+}-doped fluoride fiber system was demonstrated under cryogenic (77 K) conditions [13]. With the fiber length of 340 cm, the pump wavelength of 885 nm, and a launched pump power of 900 mW, the researchers achieved 11 mW of CW output power at 3.9 μm. Simultaneously,

λ = 1.2 µm ($^5I_6 \rightarrow {}^5I_8$ transition) radiation was emitted with 70-mW output power. This result represents the longest wavelength emitted from a fiber laser.

3.2.3 Er-based Fiber Lasers

The laser transitions between manifolds $^4I_{11/2}$ and $^4I_{13/2}$ of rare-earth cation Er^{3+} (see Figure 3.2) are suitable for tuning across the mid-IR range of 2.7–3.05 µm. There is a good overlap of the Er^{3+} absorption peak, which corresponds to the excitation of the upper laser level $^4I_{11/2}$ with the wavelength range of highly developed diode lasers emitting at ~980 nm. This makes Er-doped fiber lasers suitable candidates for efficient mid-IR sources. Operation in the mid-IR has necessitated the use of fluoride glass fiber (e.g. ZBLAN), owing to its low phonon energy. As an additional bonus, the energy transfer upconversion process in Er^{3+} ions (see Figure 3.4b) effectively depopulates the lower laser level and recycles the excitation to produce slope efficiencies as high as 35.6%, a value that exceeds the Stokes efficiency limit [14].

The first 1-W class 3-µm laser was demonstrated nearly two decades ago by Jackson et al. [15]. It was based on diode-pumped Er^{3+}, Pr^{3+}:ZBLAN fiber and emitted at 2.71 µm with the output power of 1.7 W and slope efficiency of 17.3%. Since then, Er-based fiber lasers have shown markedly improved performance. Zhu and Jain demonstrated the first 10-W-level fiber laser of the 3-µm range [16]. In their experiment, the authors used a 4-m-long heavily erbium-doped Er:ZBLAN double-clad fiber, pumped by a 975-nm laser-diode array. The output power of 9 W was obtained at 2.785 µm at the launched pump power of 42.8 W. The output, however, fluctuated showing a pulsing behavior, and at higher pump rates the authors observed optical damage of the fiber end facets [16]. This unstable behavior was attributed to the poor thermal conductivity of Er:ZBLAN, its low melting temperature, and the large amount of heat that was generated by Er^{3+} dopants in the process of optical excitation.

Tokita et al. used liquid cooling to demonstrate 24 W of power from an Er:ZBLAN fiber laser operating at λ = 2.8 µm with 975 nm diode pumping [17]. The whole fiber was soaked in a fluorocarbon liquid coolant, which circulated to maintain a constant temperature of 20 °C. The 4.2-m-long fiber had an active core diameter of 25 µm, and the maximum power of 24 W was obtained at 166 W of the diode pump [17]. Later, the same team developed a 10-W wavelength-tunable Er:ZBLAN fiber laser operating without direct liquid cooling [18]. Instead, the fiber was cooled conductively when placed between aluminum plates that were maintained at 20 °C by water cooling. Also, to remove the heat from the output fiber end, the latter was polished into a spherical shape and put in optical contact with a 2-mm-thick sapphire plate (see Figure 3.8). The pump power was 93 W at 975 nm and an external diffraction grating in the Littrow configuration was used to tune the wavelength from 2.71 to 2.88 µm.

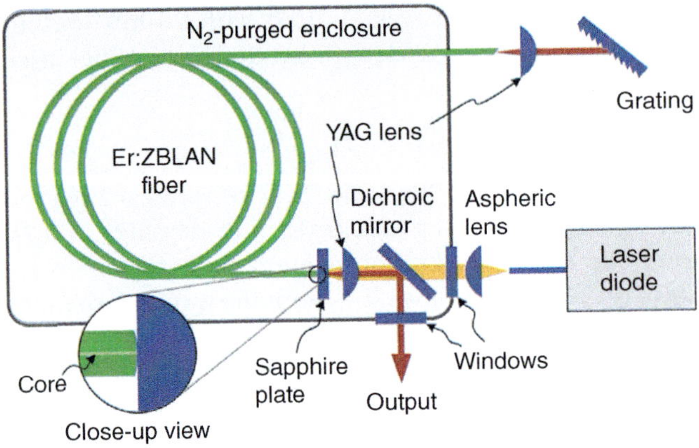

Figure 3.8 Schematic diagram of the grating-tunable Er-doped ZBLAN fiber laser pumped by a 975-nm laser diode. The inset shows how the spherical polishing of the fiber end ensures optical contact between the fiber core and the sapphire plate. A diffraction grating in the Littrow configuration was used to tune the wavelength. *Source:* reproduced from figure 1 of [18], with permission of OSA, The Optical Society.

Faucher et al. [19] obtained the output power of 20.6 W from an Er^{3+}-doped fluoride glass single-mode all-fiber laser source at $\lambda = 2.825\,\mu m$. The fiber was passively cooled by using an aluminum spool; it was 4.6-m long and had a single-mode core diameter of 16 μm. An undoped fiber with highly reflective (99.4%) Bragg grating written in it was fusion-spliced to the doped fiber and served as an end mirror. The pump source consisted of three 976-nm, 30-W fiber-coupled modules, and the slope efficiency of the laser was up to 35.4% with respect to the absorbed pump power, which is higher than the Stokes efficiency of 34.3%.

Recently, Aydin et al. reported a passively cooled cascade Er^{3+}-doped fluoride fiber laser ($\lambda = 2.825\,\mu m$) with an output power of ~13 W and a record slope efficiency of 50% (absolute optical efficiency of 37%) with respect to absorbed pump power at 976 nm [20]. The presence of an excited-state absorption band centered at 1.675 μm, between the $^4I_{13/2}$ and $^4I_{9/2}$ levels (see Figure 3.2), which partially overlaps the cascaded ($^4I_{13/2}$ to $^4I_{15/2}$) secondary emission at 1.614 μm, is believed to be responsible for the recycling of the excitation back to the upper laser level of the mid-IR transition. One of the key features of this approach was to resonate the emission at 1.614 μm, in addition to resonating the 2.8-μm emission.

Fortin et al. demonstrated a $\lambda = 2.94\,\mu m$ erbium-doped fluoride glass fiber laser delivering a record output power of 30.5 W in CW mode [21]. The laser operated on the long-wavelength transition – from the lowest Stark level of the $^4I_{11/2}$ to the highest Stark level of the $^4I_{13/2}$ manifold of the Er^{3+} ion. The passively cooled all-fiber laser cavity was formed by fiber Bragg gratings, and the laser was pumped using seven laser diodes operating around 980 nm with

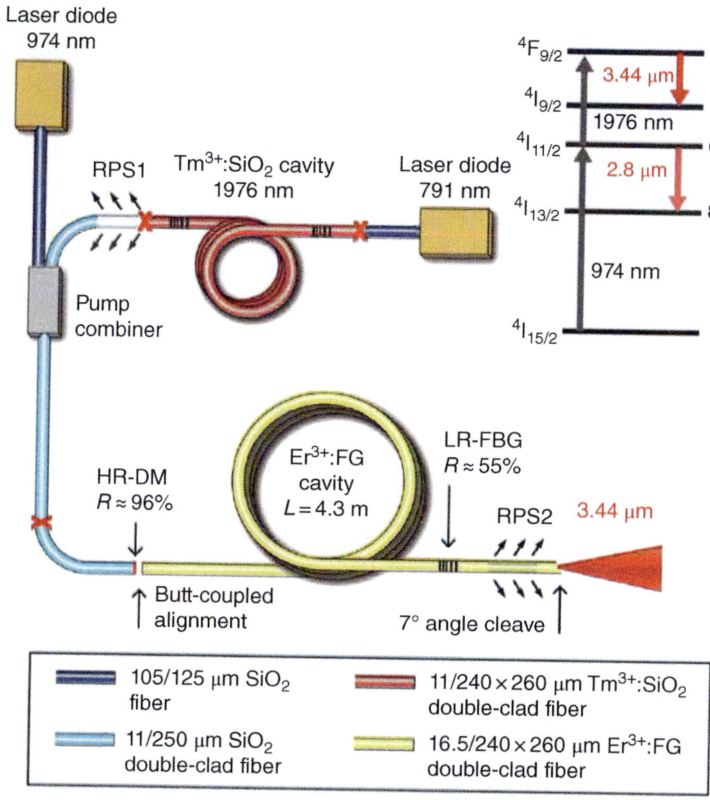

Figure 3.9 Schematic of the 3.44 μm Er-fiber laser that relies on a dual pumping scheme at 974 and 1976 nm. The inset shows a simplified energy diagram for erbium ions in fluoride glasses. A residual pump stripper (RPS) made of high-index acrylate polymer was applied to remove the remaining 791 nm pump. HR-DM, high reflection dichroic mirror; LR-FBG, low reflection fiber Bragg grating. *Source:* reproduced from figure 1 of [22], with permission of OSA, The Optical Society.

combined power of 188 W. The overall laser efficiency with respect to the launched pump power at 980 nm was 16%. The laser produced a narrow (0.15 nm) linewidth and had a single-mode beam quality with a $M^2 < 1.2$ [21]. Since the 2.94 μm wavelength is resonant with the liquid water OH absorption peak, this laser has potential to find many applications in the biomedical field.

On the longer wavelength side, a 3.44 μm all-fiber laser emitting up to 1.5 W at room temperature was demonstrated in [22]. The laser operated on the $^4F_{9/2} \rightarrow {}^4I_{9/2}$ transition of erbium-doped fluoride glasses and relied on a dual pumping scheme: at 974 and 1976 nm. In this case, a 974-nm pump was used to excite Er^{3+} ions to the long-lived level $^4I_{11/2}$, thus creating a virtual ground state, while a 1976-nm pump was used to populate the 3.5-μm transition upper level $^4F_{9/2}$ (Figure 3.9). By combining a dichroic mirror deposited on the input fiber tip and a fiber Bragg grating as an output coupler, a stable laser emission was produced with a bandwidth of less than 0.6 nm. The laser had an efficiency of 19% with respect to the launched pump power at 1976 nm and no saturation was observed, provided there is sufficient co-pumping at 974 nm.

With an improved (so-called "monolithic") design of the dual-wavelength-pumped erbium-doped fluorozirconate (Er^{3+}:ZrF_4) laser cavity, a multi-watt laser operation at 3.55 μm was demonstrated [23]. The fiber laser cavity was bounded by two fiber Bragg gratings, which allowed high-power performance without damage of the ZrF_4 fiber tip. The maximum output power and total optical efficiency were, respectively, 5.6 W and 26.4%. The laser power was the highest ever achieved from a fiber laser at this wavelength.

Based on the idea of dual-wavelength pumping, a broadly tunable long-wavelength erbium-doped Er:ZBLAN fiber laser was reported in [24]. The laser cavity consisted of a 2.8-m-long Er^{3+}-doped ZBLAN fiber gain medium with a 16-μm-diameter core and a diffraction grating. With the dual pumping at 977 and 1973 nm, the emission was centered at 3.5 μm, and the laser was tunable across 450 nm. A maximum power of 1.45 W at 3.47 μm was achieved with a 4-W pump at 977 nm and a 5.5-W pump at 1973 nm (overall optical efficiency 15%). The longest wavelength achieved was 3.78 μm, which is the longest wavelength emission from any fiber laser operating at room temperature [24].

3.2.4 Dy-based Fiber Lasers

Majewski and Jackson presented an efficient and power-scalable pump scheme for a 3-μm laser, based on dysprosium-doped fiber [25]. The authors used the free-running 2.8-μm emission from an Er^{3+}-doped fluoride fiber laser to directly excite, with a very small quantum defect of 8%, the upper laser level of the $^6H_{13/2}$–$^6H_{15/2}$ transition of the Dy^{3+} ion. The Dy^{3+} laser output at 3.04 μm was produced with a slope efficiency of 51%. When the length of the Dy^{3+}-doped fluoride fiber was extended from 92 to 140 cm, the authors observed a maximum emission wavelength of 3.26 μm, with a slope efficiency of 32% with respect to the launched pump power.

3.2.5 Raman Fiber Lasers

Fundamentally, the Raman process relies on photon–*phonon* scattering, where an optical phonon (lattice vibration) mediates the process of frequency down-conversion. The amount of red shift of the Stokes wave, with respect to the pump, depends on the frequency of the optical phonon for a given glass medium. In optical fibers, a variety of redshifted laser wavelengths extending beyond 2 μm were produced via stimulated Raman scattering. Typically, a Raman fiber laser consists of a piece of fiber enclosed between two highly reflective Bragg gratings written directly into the same fiber.

Using a germanate-based Raman gain cavity, Dianov et al. produced the output at 2.193 μm, as the fourth Stokes component, with an average power of 210 mW. A 1608-nm Er/Yb-doped fiber laser with an average power of 5 W was utilized as a pump [26]. Cumberland et al. reported Raman 2.105-μm emission wavelength from a high-concentration GeO_2 fiber Raman laser, with a maximum output of 4.6 W. The pump was a 22-W thulium-doped fiber laser at 1.938 μm [27].

In order to access longer mid-IR wavelengths, materials with extended spectral transmission windows, e.g. glasses based on fluoride, tellurite, or chalcogenide, must be considered. Fortin et al. reported a Raman laser based on a fluoride glass optical fiber [28]. The laser was pumped by a Tm^{3+}:silica CW fiber laser operating at 1940 nm with an average power of 9.6 W. A maximum output power of 580 mW was measured at 2185 nm (frequency shift of 579 cm^{-1}), corresponding to 6% conversion efficiency with respect to the launched pump power. Later, the same team demonstrated a multi-watt fluoride glass Raman fiber laser emitting at 2.23 μm. The laser was based on a nested (Raman laser inside pump laser) cavity with the pump being a Tm-doped silica fiber laser at 1.98 μm. The maximum Stokes output power recorded at 2.23 μm was as high as 3.7 W [29].

Chalcogenide glasses, arsenic sulfide (As_2S_3) and selenide (As_2Se_3) – have long been identified as ideal candidates for mid-IR Raman media because of (i) a high Raman gain coefficient (more than 50 times larger than that of a fluoride glass) and (ii) a transparency window that extends to beyond 6 and 9 μm respectively. Jackson et al. reported an As_2Se_3 chalcogenide Raman laser operating simultaneously on multiple Stokes bands. A 2.051-μm Tm^{3+}-doped silica fiber laser with a 2-W average power was employed as the pump source. The first Stokes emission was observed at 2.062 μm with an output power of 0.64 W. Two other Raman active vibrational modes of the fiber produced first Stokes outputs at 2.102 μm (0.2 W) and 2.166 μm (16 mW), respectively [30].

Bernier et al. reported the first Raman fiber laser emitting above 3 μm [31, 32]. The pump was an Er^{3+} fluoride glass fiber laser emitting at 3.005 μm (that is at the long-wavelength edge of the Er^{3+} emission band) operating in the quasi-CW mode with a pulse duration of 5 ms and a repetition rate of 20 Hz. Using a Fabry–Pérot cavity formed by a single-mode As_2S_3 chalcogenide glass fiber and fiber Bragg gratings (Figure 3.10), the authors demonstrated a Raman fiber laser action at λ = 3.34 μm. The laser output average power was 50 mW (0.6 W peak) at the launched pump average power 250 mW [31].

A long-wavelength quasi-CW fiber Raman source, operating at room temperature, was reported in [33]. The 3.77-μm output wavelength was generated by two cascades of Raman shifting; the Raman laser was based on a single-mode As_2S_3 fiber and nested Fabry–Pérot cavities formed by two pairs of fiber Bragg gratings. Similar to [32], the pump was an Er^{3+}:fluoride glass fiber laser at 3.005 μm operating in the quasi-CW mode (5-ms pulses, 20 Hz, duty cycle

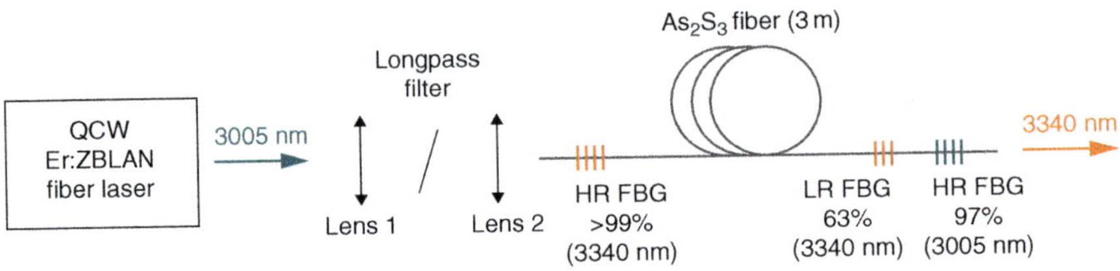

Figure 3.10 Schematic of the Raman As_2S_3 fiber laser emitting at 3.34 μm. The pump was an Er^{3+} ZBLAN fiber laser at 3.005 μm operating in the quasi-CW mode (5 ms, 20 Hz). FBG, fiber Bragg grating. *Source:* reproduced from figure 9 of [32], with permission of SPIE.

10%). The first and second Stokes orders were at 3.345 and 3.766 μm, respectively. The maximum average output power of 9 mW was measured at 3.766 μm for an averaged launched pump power of 371 mW, corresponding to a peak power of 112 mW [33].

3.3 Q-switched Mid-IR Fiber Lasers

Mid-IR fiber lasers operating in the Q-switched mode with microsecond to nanosecond pulse duration are desirable for such applications as light detection and ranging (LIDAR), pumping longer-wavelength mid-IR optical parametric oscillations (OPOs), supercontinuum generation, polymer processing, and for laser surgery where efficient and painless tissue removal can be achieved due to fast energy deposition.

In the 2-μm region, both Tm- and Ho-fiber lasers are suitable for pulsed-mode operation. Eichhorn and Jackson demonstrated a Q-switched Tm^{3+}-doped fiber laser providing multi-watt average power at $\lambda = 1.98$ μm [34]. The silica double-clad fiber was 2.3 m long with a core diameter of 20 μm and was diode-pumped at 792 nm. The laser was Q-switched with an acousto-optic modulator (AOM) and generated up to 30 W of average power at 110 kHz repetition rate, with 270 μJ pulse energy, pulse duration 41 ns, and peak power 6.6 kW. The maximum power, according to the authors, was limited by the buildup of amplified spontaneous emission (ASE). With a diffraction grating in the resonator, the laser was tuned from 1.93 to 2.05 μm with a pulse width of 41–50 ns at 125 kHz repetition rate [34]. Kadwani et al. reported an acousto-optically Q-switched oscillator based on Tm-doped photonic crystal fiber (PCF). The pump was a 100-W diode laser at 793 nm, and the active fiber had a large mode-field area >1000 μm^2. The laser maintained single-mode beam quality in a polarized beam and provided 435-μJ, 49-ns pulses at 10 kHz repetition rate at 2 μm wavelength, corresponding to an average power of 4.4 W and peak power as high as 8.9 kW [35].

A pulsed-mode operation with an acousto-optic Q-switching of a Tm^{3+}, Ho^{3+}-codoped double-clad silica fiber laser was reported by Eichhorn and Jackson [36]. Here, thulium provides efficient excitation using standard 790-nm laser diodes, while a resonant energy transfer process efficiently populates the Ho^{3+} ions. The fiber was symmetrically diode-pumped from two sides at 792 nm, with up to 30 W per each side. With the use of an intracavity diffraction grating and by taking advantage of both Tm^{3+} and Ho^{3+} emission bands, the laser was tuned in the range 1.95–2.13 μm. In the Ho^{3+} band, the shortest pulses were reached at $\lambda = 2.07$ μm at 20 kHz repetition frequency with pulse duration of 58 ns and pulse energy of 250 μJ, corresponding to an average power of 5 W. In the Tm^{3+} emission band, at $\lambda = 2.02$ μm, the maximum average power was 15 W at 100 kHz repetition frequency (pulse energy 150 μJ) [36]. A high average power nanosecond system based on thulium-doped fiber amplifier with 793-nm diode pump was achieved in a two-stage oscillator–amplifier setup [37]. The pulses with variable duration between 4 and 72 ns were generated by mode-locking a long-cavity low repetition rate oscillator with the aid of a nonlinear optical loop mirror. At 1.07 MHz repetition frequency and 72 ns pulse duration, 100.4 W of average power was produced at $\lambda \approx 2$ μm.

A number of groups achieved Q-switched fiber laser operation in the 3-μm band. Hu et al. demonstrated a Q-switched laser based on codoped Ho^{3+}, Pr^{3+} fiber operating at 2.87 μm. (Codoping Ho with Pr ions allows de-excitation of the $^5I_7$ state in holmium, providing high slope efficiencies.) With an acousto-optical Q-switching (Figure 3.11), the authors achieved 78-ns-long pulses at 120 kHz, with an average power of 0.72 W and slope efficiency of 20% with respect to the launched pump power at 1150 nm [38].

Tokita et al. demonstrated an impressive result in terms of 3-μm fiber laser peak and average power [39]. A diode-pumped 2.8 μm laser oscillator, based on 35-μm-core Er-doped ZBLAN fiber, was Q-switched with an AOM. At a repetition rate of 120 kHz and a maximum 975-nm diode pump power of 75 W, the pulse energy reached 100 μJ at a pulse duration of 90 ns, corresponding to the average output power of 12.4 W (0.9 kW peak).

Compared to actively Q-switched fiber lasers, passively Q-switched lasers have the advantages of low cost and compactness. Several passively Q-switched

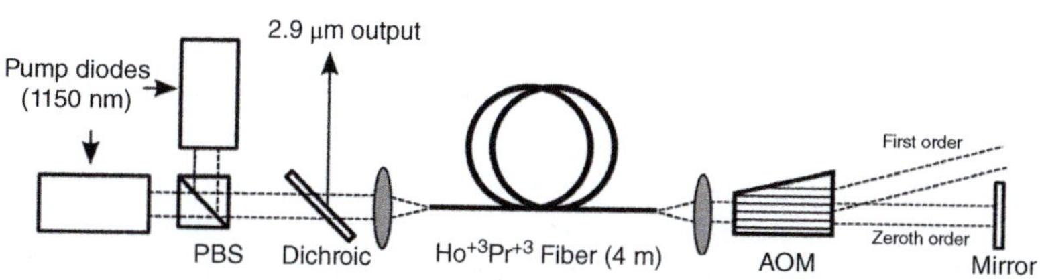

Figure 3.11 Schematic of a 2.87-μm acousto-optically Q-switched $Ho^{3+}Pr^{3+}$ fiber laser. *Source:* reproduced from figure 1 of [38], with permission of OSA, The Optical Society.

fiber lasers around 3 μm with ~μs pulse duration have been reported recently, using different saturable absorbers: graphene [40, 41], Fe^{2+}:ZnSe [41], semiconductor saturable absorber mirror (SESAM) based on indium arsenide (InAs) [42], and using a topological insulator Bi_2Te_3 [43]. (Topological insulators represent a novel class of so-called Dirac materials. In the bulk state, they have a narrow band gap while at the surface they are in the gapless metallic state and saturable absorption is observed in a broadband spectral range via the Pauli blocking effect.) The average output power in [42, 43] exceeded 300 mW at repetition frequencies around 50–100 kHz (see Table 3.1).

3.4 Mode-locked Mid-IR Fiber Lasers

Fiber-based mode-locked lasers are desirable sources of ultrashort pulses because of their compactness and environmental reliability, as has already been demonstrated by their near-IR counterparts. Comparatively large gain bandwidths of fiber lasers, due to the amorphous nature of the glass host, typically allow achieving sub-100-fs pulse durations. Mode-locked lasers in the mid-IR can further fuel the ultrafast laser science and offer a wide variety of promising applications, such as generation of supercontinuum and optical frequency combs, and precision laser surgery, to name a few.

Ultrafast fiber lasers are usually based on passively mode-locked oscillators. While erbium at $\lambda = 1.55$ μm and ytterbium at $\lambda = 1.05$ μm silica fiber lasers still dominate in the commercial realm, enormous progress took place recently in the development of thulium (Tm) and holmium (Ho) doped fiber lasers emitting near 2.0 μm and Er-fluoride fiber lasers emitting near 3.0 μm [52].

Tm-doped silica-fiber-based 2-μm lasers are the most mature among the mid-IR mode-locked fiber lasers. In the early stage of their development, passively mode-locked femtosecond Tm-doped fiber oscillators were demonstrated in the mid-1990s. For example, Nelson et al. [53] used additive-pulse mode-locking based on nonlinear polarization evolution to obtain tunable 1.8–1.9 μm pulses in the soliton regime, whereas Sharp et al. [54] reported a mode-locked thulium-doped silica fiber laser using an InGaAs saturable absorber. Later, watt-level femtosecond Tm-fiber lasers were developed. For example, by amplification of a Raman-shifted output of a 1.56-μm Er-doped fiber laser in a Tm-doped fiber having a 25-μm-diameter core, Imeshev and Fermann obtained 108-fs pulses at 2 μm wavelength with an average power as high as 3.1 W, pulse energy of 31 nJ, and peak power of 230 kW [44]. Subsequently, stable 2-μm Tm-fiber-based oscillators were developed using SESAM combined with nonlinear polarization evolution for mode-locking mechanism. After amplification in a dispersion-compensated Tm-fiber amplifier, high-power (>2.5 W) sub-70-fs pulses at ~2 μm were generated [45, 55]. These developments have also led to creating fully stabilized optically referenced frequency combs discussed in Chapter 6.

Table 3.1 Summary of mid-IR fiber lasers.

Dopant/fiber	Wavelength (μm)	Laser characteristics	Ref.
Continuous wave, CW			
Tm^{3+}-silica	2.045	Ave. power 1 kW, pump at 790 nm	[8]
Tm^{3+}, Ho^{3+}-silica	2.105	Ave. power 83 W, pump at 793 nm	[10]
Ho^{3+}-ZBLAN	3 2.1	Cascaded laser, ave. power 0.77 W at 3 μm, and 0.24 W at 2.1 μm, pump at 1150 nm	[11]
Ho^{3+}, Pr^{3+}-ZBLAN	2.94	Ave. power 2.5 W, pump at 1150 nm	[12]
Er^{3+}-ZBLAN	2.8	Ave. power 24 W, pump at 975 nm, liquid fiber cooling	[17]
Er^{3+}-fluoride glass	2.825	Ave. power 20.6 W, pump at 980 nm (90 W)	[19]
Er^{3+}-fluoride glass	2.825	Ave. power 12.5 W, pump at 976 nm (33.5 W), cascade 1.6 μm, slope optical efficiency 50%	[20]
Er^{3+}-fluoride glass	2.94	Ave. power 30.5 W, pump at 980 nm (188 W)	[21]
Er^{3+}-fluoride glass	3.44	Ave. power 1.5 W, dual pump: at 974 and 1976 nm	[22]
Er^{3+}-fluoride glass	3.55	Ave. power 5.6 W, dual pump: at 974 and 1976 nm, total optical efficiency 26.4%	[23]
Er^{3+}-ZBLAN	3.33–3.78	Grating-tunable, ave. power 1.45 W @ 3.47 μm dual pump: at 977 and 1973 nm	[24]
Dy^{3+}-ZBLAN	3.04	Ave. power 80 mW, pump at 2.8 μm (300 mW)	[25]
Raman lasers			
Fluoride glass	2.23	Ave. power 3.7 W, nested cavity, pump at 1.98 μm	[29]
As_2S_3 glass	3.34	Quasi-CW mode (5 ms, 20 Hz), ave. power 50 mW, pump at $\lambda = 3.005$ μm (250 mW)	[31]
As_2S_3 glass	3.77	Quasi-CW mode (5 ms, 20 Hz), ave. power 9 mW, pump at $\lambda = 3.005$ μm (371 mW)	[33]
Q-switched			
Tm^{3+}-silica	1.98	Ave. power 30 W (peak 6.6 kW), AOM Q-sw, 41 ns, 270 μJ, 110 kHz	[34]
Tm^{3+}-silica	1.95	Ave. power 4.4 W (peak 8.9 kW), AOM Q-sw, 49 ns, 435 μJ, 10 kHz	[35]

(*Continued*)

Table 3.1 (Continued)

Dopant/fiber	Wavelength (μm)	Laser characteristics	Ref.
Tm^{3+}, Ho^{3+}-silica	2.07	Ave. power 5 W, AOM Q-sw, 58 ns, 250 μJ, 20 kHz	[34]
Tm^{3+}-silica	~2	Ave. power 100 W, nonlinear optical loop mirror, oscillator–amplifier, 72 ns, 94 μJ, 1 MHz	[37]
Ho^{3+}, Pr^{3+}-ZBLAN	2.87	Ave. power 0.72 W, AOM Q-sw, 78 ns, 6 μJ, 120 kHz	[38]
Er^{3+}-ZBLAN	2.8	Ave. power 12.4 W, AOM Q-sw, 90 ns, 103 μJ, 120 kHz	[39]
Ho^{3+}-ZBLAN	2.97	Ave. power 317 mW, SESAM, 1.68 μs, 6.65 μJ, 48 kHz	[42]
Ho^{3+}-ZBLAN	2.98	Ave. power 327 mW, Bi_2Te_3 topological insulator, 1.37 μs, 4 μJ, 82 kHz	[43]
Mode-locked			
Tm^{3+}-silica	1.98	Ave. power 3.1 W, Raman-shifted Er-laser and Tm-amplifier, 108 fs, 100 MHz	[44]
Tm^{3+}-silica	1.94–1.97	Ave. power 2.5 W, SESAM mode-locked oscillator, 100 fs, 418 MHz	[45]
Er^{3+}-ZBLAN	2.8	Ave. power 440 mW, SESAM and a fiber Bragg grating, 60 ps, 52 MHz	[46]
Ho^{3+}, Pr^{3+}-ZBLAN	2.9	Ave. power 70 mW, InAs saturable absorber, 6 ps, 25 MHz	[47]
Er^{3+}-ZBLAN	2.8	Ave. power 206 mW, nonlinear polarization rotation, 497 fs, 57 MHz, peak 6.4 kW	[48]
Er^{3+}-ZBLAN	2.8	Ave. power 40 mW, nonlinear polarization rotation, 207 fs, 55 MHz, peak 3.5 kW	[49]
Er^{3+}-ZBLAN	2.8	Ave. power 676 mW, nonlinear polarization rotation, 270 fs, 97 MHz	[50]
Er^{3+}-ZBLAN	2.8–3.6	Ave. power ~2 W at 3.4 μm, Raman self-frequency shifted solitons, 160 fs, pump at 2.8 μm	[51]

A mode-locked soliton fiber laser based on Ho^{3+}-doped glass (λ = 2.08 μm) was demonstrated with graphene as a saturable absorber [56]. The laser was pumped by a CW Tm-fiber laser (0.5 W at 1950 nm) and produced 811-fs duration pulses at 34 MHz repetition rate and 44 mW average power.

Based on a variety of saturable absorbers, passively mode-locked lasers in the 3-μm range were also reported, with pulse durations ~20 ps and average power ~100 mW. These include an Er^{3+}-doped ZBLAN fiber laser at 2.8 μm, passively mode-locked by a Fe^{2+}:ZnSe crystal inside the laser cavity [57] and Ho^{3+}, Pr^{3+} codoped ZBLAN fiber laser at 2.87 μm, which used an InAs-based SESAM as passive mode-locker [58]. Yet, the above lasers exhibited "pulsing" – both Q-switching and mode-locking behavior simultaneously.

Haboucha et al. reported a truly continuous mode-locking from an Er^{3+}-ZBLAN glass fiber laser operating at 2.8 μm by using SESAM in conjunction with a fiber Bragg grating inside a linear cavity [46]. The Bragg grating allowed for the generation of a stable, self-starting pulse train with a repetition rate of 52 MHz, pulse duration of 60 ps, and an average power of 440 mW. In parallel, Hu et al. demonstrated a stable mode-locked fiber ring laser near 2.9 μm based on a Ho^{3+}, Pr^{3+}-codoped ZBLAN fiber using an InAs saturable absorber. This configuration enabled the generation of 6-ps pulses at 25 MHz repetition rate with an average power of 70 mW [47].

Most recently, two groups have independently demonstrated the possibility of generating femtosecond pulses from an Er^{3+}-doped fluoride (ZBLAN) fiber at 2.8 μm based on nonlinear polarization rotation [48, 49]. This technique relies upon the high Kerr nonlinearity of the fluoride fiber ($n_2 = 2.1 \times 10^{-20}\,m^2/W$) to create an intensity-dependent rotation of the polarization state inside the fiber, which, when combined with polarizing optics (quarter wave plate, half wave plate, and an optical isolator), makes an effective saturable absorber for mode-locking. Hu and colleagues were able to produce 497-fs pulses at 2.8 μm with 206 mW average power at 57 MHz repetition rate (Figure 3.12) [48]. Similarly, Duval et al. [49] reported a passively mode-locked Er-ZBLAN fiber ring laser based on nonlinear polarization evolution. The laser generated 207 fs pulses at λ = 2.8 μm with a repetition rate of 55 MHz at an average power of 40 mW. Since ZBLAN fiber has anomalous dispersion at 2.8 μm, both of these lasers were reported to operate in the soliton regime. By varying the Er^{3+}:fluoride fiber length and the output coupling, Duval et al. were able to scale up the 2.8-μm mode-locked laser power. They demonstrated stable 270-fs pulses with the repetition rate of 96.6 MHz and the average power of 675 mW (pulse energy 7 nJ, peak power 23 kW) [50].

The same team reported on a tunable (2.8–3.6 μm) and high-power (>1 W) ultrafast fiber laser system based on erbium-doped zirconium fluoride glass [51]. Tuning of the central wavelength of the pulses was achieved through a process known as soliton Raman self-frequency shift, which is facilitated by the anomalous dispersion of fluoride glass fibers in this spectral range. An ultrafast fiber oscillator (the pump) operating at λ = 2.8 μm and a repetition rate of 58 MHz was similar to the one described in [50]. It was followed by a second erbium-doped fluoride glass fiber, a portion of which (L = 1.25 m) served as an amplifier, while the remaining unpumped segment (L = 8 or 22 m) acted as a passive fiber to shift the amplified soliton to longer wavelengths. As the amplifier pump power was increased, the spectrum of a clean and isolated soliton

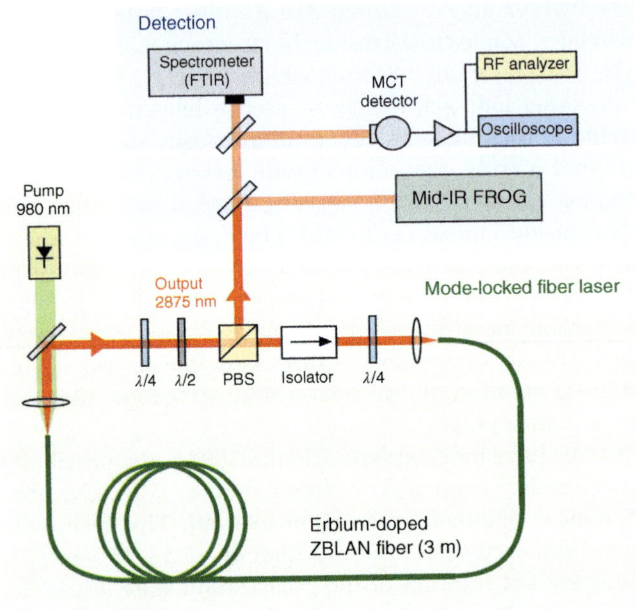

Figure 3.12 Experimental setup for generating mode-locked pulses from an Er^{3+} ZBLAN fiber laser at 2.8 μm. A ring cavity is formed with a free-space section containing polarization optics to provide the nonlinear polarization evolution mode-locking. The pulses were analyzed in the time domain, using a mid-IR frequency-resolved optical gating (FROG) setup. *Source:* reproduced from figure 1 of [48], with permission of OSA, The Optical Society.

continuously shifted toward longer wavelengths in the passive portion of the fiber, reaching a maximum central wavelength of 3.4 μm in the 8-m fiber and 3.59 μm in the 22-m fiber. Overall, high-energy Raman soliton pulses tunable from 2.8 to 3.6 μm were generated. For example, at $\lambda = 3.4\,\mu m$, 160-fs pulses were produced with the pulse energy of 37 nJ (200 kW peak), corresponding to an average output power >2 W [51].

The main results for mid-IR fiber lasers are presented in Table 3.1.

3.5 Summary

So far, the most successful fiber lasers in the mid-IR are based on the Tm^{3+} ($^3H_4 \rightarrow {}^3H_6$) laser transition, the Er^{3+} ($^4I_{11/2} \rightarrow {}^4I_{13/2}$) transition, and the Ho^{3+} ($^5I_6 \rightarrow {}^5I_7$) transition. They cover, respectively, 1.9–2.1, 2.7–3.0, and 2.8–3.05 μm spectral ranges.

Other wavelengths can be reached via dual-wavelength pumping (Er^{3+}, 3.44–3.55 μm), Raman effect (chalcogenide glass, 3.34 μm), or via cryogenic cooling (Ho^{3+}, 3.9 μm).

In terms of the output power, Tm-doped fibers enable getting 1-kW CW power in a single-spatial mode at 2.045 μm with optical-to-optical conversion efficiency higher than that dictated by the quantum defect, thanks to the "two-for-one" cross-relaxation process. At longer wavelengths, achieving high-power operation is more challenging. The main cause of power decline is the increase of quantum defect since the most efficient high-power laser diodes typically emit in the near-IR. Thus, there is more energy dissipation, and created heat becomes an ever-increasing fraction of the absorbed pump energy as the wavelength increases. Nevertheless, based on the 3-μm transition in erbium, benefiting from "two-for-one" cross-relaxation process, more than 30 W of average power has been obtained at 2.94 μm with passively cooled Er^{3+}-doped fluoride glass fiber.

Mid-IR fiber lasers operating in the CW, Q-switched, and mode-locked modes open up a plethora of new applications in biomedicine, frequency combs generation, and in nonlinear optics and spectroscopy. For further reading, excellent reviews on mid-IR fiber lasers can be found in [1, 32].

References

1 Jackson, S.D. (2012). Towards high-power mid-infrared emission from a fibre laser. *Nat. Photonics* 6: 423.

2 Hanna, D.C., Percival, R.M., Smart, R.G., and Tropper, A.C. (1990). Efficient and tunable operation of a Tm-doped fiber laser. *Opt. Commun.* 75: 283.

3 Jackson, S.D. and King, T.A. (1998). High-power diode-cladding-pumped Tm-doped silica fiber laser. *Opt. Lett.* 23: 1462.

4 Hayward, R.A., Clarkson, W.A., Turner, P.W., Nilsson, J., Grudinin, A.B., and Hanna, D.C. (2000). Efficient cladding-pumped Tm-doped silica fiber laser with high power singlemode output at 2 μm. *Electron. Lett.* 36: 711.

5 Moulton, P.F., Rines, G.A., Slobodtchikov, E.V., Wall, K.F., Frith, G., Samson, B., and Carter, A.L.G. (2009). Tm-doped fiber lasers: fundamentals and power scaling. *IEEE J. Sel. Top. Quant. Electron.* 15: 85.

6 Clarkson, W.A., Barnes, N.P., Turner, P.W., Nilsson, J., and Hanna, D.C. (2002). High-power cladding-pumped Tm-doped silica fiber laser with wavelength tuning from 1860 to 2090 nm. *Opt. Lett.* 27: 1989.

7 Jackson, S.D. (2004). Cross relaxation and energy transfer upconversion processes relevant to the functioning of 2 μm Tm^{3+}-doped silica fibre lasers. *Opt. Commun.* 230: 197.

8 Ehrenreich, T., Leveille, R., Majid, I., Tankala, K., Rines, G., and Moulton, P. (2010). 1 kW, all-glass Tm:fiber laser. *Proc. SPIE* 7580: 7580–7112.

9 Jackson, S.D. (2006). Midinfrared holmium fiber lasers. *IEEE J. Quant. Electron.* 42: 187.
10 Jackson, S.D., Sabella, A., Hemming, A., Bennetts, S., and Lancaster, D.G. (2007). High-power 83 W holmium-doped silica fiber laser operating with high beam quality. *Opt. Lett.* 32: 241.
11 Li, J., Hudson, D.D., and Jackson, S.D. (2011). High-power diode-pumped fiber laser operating at 3 μm. *Opt. Lett.* 36: 3642.
12 Jackson, S.D. (2009). High-power and highly efficient diode-cladding pumped holmium-doped fluoride fiber laser operating at 2.94 μm. *Opt. Lett.* 34: 2327.
13 Schneide, J., Carbonnier, C., and Unrau, U.B. (1997). Characterization of a Ho^{3+}-doped fluoride fiber laser with a 3.9 μm emission wavelength. *Appl. Opt.* 36: 8595.
14 Faucher, D., Bernier, M., Caron, N., and Vallée, R. (2009). Erbium-doped all-fiber laser at 2.94 μm. *Opt. Lett.* 34: 3313.
15 Jackson, S.D., King, T.A., and Pollnau, M. (1999). Diode-pumped 1.7-W erbium 3-μm fiber laser. *Opt. Lett.* 24: 1133.
16 Zhu, X. and Jain, R. (2007). 10-W-level diode-pumped compact 2.78 μm ZBLAN fiber laser. *Opt. Lett.* 32: 26.
17 Tokita, S., Murakami, M., Shimizu, S., Hashida, M., and Sakabe, S. (2009). Liquid-cooled 24 W mid-infrared Er:ZBLAN fiber laser. *Opt. Lett.* 34: 3062.
18 Tokita, S., Hirokane, M., Murakami, M., Shimizu, S., Hashida, M., and Sakabe, S. (2010). Stable 10 W Er:ZBLAN fiber laser operating at 2.71–2.88 μm. *Opt. Lett.* 35: 3943.
19 Faucher, D., Bernier, M., Androz, G., Caron, N., and Vallée, R. (2011). 20 W passively cooled single-mode all-fiber laser at 2.8 μm. *Opt. Lett.* 36: 1104.
20 Aydin, Y.O., Fortin, V., Maes, F., Jobin, F., Jackson, S.D., Vallée, R., and Bernier, M. (2017). Diode-pumped mid-infrared fiber laser with 50% slope efficiency. *Optica* 4: 235.
21 Fortin, V., Bernier, M., Bah, S.T., and Vallée, R. (2015). 30 W fluoride glass all-fiber laser at 2.94 μm. *Opt. Lett.* 40: 2882.
22 Fortin, V., Maes, F., Bernier, M., Bah, S.T., D'Auteuil, M., and Vallée, R. (2016). Watt-level erbium-doped all-fiber laser at 3.44 μm. *Opt. Lett.* 41: 559.
23 Maes, F., Fortin, V., Bernier, M., and Vallée, R. (2017). 5.6 W monolithic fiber laser at 3.55 μm. *Opt. Lett.* 42: 2054.
24 Henderson-Sapir, O., Jackson, S.D., and Ottaway, D.J. (2016). Versatile and widely tunable mid-infrared erbium doped ZBLAN fiber laser. *Opt. Lett.* 41: 1676.
25 Majewski, M.R. and Jackson, S.D. (2016). Highly efficient mid-infrared dysprosium fiber laser. *Opt. Lett.* 41: 2173.
26 Dianov, E.M., Bufetov, I.A., Mashinsky, V.M., Shubin, A.V., Medvedkov, O.I., Rakitin, A.E., Melkumov, M.A., Khopin, V.F., and Gur'yanov, A.N. (2005). Raman fiber lasers based on heavily GeO_2-doped fibers. *Quantum Electron.* 35: 435.

27 Cumberland, B.A., Popov, S.V., Taylor, J.R., Medvedkov, O.I., Vasiliev, S.A., and Dianov, E.M. (2007). 2.1 μm continuous-wave Raman laser in GeO_2 fiber. *Opt. Lett.* 32: 1848.

28 Fortin, V., Bernier, M., Carrier, J., and Vallée, R. (2011). Fluoride glass Raman fiber laser at 2185 nm. *Opt. Lett.* 36: 4152.

29 Fortin, V., Bernier, M., Faucher, D., Carrier, J., and Vallée, R. (2012). 3.7 W fluoride glass Raman fiber laser operating at 2231 nm. *Opt. Express* 20: 19412.

30 Jackson, S.D. and Anzueto-Sánchez, G. (2006). Chalcogenide glass Raman fiber laser. *Appl. Phys. Lett.* 88: 221106.

31 Bernier, M., Fortin, V., Caron, N., El-Amraoui, M., Messaddeq, Y., and Vallée, R. (2013). Mid-infrared chalcogenide glass Raman fiber laser. *Opt. Lett.* 38: 127.

32 Fortin, V., Bernier, M., Caron, N., Faucher, D., El Amraoui, M., Messaddeq, Y., and Vallée, R. (2013). Towards the development of fiber lasers for the 2 to 4 μm spectral region. *Opt. Eng.* 52: 054202.

33 Bernier, M., Fortin, V., El-Amraoui, M., Messaddeq, Y., and Vallée, R. (2014). 3.77 μm fiber laser based on cascaded Raman gain in a chalcogenide glass fiber. *Opt. Lett.* 39: 2052.

34 Eichhorn, M. and Jackson, S.D. (2007). High-pulse-energy actively Q-switched Tm^{3+}-doped silica 2 μm fiber laser pumped at 792 nm. *Opt. Lett.* 32: 2780.

35 Kadwani, P., Modsching, N., Sims, R.A., Leick, L., Broeng, J., Shah, L., and Richardson, M. (2012). Q-switched thulium-doped photonic crystal fiber laser. *Opt. Lett.* 37: 1664.

36 Eichhorn, M. and Jackson, S.D. (2008). High-pulse-energy, actively Q-switched Tm^{3+},Ho^{3+}-codoped silica 2-μm fiber laser. *Opt. Lett.* 33: 1044.

37 Zhao, J., Ouyang, D., Zheng, Z., Liu, M., Ren, X., Li, C., Ruan, S., and Xie, W. (2016). 100 W dissipative soliton resonances from a thulium-doped double-clad all-fiber-format MOPA system. *Opt. Express* 24: 12072.

38 Hu, T., Hudson, D.D., and Jackson, S.D. (2012). Actively *Q*-switched 2.9 μm $Ho^{3+}Pr^{3+}$-doped fluoride fiber laser. *Opt. Lett.* 37: 2145.

39 Tokita, S., Murakami, M., Shimizu, S., Hashida, M., and Sakabe, S. (2011). 12 W Q-switched Er:ZBLAN fiber laser at 2.8 μm. *Opt. Lett.* 36: 2812.

40 Wei, C., Zhu, X., Wang, F., Xu, Y., Balakrishnan, K., Song, F., Norwood, R.A., and Peyghambarian, N. (2013). Graphene Q-switched 2.78 μm Er^{3+}-doped fluoride fiber laser. *Opt. Lett.* 38: 3233.

41 Zhu, G., Zhu, X., Balakrishnan, K., Norwood, R.A., and Peyghambarian, N. (2013). Fe^{2+}:ZnSe and graphene Q-switched singly Ho^{3+}-doped ZBLAN fiber lasers at 3 μm. *Opt. Mater. Express* 3: 1365.

42 Li, J.F., Luo, H.Y., He, Y.L., Liu, Y., Zhang, L., Zhou, K.M., Rozhin, A.G., and Turistyn, S.K. (2014). Semiconductor saturable absorber mirror passively Q-switched 2.97 μm fluoride fiber laser. *Laser Phys. Lett.* 11: 065102.

43 Li, J., Luo, H., Wang, L., Zhao, C., Zhang, H., Li, H., and Liu, Y. (2015). 3-μm mid-infrared pulse generation using topological insulator as the saturable absorber. *Opt. Lett.* 40: 3659.

44 Imeshev, G. and Fermann, M.E. (2005). 230-kW peak power femtosecond pulses from a high power tunable source based on amplification in Tm-doped fiber. *Opt. Express* 13: 7424.

45 Lee, K.F., Mohr, C., Jiang, J., Schunemann, P.G., Vodopyanov, K.L., and Fermann, M.E. (2015). Midinfrared frequency comb from self-stable degenerate GaAs optical parametric oscillator. *Opt. Express* 23: 26596.

46 Haboucha, A., Fortin, V., Bernier, M., Genest, J., Messaddeq, Y., and Vallée, R. (2014). Fiber Bragg grating stabilization of a passively mode-locked 2.8 μm Er^{3+}: fluoride glass fiber laser. *Opt. Lett.* 39: 3294.

47 Hu, T., Hudson, D.D., and Jackson, S.D. (2014). Stable, self-starting, passively mode-locked fiber ring laser of the 3 μm class. *Opt. Lett.* 39: 2133.

48 Hu, T., Jackson, S.D., and Hudson, D.D. (2015). Ultrafast pulses from a mid-infrared fiber laser. *Opt. Lett.* 40: 4226.

49 Duval, S., Bernier, M., Fortin, V., Genest, J., Piché, M., and Vallée, R. (2015). Femtosecond fiber lasers reach the mid-infrared. *Optica* 2: 623.

50 Duval, S., Olivier, M., Fortin, V., Bernier, M., Piché, M., and Vallée, R. (2016). 23-kW peak power femtosecond pulses from a mode-locked fiber ring laser at 2.8 μm. *Proc. SPIE* 9728: 972 802.

51 Duval, S., Gauthier, J.-C., Robichaud, L.-R., Paradis, P., Olivier, M., Fortin, V., Bernier, M., Piché, M., and Vallée, R. (2016). Watt-level fiber-based femtosecond laser source tunable from 2.8 to 3.6 μm. *Opt. Lett.* 41: 5294.

52 Fermann, M.E. and Hartl, I. (2013). Ultrafast fibre lasers. *Nat. Photonics* 7: 868.

53 Nelson, L.E., Ippen, E.P., and Haus, H.A. (1995). Broadly tunable sub-500 fs pulses from an additive-pulse mode-locked thulium-doped fiber ring laser. *Appl. Phys. Lett.* 67: 19.

54 Sharp, R.C., Spock, D.E., Pan, N., and Elliot, J. (1996). 190-fs passively mode-locked thulium fiber laser with a low threshold. *Opt. Lett.* 21: 881.

55 Bethge, J., Jiang, J., Mohr, C., Fermann, M., and Hartl, I. (2012). Optically referenced Tm-fiber-laser frequency comb, Technical Digest. In: *Advanced Solid State Photonics*. Washington, DC: Optical Society of America, Paper AT5A.3.

56 Sotor, J., Pawliszewska, M., Sobon, G., Kaczmarek, P., Przewolka, A., Pasternak, I., Cajzl, J., Peterka, P., Honzátko, P., Kašík, I., Strupinski, W., and Abramski, K. (2016). All-fiber Ho-doped mode-locked oscillator based on a graphene saturable absorber. *Opt. Lett.* 41: 2592.

57 Wei, C., Zhu, X., Norwood, R.A., and Peyghambarian, N. (2012). Passively continuous-wave mode-locked Er^{3+}-doped ZBLAN fiber laser at 2.8 μm. *Opt. Lett.* 37: 3849.

58 Li, J., Hudson, D.D., Liu, Y., and Jackson, S.D. (2012). Efficient 2.87 μm fiber laser passively switched using a semiconductor saturable absorber mirror. *Opt. Lett.* 37: 3747.

4

Semiconductor Lasers

Semiconductor mid-IR lasers are directly driven by a current source, hence they represent the most desirable and compact form of laser. Semiconductor lasers operating at room temperature (RT) have important applications in spectroscopy, remote trace-gas monitoring, secure communications, and infrared countermeasures, and are used as pump sources for frequency converters based on microresonators and nonlinear waveguides. In this chapter, we will look at four main types of semiconductor lasers: heterojunction lasers, quantum cascade lasers (QCLs), interband cascade lasers (ICLs), and optically pumped semiconductor disk lasers (OPSDLs).

4.1 Heterojunction Mid-IR Lasers

The emission wavelength of the interband heterojunction diode laser is determined primarily by the bandgap energy of the material used for the active layer (Figure 4.1). Semiconductor lasers operating in the mid-IR have been around for a long time – the first mid-IR semiconductor laser was demonstrated by Melngailis in 1963 using InAs, a III–V semiconductor [1]. The InAs laser operated only under pulsed conditions and at cryogenic temperatures. It was soon discovered, however, that IV–VI compound lead-salt semiconductors such as PbS, PbSn, and PbSe were more promising for the mid-IR laser action [2]. Indeed, lead-salt lasers have been used extensively for high-precision spectroscopy because they can cover a wide emission wavelength range between 3 and 30 μm, provided that one uses a large set of individual lasers. Lead-salt lasers can operate continuous wave (CW) and can be tuned in a limited wavelength range by adjusting the temperature [3]. Nonetheless, little progress has been made since their invention, and the use of lead-salt lasers has been mostly confined to scientific applications because they operate only at low temperatures, typically around the liquid-nitrogen temperature (77 K). These lasers are outside the scope of this book; however, there are several excellent reviews on the topic, such as [3, 4].

Laser-based Mid-infrared Sources and Applications, First Edition. Konstantin L. Vodopyanov.

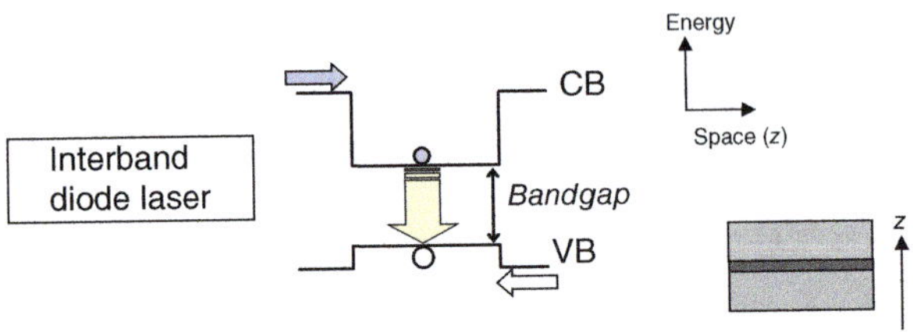

Figure 4.1 Energy-level diagram of a type-I interband diode laser. The emission wavelength is determined primarily by the bandgap of the material used for the active layer. CB, conduction band; VB, valence band.

From their first demonstration in 1963, progress in the area of mid-IR heterojunction lasers was not as fast as the development of their near-IR counterparts. There are at least two reasons for that.

- Narrow-gap semiconductors are energetically predisposed to nonradiative Auger decay rather than radiative recombination. (Auger recombination is an intrinsic three-body process where an electron and a hole recombine and the resulting energy is transferred to a third carrier, either an electron or a hole.)
- Internal losses due to free carrier absorption increase rapidly with increasing wavelength.

Thus, the level of difficulty in achieving high performance in the mid-IR tends to increase monotonically with increasing wavelength [5].

4.1.1 GaSb-based Diode Lasers

Starting from the 1980s, there have been renewed efforts, among the scientific community, to use III–V compound semiconductors, mostly GaSb-based materials, for mid-IR lasers [6]. Injection lasers based on GaInAsSb active layer and AlGaAsSb barriers (also referred to as GaSb-based lasers) appear to be very well suited for mid-IR generation. Theoretically, the bandgap of the GaInAsSb material grown lattice-matched on a GaSb substrate allows an emission wavelength between 1.7 and 4.2 μm [5]. The first RT operation of a diode laser based on GaInAsSb semiconductor material was reported by Bochkarev et al. [6]. A double GaAlSbAs/InGaSbAs/GaAlSbAs heterostructure was grown by liquid-phase epitaxy on a p-type GaSb substrate. Pulsed lasing was observed in the wavelength range 1.9–2.3 μm. Subsequently, CW operation at RT was achieved at $\lambda \approx 2\,\mu m$ [7] and at 2.2–2.4 μm [8].

A large leap in GaInAsSb laser performance occurred when quantum well (QW) active regions were employed to realize a much higher gain per injected carrier than in the bulk double-heterostructure devices. The first GaInAsSb/AlGaAsSb QW laser reported by Choi and Eglash was grown by molecular beam epitaxy (MBE); it had a low CW threshold current density (260 A/cm^2) and emitted up to 190 mW of CW output per laser facet at $\lambda = 2.1\,\mu m$ at RT [9]. At longer wavelengths, Lee et al. demonstrated pulsed operation at 2.7–2.8 μm in a GaInAsSb/AlGaAsSb QW laser. At 15 °C, the maximum average output power was 30 mW and the laser operated up to 60 °C [10].

With a new design of InGaAsSb/AlGaAsSb QW diode laser heterostructure consisting of two QWs, CW RT lasing with the wavelength up to 2.7 μm became possible [11]. A double QW structure was chosen in this case to avoid fast gain saturation, previously observed with a single QW active region. Heavily strained (by increasing the *In* composition) QWs were placed inside a waveguide laser structure (Figure 4.2a). Very low threshold current density

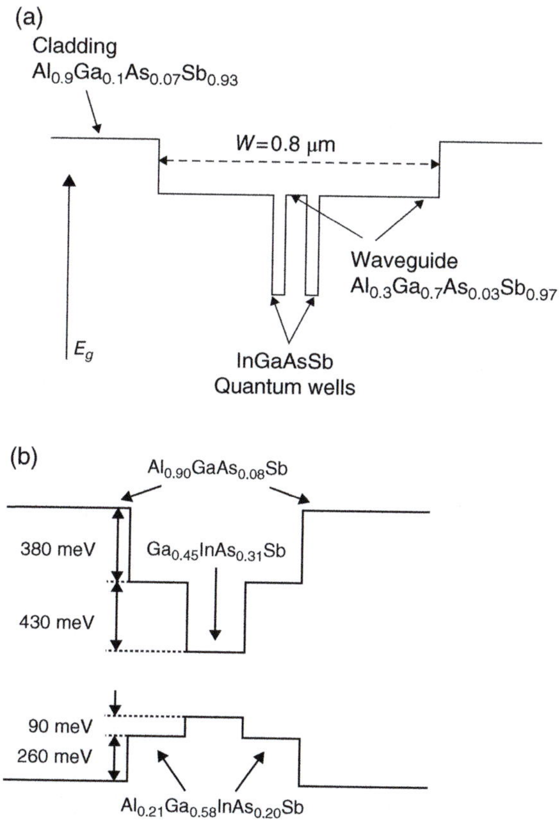

Figure 4.2 (a) Band structure of an InGaAsSb–AlGaAsSb quantum-well diode laser consisting of two QWs with separate confinement. *Source:* reproduced from figure 1 of [11], with permission of IEEE. (b) Band structure of a diode laser using a GaInAsSb QW and quinternary AlGaInAsSb barriers. *Source:* reproduced from figure 1 of [12], with permission of AIP, American Institute of Physics.

($300\,A/cm^2$) and high CW output powers (>100 mW) were obtained from devices operating in the whole 2.3–2.6-μm wavelength range [11]. Creating strain in the QW structure has beneficial influence on the operating characteristics of InGaAsSb/AlGaAsSb lasers. Either the compressive or the tensile strain of the QW region reduces the valence-band density of states, thereby lowering the threshold current since quasi-Fermi-level separation required for population inversion is achieved at lower carrier concentration. Strain also substantially reduces the Auger coefficients in InGaAsSb/AlGaAsSb QWs [5, 13].

As a result of design improvement (e.g. creating structures with increased QW compressive strain), it became possible, by the year 2007, to produce more than 1 W of CW output power at (or close to) room temperature from GaSb-based type-I QW diode lasers, in the whole range of 2–2.5 μm [14–17] (see Table 4.1). Kelemen et al. demonstrated linear GaInSb/AlGaAsSb laser arrays consisting of 19 emitters having resonator lengths between 1.0 and 1.5 mm and stripe widths between 90 and 200 μm [18]. The arrays emitted at different wavelengths between 1.8 and 2.3 μm and produced 20 W at 1.9 μm and 16 W at 2.2 μm in CW mode at 20 °C, with wall-plug efficiency (WPE) of up to 30%. Fiber-coupled bar-based laser modules have also been realized in this work.

By carefully controlling the compressive strain in the active layer, Lin et al. [30] pushed the GaSb laser operation to longer wavelengths. The authors demonstrated CW operation at 2.24–3.04 μm using GaInAsSb/AlGaAsSb double-QW ridge-waveguide lasers operating at temperatures up to 20 °C. Still, the threshold increased strongly with wavelength due to Auger recombination. For example, at $\lambda = 3.04$ μm, the threshold was three times that for a 2.24-μm device. GaSb-based diode lasers operating near 3 μm with heavily strained type-I QW with improved carrier confinement were reported by Shterengas et al. [20]. The devices had CW output powers of more than 200 mW at 3 μm at −23 °C (a temperature accessible with thermoelectric coolers) and more than 80 mW at 3.1 μm at 12 °C (Figure 4.3).

Introducing the quinternary AlGaInAsSb compound as a new barrier material for GaInAsSb QW laser diodes (Figure 4.2b) allowed the improvement of hole confinement in the GaInAsSb QWs and extended the emission wavelength to 3.26 μm, with ridge-waveguide lasers working in pulsed operation up to 50 °C [12]. Additional information on high-power RT CW GaSb-based type-I QW lasers operating at $\lambda = 1.95$–2.3 μm can be found in [31–34] and on lasers operating at $\lambda > 2.5$ μm in [35, 36].

Both the power conversion efficiency and the output power level of GaSb-based diode lasers that emit near or above 3 μm at RT were dramatically improved when cascade pumping schemes were introduced [21, 22, 37]. In such a cascade system the laser heterostructure has three single-QW GaInAsSb gain stages connected in series by means of GaSb/AlSb/InAs tunnel junctions followed by InAs/AlSb electron injectors. This provides (although at the

Table 4.1 Summary of GaSb-based injection lasers operating at or near room temperature (RT).

GaSb laser structure	Wavelength (μm)	Laser characteristics	Ref.
GaInAsSb/AlGaAsSb type I QW laser	2.1	190 mW CW at RT	[9]
GaInSb QW laser with broad-area 1000 × 150 μm^2	2.0	1.96 W CW at 16 °C	[14]
InGaAsSb/AlGaAsSb QW laser	2.3	1.15 W CW at 18 °C	[15]
InGaAsSb/AlGaAsSb compressively strained (1.6%) double QW laser	2.4	1.05 W, CW at RT	[16]
InGaAsSb/AlGaAsSb double QW laser	2.5	1 W, CW at 12 °C	[17]
GaInSb/AlGaAsSb linear laser array, 1-cm wide, 19 emitters	1.8–2.3	20 W CW at 1.9 μm; 16 W CW at 2.2 μm at 20 °C; wall-plug efficiency 30%	[18]
GaInAsSb/AlGaAsSb double QW	2.7–2.8	500 mW CW at 2.7 μm, 160 mW at 2.8 μm at 16 °C	[19]
GaInAsSb QW, heavily strained	3–3.1	200 mW at 3 μm at −23 °C, 80 mW at 3.1 μm at 12 °C	[20]
GaInAsSb/AlGaInAsSb triple QW	3.26	Pulsed (500 ns, 20 kHz), 10 mW at 20 °C, 1 mW at 50 °C	[12]
GaInAsSb QW, three-stage cascade pumping	3–3.2	960 mW CW at 3 μm, 500 mW at 3.1–3.2 μm at 17 °C	[21, 22]
GaInAsSb QW, narrow-ridge, two-stage cascade	3	107 mW CW at 17 °C in nearly diffraction-limited beam	[23]
DFB lasers			
GaInAsSb double QW, DFB ridge waveguide	2.97–3.02	3.6 mW single longitudinal mode (SLM) at 20 °C	[24]
GaInAsSb double QW, DFB ridge waveguide	3.06–3.065	6 mW SLM at 20 °C	[25]
GaInAsSb QW external-cavity laser with diffraction grating	3.18–3.24	1.8 mW SLM at 10 °C	[26]
GaInAsSb, cascade pumping, laterally coupled distributed feedback (LC-DFB)	2.65	25 mW SLM at 20 °C	[27]
GaInAsSb, cascade pumping, LC-DFB	2.89–2.9	13 mW SLM at 20 °C	[28]
GaInAsSb, cascade pumping, LC-DFB	3.27	15 mW SLM at 17 °C 40 mW SLM at −20 °C (tuning 0.27 nm/K)	[29]

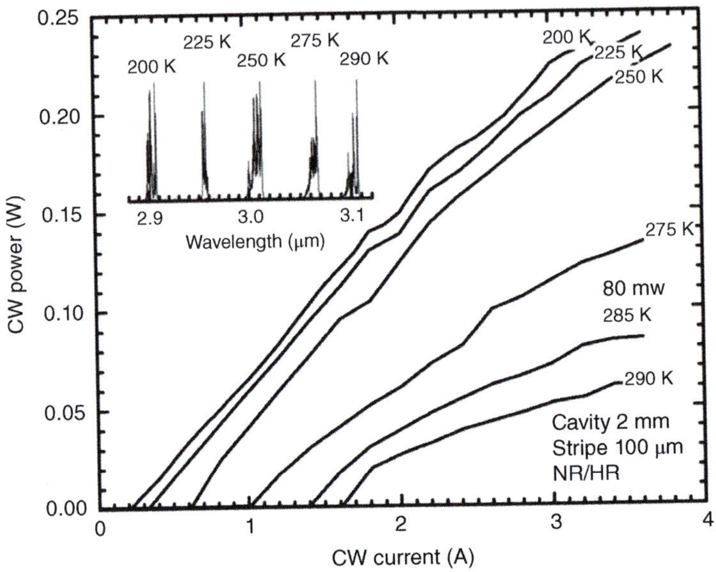

Figure 4.3 CW light-current characteristics of a 100-μm-wide, 2-mm-long laser structure based on compressively strained (1.8%) InGaAsSb quantum wells, measured in the temperature range 200–290 K. The inset shows the laser spectrum near threshold at different temperatures. *Source:* reproduced from figure 1 of [20], with permission of AIP, American Institute of Physics.

expense of higher driving voltage) carrier recycling between gain stages, and improves overall laser efficiency and the output power level. The devices demonstrated record low CW threshold current densities ~100 A/cm^2 and power conversion efficiencies of 16% at 17 °C. Three-stage cascade lasers with 100-μm-wide aperture demonstrated CW output power of 960 mW at $\lambda = 3$ μm and 500 mW at 3.1–3.2 μm.

To achieve diffraction-limited single spatial mode performance of mid-IR GaSb-based QW diode lasers, Liang et al. used a shallow narrow ridge-waveguide-type structure [23]. The laser used a two-stage cascade gain region and emitted at $\lambda \approx 3$ μm. The 2-mm-long laser devices with a ridge width of 5 μm (Figure 4.4) generated 107 mW of CW output power at 17 °C in nearly diffraction-limited beam.

4.1.2 Distributed Feedback GASb-based Lasers

Several applications, for example, molecular spectroscopy, require narrow-linewidth single longitudinal mode (SLM) laser emission. Narrow-linewidth distributed feedback (DFB) lasers based on GaInAsSb QWs and GaSb barriers and operating near 3 μm were fabricated using lateral chromium gratings as

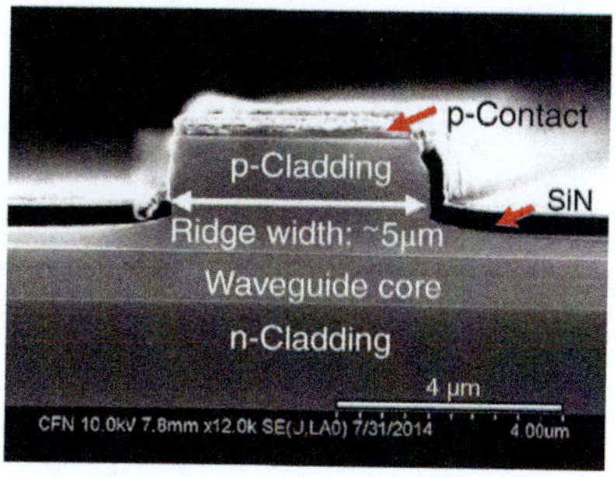

Figure 4.4 Scanning electron microscope image of the facet of a narrow-ridge (~5 µm) waveguide GaSb-based type-I quantum-well diode laser operating at $\lambda \approx 3\,\mu m$. *Source:* reproduced from figure 1 of [23], with permission of IEEE.

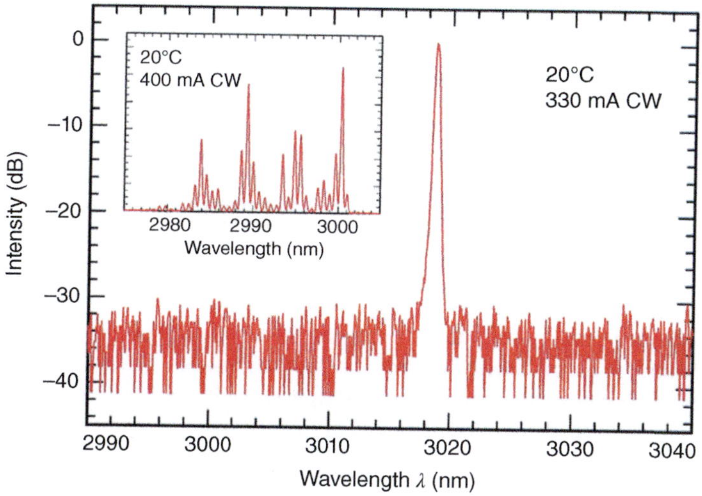

Figure 4.5 Emission spectrum of the DFB laser operating at a wavelength of 3.019 µm in CW mode at room temperature. The inset shows the emission spectrum of a reference laser (no DFB grating). *Source:* reproduced from figure 4 of [24], with permission of AIP, American Institute of Physics.

a wavelength selective element [24]. The maximum output power in the single-mode regime was 3.6 mW at 20 °C. Figure 4.5 presents the emission spectrum of such a DFB laser that shows single-mode emission at 3.019 µm with a side-mode suppression ratio of more than 30 dB. The corresponding laser spectrum of a ridge-waveguide reference sample with no grating is shown as an inset. By varying the metal grating period between 405 and 410 nm, the single-mode emission wavelength of the laser structure was varied from 2.97 to 3.02 µm. Similarly, Belahsene et al. achieved single-mode CW laser operation suitable for gas sensing in a ridge-waveguide double-QW GaInAsSb laser that had a DFB metal grating deposited next to its sidewalls. The single-mode laser was tunable from 3.060 to 3.065 µm by changing the pump current with a maximum power of 6 mW at 20 °C [25].

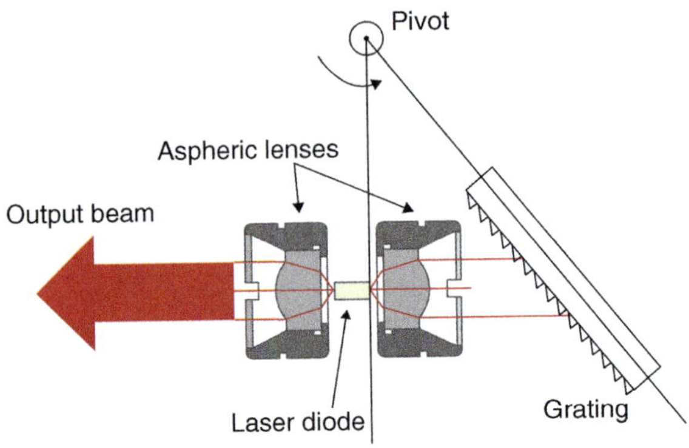

Figure 4.6 Schematic of an external-cavity GaSb laser in a modified Littrow configuration. *Source:* reproduced from figure 1 of [26], with permission of Institution of Engineering and Technology (IET).

Using an external-cavity laser, Gupta et al. [26] produced narrow-band tunable output around 3.2 μm. The laser used 17-nm InGaAsSb compressively strained QWs and 30-nm AlInGaAsSb barriers. The external cavity contained a diffraction grating in a Littrow configuration (Figure 4.6). In CW operation at 10 °C, the laser produced 1.8 mW of single-mode output power with a tuning range of 60 nm (3.180–3.241 μm). Additional fine-tuning was achieved using temperature (at a rate of $-0.26\,cm^{-1}/K$) and diode current tuning (at a rate of $-0.0165\,cm^{-1}/mA$). This device is useful for molecular spectroscopy, since it can be tuned near the fundamental vibrational absorption features of hydrocarbons.

Recent development of a new type of GaSb-based DFB lasers – laterally coupled distributed feedback (LC-DFB) ridge laser diode lasers (Figure 4.7) that minimize the intracavity loss [38, 39] – resulted in large improvement of the output power of narrow-linewidth lasers operating between 2.65 and 3.4 μm [27–29, 40–42] (see Table 4.1 for main results).

Overall, contemporary GaSb-based type-I QW diode lasers can operate in CW regime at RT with up to 3.44 μm wavelength [13]. Once the lack of sufficient hole confinement was identified as the main mechanism limiting RT performance of such lasers, the new barrier quaternary alloys were introduced leading to record-low threshold current densities (~100 A/cm^2 near 2 μm and ~200 A/cm^2 near 3 μm), comparable to that of the best GaAs near-IR lasers. Also, optimization of the laser heterostructure design has led to the reduction of the role of Auger recombination resulting in a remarkable enhancement of device efficiency and lasing threshold. Further improvement of the output power levels was demonstrated recently by introducing a cascaded pumping scheme, in the whole spectral range from 1.9 to 3.3 μm (2 W per 100 μm stripe near 2 μm, 960 mW near 3 μm, 500 mW near 3.15 μm, and 360 mW near 3.25 μm). Still, these devices exhibit strong temperature dependence of their performance.

The main results for GaSb-based diode lasers are presented in Table 4.1.

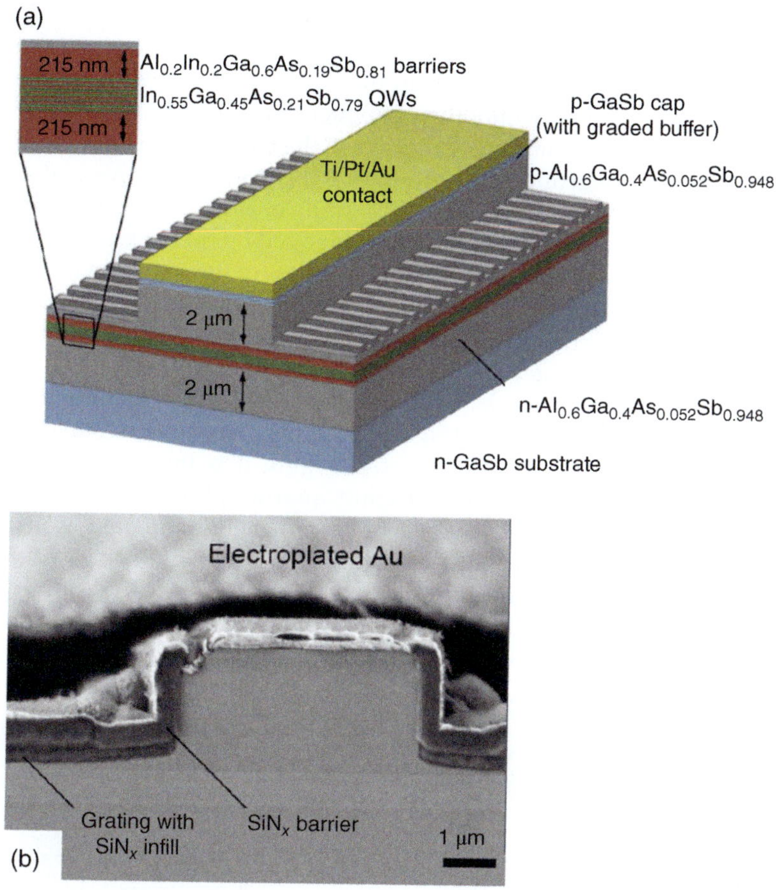

Figure 4.7 Schematic of an index-coupled distributed-feedback diode laser at $\lambda = 2.65$ μm. The laser employs InGaAsSb/AlInGaAsSb multi-quantum-well structures grown by molecular beam epitaxy on GaSb. Single-longitudinal-mode emission is achieved using laterally coupled second-order Bragg gratings (LC-DFB) etched alongside a narrow-ridge waveguide. Index modulation, instead of loss modulation with metal gratings, results in higher output power. (a) The LC-DFB laser structure (SiN_x barrier layer and electroplated Au contact are not shown). (b) Scanning electron micrograph of a fabricated laser ridge. *Source:* reproduced from figure 1 of [27], with permission of OSA, The Optical Society.

4.2 Quantum Cascade Lasers

First operated at Bell Labs by Faist et al. [43], the quantum cascade lasers (QCLs) use a totally different operating principle, as compared to conventional inter-band heterojunction semiconductor lasers, by relying on a process for light emission that is independent of the semiconductor bandgap. Instead of using electrons at the bottom of the conduction band and holes at the top of the valence band, which recombine to produce light of frequency $\nu \approx E_g/h$, where E_g is the energy bandgap and h is Planck's constant, QCLs use only one type of charge carriers – electrons – that undergo quantum jumps between energy levels E_n and E_{n-1} to create a laser photon of frequency $\nu = (E_n - E_{n-1})/h$. These energy levels (Figure 4.8a) do not exist naturally in the constituent materials of the active region but are artificially created by structuring the active region into QWs of nanometer thickness [5, 45]. The motion of electrons perpendicular to the layer interfaces is quantized and characterized by energy levels (subbands) whose difference is determined by the thickness of the wells and by the height of the energy barriers separating them. As a result, QCL wavelengths can be tailored over a wide range: 3–24 μm in the mid-IR and 64–225 μm in the far-IR (terahertz) spectral range.

In QCLs, unlike in a laser diode, an electron remains in the conduction band after emitting a laser photon (Figure 4.8a). The electron is recycled because after emitting a photon in an active region it is injected into an adjacent identical active region, where it emits another photon, and so the process continues. To achieve this process of cascaded emission of photons, active regions are alternated with doped electron injectors, and a bias voltage is applied. The active-region-injector stages of QCLs give rise to an energy staircase in which photons are emitted at each of the steps. Each step

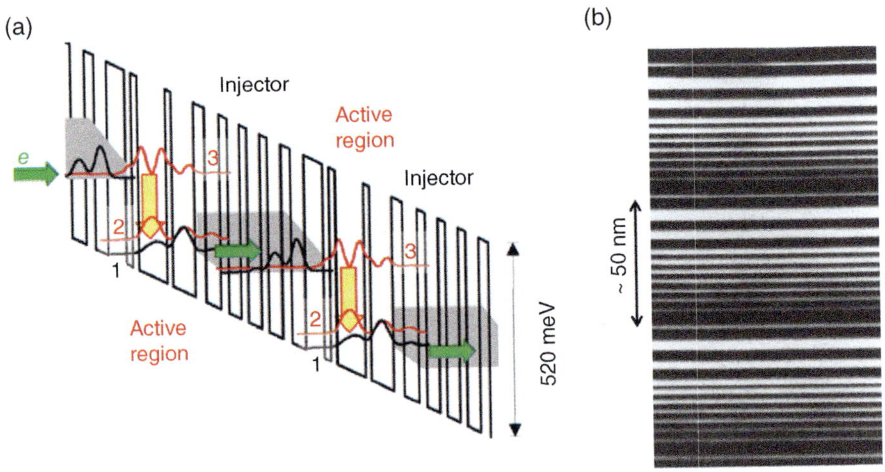

Figure 4.8 (a) Energy diagram of two stages of a quantum cascade laser emitting at $\lambda = 7.5\,\mu m$ when electric field ~50 kV/cm is applied, showing the energy levels and electron wavefunction probability distributions. Each stage (injector plus active region) is ~50 nm thick. The energy wells and barriers are made, respectively, of GaInAs and AlInAs alloys, lattice-matched to InP. (b) Transmission electron microscope (TEM) micrograph of a portion of an exemplary QCL structure. The black and white regions represent the barriers and wells, respectively. *Source:* reproduced from figure 1 of [44], with permission of SPIE.

contributes to an optical gain and, for lasers designed to emit in the mid-IR range, the number of stages typically ranges from 20 to 50 [44, 45]. The active region contains three quantized states, and the laser transition is defined by the energy difference between states 3 and 2 in Figure 4.8a, which is determined by the thickness of the QWs. To get population inversion between levels 3 and 2, one needs to have the lifetime of level 3 to be longer than that of state 2. To achieve this condition, the lowest level 1 is positioned about an optical phonon energy (≈34 meV) below level 2; this ensures that electrons rapidly scatter from level 2 to level 1 by emission of an optical phonon. Because of its resonant nature, this process is very fast: the 2–1 relaxation time is on the order of 0.1–0.2 ps. Electrons in level 3 have instead a longer lifetime, on the order of 1 ps or more. Finally, electrons are injected into the upper laser level of the next QCL stage by a process known as resonant tunneling, which ensures highly selective injection when the applied voltage is increased above a certain value. Shown in Figure 4.8b is a transmission electron micrograph of the layer structure of an exemplary QCL. The dark layers correspond to the AlInAs barrier layers, while the bright ones to GaInAs wells.

Thus, interband and intersubband semiconductor lasers are radically different devices [45] with the main dissimilarities listed below.

In an interband laser:

- The photon energy is determined by the semiconductor bandgap.
- The upper-state lifetime and oscillator strength depend on a semiconductor material.
- The laser upper-state lifetime is measured in nanoseconds.
- The population inversion is obtained through injecting electrons and holes.

In an intersubband laser:

- The photon energy is determined by the QW thickness.
- The upper- and lower-state lifetimes and oscillator strengths can be engineered.
- The laser upper-state lifetime is short, typically 1–3 ps.
- The population inversion is established by pump current and via a suitable design of the active region.

The first QCL was fabricated in the InGaAs/InAlAs material system lattice-matched to an InP substrate [43]. In 1998, Sirtori et al. demonstrated a QCL based on a different material system, GaAs/AlGaAs [46], proving that the QCL concept is not restricted to one material system. Later, QCLs based on an InAs/AlSb material system were implemented for the development of short-wavelength ($\lambda < 3\,\mu m$) QCLs [47, 48] (see Section 4.2.4).

4.2.1 High Power and High Efficiency QCLs

The holy grail of QCL was to make them work (i) at RT, (ii) in a CW regime, (iii) with high (more than 10%) WPE, and (iv) more than 1 W of average power. The first QCL operated in the pulsed mode at a wavelength of 4.2 μm with much less than a milliwatt average power and at cryogenic temperatures below $T = 90\,\text{K}$ [43]. In 1996, an RT QCL operating in pulsed mode [49] and in 2002, an RT QCL operating in the CW mode [50] were demonstrated. In the latter case, the gain region consisting of a lattice-matched InGaAs/InAlAs core was vertically and laterally buried within InP cladding layers. This geometry greatly improved the heat transport by allowing heat flow from all sides of the active region. The produced CW output power was 17 mW at 292 K. The emission wavelength was $\lambda = 9.1\,\mu\text{m}$.

Subsequently, through optimization of material and processing technology, QCL power efficiency and output power were dramatically enhanced. In 2008, the Razeghi team at Northwestern University demonstrated 2.5 W CW output power at close to RT ($T = 288\,\text{K}$) with a WPE of 12.5% at $\lambda = 4.6\,\mu\text{m}$ [51]. The authors used an uncoated GaInAs/AlInAs QCL device with the ridge width 10.6 μm and cavity length 4.8 mm that was bonded epilayer-down to a diamond submount. At cryogenic ($T = 80\,\text{K}$) temperature and similar design (GaInAs/AlInAs heterostructure, 19.7-μm-wide, 5-mm-long, ridge-waveguide design) the team demonstrated an average CW power of 10 W at $\lambda = 4.6\,\mu\text{m}$, with the WPE reaching 33% (at 7-W output). Exemplary power–current–voltage curves for cryogenic temperatures ranging from 80 to 240 K are shown in Figure 4.9.

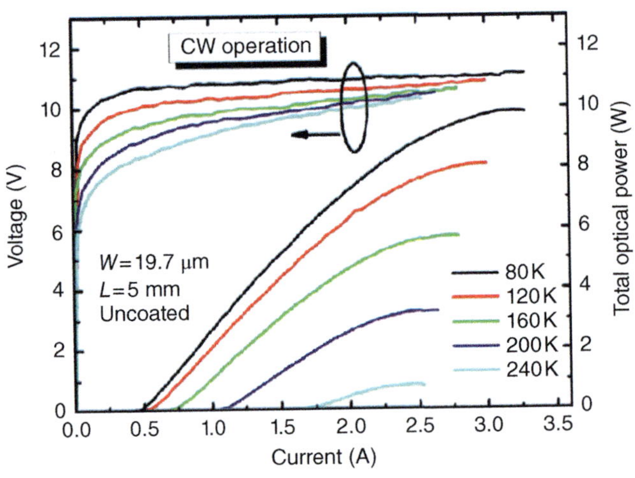

Figure 4.9 Power–current–voltage behavior of an uncoated QCL operating in the CW mode at temperatures ranging from 80 to 240 K. The optical power was measured from one facet and doubled to give the total power that reached 10 W at 80 K. The maximum wall-plug efficiency reached 33%. *Source:* reproduced from figure 1a of [51], with permission of AIP, American Institute of Physics.

By 2011, the maximum RT CW power from a single-facet GaInAs/AlInAs/InP QCL device reached 5.1 W at λ = 4.9 μm (Bai et al. [52]). The core of a buried-ridge QCL device consisted of 40 stages with a ridge width of 8 μm and laser cavity length 5 mm. At RT (298 K) and CW operation the WPE reached 21%, while in the pulsed operation it was as high as 27%.

Lyakh et al. reported fully hermetically packaged (Figure 4.10) QCLs using strain-balanced InP-based structures grown by metal–organic chemical vapor deposition (MOCVD) emitting high power (>1.5 W) at 4.6 μm at RT [53]. Subsequently, the same team reported a QCL emitting at 4.7 μm that generated over 4.5 W of optical power in the CW regime and fundamental transverse TM_{00} mode at T = 283 K. A buried strain-balanced $Al_{0.78}In_{0.22}As/In_{0.72}Ga_{0.28}As$/InP heterostructure consisted of a 10-mm-long, 7.5-μm-wide central laser section that was tapered to 20 μm width at the laser facets [54, 55]. The maximum WPE of 16.3% for this laser was reached at an output power of 4 W [55]. Figure 4.11 plots the main characteristics of such a laser.

On the longer wavelength side, a power-efficient QCL at λ = 7.1 μm with strain-balanced active region and low voltage defect (electron energy difference between the lower laser level of one gain stage and the upper laser level of the next stage) was reported [56]. Strain in the active region was used to increase the conduction-band offset and hence improve carrier confinement. The maximum WPE at RT (T = 293 K) was 19% in pulsed mode and 10% in CW mode, with up to 1.4 W of measured CW power. Lyakh et al. reported a multiwatt RT QCL at λ = 9.2 μm based on a strain-balanced AlInAs/InGaAs/InP structure grown by MBE, in buried heterostructure geometry.

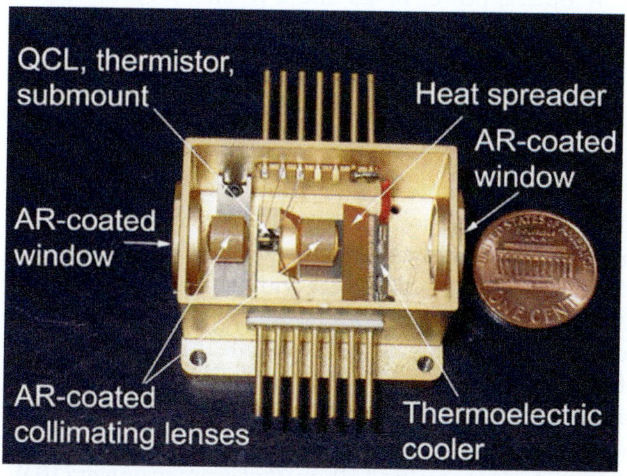

Figure 4.10 Hermetic butterfly-style package of a high-power (>1.5 W) strain-balanced InP-based QCL operating at 4.6 μm at room temperature. The top lid is removed for clarity. *Source:* reproduced from figure 4 of [53], with permission of AIP, American Institute of Physics.

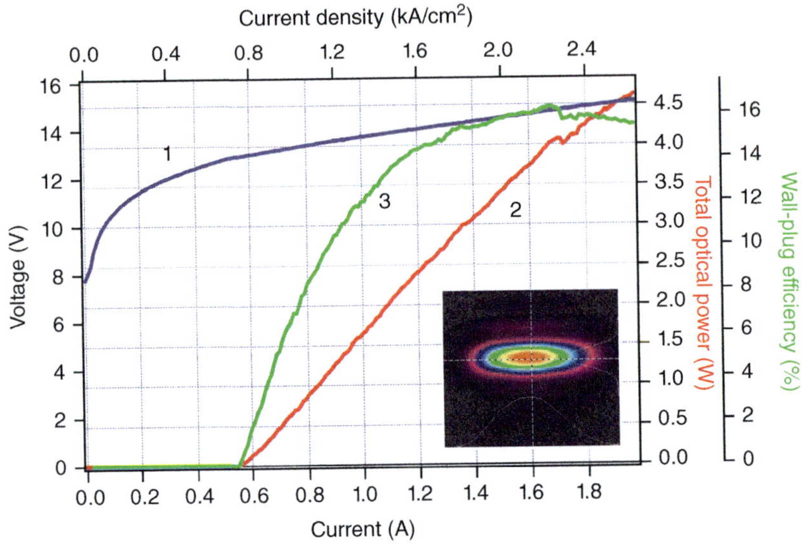

Figure 4.11 Pump current dependence of voltage (curve 1), total optical power (curve 2), and wall-plug efficiency (curve 3) of a room-temperature (283 K) hermetically packaged 4.7-μm QCL mounted on an AlN/SiC submount. The inset shows a fundamental transverse TM_{00} beam profile. *Source:* adapted from figure 3 of [55], with permission of OSA, The Optical Society.

Maximum pulsed and CW optical power of 4.5 and 2 W, and WPE of 16 and 10%, respectively, were demonstrated for a laser having 3-mm-long and 10-μm-wide active region, mounted epi-side down on an AlN/SiC composite submount [57].

One of the highest QCL power efficiencies was achieved at cryogenic (T = 80 K) temperature and pulsed-mode operation (5 kHz repetition rate and 100 ns pulse width) using a 13.6-μm-wide and 2.9-mm-long ridge waveguide based on an InGaAs/AlInAs/InP material system with a special design that uses an ultra-strong coupling between the injector and active multiple quantum well (MQW) regions. The WPE reached 47% with 10 W of peak optical output power (5 mW average power) at λ = 4.5 μm [58]. The absolute highest wall plug QCL efficiency reported to date, 53% – which means that more light is produced than heat from the power supply – was achieved at T = 40 K in a pulsed mode with a repetition rate 250 kHz and pulse width 200 ns (duty cycle 5%). The QCL was based on a buried-ridge GaInAs/AlInAs/InP structure with a single-well injector design. The laser emitted 500 mW of average power (10 W peak), and operated at λ = 5 μm [59].

4.2.2 Single-mode Distributed Feedback (DFB) QCLs

Fabry–Pérot laser is the simplest of QCLs. An optical waveguide is first fabricated out of the quantum cascade material to form the gain medium. The ends of the device are then cleaved to form a Fabry–Pérot resonator. The residual reflectivity on the cleaved facets from the semiconductor-to-air interface is sufficient to make the laser operate. Although Fabry–Pérot QCLs are capable of producing high powers, they are typically multi-longitudinal-mode at higher operating currents. The spectrum of QCL in such a cavity design is broad (10–50 cm^{-1}) and depends on driving current and operating temperature (curve 1 in Figure 4.12). Yet, mid-IR sensor applications require a tunable source with a narrow linewidth, smaller than the pressure-broadened absorption linewidth of gases at atmospheric pressure (typically 0.1–0.2 cm^{-1}).

In a DFB QCL [60] a grating is built on top of the laser waveguide to prevent it from emitting at other than the desired wavelength. The grating of periodicity Λ creates modulation of the real and imaginary part of the effective index (n_{eff}) of the laser mode. This forces single-mode operation of the laser, even at high operating currents, with the wavelength (λ_B) that is determined by the Bragg reflection condition $\lambda_B = 2n_{eff}\Lambda$. Tuning of DFB lasers is achieved by varying the temperature, which changes the effective refractive index n_{eff} and therefore shifts the wavelength at which the Bragg condition is attained. Faist

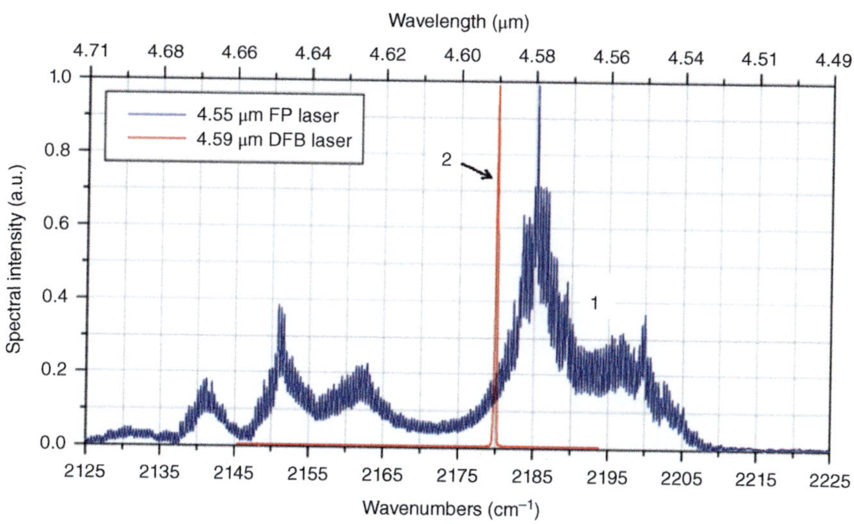

Figure 4.12 Fabry–Pérot versus DFB laser comparison near $\lambda = 4.6\,\mu m$. Fabry–Pérot lasers (curve 1) have broadband emission, while DFB lasers (curve 2) emit at a well-defined wavelength. *Source:* reproduced from www.thorlabs.com, with permission of Thorlabs.

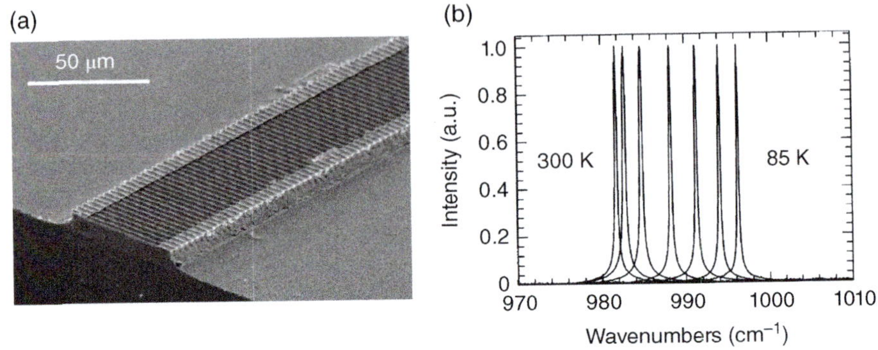

Figure 4.13 (a) Scanning electron microscopy picture of a DFB quantum cascade laser. The central top part of the ridge is free of metal. (b) Lasing spectra of a 45-μm-wide and 1.2-mm-long DFB QCL at different temperatures between 85 and 300 K. *Source:* adapted from figures 1 and 4 of [61], with permission of AIP, American Institute of Physics.

et al. reported a tunable single-mode pulsed DFB laser operating near $\lambda = 8\,\mu m$ below and above RT [60]. In their design, the mode interaction with the top metal contact (and hence loss) was modulated by modulating the thickness of the heavily doped cap semiconductor layer. By changing the laser operating temperature between 80 and 315 K, the output was continuously tunable in wavelength from 7.78 to 7.93 μm (frequency tuning over $24\,cm^{-1}$). Similar results were obtained for the laser operating near 5 μm [60]. The spectrum was single mode over a temperature range 80–315 K and in this temperature interval the device tuned continuously from 5.31 to 5.38 μm (frequency tuning over $24.5\,cm^{-1}$). The reported linewidth of $0.3\,cm^{-1}$ was limited by dynamic heating during the 10-ns-long pulse.

Hofstetter et al. demonstrated a DFB QC laser at $\lambda = 10.16\,\mu m$ that emitted 80 mW of CW optical power in a single-mode operation. The holographically defined grating with 1.59 μm period was created on the top of an n-doped InGaAs cap layer [61]. Figure 4.13 shows a scanning electron microscopy picture of such a DFB QCL, as well as its lasing spectra at different temperatures between 85 and 300 K. The emission wavelength was tuned by temperature from 10.040 μm ($996\,cm^{-1}$) at 85 K to 10.183 μm ($982\,cm^{-1}$) at 300 K.

In the CW mode, DFB QCLs are characterized by a very narrow spectral linewidth: 0.1–3 MHz without frequency stabilization and <10 kHz with frequency stabilization. In the pulsed mode the linewidth is broader, ~300 MHz, due to frequency chirp from dynamic heating. Alpes Lasers manufactures a variety of DFB QC lasers operating in the pulsed or CW mode at RT, with wavelength ranging from 4.3 to 10.5 μm.[1]

1 www.alpeslasers.com.

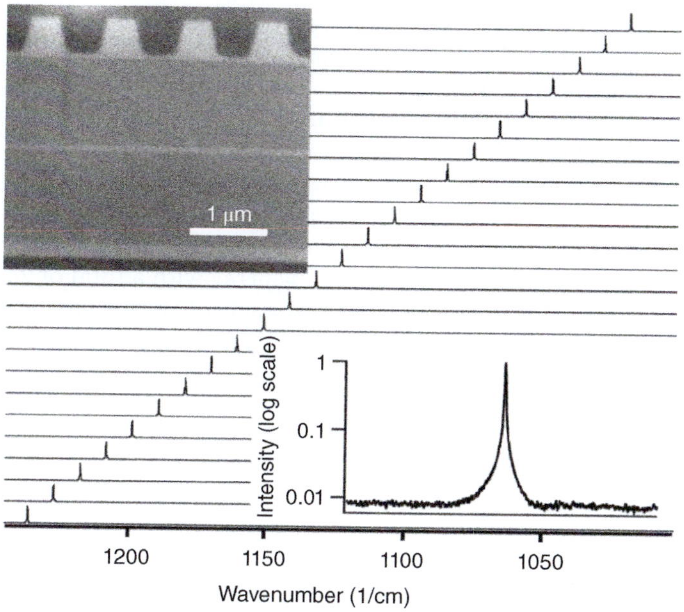

Figure 4.14 Spectra of 24 single-mode DFB lasers in one array. Laser frequencies are spaced 9.5 cm^{-1} apart and span 220 cm^{-1}. The inset at the top left is a scanning electron micrograph showing a cross section of the device, along the laser ridge. The inset at the bottom right shows a log-scale spectrum of a representative laser in the array, showing side-mode suppression >20 dB. *Source:* reproduced from figure 1 of [65], with permission of IEEE.

The Northwestern University team reported RT CW DFB QCLs with very high average power at λ = 4.6–4.8 μm [62, 63]. For example, the team demonstrated a DFB QCL with the CW output power 2.4 W at λ = 4.8 μm with WPE of 10% at T = 298 K. The laser consisted of an 11-μm-wide and 5-mm-long waveguide, and the resonator was formed by two end facets – one with high reflection and another with antireflection coating, so that the main role in achieving the laser action was played by the top surface grating. The laser showed single-longitudinal-mode operation with the side-mode suppression of 30 dB and had single-lobed far-field distribution. The laser frequency was tuned over 5 cm^{-1} by changing the pump current [63].

Several QCL applications, such as spectroscopic detection of complex organic molecules or the analysis of multicomponent gases, need broad tunability. To bypass the limitation imposed by a small tuning range of a single DFB laser (typically 5–20 cm^{-1}), Lee et al. implemented a multiwavelength QCL array [64, 65]. All DFB lasers in the array used the same gain medium with a bound-to-continuum design [66]. Due to a very broadband gain profile, the laser operation was possible between 8 and 10 μm. Figure 4.14 depicts the spectra of an array consisting of 24 single-mode DFB lasers. Laser ridges in the array were defined on top of the buried gratings and were 15 μm wide and spaced 75 μm apart. The laser frequencies were separated by a 9.5 cm^{-1} interval and span a total range of 220 cm^{-1} [65].

The current record, in terms of mid-IR spectral coverage, was achieved in [67] where a broadband heterogeneous QCL structure based on strained AlInAs/GaInAs material system with DFB grating laser array was emitting between 5.9 and 10.9 μm (~760 cm^{-1}) from a single wafer at RT. The active region consisted of six QC stages, grown on top of each other, corresponding to different central wavelengths. A DFB grating array with a total of 24 lasers was patterned by e-beam lithography, and each quantum cascade stage design and spatial arrangement of stages was engineered to exhibit a flat threshold current density across the whole range of wavelengths.

An excellent review of multiwavelength QCL arrays and their applications is given in [68].

4.2.3 Broadly Tunable QCLs with an External Cavity

An external-cavity QCL represents another scenario for achieving broadband and continuous tuning of a QCL with a narrow-line laser emission. In such a laser, one of the cleaved waveguide facets of the gain medium has an antireflection coating that prevents the lasing action on the facets. If a frequency-selective element – a diffraction grating – is included in the external cavity, it is possible to reduce the laser emission to an SLM and tune the radiation.

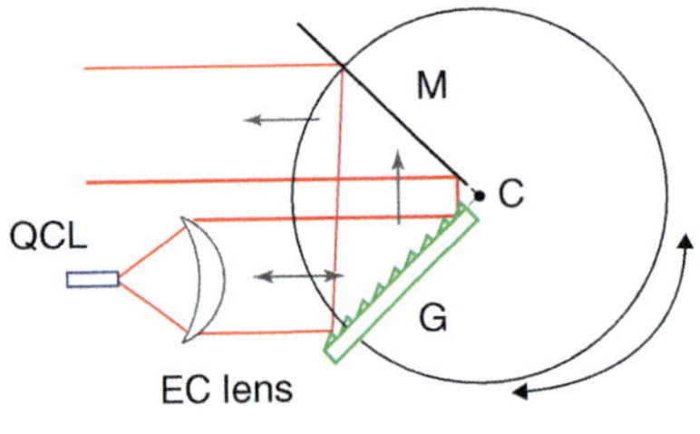

Figure 4.15 External-cavity quantum cascade laser. Frequency-selective optical feedback is provided by a diffraction grating (G) in Littrow configuration. The first-order diffracted beam is back-coupled into the laser, whereas the zero-order beam is collected through an additional mirror (M). Coarse tuning is achieved through the rotation of the grating around point C. This geometry prevents the walk-off of the outcoupled beam when the laser is tuned. *Source:* reproduced from figure 4 of [69], with permission of IOP.

Figure 4.15 shows a Littrow configuration of the external-cavity QCL [69]. The first-order diffracted beam is back-coupled into the laser, whereas the zero-order beam is collected through an additional mirror. Coarse tuning is achieved through the rotation of the grating.

Luo et al. [70] showed that for a grating-coupled quantum-cascade laser near $\lambda = 5.1$ μm operating in the pulsed mode, the output wavelength can be tuned over 140 nm at $T = 80$ K and over 127 nm at 243 K, the tuning range being limited by the gain band of the QCL at each temperature.

Gmachl et al. demonstrated an ultra-broadband QCL by a special design of an active region, namely by putting together many dissimilar optical transitions [71]. A strongly inhomogeneous system was well suited for simultaneous laser action on a continuum of wavelengths. In pulsed mode, at peak pump currents above 8 A, and at cryogenic (10–100 K) temperatures, supercontinuum emission from 6 to 8 μm wavelength was achieved [71].

For single-mode operation in an external-cavity configuration, it is desirable, however, to have a homogeneously broadened gain curve since the laser power can be concentrated in an SLM. Maulini et al. realized a heterostructure with an active region having broad gain curve with the electroluminescence spectrum being about 300 cm^{-1} wide [72], based on the so-called bound-to-continuum design. In this scheme, radiative transitions occur between a single initial state (upper state) located close to the injection barrier and a quasi-miniband of final states delocalized over the coupled QWs of a chirped superlattice [66]. Due to the fact that all the transitions share the same upper state, laser action at a particular wavelength will reduce the gain over the whole range. This reduction will be homogeneous if the populations of the lower states are negligible, which is the case in most of QCLs because of fast relaxation by emission of optical phonons. In a grating-coupled external-cavity configuration of [72], the laser was tuned between 9.11 and 10.56 μm by varying the diffraction grating angle. This tuning range corresponds to 15% of central wavelength and a frequency span of 150 cm^{-1}.

In general, QCL gain drops with increasing bandwidth and realizing both a broad gain spectrum and, at the same time, a low threshold current value seems to impose two mutually exclusive requirements. Nevertheless, because of the larger matrix element in the bound-to-continuum design, the latter does not suffer from a significantly increased threshold current density [71]. Combining broad bound-to-continuum gain regions with different central wavelengths allows even broader gain spectra. For example, using a two-stack bound-to-continuum QCL with one region centered at 8.2 μm and the other at 9.3 μm, Wittmann et al. achieved, at RT, a single-mode tuning range of 7.7–9.9 μm (frequency span 292 cm^{-1}) in pulsed mode, and 8–9.6 μm (201 cm^{-1}) in CW mode, with a maximum CW power of 135 mW [73].

Hugi et al. demonstrated a broadband QCL based on an active region incorporating five different cascade designs [74]. The gains of each cascade peaked at the wavelengths of 7.3, 8.5, 9.4, 10.4, and 11.5 μm and all cascades were based on a bound-to-continuum design with inherent broad gain spectrum. The laser operated at 15 °C at a duty cycle of 1.5% (pulse duration 15 ns, repetition rate 1 MHz). With the grating-coupled external-cavity design, it was tunable from 7.6 to 11.4 μm with an average output power of 15 mW (peak power 1 W). With a total tuning range of over 432 cm^{-1} (with the linewidth that varied between 0.12 and 2 cm^{-1}), this single QCL source covers an emission range of over 39% around the center QCL frequency.

Yao et al. used a slightly different QCL design, namely based on a "continuum-to-continuum" approach, where multiple transitions from several upper laser states to several lower laser states occur, and showed that one can realize a broad gain spectrum of 430 cm^{-1} in the 4–5 μm wavelength region. In the pulsed-mode operation (100 ns, 5 kHz) and in a Fabry–Pérot mode with no wavelength selection, these lasers have demonstrated high WPE, i.e. 20% at RT (295 K) and 40% at 80 K [75]. Fujita et al. reported a QCL device that exhibited an enormously wide electroluminescence spectrum of 600 cm^{-1} centered at 6.8 μm, based on anti-crossed dual-upper-state to multiple-lower-state design that potentially can be beneficial for broadband tuning [76]. In the anti-crossed design, the transitions take place from a specially engineered dual upper state to multiple lower states, so that numerous transition channels lead to a broad-gain spectrum. Despite the broad spectrum, the buried heterostructure laser (with no spectral selection) has shown low threshold and operated CW up to 102 °C. At RT (27 °C), the maximum CW output power reached 528 mW.

Recently, Lyakh et al. demonstrated a broadband external-cavity QCL with an ultrafast tuning capability. The emission wavelength was controlled by an intracavity acousto-optic modulator (AOM) based on a germanium crystal (Figure 4.16). By changing the AOM driving radiofrequency from 41 to 49 MHz, the laser wavelength was tuned from 8.5 to 9.8 μm (frequency tuning range 150 cm^{-1}). The measured switching time between any two wavelengths within the tuning range was extremely fast, less than 1 μs, so that spectral measurements in the 8.7–9.6 μm band took less than 20 μs. The QCL delivered 350 mW of the average power at the center of the tuning range, with a typical linewidth of 4.7 cm^{-1} [77].

An excellent review on external-cavity QCLs can be found in [69].

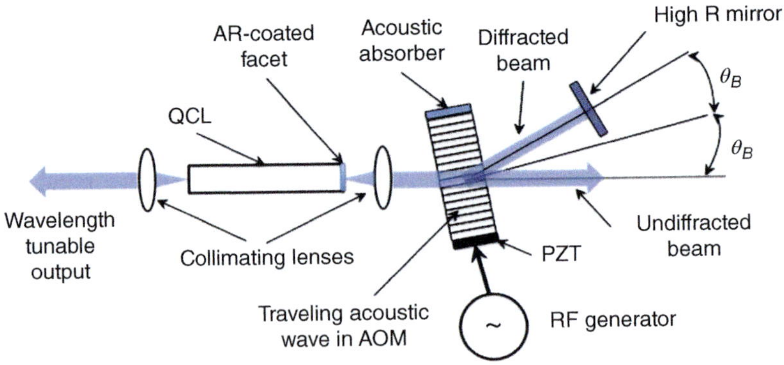

Figure 4.16 Schematic of a broadband (8.7–9.6 μm) fast-tunable external-cavity QCL based on intracavity acousto-optic modulator (AOM) in Littrow configuration. *Source:* reproduced from figure 1 of [77], with permission of AIP, American Institute of Physics.

4.2.4 Short-wavelength (<4 µm) QCLs

Operating QCLs in the atmospheric transparency window between 3 and 5 µm is desirable for many purposes, especially molecular spectroscopy and monitoring of the atmosphere. Yet, for wavelengths below 4 µm, the realization of efficient QCLs using standard material systems is challenging due to material limitations, namely because of insufficient conduction-band offset required to accommodate the laser transition inside the QW [44].

In the most widely used AlInAs/GaInAs material system, a short-wavelength QCL emitting at 3.4 µm was demonstrated using strain-compensated AlInAs/GaInAs on InP substrates [78]. Watt-level peak power RT emission at 3.3 µm was also demonstrated in a strain-compensated system based on InGaAs/InAlAs/AlAs on InP substrate, with pulsed-mode operation above 350 K [79]. An increase of the In concentration in QWs and of the Al content in the barriers made it possible to obtain a conduction-band offset of 1.2 eV in the strain-compensated $Ga_{0.21}In_{0.79}As/Al_{0.89}In_{0.11}As$ InP-based structures and demonstrate RT emission at 3.0 µm [80]. The conduction-band offset in heterostructures grown on InP can also be increased, up to 1.2–1.6 eV, by use of AlSbAs [81, 82] or AlAs-based barriers [83]. These systems allowed realizing QCLs emission at 3.1 µm [82] and 3.05 µm [83].

Lately, QCLs based on a new material system – InAs/AlSb – have brought significant progress in short-wavelength QCLs due to the high conduction-band offset of 2.1 eV [84]. (See an excellent review on InAs/AlSb QCLs in [85].) In particular, the first QCL emitting below 3 µm in an InAs/AlSb system was demonstrated by Devenson et al. [86]. The same team also reported QCLs operating at temperatures of up to 400 K at $\lambda \approx 3.3$ µm [84]. By cooling to cryogenic temperatures, wavelengths as short as 2.75 µm [47] and 2.63 µm [48] have been demonstrated. However, producing short-wavelength outputs from QCLs comes at a cost: the maximum operation temperatures of the InAs/AlSb QCLs decreased from 400 K for lasers operating at λ = 3.3 µm to 285 K for λ = 2.88 µm, and to 140 K for the devices with λ = 2.75 µm output [47].

A number of short-wavelength *single-frequency* QCLs have been realized as well. For instance, a pulsed RT single-mode DFB InAs/AlSb QCL at a wavelength 3.34 µm with a peak output power of 0.8 W was demonstrated in [87]. A record-wide (in frequency units) continuous external-cavity QCL spectral tuning of 556 cm^{-1} in the spectral region 3.28–4.01 µm was achieved by Riedi et al. [88]. In this work, a heterogeneous active region was based on the strain-compensated InGaAs/InAlAs-AlAs material system. The two active regions with center wavelengths of 3.3 and 3.7 µm were combined into a heterogeneous double-stack active region. Both stacks had bound-to-continuum transitions to intrinsically broaden the gain spectra of each stack. The spectra corresponding to the single-mode tuning regime are displayed in Figure 4.17.

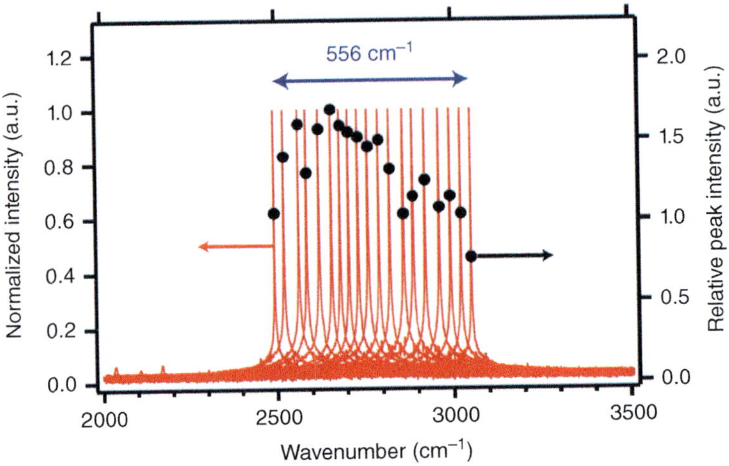

Figure 4.17 Single-mode tuning of a short-wavelength external-cavity QCL between 3.28 and 4.01 μm (2494–3050 cm^{-1}). The spectra are normalized; the black dots show relative intensities. *Source:* reproduced from figure 5 of [88], with permission of AIP, American Institute of Physics.

4.2.5 QCLs at Long (16–21 μm) Wavelengths

The InAs/AlSb material system has also been successfully utilized for long-wavelength mid-IR generation at $\lambda > 16\,\mu m$ [85]. The favorable property of InAs is that it has small electron effective mass that allows achieving high intersubband gain, even at long wavelengths. (The intersubband gain increases with decreasing effective mass [85].)

One of the main motivations for this development is the atmospheric transparency window at >16 μm that allows performing spectroscopic measurements. RT operation has been demonstrated in InAs-based QCLs utilizing metal–metal optical waveguides in the whole range between 16 and 21 μm [89–91]. Moreover, long-wavelength mid-IR single-frequency DFB QCLs based on InAs/AlSb have been realized near 17–18 μm [91] (Figure 4.18). (See also Table 4.2 for details.)

Overall, QCLs are commercially available and there are more than 20 companies that sell QCLs. For example, Pranalytica[2] makes fixed-wavelength high average power QCLs with wavelengths between 3.8 and 12 μm, with >1 W average power between 4 and 5 μm. Daylight Solutions[3] produces lasers that can cover the whole range from 4 to 12 μm (piecewise, with the tuning range of >250 cm^{-1} at different central wavelengths) based on broadly tunable external-cavity QCLs operating in the pulsed mode at RT. The average power is up to 20 mW and the linewidth is

2 http://www.pranalytica.com/index.php.
3 www.daylightsolutions.com.

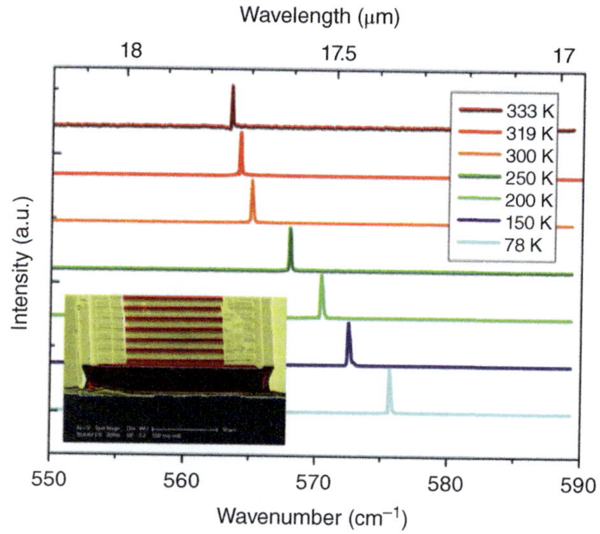

Figure 4.18 Emission spectra of a metal–metal waveguide distributed feedback InAs/AlSb QCL measured at different temperatures. The inset shows a scanning electron microscope image of the cleaved facet of the DFB laser (Ti/Au metallic layers are in yellow). The laser dimensions are 30 μm × 1 mm. *Source:* reproduced from figure 4 of [91], with permission of AIP, American Institute of Physics.

1 cm^{-1}. The CW version of a tunable external-cavity QCL has a smaller tuning range of 60 cm^{-1}, but much narrower linewidth <45 MHz and mode-hop-free performance, suitable for high-resolution spectroscopy. The central wavelengths of these lasers are between 4.4 and 10.5 μm. Block Engineering offers a QCL system with the largest gap-free tuning range $\lambda \approx 5.4$–12.8 μm. The system consists of four switchable internal laser modules covering four spectral ranges, operates at RT in a pulsed mode (pulse duration 20–500 ns, repetition rate of up to 3 MHz, duty-cycle of up to 15%), and has the spectral linewidth of 2 cm^{-1}.

Finally, in [92, 93] the reader can find excellent reviews of terahertz (far-IR) QCLs, which are beyond the scope of this book.

The main results for QCLs are presented in Table 4.2.

4.3 Interband Cascade Lasers

The interband cascade laser (ICL) is a technology that has demonstrated noteworthy performance improvements in the intermediate mid-IR wavelength region from 3 to 5 μm, thus filling the gap between injection lasers and QCLs. The concept was pioneered by Yang [94] and Meyer et al. [95, 96] and represents an elegant hybrid of the conventional diode laser and the QCL. An ICL

Table 4.2 Summary of quantum cascade lasers (QCLs).

Laser structure	Wavelength (μm)	Laser characteristics	Ref.
High-power QCLs			
Buried-ridge WG, GaInAs/AlInAs/InP	4.9	Pulsed (500 ns, 100 kHz, 5% duty-cycle), peak 10 W (T = 298 K), wall-plug efficiency (WPE) 27%, CW, 5.1 W (298 K), WPE 21%	[52]
Buried strain-balanced WG, GaInAs/AlInAs/InP	4.7	CW, 4.5 W (283 K), WPE 16.3%	[55]
Strain-balanced WG, InGaAs/AlInAs/InP	7.1	Pulsed (500 ns, 10 kHz), peak 3 W (293 K), WPE 19%, CW, 1.4 W (93 K), WPE 10%	[56]
Strain-balanced, AlInAs/InGaAs/InP	9.2	Pulsed (500 ns, 10 kHz), peak 4.5 W (293 K), WPE 16%, CW, 2 W (293 K), WPE 10%	[57]
Ridge WG, GaInAs/AlInAs/InP	4.5	Pulsed (100 ns, 5 kHz), peak 10 W, average 5 mW (80 K), WPE 47%	[58]
Buried-ridge WG, GaInAs/AlInAs/InP	5	Pulsed (200 ns, 250 kHz), peak 10 W, average 500 mW (40 K), WPE 53%	[59]
Distributed feedback, DFB			
Plasmon-enhanced waveguide; corrugated grating on the surface	7.8–7.9 5.3–5.4	Pulsed, single longitudinal mode (SLM), temperature tuned from 7.78 μm (T = 80 K) to 7.93 μm (300 K) and from 5.31 μm (80 K) to 5.38 μm (315 K), peak 60 mW (300 K)	[60]
Grating on the top of n-doped InGaAs cap layer	10.1	Pulsed, SLM, 10.04 μm (85 K) to 10.183 μm (300 K); peak 230 mW (85 K) and 80 mW (300 K)	[61]
Surface grating, epilayer-down on diamond	4.75	CW, SLM, 1.1 W (298 K), tuning 5 cm^{-1} by current and 18 cm^{-1} by temperature	[62]
Surface grating with surface-plasmon coupling	4.8	CW, SLM, 2.4 W (298 K), 10% WPE, tuning 5 cm^{-1} by current	[63]
24 single-mode DFB lasers in one array; bound-to-continuum active region	8–9.8	Pulsed (50 ns, 80 kHz) at RT, total span 220 cm^{-1}, frequency spacing 9.5 cm^{-1}, peak 100–1100 mW	[65]
24 single-mode DFB lasers in one array; heterogeneous cascade active region	5.9–10.9	Pulsed (200 ns, 50 kHz) at RT, peak 50 mW	[69]

Table 4.2 (Continued)

Laser structure	Wavelength (μm)	Laser characteristics	Ref.
InAs/AlSb DFB	3.34–3.38	Pulsed (100 ns, 1 kHz) at RT, peak 0.8 W	[87]
External cavity			
Bound-to-continuum design	9.11–10.56	Pulsed (50 ns, 500 kHz) at RT, peak 40 mW	[72]
Two bound-to-continuum active regions	7.66–9.87, (pulsed)	Pulsed (400 ns, 99 kHz) at RT, peak 0.8 W	[73]
	8–9.6 (CW)	CW at RT, up to 135 mW	
Five different bound-to-continuum cascades in one active region	7.6–11.4	Pulsed (15 ns, 1 MHz) at RT, peak 1 W, average 15 mW	[74]
Heterogeneous double-stack of two bound-to-continuum active regions	3.28–4.01	Pulsed (50 ns, 200 kHz) at RT, peak 0.8 W	[88]
Short wavelength			
Strain-compensated AlInAs/GaInAs/InP	3.4	CW 120 mW (15 K); pulsed, peak 4 mW (280 K)	[78]
Strain-compensated InGaAs/AlAsSb/InP	3.1	Pulsed (100 ns, 5 kHz), peak 1.3 W (80 K) and 8 mW (295 K)	[82]
Strain-compensated InGaAs/InAlAs/AlAs/InP	3.05	Pulsed (100 ns, 125 kHz), peak 120 mW (80 K)	[83]
InAs/AlSb	2.63–2.65	Pulsed (100 ns, 10 kHz), peak 260 mW (80 K)	[48]
Long wavelength			
InAs/AlSb, metal–metal waveguide	19 21	Pulsed mode, max $T = 291$K Pulsed mode, max $T = 250$ K	[90]
InAs/AlSb, metal–metal waveguide	17.8	Pulsed mode, max $T = 333$K Single-mode DFB, peak 15 mW (78 K)	[91]

uses a material system with type-II band alignment (also called broken-gap alignment) of the GaSb/InAs [94] or GaInSb/InAs [95] heterojunction. In the type-II alignment, the top of the valence band of GaSb (or GaInSb) is above the bottom of the conduction band of InAs, as illustrated in Figure 4.19. The optical transition occurs between an electron state confined in the InAs well layer and a hole state localized in the GaSb (or GaInSb) well layer.

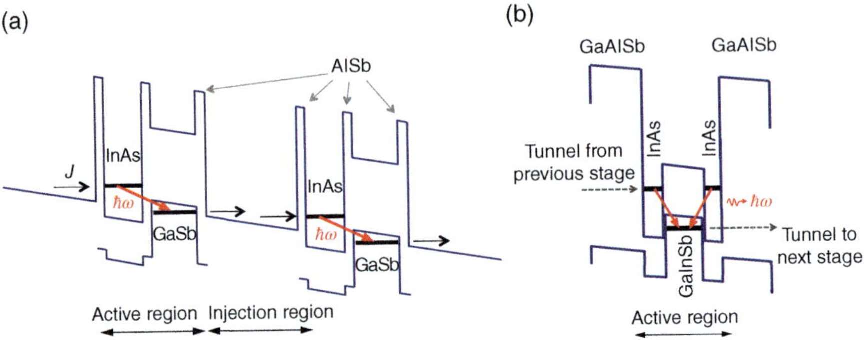

Figure 4.19 Type-II QW ICL band structures. (a) The proposed ICL structure of [94], where the optical transition occurs between an electron state confined in the InAs well layer and a hole state localized in the GaSb well layer. (b) The "W" active region ICL design of [95], where the optical transition occurs between an electron state of the InAs double well and a hole state of the GaInSb well, sandwiched between InAs layers.

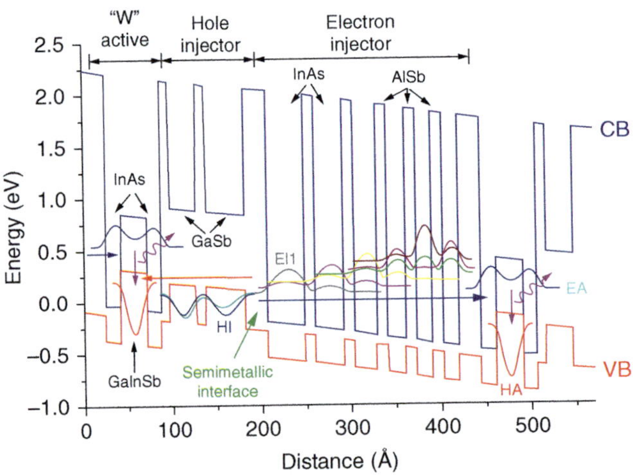

Figure 4.20 Band structure and probability densities of electron (blue) and hole (red, inverted) states in one cascade of the ICL structure under electric field bias. The active "W" region, hole injector, and electron injector are indicated. Electrons drift to the right and holes to the left. CB, conduction band; VB, valence band. *Source:* reproduced from figure 3 of [97], with permission of IEEE.

Even though the lasing transition is an interband process, multiple stages can be cascaded by energetically aligning the conduction and valence states at the structure's type-II interfaces through applying forward bias (Figure 4.20). This allows electrons in the valence band to scatter elastically to the conduction band of the next ICL active stage for recycling, with more than 10 cascading stages. Since the ICL is based on an interband transition, nonradiative relaxation between the two states is greatly suppressed, leading to longer lifetimes and hence smaller thresholds, as compared to QCLs. In fact, it has already been demonstrated that at RT, ICLs emitting in the 3–5 μm spectral range can operate continuously at drive powers more than an order of magnitude lower than state-of-the-art QCLs [98]. On the other hand, as in the case of the intersubband-based QCL, the emission photon energy is controlled almost entirely by quantum confinement rather than the energy gaps of the constituents, and can in principle be tuned anywhere from zero to about 1 eV in photon energy [95].

The first interband cascade laser (ICL) was demonstrated in 1997. It was based on an InAs/GaInSb/AlSb type-II QW structure and operated at $\lambda \approx 3.8$ μm in the pulsed mode at cryogenic temperatures from 80 to 170 K [99]. Later, the practical milestone of ambient-temperature CW operation was surpassed due to improvements in the material design and processing [100]. In 2008, the group at the Naval Research Laboratory obtained CW operation at temperatures up to 319 K at $\lambda = 3.75$ μm, with an optical power in excess of 10 mW at 300 K [101]. Also, ICL based on a narrow (3.5-μm-wide) ridge waveguide (Figure 4.21) allowed laser action up to 335 K in the continuous mode at $\lambda = 3.7$ μm [102]. Bewley et al. reported an ICL device that operates in CW mode ($\lambda = 3.9$ μm) up to a maximum temperature of 118 °C, which represents the current record for semiconductor lasers emitting in the 3–4 μm range [103].

Many ICL applications require SLM operation. In 2008, a single-mode DFB ICL was first reported that emitted 67 mW of CW output power at $\lambda = 3.27$ μm at $T = 78$ K [104]. Later, Kim et al. [105] obtained single-mode emission from an ICL that operated well above RT (up to 80 °C). The laser operated between

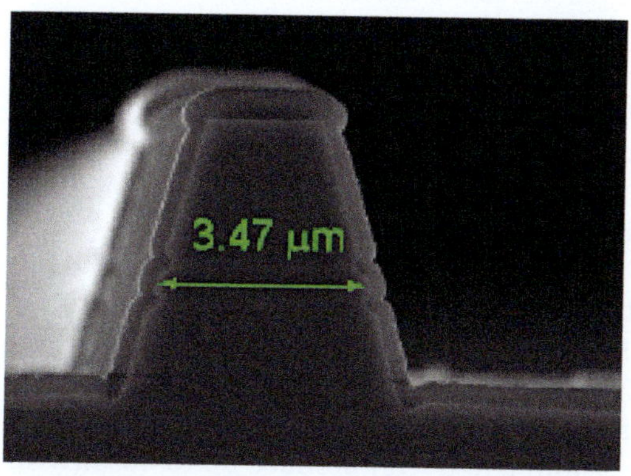

Figure 4.21 Scanning electron micrograph of a 3.5-μm-wide ridge ICL structure. *Source:* reproduced from figure 8 of [102], with permission of IOP.

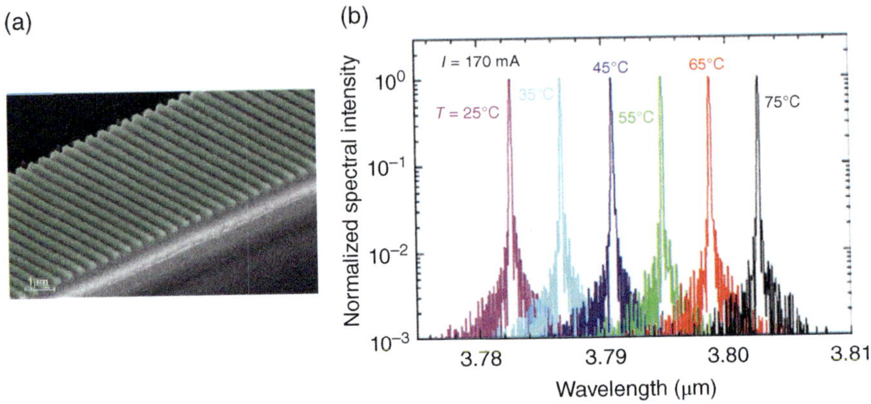

Figure 4.22 (a) Scanning electron micrograph of a ridge ICL with the first-order germanium DFB grating on top. (b) Emission spectra of a 7.4-μm-wide 2 mm-long ridge-waveguide DFB ICL at different temperatures. *Source:* reproduced from figure 3 of [105], with permission of AIP, American Institute of Physics.

3.78 and 3.8 μm and a DFB grating was formed by a patterned germanium layer on top of the ridge of the laser (Figure 4.22). The maximum single-mode output was 27 mW at 40 °C, with the minimum power required to operate the device of only 90 mW at 20 °C [105]. Most recently, a lateral-grating DFB ICL was reported, in which case metal gratings were patterned laterally to the laser ridge so that the evanescent part of the laser mode couples to the grating, leading to modulation of the losses in the resonator [106, 107]. With this approach, the ICL emitted up to 55 mW in a single spectral mode at λ = 3.6 μm at 25 °C [106]. Later, another design for SLM operation was introduced with laterally coupled gratings in a so-called "double ridge" waveguide system that minimizes optical losses. Here, the grating is placed above the active region while a second, slightly wider ridge ensures current confinement in the active region [108–110]. The authors were able to get single-mode operation at 46 °C (λ = 3.57 μm) with more than 8800 hours of lifetime with negligible degradation of the performance [108], and observed SLM emission around 6 μm in CW operation at −10 °C [110].

With DFB gratings, SLM operation was achieved beyond the ICL "sweet spot" of 3–5 μm, that is at 2.8 μm [111] and 5.2 μm [112] in CW mode at RT. It has been shown that by a combination of current and temperature tuning around RT, the ICL can be tunable over 25 nm in wavelength [112].

Close to diffraction-limited ICL performance is also desirable for practical use. Bawley et al. has shown that a tapered structure with the ridge width varying linearly from 5 to 63 μm over 4-mm length can produce up to 403 mW of CW power, with the beam propagation factor M^2 = 2.3 [113]. (For the ideal, diffraction-limited propagation M^2 = 1.) Kim et al. reported on high-brightness

ICLs emitting at $\lambda \approx 3.2\,\mu m$ and $\lambda \approx 3.45\,\mu m$. With 18-μm ridge width and 4.5-mm cavity length, the 10-stage ICL emitted 464 mW in the CW mode at 25 °C, with the beam propagation factor of $M^2 = 1.9$ [114]. Similar-design ICLs fabricated with 1-mm-long cavities exhibited CW WPEs of 18% at $T = 25\,°C$ [114]. This conversion efficiency compares favorably to 21% for the highest value ever reported for a QCL operating in CW mode at RT [52].

For high-efficiency ICLs, Canedy et al. fabricated broad-area (150-μm wide) 10-stage ICLs emitting 1 W at $\lambda = 3.0–3.2\,\mu m$ that maintain CW WPEs exceeding 40% at temperatures up to 125 K. The CW threshold current density at 80 K was only 11 A/cm^2 [115]. These results demonstrate that at cryogenic temperatures ICLs can convert electrical to optical energy nearly as efficiently as the best intersubband-based QCLs.

On the long-wavelength side, ICLs operated in CW mode above RT up to 5.6 μm wavelength [116]. A record wavelength of 7 μm for ICLs operating at RT has been demonstrated in the pulsed mode [117].

Overall, compared to QCLs, ICLs typically have a lower output power but benefit from much smaller threshold current densities. The best low power consumption characteristics demonstrated to date with ICLs operating at RT include CW threshold drive powers as low as 29 mW and threshold current densities below 100 A/cm^2. An excellent review of ICLs can be found in [98].

The main results for ICLs are presented in Table 4.3.

Table 4.3 Summary of interband cascade lasers (ICLs).

Laser structure	Wavelength (μm)	Laser characteristics	Ref.
InAs/GaInSb/AlSb type-II QW structure	3.8	First ICL demonstrated; pulsed mode (100 ns, 1 kHz), $T = 80–170\,K$	[99]
DFB grating on sidewalls; 13.2-μm wide, 4-mm-long ridge waveguide	3.6	CW, single longitudinal mode (SLM), 55 mW (25 °C)	[106]
DFB, patterned Ge layer on top	3.78–3.8	CW, SLM, 27 mW (40 °C)	[105]
Vertical sidewall DFB gratings	5.2	CW, SLM, threshold power 138 mW (−5 °C)	[112]
10-Stage InAs/GaInSb/AlSb, 18-μm-wide ridge	3.2 and 3.45	CW, 464 mW, $M^2 = 1.9$, WPE 18% (25 °C)	[114]
10-Stage InAs/GaInSb active core, 150-μm wide WG	3.0–3.2	CW, 1 W, WPE > 40% at 80–125 K	[115]

4.4 Optically Pumped Semiconductor Disk Lasers (OPSDLs)

In contrast to edge-emitting semiconductor lasers that we looked at earlier, an optically pumped semiconductor disk laser (OPSDL) is excited in a manner similar to that of ion-doped crystalline disk lasers. These lasers are also known as the vertical-external-cavity surface-emitting lasers (VECSELs). The OPSDL chip in such a laser is formed by a sequence of epitaxially grown semiconductor layers and simultaneously acts as cavity mirror and gain region, hence it can be considered as an *active mirror*. A pump laser is focused onto the surface of the chip with a spot size typically several tens to hundreds of microns in diameter [118]. An arrangement of external mirrors and components is used to complete the laser resonator. A simple two-mirror OPSDL cavity is shown schematically in Figure 4.23. The OPSDL concept offers a number of advantages:

- One can realize, depending on the semiconductor material, fundamental emission wavelengths ranging from the visible to the mid-IR.
- Due to the mode-matching offered by the surface-emitting geometry in conjunction with the external resonator, one can produce circularly symmetric diffraction-limited output beams.
- At the same time, the pumped area can be increased to facilitate output power scaling.
- The external resonator allows flexible control of the laser properties by inserting intracavity elements. These can be elements for frequency-doubling, mode-locking with saturable absorbers, and frequency stabilization.

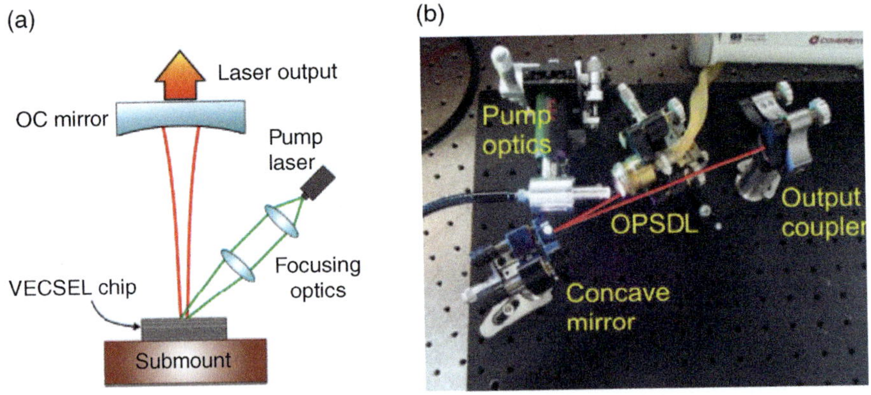

Figure 4.23 (a) A simple arrangement of a two-mirror OPSDL cavity. (b) Photograph of a three-mirror OPSDL resonator with beam paths marked. *Source:* reproduced from figures 1 and 12 of [118], with permission of Wiley.

In the near-IR, OPSDL devices with corresponding emission wavelengths near 1 μm have reached maturity due to the highly developed GaAs-based material system. For example, output powers of up to several tens of watts [119] and slope efficiencies as high as 67% [120] have been realized with GaAs-based OPSDLs with laser chips that are commonly contacted to silicon carbide or diamond heat spreaders. Since this book is about mid-IR sources, we will consider only two main classes of OPSDL: based on AlGaIn/AsSb material system, and on IV–VI compound lead-chalcogenides (PbS, PbSe, and PbTe).

4.4.1 (AlGaIn)(AsSb)-based OPSDL at $\lambda \approx 2.3\,\mu m$

Schulz et al. reported a multiwatt CW OPSDL system operating at RT near $\lambda = 2.36\,\mu m$. The authors used (AlGaIn)(AsSb)-based OPSDL structure grown by MBE on a GaSb substrate [121]. Three distinct functional regions of epitaxial layers defined the OPSDL structure are [118]: the distributed Bragg reflector (DBR), the active gain region, and the window (Figure 4.24). The DBR, or mirror, region consisted of 25 alternating quarter-wavelength-thick layer pairs having high and low refractive indices. For (AlGaIn)(AsSb)-based OPSDLs, GaSb is used as the high-index material and AlAsSb for the low-index layers. The resulting DBR reflectivity is typically ~99.8%. The active, or gain, region was grown directly on top of the DBR and consisted of 15 compressively strained 10-nm-thick GaInAsSb QWs embedded between barrier layers of AlGaAsSb – a material having a higher bandgap energy than the QWs. Typically, the QWs are positioned at the antinodes of the standing-wave optical intensity distribution within the device in order to maximize the modal gain. The OPSDL structure was completed by adding a confinement region and finally a thin capping layer. The confinement window layer is designed to have a higher bandgap energy than the barrier layers to prevent nonradiative

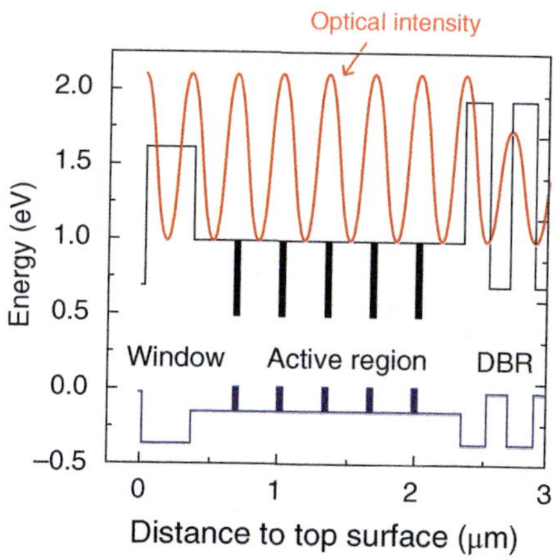

Figure 4.24 Band edges of valence and conduction bands as well as optical intensity within the active region of an (AlGaIn)(AsSb)-based OPSDL structure. DBR, distributed Bragg reflector mirror. *Source:* reproduced from figure 2 of [118], with permission of Wiley.

recombination of the photoexcited carriers at the surface of the device. For the 2.36 μm OPSDL, this was achieved using lattice-matched AlGaAsSb. Finally, a thin (a few nanometers) GaSb capping layer was grown as the topmost layer to protect the structure from the oxidation in air.

The authors of [121] employed the so-called "in-well" pumping concept, where the pump photon energy (at a wavelength of 1.96 μm) is chosen close to that of the emitted photons, leading to the pump light absorption exclusively in the active QWs instead of the surrounding barrier layers. This strategy allows minimizing quantum defect (the fractional energy difference between the pump and the emitted photons). Using a thulium-doped fiber laser as pump source at 1.96 μm, an OPSDL CW output power of 3.2 W at λ = 2.36 μm has been reached at a heat sink temperature of −15 °C. More than 2 W has been recorded at +15 °C.

4.4.2 PbS-based OPSDL at λ = 2.6–3 μm

Generally, operation of mid-IR III–V- or II–VI-based interband semiconductor lasers are restricted by Auger recombination and free carrier absorption. In contrast, the IV–VI compounds such as PbTe, PbSe, and PbS have lower Auger recombination coefficient compared to III–V and II–VI compounds, owing to the symmetric band edge structure between conduction and valence bands. Additionally, free carrier absorption is significantly reduced in the optically pumped system.

Ishida et al. achieved OPSDL operation at 2.65–3.1 μm with the output power exceeding 2 W. The OPSDL was based on PbS active layer QW structure (Figure 4.25a) [122]. The laser was pumped at 1.55 μm (pulsed excitation, 100 ns, 10 kHz). The laser structure was grown epitaxially on a BaF_2 (111) substrate and consisted of an optical excitation (OE) layer made of short-period PbSrS superlattices, active PbSrS/PbS MQW layer, and an SrS/PbSrS Bragg reflector mirror structure (Figure 4.25b). Most of the pump light is absorbed in the OE layer and generated electrons and holes diffuse into the active QW layer, which has the smallest bandgap.

Figure 4.26a plots the output spectra for the laser with temperatures ranging from −120 to +10 °C. The emission center wavelengths range from 2.65 μm near RT up to 3 μm at the lowest operation temperature of −120 °C (153 K). Figure 4.26b plots the dependence of the output power on the excitation power at −70 °C. At this temperature (with λ = 2.9 μm) the output power exceeded 2 W, and external quantum efficiency reached 16%.

4.4.3 PbSe-based OPSDL at λ = 4.2–4.8 μm

Rahim et al. demonstrated a PbSe-based OPSDL operating at λ = 4.2–4.8 μm above RT [123]. The active section consisted of a single 850-nm-thick epitaxial

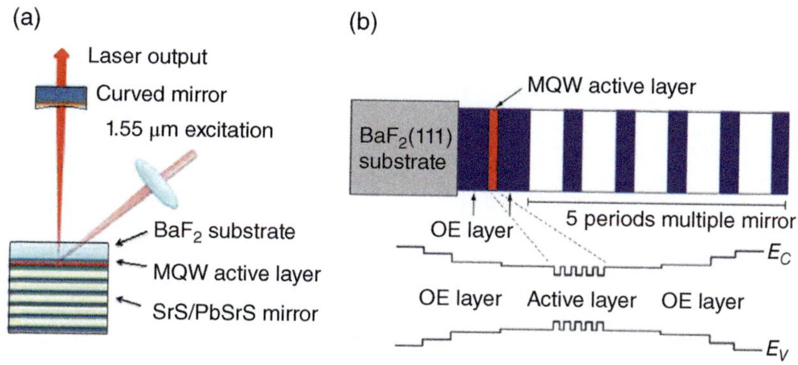

Figure 4.25 (a) Layout of the optically pumped PbS laser. (b) Laser structure built on BaF_2 substrate with optical excitation (OE) layers and MQW active layer at the center, as well as SrS/PbSrS Bragg reflector mirror. Band energy profiles for OE and MQW regions are shown at the bottom. *Source:* reproduced from figures 3 and 4 of [122], with permission of AIP, American Institute of Physics.

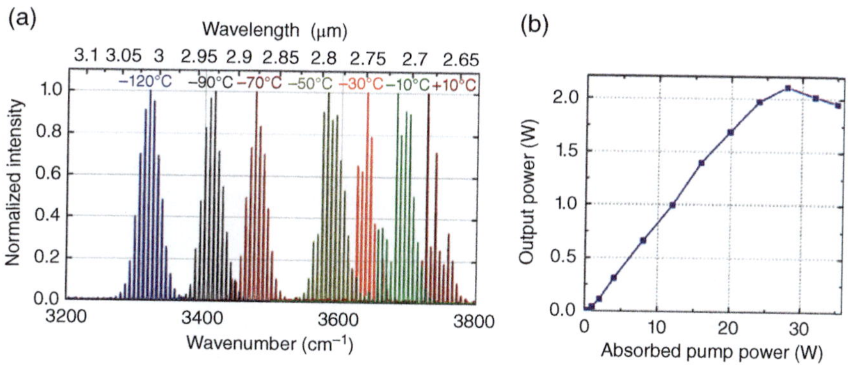

Figure 4.26 (a) Output spectra for the PbSrS/PbS MQW VECSEL from −120 to +10 °C. (b) Dependence of the output power at 2.9 μm on excitation power at −70 °C. *Source:* reproduced from figure 5 of [122], with permission of AIP, American Institute of Physics.

PbSe gain layer grown by MBE on BaF_2 substrate, followed by a $PbEuTe/BaF_2$ Bragg mirror. A thick Al layer was added to serve as a heat spreader (Figure 4.27a). The laser cavity was formed between the bottom flat Bragg mirror and the top curved mirror (Figure 4.27a).

The laser was pumped with a commercial 1.55 μm-fiber laser with 100-ns wide pulses at 10 kHz repetition rate. The cavity length was 24 mm and the diameter of the fundamental mode of about 200 μm matched the diameter of the focused pump beam. At 60 W of average pump power, the output peak power was 18 mW at heat-sink temperature of −22 °C and 6 mW at 27 °C (the average power was correspondingly 18 and 6 μW). The cavity emission

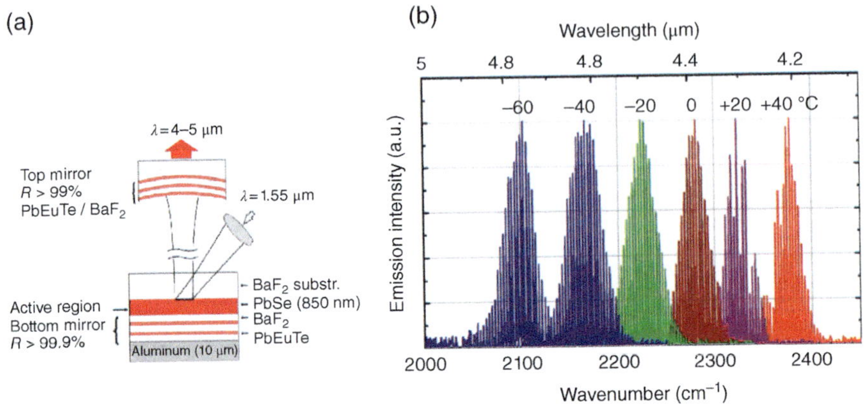

Figure 4.27 (a) Layout of the optically pumped PbSe laser. The top curved mirror uses similar Bragg layers, as the bottom mirror. (b) Normalized laser spectra at different heat-sink temperatures. *Source:* reproduced from figures 1 and 3 of [123], with permission of AIP, American Institute of Physics.

wavelength varied with temperature from 4.2 μm (at +40 °C) to 4.8 μm (at −40 °C) (Figure 4.27b). The temperature shift in the center wavelength is due to the temperature dependence of the bandgap of PbSe (0.5 meV/K).

According to the authors, low efficiency can be accounted for by a large photon energy difference between the pump and the output wavelength (quantum defect), so that 65% of the pump power was lost as heat [123].

4.4.4 PbTe-based OPSDL at $\lambda = 4.7–5.6\,\mu m$

A long-wavelength OPSDL based on PbTe active region was reported by Khiar et al. [124]. The laser produced single-frequency output with a continuous and mode-hop-free tuning range of 4.7–5.6 μm and operated at T = 100–170 K. A schematic cross section of the laser is shown in Figure 4.28a. A Si (111) wafer served as a substrate, on which a Bragg mirror was grown, consisting of four pairs of PbSrTe/EuTe layers. A 1.2-μm-thick PbTe active layer grown on top of that completed the OPSDL structure. On top, the cavity was terminated with a curved Si/SiO Bragg mirror, and the length of the cavity was kept very short, 50–100 μm, so that only one longitudinal mode was supported. The length was altered with a piezoelectric actuator. The PbTe laser was pumped with a pulsed (100 ns, 10 kHz) 1.55 μm laser. By changing the cavity length, up to 5% mode-hop-free wavelength tuning was achieved at fixed temperature. On a larger scale, the tuning was achieved by temperature (Figure 4.28b), with the tuning range that extended from 5.6 μm (T = 90 K) to 4.7 μm (T = 170 K). The output peak power was measured to be up to 100 mW (average power ~100 μW) and

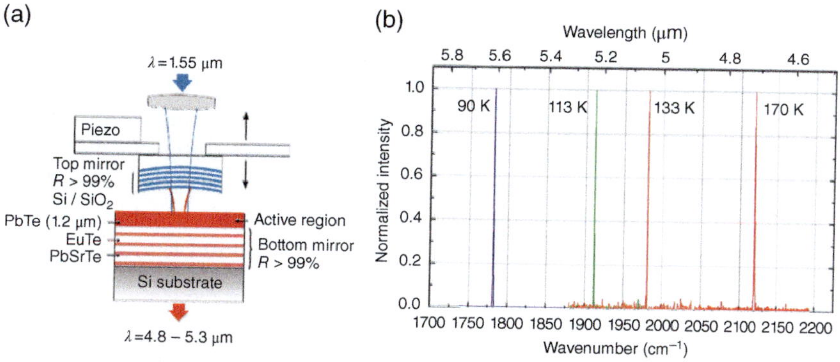

Figure 4.28 (a) Schematic cross section of a PbTe OPSDL structure, MBE grown on a Si-substrate, with a short, 50–100-μm laser cavity. (b) Normalized single-mode spectra at different temperatures and fixed cavity length. Lasing wavelengths range from 5.61 μm at 90 K to 4.71 μm at 170 K operating temperature. *Source:* reproduced from figures 1 and 4 of [124], with permission of AIP, American Institute of Physics.

Table 4.4 Summary of optically pumped semiconductor disk lasers (OPSDLs).

Active layer (/barrier)	Pump	Wavelength (μm)	Output power	Ref.
GaInAsSb/ AlGaAsSb	1.96 μm, CW, 14 W	2.36	3.2 W (−15 °C)	[121]
PbSrS/PbS	1.55 μm, pulsed, 100 ns, 10 kHz, 28 W	2.65 (RT) 2.9 (−70 °C) 3 (−120 °C)	2.2 W (−70 °C)	[122]
PbSe	1.55 μm, pulsed, 100 ns, 10 kHz, 60 W	4.2 (+40 °C) 4.8 (−40 °C)	Peak 18 mW (−22 °C) Peak 6 mW (27 °C)	[123]
PbTe	1.55 μm, pulsed, 100 ns, 10 kHz	5.6 (90 K) 4.7 (170 K)	Peak 100 mW (100 K)	[124]

the beam propagation factor $M^2 = 1.14$, that is, the laser displayed near diffraction-limited performance.

Again, low OPSDL efficiency can be accounted for by a considerable fraction (~60%) of the pump beam that does not reach the active layer due to reflections and interferences in the top Bragg mirror. Also, more than two-thirds of the pump power is wasted due to the large wavelength difference of the emitted 5-μm wavelength compared to the 1.55-μm pump wavelength.

The main results for the OPSDLs are presented in Table 4.4.

4.5 Summary

Conventional type-I mid-IR QW lasers have now reached RT CW operation, but only at wavelengths slightly exceeding 3 μm. Insufficient hole confinement appears to be responsible for much of the increasing difficulty at longer (>3.4 μm) mid-IR wavelengths. In contrast, QCLs work very well at $\lambda > 4$ μm and produce up to 10 W of average power at RT near 4.6 μm. Tunable external-cavity QCLs can provide large tunability, up to 39% around the center QCL frequency. The niche between 3 and 4 μm can be filled using a novel InAs/AlSb QCL material system. ICLs, on the other hand, operate best at 3–5 μm and can now produce 500 mW of power at RT. Their clear advantage over QCLs is much smaller threshold for laser operation enabling low energy-consuming devices for spectroscopy. OPSDLs can operate in the range from 2 to 5.6 μm and produce narrow-linewidth good-quality diffraction-limited beams. However, optical conversion efficiency of mid-IR OPSDLs does not match the high efficiency of their near-IR counterparts.

References

1 Melngailis, I. (1963). Maser action in InAs diodes. *Appl. Phys. Lett.* 2: 176.
2 Dimmock, J.O., Melngailis, I., and Strausss, A.J. (1966). Band structure and laser action in $Pb_xSn_{1-x}Te$. *Phys. Rev. Lett.* 16: 1193.
3 Linden, K.J. and Mantz, A.W. (1982). Tunable diode lasers and laser systems for the 3 to 30 μm infrared spectral region. *Proc. SPIE* 320: 109.
4 Tacke, M. (1995). New developments and application of tunable IR lead salt lasers. *Infrared Phys. Technol.* 36: 447.
5 Olafsen, L.J., Vurgaftman, I., and Meyer, J.R. (2004). Antimonide mid-IR lasers. In: *Long-Wavelength Infrared Semiconductor Lasers* (ed. H.K. Choi). Hoboken, NJ: Wiley.
6 Bochkarev, A.E., Dolginov, L.M., Drakin, A.E., Druzhininn, L.V., Eliseev, P.G., and Sverdlov, B.N. (1985). Injection InGaSbAs lasers emitting radiation of wavelengths 1.9–2.3 μm at room temperature. *Sov. J. Quantum Electron.* 15: 869.
7 Baranov, A.N., Danilova, T.N., Dzhurtanov, B.E., Imenkov, A.N., Konnikov, S.G., Litvak, A.M., Usmanskii, V.E., and Yakovlev, Yu.P. (1988). CW lasing in GaInAsSb/GaSb buried channel laser (T = 20 °C, λ = 2.0 μm). *Sov. Tech. Phys. Lett.* 14: 727.
8 Bochkarev, A.E., Dolginov, L.M., Drakin, A.E., Eliseev, P.G., and Sverdlov, B.N. (1988). Continuous-wave lasing at room temperature in InGaSbAs/GaAlSbAs injection heterostructures emitting in the spectral range 2.2–2.4 μm. *Sov. J. Quantum Electron.* 18: 1362.
9 Choi, H.K. and Eglash, S.J. (1992). High-power multiple-quantum-well GaInAsSb/AlGaAsSb diode lasers emitting at 2.1 μm with low threshold current density. *Appl. Phys. Lett.* 61: 1154.

10 Lee, H., York, P.K., Menna, R.J., Martinelli, R.U., Garbuzov, D.Z., Narayan, S.Y., and Connolly, J.C. (1995). Room-temperature 2.78 μm AlGaAsSb/InGaAsSb quantum well lasers. *Appl. Phys. Lett.* 66: 1942.

11 Garbuzov, D.Z., Lee, H., Khalfin, V., Martinelli, R., Connolly, J.C., and Belenky, G.L. (1999). Operation of InGaAsSb–AlGaAsSb broad waveguide SCH-QW diode lasers. *IEEE Photon. Technol. Lett.* 11: 794.

12 Grau, M., Lin, C., Dier, O., Lauer, C., and Amann, M.-C. (2005). Room-temperature operation of 3.26 μm GaSb-based type-I lasers with quinternary AlGaInAsSb barriers. *Appl. Phys. Lett.* 87: 241104.

13 Belenky, G., Shterengas, L., Kipshidze, G., and Hosoda, T. (2011). Type-I diode lasers for spectral region above 3 μm. *IEEE J. Sel. Topics Quantum Electron.* 17: 1426.

14 Rattunde, M., Schmitz, J., Kaufel, G., Kelemen, M., Weber, J., and Wagner, J. (2006). GaSb-based 2.X μm quantum-well diode lasers with low beam divergence and high output power. *Appl. Phys. Lett.* 88: 081115.

15 Donetsky, D., Kipshidze, G., Shterengas, L., Hosoda, T., and Belenky, G. (2007). 2.3 μm type-I quantum well GaInAsSb/AlGaAsSb/GaSb laser diodes with quasi-CW output power of 1.4 W. *Electron. Lett.* 43: 810.

16 Shterengas, L., Belenky, G., Kisin, M.V., and Donetsky, D. (2007). High power 2.4 μm heavily strained type-I quantum well GaSb-based diode lasers with more than 1 W of continuous wave output power and a maximum power-conversion efficiency of 17.5%. *Appl. Phys. Lett.* 90: 011119.

17 Kim, J.G., Shterengas, L., Martinelli, R.U., Belenky, G.L., Garbuzov, D.Z., and Chan, W.K. (2002). Room-temperature 2.5 mm InGaAsSb/AlGaAsSb diode lasers emitting 1 W continuous waves. *Appl. Phys. Lett.* 81: 3146.

18 Kelemen, M.T., Gilly, J., Moritz, R., Schleife, J., Fatscher, M., Kaufmann, M., Ahlert, S., and Biesenbach, J. (2010). Diode laser arrays for 1.8 to 2.3 μm wavelength range. *Proc. SPIE* 7686: 76860N.

19 Kim, J.G., Shterengas, L., Martinelli, R.U., and Belenky, G.L. (2003). High-power room-temperature continuous wave operation of 2.7 and 2.8 μm In(Al)GaAsSb/GaSb diode lasers. *Appl. Phys. Lett.* 83: 1926.

20 Shterengas, L., Belenky, G., Kipshidze, G., and Hosoda, T. (2008). Room temperature operated 3.1 μm type-I GaSb-based diode lasers with 80 mW continuous-wave output power. *Appl. Phys. Lett.* 92: 171111.

21 Shterengas, L., Liang, R., Kipshidze, G., Hosoda, T., Belenky, G., Bowman, S.S., and Tober, R.L. (2014). Cascade type-I quantum well diode lasers emitting 960 mW near 3 μm. *Appl. Phys. Lett.* 105: 161112.

22 Hosoda, T., Wang, M., Shterengas, L., Kipshidze, G., and Belenky, G. (2015). Three stage cascade diode lasers generating 500 mW near 3.2 μm. *Appl. Phys. Lett.* 107: 111106.

23 Liang, R., Hosoda, T., Shterengas, L., Stein, A., Lu, M., Kipshidze, G., and Belenky, G. (2015). Narrow ridge $\lambda \approx 3$-μm cascade diode lasers with output power above 100 mW at room temperature. *IEEE Photon. Technol. Lett.* 27: 2425.

24 Lehnhardt, T., Hümmer, M., Rößner, K., Müller, M., Höfling, S., and Forchel, A. (2008). Continuous wave single mode operation of GaInAsSb/GaSb quantum well lasers emitting beyond 3 μm. *Appl. Phys. Lett.* 92: 183508.

25 Belahsene, S., Naehle, L., Fischer, M., Koeth, J., Boissier, G., Grech, P., Narcy, G., Vicet, A., and Rouillard, Y. (2010). Laser diodes for gas sensing emitting at 3.06 μm at room temperature. *IEEE Photon. Technol. Lett.* 22: 1084.

26 Gupta, J.A., Ventrudo, B.F., Waldron, P., and Barrios, P.J. (2010). External cavity tunable type-I diode laser with continuous-wave single mode operation at 3.24 μm. *Electron. Lett.* 46: 1218.

27 Briggs, R.M., Frez, C., Bagheri, M., Borgentun, C.E., Gupta, J.A., Witinski, M.F., Anderson, J.G., and Forouhar, S. (2013). Single-mode 2.65 μm InGaAsSb/AlInGaAsSb laterally coupled distributed-feedback diode lasers for atmospheric gas detection. *Opt. Express* 21: 1317.

28 Hosoda, T., Fradet, M., Frez, C., Shterengas, L., Sander, S., Forouhar, S., and Belenky, G. (2016). Laterally coupled distributed feedback cascade diode lasers emitting near 2.9 μm. *Electron. Lett.* 52: 857.

29 Liang, R., Hosoda, T., Shterengas, L., Stein, A., Lu, M., Kipshidze, G., and Belenky, G. (2014). Distributed feedback 3.27 μm diode lasers with continuous-wave output power above 15 mW at room temperature. *Electron. Lett.* 50: 1378.

30 Lin, C., Grau, M., Dier, O., and Amann, M.-C. (2004). Low threshold room-temperature continuous-wave operation of 2.24–3.04 μm GaInAsSb/AlGaAsSb quantum-well lasers. *Appl. Phys. Lett.* 84: 5088.

31 Shterengas, L., Belenky, G.L., Gourevitch, A., Donetsky, D., Kim, J.G., Martinelli, R.U., and Westerfeld, D. (2004). High power 2.3-μm GaSb-based linear laser array. *IEEE Photon. Technol. Lett.* 16: 2218.

32 Chen, J., Kipshidze, G., and Shterengas, L. (2010). Diode lasers with asymmetric waveguide and improved beam properties. *Appl. Phys. Lett.* 96: 241111.

33 Kipshidze, G., Hosoda, T., Sarney, W.L., Shterengas, L., and Belenky, G. (2011). High-power 2.2-μm diode lasers with metamorphic arsenic-free heterostructures. *IEEE Photon. Technol. Lett.* 23: 317.

34 Liang, R., Chen, J., Kipshidze, G., Westerfeld, D., Shterengas, L., and Belenky, G. (2011). High-power 2.2-μm diode lasers with heavily strained active region. *IEEE Photon. Technol. Lett.* 23: 603.

35 Shterengas, L., Belenky, G.L., Kim, J.G., and Martinelli, R.U. (2004). Design of high-power room-temperature continuous-wave GaSb-based type-I quantum-well lasers with lambda >2.5 μm. *Semicond. Sci. Technol.* 19: 655.

36 Shterengas, L., Kipshidze, G., Hosoda, T., Chen, J., and Belenky, G. (2009). Diode lasers emitting at 3 μm with 300 mW of continuous-wave output power. *Electron. Lett.* 45: 942.

37 Hosoda, T., Feng, T., Shterengas, L., Kipshidze, G., and Belenky, G. (2016). High power cascade diode lasers emitting near 2 μm. *Appl. Phys. Lett.* 108: 131109.

38 Forouhar, S., Briggs, R.M., Frez, C., Franz, K.J., and Ksendzov, A. (2012). High-power laterally coupled distributed feedback GaSb-based diode lasers at 2 μm wavelength. *Appl. Phys. Lett.* 100: 031107.

39 Ksendzov, A., Forouhar, S., Briggs, R.M., Frez, C., Franz, K.J., and Bagheri, M. (2012). Linewidth measurement of high power diode laser at 2 μm for carbon dioxide detection. *Electron. Lett.* 48: 520.

40 Naehle, L., Belahsene, S., von Edlinger, M., Fischer, M., Boissier, G., Grech, P., Narcy, G., Vicet, A., Rouillard, Y., Koeth, J., and Worschech, L. (2011). Continuous-wave operation of type-I quantum well DFB laser diodes emitting in 3.4 μm wavelength range around room temperature. *Electron. Lett.* 47: 46.

41 Naehle, L., Zimmermann, C., Belahsene, S., Fischer, M., Boissier, G., Grech, P., Narcy, G., Lundqvist, S., Rouillard, Y., Koeth, J., Kamp, M., and Worschech, L. (2011). Monolithic tunable GaSb-based lasers at 3.3 μm. *Electron. Lett.* 47: 1092.

42 Gupta, J.A., Bezinger, A., Barrios, P.J., Lapointe, J., Poitras, D., and Waldron, P. (2012). High-resolution methane spectroscopy using InGaAsSb/AlInGaAsSb laterally-coupled index-grating distributed feedback laser diode at 3.23 μm. *Electron. Lett.* 48: 396.

43 Faist, J., Capasso, F., Sivco, D.L., Sirtori, C., Hutchinson, A.L., and Cho, A.Y. (1994). Quantum cascade laser. *Science* 264: 553.

44 Capasso, F. (2010). High-performance midinfrared quantum cascade lasers. *Opt. Eng.* 49: 111102.

45 Faist, J. (2013). *Quantum Cascade Lasers*. Oxford: Oxford University Press.

46 Sirtori, C., Kruck, P., Barbieri, S., Collot, P., Nagle, J., Beck, M., Faist, J., and Oesterle, U. (1998). GaAs/Al_xGa_{1-x} as quantum cascade lasers. *Appl. Phys. Lett.* 73: 3486.

47 Devenson, J., Teissier, R., Cathabard, O., and Baranov, A.N. (2007). InAs/AlSb quantum cascade lasers emitting at 2.75–2.97 μm. *Appl. Phys. Lett.* 91: 251102.

48 Cathabard, O., Teissier, R., Devenson, J., Moreno, J.C., and Baranov, A.N. (2010). Quantum cascade lasers emitting near 2.6 μm. *Appl. Phys. Lett.* 96: 141110.

49 Faist, J., Capasso, F., Sirtori, C., Sivco, D.L., Baillargeon, J.N., Hutchinson, A.L., Chu, S.-N.G., and Cho, A.Y. (1996). High power mid-infrared (λ~5 μm) quantum cascade lasers operating above room temperature. *Appl. Phys. Lett.* 68: 3680.

50 Beck, M., Hofstetter, D., Aellen, T., Faist, J., Oesterle, U., Ilegems, M., Gini, E., and Melchior, H. (2002). Continuous wave operation of a mid-infrared semiconductor laser at room temperature. *Science* 295: 301.

51 Bai, Y., Slivken, S., Darvish, S.R., and Razeghi, M. (2008). Room temperature continuous wave operation of quantum cascade lasers with 12.5% wall plug efficiency. *Appl. Phys. Lett.* 93: 021103.

52 Bai, Y., Bandyopadhyay, N., Tsao, S., Slivken, S., and Razeghi, M. (2011). Room temperature quantum cascade lasers with 27% wall plug efficiency. *Appl. Phys. Lett.* 98: 181102.

53 Lyakh, A., Pflügl, C., Diehl, L., Wang, Q.J., Capasso, F., Wang, X.J., Fan, J.Y., Tanbun-Ek, T., Maulini, R., Tsekoun, A., Go, R., and Patel, C.K.N. (2008). 1.6 W high wall plug efficiency, continuous-wave room temperature quantum cascade laser emitting at 4.6 μm. *Appl. Phys. Lett.* 92: 111110.

54 Lyakh, A., Maulini, R., Tsekoun, A., Go, R., Pflügl, C., Diehl, L., Wang, Q.J., Capasso, F., and Patel, C.K.N. (2009). 3 W continuous-wave room temperature single-facet emission from quantum cascade lasers based on nonresonant extraction design approach. *Appl. Phys. Lett.* 95: 141113.

55 Lyakh, A., Maulini, R., Tsekoun, A., Go, R., and Patel, C.K.N. (2012). Tapered 4.7 μm quantum cascade lasers with highly strained active region composition delivering over 4.5 watts of continuous wave optical power. *Opt. Express* 20: 4382.

56 Maulini, R., Lyakh, A., Tsekoun, A., Kumar, C., and Patel, N. (2011). λ~7.1 μm quantum cascade lasers with 19% wallplug efficiency at room temperature. *Opt. Express* 19: 17203.

57 Lyakh, A., Maulini, R., Tsekoun, A., Go, R., and Patel, C.K.N. (2012). Multiwatt long wavelength quantum cascade lasers based on high strain composition with 70% injection efficiency. *Opt. Express* 20: 24272.

58 Liu, P.Q., Hoffman, A.J., Escarra, M.D., Franz, K.J., Khurgin, J.B., Dikmelik, Y., Wang, X., Fan, J.-Y., and Gmachl, C.F. (2010). Highly power-efficient quantum cascade lasers. *Nat. Photon.* 4: 95.

59 Bai, Y., Slivken, S., Kuboya, S., Darvish, S.R., and Razeghi, M. (2010). Quantum cascade lasers that emit more light than heat. *Nat. Photon.* 4: 99.

60 Faist, J., Gmachl, C., Capasso, F., Sirtori, C., Sivco, D.L., Baillargeon, J.N., and Cho, A.Y. (1997). Distributed feedback quantum cascade lasers. *Appl. Phys. Lett.* 70: 2670.

61 Hofstetter, D., Faist, J., Beck, M., and Müller, A. (1999). Demonstration of high-performance 10.16 μm quantum cascade distributed feedback lasers fabricated without epitaxial regrowth. *Appl. Phys. Lett.* 75: 665.

62 Lu, Q.Y., Bai, Y., Bandyopadhyay, N., Slivken, S., and Razeghi, M. (2010). Room-temperature continuous wave operation of distributed feedback quantum cascade lasers with watt-level power output. *Appl. Phys. Lett.* 97: 231119.

63 Lu, Q.Y., Bai, Y., Bandyopadhyay, N., Slivken, S., and Razeghi, M. (2011). 2.4 W room temperature continuous wave operation of distributed feedback quantum cascade lasers. *Appl. Phys. Lett.* 98: 181106.

64 Lee, B.G., Belkin, M.A., Audet, R., MacArthur, J., Diehl, L., Pflügl, C., Capasso, F., Oakley, D.C., Chapman, D., Napoleone, A., Bour, D., Corzine, S., Höfler, G., and Faist, J. (2007). Widely tunable single-mode quantum cascade laser source for mid-infrared spectroscopy. *Appl. Phys. Lett.* 91: 231101.

65 Lee, B.G., Zhang, H.A., Pfluegl, C., Diehl, L., Belkin, M.A., Fischer, M., Wittmann, A., Faist, J., and Capasso, F. (2009). Broadband distributed-feedback quantum cascade laser array operating from 8.0 to 9.8 μm. *IEEE Photon. Technol. Lett.* 21: 914.

66 Faist, J., Beck, M., Aellen, T., and Gini, E. (2001). Quantum cascade lasers based on a bound-to-continuum transition. *Appl. Phys. Lett.* 78: 147.
67 Bandyopadhyay, N., Chen, M., Sengupta, S., Slivken, S., and Razeghi, M. (2015). Ultra-broadband quantum cascade laser, tunable over 760 cm^{-1}, with balanced gain. *Opt. Express* 23: 21159.
68 Rauter, P. and Capasso, F. (2015). Multi-wavelength quantum cascade laser arrays. *Laser Photon. Rev.* 9: 452.
69 Hugi, A., Maulini, R., and Faist, J. (2010). External cavity quantum cascade laser. *Semicond. Sci. Technol.* 25: 083001.
70 Luo, G., Peng, C., Le, H.Q., Pei, S.-S., Lee, H., Hwang, W.-Y., Ishaug, B., and Zheng, J. (2002). Broadly wavelength-tunable external cavity mid-infrared quantum cascade lasers. *IEEE J. Quantum Electron.* 38: 486.
71 Gmachl, C., Sivco, D.L., Colombelli, R., and Capasso, F. (2002). Ultra-broadband semiconductor laser. *Nature* 415: 883.
72 Maulini, R., Beck, M., Faist, J., and Gini, E. (2004). Broadband tuning of external cavity bound-to-continuum quantum-cascade lasers. *Appl. Phys. Lett.* 84: 1659.
73 Wittmann, A., Hugi, A., Gini, E., Hoyler, N., and Faist, J. (2008). Heterogeneous high-performance quantum-cascade laser sources for broad-band tuning. *IEEE J. Quantum Electron.* 44: 1083.
74 Hugi, A., Terazzi, R., Bonetti, Y., Wittmann, A., Fischer, M., Beck, M., Faist, J., and Gini, E. (2009). External cavity quantum cascade laser tunable from 7.6 to 11.4 μm. *Appl. Phys. Lett.* 95: 061103.
75 Yao, Y., Wang, X., Fan, J.-Y., and Gmachl, C.F. (2010). High performance "continuum-to-continuum" quantum cascade lasers with a broad gain bandwidth of over 400 cm^{-1}. *Appl. Phys. Lett.* 97: 081115.
76 Fujita, K., Furuta, S., Sugiyama, A., Ochiai, T., Ito, A., Dougakiuchi, T., Edamura, T., and Yamanishi, M. (2011). High performance quantum cascade lasers with wide electroluminescence (600 cm^{-1}), operating in continuous-wave above 100 °C. *Appl. Phys. Lett.* 98: 231102.
77 Lyakh, A., Barron-Jimenez, R., Dunayevskiy, I., Go, R., Kumar, C., and Patel, N. (2015). External cavity quantum cascade lasers with ultra-rapid acousto-optic tuning. *Appl. Phys. Lett.* 106: 141101.
78 Faist, J., Capasso, F., Sivco, D.L., Hutchinson, A.L., Chu, S.N.G., and Cho, A.Y. (1998). Short wavelength ($\lambda \sim 3.4$ μm) quantum cascade laser based on strained compensated InGaAs/AlInAs. *Appl. Phys. Lett.* 72: 680.
79 Bismuto, A., Beck, M., and Faist, J. (2011). High power Sb-free quantum cascade laser emitting at 3.3 μm above 350 K. *Appl. Phys. Lett.* 98: 191104.
80 Bandyopadhyay, N., Bai, Y., Tsao, S., Nida, S., Slivken, S., and Razeghi, M. (2012). Room temperature continuous wave operation of $\lambda \sim 3$–3.2 μm quantum cascade lasers. *Appl. Phys. Lett.* 101: 241110.
81 Revin, D.G., Cockburn, J.W., Steer, M.J., Airey, R.J., Hopkinson, M., Krysa, A.B., Wilson, L.R., and Menzel, S. (2007). InGaAs/AlAsSb/InP quantum cascade lasers operating at wavelengths close to 3 μm. *Appl. Phys. Lett.* 90: 021108.

82 Zhang, S.Y., Revin, D.G., Cockburn, J.W., Kennedy, K., Krysa, A.B., and Hopkinson, M. (2009). λ~ 3.1 μm room temperature InGaAs/AlAsSb/InP quantum cascade lasers. *Appl. Phys. Lett.* 94: 031106.
83 Semtsiv, P., Wienold, M., Dressler, S., and Masselink, W.T. (2007). Short wavelength (λ ≈ 3.05 μm) InP-based strain-compensated quantum cascade laser. *Appl. Phys. Lett.* 90: 051111.
84 Devenson, J., Barate, D., Cathabard, O., Teissier, R., and Baranov, A.N. (2006). Very short wavelength (λ = 3.1–3.3 μm) quantum cascade lasers. *Appl. Phys. Lett.* 89: 191115.
85 Baranov, A.N. and Teissier, R. (2015). Quantum cascade lasers in the InAs/AlSb material system. *IEEE J. Sel. Topics Quantum Electron.* 21: 1200612.
86 Devenson, J., Teissier, R., Cathabard, O., and Baranov, A.N. (2007). InAs/AlSb quantum cascade lasers emitting below 3 μm. *Appl. Phys. Lett.* 90: 111118.
87 Cathabard, O., Teissier, R., Devenson, J., and Baranov, A.N. (2009). InAs-based distributed feedback quantum cascade lasers. *Electron. Lett.* 45: 1028.
88 Riedi, S., Hugi, A., Bismuto, A., Beck, M., and Faist, J. (2013). Broadband external cavity tuning in the 3–4 μm window. *Appl. Phys. Lett.* 103: 031108.
89 Bahriz, M., Lollia, G., Baranov, A.N., Laffaille, P., and Teissier, R. (2013). InAs/AlSb quantum cascade lasers operating near 20 μm. *Electron. Lett.* 49: 1238.
90 Chastanet, D., Lollia, G., Bousseksou, A., Bahriz, M., Laffaille, P., Baranov, A.N., Julien, F., Colombelli, R., and Teissier, R. (2014). Long-infrared InAs-based quantum cascade lasers operating at 291 K (λ = 19 μm) with metal–metal resonators. *Appl. Phys. Lett.* 104: 021106.
91 Chastanet, D., Bousseksou, A., Lollia, G., Bahriz, M., Julien, F.H., Baranov, A.N., Teissier, R., and Colombelli, R. (2014). High temperature, single mode, long infrared (λ =17.8 μm) InAs-based quantum cascade lasers. *Appl. Phys. Lett.* 105: 111118.
92 Kohler, R., Tredicucci, A., Beltram, F., Beere, H.E., Linfield, E.H., Davies, A.G., Ritchie, D.A., Iotti, R.C., and Rossi, F. (2002). Terahertz semiconductor-heterostructure laser. *Nature* 417: 156.
93 Williams, B.S. (2007). Terahertz quantum-cascade lasers. *Nat. Photon.* 1: 517.
94 Yang, R.Q. (1995). Infrared laser based on intersubband transitions in quantum wells. *Superlattices Microstruct.* 17: 77.
95 Meyer, J.R., Hoffman, C.A., Bartoli, F.J., and Ram-Mohan, L.R. (1995). Type-II quantum-well lasers for the mid-wavelength infrared. *Appl. Phys. Lett.* 67: 757.
96 Meyer, J.R., Vurgaftman, I., Yang, R.Q., and Ram-Mohan, L.R. (1996). Type-II and type-I interband cascade lasers. *Electron. Lett.* 32: 45.
97 Vurgaftman, I., Bewley, W.W., Canedy, C.L., Kim, C.S., Kim, M., Lindle, J.R., Merritt, C.D., Abell, J., and Meyer, J.R. (2011). Mid-IR type-II interband cascade lasers. *IEEE J. Sel. Topics Quantum Electron.* 17: 1435.
98 Vurgaftman, I., Weih, R., Kamp, M., Meyer, J.R., Canedy, C.L., Kim, C.S., Kim, M., Bewley, W.W., Merritt, C.D., Abell, J., and Höfling, S. (2015). Interband cascade lasers. *J. Phys. D Appl. Phys.* 48: 123001.

99 Lin, C.-H., Yang, R.Q., Zhang, D., Murry, S.J., Pei, S.S., Allerman, A.A., and Kurtz, S.R. (1997). Type-II interband quantum cascade laser at 3.8 μm. *Electron. Lett.* 33: 598.

100 Vurgaftman, I., Bewley, W.W., Canedy, C.L., Kim, C.S., Kim, M., Merritt, C.D., Abell, J., Lindle, J.R., and Meyer, J.R. (2011). Rebalancing of internally generated carriers for mid-infrared interband cascade lasers with very low power consumption. *Nat. Commun.* 2: 585.

101 Kim, M., Canedy, C.L., Bewley, W.W., Kim, C.S., Lindle, J.R., Abell, J., Vurgaftman, I., and Meyer, J.R. (2008). Interband cascade laser emitting at $\lambda = 3.75$ μm in continuous wave above room temperature. *Appl. Phys. Lett.* 92: 191110.

102 Vurgaftman, I., Canedy, C.L., Kim, C.S., Kim, M., Bewley, W.W., Lindle, J.R., Abell, J., and Meyer, J.R. (2009). Mid-infrared interband cascade lasers operating at ambient temperatures. *N. J. Phys.* 11: 125015.

103 Bewley, W.W., Canedy, C.L., Kim, C.S., Kim, M., Merritt, C.D., Abell, J., Vurgaftman, I., and Meyer, J.R. (2012). High-power room-temperature continuous-wave mid-infrared interband cascade lasers. *Opt. Express* 20: 20894.

104 Kim, C.S., Kim, M., Bewley, W.W., Lindle, J.R., Canedy, C.L., Nolde, J.A., Larrabee, D.C., Vurgaftman, I., and Meyer, J.R. (2008). Broad-stripe, single-mode, mid-IR interband cascade laser with photonic-crystal distributed-feedback grating. *Appl. Phys. Lett.* 92: 071110.

105 Kim, C.S., Kim, M., Abell, J., Bewley, W.W., Merritt, C.D., Canedy, C.L., Vurgaftman, I., and Meyer, J.R. (2012). Midinfrared distributed-feedback interband cascade lasers with continuous-wave single-mode emission to 80 °C. *Appl. Phys. Lett.* 101: 061104.

106 Vurgaftman, I., Bewley, W.W., Canedy, C.L., Kim, C.S., Kim, M., Merritt, C.D., Abell, J., and Meyer, J.R. (2013). Interband cascade lasers with low threshold powers and high output powers. *IEEE J. Sel. Topics Quantum Electron.* 19: 1200210.

107 Weih, R., Nähle, L., Höfling, S., Koeth, J., and Kamp, M. (2014). Single mode interband cascade lasers based on lateral metal gratings. *Appl. Phys. Lett.* 105: 071111.

108 Forouhar, S., Borgentun, C., Frez, C., Briggs, R.M., Bagheri, M., Canedy, C.L., Kim, C.S., Kim, M., Bewley, W.W., Merritt, C.D., Abell, J., Vurgaftman, I., and Meyer, J.R. (2014). Reliable mid-infrared laterally-coupled distributed-feedback interband cascade lasers. *Appl. Phys. Lett.* 105: 051110.

109 Borgentun, C., Frez, C., Briggs, R.M., Fradet, M., and Forouhar, S. (2015). Single-mode high-power interband cascade lasers for mid-infrared absorption spectroscopy. *Opt. Express* 23: 2446.

110 Dallner, M., Scheuermann, J., Nähle, L., Fischer, M., Koeth, J., Höfling, S., and Kamp, M. (2015). InAs-based distributed feedback interband cascade lasers. *Appl. Phys. Lett.* 107: 181105.

111 Scheuermann, J., Weih, R., von Edlinger, M., Nähle, L., Fischer, M., Koeth, J., Kamp, M., and Höfling, S. (2015). Single-mode interband cascade lasers emitting below 2.8 μm. *Appl. Phys. Lett.* 106: 161103.

112 von Edlinger, M., Scheuermann, J., Weih, R., Zimmermann, C., Nähle, L., Fischer, M., Koeth, J., Höfling, S., and Kamp, M. (2014). Monomode interband cascade lasers at 5.2 μm for nitric oxide sensing. *IEEE Photon. Technol. Lett.* 26: 480.

113 Bewley, W.W., Kim, C.S., Canedy, C.L., Merritt, C.D., Vurgaftman, I., Abell, J., Meyer, J.R., and Kim, M. (2013). High-power, high-brightness continuous-wave interband cascade lasers with tapered ridges. *Appl. Phys. Lett.* 103: 111111.

114 Kim, M., Bewley, W.W., Canedy, C.L., Kim, C.S., Merritt, C.D., Abell, J., Vurgaftman, I., and Meyer, J.R. (2015). High-power continuous-wave interband cascade lasers with 10 active stages. *Opt. Express* 23: 9664.

115 Canedy, C.L., Kim, C.S., Merritt, C.D., Bewley, W.W., Vurgaftman, I., Meyer, J.R., and Kim, M. (2015). Interband cascade lasers with >40% continuous-wave wallplug efficiency at cryogenic temperatures. *Appl. Phys. Lett.* 107: 121102.

116 Bewley, W.W., Canedy, C.L., Kim, C.S., Kim, M., Merritt, C.D., Abell, J., Vurgaftman, I., and Meyer, J.R. (2012). Continuous-wave interband cascade lasers operating above room temperature at $\lambda = 4.7–5.6$ μm. *Opt. Express* 20: 3235.

117 Dallner, M., Hau, F., Hofling, S., and Kamp, M. (2015). InAs-based interband-cascade-lasers emitting around 7 μm with threshold current densities below 1 kA/cm^2 at room temperature. *Appl. Phys. Lett.* 106: 041108.

118 Schulz, N., Hopkins, J.-M., Rattunde, M., Burns, D., and Wagner, J. (2008). High-brightness long-wavelength semiconductor disk lasers. *Laser Photon. Rev.* 2: 160.

119 Chilla, J., Butterworth, S., Zeitschel, A., Charles, J., Caprara, A.L., Reed, M.K., and Spinelli, L. (2004). High power optically pumped semiconductor lasers. *Proc. SPIE* 5332: 143.

120 Beyertt, S.-S., Brauch, U., Demaria, F., Dhidah, N., Giesen, A., Kuebler, T., Lorch, S., Rinaldi, F., and Unger, P. (2007). Efficient gallium-arsenide disk laser. *IEEE J. Quantum Electron* 43: 869.

121 Schulz, N., Rattunde, M., Ritzenthaler, C., Roesener, B., Manz, C., Koehler, K., and Wagner, J. (2007). Resonant optical in-well pumping of an (AlGaIn)(AsSb)-based vertical-external-cavity surface-emitting laser emitting at 2.35 μm. *Appl. Phys. Lett.* 91: 091113.

122 Ishida, A., Sugiyama, Y., Isaji, Y., Kodama, K., Takano, Y., Sakata, H., Rahim, M., Khiar, A., Fill, M., Felder, F., and Zogg, H. (2011). 2 W high efficiency PbS mid-infrared surface emitting laser. *Appl. Phys. Lett.* 99: 121109.

123 Rahim, M., Khiar, A., Felder, F., Fill, M., and Zogg, H. (2009). 4.5-μm wavelength vertical external cavity surface emitting laser operating above room temperature. *Appl. Phys. Lett.* 94: 201112.

124 Khiar, A., Rahim, M., Fill, M., Felder, F., Hobrecker, F., and Zogg, H. (2010). Continuously tunable monomode mid-infrared vertical external cavity surface emitting laser on Si. *Appl. Phys. Lett.* 97: 151104.

5

Mid-IR by Nonlinear Optical Frequency Conversion

This chapter reviews devices based on nonlinear optical (NLO) frequency downconversion that operate in the 2–20 μm mid-IR region. It discusses the principles of nonlinear frequency conversion, gives a brief assessment of existing and emerging NLO materials suitable for mid-IR, and reviews current approaches for obtaining tunable mid-IR output. We start first with the techniques that involve three-wave processes, namely those based on the second-order nonlinearity $\chi^{(2)}$.

5.1 Two Approaches to Frequency Downconversion Using Second-order Nonlinearity

There are two basic techniques (Figure 5.1) for achieving broadly tunable mid-IR output via frequency downconversion in a three-wave process that relies upon second-order nonlinearity $\chi^{(2)}$ of optical materials, namely (i) using difference frequency generation (DFG) and (ii) using optical parametric oscillators (OPOs). The OPO variations include traveling-wave optical parametric generators (OPGs) and optical parametric amplifiers (OPAs).

While both DFG and OPO are based on three-photon processes, the main distinction between them is that in the former case one needs two pump sources (and at least one of them has to be tunable to achieve mid-IR tunability) while in the latter case one uses just one pump laser source (and a "seed" wave in the case of an OPA).

In both DFG and OPO processes, the photon energy conservation dictates that

$$\omega_3 = \omega_2 + \omega_1, \tag{5.1}$$

while the photon momentum conservation (also referred to as phase matching) requires that

Laser-based Mid-infrared Sources and Applications, First Edition. Konstantin L. Vodopyanov.

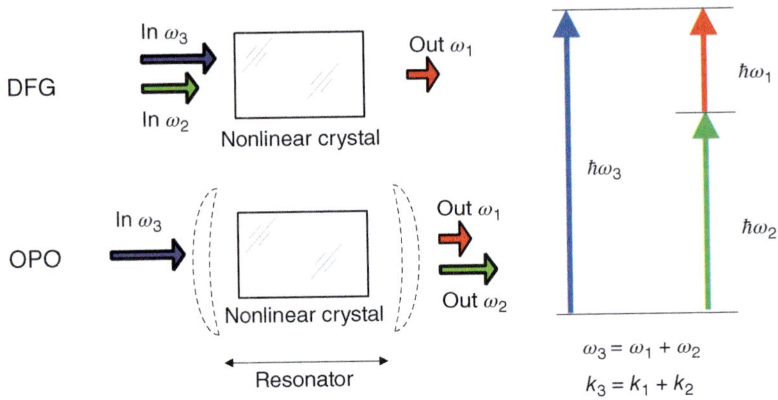

Figure 5.1 DFG and OPO – the two main techniques for using second-order nonlinearity of the materials for achieving tunable mid-IR output.

$$k_3 = k_2 + k_1. \tag{5.2}$$

Here, ω_3, ω_2, and ω_1 are the so-called *pump*, *signal*, and *idler* angular frequencies, such that $\omega_3 > \omega_2 > \omega_1$, and k_3, k_2, and k_1 are, correspondingly, their wave vectors (such that their modules are $|k_i| = \omega_i n_i / c$, where $i = 1, 2, 3$ and n_i is a refractive index).

The momentum conservation condition is never satisfied in *isotropic* media (at least in the transparency range, where normal dispersion applies, i.e. refractive index n increases with the optical frequency ω). However, it can be fulfilled in birefringent crystals where one can use orthogonal polarizations so that the wavelength dispersion is compensated by polarization dispersion (so-called *birefringent* or *angular* phase matching).

Alternatively, the photon momentum conservation can be satisfied in so-called quasi-phase-matched (QPM) crystals. Quasi-phase-matching is obtained by periodic inversion of the crystalline orientation along the optical path. This periodicity of crystal orientation, and associated reversal of the sign of the optical nonlinearity, compensates for the wave-vector mismatch because the relative phase is corrected at regular intervals [1].

The period of QPM grating Λ inside the crystal (Figure 5.2) should be such that

$$2\pi / \Lambda = k_3 - k_2 - k_1 \tag{5.3}$$

In this case, an additional k-vector associated with orientation reversal periodicity ($2\pi/\Lambda$) compensates the mismatch between participating waves $\Delta k = k_3 - k_2 - k_1$.

Quasi-phase-matching can be illustrated as follows: suppose one needs to drive a car from point A to point B, but the wheels of the car cannot be directed straight, but rather at +45° or −45°. In this case, the strategy is

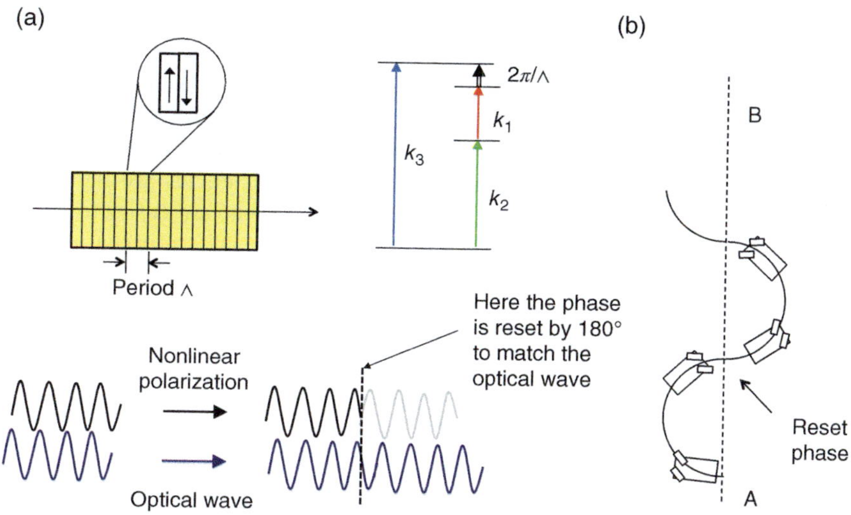

Figure 5.2 Quasi-phase-matching. (a) In quasi-phase-matched (QPM) crystals with periodically flipped sign of the nonlinearity, the artificially created grating of crystal orientation reversal compensates for the wave-vector mismatch. (b) Quasi-phase-matching is similar to the situation when one needs to drive a car from point A to point B, but the wheels of the car cannot be put straight, but rather at either +45° or at −45°. In this case, the strategy is to start going in the direction of 90° to the target and each time the car makes a semicircle, flip the wheels from +45° to −45°, or vice versa.

as follows: one starts going in the direction of 90° to the target and when the car makes a semicircle, steers the wheels from −45° to +45° position (or vice versa). The average travel distance will be increased by a factor of $\pi/2$, as compared to the straight line, the gas mileage will be reduced correspondingly by $2/\pi$ and that is exactly the factor by which the effective nonlinear coefficient is reduced in the QPM scenario, as compared to the true phase-matched case. The key point is to flip the steering wheel at the right moment. Similarly, in the case of QPM crystals, the period of flipping the crystal orientation Λ needs to be precisely adjusted for a given three-wave interaction.

Unlike birefringent phase matching, quasi-phase-matching allows (i) to access the highest available nonlinear coefficient, like d_{33} in lithium niobate, and (ii) eliminates spatial (birefringent) walk-off of the beams with orthogonal polarizations. This allows tight, so-called noncritical, beam focusing. A typical example of QPM crystal is periodically poled lithium niobate (PPLN).

5.1.1 Difference Frequency Generation

DFG allows achieving compact mid-IR sources based on frequency mixing the outputs of a pair of well-developed near-IR lasers. DFG is a cavityless, single-pass wavelength conversion scheme. It only requires the input laser beams to be spatially and temporally overlapped within the nonlinear crystal. For example, lithium niobate ($LiNbO_3$) is a low-cost robust material, and periodic poling allows it to be tailored to phase-match practically any three-wave DFG interaction, limited only by the $LiNbO_3$ absorption edge near $\lambda \approx 5\,\mu m$ [2]. Both waveguide and bulk interaction geometries in PPLN have been used to mix near-IR wavelengths to the mid-IR. While DFG conversion efficiencies are higher in waveguides, bulk materials can handle higher power levels, are easier to fabricate, and are far less sensitive to the beam alignment. The DFG source tuning is achieved by changing the wavelength of either one of the near-IR lasers, or both of them.

In the limit of small conversion efficiency and perfect phase matching between the interacting waves, the DFG power P_1 is expressed by the following product:

$$P_1 = \eta \times P_2 P_3, \tag{5.4}$$

where P_2 and P_3 are the powers of the two pump lasers (the one with a shorter wavelength, P_3, is called the *pump* and another, P_2, is called the *signal*). Thus, P_1 is linear with respect to both P_2 and P_3 and one can introduce a normalized conversion efficiency η (in % per W) that depends on the effective nonlinear coefficient (d_{eff}), the output mid-IR frequency (ω_1), focusing strength, and interaction length (L) in the following way [3]:

$$\eta \propto \omega_1^2 \left(\frac{d_{eff}^2}{n^3} \right) \frac{L^2}{\text{area}}. \tag{5.5}$$

Here, n is the average refractive index, and “area” is the effective cross section – a measure of how tightly the interacting beams are focused inside the crystal (the effective cross section can be very small in waveguides). The important term that determines DFG efficiency, d_{eff}^2/n^3, is referred to as the nonlinear optical figure of merit (NLO FOM) of a crystal. One can also see that the DFG conversion efficiency drastically drops on increasing the mid-IR wavelength, due to the ω_1^2 term.

5.1.2 Optical Parametric Oscillators (OPOs)

Optical parametric oscillators offer extremely wide tunability, intrinsically limited only by the material transparency, require only a single-pump laser, and typically have very high conversion efficiency from the pump, much larger than that of DFG. OPOs and other parametric devices such as OPGs and OPAs are the sources of choice when one needs the broadest continuous tunability (up to two or three octaves in frequency), high peak (>1 kW) or average (>1 W) power, and high (>50%) quantum conversion efficiency. This comes at a cost of the need for a resonant cavity (typically resonant for the signal wave), or having a high peak power (>1 MW) pump in the case of single-pass devices such as OPG or OPA. Also, OPOs have a distinct oscillation threshold, while DFG devices do not (one always gets an output from a DFG system).

Parametric frequency downconversion in an OPO can be regarded as the inverse process of sum-frequency generation with an NLO crystal that can be viewed as a catalyst that promotes decay of the pump photon into two smaller photons. The OPO tuning is achieved by changing the phase-matching condition that in turn changes the ratio between the signal and idler photon energies, and thus tunes the output frequency. This can be accomplished by:

- rotating the crystal (for birefringent NLO materials)
- changing the inversion period (for QPM crystals)
- changing the temperature of the crystal
- tuning the pump wavelength.

Figure 5.3 depicts four basic types of optical parametric devices, namely (i) continuous wave (CW), (ii) pulsed nanosecond, (iii) synchronously pumped ultrafast, and (iv) traveling-wave device such as OPG and OPA. Since parametric gain depends on the instantaneous intensity (power density) of the pump, CW OPOs (Figure 5.3a) require a high average power (typically >1 W) to reach the threshold; also, since in CW OPOs the gain is small, one requires a low-loss cavity. In contrast, with nanosecond pumping (Figure 5.3b), the requirements for cavity loss are much less demanding (e.g. a simple "plano-plano" short cavity is sufficient); the OPO threshold in this case can be achieved with modest pump pulse energies, in the microjoule to millijoule range, with correspondingly modest average pump powers [4]. With pico- or femtosecond pumping (Figure 5.3c), it is necessary to match the OPO roundtrip time to the repetition period of a pump laser – a synchronous pumping (or sync-pumping). In this case, the OPO pulse after a full roundtrip will be amplified again by the next incoming pump pulse. Due to the short duration of mode-locked pump pulses, the OPO threshold can be low in terms of the average power, around 100 mW or less. Finally, OPAs and OPGs are single-pass devices that do not require a cavity (Figure 5.3d). As a pump, they typically use short-duration (ps- or fs-range) intense pulses with the input pump power density of at least 100 MW/cm^2. The difference between an OPA and an OPG is that, in the former case, there is a weak "seed" pulse (characteristically at the *signal* wave frequency ω_2) that is injected along with a strong pump at ω_3, while in the OPG it is the quantum noise that serves as a seed for parametric amplification. In both cases, a single-pass gain in an NLO crystal is very high (~10^6 or more, an exponential process) and a substantial fraction (>10%) of the pump is converted into the signal and the idler waves.

For perfect phase matching between the interacting waves and in the limit of no pump depletion, a single-pass parametric gain for a seed pulse P_{in} (at either signal or idler wave) in the presence of a pump field at ω_3 is expressed [3] as

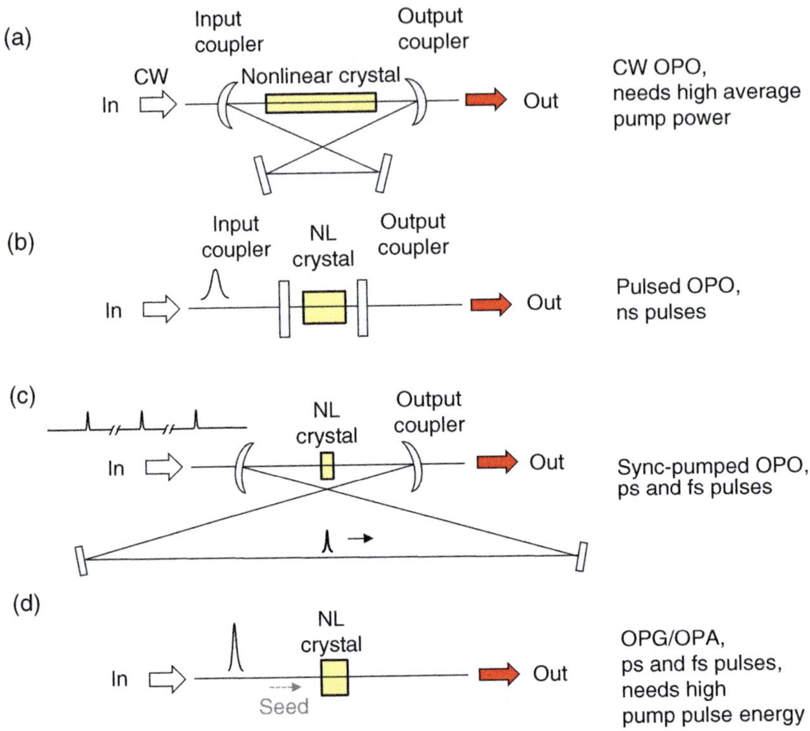

Figure 5.3 Four basic types of optical parametric devices. (a) Continuous-wave OPO. (b) Pulsed nanosecond OPO. (c) Sync-pumped ultrafast OPO. (d) Traveling-wave device such as OPG and OPA.

$$G = \frac{P_{out}}{P_{in}} = \cos h^2(\Gamma L), \tag{5.6}$$

where L is the length of a nonlinear crystal and Γ is the gain increment given by:

$$\Gamma^2 = \left(\frac{d_{eff}^2}{n^3}\right)\frac{2\omega_1\omega_2 I_{pump}}{\varepsilon_0 c^3} = \left(\frac{d_{eff}^2}{n^3}\right)\frac{8\pi^2 I_{pump}}{\lambda_1\lambda_2\varepsilon_0 c} \tag{5.7}$$

Here, I_{pump} is the pump laser intensity (power density), ω_1 and ω_2 are idler and signal frequencies, λ_1 and λ_2 are corresponding wavelengths (such that $\omega_i = 2\pi c/\lambda_i$), d_{eff} is the effective nonlinear coefficient, and n is the average refractive index. Again, the important term, which determines the gain, is the NLO FOM $\left(d_{eff}^2 / n^3\right)$.

In fact, a single-pass parametric gain varies drastically with pump intensity. For example, in the low-gain limit ($\Gamma L \ll 1$),

$$G = \cosh^2(\Gamma L) \approx 1 + (\Gamma L)^2 \tag{5.8}$$

so that the incremental *increase* in power is

$$\frac{(P_{out} - P_{in})}{P_{in}} = G - 1 = (\Gamma L)^2, \tag{5.9}$$

and is proportional to the pump power density. On the other hand, in the high-gain limit ($\Gamma L \gg 1$), the cosh function is approximated by an exponent and

$$G = \cosh^2(\Gamma L) \approx \frac{1}{4} e^{2\Gamma L} \tag{5.10}$$

so that G grows exponentially with the pump field.

As a numerical example, consider PPLN with an effective NLO coefficient d_{eff} = 14 pm/V suitable for mid-IR range [5], with λ_3 = 1.06 μm (pump wavelength), λ_2 = 1.57 μm (signal), and λ_1 = 3.3 μm (idler). At a CW pump power of 10 W, a beam diameter of 100 μm (assume for simplicity a flat-top uniform intensity distribution for all three waves), which corresponds to a pump power density of 0.13 MW/cm^2, and for a 30-mm-long crystal, we get from (5.8) to (5.9) an incremental increase in power of only 13% per single pass. For ultrafast pump pulses, though, the situation drastically changes. At 1-ps pulse duration, 100-μm flat-top beam diameter, at only 5-mm-long crystal, and a modest pulse energy of 1 μJ, corresponding to the pump power density of 13 GW/cm^2, a single-pass parametric gain reaches $G = 1.7 \cdot 10^{16}$.

5.1.3 Brief Review of $\chi^{(2)}$ Nonlinear Crystals for Mid-IR

Table 5.1 compares linear and NLO properties of selected $\chi^{(2)}$ NLO crystals that are most suitable for mid-IR applications. The crystals are divided into three categories in this table:

- Periodically poled (PP) QPM oxides, such as lithium niobate ($LiNbO_3$) or potassium titanyl phosphate ($KTiOPO_4$).
- Birefringent crystals, some of which have been known for quite a long time (e.g. AGS, AGSe, ZGP, CGA, CdSe, and GaSe) [12], while the others have been developed quite recently, such as cadmium silicon phosphide (CSP).
- Newly developed QPM semiconductors with periodic inversion of crystalline orientation, such as orientation-patterned gallium arsenide (OP-GaAs), gallium phosphide (OP-GaP), zinc selenide (OP-ZnSe), and gallium nitride (OP-GaN).

5.1.3.1 Periodically Poled Oxides

PP oxides achieved their maturity in the mid-1990s and are widely used because of their comparatively low cost and high nonlinear coefficient d_{33}. The advent of the first practical QPM material – PPLN – with the successful implementation of quasi-phase-matching by periodic inversion of its ferroelectric domains, represented a new paradigm in NLO materials development [1, 17]. No longer constrained by the strict requirements of birefringence phase matching, QPM enabled the use of a much broader range of materials, provided that a mechanism existed for creating the alternating QPM domain structure. In ferroelectric oxides, this was achieved by applying lithographic electrodes, followed by electric field poling of periodic domains of alternating polarity. QPM offered the advantages of noncritical phase matching, polarization flexibility, and engineerable functionality.

Potassium titanyl phosphate, $KTiOPO_4$ (KTP), and its isomorphs, belong to another class of crystals that are quite attractive for periodic poling [7]. KTP possesses high laser damage threshold, and has low susceptibility for the so-called photorefractive effect. (The photorefractive effect is an unwanted altering of the refractive index in the crystal caused by parasitically generated visible light, characteristic of PPLN and PPLT.) For KTP, the nonlinear coefficient d_{33} is about 2/3 of that of $LiNbO_3$. Periodic poling of the isomorphs of KTP like $KTiOAsO_4$ (KTA) and $RbTiOAsO_4$ (RTA) has also been successfully implemented. The KTA and RTA crystals have been widely used in mid-IR OPOs, primarily because of their slightly wider infrared transmission.

Generally, ferroelectric NLO oxide crystals such as lithium niobate, lithium tantalite, or KTP and its analogs limit the longwave operation to 4–5 μm due to the onset of multi-phonon absorption.

5.1.3.2 Birefringent Crystals

Among birefringent NLO crystals for mid-IR listed in Table 5.1, ZGP is currently the most robust and is the material of choice for 2-μm-pumped optical parametric oscillators. ZGP has a very high nonlinear coefficient of 75 pm/V (see Table 5.1) and the highest thermal conductivity (35 W/mK) of any bulk birefringent crystal that is phase-matchable for 2-μm pump. ZGP allows achieving OPO tunability from 2.5 to >10 μm. Improved polishing and antireflection coatings enable laser damage thresholds in excess of 4 J/cm^2 for pulsed (20 ns, 10 kHz) illumination. Very high mid-IR (3–5 μm) output powers in excess of 30 W have been achieved with ZGP in the pulse-periodic mode [18].

$CdSiP_2$ (CSP) is a new NLO crystal suitable for laser frequency conversion in the 1–7 μm spectral range [19]. The material is particularly promising due to its high nonlinear coefficient of 84.5 pm/V [9], which is the largest of any new phase-matchable inorganic crystal grown in the past four decades. Most importantly, with its bandgap of 2.45 eV, CSP allows 1.064-μm OPO pumping

Table 5.1 Linear and nonlinear optical properties of selected second-order nonlinear crystals suitable for mid-IR applications.

Crystal	Transparency range (μm)	d_{eff} (pm/V)[a]	Ave. ref. index	NLO FOM $\left(d_{eff}^2 / n^3\right)$ with respect to PPLN	Ref.
Periodically poled oxides					
PP LN ($LiNbO_3$)	0.4–5.5	$(2/\pi)d_{33} = (2/\pi)\cdot 22 = 14$	2.12	1	[5]
PP LT ($LiTaO_3$)	0.35–4.5	$(2/\pi)d_{33} = (2/\pi)\cdot 10.7 = 6.8$	2.11	0.24	[6]
PP KTP ($KTiOPO_4$)	0.35–4.3	$(2/\pi)d_{33} = (2/\pi)\cdot 16.9 = 10.8$	1.8	1.0	[7]
PP KTA ($KTiOAsO_4$)	0.35–5.3	$(2/\pi)d_{33} = (2/\pi)\cdot 16.2 = 10.3$	1.8	0.9	[7]
PP RTA ($RbTiOAsO_4$)	0.35–5.3	$(2/\pi)d_{33} = (2/\pi)\cdot 15.8 = 10.1$	1.8	0.9	[7]
Birefringent					
AGS ($AgGaS_2$)	0.47–13	$d_{36} = 12$	2.4	0.5	[8]
AGSe ($AgGaSe_2$)	0.71–19	$d_{36} = 33$	2.65	2.8	[8]
CSP ($CdSiP_2$)	0.65–7	$d_{36} = 84.5$	3.05	12.2	[9]
LIS ($LiInS_2$)	0.4–12	$d_{31} = 7.2; d_{24} = 5.9$	2.1	0.23	[10]
LGS ($LiGaS_2$)	0.33–11.6	$d_{31} = 5.7; d_{24} = 5.2$	2.1	0.16	[11]
CdSe	0.75–25	$d_{31} = 18$	2.46	1.1	[8]
ZGP ($ZnGeP_2$)	0.74–12	$d_{36} = 75$	3.13	8.9	[12]
GaSe	0.65–20	$d_{22} = 54$	2.73	7	[12]
CGA ($CdGeAs_2$)	2.4–18	$d_{36} = 236$	3.6	65	[12]

(Continued)

Table 5.1 (Continued)

Crystal	Transparency range (μm)	d_{eff} (pm/V)[a]	Ave. ref. index	NLO FOM $\left(d_{eff}^2 / n^3\right)$ with respect to PPLN	Ref.
Orientation patterned, cubic					
OP-GaAs	0.9–17	$(2/\pi)\sqrt{4/3}\ d_{14} = (2/\pi)\sqrt{4/3} \cdot 94 = 69.1$	3.3	6.5	[13]
OP-GaP	0.57–13	$(2/\pi)\sqrt{4/3}\ d_{14} = (2/\pi)\sqrt{4/3} \cdot 37 = 27.2$	3.05	1.27	[6]
OP-GaP	0.57–13	$(2/\pi)\sqrt{4/3}\ d_{14} = (2/\pi)\sqrt{4/3} \cdot 35 = 25.7$	3.05	1.13	[b]
OP-ZnSe	0.55–20	$(2/\pi)\sqrt{4/3}\ d_{14} = (2/\pi)\sqrt{4/3} \cdot 30 = 22.1$	2.44	1.6	[14][c]
OP-ZnSe	0.55–20	$(2/\pi)\sqrt{4/3}\ d_{14} = (2/\pi)\sqrt{4/3} \cdot 20 = 14.7$	2.44	0.7	[d]
Orientation patterned, hexagonal					
OP-GaN	0.37–7	$(2/\pi)d_{33} = (2/\pi)\cdot 16.5 = 12.1$	2.3	0.6	[15]

[a] The $(2/\pi)$ reduction factor accounts for the effect of quasi-phase-matching [1]. For cubic-symmetry crystals (e.g. GaAs), there is an additional $\sqrt{4/3}$ factor, which appears when d_{eff} is maximized (for example, when all polarizations are along <111>) and $d_{eff} = \sqrt{4/3}\ d_{14}$. Listed here are nonlinear coefficients that are best suited for mid-IR. They might be lower than those for the visible and near-IR because of the dispersion [16].

[b] Mid-IR d_{14} coefficient of GaP was measured in my group at University of Central Florida, to be 35 ± 2 pm/V, via direct comparison of SHG efficiency between quasi-phase-matched OP-GaP and OP-GaAs crystals in a frequency-doubling process using nanosecond pulses at $\lambda = 4.7\,\mu m$, and referencing to the known d_{14} of GaAs (94 pm/V).

[c] d_{14} of ZnSe was measured in this work via SHG of 1.3 μm.

[d] Mid-IR d_{14} coefficient of ZnSe was measured in my group at University of Central Florida, to be 20 ± 2 pm/V, via direct comparison of SHG efficiency between a single-coherence-length (~100 μm) 110-cut ZnSe and a single-coherence-length (~45 μm) 110-cut GaAs in a frequency-doubling process using nanosecond pulses at $\lambda = 4.7\,\mu m$, and referencing to the known d_{14} of GaAs (94 pm/V).

without the onset of two-photon absorption. Its thermal conductivity (13.6 W/mK) is higher than that of existing longwave infrared NLO crystals suitable for 1-μm pump, most notable $AgGaS_2$ that has thermal conductivity of only 1.4 W/mK.

Gallium selenide (GaSe) deserves a special place in the family of birefringent NLO crystals. It is a two-dimensional III–VI semiconductor with layered structure and weak (van der Waals) interlayer coupling. Although the crystal's mechanical softness makes it difficult to cut and polish it along arbitrary directions (GaSe can only be cleaved perpendicular to its *c*-axis, along 001 planes), the crystal has unique advantages including: enormous birefringence ($\Delta n \sim 0.3$) that makes it phase-matchable for numerous three-wave processes (including THz generation), broad transparency range (0.65–20 μm) with extremely low optical losses (<0.1 cm^{-1} at 1–15 μm), large NLO coefficient (54 pm/V), and large bandgap (2 eV) that minimizes two-photon absorption at $\lambda > 1.25$ μm. Here is a tribute to GaSe crystal – a list of the most prominent NLO applications of GaSe in the mid-IR as well as in the longwave IR (terahertz) spectral regions:

- broadband frequency doubling of 6–12 μm radiation [20]
- DFG in the CW regime (8.8–15 μm) [21]
- DFG with picosecond pulses (4–18 μm) [22]
- DFG with femtosecond pulses (3–20 μm) [23]
- optical parametric generation (OPG) with continuous tunability from 3 to 19 μm [24]
- generation of frequency combs in the 4–17 μm spectral range via DFG [25–27]
- generation of single-cycle electromagnetic transients (center wavelength from 7 to 3000 μm) via optical rectification [28, 29]
- generation of broadly tunable (0.2–5.3 THz) terahertz radiation by nanosecond DFG [30]
- high (up to 23rd order) harmonic generation from 10-μm mid-IR pulses [31]
- coherent electro-optic detection of mid-IR and THz waves [32]

5.1.3.3 Emerging QPM Nonlinear Optical Materials

Zinc blende semiconductors such as GaAs and GaP are particularly appealing for mid-IR nonlinear-optical frequency conversion – due to their deep infrared transparency, high second-order nonlinearity, superior thermal conductivity, and high surface damage threshold. Their cubic crystal structure lacks birefringence; therefore, they require spatial modulation of the optical nonlinearity via quasi-phase-matching. Zinc-blende-type semiconductors are not ferroelectric and no techniques analogous to electric field poling in $LiNbO_3$ exist for inducing domain flipping in already grown crystal.

OP-GaAs was the first realization of a practical QPM semiconductor. This was done using all-epitaxial processing pioneered independently at Stanford University [33] and at the University of Tokyo [34]. The present approach for making OP-GaAs relies on polar-on-nonpolar molecular beam epitaxy (MBE) whereby a thin Ge layer is deposited on a GaAs substrate and the subsequent GaAs layer – under the proper growth conditions – has an inverted orientation relative to the substrate. This layer is then lithographically patterned, etched back to the original substrate, and then simultaneously regrown with the opposite substrate polarity – first by MBE and then by high-growth-rate hydride vapor phase epitaxy (HVPE) – to produce thick (>500 μm) QPM structures suitable for bulk NLO applications [18, 35–37]. As we will see further in this chapter, numerous mid-IR devices based on the OP-GaAs have been demonstrated from CW to femtosecond in the time domain, and from mid-IR to THz in frequency.

Although the OP-GaAs exhibited outstanding performance with 2 μm pumping, it cannot be pumped at wavelengths below 1.8 μm due to the onset of two-photon absorption and the resulting accumulation of free carriers. By contrast, gallium phosphide (GaP) – another III–V zinc blende semiconductor – has wider bandgap and small two-photon absorption at 1 μm. The growth and processing of OP-GaP was a direct extension of the OP-GaAs process, where phosphorus was substituted for arsenic and silicon for germanium as the lattice-matched nonpolar layer. As a result, the usable OP-GaP samples with up to 1-mm QPM layer thickness were grown at BAE Systems and a number of mid-IR devices ranging from CW to ultrafast were demonstrated [18].

5.1.3.3.1 Zinc Selenide as an NLO Candidate

ZnSe, a II–VI semiconductor with zinc blende symmetry, is an excellent candidate for optical frequency conversion because of its high (~20 pm/V) nonlinear susceptibility, high optical damage threshold, good mechanical properties, and outstanding transparency range, from 0.55 to 20 μm. Kanner et al. reported the first results on the growth of QPM OP-ZnSe [38]. In order to achieve orientation patterning, ZnSe films were grown on patterned GaAs templates. Because ZnSe has the same (zinc blende) crystal structure as GaAs, and is almost lattice-matched to GaAs, it became possible to grow thick (>750 μm) films of ZnSe on top of OP-GaAs template and preserve its orientation patterning. Initial results on frequency doubling of 1.6-μm radiation as well as CO_2 laser radiation, and DFG near 9 μm have been demonstrated [38].

5.1.3.3.2 Gallium Nitride as an NLO Candidate

A III–V semiconductor GaN has a wurtzite crystal structure. Its diagonal nonlinear-optical coefficient ($d_{33} \sim 16$ pm/V) is of a similar magnitude to that of PPLN, but GaN has broader transparency range, from 0.37 to 7 μm. Because of its uniquely high thermal conductivity (220 W/mK), high laser damage threshold, wide bandgap (3.4 eV), and thus the ability to use standard 1-μm lasers as a pump, it is well suited for high-power mid-IR NLO applications [15, 39, 40]. Although GaN has not yet reached maturity as an NLO material, several groups have demonstrated periodic inversion of the GaN polarity to achieve quasi-phase-matching (with GaN grown on both sapphire and GaN substrates) [41–44]. Also, second harmonic generation from 1.66 μm was demonstrated in QPM GaN [41].

Petrov [11] published a detailed overview of the available non-oxide NLO materials, emphasizing most recent developments of both birefringent (angle-tuned) and QPM crystals. Also, a review paper by Schunemann et al. [18] highlights advances in the growth of the birefringent crystals $ZnGeP_2$ and $CdSiP_2$, as well as all-epitaxial growth of orientation-patterned semiconductors gallium arsenide (OP-GaAs) and gallium phosphide (OP-GaP).

5.2 Continuous-wave (CW) Regime

5.2.1 DFG of CW Radiation

Specific advantages of the DFG approach for generating mid-IR include: readily available room-temperature single-frequency near-IR "pump" and "signal" laser sources, no need for cryogenic cooling, and high mid-IR beam quality. For example, the outputs of well-developed telecom-range narrow-linewidth diode- or fiber-lasers can be fiber-coupled and mixed in a nonlinear crystal. In addition, coherence properties of the DFG output are inherited from those of the pump lasers. Therefore, narrow-linewidth (1 kHz to 1 MHz) mid-IR output can be produced via DFG using appropriate pump and signal inputs.

The best DFG results, in terms of efficiency, are usually obtained in QPM crystals. In the early work on using PP crystals for frequency conversion, Sanders et al. [45] demonstrated a broadly tunable mid-IR source based on mixing the outputs of two wavelength-tunable single-spatial-mode laser diodes at around 780 and 980 nm in bulk PPLN crystal fabricated by electric field poling of a 0.5-mm thick z-cut $LiNbO_3$ wafer. Coherent mid-IR radiation was generated in a 7.8-mm-long crystal over the range 3.6–4.3 μm with mid-IR power of 6 μW. A more efficient PPLN DFG system (Figure 5.4) was based upon mixing the outputs of a distributed feedback (DFB) diode laser at 1562 nm amplified in an erbium fiber amplifier (EFA) to increase the optical power to 500 mW (with 360-kHz linewidth) and a DFB fiber laser at 1083 nm amplified in an ytterbium fiber amplifier (YFA) to raise the power to 800 mW (100-kHz linewidth). Tunable narrow-linewidth output with 0.4-mW power was demonstrated near 3.5-μm wavelength, corresponding to DFG normalized conversion efficiency η = 0.1%/W [46, 47]. Also, the authors performed sensitive and selective spectroscopic detection of formaldehyde with this DFG source near 3.53 μm.

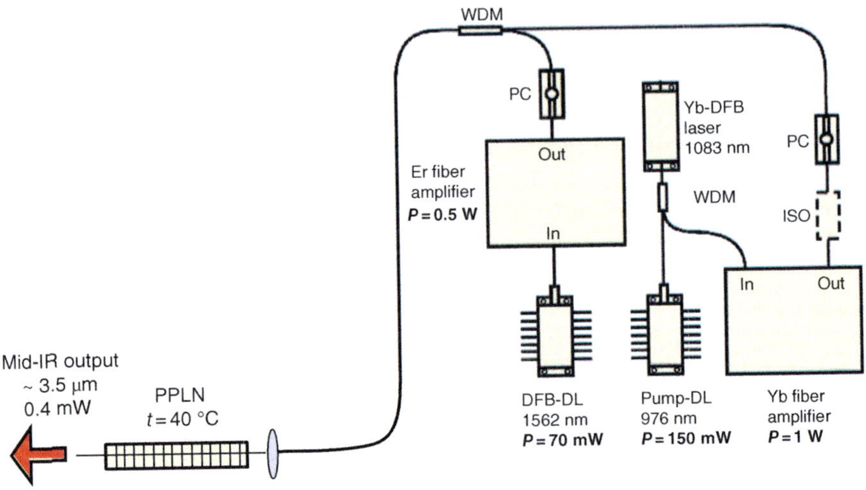

Figure 5.4 Experimental setup for the 3.5-μm DFG source based on mixing the amplified outputs of two DFB lasers in PPLN. DL, diode laser; ISO, optical isolator; PC, polarization controller; WDM, wavelength division multiplexer. *Source:* based on figure 3 of [46], with permission of OSA, The Optical Society.

As much as 3.5 W of CW power at 3.4 μm (linewidth ~1.5 nm) was obtained in a 5-cm long PPLN crystal by difference frequency mixing of 1.064 and 1.55 μm linearly polarized fiber lasers, with correspondingly 43.3 and 31 W of CW power [48]. This is one of the highest CW DFG powers demonstrated so far.

In an alternative DFG approach utilizing a ridge-type PPLN waveguide, the conversion efficiency can be more than 100 times higher than in bulk PPLN crystals, resulting in tens of milliwatts of DFG power [49–52]. This improvement comes from the fact that the beam size is no longer limited by diffraction and is kept at a very small size over the whole length of the crystal. Using direct-bonded PPLN with an integral ridge waveguide structure, an 11-mW pump wave (1047 nm), and a 66-mW signal wave (1550 nm), Tadanaga and coworkers at NTT corporation produced 0.26 mW of radiation at 3.3 μm with a normalized conversion efficiency of η = 40%/W [49]. Asobe et al. [52] achieved highly efficient DFG using a QPM Zn:$LiNbO_3$ waveguide fabricated by direct bonding. The DFG performance was measured using a 1.064 mm pump generated with a laser diode and amplified in an ytterbium-doped fiber amplifier (YDFA) and a signal generated by a 1.55-μm external cavity tunable laser diode and an erbium-doped fiber amplifier (EDFA). CW signal and pump powers of 558 and 444 mW, respectively, were injected into the ridge waveguide (11 μm thick, 17 μm wide, and 38 mm long) and a 65-mW mid-IR output was obtained at 3.4 μm, corresponding to the normalized conversion efficiency η = 26%/W. The reported DFG source could be tunable over 10 nm, determined by its phase-matching bandwidth, by scanning the signal wavelength [52].

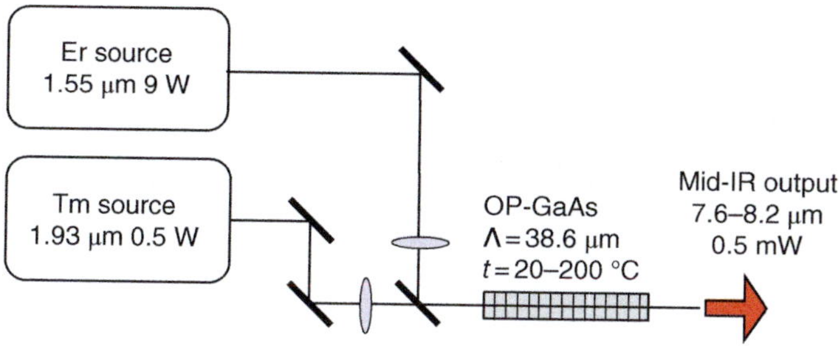

Figure 5.5 Schematic of the experimental DFG setup based on OP-GaAs to produce 0.5-mW 7.6–8.2 μm output. *Source:* based on figure 1 of [55], with permission of OSA, The Optical Society.

Development of new QPM nonlinear crystals with deeper IR transparency, such as GaAs, potentially allows extending tunability range of DFG sources to $\lambda > 10\,\mu m$. DFG of 8-μm radiation using orientation-patterned GaAs was first demonstrated by Levi et al. in a 19-mm-long OP-GaAs crystal [53]. The pump source – a fiber-coupled CW DFB laser diode (1.306–1.314 μm, 3.3 mW) – was mixed with a signal source – an external-cavity diode laser (1.51–1.58 μm), amplified in an EDFA to 787 mW – to generate 7.9-μm idler radiation with 38-nW power. Later, the same team produced several microwatts of the DFG output in OP-GaAs that was tunable in the broad range of 7–9 μm (span 300 cm^{-1}) by mixing the outputs of tunable 1.3-μm (80 mW) and fixed-wavelength 1.55-μm (2 W) lasers [54]. The authors also used the DFG system to perform cavity ring-down spectroscopy of N_2O gas. In terms of the average DFG power, the above result was improved by Vasilyev et al. [55], who generated a tunable (7.6–8.2 μm) single-frequency DFG output with 0.5-mW power from a 33 mm-long OP-GaAs crystal. The DFG source was pumped by a tunable single-frequency external-cavity diode laser, amplified by an EDFA (9 W at 1.55 μm), and a custom Tm-doped fiber laser (0.5 W at 1.93 μm, Figure 5.5). The DFG output wavelength was tuned by simultaneous tuning of the diode laser wavelength and of the OP-GaAs crystal temperature. The authors tested spectroscopic capabilities of their DFG source by measuring methane absorption spectra near 7.65 μm [55].

Thus, difference frequency mixing of telecom sources in QPM crystals such as PPLN and OP-GaAs has been shown to be an effective method for the generation of broadly tunable spectroscopic-grade output in the mid-IR. The main results on CW DFG are summarized in Table 5.2.

5.2.2 CW OPOs

Due to a non-dissipative nature of the parametric conversion process, optical parametric oscillators can be very efficient converters from the near-IR to mid-IR, mostly limited by the Stokes (quantum defect) limit – e.g. by the photon

Table 5.2 Summary of CW DFG in QPM crystals.

DFG structure	DFG wavelength (μm)	Pump and signal	DFG CW power and normalized efficiency	Ref.
PPLN, bulk				
PPLN, $L = 7.8$ mm	3.6–4.3	780 nm (180 mW) 980 nm (400 mW)	6 μW, $\eta = 0.008\%/W$	[45]
PPLN, $L = 50$ mm	3.5	1083 nm (800 mW) 1562 nm (500 mW)	0.4 mW, $\eta = 0.1\%/W$	[46, 47]
PPLN, $L = 50$ mm	3.4	1064.6 nm (43.3 W) 1549.8 nm (31 W)	3.55 W, $\eta = 0.26\%/W$	[48]
PPLN, waveguides				
PPLN ridge WG, direct bonded, $L = 50$ mm	3.2	1047 nm (11 mW) 1550 nm (66 mW)	260 μW, $\eta = 40\%/W$	[49]
PPLN ridge WG, direct bonded, $L = 50$ mm	3.36	1064 nm (46 mW) 1558 nm (7 mW)	146 μW, $\eta = 45\%/W$	[51]
PPLN ridge WG, direct bonded, $L = 50$ mm	3.52	1083 nm (318 mW) 1562 nm (503 mW)	15 mW, $\eta = 9.4\%/W$	[50]
Zn:PPLN ridge WG, direct bonded, $L = 38$ mm	3.4	1064 nm (444 mW) 1550 nm (558 mW)	65 mW, $\eta = 26\%/W$	[52]
OP-GaAs, bulk				
OP-GaAs, $L = 19$ mm	7.9	~1300 nm (11 mW) ~1550 nm (66 mW)	38 nW, $\eta = 0.005\%/W$	[53]
OP-GaAs, $L = 19$ mm	7–9	~1300 nm (80 mW) ~1550 nm (2 W)	A few microwatts, $\eta = 0.002\%/W$	[54]
OP-GaAs, $L = 33$ mm	7.6–8.2	~1550 nm (9 W) ~1930 nm (0.5 W)	0.5 mW, $\eta = 0.003\%/W$	[55]

energy difference between the pump and the idler wave. The OPO gain originates from coherently driven nonlinear polarization, and its long-wavelength tuning cutoff is set only by the transparency of the nonlinear medium (OPOs can produce an idler wave even in the THz range, below the band of phonon resonances [56]).

In an optical parametric oscillator, at least one of the two generated waves (signal or idler) resonates in a cavity. Typically, it is the shorter-wavelength “signal” wave (this scenario is easier to implement from a practical point of view). The oscillation starts from quantum noise if the gain for the resonant field, provided by the incident pump light, overcomes the round-trip loss. As compared to singly resonant OPOs (SROs), much lower pump thresholds can be achieved in doubly resonant (with signal and idler or signal and pump resonating) or even triply resonant (signal, idler, and pump resonating) OPOs. However, the latter two versions are more difficult to implement because of tight constrains, e.g. on the cavity length, imposed by the requirement for multiple resonances. The first operation of a CW OPO was demonstrated in a cavity that was resonant for both signal and idler waves [57].

Fabrication of PPLN in the early 1990s tremendously stimulated development of OPOs. In 1996, Bosenberg et al. demonstrated the first CW QPM OPO, which promptly revealed the exceptional performance of QPM materials. The OPO was resonant at a signal wave in a two-mirror linear [58] or in a four-mirror ring bow-tie cavity configuration [59] and was based on a 50-mm-long PPLN crystal as a nonlinear medium. The pump source for this device was a 1064-nm Nd:YAG laser with 13.5 W of available power. The oscillation threshold power was 2.9 W in a linear cavity and 3.6 W (pump power density ~55 kW/cm^2) in a ring cavity (Figure 5.6a). The milestones of their research were (i) obtaining the idler output power of 3.55 W at λ = 3.25 μm, which corresponds to 80% of the quantum-limited performance, (ii) observing 93% pump depletion (Figure 5.6b), (iii) achieving the OPO idler tuning over the 3.24–3.95 μm spectral range by using PPLN crystal with multiple grating periods, and (iv) maintaining single-longitudinal-mode oscillation in a ring cavity containing an intracavity etalon.

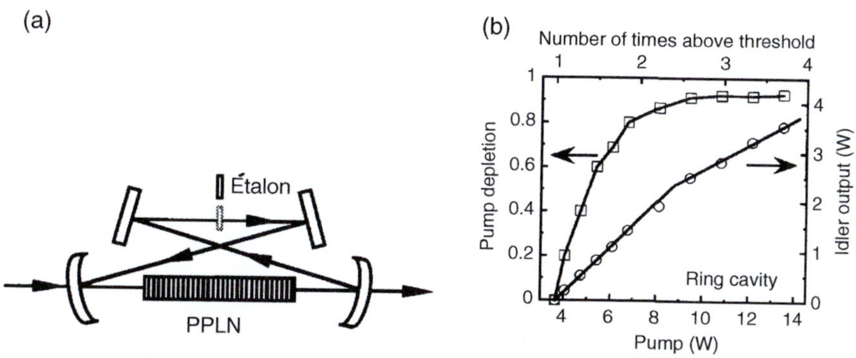

Figure 5.6 (a) Schematic of a ring-cavity CW OPO based on PPLN crystal [59]. The two curved mirrors had 100-mm radius of curvature, and the remaining two mirrors were flat. The PPLN crystal was 50 mm long and had a grating period of 29.75 μm. (b) Pump depletion and idler output versus pump input for the ring cavity operating at an idler wavelength of 3.25 μm. *Source:* reproduced from figures 1 and 2 of [59], with permission of OSA, The Optical Society.

Kumar et al. demonstrated a high-power CW optical parametric oscillator that was pumped by a single-frequency ytterbium fiber laser (1064 nm, maximum power 28.6 W) and produced the total output power of 17.5 W, out of which 9.8 W were in the near-IR "signal" at $\lambda_s = 1.63\,\mu m$ and 7.7 W in the "idler" wave at $\lambda_i = 3.07\,\mu m$ [60]. Figure 5.7 shows the experimental setup and how the signal/idler outputs change with the signal-wave outcoupling, while the inset shows variation of the OPO threshold pump power. The maximum overall power of 17.5 W was reached at the optimal output coupling of ~3.8%, corresponding to the OPO conversion efficiency of 61%. Under this condition the pump depletion reached 69.4%. In the absence of the output coupling for the signal wave, 8.6 W of the idler power was generated at $\lambda = 3.06\,\mu m$, with pump-to-idler conversion efficiency of 30%, and pump threshold of 3.6 W. In

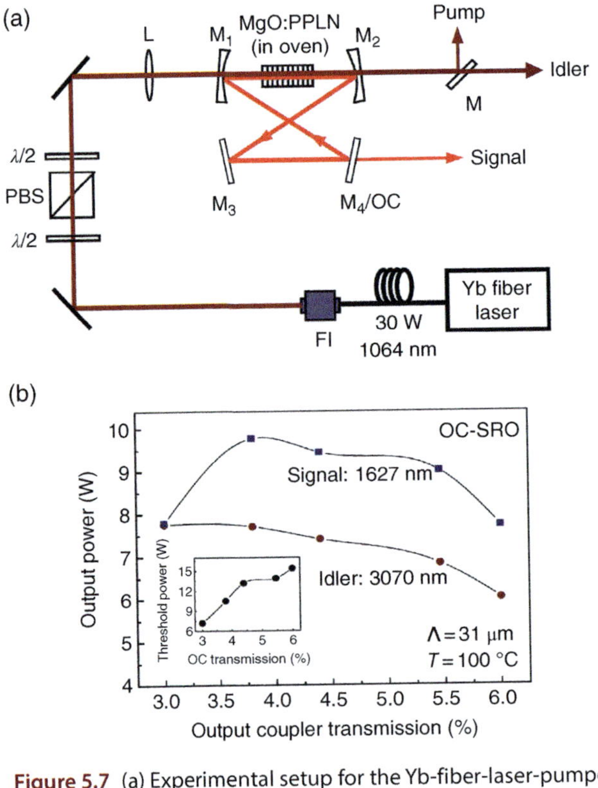

Figure 5.7 (a) Experimental setup for the Yb-fiber-laser-pumped ring-cavity PPLN OPO with a total (signal + idler) power of 17.5 W. FI, Faraday isolator; λ/2, half-wave plate; PBS, polarizing beamsplitter; L, lens; M1–M4, dichroic mirrors. The mirror M4 was used as an output coupler (OC) for the resonating signal wave across 1.6–1.7 μm. (b) Variation of the extracted signal (1.63 μm) and idler (3.07 μm) power with OC transmission. Inset: Threshold pump power versus OC transmission. *Source:* reproduced from figures 1 and 2 of [60], with permission of Springer.

the idler branch, the OPO was tunable from 2.8 to 3.2 μm by changing the PPLN temperature [60].

The same research team demonstrated a high-power CW OPO based on periodically poled lithium tantalate (PPLT), $LiTaO_3$, crystal. Despite its twice smaller nonlinear coefficient, as compared to the lithium niobate (see Table 5.1), the crystal has an increased resistance to photorefractive damage and higher thermal conductivity, along with increased optical damage threshold, thus making it attractive for multi-watt mid-IR generation. The OPO pump source was a single-frequency Yb fiber laser delivering up to 30 W at 1064 nm. A 30-mm-long MgO-doped stoichiometric PP $LiTaO_3$ (MgO:sPPLT) was used in this setup as a gain element [61]. The singly resonant OPO (SRO) was tunable from 3.03 to 3.46 μm and generated 5.5 W of CW power at 3.1–3.3 μm with the pump depletion up to 64%.

Henderson et al. reported a broadly tunable single-frequency (<1 MHz linewidth) CW OPO [62]. The OPO had a four-mirror ring cavity, utilized an 80-mm-long PPLN crystal, and was driven by a narrow-linewidth (50 kHz) all-fiber pump source at 1083 nm. The oscillation threshold as low as 780 mW was measured at λ = 2.8 μm idler wave; the maximum pump depletion was 85% at a pump power of 2.8 W. At this input level, 750 mW of idler was produced at λ = 2.8 μm in a near-diffraction-limited beam. The overall tunability was 2.65–3.2 μm achieved by changing the QPM period of PPLN containing multiple grating periods. By fine-tuning only the pump wavelength, mode-hope free tuning range of 60 GHz ($\Delta\lambda \sim 2$ nm) was achieved [62]. More results on broadly (>1000 nm) and continuously tunable mid-IR single longitudinal mode (SLM) CW optical parametric oscillators based on PPLN can be found in [63–66].

Several groups reported on widely tunable CW single-frequency OPOs where the pump wave resonates, in addition to the OPO signal (or idler) wave. This pump-enhanced SRO configuration is labeled as PE-SRO [67–70]. Rihan et al. demonstrated a PE-SRO pumped by a single-frequency Ti:sapphire laser delivering 760 mW at 795 nm with an octave-wide tunability of the idler wave (1.7–3.5 μm) with 20–50 mW of the output power. Thanks to the pump enhancement, a threshold as low as 110 mW was achieved [71] (see Table 5.3).

Until recently, the long wavelength range of CW OPOs was limited to around 5 μm – a limitation set by multi-phonon absorption in oxide crystals such as lithium niobate, lithium tantalate, or potassium titanyl phosphate [74]. As an attempt to extend this limit to longer wavelengths, Schunemann and coauthors employed OP-GaAs as a nonlinear medium and made the first experimental demonstration of CW OPO operation in GaAs, as well as the first CW OPO pumped at a laser wavelength longer than 1.55 μm [72]. The OPO was emitting near degeneracy, at $\lambda_s \sim 3.8$ μm and $\lambda_i \sim 4.7$ μm. The OP-GaAs sample, 40-mm-long, with 1.7-mm-thick QPM grating structure, and QPM period 63.5 μm, was grown using low-pressure HVPE with the bulk absorption loss of 0.004 cm^{-1} at 2.4 μm. The OPO pump was a Ho:YAG oscillator–amplifier system at

Table 5.3 Summary of CW mid-IR OPOs.

OPO crystal	Pump laser	Wavelength tuning (μm)	Other parameters	Ref.
PPLN $L = 50$ mm	1.064 μm 13.5 W	3.24–3.95	SRO, idler power 3.55 W at 3.25 μm, threshold 3.6 W (ring cavity), quantum efficiency 80%, pump depletion 93%	[58, 59]
PPLN $L = 50$ mm	1.064 μm 28.6 W	2.8–3.2	SRO, idler power 8.6 W at 3.06 μm, threshold 3.6 W, pump-to-idler conversion 30%, pump depletion 79%	[60]
PPLT $L = 30$ mm	1.064 μm 30 W	2.8–3.2	SRO, idler power 5.5 W at 3.1–3.3 μm, threshold 17.5 W, pump-to-idler conversion 18.5%, pump depletion 64%	[61]
PPLN $L = 80$ mm	1.083 μm 2.8 W	2.65–3.2	SRO, single frequency; idler power 750 mW at 2.8 μm; threshold 780 mW; quantum efficiency 69%; pump depletion 85%; continuous-mode-hop-free scans of 60 GHz	[62]
PPLN $L = 50$ mm	775–860 nm 6 W	2.5–4.4	SRO, single frequency; idler power 800 mW; threshold 1.5 W; pump depletion 80%; continuous-mode-hop-free scans of 40 GHz by tuning the pump wavelength	[66]
PPLN $L = 50$ mm	795 nm 760 mW	1.7–3.5	PE-SRO, single frequency; idler power 20–50 mW, threshold 110 mW	[71]
OP-GaAs $L = 40$ mm	2.09 μm 24.7 W	3.8 and 4.7	Doubly resonant, signal + idler power 5.3 W, quantum efficiency 23.6%, threshold 11.5 W	[72]
$AgGaSe_2$ whispering gallery resonator	1.57 μm 12 mW	2–8	Triply resonant, 800 μW at 2.5 μm, 10 μW at 8 μm; quantum efficiency 12%, threshold 0.5–2 mW	[73]

$\lambda = 2.09$ μm, pumped in turn by thulium fiber lasers. The OPO resonator was configured as a bow-tie ring (Figure 5.8). The maximum total output power (signal + idler) was 5.3 W at 24.7 W of pump power. The OPO threshold was 11.5 W, and maximum conversion efficiency (signal + idler) of 23.6% occurred at 1.8 times the threshold [72]. The high oscillation threshold can be

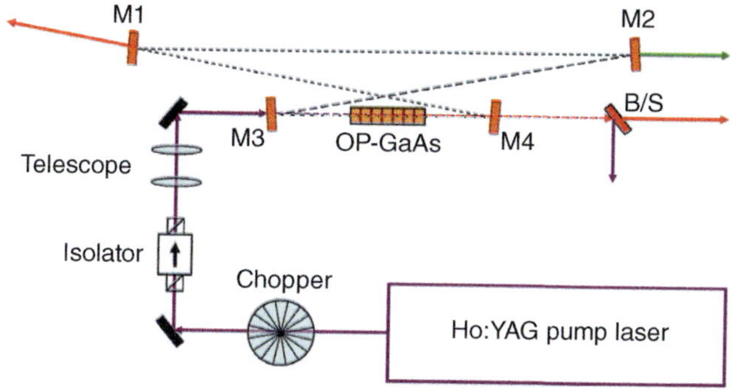

Figure 5.8 Schematic of the bow-tie CW OP-GaAs OPO system. *Source:* reproduced from figure 1 of [72], with permission of OSA, The Optical Society.

explained by at least two reasons: (i) the OPO threshold scales as the product of the signal and the idler wavelengths (see Eq. 5.7) and thus goes up when the pump wavelength is shifted to longer wavelengths, (ii) the bulk OP-GaAs losses at resonating wavelengths (3.8 and 4.7 μm) might be higher than those measured at 2.4 μm.

A totally new approach to CW OPOs is based on whispering gallery resonators (WGRs) where a millimeter-size monolithic cavity guides light via total internal reflection. With small round-trip losses for all interacting waves, a WGR OPO is intrinsically triply resonant and thresholds down to the microwatt level in lithium niobate can be reached [75–77]. Meisenheimer et al. demonstrated a CW OPO emitting mid-IR light at wavelengths up to 8 μm in a device that was based on a 3.5-mm-diameter WGR made of silver gallium selenide ($AgGaSe_2$) [73].

Employing $AgGaSe_2$ crystal significantly extends the tuning range of existing OPOs into the mid-IR (Table 5.3). The pump at $\lambda = 1.57$ μm in [73] was provided by a DFB 12-mW laser diode and coupled into the resonator via an evanescent field using a silicon prism (inset, Figure 5.9). The spatial distribution of the light field in the WGR is characterized by three mode numbers (m, p, q), one of which (m) is the longitudinal mode number, and the other two represent the transverse radial (q), and polar (p) mode numbers. Energy and angular momentum conservation laws dictate a particular relationship between these mode numbers. Figure 5.9 shows the signal output power at $\lambda_s = 2.54$ μm versus pump power dependence, where the output power of 0.8 mW was achieved at 12 mW pump. The corresponding idler power at $\lambda_i = 4.11$ μm was estimated to be 0.5 mW. For different combinations of the polar mode numbers p and crystal temperature, OPO tuning (signal and idler) in the whole range of 2–8 μm was produced with the output power ranging from 800 μW (at 2 μm) to 10 μW (at 8 μm).

Table 5.3 summarizes the main results on CW OPOs. A very good review of the progress in the development of CW optical parametric oscillators (including whispering-gallery OPO devices) can be found in [74].

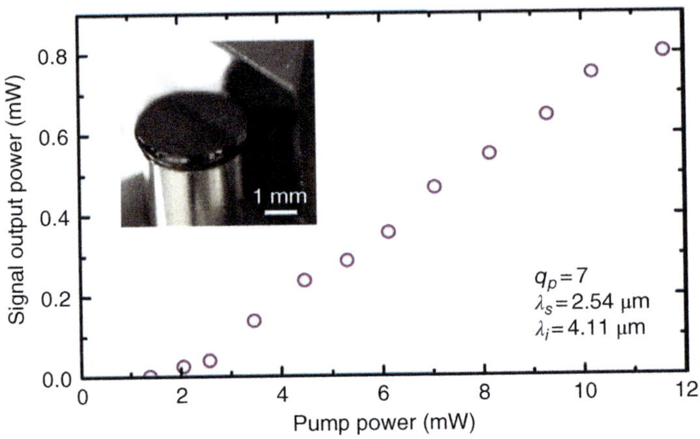

Figure 5.9 Signal output power at $\lambda_s = 2.54\,\mu m$ (with the corresponding idler wave at $\lambda_i = 4.11\,\mu m$) versus pump power for the WGR pump-mode with q-number of 7. The inset shows a microresonator and an Si prism. *Source:* reproduced from figure 3 of [73], with permission of OSA, The Optical Society.

5.3 Pulsed Regime

5.3.1 Pulsed DFG

In the pulsed regime, DFG conversion efficiency can be dramatically scaled up. As follows from (5.4), the peak power of DFG output is proportional to the product of the peak powers of the two pump beams. Hence, at the same average power, one can get much higher DFG conversion efficiency, as compared to CW regime (the enhancement factor scales as the inverse of the duty factor of the pump). In the pulsed DFG approach, it is not only that the two pump beams need to perfectly overlap in space, but they also need to overlap in time.

Since the DFG process has no threshold and requirements for optical losses in a crystal are not very demanding, the great selection of nonlinear crystals listed in Table 5.4, such as PPLN, AGS, AGSe, GaSe, and CGA, combined with a variety of pump laser sources, could be used for frequency mixing to achieve mid-IR output. For example, broadband difference frequency mixing in AGS crystal with nanosecond pulses and with wavelengths extending to $\lambda > 11\,\mu m$ has been demonstrated in a number of early works [78–81]. Seymour et al. demonstrated that the long-wavelength limit of DFG output could be extended to 18.3 μm – with detectable power even beyond the two-phonon absorption band of AGS [79]. Also, DFG from $\lambda = 2.7$ to 38.4 μm – well beyond multi-phonon infrared resonances – was reached in GaSe with nanosecond pump pulses near $\lambda \approx 1\,\mu m$ [82]. Continuously tunable 6.8–20.1-μm output has been achieved via pulsed DFG in $CdGeAs_2$ – a crystal with the highest NLO

Table 5.4 Summary of pulsed DFG sources.

DFG crystal	DFG wavelength (μm)	Pump and signal	DFG parameters	Ref.
Broadly tunable, angular phase matched				
AGS L = 1.5–1.7 mm	4.6–12	694 nm (1.4 mJ) 737–817 nm (130 μJ)	4 nJ @ 11 μm (10 ns duration)	[78]
AGS L = 2.8 mm	5.5–18.3	539–658 nm (19 μJ) 555–747 nm (23 μJ)	16 nJ @ 11.8 μm (4 ns duration)	[79]
AGS L = 10 mm	5–11	1.064 μm (120 mJ) 1.18–1.35 μm (~12 mJ)	2 μJ @ 6 μm (12 ns, 10 Hz)	[80]
AGS L = 10 mm	5–12	1.76–2.01 μm (18 mJ) 2.26–2.7 μm (17 mJ)	96 μJ @ 7.5 μm (8 ns, 30 Hz)	[81]
GaSe L = 20 mm	2.7–38.4	1.064 μm (6 mJ) 1.09–1.75 μm (3–5 mJ)	12 μJ @ 5.9 μm (5 ns, 10 Hz)	[82]
CGA L = 5.8 mm	6.8–20.1	4–5 μm (5 μJ) 6.5–9.5 μm (1.5 μJ)	0.3 μJ @ 10–13 μm (0.1 ns, 3 Hz)	[83]
High energy per pulse				
KTA L = 15 mm	3–5.3	785–886 nm (70 mJ) 1.064 μm (170 mJ)	0.5 mJ @ 4.4 μm, 0.3 mJ @ 5 μm (2 ns, 10 Hz)	[84]
KTA six crystals in series, total L = 60 mm	3.14–4.81	1.065 μm (50 mJ) 1.37–1.61 μm (4 mJ)	1 mJ @ 3.5 μm, 0.4 mJ @ 4.5 μm, (2 ns, 20 Hz)	[85]
High repetition rate				
PPLN L = 50 mm	3.52	1.064 μm (19 μJ, 7.7 W ave.) 1.525 μm (10 mW CW)	1 W ave. power (2.5 ns, 400 kHz), 1 GHz linewidth, conv. eff. from pump 13.4%	[86]
PPLN L = 50 mm	3.2–5.7	1.064 μm (30 μJ, 252 mW ave.) 1.5–1.6 μm (8 mW CW)	0.2–14 mW ave. (6 ns, 8.4 kHz), linewidth 154 MHz @ 2 mW, 195 GHz @ 14 mW	[87]

coefficient (236 pm/V) among all $\chi^{(2)}$ materials in practical use. For a pump, the authors used the signal (4–5 μm) and the idler (6.5–9.5 μm) outputs of a $ZnGeP_2$ OPG [83] (see Table 5.4).

High, millijoule-level, mid-IR DFG pulse energies could be obtained at low (~10 Hz) repetition rates. For example, Kung produced continuously tunable mid-IR radiation from 3.0 to 5.3 μm by difference frequency mixing in a KTA crystal using amplified Ti:sapphire laser at 785–886 nm (pulse energy 60–70 mJ) and an Nd:YAG laser at 1064 nm (pulse energy 170 mJ) at 10 Hz repetition rate [84]. An energy of over 0.5 mJ in a 2-ns pulse (30-GHz linewidth) was achieved at λ = 4.4 μm, providing peak power of 250 kW. Similarly, Miyamoto et al. reported on the generation of nanosecond mid-IR pulses with high energy (0.4–1 mJ/pulse), large tunability (from 3.1 to 4.8 μm), and narrow linewidth (1.4 GHz) using DFG in KTA. Five or six KTA crystals in series with the total length of 50–60 mm were used in this work [85]. DFG was obtained by mixing 50-mJ pump pulses at 1065 nm with wavelength-tunable (1368–1611 nm) 4-mJ signal pulses. A narrow linewidth and good frequency reproducibility of the DFG output were confirmed by observing a ro-vibrational absorption line of CO gas at 4.587 μm.

For DFG at high (>1 kHz) repetition rates in the 2–5 μm range – a region suitable for molecular spectroscopy – PPLN is one of the most appropriate materials due to its high conversion efficiency and commercial availability of crystals that can exceed 50 mm in length. Belden et al. generated >1 W of average DFG power in a narrow-linewidth output at 3.52 μm using a 5 cm-long PPLN crystal [86]. The pump was an all-fiber laser source, producing 2.5-ns pulses at 1064 nm at 400 kHz repetition rate, and 7.7 W average power, while the "signal" beam was from a fiber-coupled DFB diode laser emitting 10 mW of CW power at 1525 nm. One should notice that the peak pump power was about million times stronger than the "signal" and the DFG operation was closer to the regime, characteristic of the OPA – with the "signal" beam serving as a seed. The threshold for the onset of the exponential behavior corresponding to (5.10) was at 2.5 W of pump power. Overall, the 1.064–3.52-μm power conversion efficiency was as high as 13.4% (quantum efficiency 44%).

A 3–8 kHz repetition rate DFG system based on PPLN crystal with the average power of 0.2–14 mW designed for gas sensing via photoacoustic spectroscopy in the wavelength range 3.2–3.7 μm is described in [87]. The pump laser was a diode-pumped pulsed high peak power passively Q-switched Nd:YAG laser in a nonplanar ring oscillator (NPRO) configuration. This laser set the pulse repetition rate, pulse duration, and the linewidth (154 MHz) for the generated DFG output. The "signal" was a CW external cavity diode laser tunable between 1500 and 1600 nm, which set the DFG tuning range. As in [86], this DFG source was based on a combination of a pulsed pump laser and a CW signal laser. In this configuration, the exponential OPA process was considerably suppressed at low (<2 mW) output DFG power. At the higher average power (up to 14 mW), the OPA regime became dominant and the linewidth was broadened by more than 1000 times, to 195 GHz [87].

Table 5.4 gives a summary of pulsed DFG sources.

The reader can find more information on pulsed DFG in the reviews [11, 87].

5.3.2 Pulsed OPOs

To achieve OPO action with nanosecond (*ns*) pump pulses, a simple short-length linear cavity consisting of two flat mirrors might be sufficient. The threshold of a pulsed SRO (resonating at either signal wave or idler wave) is determined by the two main factors:

- Parametric gain should compensate the round-trip loss in the cavity at the resonating wave.
- Parametric gain should be high enough to overcome the so-called "buildup" loss. The latter is related to the fact that for an *ns* pump, the intracavity OPO power needs to build up in the OPO cavity in a limited amount of time (typically 10–100 round trips) – from quantum noise to a detectable value [88].

Interestingly, the loss in the cavity plays a secondary role in determining the threshold of a pulsed OPO. In fact, the equivalent quantum noise "seed" power of an OPO is given by Yariv [3] and is expressed as the following product: *photon energy* × *bandwidth*. For a wavelength of ~3 μm (photon energy = $6.6{\cdot}10^{-20}$ J) and an OPO bandwidth of ~1 cm^{-1} ($3{\cdot}10^{10}$ Hz), one gets the quantum noise seed power of ~2 nW. Assuming detectable *peak* power of ~1 W, one needs the total buildup of intracavity intensity of about 10^9. With typically ~30 roundtrips in a pulsed ns OPO, this means that a single-pass OPO gain factor in (5.8) should be $G \approx 2$, which is larger than a typical intracavity loss for the resonant wave.

5.3.2.1 Broadly Tunable Pulsed OPOs

Shortly after the invention of the laser, pulsed OPOs with their capability of producing broadband mid-IR tuning, reaching several octaves, became important spectroscopic tools. As early as 1973, Hanna et al. had demonstrated an optical parametric oscillator based on a proustite (Ag_3AsS_3) crystal pumped by a Q-switched neodymium (1.065 μm) laser that was tuned over a substantial (1.22–8.5 μm) portion of the infrared [89].

Two decades later, owing to the development of the poling technology in ferroelectrics, PPLN became one of the most widely used OPO nonlinear materials for generating tunable light, especially in the 1–5 μm spectral region. High nonlinear coefficient, good optical quality, and low absorption allowed dramatic reduction in OPO pump threshold and improvement of the conversion efficiency. The first nanosecond OPO based on PPLN crystal was reported by Myers et al. [90]. The authors used a 15-mm-long, 0.5-mm-thick PPLN crystal with a 31-μm QPM grating period, pumped by a Q-switched diode-pumped 1.064-μm Nd:YAG laser with 7–20 ns pulse duration and 100 Hz to 10 kHz repetition rate. The OPO resonator was a linear cavity with mirrors selected to resonate at the signal wave. With the focused pump laser beam size of 47 μm, the OPO pump threshold was 12 μJ, and the output was continuously tunable from 1.66 to 2.95 μm, with the crystal temperature varying from room temperature to 180 °C. To extend the tuning range of the PPLN OPO, the same team reported an OPO that used multiple grating sections on a single PPLN chip. The PPLN chip was 26-mm long and consisted of 25 0.5-mm-wide gratings with QPM periods from 26 to 32 μm in 0.25-μm steps (Figure 5.10a) [91]. The oscillation threshold was as low as 6 μJ with a 7-ns pump pulse (energy fluence 0.09 J/cm^2). For tuning, the PPLN crystal was translated across the beam so that it interacted with different grating sections, and a tunable output of 1.36–4.83 μm was achieved. The OPO tuning curve with respect to the grating period is shown in Figure 5.10b. At a pump repetition rate of 1 kHz and the average pump power 100 mW (100 μJ/pulse), the output power at 4 and 4.83 μm amounted to 6 and 2 mW correspondingly.

Silver gallium sulfide, $AgGaS_2$ (AGS), is one of the few nonlinear materials that can be conveniently pumped by commercial 1-μm lasers and provide very large tunability, well over an octave, with the idler wavelength stretching beyond 10 μm. The first OPO based on AGS crystal was reported in 1984 by Fan et al. [92]. A type-I angle-tuned OPO was pumped by a Q-switched Nd:YAG laser (τ = 18 ns) and tuned from 1.4 to 4 μm. Subsequently, Vodopyanov et al. demonstrated an AGS OPO with even larger tuning range of 3.9–11.3 μm [93].

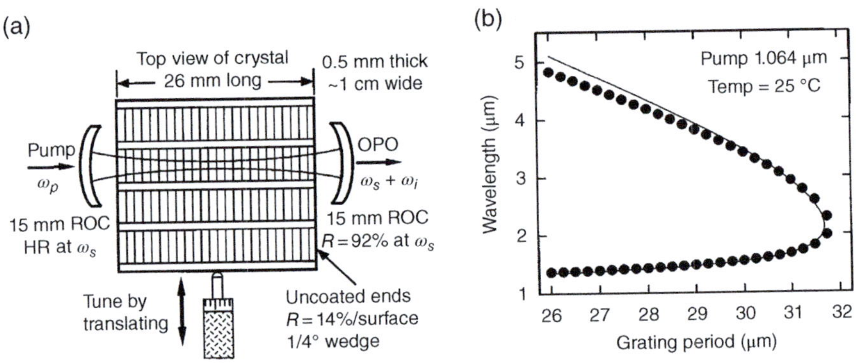

Figure 5.10 (a) Experimental setup for the pulsed OPO using multi-grating PPLN crystal. For tuning, the PPLN crystal was translated through the resonator so the pump beam interacted with different grating sections. (b) The OPO wavelength tuning as a function of grating period, achieved by translation of the PPLN crystal through different grating sections. *Source:* reproduced from figures 2 and 3 of [91], with permission of OSA, The Optical Society.

The singly resonant angle-tuned OPO was formed by two flat mirrors and used a 20-mm AGS crystal cut at 45.1° to the optical z-axis for the type-II phase matching. The crystal was pumped by 1.06-μm pulses from a Nd:YAG laser with 12–100 ns pulse duration and yielded up to 0.37 mJ idler wave pulse energy at λ = 6 μm at 15 mJ pump. In the configuration with the recycling of the pump and idler beams in the second pass, the OPO pump threshold was 85 μJ. The OPO linewidth was ~1 cm^{-1} and the quantum conversion efficiency reached 22%. This device, with its tuning range of 3.9–11.3 μm (Figure 5.11), can be considered to be one of the longest wavelength OPOs pumped at 1 μm (only surpassed by a BGSe crystal, see later in this section). Nevertheless, the practical use of AGS is limited because of its comparatively low surface damage threshold (~0.2 J/cm^2 for ns pulses).

Because of its longwave IR cutoff (19 μm), silver gallium selenide, $AgGaSe_2$, has been successful in providing longwave mid-IR OPO tunability. As opposed to AGS, it is not phase-matchable (at least using angular phase matching) for 1-μm pumping; however, it phase matches with a variety of pump sources at λ > 1.3 μm. The first operation of an AGSe OPO was reported by Eckardt et al. with continuous tuning ranges of 6.7–6.9 μm (1.34-μm neodymium laser pump) and 2.65–9.02 μm (2.05-μm holmium laser pump) [94]. Quarles et al. demonstrated the full continuous tuning range between 2.49 and 12.05 μm with a 2.05-μm holmium-laser pump, with a single angle-tuned AGSe crystal, and with the output OPO energy up to a few millijoules [95]. Continuously tunable IR output in the range 6.1–14.1 μm has been demonstrated by Chandra et al. [96] in a cascaded OPO based on angle-tuned AGSe crystal pumped by 1.57-μm pulses (the output of a KTP OPO). Energies of up to 1.2 mJ/pulse at 9 μm idler wave (quantum efficiency 23%) with bandwidths of 5 cm^{-1} were obtained using a 35-mm-long AGSe crystal and type-I phase matching. The main limitation was the surface damage of the AGSe crystal starting at ~20 MW/cm^2 intensity (0.12 J/cm^2 fluence).

An extremely high nonlinear-optical coefficient of a $ZnGeP_2$ (ZGP) crystal d_{NL} = 75 pm/V, with its NLO FOM d^2/n^3 (n is refractive index) almost nine

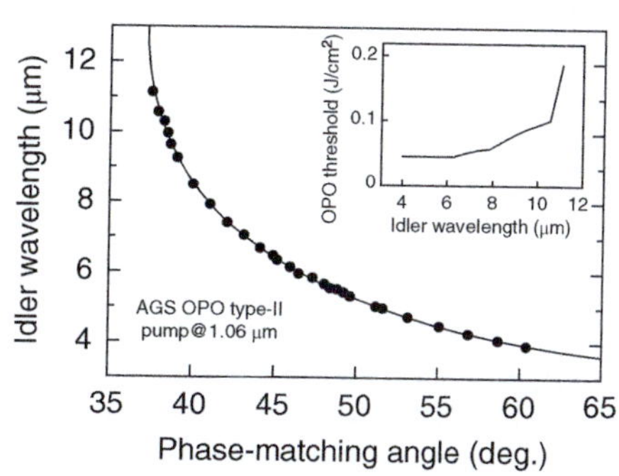

Figure 5.11 AGS type-II OPO angular tuning curve (idler wave only). Solid line – theoretical tuning curve. Inset: OPO threshold fluence as a function of the idler wavelength. *Source:* reproduced from figure 2 of [93], with permission of AIP, American Institute of Physics.

times that of PPLN, combined with good optical, mechanical, thermal, and surface damage properties, favors a variety of nonlinear-optical applications in the 2–12 μm range, including efficient high average power, high pulse energy, as well as broadly tunable OPOs. Despite its large (2 eV) bandgap, ZGP has a residual absorption tail in the near-IR region. As a result, for efficient performance, the ZGP pump wavelength should be chosen at 2 μm or above. For example, 2-μm holmium or 3-μm erbium lasers are suitable candidates for this purpose. A ZGP OPO with 3.8–12.4 μm mid-IR tunability and >1 mJ idler pulse energy was demonstrated in [97, 98]. A pump source was a Q-switched erbium laser with $\lambda = 2.8$ μm (Er, Cr:YSGG) or 2.93 μm (Er, Cr, Tm:YAG), with 100-ns pulse duration, 10-Hz repetition rate, and 10-mJ energy. An AR-coated ZGP crystal was 20-mm long and was cut for type-I ($\theta_0 = 49.5°$) or for type-II ($\theta_0 = 70°$) phase matching. The lowest OPO threshold was obtained in a flat–flat double-pass OPO cavity configuration shown in the inset of Figure 5.12a. The OPO output was continuously tunable, via crystal angle tuning, from 3.8 to 12.4 μm (type-I phase matching, Figure 5.12a) and from 4 to 10 μm (type-II phase matching, Figure 5.12b) with a linewidth of 2–3 cm^{-1}. Figure 5.12c shows the dependence of the type-I OPO idler energy at $\lambda = 8.1$ μm as a function of the pump energy, where a 1-mJ pulse energy was reached at 10.5 mJ of the pump. The OPO pumping threshold was less than 1 mJ/pulse in the whole 4–12 μm range and the quantum conversion efficiency reached 35%.

High quadratic nonlinearity of ZGP made it possible to demonstrate a record-low-threshold OPO with a wide tunability. The OPO was a tandem system, where a noncritically phase-matched (NCPM) ZGP OPO was pumped by the idler wave output of a PPLN OPO, pumped in turn by a Nd:YAG laser (1.6 mJ, 20 ns, 1 kHz) [99]. The singly resonant ZGP OPO (Figure 5.13a) contained a 24-mm $\theta = 90°$-cut ZGP crystal and was formed by a concave output coupler mirror M3 (highly reflective at signal and transmissive at the idler and the pump) and a gold reflector mirror M4, which was deposited directly into the polished flat surface of the ZGP crystal (the front surface of ZGP was AR-coated). Tuning the PPLN OPO pump wavelength in the range 2.3–3.7 μm resulted in tuning the ZGP OPO output from 3.7 to 10.2 μm. The OPO tuning curves (versus pump wavelength) are shown in Figure 5.13b. Solid lines on this figure correspond to a theoretical prediction based on known dispersion data. The ZGP OPO idler output amounted to 25 μJ at 1 kHz and $\lambda = 7$ μm corresponding to photon conversion efficiency of 30%. The total conversion from the 1.064 μm laser to the 7 μm output amounted to 1.5%. At a pump beam size of $w_0 = 125$ μm (at $\lambda = 3.1$ μm), which is close to a confocal focusing condition, the OPO pump threshold was remarkably low, 2 μJ [99]. This is the lowest OPO threshold reported so far for a singly resonant pulsed OPO.

The first optical parametric oscillator based on gallium arsenide was demonstrated in [100, 101]. The OPO utilized a QPM OP-GaAs crystal, 11-mm long, 5-mm wide, and 0.5-mm thick, with a domain reversal period of 61.2 μm.

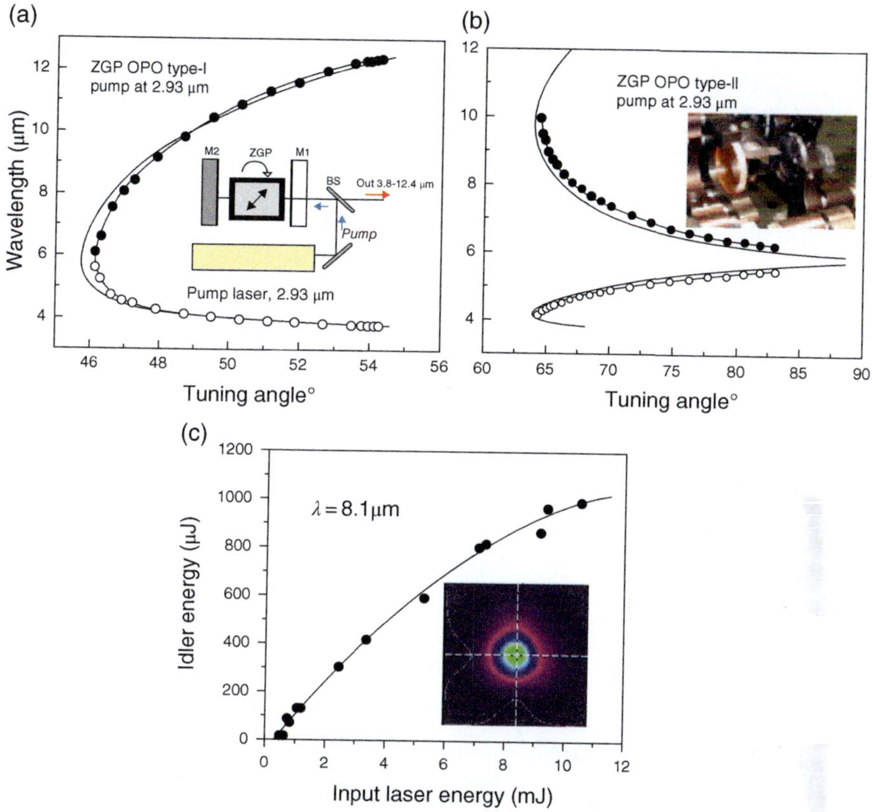

Figure 5.12 ZGP OPO with a wide (3.8–12.4 μm) mid-IR tunability and more than 1 mJ pulse energy. (a) Angular tuning curve for the type-I OPO. Inset: OPO schematic. The front OPO mirror M1 was transmissive for the pump and the idler and highly reflective (98%) for the signal. A gold rear mirror M2 was highly reflective ($R > 98\%$) for the pump, signal, and idler. Thus, the signal wave resonated, while the pump and the idler were double-passed. A dichroic beam splitter (BS) separated the incoming pump beam from the outcoming idler. (b) Angular tuning curve for the type-II OPO. (c) Dependence of the type-I OPO idler pulse energy as a function of the pump. The inset shows the OPO idler beam far-field intensity distribution (the corresponding beam quality factor $M^2 \approx 1.5$). *Source:* reproduced from figures 2 and 3 of [97], with permission of OSA, The Optical Society.

Tunable (1.75–2 μm) pulses from a Nd:YAG-laser-pumped PPLN OPO (10 Hz, 0–70 μJ, 6 ns) were used as a pump. A 13-mm-long OP-GaAs OPO cavity (Figure 5.14) was formed by two flat mirrors M4–M5. The input–output mirror M4 was reflective at the signal wavelength and transmissive at the pump and the idler. In a single-pass arrangement, mirror M5 was dielectric, identical to M4, while in a two-pass arrangement, a flat gold mirror was used as M5 to reflect all three waves. Thus, the signal wave was resonated, whereas the pump and idler

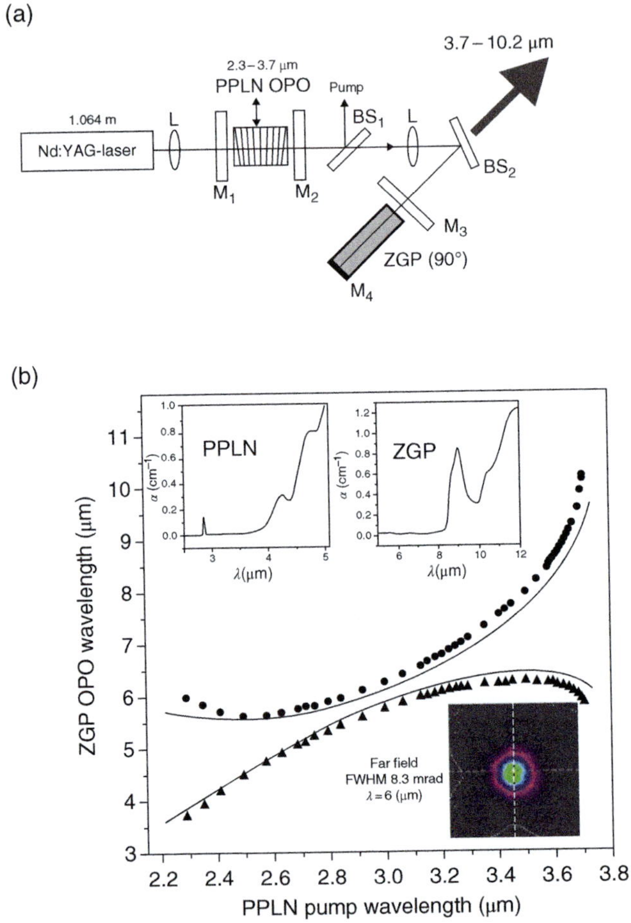

Figure 5.13 (a) Schematic of the tandem noncritically phase-matched ZGP OPO. M1–M4, OPO mirrors; L, infrared focusing lenses; BS, beamsplitters. (b) Tuning curve for ZGP OPO, as a function of pump wavelength. The insets show longwave absorption coefficients for PPLN and ZGP and also the far-field beam profile for the idler beam at $\lambda = 6\,\mu m$. *Source:* reproduced from figures 1 and 2 of [99], with permission of OSA, The Optical Society.

waves were recycled to have a second pass before leaving the cavity. Because of the symmetry of GaAs, the output polarizations of the OP-GaAs OPO were orthogonal to that of the pump and were extracted by use of a ZnSe plate at Brewster's angle (Figure 5.14). The lowest OPO pump threshold (16 μJ) and the highest output were obtained in a two-pass arrangement in the GaAs OPO. Figure 5.15a shows the OP-GaAs OPO tuning curve (with respect to the pump wavelength). The OP-GaAs crystal allowed mid-IR tuning between 2 and 11 μm, which was limited by (i) the spectral range of the OPO mirrors and (ii) the onset

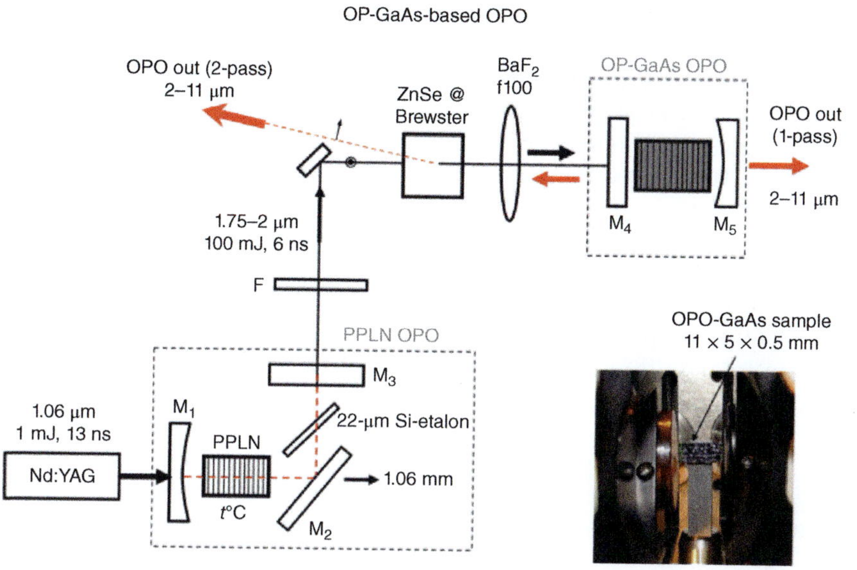

Figure 5.14 Schematic of the OP-GaAs OPO. Tunable, 1.75–2-μm, signal pulses from a three-mirror PPLN OPO were used as a pump (a 22-μm-thick uncoated silicon etalon was used inside PPLN OPO cavity to reduce its linewidth to 5 cm^{-1}). M1–M3, PPLN OPO mirrors; M4–M5, GaAs OPO mirrors. The filter F was used to block both the PPLN idler wave (2.3–2.6 μm) and the 1.06-μm pump. *Source:* reproduced from figure 1 of [100], with permission of OSA, The Optical Society.

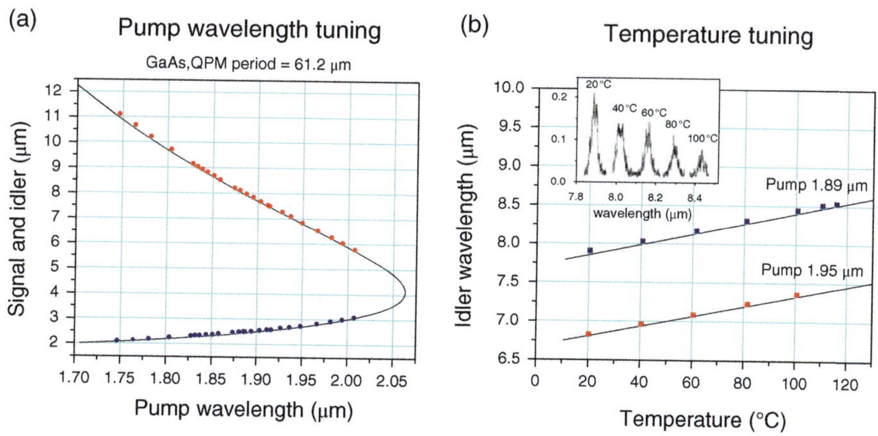

Figure 5.15 (a) OP-GaAs OPO tuning curve ($t = 20$ °C) with respect to the pump wavelength. (b) OP-GaAs OPO temperature-tuning curves for two selected pump wavelengths. Solid lines are calculated, based on data from [102]. Inset shows OPO spectral line shapes at different GaAs temperatures (pump 1.89 μm). *Source:* reproduced from figures 3 and 4 of [101], with permission of SPIE.

of two-photon absorption at <1.75 μm pump wavelength. Temperature tuning curves for two selected pump wavelengths are represented in Figure 5.15b. At the pump energy around 40–50 μJ, the OPO idler output was 3 μJ (λ = 7.9 μm), corresponding to the quantum conversion efficiency of 26%.

The broadest continuous mid-IR tunability from a GaAs system was reported in [103]. With the choice of the pump near $\lambda \approx 3$ μm (from a PPLN OPO), the authors were able to achieve tunable output in the whole range of 4–14.2 μm with a linewidth of 2–6 cm^{-1}, using a single OP-GaAs crystal with a domain reversal period of 150 μm (Figure 5.16). The OPO output was tuned using (i) an intracavity diffraction grating and (ii) fine adjustment (within 160 nm) of the pump wavelength near 3 μm. The OPO was operating at 2 kHz repetition rate. Its threshold, in terms of pump pulse energy, was approximately 25 μJ and the output average power reached 14 mW at 6-μm wavelength [103].

CSP has a transparency window extending from 1 to 6.5 μm [104]. Prominently, when pumped at 1.064 μm, CSP allows parametric generation under noncritical (θ = 90°) phase matching with a large nonlinearity (see Table 5.1),

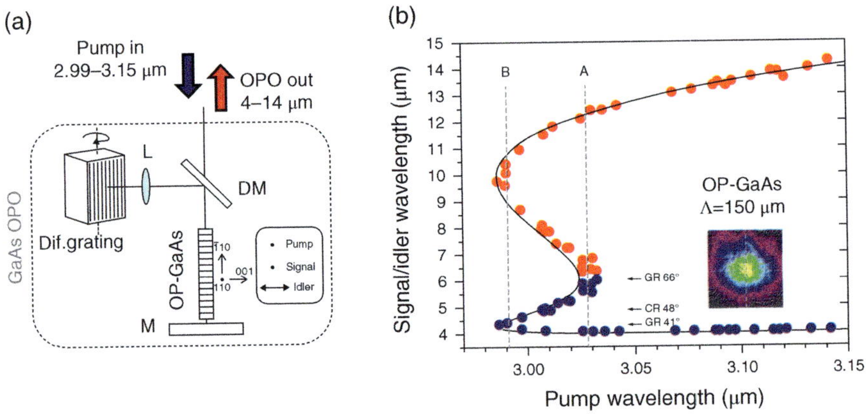

Figure 5.16 (a) Schematic of the 4–14 μm OP-GaAs OPO. Tunable pump pulses (near ~3 μm) are directed into an OP-GaAs crystal. The L-shaped OPO cavity is formed by: (i) diffraction grating, (ii) dichroic mirror DM that reflects ($R > 98\%$) the signal at 4–6 μm and transmits the 3-μm pump and idler at 6–14 μm, and (iii) metallic mirror M. The lens L is introduced to make a stable cavity and expand the beam at the grating. (b) The OPO tuning curve. The wavelength was tuned by (i) an intracavity diffraction grating and by (ii) fine adjustment (within 160 nm) of the pump wavelength. Solid line: theoretical tuning curve. The vertical dashed lines "A" and "B" represent turning points of the tuning curve. Also shown are diffraction grating angles for several resonating signal wavelengths. The inset shows a far-field beam profile at $\lambda \approx 8$ μm. *Source:* reproduced from figure 1 of [103], with permission of OSA, The Optical Society.

providing mid-IR idler radiation near 6 μm, which is important for medical applications. The main problem yet to be solved with this crystal is the residual absorption close to the bandgap, which is not intrinsic. For example, at 1.064 μm, the bulk absorption is ~0.2 cm^{-1}. The first OPO operation of CSP was demonstrated by Petrov et al. in the nanosecond noncritical (θ = 90°) singly resonant regime, pumped at 1.064 μm [105]. The CSP sample had a length of 8 mm and the OPO cavity consisted of two plane mirrors separated by 9.5 mm. With the signal/idler wavelengths of 1.285/6.193 μm, the pump threshold was 1.8 mJ. The maximum idler energy was 0.47 mJ (10 Hz repetition rate), at an incident pump energy of 21.4 mJ. This gives an idler conversion efficiency of 2.2% and quantum conversion efficiency of 12.8%. The CSP damage threshold (0.22 J/cm^2), in terms of pulse energy fluence, was similar to that of $AgGaS_2$.

Marchev et al. used a 21.4-mm-long θ = 90°-cut CSP crystal pumped by 8-ns pulses at 1.064 μm in an OPG arrangement (without a cavity but with weak residual reflections from the parallel crystal surfaces), in a double-pass configuration for pump, signal, and idler (Figure 5.17) [106]. A 45° ZnSe bending mirror (BM) was used for separating the pump radiation from the signal and idler outputs, and a silver-coated mirror – to retroreflect all the three waves for a second pass. The pumping threshold for the device was 213 μJ (0.23 MW/cm^2 on-axis intensity). At the maximum pump energy of 12 mJ (12.7 MW/cm^2), the total output energy exceeded 4 mJ, from which 3.64 mJ

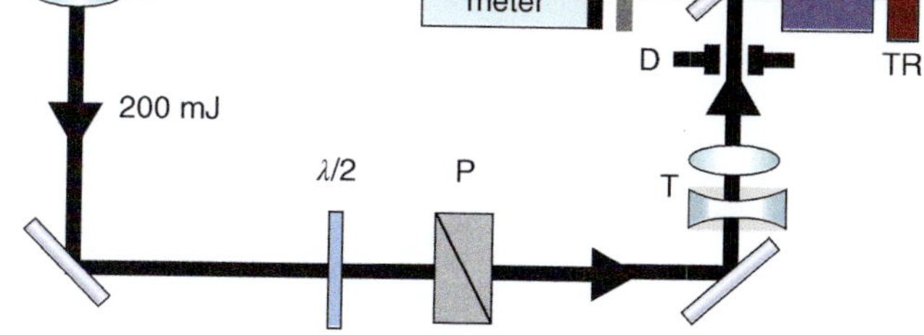

Figure 5.17 Experimental setup of the CSP-based OPG device. T, telescope; D, diaphragm; BM, bending mirror; TR, total reflector (metal mirror); F, long-pass filter; P, polarizer; λ/2, half-wave plate; D and DD, diaphragms. *Source:* reproduced from figure 1 of [106], with permission of OSA, The Optical Society.

were at 1.288 μm (signal) and 0.52 mJ at 6.125 μm (idler), with a quantum conversion efficiency of 34.7%. The beam propagation parameter M^2 for the idler wave was between 7 and 8. At a repetition rate of 100 Hz, the average idler power was 52.3 mW at 6 μm.

An extra-wide mid-IR tunability (2.7–17 μm) under 1.064 μm pumping was demonstrated in a newly developed chalcogenide crystal BGSe ($BaGa_4Se_7$) with monoclinic symmetry [107]. The crystal has an excellent optical quality and is transparent from 0.8 to 15 μm (at 0.3 cm^{-1} absorption level). A singly resonant nanosecond OPO had a linear cavity consisting of a flat input–output coupler and a flat gold-coated total rear reflector, which ensures recycling of the pump and a double pass for the nonresonant idler prior to its extraction from the cavity. Pumping via a 45° ZnSe BM, highly transmitting for the signal and idler, ensured separation of the input and output waves. The pump source was a diode-pumped Nd:YAG master oscillator–power amplifier (MOPA) system operating at 10 Hz, with 8 ns pulse duration and 63 mJ pulse energy. The OPO angle tuning curves for type-I and type-II phase matching are shown in Figure 5.18. The best results at 63-mJ pump include: unprecedented tuning range, from 2.7 to 17 μm achieved with a single crystal cut; 4.7-mJ pulse energy obtained for the λ = 5.3 μm idler wave (pump-to-idler conversion efficiency of 7.5%); and 3.7-mJ idler output at λ = 7.2 μm (pump-to-idler conversion efficiency of 5.9%; quantum conversion efficiency of 40%) [107].

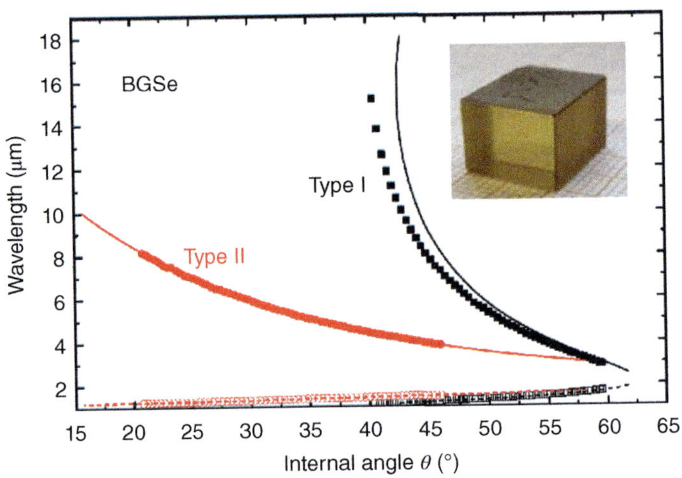

Figure 5.18 Experimental angle tuning curves for type I and type II BGSe-based OPO pumped at 1.064 μm. Solid curves: calculated. *Source:* reproduced from figure 3 of [107], with permission of OSA, The Optical Society.

5.3.2.2 Narrow-linewidth Pulsed OPOs

Because of the limited buildup time in pulsed OPOs, it is more challenging to achieve a narrow-linewidth output in such devices, as compared to their CW counterparts. Richman et al. have developed a PPLN-based pulsed mid-IR OPO that resonates the signal wave and uses just one intracavity etalon to restrict lasing to an SLM of the resonator cavity [108]. The OPO ring resonator (Figure 5.19) consisted of three mirrors. The two dichroic mirrors on either side of the PPLN crystal (25 mm long, 0.5 mm high, and 19 mm wide) transmitted the pump and idler beams and reflected the signal beam. The third mirror outcoupled 10% of the signal beam. The airspace etalon was made of one convex and one concave mirror, with a 95% reflective coating for the signal wave and each with a curvature that matches the phase-front curvature of the ring-cavity TEM_{00} mode. The etalon with mirror spacing of 357 μm (free spectral range $14\,cm^{-1}$) had a finesse of 60 and the insertion loss of 30%. The pump was a narrow-linewidth injection-seeded Q-switched laser at 1.06 μm and 1-kHz repetition rate, and its output was focused to a 240-μm spot size in the PPLN crystal.

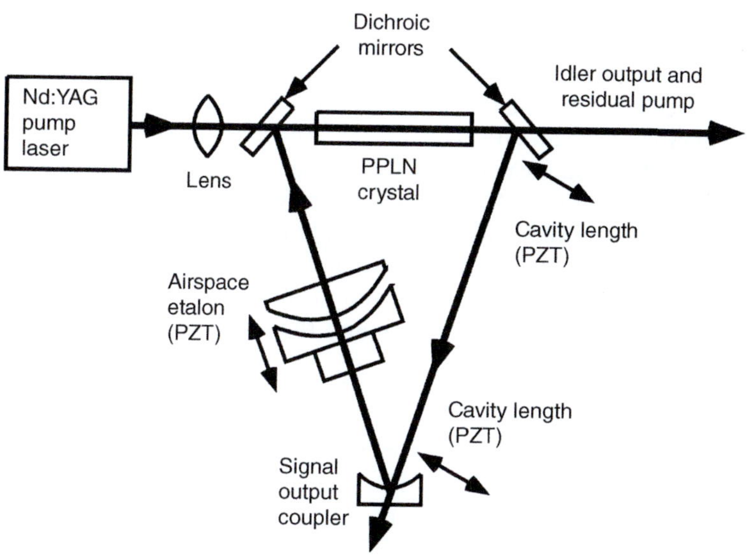

Figure 5.19 Schematic of a pulsed single-longitudinal-mode tunable PPLN OPO. The ring resonator consists of three mirrors. The two flat mirrors on either side of the PPLN crystal transmit the pump and idler beams and reflect the signal beam. The third (curved) mirror outcouples 10% of the signal beam for spectral characterization. The airspace etalon is made of two lenses, each with a 95% reflective coating and each with a curvature that nearly matches the phase-front curvature of the ring-cavity TEM_{00} mode. The right-hand dichroic mirror and the curved mirror are on a single translation stage driven by a PZT to control the cavity length. *Source:* reproduced from figure 1 of [108], with permission of OSA, The Optical Society.

Both the etalon mirror spacing and the OPO resonator length were adjusted by piezoelectric translators (PZTs), so that it was possible to tune the frequency continuously over 10 cm^{-1} without using motorized parts. Translation of a multi-grating PPLN wafer allowed access to any wavelength from 1.45 to 1.8 μm (signal) and from 2.6 to 4 μm (idler). Up to 18 μJ/pulse in the idler beam and up to 15 μJ/pulse in the signal beam were produced with only 200-μJ pump energy. The measured single-mode OPO linewidth was 0.005 cm^{-1}.

Ganikhanov et al. demonstrated a narrow-linewidth and broadly tunable output from a singly resonant (idler wave) nanosecond OPO based on type-II ZGP crystal pumped at 2.55 μm. With the OPO cavity containing a diffraction grating and an Si etalon, the authors achieved mid-IR tunability from 3.7 to 8 μm with the output energies 10–200 μJ/pulse, and with the output linewidth of 0.1 cm^{-1}, corresponding to three axial cavity modes [109]. It was also demonstrated that a narrow-linewidth idler OPO output can be achieved even using a comparatively broadband pump.

The concept of a doubly resonant nested-cavity OPO enables achieving an SLM emission from a pulsed OPO without using any intracavity etalons or resorting to injection seeding. In addition, one can achieve very low oscillation thresholds of a few microjoules [110, 111]. Traditionally it was understood that a doubly resonant condition precludes continuous frequency tuning, since it is satisfied only at particular wavelengths. A nested-cavity doubly resonant OPO solves this problem. In such an OPO (Figure 5.20) the inner mirrors (M2, M3) are deposited onto the nonlinear crystal faces whereas the external mirrors (M1, M4) are mounted on two PZT actuators for fine-tuning of the lengths, such that the signal wave oscillates between M1 and M3, while the idler between M2 and M4 (alternatively, it can be a three-mirror design). The mode spacing is chosen to be slightly different for the signal and the idler, so that only one pair of signal–idler modes oscillates. By fine-tuning the PZT actuators, tunable

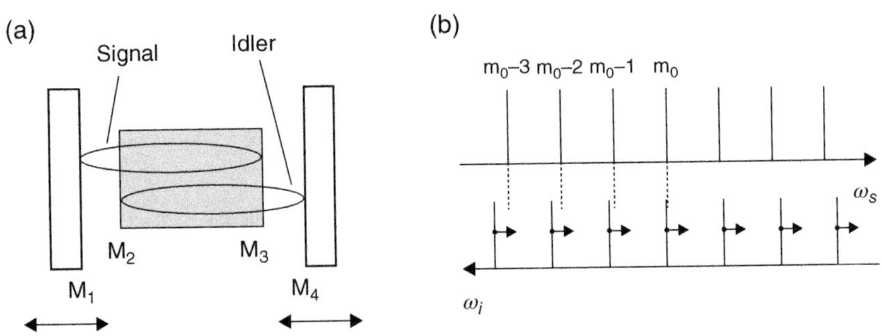

Figure 5.20 (a) Schematic of a nested dual-cavity doubly resonant OPO for SLM operation. M1–M3 form the "signal" cavity, and M2–M4 form the "idler" cavity. (b) Mode picture for the signal and idler waves. Since the mode spacing is different for the signal and the idler waves, only one pair of signal–idler modes oscillates. *Source:* reproduced from figure 16 of [110], with permission of OSA, The Optical Society.

mode-hop-free SLM oscillation can be achieved over >100 GHz range. Low-threshold nested-cavity OPOs based on PPLN crystals pumped by compact passively Q-switched single-frequency microlasers of fiber-based lasers operating at high (4.8–100 kHz) repetition rates have been demonstrated and used for various spectroscopic applications in the range from 2 to 4.3 μm [112, 113].

Clément et al. reported a nanosecond doubly resonant nested-cavity OPO based on 10-mm-long OP-GaAs emitting a tunable single-frequency radiation in the longwave IR region [114]. The OPO was pumped by an SLM Tm:YAP microlaser emitting pulses at λ = 1.94 μm, with a maximum output energy of 170 μJ, pulse duration of 36 ns, and a repetition rate of 100 Hz. The OPO pump threshold energy was 10 μJ. With the OP-GaAs QPM period of 72.6 μm, SLM wavelength tuning over the range of 10.3–10.9 μm was obtained by varying the crystal temperature. Based on this device, a differential absorption LIDAR was demonstrated by carrying out detection of ammonia vapor around λ = 10.4 μm [114].

The summary of pulsed narrow-band OPOs can be found in Table 5.5.

Table 5.5 Summary of pulsed nanosecond OPOs.

OPO crystal	OPO wavelength (μm)	Pump	OPO parameters	Ref.
Broadly tunable				
PPLN	1.36–4.83	1.064 μm (100 μJ, 7 ns, 1 kHz)	6 μJ @ 4 μm; 2 μJ @ 4.8 μm	[91]
AGS	3.9–11.3	1.064 μm (15 mJ, 12 ns, 10 Hz)	0.37 mJ @ 6 μm	[93]
AGSe	6.1–14.1	1.57 μm (30 mJ, 6 ns, 5 Hz)	1.2 mJ @ 9 μm	[96]
ZGP	3.8–12.4	2.93 μm (10 mJ, 100 ns, 10 Hz)	1.2 mJ @ 6.6 μm; 1 mJ @ 8.1 μm	[97]
OP-GaAs	2–11	1.75–2 μm (60 μJ, 6 ns, 10 Hz)	3 μJ @ 7.9 μm	[101]
OP-GaAs	4–14	~3 μm (120 μJ, 20 ns, 2 kHz)	7 μJ @ 6 μm	[103]
BGSe	2.7–17	1.064 μm (63 mJ, 8 ns, 10 Hz)	4.7 mJ @ 5.3 μm; 3.7 mJ @ 7.2 μm	[107]
Narrow linewidth				
PPLN	~3	1.064 μm (200 μJ, 10 ns, 1 kHz)	18 μJ, 0.01 cm^{-1}	[108]
PPLN	3.8–4.3	1.064 μm (16 μJ, 9.8 ns, 4.8 kHz)	Nested-cavity SLM OPO, 0.3 μJ @ 3.9 μm	[112]

(*Continued*)

Table 5.5 (Continued)

OPO crystal	OPO wavelength (μm)	Pump	OPO parameters	Ref.
PPLN	3.3–3.5	1.064 μm (4–10 μJ, 1 μs, 40–100 kHz)	Nested-cavity SLM OPO, 0.5 μJ @ 3.5 μm	[113]
ZGP	3.7–8	2.55 μm (12 mJ, 12 ns, 10 Hz)	10 μJ @ 3.7 μm, 0.1 cm^{-1}; 10 μJ @ 8 μm, 0.1 cm^{-1}	[109]
OP-GaAs	10.3–10.9	1938.5 μm (170 μJ, 36 ns, 100 Hz)	Nested-cavity SLM OPO, 2 μJ/pulse	[114]
High average power				
PPKTP	1.72 and 2.76	1.064 μm (7.2 W, 20 kHz, 5.8 ns)	2 W (sig + idler)	[7]
KTP	2.13	1.064 μm (135 W, 20 kHz, 40 ns)	53 W in two polarizations	[115]
ZGP	3.67 and 4.67	2.05 μm (20.1 W, 10 kHz, 11 ns)	10.1 W (sig + idler), conv. eff. 50.2%	[116]
ZGP	3.7–4.1 4.4–4.8	2.13 μm (25 W, 20 kHz, 40 ns)	14 W (sig + idler)	[115]
ZGP	3.8 and 4.6	2.1 μm (37.7 W, 32 ns, 45 kHz)	22 W (sig + idler), $M^2 \approx 1.4$, conv. eff. 58% (slope 75%)	[117]
ZGP	4.3	2 μm (7.2 W, 5 kHz)	2.53 W, cascaded intracavity OPO	[118]
ZGP	Broadband 3.5–5 μm	2.09 μm (43 W, 50 ns, 35 kHz)	27 W (slope eff. 67%); 99 W (25% duty cycle)	[119]
OP-GaAs	Broadband 3.5–5 μm	2.09 μm (6.1 W, 65 ns, 20 kHz)	2.85 W ave. power (sig + idler)	[120]
OP-GaAs	3.6 and 4.4	≈2 μm, 60 W ave. power	18 W ave. power (sig + idler)	[18]
OP-GaAs	10.6	1.95 μm (12 W, 160 ns, 50 kHz)	800 mW ave. power, conv. eff. 6.8%, quant. eff. 36.8%	[121]
High pulse energy				
KTP L = 15 mm	2.6–3.2	1.064 μm (145 mJ, 5 ns, 10 Hz)	17 mJ @ 2.9 μm	[122]
$LiNbO_3$ L = 50 mm	2.13	1.064 μm (600 mJ, 15 ns)	300 mJ	[123]
PPLN	1.89 and 2.43	1.064 μm (196 mJ, 10 ns, 30 Hz)	124 mJ (sig + idler)	[124]
PPLT L = 40 mm	1.84 and 2.52		118 mJ (sig + idler)	

Table 5.5 (Continued)

OPO crystal	OPO wavelength (μm)	Pump	OPO parameters	Ref.
ZGP $L = 25\,mm$	6.9–9.9	2.8 μm (25 mJ,50 ns, 10 Hz)	2.4 mJ @ 6.9 μm, 0.7 mJ @ 9.9 μm	[125]
ZGP RISTRA $L = 10\,mm$	3.4	2.05 μm (55 mJ,14 ns, 500 Hz)	10 mJ, near-diffraction-limited	[126]
ZGP RISTRA	Broadband 3.5–5 μm	2.053 μm (45.6 mJ,100 Hz)	23.8 mJ (sig + idler)	[127]
ZGP MOPA	Broadband 3.5–5 μm	2.05 μm (500 mJ,15 ns, 1 Hz)	212 mJ (sig + idler)	[128]

5.3.2.3 High Average Power OPOs

5.3.2.3.1 *Based on PPKTP and KTP*

Among the family of PP oxide crystals, large-aperture PPKTP crystals are well suited for generating average powers exceeding 1 W in the pulsed nanosecond regime. Peltz et al. have reported a high average power OPO using a 3-mm-thick PPKTP crystal [7]. A diode-pumped $Nd:YVO_4$ laser system (1.064 μm, 5.8 ns, 10–20 kHz) served as a pump. At a repetition rate of 20 kHz and 7.2 W pump power, the total signal + idler output power (at correspondingly 1.72 and 2.76 μm) reached 2 W.

Bulk KTP crystals enable much higher average powers, as compared to PPKTP, due to their larger apertures. Cheung et al. [115] used a diode array-pumped Nd:YAG MOPA system with the repetition rate of 20 kHz, pulse duration of 40 ns, and the average power of 135 W, as a pump source for an OPO based on bulk KTP crystals. A degenerate OPO ($\lambda \approx 2.13$ μm) used six KTP crystals, $3 \times 3 \times 6$ mm in size, cut for type-II phase matching in a walk-off compensated configuration. The average power of 53 W (in two orthogonal polarizations) has been achieved with 43% conversion efficiency.

5.3.2.3.2 *Based on ZGP*

The ZGP crystal is extremely suitable for scaling to high average OPO powers in the longwave mid-IR range, since in addition to high quadratic nonlinearity it exhibits excellent thermal and mechanical properties. Typically, high-power ZGP OPOs are based on frequency downconversion of 2-μm lasers.

Wu et al. [118] demonstrated a coupled tandem approach, where a 2-μm output of a KTP OPO pumped a ZGP OPO. Not only was the ZGP OPO placed within the cavity of the KTP OPO, but also the KTP OPO was placed within the cavity of a diode-pumped Nd:YALO laser (Figure 5.21). The Nd:YALO laser was able to generate ~58 W of the output power at $\lambda \sim 1$ μm at a pulse

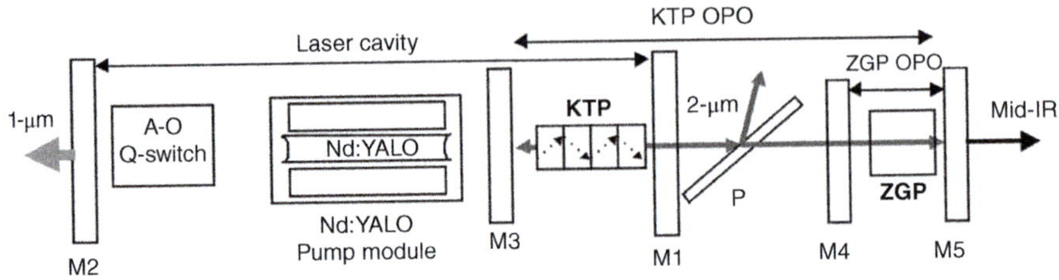

Figure 5.21 Schematic diagram of the coupled tandem KTP OPO–ZGP OPO. The Nd:YALO laser cavity (the pump) was formed by a high reflector M1 and an output coupler M2. The KTP OPO utilized four type-II, 51°-cut, diffusion-bonded walk-off-compensated KTP crystals (5 × 5 × 8 mm), placed between mirrors M3 and M1. A ZGP OPO cavity was formed by mirrors M4 and M5 and used type I, 53°-cut ZGP with 5 × 5 × 10 mm dimensions. *Source:* reproduced from figure 1 of [118], with permission of OSA, The Optical Society.

repetition rate of 5 kHz. The output from the ZGP OPO was tunable over the range of 3–6 μm, with the maximum output power of 2.53 W achieved at 4.3 μm, at approximately 580 W of diode pump power.

The first ZGP OPO with an average output power exceeding 10 W was reported by Budni et al. [116]. The OPO used a 14-mm-long type-I ZGP crystal and was pumped at 10 kHz by 11-ns, λ = 2.05 μm pulses from a diode-pumped Q-switched Ho,Tm:YLF-laser operating at $T = 77$ K. At the maximum pump drive level of 20.1 W incident onto the ZGP crystal, the output average power (signal at 3.67 μm plus idler at 4.67 μm) reached 10.1 W, corresponding to a conversion efficiency of 50.2%. It is important that the operation of this OPO at maximum power level was well below the ZGP damage threshold.

Cheung et al. used the high-power 2.13-μm output obtained from the KTP OPO described above for pumping a ZGP OPO. A polarizer was used to separate the two orthogonally polarized 2.13-μm beams from the KTP OPO (approximately 25 W in each beam), and each beam was sent into a separate ZGP OPO producing broadband output in the 3–5 μm band. With two ZGP OPOs running simultaneously, 13 and 11 W were obtained with the beam quality factor $M^2 \approx 4$ [115].

Lippert et al. have demonstrated a 22-W output from a single ZGP OPO [117]. The OPO was pumped by a pulsed 2.1-μm Ho:YAG laser (pumped in turn by a Tm-fiber laser) with 32-ns pulse duration and 45-kHz repetition frequency. An innovative V-shaped three-mirror ring OPO design was used to achieve an excellent beam quality. The ring is in the noncritical plane of the crystal, as shown in Figure 5.22. Compared to other ring resonators, the V-shaped three-mirror ring cavity has the advantages of short round-trip time and also that a single ZGP crystal can be used for two passes. The crystal was tuned by rotation in the orthogonal plane. With 37.7 W of Ho:YAG pump, the authors obtained 22 W of average power in the 3–5 μm range (3.8-μm signal plus 4.6-μm idler), with a beam quality factor of $M^2 \approx 1.4$. The absolute conversion efficiency was 58%, and the slope efficiency was 75%. The output

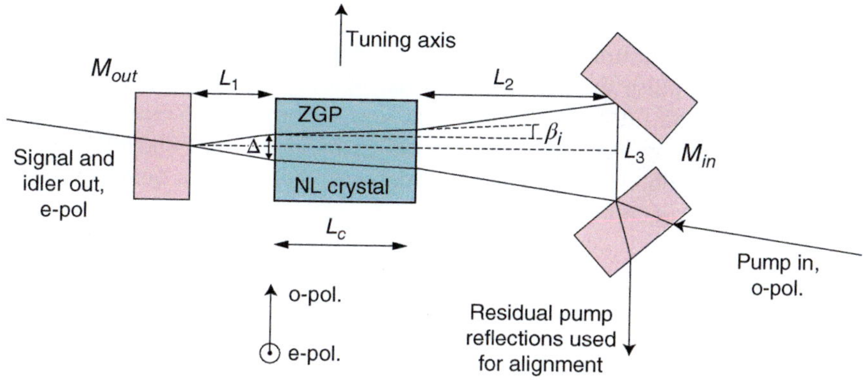

Figure 5.22 V-shaped 3-mirror ring resonator of a high-power ZGP OPO with two passes through the same crystal and angle tuning about an axis in the plane of the ring. *Source:* reproduced from figure 1 of [117], with permission of OSA, The Optical Society.

slope showed no sign of roll-off at high power, which indicates that the high efficiency can be maintained at even higher pump power.

Hemming et al. have generated the highest reported output power from a mid-IR ZGP OPO. The pump was a thulium-fiber-laser-pumped Q-switched Ho:YAG laser (2.09 μm) operating at 35-kHz pulse repetition rate, with up to 60 W of output power. The system produced 27 W of the output power in the 3–5 μm band with the beam quality factor $M^2 = 4$ when operating continuously in a repetitively Q-switched mode. However, when the OPO operated at a duty cycle of 25%, the average output power during the "open" cycle reached 99 W [119].

5.3.2.3.3 *Based on OP-GaAs*

In the early work on OP-GaAs, Kieleck et al. reported a high-efficiency high repetition rate mid-IR OPO utilizing OP-GaAs pumped by a 2.09-μm Ho:YAG laser [120]. The OP-GaAs crystal was 20 mm long, 5 mm wide, and 450 μm thick, and had 63-μm grating period. Up to 2.85 W are obtained in the 3–5 μm band for 6.1 W of pump power at 20 kHz repetition rate, corresponding to an optical-to-optical conversion efficiency of 46.5%, with an OPO pump threshold of 1 W. According to the authors, the OP-GaAs OPO performance, in terms of the average power, was only limited by the thickness of the crystal. Thanks to the further progress in all-epitaxial processing of OP-GaAs, namely increasing crystal's thickness in the direction of the epitaxial growth and improving the quality of orientation patterning, it became possible to demonstrate 18 W of the average power (signal + idler) from an OP-GaAs OPO pumped by a pulsed high repetition rate Tm-fiber laser ($\lambda \approx 2$ μm), with the signal and idler wavelengths of 3.6 and 4.4 μm, respectively [18].

On a longer-wavelength side, an optical parametric oscillator based on OP-GaAs generating an idler output at around 10.6 μm with relatively high average power was reported by Wueppen et al. [121]. The system used an SLM (spectral width <100 MHz) thulium fiber laser at 1.95-μm wavelength, 50-kHz repetition rate, and 150-ns pulse duration as a pump. With a signal-resonant bow-tie OPO cavity and an OP-GaAs crystal that was 40 mm long, 5 mm wide, and 1.3 mm thick, the average idler output power of more than 800 mW (pulse energy 16 μJ) was obtained at 10.6 μm, which corresponds to a quantum conversion efficiency of 36.8% [121].

5.3.2.4 High Pulse Energy OPOs

Large-aperture bulk oxide crystals such as lithium niobate and its family, as well as KTP and its family are well suited for producing high pulse energies. Vysniauskas et al. [122] demonstrated a high-energy OPO based on angular-tuned KTA crystal. The OPO was pumped by a Q-switched Nd:YAG laser with 5 ns pulse duration and 10 Hz repetition rate. The pump beam had a diameter of 4 mm and was double-passed through the 15-mm-long KTA crystal. The OPO was singly resonant for the idler wave and had an unstable resonator cavity to improve the spatial beam distribution. The idler beam tunability over the 2.6–3.2 μm range was achieved with the pulse energy of 17 mJ at 2.9 μm for a pump pulse energy of 145 mJ. Even higher mid-IR OPO output energies were reported by Mennerat and Kupecek [123]. An OPO was based on a 50 mm-long, 15 mm-aperture, 47°-cut $LiNbO_3$ crystal and was pumped with 15-ns pulses at 1.064 μm. The maximum output energy of 300 mJ (signal + idler) was achieved near OPO degeneracy (2.13 μm) at a 600 mJ pump.

Scaling a nanosecond OPO to high pulse energies entails increasing the beam diameters to avoid surface damage. The OPO cavity, however, should be kept short for high efficiency and low threshold. As a result, the cavity Fresnel-number ($D^2/\lambda L$, where D is the beam diameter and L is the cavity length) increases, which worsens the spatial coherence across the beam. Hansson et al. [129] used an unstable OPO resonator to improve the beam quality of a PPRTA-based OPO (Figure 5.23). An unstable resonator effectively filters out

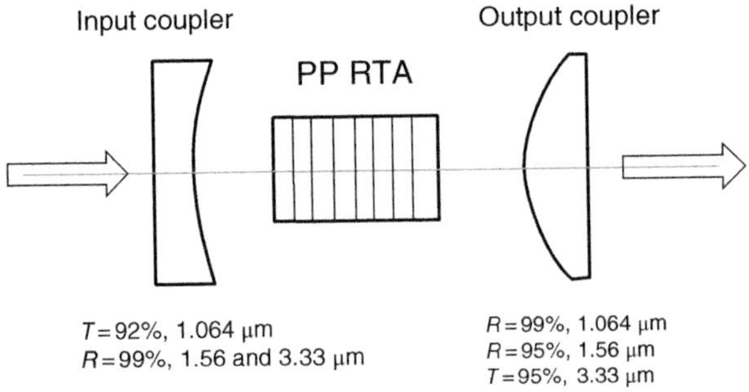

Figure 5.23 PP RTA-based OPO with an unstable resonator cavity. *Source:* reproduced from figure 1 of [129], with permission of OSA, The Optical Society.

the OPO modes with high spatial frequency components by a combination of laser-mode magnification and feedback of only the lowest-order spatial modes. The authors demonstrated an improvement in M^2 – the beam quality parameter – by a factor of 3 for a λ = 3.3-μm OPO idler wave, as compared to that of a plane-parallel OPO resonator.

Over the past decade, QPM ferroelectric PPLN and PPLT crystals with large thickness along the poling z-axis, exceeding 5 mm, have become available, thanks to the maturity of high-voltage poling technology [124, 130, 131]. This progress allowed producing high-pulse-energy outputs from OPOs based on QPM crystals, while benefiting from their engineerable phase matching and access to the highest nonlinear coefficients. Ishizuki et al. produced high-energy output from an OPO based on congruent MgO-doped PPLT (MgO–PPLT) crystal [124]. The authors used 196-mJ pump pulses at 1064 nm with 10-ns duration at 30 Hz repetition rate and generated 118 mJ total output in the OPO signal (1.84 μm) plus idler (2.52 μm) waves. Such high pulse energies from a QPM crystal became possible because of the availability of large-aperture (5 × 16 mm) PPLT structure with 40-mm length [124]. The same team also demonstrated an OPO based on large-aperture MgO-doped PPLN (MgO–PPLN) crystal using the same setup, including OPO mirrors, cavity, and the pump source. With the wavelength of signal and idler waves of 1.89 and 2.43 μm, the total OPO output (signal + idler) was measured to be a 124 mJ/pulse at 193-mJ pump pulse energy [124]. The authors argue that despite its lower second-order nonlinearity, lithium tantalate has higher thermal conductivity and shorter absorption edge in the UV region, compared to lithium niobate, and produces similar OPO outputs in terms of energy per pulse.

OPOs based on the ZGP crystal are widely used for generating tunable IR radiation in the whole of the 2.5–12-μm region. ZGP crystal possesses a set of unique properties such as high nonlinearity (d_{36} = 75 ± 8 pm/V), high thermal conductivity of 35–36 W/(mK) (higher than YAG crystal), and relatively high hardness 5.5 (Mohs scale). Hence, it is a remarkably promising material for generating high-energy mid-IR pulses. Allik et al. [125] reported a high-energy-per-pulse ZGP optical parametric oscillator pumped by a 2.8 μm Er,Cr:YSGG laser with a 10-Hz repetition rate, 25-mJ energy, and 50-ns duration. A 25-mm-long ZGP crystal was cut at θ = 65° for type-II phase matching. The OPO yielded idler output in the forward direction of 0.7–2.4 mJ/pulse, in the wavelength range 6.9–9.9 μm. The quantum conversion efficiency reached 29% at the OPO linewidth of typically 4 cm^{-1}.

By exploring a nonplanar OPO cavity containing a ZGP crystal, Dergachev et al. reported on a near-diffraction-limited output at the OPO signal wave at λ = 3.4 μm with pulse energy of 10 mJ at a repetition rate of 500 Hz [126]. As a pump source, the authors utilized a 2-μm, Ho:YLF MOPA system producing >55 mJ pulse energy at 500 Hz. (The Ho:YLF MOPA was pumped in turn by a 100-W Tm-fiber laser at 1940 nm.) The OPO used a 10-mm-long ZGP crystal, which was AR-coated to minimize reflections at the pump, signal, and idler wavelengths. The OPO resonator, shown in Figure 5.24, was based on a four-mirror nonplanar image-rotating ring cavity, known as Rotated Image Singly

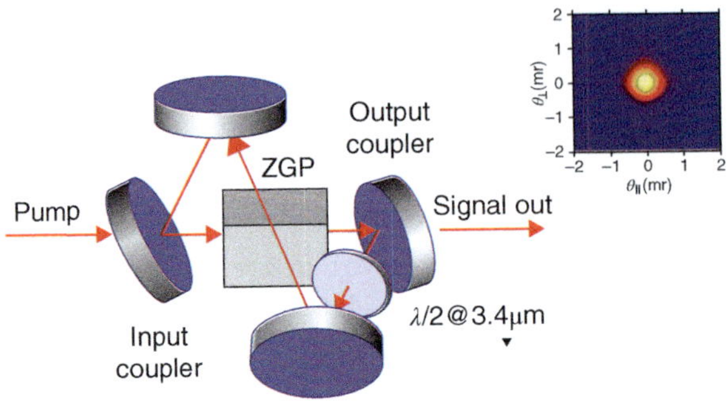

Figure 5.24 Geometric configuration of a nonplanar cavity for a high-pulse-energy Rotated Image Singly Resonant Twisted RectAngle (RISTRA) ZGP-based OPO. The inset shows the far-field beam profile. *Source:* adapted from figure 4a of [126], with permission of OSA, The Optical Society.

Resonant Twisted RectAngle (RISTRA) [132]. This cavity design was selected because the intracavity image rotation generates highly symmetric, high-quality output beams under operating conditions that would otherwise lead to very poor beam quality. Image rotation is particularly effective for improving beam quality in high-energy ns OPOs, where the ratio of beam diameter to cavity length results in very large Fresnel numbers.

Using a ZGP-based RISTRA OPO, Stöppler et al. demonstrated an overall OPO pulse energy of 23.8 mJ in the broadband 3–5 μm output. As a pump, the authors used 45.6-mJ pulses from a fiber-pumped oscillator–amplifier system based on Ho^{3+}:$LuLiF_4$ active medium (λ = 2.053 μm), at a repetition rate of 100 Hz [127]. Haakestad et al. reported broadband (3.5–5 μm) energetic pulses produced via nonlinear conversion in a near-degenerate ZGP-based MOPA system, pumped by 0.5-J pulses from a Q-switched cryogenic Ho:YLF oscillator at 2.05 μm [128]. A singly resonant master OPO had a V-shaped three-mirror ring resonator and contained two ZGP crystals, while the power amplifier used from one to three large-aperture ZGP crystals (Figure 5.25). Pulses with up to 212 mJ energy at 1 Hz repetition rate were obtained, with pulse duration of 15 ns and beam quality M^2 = 3. These are the highest-energy pulses generated in this wavelength region by a nanosecond solid-state laser source. The main advantages of the V-shaped three-mirror ring-cavity OPO used in this work are: (i) two-pass pumping without pump feedback into the laser, (ii) simple alignment, (iii) compactness, and (iv) reduced fluence at crystal surfaces because the forward and backward propagating beams were not overlapping.

5.3.2.5 Waveguide OPOs

Singly resonant pulsed OPO operation ($\tau \approx$ 100 ns) has been demonstrated in a PPLN-based waveguide. The OPO was pumped at 760 nm and had a very low oscillation threshold, 1.6 W in terms of the peak power (pump pulse energy 0.16 μJ) [133]. The idler tuning range of 1.18–2.080 μm was achieved by tuning

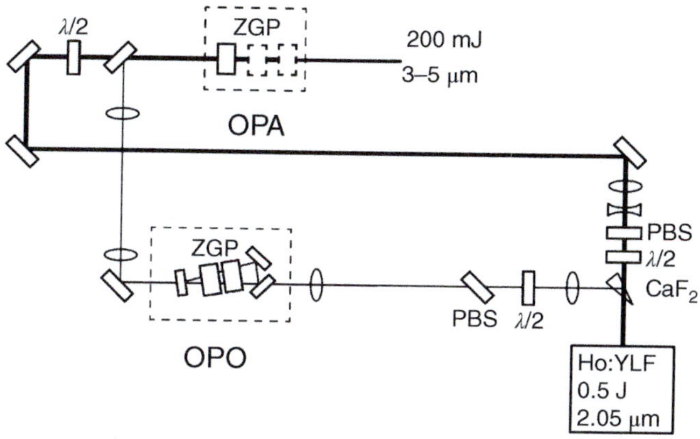

Figure 5.25 Schematic of the ZGP-based master oscillator–power amplifier system. A Q-switched cryogenic Ho:YLF oscillator provided 0.5-J pulses at $\lambda = 2.05\,\mu m$ and 1 Hz repetition rate. The pump beam for the OPO was obtained using one reflection from a CaF_2 wedge, while the transmitted pump was directed to the OPA. The ring-cavity OPO used two 6-mm-long ZGP crystals; the number of ZGP crystals in the OPA varied between one and three. PBS, polarizing beam splitter; $\lambda/2$, half-wave plate at 2.05 μm. *Source:* reproduced from figure 1 of [128], with permission of OSA, The Optical Society.

the pump wavelength from 756 to 772 nm, with the idler peak power of 220 mW (corresponding to the pulse energy 22 nJ).

The first demonstration of the OPO operation in a GaAs-based waveguide was reported by Oron et al. [134]. A 13-mm-long waveguide was based on an OP-GaAs with $12 \times 3\,\mu m^2$ cross section. A monolithic OPO cavity was formed by dielectric facet coating. With a pulsed pump near 2 μm (25 ns, 10 kHz), the OPO peak threshold power was 7 W (pulse energy 175 nJ). At the pump peak power of 11.6 W, the overall OPO output peak power of 0.6 W (pulse energy 15 nJ) was generated in the signal (3.6 μm) plus idler (4.5 μm) waves. In a quasi-CW pump mode (duty factor 5%, chopping rate 1 kHz), the OPO pump threshold was 5.7 W. At the maximum available pump power of 6.6 W, the detected OPO signal power was a few milliwatts. The authors explain the low power output by the fact that the pump power was too close to the OPO threshold to enable efficient energy conversion [134].

The results on the pulsed nanosecond OPOs are summarized in Table 5.5.

5.4 Regime of Ultrashort (ps and fs) Pulses

5.4.1 Ultrafast DFG

Downconversion of near-IR sources by DFG remains one of the most common approaches for getting ultrafast pulses in the mid-IR. DFG is attractive because it offers a straightforward solution with a single-pass geometry. In the early work on ultrafast DFG, Dahinten et al. generated nearly bandwidth-limited mid-IR pulses of 1 ps duration, tunable between 4 and 18 μm [22]. The mid-IR output was generated via difference frequency mixing of Nd:glass laser pulses and pulses from an infrared dye laser, pumped by the same Nd laser. Tuning between 4 and 18 μm was achieved by various combinations of laser dyes and nonlinear crystals ($AgGaS_2$ and GaSe). The energy of the mid-IR pulses reached a few microjoules, with the photon conversion efficiency ~2%. Ehret and Schneider generated mid-IR via difference frequency mixing of the signal and idler output of an optical parametric oscillator, pumped by a Ti:sapphire laser (λ = 815 nm) at a repetition rate of 76 MHz. $AgGaS_2$ (2-mm thick) and GaSe (1-mm thick) crystals were applied as nonlinear media [135]. GaSe had a larger tuning range and was more efficient in the whole spectral range than $AgGaS_2$. The average IR power at 8.5 μm was 2 mW with GaSe crystal and 1.3 mW with $AgGaS_2$. Kaindl et al. reported on a mid-IR light source that provides femtosecond pulses on a microjoule energy scale, broadly tunable in the 3–20-μm wavelength range with pulse durations as short as 50 fs at 5 μm [23]. The pulses were generated by phase-matched difference-frequency mixing in a 1-mm-thick GaSe using near-IR signal and idler pulses from a parametric device based on a 1-kHz Ti:sapphire amplifier system.

In the past decade, an approach based on generating mid-IR ultrafast pulses via DFG with well-established ultrafast Yb, Er, and Tm-doped fibers has gained much popularity. Erny et al. reported on DFG of mid-IR femtosecond pulses tunable in the 3.2–4.8 μm range from a two-branch mode-locked Er-doped fiber source operating at a repetition rate of 82 MHz. DFG was achieved via nonlinear mixing, in a 2-mm-thick MgO:PPLN crystal, of 170-mW, 65-fs pump pulses at a fixed wavelength of 1.58 μm with 11.5-mW, 40-fs pulses that were tunable between 1.05 and 1.18 μm and were produced in a highly nonlinear fiber with a core diameter of 3.7 μm [136]. The average DFG power of 1.07 mW, produced at 3.6 μm, corresponds to the quantum efficiency of ~30%, if counted with respect to the shorter-wave pump component.

Winters et al. presented an approach to femtosecond mid-IR DFG that was based on frequency redshifted solitons and where (as in the previous example) the pump and signal DFG pulses were derived from the same 1.55-μm Er-fiber laser [137]. A spectrally shifted pulse was created through intra-pulse Raman scattering of femtosecond optical solitons propagating in anomalous-dispersion single-mode polarization-maintaining optical fiber (L = 25 m, mode-field diameter 10.5 μm). The soliton redshift had linear dependence on the injected pump power (Figure 5.26a); thus, the authors were able to generate a signal pulse that was detuned from the pump by a controllable amount, up to more than 1000 cm^{-1}. This allowed generating mid-IR pulses over the range of 9.7–14.9 μm using a 1-mm-thick GaSe or a 5-mm-thick $AgGaSe_2$ crystal (Figure 5.26b). According to the authors, $AgGaSe_2$ produced higher DFG output powers than GaSe (possibly because of larger crystal length). Yao et al.

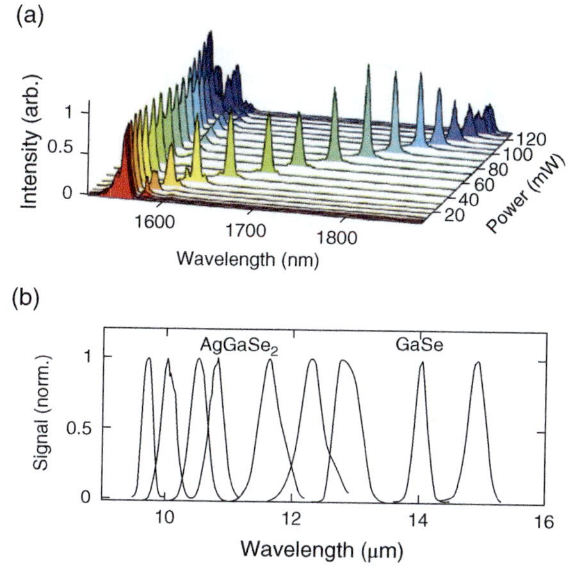

Figure 5.26 (a) Spectra of the frequency-shifted solitons, along with the pump spectra, obtained through intra-pulse Raman scattering in a single-mode fiber at different launched 1.55-μm pump powers. (b) Normalized DFG spectra obtained with $AgGaSe_2$ and GaSe crystals. *Source:* reproduced from figures 1 and 3 of [137], with permission of OSA, The Optical Society.

performed DFG by mixing the outputs from two photonic crystal fibers (PCF) pumped by the same Yb-fiber laser (1.035 μm, 1.3 W, 300 fs, 40 MHz). Facilitated by self-phase modulation, the output spectrum of the first PCF possesses two dominant outermost peaks that can be extended to 970 and 1092 nm. Spectral tuning was realized by varying the coupled-in power, which alters the wavelength. The second PCF with two closely spaced zero dispersion wavelengths around the laser wavelength was used to generate intense Stokes pulses between 1.24 and 1.26 μm. The two sets of pulses were mixed in an $AgGaS_2$ crystal, which resulted in producing mid-IR pulses tunable from 4.2 to 9 μm with a maximum average power of 640 μW at 4.5 μm [138].

Through DFG in a 2-mm-long fanned-out OP-GaAs, Phillips et al. demonstrated a mid-IR output tunable from 6.7 to 12.7 μm. A Tm-doped-fiber oscillator–amplifier system was used to generate 150-fs pulses at $\lambda \approx 1.95$ μm. These pulses were utilized to generate a comparatively broadband output centered at $\lambda \approx 2.5$ μm, through Raman soliton self-frequency shift in a fluoride fiber, followed by mid-IR DFG in the OP-GaAs crystal [139]. The concept is illustrated in Figure 5.27. By lateral translation of the fan-out OP-GaAs (and thus changing the period of the quasi-phase-matching grating), a tuning range of 6.7–12.7 μm was achieved with up to 1.3 mW of the average power at ~9 μm.

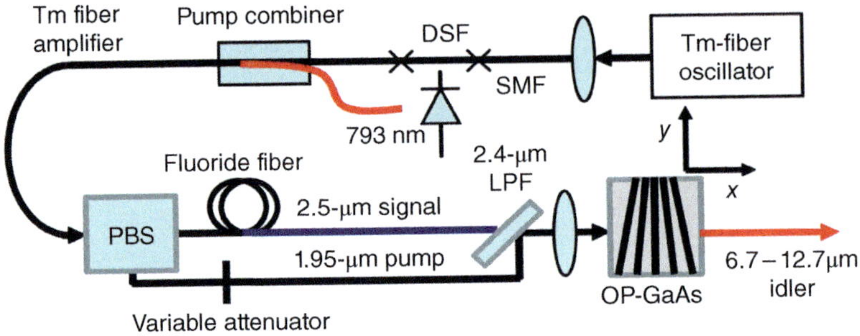

Figure 5.27 Setup of a Tm-fiber laser-based tunable 6.7–12.7 μm DFG system. After the Tm amplifier, the pulses from one arm were coupled into a single-mode fluoride fiber in order to facilitate Raman soliton self-frequency shift to ~2.5 μm. Subsequently, the pulses were recombined with the second-arm 1.95-μm beam in the OP-GaAs to produce DFG. SMF, single-mode fiber; DSF, dispersion-shifted fiber; PBS, polarizing beam splitter; LPF, long-pass filter. *Source:* reproduced from figure 1 of [139], with permission of OSA, The Optical Society.

A high-power mid-IR DFG source was reported by Zhu et al. [140]. The researchers used intense ultrashort pulses centered at ~1.05 and 1.55 μm, seeded by a common 250-MHz mode-locked Er-fiber oscillator. Half of the output from the oscillator was amplified by an EDFA to about 450 mW and coupled into a highly nonlinear fiber to broaden the spectrum and to generate 1.05-μm pulses through the process of self-phase modulation. These seed pulses were amplified by a YDFA to 1.2 W. The other half of the output from the Er-fiber oscillator was amplified by the second EDFA to about 450 mW (center wavelength 1.55 μm, pulse duration 60 fs). The DFG occurred in a MgO:PPLN crystal, and up to 120 mW of the output power in a broad spectrum spanning from 2.9 to 3.6 μm was produced [140]. Based on a similar approach, even higher DFG powers were produced near 3 μm wavelength, namely 150 mW in a broadband output with a spectral bandwidth from 2.7 to 3.45 μm [141]. By frequency mixing amplified 1.05- and 1.55-μm fs pulses seeded by the same oscillator (the average power after amplification was correspondingly 4 W and 140 mW) and by using a 3-mm-long MgO:PPLN crystal, Cruz et al. produced a DFG output at 100 MHz repetition rate and center wavelength tunable from 2.6 to 5.2 μm. The highest DFG average power of 500 mW was reported for an output with an instantaneous span of 2.8–3.5 μm [142].

At longer waves, a high-power ultrafast DFG system was employed by Beutler et al. [143]. The authors used an $AgGaSe_2$ crystal for DFG between the signal and idler waves of an OPO, which was synchronously pumped by a Yb-fiber oscillator–amplifier system (1.032 μm, 7.8 W). In the picosecond regime, continuous 5–18 μm tuning was achieved at 80 MHz repetition rate with average power of 140 mW at 6 μm. In the femtosecond regime at 53 MHz repetition rate, similar tunability (5–17 μm) was achieved, with the average power of 69 mW at 6 μm.

5.4.2 Intra-pulse DFG (Optical Rectification)

Mid-IR radiation can be generated by mixing spectral components within the same broadband near-IR pulse. This process is referred to as an optical rectification, or self-mixing, or intrapulse DFG. In the time domain, a few-cycle pulse creates nonlinear polarization, resulting in forward emission of long-wavelength light having an optical period on the order of the pump pulse duration. In the frequency domain, this corresponds to DFG between the spectral components within the same pump pulse [144, 145]. Thus, there is no need for a second near-IR pulse. However, the shortest mid-IR wavelength that can be produced via optical rectification is limited by the spectral span of the pump pulse, therefore extremely short (few-optical-cycle) pulses with 10–20 fs pulse duration are typically used for mid-IR optical rectification.

Using phase-matched optical rectification in GaSe crystal with broadband 20-fs pulses from a mode-locked Ti:sapphire oscillator, Kaindl et al. generated femtosecond pulses tunable from 9 to 18 μm at 88-MHz repetition rate [146]. Direct measurements of the spectrum and the pulse duration at center $\lambda \approx 11.5\,\mu m$ demonstrated nearly bandwidth-limited pulses of 140-fs duration. Huber et al. produced bandwidth-limited infrared pulses as short as 50 fs using phase-matched optical rectification of 10-fs 780-nm laser pulses in a thin (90-μm) GaSe crystal. The central frequency of the transients was continuously tunable by changing the GaSe phase-matching angle over a wide interval of wavelengths extending from 7 μm all the way to the far-IR domain, $\lambda = 3000\,\mu m$ [28].

Recently, intra-pulse DFG using 1.03-μm pump pulses was employed to generate high-average-power mid-IR pulses. The pump system included a 90-W average power thin-disc Yb:YAG oscillator followed by a nonlinear compression stage to obtain 50 W of the average power in 19-fs-long pulses at 100-MHz repetition rate. A 1-mm-thick $LiGaS_2$ (LGS) crystal was used to generate mid-IR radiation via self-mixing. The generated spectrum spanned from 6.7 to 18 μm (at −30 dB level) [147]. The duration of the mid-IR pulse, measured by electro-optical sampling, was 66 fs, which corresponds to about two cycles of the mid-IR electric field at the center wavelength of ~11 μm. The generated average mid-IR power was 100 mW, corresponding to ~0.1% conversion efficiency from the original (uncompressed) 1.03-μm pump.

Using another high-power laser source – a mode-locked Ho:YAG thin-disk oscillator ($\lambda \approx 2\,\mu m$) with 18.7 W average power and a repetition rate of 77 MHz, whose output pulses were subsequently compressed to 15 fs duration in a nonlinear fiber – Zhang et al. demonstrated, via intra-pulse DFG in a GaSe crystal, a broadband longwave mid-IR output with the spectral span of 4.4–20 μm (at −30 dB level), with an average power of 24 mW [148].

Vasilyev et al. [149] showed that relatively long, $\lambda \sim 2.5\,\mu m$, central wavelength of a few-cycle Cr^{2+}:ZnS driving source (20-fs pulse duration, 6 W average power, 78 MHz repetition rate) enabled the use of a highly nonlinear ZGP crystal for IDFG with high conversion efficiency (>3%) and output power of 148 mW, with a spectral span of 5.8–12.5 μm. An even broader spectrum (although at a smaller, 13 mW, output power) was achieved in GaSe crystal with the same pump: 4.3–16.6 μm for type I and 5.8–17.6 μm for type-II phase matching.

The main results on ultrafast DFG (including IDFG) are summarized in Table 5.6. Additional information can be found in the comprehensive review [11] and also in Section 6.2.3.

Table 5.6 Summary of the ultrafast downconversion systems.

Nonlinear crystal	Tuning range (μm)	Pump	Output parameters	Ref.
Ultrafast DFG				
$AgGaS_2$; GaSe	4–10; 6–18	1064 μm, 1 mJ, 2 ps and 1.1–1.4 μm, 1 Hz	1 nJ to 3 μJ/pulse, 1 ps, quant. eff. 2%	[22]
GaSe	5.3–18	1.35–1.6 μm, 300 mW, and 2.05–1.65 μm, 230 mW, 76 MHz	2 mW at 8.5 μm, 300 fs, quant. eff. 3.3%	[135]
GaSe	3–20	1.26–1.54 μm, 50 μJ and 2.2–1.66 μm, 30 μJ, 1 kHz	1 μJ, 54 fs at 5.5 μm, quant. eff. 10%	[23]
PPLN	3.2–4.8	1.58 μm, 170 mW, 65 fs and 1.05–1.18 μm, 11.5 mW, 40 fs, 82 MHz	1.07 mW at 3.6 μm, quant. eff. 30%	[136]
$AgGaSe_2$	9.7–14.9	1.55 μm, 135 mW and 1.7–1.85 μm, 37 MHz	1.5 μW, 420 fs	[137]
$AgGaS_2$	4.2–9	0.97–1.092 μm, and 1.24–1.26 μm, 50 mW, 300 fs, 40 MHz	640 μW at 4.5 μm	[138]
OP-GaAs	6.7–12.7	1.95 μm, 430 mW, 150 fs and ~2.5 μm, 30 mW, 72 MHz	1.3 mW at ~9 μm	[139]
GaSe	8–14	1.55 μm, 550 mW, 50 fs and 1.76–1.93 μm, 100–250 mW, 84 fs, 250 MHz	4 mW at 7.8 μm	[150]

Table 5.6 (Continued)

Nonlinear crystal	Tuning range (μm)	Pump	Output parameters	Ref.
GaSe	4–17	1.55 μm, 360 mW, 100 fs and SC 1.7–2.3 μm, 160 mW, 40 MHz	1 mW	[27]
GaSe $AgGaSe_2$	10.5–16.5	1.56–1.62 μm and 1.8–1.86 μm, total 1.45 W, 250 fs, 42 MHz	4.3 mW at 13 μm	[151]
GaSe	16–20	1.04 μm, 1.1 W and 1.105 μm, 100 mW, 500 fs, 50 MHz	1.5 mW at 18 μm	[152]
PPLN	2.9–3.6	1.05 μm, 1.2 W, 90 fs and 1.55 μm, 450 mW, 60 fs, 250 MHz	120 mW, broadband	[140]
PPLN	2.7–3.45	1.048 μm, 4 W, 210 fs and 1.57 μm, 140 mW, 60 fs, 100 MHz	150 mW, broadband	[141]
PPLN	2.7–4.2	1.04 μm, 2.3 W, 150 fs and 1.3–1.7 μm, 125 MHz	237 mW at ~3.3 μm	[153]
PPLN	2.6–5.2	1.048 μm, 4 W, 210 fs and 1.57 μm, 140 mW, 60 fs, 100 MHz	500 mW at ~3.15 μm	[142]
$AgGaSe_2$	5–18	1.38–1.98 μm, 2 W and 4.1–2.2 μm, 1.3 W, 2.1–2.6 ps, 80 MHz	140 mW at 6 μm	[143]
Ultrafast intra-pulse DFG (optical rectification)				
GaSe	9–18	0.83 μm, 100 mW, 20 fs,	Ave. power 1 μW, 140 fs at 11.5 μm	[146]
GaSe	Broadband 9–12	0.8 μm, 500 mW, 10 fs	10 μW	[154]
LGS	Broadband 6.7–18	1.03 μm, 50 W, 19 fs, 100 MHz	103 mW	[147]
GaSe	Broadband 4.4–20	2 μm, 18.7 W, 15 fs, 77 MHz	24 mW	[148]
GaSe	Broadband 6–18	2 μm, 32 W, 16 fs, 1.25 MHz	450 mW	[155]
GaSe	Broadband 4.3–16.6; 5.8–17.6	2.5 μm, 5.9 W, 20 fs, 78 MHz	13 mW	[149]

(*Continued*)

Table 5.6 (Continued)

Nonlinear crystal	Tuning range (μm)	Pump	Output parameters	Ref.
ZGP	Broadband 5.8–12.5	2.5 μm, 4.5 W, 20 fs, 78 MHz	148 mW	[149]
Picosecond OPOs				
KTP	1.54 and 3.47	1.064 μm, 29 W, 7 ps, 83 MHz	6.4 W at 3.47 μm quant. eff. 72.5%	[156]
PPLN	2.13; 2.32	1.064 μm, 30.8 W, 37 ps, 103 MHz	20.2 W at 2.13 μm 10.5 W at 2.32 μm quant. eff. 76%	[157]
PPLN	3.06–4.16	1.064 μm, 16 W, 21 ps, 81 MHz	4.6 W at 3.33 μm quant. eff. 90%	[158]
PPLN	2.3–3.5	1.035 μm, 11 W, 150 ps, 1 MHz	1.5 W (1.5 μJ at 1 MHz) quant. eff. 43%	[159]
CSP	6.09–6.58	1.064 μm, 600 mW; 1-μs macropulses at 20 Hz filled with 8.6-ps micropulses at 450 MHz	30 mW (1.5 mJ in macropulse) quant. eff. 29.5%	[160]
Femtosecond OPOs				
PPLN	1.7–5.4	790–815 nm, 850 mW, 90 ps, 81 MHz	20 mW at 5.4 μm	[161]
PPLN	2.18–3.73	790 nm, 1 W, 20 fs, 100 MHz	33 mW at 3.72 μm, min. duration 33 fs at 2.7 μm	[162]
$AgGaSe_2$	4.1–7.9	1.55 μm, 400 mW, 120 fs, 82 MHz	35 mW at 4.55 μm, 22 mW at 5.25 μm	[163]
CSP	6.54–7.19	1.029 μm, 3.7 W, 560 fs, 43 MHz	110 mW at 7.05 μm, quant. eff. 20%	[164]
CSP	6.32–7.06	1.015–1.074 μm, 0.7–1 W, 140 fs, 80 MHz	32 mW at 6.8 μm, quant. eff. 22%	[165]
OP-GaP	5–12	1.04 μm, 150 fs, 101 MHz	55 mW (5.4 μm), 7.5 mW (11.8 μm)	[166]

Table 5.6 (Continued)

Nonlinear crystal	Tuning range (μm)	Pump	Output parameters	Ref.
OPGs				
AGS	1.2–10	1.064 μm (10 mJ, 20 ps)	Quant. eff. 0.1–10%, threshold 3 GW/cm^2	[167]
ZGP	3.9–10	2.8 μm (3 mJ, 100 ps, 3 Hz)	Quant. eff. 18%, threshold 0.09 GW/cm^2	[168]
ZGP	5–11	3.15 μm (60 μJ, 2.7 ps, 10 Hz)	Quant. eff. 20%, threshold 0.1 GW/cm^2	[169]
CdSe	3.6–4.38–13	2.8 μm (3 mJ, 100 ps, 3 Hz)	Quant. eff. 10%, threshold 0.47 GW/cm^2	[170]
GaSe	3–19	2.8 μm (3 mJ, 100 ps, 3 Hz)	Quant. eff. 5%, threshold 1.1 GW/cm^2	[168]
CSP	6.15–6.73	1.064 μm (2.1 mJ, 20 ps, 5 Hz)	Quant. eff. 8.6%, threshold 0.4 GW/cm^2	[171]
OPAs				
KTP, OPCPA	3.9	1.03 μm (250 mJ, 70 ps, 20 Hz)	8 mJ, 83 fs	[172]
ZGP two-stage OPA	5	2.05 μm (1.6 mJ, 60 fs, 100 Hz)	0.2 mJ, 450 fs	[173]
ZGP, OPCPA	7	2 μm (40 mJ, 70 ps, 100 Hz)	0.2 mJ, 180 fs	[174]

5.4.3 Ultrafast OPOs

5.4.3.1 Picosecond Mode

The use of ultrashort pump pulses with an SRO operating in the synchronously pumped regime (that is, when the cavity length of the OPO exactly matches the repetition rate of the pump) can dramatically reduce the OPO threshold, down to <100 mW, in terms of the average pump power, thanks to the high peak power of ultrashort pulses that effectively determines the parametric gain [175].

In the early work on mid-IR OPOs, an efficient and highly stable optical parametric oscillator with both high peak and high average power based on a KTA crystal synchronously pumped by a diode-pumped mode-locked $Nd:YVO_4$ oscillator–amplifier system was achieved by Ruffing et al. [156]. The pump laser delivered 7-ps-long pulses at 1.064 μm with a repetition rate of 83 MHz and an average power of 29 W. The synchronously pumped OPO had a folded signal-resonant linear resonator (Figure 5.28) and contained a 15-mm-long KTA crystal, cut for type-II noncritical phase matching. The cavity-length detuning tolerance was a few 100 μm. The OPO operated at a fixed signal–idler pair of wavelengths; the average power for the signal wave (1.54 μm) was 14.6 W, while for the idler wave (3.47 μm), the power was 6.4 W. The total OPO output of 21 W corresponds to the optical conversion efficiency of 72.5%.

By the use of a 50-mm-long MgO-doped PPLN nonlinear crystal and an $Nd:YVO_4$ mode-locked laser as a pump (pulse duration 37 ps, repetition rate 103 MHz, average power 30.8 W), Qin et al. obtained 20.2 W of average power at 2.128 μm from a picosecond optical parametric oscillator operating at degeneracy [157]. The OPO pulse duration was measured to be 29 ps, and the optical conversion efficiency was 66%. In the nondegenerate mode of operation, a signal (1.97 μm) power of 12.8 W and an idler (2.32 μm) power of 10.5 W were generated, corresponding to a conversion efficiency of as high as 76%.

An elegant way of generating high-energy, mid-IR picosecond pulses at 1 MHz repetition rate was demonstrated in [159]. The authors used an OPO cavity with a length that was a small fraction (1/193) of that required to match the pump repetition rate of 1 MHz. The OPO, based on an MgO-doped PPLN

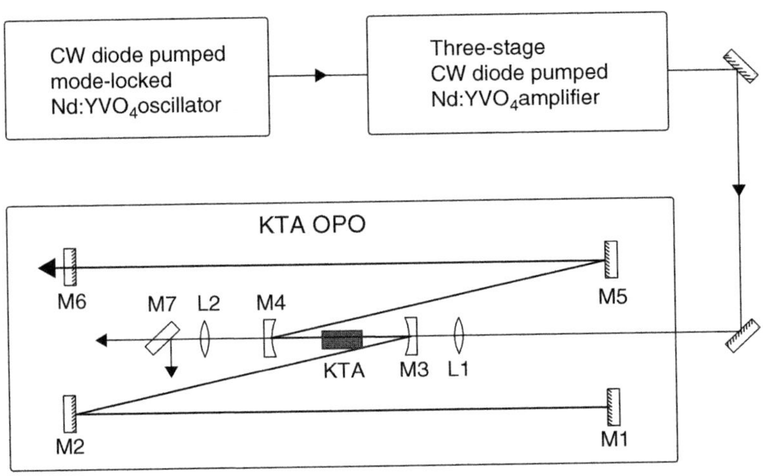

Figure 5.28 A picosecond optical parametric oscillator based on a KTA crystal, synchronously pumped by a mode-locked $Nd:YVO_4$ oscillator–amplifier system. *Source:* reproduced from figure 1 of [156], with permission of Springer.

crystal, was pumped by a fiber MOPA system delivering 11-μJ, 150-ps pulses at 1.035 μm. For the 1.55-m-long OPO cavity, the resonating near-IR signal pulses had a repetition rate that was the 193rd harmonic of the 1-MHz pump. In contrast, the repetition frequency of the nonresonant idler wave, tunable from 2.3 to 3.5 μm, was the same as that of the pump, since it was only generated in the presence of both the pump and the signal pulses. The authors demonstrated OPO idler pulses with energies as high as 1.5 μJ, with the quantum efficiency up to 43%.

5.4.3.2 Femtosecond Mode

In 1997, Burr et al. demonstrated an efficient high-repetition-rate femtosecond optical parametric oscillator that was broadly tunable in the mid-IR [161, 176]. The ring-cavity OPO in [161] was based on PPLN and was synchronously pumped at a repetition rate of 81 MHz by a mode-locked Ti:sapphire laser with 90-fs duration, average power of 850 mW, and tunable (790–815 nm) central wavelength. A small noncollinear angle between the pump and the signal was introduced (Figure 5.29), so that the long-wavelength idler beam was extracted at an angle to the resonating signal wave. The OPO cavity contained a prism sequence for group velocity dispersion compensation. (In the case of the femtosecond pump, the OPO spectral span is large enough to worry about dispersion of the roundtrip travel time for different spectral components of the resonating wave.) The noncollinear angle between pump and signal, measured internal to the crystal, varied from 1.0° to 1.6°, and PPLN length varied between 0.3 and 0.8 mm, so that the idler was tunable from 1.7 μm to beyond 5.4 μm, with a maximum average power of 200 mW near degeneracy (~1.6 μm), and more than 20 mW at $\lambda = 5.4$ μm. Interferometric autocorrelation (Figure 5.29) was used to characterize the mid-IR idler pulses, which typically had duration of 125 fs. Near degeneracy, the OPO had a pumping threshold as low as 65 mW of average pump power and the pump depletion was as high as 85%.

A PPLN-based optical parametric oscillator delivering extremely short (few-optical-cycle) mid-IR pulses was reported in [162]. The OPO was synchronously pumped by 20-fs Ti:sapphire laser pulses. By using a PPLN crystal length as short as 250 μm, and by careful dispersion management, near-transform-limited, few-cycle idler pulses tunable across the 2.18–3.73 μm range have been generated, with 33-fs duration at $\lambda = 2.7$ μm.

Using $AgGaSe_2$, a crystal with much deeper IR cutoff, as compared to PPLN, Marzenell et al. demonstrated operation of an ultrafast OPO that was tunable from 2 to 8 μm with the exception of a small gap near 3–4 μm [163]. To ensure a reduced two-photon absorption in the $AgGaSe_2$ crystal, the OPO was pumped at 1.55 μm (the output of a Ti:sapphire laser-pumped OPO). Femtosecond pulses were generated between 1.93 and 2.49 μm (signal wave), and between 4.1 and 7.9 μm (idler wave) with the average output power of

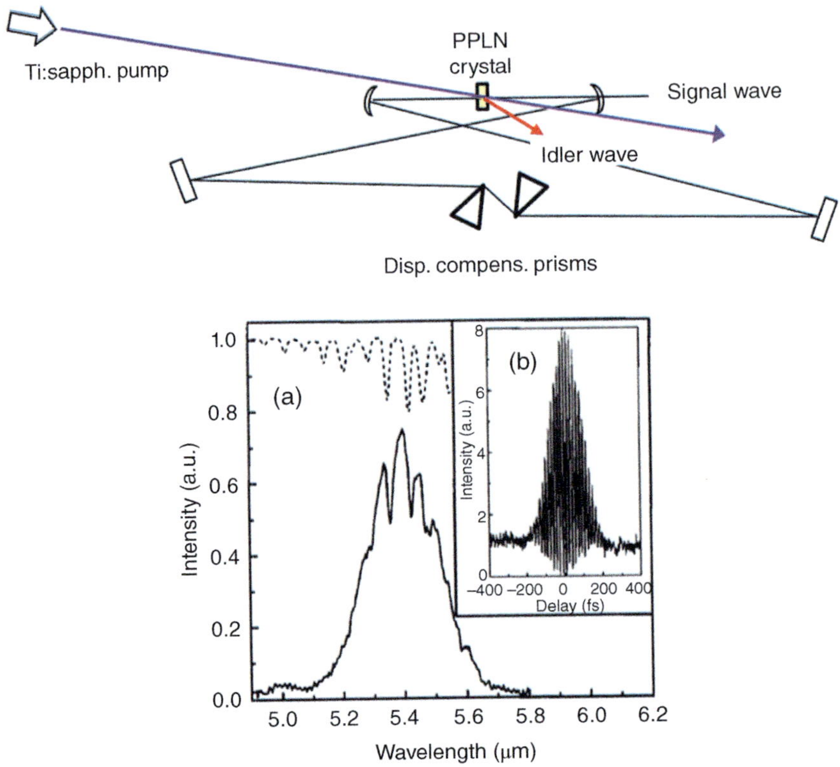

Figure 5.29 Femtosecond mid-IR PPLN OPO synchronously pumped by a Ti:sapphire laser. A prism sequence was used for group velocity dispersion compensation. The inset shows: (a) idler spectrum (solid curve) and background transmission in air (dashed curve); (b) interferometric autocorrelation of the idler at $\lambda \sim 5.4\,\mu m$. *Source:* the inset is reproduced from figure 2 of [161], with permission of OSA, The Optical Society.

67 mW (at 2.35 μm) and 35 mW (at 4.55 μm). The signal and idler pulse durations were measured to be, respectively, 230–520 and 300–640 fs. For the low-loss cavity (small output coupling), the OPO threshold was <100 mW.

Kumar et al. reported a femtosecond OPO, based on a CSP crystal. The OPO was pumped near 1-μm wavelength and delivered idler wave tunability in the range 6.3–7.2 μm [164, 165], which is important for spectroscopic and also medical applications (6.45 μm coincides with the absorption peak of amide II protein of human tissue). The maximum OPO average power reached 100 mW at $\lambda = 7.05\,\mu m$.

Maidment et al. reported a broadly tunable femtosecond optical parametric oscillator based on a new semiconductor nonlinear material, OP-GaP, which enables the production of femtosecond pulses covering a large portion (5–12 μm)

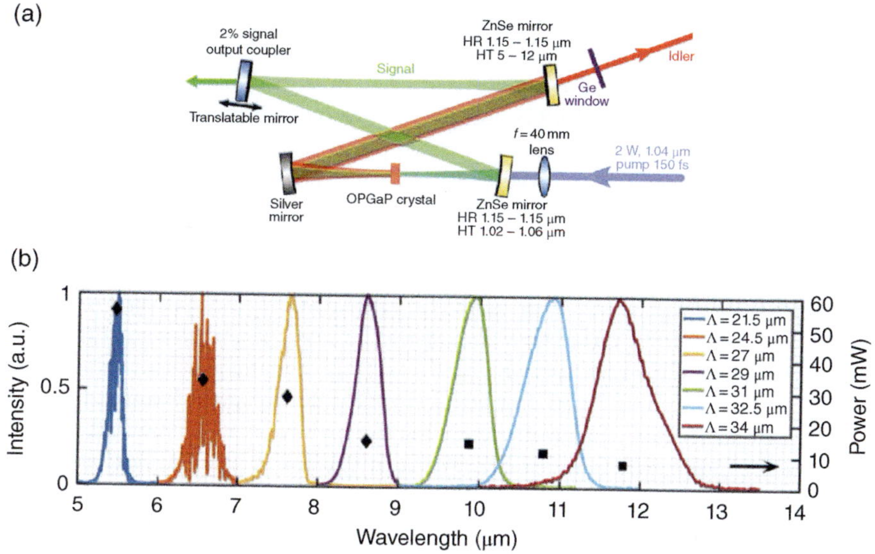

Figure 5.30 (a) Schematic of the synchronously pumped singly resonant femtosecond OPO based on the OP-GaP crystal. (b) Idler spectra obtained with seven orientation-reversal periods of the OP-GaP. Diamonds and squares indicate the measured average power. *Source:* reproduced from figures 3 and 4 of [166], with permission of OSA, The Optical Society.

of the long-wavelength mid-IR region [166]. The pump source was a 1.04-μm Yb laser delivering 150-fs pulses at 101 MHz repetition frequency. To obtain broad spectral coverage, seven 1-mm-long OP-GaP crystals, with orientation-reversal periods variable from $\Lambda = 21.5$ to $\Lambda = 34.0\,\mu m$ were used in this experiment. The idler wave spectra, centered from 5.4 to 11.8 μm, are presented in Figure 5.30, together with the average power data for each crystal. The output average power was 55 mW at 5.4 μm and 7.5 mW at 11.8 μm [166].

The main results on ultrafast OPOs are summarized in Table 5.6. Some additional information on ultrafast optical parametric sources can be found in the reviews [11, 177], and also in Sections 6.2.4 and 6.2.5.

5.4.4 Ultrafast OPGs

The principle of operation of a traveling-wave "superfluorescent" OPG is based on a single-pass high-gain ($>10^{10}$) amplification of quantum noise in a nonlinear crystal pumped by intense short laser pulses.

The main characteristics of the traveling-wave OPGs are:

- Simplicity (no cavity needed).
- Ability to produce high peak power outputs (>1 MW) in the form of a single pulse.

- Broad tunability, restricted only by the phase matching and crystal transparency.
- No buildup time. This allows generating synchronized, independently tunable pulses from different OPGs pumped by the same laser, which is attractive for time-resolved pump-probe spectroscopy.
- High pump power density, typically >1 GW/cm_2, is needed to achieve an OPG threshold.

In their early work, Elsaesser et al. reported a traveling-wave OPG operating over the whole range of 1.2–8 μm based on proustite (Ag_3AsS_3) crystal pumped by single pulses of Nd:YAG laser radiation (λ = 1.06 μm, pulse duration 21 ps) with a threshold pump intensity 6 GW/cm^2 [178]. The energy conversion efficiency amounted to 10^{-4}–10^{-2}, and the OPG-pulse spectral bandwidth was 10–40 cm^{-1}, with OPG-pulse duration being 8 ps. The OPG conversion efficiency was significantly improved (to 10^{-3}–10^{-1}) when AGS crystals were used. Two AGS crystals of 1.5 and 3 cm in length were placed in series and were pumped by picosecond Nd:YAG laser radiation [167]. The output was tunable in the range of 1.2–10 μm and the pumping threshold was 3 GW/cm^2.

Subsequently, using λ = 2.8-μm 100-ps pump pulses from a mode-locked Er,Cr:YSGG laser, traveling-wave OPGs with a number of crystals (ZGP, CdSe, and GaSe) were demonstrated [24, 168, 170, 179]. The ZGP crystal has shown the best performance in terms of the smallest pump threshold, which was only 0.09 GW/cm^2 for a 4-cm-long crystal, with the OPG tunability from 3.9 to 10 μm and quantum conversion efficiency that reached 18% (see Table 5.6) [168]. Using the same pump source and a CdSe crystal, the tuning range was 3.6–4.3 μm (signal wave) and 8–13 μm (idler wave) [170]. It has also been shown that in the important spectral range of 8–12 μm, CdSe was superior to ZGP in GaSe, in the sense of larger conversion efficiency and narrower linewidth.

When pumped by λ = 2.8-μm pulses, an OPG based on GaSe crystal holds the record for the broadest continuous tunability from 3.3 to 19 μm. This was achieved in a double-pass OPG geometry with a single z-cut angle-tuned GaSe crystal [168]. Despite the fact that GaSe can be cleaved only along the (001) plane (z-cut, $\theta = 0°$), its extremely large birefringence ($\Delta n \sim 0.35$) allows almost any conceivable three-wave interaction in its transparency range to be phase-matched. The GaSe crystal was 14-mm long and an elliptical focusing with an aspect ratio 1:20 was used for the 2.8-μm 100-ps 3-mJ pump beam – in order to keep the beam size sufficiently large in the walk-off plane (the walk-off amounted to 0.8 mm for the internal phase-matching angle of θ = 12°). The OPG threshold intensity was 1.1 GW/cm^2. At the pump intensity ~5 GW/cm^2, the quantum conversion efficiency was 5% in the 4–11 μm range (and declined at longer wavelengths) [168, 179]. Figure 5.31 shows experimental tuning curves for the type I and type II GaSe OPG.

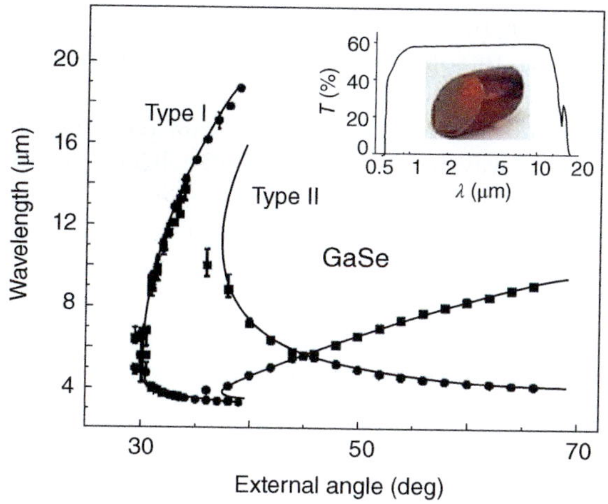

Figure 5.31 OPG tuning curves obtained with a single z-cut GaSe crystal pumped at 2.8 μm. Vertical bars represent experimental linewidths. Solid curves – calculated tuning curves. Inset: transmission spectrum of an uncoated crystal. *Source:* reproduced from figure 3 of [179], with permission of OSA, The Optical Society.

The main disadvantage of OPGs is high divergence – the result of the lack of the optical cavity. For example, both collinear and noncollinear three-wave interactions may occur in the same pumping geometry. This can also result in spectral broadening (especially for type-I phase matching). A solution to this problem, however, may be a double-crystal or a double-pass scheme with a single crystal, where the second pass filters out the off-axis components of the OPG beam produced in the first pass. Very often an OPG is used as a "seed" source for a high-power ultrafast OPA system, as in [173].

The summary of the main OPG results can be found in Table 5.6.

5.4.5 Ultrafast OPAs

Ultrafast pulses of mid-IR coherent radiation play an increasingly important role in such fields as strong-field physics and generation of coherent X-rays by high harmonic generation. An OPA approach allows producing high peak power (up to >100 GW) femtosecond pulses with superior spatial, temporal, and spectral characteristics. This is especially true when optical parametric chirped pulse amplification (OPCPA) is exploited. The concept of chirped-pulse amplification was first developed for laser amplifiers, but it was soon realized that it is also suitable for OPAs. First, the seed pulse is temporally stretched – to 100-ps–1-ns duration. This makes it possible to apply much higher pump energies from comparatively simple Q-switched lasers operating in the nanosecond regime and obtain much higher amplified pulse energies.

After optical parametric amplification, the chirped mid-IR pulse is compressed (e.g. with a diffraction grating pair), in some occasions down to few optical cycles.

Andriukaitis et al. demonstrated a compact 20-Hz repetition-rate mid-IR OPCPA system operating at a central wavelength of 3.9 μm and delivering 8-mJ pulses of 83 fs pulse duration (<7 optical cycles) [172]. The three OPAs in series were based on KTP crystals and were driven by 190-fs pulses at 1030 nm. The OPA seed pulse was generated via white-light generation in a 6-mm-long YAG crystal using a 2-μJ portion of the pump. The energy in the signal wave (at ~1.4 μm) after the final OPA stage was 65 μJ. Finally, the OPCPA section, consisting of two stages using 10-mm-long KTA crystals, pumped by an Nd system with 70-ps-long pulses with energies 250 mJ, produced 22-mJ signal (at 1.4 μm) and 13-mJ idler (at 3.9 μm) pulses. Subsequently, the idler pulses were compressed to an 83-fs duration [172].

Sanchez et al. used an OPCPA architecture to yield <8 optical-cycle duration pulses at a longer (7-μm) central wavelength at 100 Hz repetition rate [174]. The mid-IR seed at $\lambda = 7$ μm for the OPA chain was generated in a CSP crystal through DFG of the femtosecond outputs of two mutually coherent Er- and Tm-fiber lasers. An OPCPA was pumped by optically synchronized Ho:YLF laser pulses at 2 μm (40 mJ, 11 ps). The stretched, picojoule-level seed pulses (of 6-ps duration) were amplified in a chain of three consecutive noncollinear ZGP OPA stages (with ZGP crystal lengths of 5, 5, and 3 mm respectively) and subsequently compressed to get 7-μm pulses with 0.2-mJ energy and 180-fs duration. In addition, the system produced intrinsically phase-stable pulses – in terms of their carrier-envelope phase [174].

The other selected results of ultrafast OPAs are listed in Table 5.6.

5.5 Raman Frequency Converters

A Raman frequency converter is a coherent light source where an amplifier medium is based on stimulated Raman scattering (SRS), a parametric third-order NLO process [180]. SRS has been successfully utilized for several decades as a means to generate longer wavelengths from relatively mature laser sources [181]. The Raman-active medium can be either a bulk crystal, an optical fiber, a waveguide, a liquid, or a gas. Raman frequency conversion can be achieved both as a single-pass super-radiant SRS emission or as output from an oscillator (often referred to as a "Raman laser"). In fibers or waveguides, the long interaction length makes it easy to exceed the oscillation threshold, even in the CW mode [182]. Thanks to a broad Raman gain contour in glasses and other solids, Raman oscillators can be made tunable over hundreds of wavenumbers, even with a fixed-wavelength pump. Also, the ability to cascade several Raman Stokes orders in the same device opens up the possibility of obtaining mid-wavelengths from a significantly shorter wavelength pump.

5.5.1 Crystalline Raman Converters

In general, mid-IR Raman conversion is challenging, as compared to the visible and near-IR range – due to a reduced Raman gain, which is inversely proportional to the wavelength [180], and also due to limited IR transparency of materials. Currently, the most common Raman crystals include $SrWO_4$, $KGdWO_4$, $BaWO_4$, $PbWO_4$, $Ba(NO_3)_2$, YVO_4, $GdVO_4$, $LiIO_3$, and diamond. Among these crystals, $BaWO_4$ is particularly attractive due to its respectable optical properties and high Raman gain. When pumped with 10-ns pulses at $\lambda = 1.56\,\mu m$, multistage SRS to the first, second, third, and fourth Stokes-shifted components with the new wavelengths of, respectively, 1.8, 2.2, 2.75, and 3.7 μm were observed, with pump thresholds in the range of 2.7–8 mJ [183]. Also, using $BaWO_4$ crystal, a diode-end-pumped actively Q-switched intracavity Raman laser with Tm,Ho:$GdVO_4$ (2.053 μm) laser gain medium has been demonstrated [184]. At the first Stokes wavelength of 2.53 μm, the average output power of 186 mW (1 kHz, 7.8 ns) was obtained at an 802-nm diode pump power of 2.8 W, corresponding to a diode-to-Stokes optical conversion efficiency of 6.6%. The Raman lasing threshold was 2 W with respect to the diode power.

5.5.2 Fiber Raman Converters

Single-pass cascaded Raman generation was demonstrated by Rakich et al. [185] as a practical and efficient means of power transfer from telecommunications wavelengths to the mid-IR through use of conventional silica fibers and amplifiers. Silica fibers were shown to facilitate 37% efficient Raman power conversion from 1.53 to 2.15 μm and 16% from 1.53 to 2.41 μm band, using 2-ns 680-kHz-repetition-rate pulses from an all-fiber laser source. At high powers, anomalous dispersion and Kerr nonlinearities give rise to modulation instabilities, yielding significant spectral broadening and pulse distortion. For this reason, normal dispersion fibers are more desirable for controlled and maximally efficient cascaded Raman at the nanosecond time scale. The authors in [185] have shown that normal fiber dispersion can be obtained over the entire silica transparency window through proper choice of fiber cutoff wavelength λ_c and numerical aperture. The spectral evolution, with the pump power increase, of the resulting cascaded Raman process is summarized by the intensity map of Figure 5.32. As the laser power is increased, significant and controlled spectral redshifts take place through higher-order cascaded Raman power transfer, where the fundamental 1.531-μm pump wavelength is shifted in 14.7-THz (490-cm^{-1}) steps to 1.64, 1.78, 1.94, 2.14, and 2.41 μm wavelength bands. While

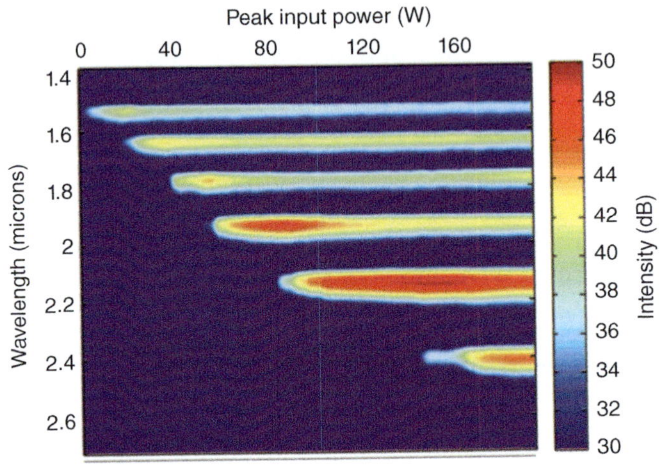

Figure 5.32 Cascaded Raman process in a silica fiber: an intensity map shows the measured fraction of the pump power (1.531 μm) and the Stokes orders at 1.64, 1.78, 1.94, 2.14, and 2.41 μm, as the pump laser power is increased. *Source:* reproduced from figure 3a of [185], with permission of OSA, The Optical Society.

the spectral width of each successive order does broaden, the generated spectral bands are very clean, showing negligible power transfer to continuum. Despite the high material losses of silica at 2.41 μm, a strong fifth Raman order is formed, producing significant (16%) power transfer from the pump.

More information on fiber-based Raman converters can be found in Section 3.2.5.

5.5.3 Silicon Raman Converters

With the Raman frequency shift of 520 cm^{-1}, silicon Raman lasers are considered as potential candidates for covering the technologically important mid-IR region of 2.2–6.5 μm, that is below the multiphonon absorption in silicon (it becomes significant above 6.5 μm). High Raman gain coefficient combined with good crystal quality, high thermal conductivity, and high optical damage threshold renders silicon a very attractive Raman medium, even when compared to the very best Raman crystals [186, 187]. Silicon-based photonic devices have an additional advantage due to their compatibility with the manufacturing infrastructure of silicon electronics, which offers the possibility of creating integrated low-loss waveguides, microcavities, and resonators.

The first silicon Raman laser was reported in the near-IR by Boyraz and Jalali [188]. The laser consisted of a silicon gain medium – a 2-cm-long rib waveguide – incorporated into a fiber loop cavity that was synchronously pumped with 30-ps 1540-nm pulses at a 25 MHz repetition rate and produced output pulses at the Stokes wavelength of 1675 nm. A lasing threshold was at 9 W of peak power, and the slope efficiency was 8.5%. This result was followed by the demonstration of the first CW silicon Raman laser (λ_{pump} = 1550 nm, λ_{Raman} = 1686 nm), with the pump threshold of 180 mW [189]. Here, the authors incorporated reverse biased PIN diodes along the waveguides to remove free carriers created by two-photon absorption in silicon, resulting in additional severe losses. Takahashi et al. reported a CW near-IR Raman silicon laser (λ_{pump} = 1428 nm, λ_{Raman} = 1543 nm) using a high-quality-factor nanocavity, yielding a device with a cavity size of less than 10 μm and a very low lasing threshold of 1 μW [190]. Using a silicon rib waveguide in a ring-cavity "racetrack" configuration, Rong et al. obtained, with a 1550-nm pump, cascaded Raman lasing at 1686 nm (first Stokes) and 1848 nm (second Stokes), with the threshold of appearance of the second Stokes of 120 mW, with respect to the coupled pump power, and with an output power at the second Stokes of 5 mW [191].

In the mid-IR, Raghunathan et al. demonstrated a silicon Raman amplifier, where an input signal at 3.39 μm wavelength was amplified by 12 dB in a 2.5-cm-long bulk silicon sample. The sample was pumped with 5-ns-long pulses at 2.88 μm with a peak pump intensity of 217 MW/cm^2 (close to the damage threshold of the sample surface) [192]. Parametric amplification and parametric generation of mid-IR light based on another third-order nonlinear effect, namely four-wave mixing (FWM) based on Kerr nonlinearity, were demonstrated independently by two groups using silicon waveguides and pump wavelengths near one-half the bandgap energy ($E \approx 0.55$ eV, $\lambda \approx 2.2$ μm), so that there is no parasitic two-photon absorption [193, 194]. In [193], a parametric gain of 25.4 dB was obtained in a 4-mm-long silicon chip using 2-ps, 28-W peak power pulses.

Using a silicon microresonator, Griffith et al. demonstrated near-octave spanning, coherent mid-IR frequency comb generation based on the interplay between FWM and coherent generation of Stokes and anti-Stokes frequencies via SRS in silicon [195]. The silicon microresonator was pumped with a monochromatic OPO source at 2.6 μm with 180 mW of power and produced a broadband output spanning 2.46–4.28 μm. The authors of [195] note that silicon suffers from three-photon absorption (3PA) in the 2.2–3.3 μm wavelength range, and the generated photocarriers with long lifetimes can cause significant mid-IR absorption. To mitigate this, the silicon microresonators were embedded in an integrated PIN diode to enable extraction of the generated free carriers. When a reverse-bias voltage of −12 V was applied to the PIN junction, carriers were swept out of the diode depletion region, thus considerably improving the device performance.

5.5.4 Diamond Raman Converters

Diamond has several properties well suited for nonlinear mid-IR generation including wide transparency window, exceptional thermal conductivity, high damage threshold, high Raman gain coefficient, and an absence, due to its crystal symmetry, of the first-order IR lattice absorption. The large bandgap of diamond (5.4 eV) ensures that there is no two-photon absorption and a large Raman shift frequency (1332 cm^{-1}) allows a big wavelength shift from the pump wavelength. However, there are two mid-IR absorption bands in bulk diamond. The three-phonon absorption band extends from 2.5 to 3.75 μm and the stronger two-phonon band is prominent from 3.75 to 6 μm.

Sabella et al. reported a pulsed mid-IR diamond Raman laser with the output that was tuned from 3.38 to 3.80 μm (that is, in the local minima between the two multi-phonon absorption bands) [196]. The 21-mm-long cavity contained an 8-mm-long uncoated diamond crystal, grown by chemical vapor deposition (CVD). As a pump, the authors used a tunable OPO in the vicinity of 2.5 μm with up to 1 mJ/pulse (10 Hz, 5 ns) and a linewidth of 0.55 cm^{-1}, which was less than the Raman linewidth of diamond (1.5 cm^{-1}). The Raman laser tuning of 3.38–3.80 μm was obtained through varying the pump wavelength in the range 2.33–2.52 μm. The highest pulse energy obtained was 80 μJ at 3.7 μm, which corresponds to 15% conversion efficiency (22% quantum efficiency) with respect to the pump entering the crystal, with a threshold near 250 μJ/pulse.

Latawiec et al. demonstrated a CW diamond Raman laser operating near 2-μm wavelength. The authors used a high-quality-factor (Q) "racetrack" diamond microresonator (path length 600 μm, waveguide width 800 nm, height 700 nm) embedded in silica on a silicon chip. By tuning the telecom-range pumping wavelength ($\lambda \sim 1.6$ μm), the output was discretely tunable over a ~ 100 nm bandwidth around the Stokes wavelength of 2 μm with the output power of 250 μW. The CW operation pump threshold power was 85 mW [197].

5.5.5 Other Raman Converters

Whispering gallery mode (WGM) optical microcavities, such as microspheres, microdisks, and microtoroids, offer good possibilities for SRS emission, thanks to their high quality factors and small mode volume. In the near-IR range, SRS threshold powers as low as 74 and 15 μW were observed, respectively in silica microtoroids [198] and CaF_2 microdisks [199]. Chalcogenide glasses such as As_2S_3 and As_2Se_3 are very advantageous for SRS emission. As_2S_3 has a high Raman gain coefficient, almost 100 times that of silica, and has a transparency window that extends up to 6 μm. Vanier et al. presented the first observation of SRS in high-quality-factor 40-μm-diameter As_2S_3 microspheres fabricated using CO_2 laser melting [200]. Raman laser emission (in both forward and backward direction) was observed with the coupled 1550-nm pump power threshold of 13 μW, with internal conversion efficiency of 10.7%. Although the Raman emission wavelength was still in the near-IR (at 1640.7 nm), considering a respectable mid-IR transparency of As_2S_3, microspheres made of this material can be regarded as good candidates for mid-IR Raman generation, e.g. via cascaded SRS process.

Using stimulated backward Raman scattering in solid para-hydrogen at $T = 4\,K$ (Raman shift $4149.7\,cm^{-1}$, linewidth 7 MHz), Kuyanov et al. developed a single-pass Raman converter operating in the mid-IR range with continuous tunability of 4.4–8 μm [201]. The device was pumped by a tunable near-IR optical parametric oscillator (OPO) with an output energy of 20 mJ, 7-ns pulse duration, and 20-Hz repetition rate. The output wavelength tuning (4.4–8 μm, spectral linewidth $0.4\,cm^{-1}$) was achieved by tuning the pump in the range 1.56–1.85 μm. The mid-IR output energies varied from 1.7 mJ at 4.4 μm to 120 μJ at 8 μm, which corresponds to quantum efficiencies of 53 and 8%, respectively. The threshold for the SRS generation was between 2 and 8 mJ; also, the backward-propagating beam contained the dominant part of the SRS radiation.

Gladyshev et al. demonstrated a Raman source at $\lambda = 4.4\,\mu m$ based on silica hollow core fiber (HCF) filled with hydrogen gas [202]. The hollow core diameter was 77 μm and the fiber cladding was formed by 10 non-contacting silica capillaries with 1.15-μm-thick walls (a "revolver" design, Figure 5.33). The calculated optical losses for the fundamental modes for the pump (1.56 μm) and the Stokes (4.4 μm) wavelengths were, respectively, 0.0025 and 0.92 dB/m. A 15-m segment of the revolver HCF was filled with molecular hydrogen at a pressure of 30 atm. The fiber ends were hermetically glued into miniature gas cells with sapphire windows for light in- and outcoupling. Under pumping by a pulsed erbium fiber laser (1.56 μm, 2 ns, 25 kHz, 1.2 W) in a single-pass

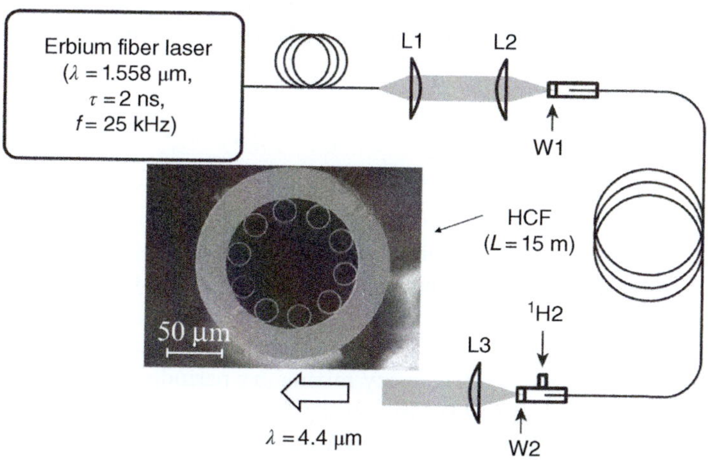

Figure 5.33 Schematic of the 4.4-μm Raman source based on a silica hollow core fiber (HCF). L1, L2, incoupling fused silica lenses; W1, W2, sapphire windows of the gas cells; L3, ZnSe collimating lens. The inset shows an electron-microscope image of the fiber cross section. *Source:* reproduced from figure 1 of [202], with permission of Turpion, Ltd and Kvantovaya Elektronika.

geometry, the Raman quantum conversion efficiency reached 15%, corresponding to 30 mW of the average power (0.6 kW peak) at $\lambda = 4.4\,\mu m$ [202].

5.6 Summary

Difference frequency generators conveniently convert radiation of near-IR lasers to the mid-IR, preserving coherence properties of the pump (e.g. narrow linewidth) and are widely used in spectroscopy owing to the broad and continuous tuning. The main drawback of DFG sources is typically low conversion efficiency and also the need for two driving laser sources. OPOs, on the other hand, need only one pump laser, are more efficient (the quantum conversion efficiency from the pump can be as high as 80%), and can produce high average power (>10 W) or high pulse energy (>100 mJ). However, this comes at the expense of the requirement for a resonant cavity. As for nonlinear crystals, QPM materials, such as PPLN and PPLT, as well as PPKTP crystal and its family, serve as a workhorse for frequency conversion in the whole wavelength range of 1–5 μm, thanks to their robustness, high nonlinearity, and affordable cost. OP-GaAs and OP-GaP, grown by epitaxy, are remarkably robust and are being used in a number of longwave (>5 μm) applications. OP-GaAs serves (similar to bulk ZGP) in many high-power OPO applications in the whole range between 3 and 10 μm. The newly developed nonlinear crystal CSP allows direct downconversion from a 1-μm pump to the 6-μm output under noncritical phase matching. Also, Raman frequency converters based on the well-developed silicon technology have the potential for covering the technologically important region of 2–6.5 μm.

References

1 Fejer, M.M., Magel, G.A., Jundt, D.H., and Byer, R.L. (1992). Quasi-phase-matched 2nd harmonic-generation: tuning and tolerances. *IEEE J. Quantum Electron.* 28: 2631.

2 Lim, E., Hertz, H., Bortz, M., and Fejer, M. (1991). Infrared radiation generated by quasi-phasematched difference-frequency-mixing in a periodically-poled lithium niobate waveguide. *Appl. Phys. Lett.* 59: 2207.

3 Yariv, A. (1989). *Quantum Electronics*, 3e. New York: Wiley.

4 Vodopyanov, K.L. (2003). Pulsed mid-IR optical parametric oscillators. In: *Solid-State Mid-Infrared Laser Sources* (ed. I.T. Sorokina and K.L. Vodopyanov), Topics Appl. Phys., vol. 89. Berlin: Springer.

5 Myers, L.E. and Bosenberg, W.R. (1997). Periodically poled lithium niobate and quasi-phase-matched optical parametric oscillators. *IEEE J. Quantum Electron.* 33: 1663.

6 Shoji, I., Kondo, T., Kitamoto, A., Shirane, M., and Ito, R. (1997). Absolute scale of second-order nonlinear-optical coefficients. *J. Opt. Soc. Am. B* 14: 2268.

7 Peltz, M., Bäderl, U., Borsutzky, A., Wallenstein, R., Hellström, J., Karlsson, H., Pasiskevicius, V., and Laurell, F. (2001). Optical parametric oscillators for high pulse energy and high average power operation based on large aperture periodically poled KTP and RTA. *Appl. Phys. B* 73: 663.

8 Roberts, D.A. (1992). Simplified characterization of uniaxial and biaxial nonlinear optical crystals: a plea for standardization of nomenclature and conventions. *IEEE J. Quantum Electron.* 28: 2057.

9 Petrov, V., Noack, F., Tunchev, I., Schunemann, P., and Zawilski, K. (2009). The nonlinear coefficient d_{36} of $CdSiP_2$. *Proc. SPIE* 7197: 719721.

10 Fossier, S., Salaün, S., Mangin, J., Bidault, O., Thenot, I., Zondy, J.J., Chen, W., Rotermund, F., Petrov, V., Petrov, P., Henningsen, J., Yelisseyev, A., Isaenko, L., Lobanov, S., Balachninaite, O., Slekys, G., and Sirutkaitis, V. (2004). Optical, vibrational, thermal, electrical, damage and phase-matching properties of lithium thioindate. *J. Opt. Soc. Am. B* 21: 1981.

11 Petrov, V. (2015). Frequency down-conversion of solid-state laser sources to the mid-infrared spectral range using non-oxide nonlinear crystals. *Prog. Quantum Electron.* 42: 1.

12 Nikogosyan, D.N. (1997). *Properties of Optical and Laser-Related Materials.* Chichester, NY: Wiley.

13 Skauli, T., Vodopyanov, K.L., Pinguet, T.J., Schober, A., Levi, O., Eyres, L.A., Fejer, M.M., Harris, J.S., Gerard, B., Becouarn, L., Lallier, E., and Arisholm, G. (2002). Measurement of the nonlinear coefficient of orientation-patterned GaAs and demonstration of highly efficient second-harmonic generation. *Opt. Lett.* 27: 628.

14 Wagner, H.P., Kühnelt, M., Langbein, W., and Hvam, J.M. (1998). Dispersion of the second-order nonlinear susceptibility in ZnTe, ZnSe, and ZnS. *Phys. Rev. B* 58: 494.

15 Zhang, H.Y., He, X.H., Shih, Y.H., Schurman, M., Feng, Z.C., and Stall, R.A. (1996). Study of nonlinear optical effects in GaN:Mg epitaxial film. *Appl. Phys. Lett.* 69: 2953.

16 Miller, R.C. (1964). Optical second harmonic generation in piezoelectric crystals. *Appl. Phys. Lett.* 5: 17.

17 Byer, R.L. (1997). Quasi-phasematched nonlinear interactions and devices. *J. Nonlinear Opt. Phys. Mater.* 6: 549.

18 Schunemann, P.G., Zawilski, K.T., Pomeranz, L.A., Creeden, D.J., and Budni, P.A. (2016). Advances in nonlinear optical crystals for mid-infrared coherent sources. *J. Opt. Soc. Am.* 33: D36.

19 Schunemann, P.G. and Zawilski, K.T. (2013). Nonlinear optical $CdSiP_2$ crystal and producing method and devices therefrom. U.S. patent 8,379,296, 19 February 2013.

20 Auerhammer, J.M. and Eliel, E.R. (1996). Frequency doubling of mid-infrared radiation in gallium selenide. *Opt. Lett.* 21: 773.

21 Eckhoff, W.C., Putnam, R.S., Wang, S.X., Curl, R.F., and Tittel, F.K. (1996). A continuously tunable long-wavelength cw IR source for high-resolution spectroscopy and trace-gas detection. *Appl. Phys. B* 63: 437.
22 Dahinten, T., Plodereder, U., Seilmeier, A., Vodopyanov, K.L., Allakhverdiev, K.R., and Ibragimov, Z.A. (1993). Infrared pulses of 1 picosecond duration tunable between 4 μm and 18 μm. *IEEE J. Quantum Electron.* 29: 2245.
23 Kaindl, R.A., Wurm, M., Reimann, K., Hamm, P., Weiner, A.M., and Woerner, M. (2000). Generation, shaping, and characterization of intense femtosecond pulses tunable from 3 to 20 μm. *J. Opt. Soc. Am. B* 17: 2086.
24 Vodopyanov, K.L. (1993). Parametric generation of tunable infrared radiation in $ZnGeP_2$ and GaSe pumped at 3 μm. *J. Opt. Soc. Am. B* 10: 1723.
25 Keilmann, F., Gohle, C., and Holzwarth, R. (2004). Time-domain mid-infrared frequency-comb spectrometer. *Opt. Lett.* 29: 1542.
26 Ruehl, A., Gambetta, A., Hartl, I., Fermann, M.E., Eikema, K.S.E., and Marangoni, M. (2012). Widely-tunable mid-infrared frequency comb source based on difference frequency generation. *Opt. Lett.* 37: 2232.
27 Keilmann, F. and Amarie, S. (2012). Mid-infrared frequency comb spanning an octave based on an Er fiber laser and difference-frequency generation. *J. Infrared Milli. Terahz Waves* 33: 479.
28 Huber, R., Brodschelm, A., Tauser, F., and Leitenstorfer, A. (2000). Generation and field-resolved detection of femtosecond electromagnetic pulses tunable up to 41 THz. *Appl. Phys. Lett.* 76: 3191.
29 Reimann, K., Smith, R.P., Weiner, A.M., Elsaesser, T., and Woerner, M. (2003). Direct field-resolved detection of terahertz transients with amplitudes of megavolts per centimeter. *Opt. Lett.* 28: 471.
30 Shi, W., Ding, Y.J., Fernelius, N., and Vodopyanov, K.L. (2002). Efficient, tunable, and coherent 0.18–5.27-THz source based on GaSe crystal. *Opt. Lett.* 27: 1454.
31 Schubert, O., Hohenleutner, M., Langer, F., Urbanek, B., Lange, C., Huttner, U., Golde, D., Meier, T., Kira, M., Koch, S.W., and Huber, R. (2014). Sub-cycle control of terahertz high-harmonic generation by dynamical Bloch oscillations. *Nat. Photon.* 8: 119.
32 Kubler, C., Huber, R., Tubel, S., and Leitenstorfer, A. (2004). Ultrabroadband detection of multi-terahertz field transients with GaSe electro-optic sensors: approaching the near infrared. *Appl. Phys. Lett.* 85: 3360.
33 Ebert, C.B., Eyres, L.A., Fejer, M.M., and Harris, J.S. (1999). MBE growth of antiphase GaAs films using GaAs/Ge/GaAs heteroepitaxy. *J. Cryst. Growth* 201: 187.
34 Koh, S., Kondo, T., Ebihara, M., Ishiwada, T., Sawada, H., Ichinose, H., Shoji, I., and Ito, R. (1999). GaAs/Ge/GaAs sublattice reversal epitaxy on GaAs (100) and (111) substrates for nonlinear optical devices. *Jpn. J. Appl. Phys.* 38: L508.
35 Eyres, L.A., Tourreau, P.J., Pinguet, T.J., Ebert, C.B., Lallier, E., Harris, J.S., Fejer, M.M., and Gerard, B. (2000). Quasi-phasematched frequency conversion

in thick all-epitaxial, orientation-patterned GaAs films, Technical Digest. In: *Advanced Solid State Lasers*. Washington, DC: Optical Society of America, Paper TuA2.

36 Lynch, C., Bliss, D.F., Zens, T., Lin, A., Harris, J.S., Kuo, P.S., and Fejer, M.M. (2008). Growth of mm-thick orientation-patterned GaAs for IR and THz generation. *J. Cryst. Growth* 310: 241.

37 Schunemann, P.G. and Setzler, S.D. (2011). Future directions in quasi-phasematched semiconductors for mid-infrared lasers. *Proc. SPIE* 7917: 79171F.

38 Kanner, G.S., Marable, M.L., Singh, N.B., Berghmans, A., Kahler, D., Wagner, B., Lin, A., Fejer, M.M., Harris, J.S., and Schepler, K.L. (2009). Optical probes of orientation-patterned ZnSe quasi-phase-matched devices. *Opt. Eng.* 48: 114201.

39 Miragliotta, J., Wickenden, D.K., Kistenmacher, T.J., and Bryden, W.A. (1993). Linear- and nonlinear-optical properties of GaN thin films. *J. Opt. Soc. Am. B* 10: 1447.

40 Abe, M., Sato, H., Shoji, I., Suda, J., Yoshimura, M., Kitaoka, Y., Mori, Y., and Kondo, T. (2010). Accurate measurement of quadratic nonlinear-optical coefficients of gallium nitride. *J. Opt. Soc. Am. B* 27: 2026.

41 Chowdhury, A., Ng, H.M., Bhardwaj, M., and Weimann, N.G. (2003). Second-harmonic generation in periodically poled GaN. *Appl. Phys. Lett.* 83: 1077.

42 Mita, S., Collazo, R., and Sitar, Z. (2009). Fabrication of a GaN lateral polarity junction by metalorganic chemical vapor deposition. *J. Cryst. Growth* 311: 3044.

43 Katayama, R., Kuge, Y., Onabe, K., Matsushita, T., and Kondo, T. (2006). Complementary analyses on the local polarity in lateral polarity-inverted GaN heterostructure on sapphire (0001) substrate. *Appl. Phys. Lett.* 89: 231910.

44 Hite, J., Twigg, M., Mastro, M., Freitas, J. Jr., Meyer, J., Vurgaftman, I., O'Connor, S., Condon, N., Kub, F., Bowman, S., and Eddy, C. Jr. (2012). Development of periodically oriented gallium nitride for non-linear optics. *Opt. Mater. Express* 2: 1203.

45 Sanders, S., Lang, R.J., Myers, L.E., Fejer, M.M., and Byer, R.L. (1996). Broadly tunable mid-IR radiation source based on difference frequency mixing of high power wavelength-tunable laser diodes in bulk periodically poled $LiNbO_3$. *Electron. Lett.* 32: 218.

46 Weibring, P., Richter, D., Walega, J.G., and Fried, A. (2007). First demonstration of a high performance difference frequency spectrometer on airborne platforms. *Opt. Express* 15: 13476.

47 Weibring, P., Richter, D., Fried, A., Walega, J.G., and Dyroff, C. (2006). Ultra-high-precision mid-IR spectrometer II: system description and spectroscopic performance. *Appl. Phys. B* 85: 207.

48 Guha, S., Barnes, J.O., and Gonzalez, L.P. (2014). Multiwatt-level continuous-wave midwave infrared generation using difference frequency mixing in periodically poled MgO-doped lithium niobate. *Opt. Lett.* 39: 5018.

49 Tadanaga, O., Yanagawa, T., Nishida, Y., Miyazawa, H., Magari, K., Asobe, M., and Suzuki, H. (2006). Efficient 3-μm difference frequency generation using direct-bonded quasi-phase-matched $LiNbO_3$ ridge waveguides. *Appl. Phys. Lett.* 88: 061101.

50 Richter, D., Weibring, P., Fried, A., Tadanaga, O., Nishida, Y., Asobe, M., and Suzuki, H. (2007). High-power, tunable difference frequency generation source for absorption spectroscopy based on a ridge waveguide periodically poled lithium niobate crystal. *Opt. Express* 15: 564.

51 Denzer, W., Hancock, G., Hutchinson, A., Munday, M., Peverall, R., and Ritchie, G.A.D. (2007). Mid-infrared generation and spectroscopy with a PPLN ridge waveguide. *Appl. Phys. B* 86: 437.

52 Asobe, M., Tadanaga, O., Yanagawa, T., Umeki, T., Nishida, Y., and Suzuki, H. (2008). High-power mid-infrared wavelength generation using difference frequency generation in damage-resistant $Zn:LiNbO_3$ waveguide. *Electron. Lett.* 44: 288.

53 Levi, O., Pinguet, T.J., Skauli, T., Eyres, L.A., Parameswaran, K.R., Harris, J.S. Jr., Fejer, M.M., Kulp, T.J., Bisson, S.E., Gerard, B., Lallier, E., and Becouarn, L. (2002). Difference frequency generation of 8-μm radiation in orientation-patterned GaAs. *Opt. Lett.* 27: 2091.

54 Bisson, S.E., Kulp, T.J., Levi, O., Harris, J.S., and Fejer, M.M. (2006). Long-wave IR chemical sensing based on difference frequency generation in orientation-patterned GaAs. *Appl. Phys. B* 85: 199.

55 Vasilyev, S., Schiller, S., Nevsky, A., Grisard, A., Faye, D., Lallier, E., Zhang, Z., Boyland, A.J., Sahu, J.K., Ibsen, M., and Clarkson, W.A. (2008). Broadly tunable single-frequency cw mid-infrared source with milliwatt-level output based on difference-frequency generation in orientation-patterned GaAs. *Opt. Lett.* 33: 1413.

56 Sowade, R., Breunig, I., Mayorga, I.C., Kiessling, J., Tulea, C., Dierolf, V., and Buse, K. (2009). Continuous-wave optical parametric terahertz source. *Opt. Express* 17: 22303.

57 Smith, R.G., Geusic, J.E., Levinstein, H., Rubin, J.J., Singh, S., and Van Uiter, L. (1968). Continuous optical parametric oscillation in $Ba_2NaNb_5O_{15}$. *Appl. Phys. Lett.* 12: 308.

58 Bosenberg, W.R., Drobshoff, A., Alexander, J.I., Myers, L.E., and Byer, R.L. (1996). Continuous-wave singly resonant optical parametric oscillator based on periodically poled $LiNbO_3$. *Opt. Lett.* 21: 713.

59 Bosenberg, W.R., Drobshoff, A., Alexander, J.I., Myers, L.E., and Byer, R.L. (1996). 93% pump depletion, 3.5-W continuous-wave, singly resonant optical parametric oscillator. *Opt. Lett.* 21: 1336.

60 Kumar, S.C., Das, R., Samanta, G., and Ebrahim-Zadeh, M. (2011). Optimally-output-coupled, 17.5 W, fiber-laser-pumped continuous-wave optical parametric oscillator. *Appl. Phys. B* 102: 31.

61 Kumar, S.C. and Ebrahim-Zadeh, M. (2011). High-power, continuous-wave, mid-infrared optical parametric oscillator based on MgO:sPPLT. *Opt. Lett.* 36: 2578.

62 Henderson, A. and Stafford, R. (2006). Low threshold, singly-resonant CW OPO pumped by an all-fiber pump source. *Opt. Express* 14: 767.

63 Popp, A., Müller, F., Kühnemann, F., Schiller, S., von Basum, G., Dahnke, H., Hering, P., and Mürtz, M. (2002). Ultra-sensitive mid-infrared cavity leak-out spectroscopy using a cw optical parametric oscillator. *Appl. Phys. B* 75: 751.

64 Ngai, A.K.Y., Persijn, S.T., von Basum, G., and Harren, F.J.M. (2006). Automatically tunable continuous-wave optical parametric oscillator for high-resolution spectroscopy and sensitive trace-gas detection. *Appl. Phys. B* 85: 173.

65 Lindsay, I., Adhimoolam, B., Groß, P., Klein, M., and Boller, K. (2005). 110 GHz rapid, continuous tuning from an optical parametric oscillator pumped by a fiber-amplified DBR diode laser. *Opt. Express* 13: 1234.

66 Siltanen, M., Vainio, M., and Halonen, L. (2010). Pump-tunable continuous-wave singly resonant optical parametric oscillator from 2.5 to 4.4 μm. *Opt. Express* 18: 14087.

67 Scheidt, M., Beier, B., Boller, K.-J., and Wallenstein, R. (1997). Frequency-stable operation of a diode-pumped continuous-wave $RbTiOAsO_4$ optical parametric oscillator. *Opt. Lett.* 22: 1287.

68 Turnbull, G.A., McGloin, D., Lindsay, I.D., Ebrahimzadeh, M., and Dunn, M.H. (2000). Extended mode-hop-free tuning by use of a dual-cavity, pump-enhanced optical parametric oscillator. *Opt. Lett.* 25: 341.

69 Lindsay, I.D., Petriokes, C., Dunn, M.H., and Ebrahimzadeh, M. (2001). Continuous-wave pump-enhanced singly resonant optical parametric oscillator pumped by an extended-cavity diode laser. *Appl. Phys. Lett.* 78: 871.

70 Stothard, D.J.M., Lindsay, I.D., and Dunn, M.H. (2004). Continuous-wave pump-enhanced optical parametric oscillator with ring resonator for wide and continuous tuning of single-frequency radiation. *Opt. Express* 12: 502.

71 Rihan, A., Andrieux, E., Zanon-Willette, T., Briaudeau, S., Himbert, M., and Zondy, J.-J. (2011). A pump-resonant signal-resonant optical parametric oscillator for spectroscopic breath analysis. *Appl. Phys. B* 102: 367.

72 Schunemann, P.G., Pomeranz, L.A., Setzler, S.D., Jones, C.W., and Budni, P.A. (2013). CW mid-IR OPO based on OP-GaAs, Technical Digest. In: *The European Conference on Lasers and Electro-Optics*. Washington, DC: Optical Society of America, Paper JSII_2_3.

73 Meisenheimer, S.-K., Fürst, J.U., Buse, K., and Breunig, I. (2017). Continuous-wave optical parametric oscillation tunable up to an 8 μm wavelength. *Optica* 4: 189.

74 Breunig, I., Haertle, D., and Buse, K. (2011). Continuous-wave optical parametric oscillators: recent developments and prospects. *Appl. Phys. B* 105: 99.

75 Fürst, J.U., Strekalov, D.V., Elser, D., Aiello, A., Andersen, U.L., Marquardt, C., and Leuchs, G. (2010). Low-threshold optical parametric oscillations in a whispering gallery mode resonator. *Phys. Rev. Lett.* 105: 263904.

76 Beckmann, T., Linnenbank, H., Steigerwald, H., Sturman, B., Haertle, D., Buse, K., and Breunig, I. (2011). Highly tunable low-threshold optical parametric oscillation in radially poled whispering gallery resonators. *Phys. Rev. Lett.* 106: 143903.

77 Werner, C.S., Buse, K., and Breunig, I. (2015). Continuous-wave whispering-gallery optical parametric oscillator for high-resolution spectroscopy. *Opt. Lett.* 40: 772.

78 Hanna, D.C., Rampal, V.V., and Smith, R.C. (1973). Tunable infrared down-conversion in silver thiogallate. *Opt. Commun.* 8: 151.

79 Seymour, R.J. and Zernike, F. (1976). Infrared radiation tunable from 5.5 to 18.3 μm generated by mixing in $AgGaS_2$. *Appl. Phys. Lett.* 29: 705.

80 Kato, K. (1984). High-power difference-frequency generation at 5–11 μm in $AgGaS_2$. *IEEE J. Quantum Electron.* QE-20: 698.

81 Haidar, S., Nakamura, K., Niwa, E., Masumoto, K., and Ito, H. (1999). Mid-infrared (5–12-μm) and limited (5.5–8.5-μm) single-knob tuning generated by difference-frequency mixing in single-crystal $AgGaS_2$. *Appl. Opt.* 38: 1798.

82 Shi, W. and Ding, Y. (2004). A monochromatic and high-power terahertz source tunable in the ranges of 2.7–38.4 and 58.2–3540 μm for variety of potential applications. *Appl. Phys. Lett.* 84: 1635.

83 Vodopyanov, K.L. and Schunemann, P.G. (1998). Efficient difference frequency generation of 7–20-μm radiation in $CdGeAs_2$. *Opt. Lett.* 23: 1096.

84 Kung, A.H. (1994). Narrowband mid-infrared generation using $KTiOAsO_4$. *Appl. Phys. Lett.* 65: 1082.

85 Miyamoto, Y., Hara, H., Masuda, T., Hiraki, T., Sasao, N., and Uetake, S. (2017). Injection-seeded tunable mid-infrared pulses generated by difference frequency mixing. *Jpn. J. Appl. Phys.* 56: 032101.

86 Belden, P., Chen, D.W., and Di Teodoro, F. (2015). Watt-level, gigahertz-linewidth difference-frequency generation in PPLN pumped by a nanosecond-pulse fiber laser source. *Opt. Lett.* 40: 958.

87 Fischer, C. and Sigrist, M.W. (2003). Mid-IR difference frequency generation. In: *Solid-state Mid-infrared Laser Sources* (ed. I.T. Sorokina and K.L. Vodopyanov). Topics Appl. Phys., vol. 89. Berlin: Springer.

88 Brosnan, S.J. and Byer, R.L. (1979). Optical parametric oscillator threshold and linewidth studies. *IEEE J. Quantum Electron.* QE-15: 415.

89 Hanna, D.C., Luther-Davies, B., and Smith, R.C. (1973). Singly resonant proustite parametric oscillator tuned from 1.22 to 8.5 μm. *Appl. Phys. Lett.* 22: 440.

90 Myers, L.E., Eckardt, R.C., Fejer, M.M., Byer, R.L., Bosenberg, W.R., and Pierce, J.W. (1995). Quasi-phase-matched optical parametric oscillators in bulk periodically poled $LiNbO_3$. *J. Opt. Soc. Am. B* 12: 2102.

91 Myers, L.E., Eckardt, R.C., Fejer, M.M., Byer, R.L., and Bosenberg, W.R. (1996). Multi-grating quasi-phase-matched optical parametric oscillators in periodically poled $LiNbO_3$. *Opt. Lett.* 21: 591.

92 Fan, Y.X., Eckardt, R.L., Byer, R.L., Route, R.K., and Feigelson, R.S. (1984). $AgGaS_2$ infrared parametric oscillator. *Appl. Phys. Lett.* 45: 313.

93 Vodopyanov, K.L., Maffetone, J.P., Zwieback, I., and Ruderman, W. (1999). $AgGaS_2$ optical parametric oscillator continuously tunable from 3.9 to 11.3 μm. *Appl. Phys. Lett.* 75: 1204.

94 Eckardt, R.C., Fan, Y.X., Byer, R.L., Marquardt, C.L., Storm, M.E., and Esterowitz, L. (1986). Broadly tunable infrared parametric oscillator using $AgGaSe_2$. *Appl. Phys. Lett.* 49: 608.

95 Quarles, G.J., Marquardt, C.L., and Esterowitz, L. (1990). 2-μm pumped $AgGaSe_2$ with continuous tuning 2.49–12.05 μm. In: Proc. LEOS. Boston, MA, Paper ELT7.1.

96 Chandra, S., Allik, T.H., Catella, G., Utano, R., and Hutchinson, J.A. (1997). Continuously tunable, 6–14 μm silver-gallium selenide optical parametric oscillator pumped at 1.57 μm. *Appl. Phys. Lett.* 71: 584.

97 Vodopyanov, K.L., Ganikhanov, F., Maffetone, J.P., Zwieback, I., and Ruderman, W. (2000). $ZnGeP_2$ optical parametric oscillator with 3.8–12.4-μm tunability. *Opt. Lett.* 25: 841.

98 Vodopyanov, K.L. (2001). OPOs target the longwave infrared. *Laser Focus World* 37: 225.

99 Vodopyanov, K.L. and Schunemann, P.G. (2003). Broadly tunable noncritically phase-matched ZGP optical parametric oscillator with a 2-microjoule pump threshold. *Opt. Lett.* 28: 441.

100 Vodopyanov, K.L., Levi, O., Kuo, P.S., Pinguet, T.J., Harris, J.S., Fejer, M.M., Gerard, B., Becouarn, L., and Lallier, E. (2004). Optical parametric oscillation in quasi-phase-matched GaAs. *Opt. Lett.* 29: 1912.

101 Vodopyanov, K.L., Levi, O., Kuo, P.S., Pinguet, T.J., Harris, J.S., Fejer, M.M., Gerard, B., Becouarn, L., and Lallier, E. (2004). Optical parametric oscillator based on orientation-patterned GaAs. *Proc. SPIE* 5620: 63–69.

102 Skauli, T., Kuo, P.S., Vodopyanov, K.L., Pinguet, T.J., Levi, O., Eyres, L.A., Harris, J.S., and Fejer, M.M. (2003). Determination of GaAs refractive index and its temperature dependence, with application to quasi-phasematched nonlinear optics. *J. Appl. Phys.* 94: 6447.

103 Vodopyanov, K.L., Makasyuk, I., and Schunemann, P.G. (2014). Grating tunable 4–14 μm GaAs optical parametric oscillator pumped at 3 μm. *Opt. Express* 22: 4131.

104 Zawilski, K.T., Schunemann, P.G., Pollak, T.C., Zelmon, D.E., Fernelius, N.C., and Hopkins, F.K. (2010). Growth and characterization of large $CdSiP_2$ single crystals. *J. Cryst. Growth* 312: 1127.

105 Petrov, V., Schunemann, P.G., Zawilski, K.T., and Pollak, T.M. (2009). Noncritical singly resonant optical parametric oscillator operation near 6.2 μm based on a $CdSiP_2$ crystal pumped at 1064 nm. *Opt. Lett.* 34: 2399.

106 Marchev, G., Tyazhev, A., Petrov, V., Schunemann, P.G., Zawilski, K.T., Stöppler, G., and Eichhorn, M. (2012). Optical parametric generation in $CdSiP_2$ at 6.125 μm pumped by 8 ns long pulses at 1064 nm. *Opt. Lett.* 37: 740.

107 Kostyukova, N.Y., Boyko, A.A., Badikov, V., Badikov, D., Shevyrdyaeva, G., Panyutin, V., Marchev, G.M., Kolker, D.B., and Petrov, V. (2016). Widely tunable in the mid-IR $BaGa_4Se_7$ optical parametric oscillator pumped at 1064 nm. *Opt. Lett.* 41: 3667.

108 Richman, B.A., Aniolek, K.W., Kulp, T.J., and Bisson, S.E. (2000). Continuously tunable, single-longitudinal-mode, pulsed mid-infrared optical parametric oscillator based on periodically poled lithium niobate. *J. Opt. Soc. Am. B* 17: 1233.

109 Ganikhanov, F., Caughey, T., and Vodopyanov, K.L. (2001). Narrow-linewidth middle-infrared $ZnGeP_2$ optical parametric oscillator. *J. Opt. Soc. Am. B* 18: 818.

110 Scherrer, B., Ribet, I., Godard, A., Rosencher, E., and Lefebvre, M. (2000). Dual-cavity doubly resonant optical parametric oscillators: demonstration of pulsed single-mode operation. *J. Opt. Soc. Am. B* 17: 1716.

111 Drag, C., Desormeaux, A., Lefebvre, M., and Rosencher, E. (2002). Entangled cavity optical parametric oscillator for mid-infrared pulsed single-longitudinal-mode operation. *Opt. Lett.* 27: 1238.

112 Hardy, B., Berrou, A., Guilbaud, S., Raybaut, M., Godard, A., and Lefebvre, M. (2011). Compact, single-frequency, doubly resonant optical parametric oscillator pumped in an achromatic phase-adapted double-pass geometry. *Opt. Lett.* 36: 678.

113 Barria, J.B., Roux, S., Dherbecourt, J.-B., Raybaut, M., Melkonian, J.-M., Godard, A., and Lefebvre, M. (2013). Microsecond fiber laser pumped, single-frequency optical parametric oscillator for trace gas detection. *Opt. Lett.* 38: 2165.

114 Clément, Q., Melkonian, J.-M., Dherbecourt, J.-B., Raybaut, M., Grisard, A., Lallier, E., Gérard, B., Faure, B., Souhaité, G., and Godard, A. (2015). Longwave infrared, single-frequency, tunable, pulsed optical parametric oscillator based on orientation-patterned GaAs for gas sensing. *Opt. Lett.* 40: 2676.

115 Cheung, E., Palese, S., Injeyan, H., Hoefer, C., Hilyard, R., Komine, H., Berg, J., and Bosenberg, W. (1999). High power conversion to mid-IR using KTP and ZGP OPOs, Technical Digest. In: *Advanced Solid-State Lasers*. Washington, DC: Optical Society of America, Paper WC1.

116 Budni, P.A., Pomeranz, L.A., Lemons, M.L., Schunemann, P.G., Pollak, T.M., and Chicklis, E.P. (1998). 10 W mid-IR holmium pumped $ZnGeP_2$ OPO, Technical Digest. In: *Advanced Solid State Lasers*. Washington, DC: Optical Society of America, Paper FC1.

117 Lippert, E., Fonnum, H., Arisholm, G., and Stenersen, K. (2010). A 22-watt mid-infrared optical parametric oscillator with V-shaped 3-mirror ring resonator. *Opt. Express* 18: 26475.

118 Wu, R., Lai, K.S., Lau, W.E., Wong, H.F., Lim, Y.L., Lim, K.W., and Chia, L.C.L. (2002). A novel laser integrated with a coupled tandem OPO configuration, Technical Digest. In: *Conference on Lasers and Electro-Optics*. Washington, DC: Optical Society of America, Paper CTuD6.

119 Hemming, A., Richards, J., Davidson, A., Carmody, N., Bennetts, S., Simakov, N., and Haub, J. (2013). 99-W mid-IR operation of a ZGP OPO at 25% duty cycle. *Opt. Express* 21: 10062.

120 Kieleck, C., Eichhorn, M., Hirth, A., Faye, D., and Lallier, E. (2009). High-efficiency 20–50 kHz mid-infrared orientation-patterned GaAs optical parametric oscillator pumped by a 2-μm holmium laser. *Opt. Lett.* 34: 262.

121 Wueppen, J., Nyga, S., Jungbluth, B., and Hoffmann, D. (2016). 1.95 μm-pumped OP-GaAs optical parametric oscillator with 10.6 μm idler wavelength. *Opt. Lett.* 41: 4225.

122 Vysniauskas, G., Burns, D., and Bente, E. (2002). Development of a nanosecond high energy KTA OPO system operating 2.9 μm, Technical Digest. In: *Conference on Lasers and Electro-Optics*. Washington, DC: Optical Society of America, Paper CWA28.

123 Mennerat, G. and Kupecek, P. (1998). High-energy narrow-linewidth tunable source in the mid infrared, Technical Digest. In: *Advanced Solid-State Lasers*. Washington, DC: Optical Society of America, Paper FC13.

124 Ishizuki, H. and Taira, T. (2010). High energy quasi-phase matched optical parametric oscillation using mg-doped congruent $LiTaO_3$ crystal. *Opt. Express* 18: 253.

125 Allik, T.H., Chandra, S., Rines, D.M., Schunemann, P.G., Hutchinson, J.A., and Utano, R. (1997). Tunable 7–12-μm optical parametric oscillator using a Cr,Er:YSGG laser to pump CdSe and $ZnGeP_2$ crystals. *Opt. Lett.* 22: 597.

126 Dergachev, A., Armstrong, D., Smith, A., Drake, T., and Dubois, M. (2007). 3.4-μm ZGP RISTRA nanosecond optical parametric oscillator pumped by a 2.05-μm ho:YLF MOPA system. *Opt. Express* 15: 14404.

127 Stöppler, G., Schellhorn, M., and Eichhorn, M. (2013). Ho^{3+}:LLF MOPA pumped RISTRA ZGP OPO at 3–5 μm. *Proc. SPIE* 8604: 86040I.

128 Haakestad, M., Fonnum, H., and Lippert, E. (2014). Mid-infrared source with 0.2 J pulse energy based on nonlinear conversion of Q-switched pulses in $ZnGeP_2$. *Opt. Express* 22: 8556.

129 Hansson, G., Karlsson, H., and Laurell, F. (2001). Unstable resonator optical parametric oscillator based on quasi-phase-matched $RbTiOAsO_4$. *Appl. Opt.* 40: 5446.

130 Ishizuki, H. and Taira, T. (2005). High-energy quasi-phase-matched optical parametric oscillation in a periodically poled MgO:$LiNbO_3$ device with a 5 mm × 5 mm aperture. *Opt. Lett.* 30: 2918.

131 Saikawa, J., Miyazaki, M., Fujii, M., Ishizuki, H., and Taira, T. (2008). High-energy, broadly tunable, narrow-bandwidth mid-infrared optical parametric system pumped by quasi-phase-matched devices. *Opt. Lett.* 33: 1699.

132 Smith, A.V. and Armstrong, D.J. (2002). Nanosecond optical parametric oscillator with 90° image rotation: design and performance. *J. Opt. Soc. Am. B* 19: 1801.

133 Arbore, M.A. and Fejer, M.M. (1997). Singly resonant optical parametric oscillation in periodically poled lithium niobate waveguides. *Opt. Lett.* 22: 151.

134 Oron, M.B., Blau, P., Pearl, S., and Katz, M. (2012). Optical parametric oscillation in orientation patterned GaAs waveguides. *Proc. SPIE* 8240: 82400C-11.

135 Ehret, S. and Schneider, H. (1998). Generation of subpicosecond infrared pulses tunable between 5.2 μm and 18 μm at a repetition rate of 76 MHz. *Appl. Phys. B* 66: 27.

136 Erny, C., Moutzouris, K., Biegert, J., Kühlke, D., Adler, F., Leitenstorfer, A., and Keller, U. (2007). Mid-infrared difference-frequency generation of ultrashort pulses tunable between 3.2 and 4.8 μm from a compact fiber source. *Opt. Lett.* 32: 1138.

137 Winters, D.G., Schlup, P., and Bartels, R.A. (2010). Subpicosecond fiber-based soliton-tuned mid-infrared source in the 9.7–14.9 μm wavelength region. *Opt. Lett.* 35: 2179.

138 Yao, Y. and Knox, W.H. (2013). Broadly tunable femtosecond mid-infrared source based on dual photonic crystal fibers. *Opt. Express* 21: 26612.

139 Phillips, C.R., Jiang, J., Mohr, C., Lin, A.C., Langrock, C., Snure, M., Bliss, D., Zhu, M., Hartl, I., Harris, J.S., Fermann, M.E., and Fejer, M.M. (2012). Widely tunable midinfrared difference frequency generation in orientation-patterned GaAs pumped with a femtosecond Tm-fiber system. *Opt. Lett.* 37: 2928.

140 Zhu, F., Hundertmark, H., Kolomenskii, A.A., Strohaber, J., Holzwarth, R., and Schuessler, H.A. (2013). High-power mid-infrared frequency comb source based on a femtosecond Er:fiber oscillator. *Opt. Lett.* 38: 2360.

141 Meek, S.A., Poisson, A., Guelachvili, G., Hänsch, T.W., and Picqué, N. (2014). Fourier transform spectroscopy around 3 μm with a broad difference frequency comb. *Appl. Phys. B* 114: 573.

142 Cruz, F.C., Maser, D.L., Johnson, T., Ycas, G., Klose, A., Giorgetta, F.R., Coddington, I., and Diddams, S.A. (2015). Mid-infrared optical frequency combs based on difference frequency generation for molecular spectroscopy. *Opt. Express* 23: 26814.

143 Beutler, M., Rimke, I., Büttner, E., Farinello, P., Agnesi, A., Badikov, V., Badikov, D., and Petrov, V. (2015). Difference-frequency generation of ultrashort pulses in the mid-IR using Yb-fiber pump systems and $AgGaSe_2$. *Opt. Express* 23: 2730.

144 Bonvalet, A., Joffre, M., Martin, J.L., and Migus, A. (1995). Generation of ultrabroadband femtosecond pulses in the mid-infrared by optical rectification of 15 fs light pulses at 100 MHz repetition rate. *Appl. Phys. Lett.* 67: 2907.

145 Nahata, A., Weling, A.S., and Heinz, T.F. (1996). A wideband coherent terahertz spectroscopy system using optical rectification and electro-optic sampling. *Appl. Phys. Lett.* 69: 2321.

146 Kaindl, R.A., Smith, D.C., Joschko, M., Hasselbeck, M.P., Woerner, M., and Elsaesser, T. (1998). Femtosecond infrared pulses tunable from 9 to 18 μm at an 88-MHz repetition rate. *Opt. Lett.* 23: 861.

147 Pupeza, I., Sánchez, D., Zhang, J., Lilienfein, N., Seidel, M., Karpowicz, N., Paasch-Colberg, T., Znakovskaya, I., Pescher, M., Schweinberger, W., Pervak, V., Fill, E., Pronin, O., Wei, Z., Krausz, F., Apolonski, A., and Biegert, J. (2015). High-power sub-two-cycle mid-infrared pulses at 100 MHz repetition rate. *Nat. Photon.* 9: 721.

148 Zhang, J., Mak, K., Nagl, N., Seidel, M., Bauer, D., Sutter, D., Pervak, V., Krausz, F., and Pronin, O. (2018). Multi-mW, few-cycle mid-infrared continuum spanning from 500 to 2250 cm^{-1}. *Light Sci. Appl.* 7: 17180.

149 Vasilyev, S., Moskalev, I.S., Smolski, V.O., Peppers, J.M., Mirov, M., Muraviev, A.V., Zawilski, K., Schunemann, P.G., Mirov, S.B., Vodopyanov, K.L., and Gapontsev, V.P. (2019). Super-octave longwave mid-infrared coherent transients produced by optical rectification of few-cycle 2.5-μm pulses. *Optica* 6: 111.

150 Gambetta, A., Coluccelli, N., Cassinerio, M., Gatti, D., Laporta, P., Galzerano, G., and Marangoni, M. (2013). Milliwatt-level frequency combs in the 8–14 μm range via difference frequency generation from an Er:fiber oscillator. *Opt. Lett.* 38: 1155.

151 Hegenbarth, R., Steinmann, A., Sarkisov, S., and Giessen, H. (2012). Milliwatt-level mid-infrared (10.5–16.5 μm) difference frequency generation with a femtosecond dual-signal-wavelength optical parametric oscillator. *Opt. Lett.* 37: 3513.

152 Hajialamdari, M. and Strickland, D. (2012). Tunable mid-infrared source from an ultrafast two-color Yb:fiber chirped-pulse amplifier. *Opt. Lett.* 37: 3570.

153 Soboń, G., Martynkien, T., Mergo, P., Rutkowski, L., and Foltynowicz, A. (2017). High-power frequency comb source tunable from 2.7 to 4.2 μm based on difference frequency generation pumped by an Yb-doped fiber laser. *Opt. Lett.* 42: 1748.

154 Schliesser, A., Brehm, M., and Keilmann, F. (2005). Frequency-comb infrared spectrometer for rapid, remote chemical sensing. *Opt. Express* 13: 9029.

155 Gaida, C., Gebhardt, M., Huermann, T., Stutzki, F., Jauregui, C., Antonio-Lopez, J., Schülzgen, A., Amezcua-Correa, R., Tünnermann, A., Pupeza, I., and Limpert, J. (2018). Watt-scale super-octave mid-infrared intrapulse difference frequency generation. *Light Sci. Appl.* 7: 94.

156 Ruffing, B., Nebel, A., and Wallenstein, R. (1998). All-solid-state cw mode-locked picosecond $KTiOAsO_4$ (KTA) optical parametric oscillator. *Appl. Phys. B* 67: 537.

157 Qin, Z., Xie, G., Ge, W., Yuan, P., and Qian, L. (2015). Over 20-W mid-infrared picosecond optical parametric oscillator. *IEEE Photon. J.* 7: 1400506.

158 Kokabee, O., Esteban-Martin, A., and Ebrahim-Zadeh, M. (2010). Efficient, high-power, ytterbium-fiber-laser-pumped picosecond optical parametric oscillator. *Opt. Lett.* 35: 3210.

159 Xu, L., Chan, H.-Y., Alam, S.-U., Richardson, D.J., and Shepherd, D.P. (2015). Fiber-laser-pumped, high-energy, mid-IR, picosecond optical parametric oscillator with a high-harmonic cavity. *Opt. Lett.* 40: 3288.
160 Kumar, S.C., Agnesi, A., Dallocchio, P., Pirzio, F., Reali, G., Zawilski, K.T., Schunemann, P.G., and Ebrahim-Zadeh, M. (2011). Compact, 1.5 mJ, 450 MHz, $CdSiP_2$ picosecond optical parametric oscillator near 6.3 μm. *Opt. Lett.* 36: 3236.
161 Burr, K.C., Tang, C.L., Arbore, M.A., and Fejer, M.M. (1997). Broadly tunable mid-infrared femtosecond optical parametric oscillator using all-solid-state-pumped periodically poled lithium niobate. *Opt. Lett.* 22: 1458.
162 Kumar, S.C., Esteban-Martin, A., Ideguchi, T., Yan, M., Holzner, S., Hänsch, T.W., Picqué, N., and Ebrahim-Zadeh, M. (2014). Few-cycle, broadband, mid-infrared optical parametric oscillator pumped by a 20-fs Ti:sapphire laser. *Laser Photon. Rev.* 91: 86.
163 Marzenell, S., Beigang, R., and Wallenstein, R. (1999). Synchronously pumped femtosecond optical parametric oscillator based on $AgGaSe_2$ tunable from 2 μm to 8 μm. *Appl. Phys. B* 69: 423.
164 Kumar, S.C., Krauth, J., Steinmann, A., Zawilski, K.T., Schunemann, P.G., Giessen, H., and Ebrahim-Zadeh, M. (2015). High-power femtosecond mid-infrared optical parametric oscillator at 7 μm based on $CdSiP_2$. *Opt. Lett.* 40: 1398.
165 Kumar, S.C., Esteban-Martin, A., Santana, A., Zawilski, K.T., Schunemann, P.G., and Ebrahim-Zadeh, M. (2016). Pump-tuned deep-infrared femtosecond optical parametric oscillator across 6–7 μm based on $CdSiP_2$. *Opt. Lett.* 41: 3355.
166 Maidment, L., Schunemann, P.G., and Reid, D.T. (2016). Molecular fingerprint-region spectroscopy from 5 to 12 μm using an orientation-patterned gallium phosphide optical parametric oscillator. *Opt. Lett.* 41: 4261.
167 Elsaesser, T., Seilmeier, A., Kaiser, W., Koidl, P., and Brandt, G. (1984). Parametric generation of tunable picosecond pulses in the medium infrared using $AgGaS_2$ crystals. *Appl. Phys. Lett.* 44: 383.
168 Vodopyanov, K.L. and Chazapis, V. (1997). Extra-wide tuning range optical parametric generator. *Opt. Commun.* 135: 98.
169 Petrov, V., Tanaka, Y., and Suzuki, T. (1997). Parametric generation of 1-ps pulses between 5 and 11 μm with a $ZnGeP_2$ crystal. *IEEE J. Quantum Electron.* 33: 1749.
170 Vodopyanov, K.L. (1998). Megawatt peak power 8–13 μm CdSe optical parametric generator pumped at 2.8 μm. *Opt. Commun.* 150: 210.
171 Kumar, S.C., Jelínek, M., Baudisch, M., Zawilski, K.T., Schunemann, P.G., Kubeček, V., Biegert, J., and Ebrahim-Zadeh, M. (2012). Tunable, high-energy, mid-infrared, picosecond optical parametric generator based on $CdSiP_2$. *Opt. Express* 20: 15703.

172 Andriukaitis, G., Balčiūnas, T., Ališauskas, S., Pugžlys, A., Baltuška, A., Popmintchev, T., Chen, M.-C., Murnane, M.M., and Kapteyn, H.C. (2011). 90 GW peak power few-cycle mid-infrared pulses from an optical parametric amplifier. *Opt. Lett.* 36: 2755.

173 Wandel, S., Lin, M.W., Yin, Y., Xu, G., and Jovanovic, I. (2016). Parametric generation and characterization of femtosecond mid-infrared pulses in $ZnGeP_2$. *Opt. Express* 24: 5287.

174 Sanchez, D., Hemmer, M., Baudisch, M., Cousin, S.L., Zawilski, K., Schunemann, P., Chalus, O., Simon-Boisson, C., and Biegert, J. (2016). 7μm, ultrafast, sub-millijoule-level, mid-infrared optical parametric chirped pulse amplifier pumped at 2 μm. *Optica* 3: 147.

175 McCarthy, M.J. and Hanna, D.C. (1993). All-solid-state synchronously pumped optical parametric oscillator. *J. Opt. Soc. Am. B* 10: 2180.

176 Burr, K.C., Tang, C.L., Arbore, M.A., and Fejer, M.M. (1997). High-repetition-rate femtosecond optical parametric oscillator based on periodically poled lithium niobate. *Appl. Phys. Lett.* 70: 3341.

177 Kumar, S.C., Schunemann, P.G., Zawilski, K.T., and Ebrahim-Zadeh, M. (2016). Advances in ultrafast optical parametric sources for the mid-infrared based on $CdSiP_2$. *J. Opt. Soc. Am. B* 33: D44.

178 Elsaesser, T., Seilmeier, A., and Kaiser, W. (1983). Parametric generation of tunable picosecond pulses in proustite between 1.2 and 8 μm. *Opt. Commun.* 44: 293.

179 Vodopyanov, K.L. (1999). Mid-IR optical parametric generator with extrawide (3–19 μm) tunability: applications for spectroscopy of two-dimensional electrons in quantum wells. *J. Opt. Soc. Am. B* 16: 1579.

180 Shen, Y.R. (2003). *The Principles of Nonlinear Optics*. Hoboken, NJ: Wiley.

181 Stolen, R.H., Ippen, E.P., and Tynes, A.R. (1972). Raman oscillation in glass optical waveguide. *Appl. Phys. Lett.* 20: 62.

182 Agrawal, G.P. (2013). *Nonlinear Fiber Optics*, 5e. Oxford: Academic Press.

183 Basiev, T.T., Basieva, M.N., Doroshenko, M.E., Fedorov, V.V., Osiko, V.V., and Mirov, S.B. (2006). Stimulated Raman scattering in mid IR spectral range 2.31–2.75–3.7 μm in $BaWO_4$ crystal under 1.9 and 1.56 μm pumping. *Laser Phys. Lett.* 3: 17.

184 Zhao, J., Zhang, X., Guo, X., Bao, X., Li, L., and Cui, J. (2013). Diode-pumped actively Q-switched Tm, Ho:$GdVO_4$/$BaWO_4$ intracavity Raman laser at 2533 nm. *Opt. Lett.* 38: 1206.

185 Rakich, P.T., Fink, Y., and Soljacic, M. (2008). Efficient mid-IR spectral generation via spontaneous fifth-order cascaded-Raman amplification in silica fibers. *Opt. Lett.* 33: 1690.

186 Jalali, B., Raghunathan, V., Shori, R., Fathpour, S., Dimitropoulos, D., and Stafsudd, O. (2006). Prospects for silicon mid-IR Raman lasers. *IEEE J. Selected Topics Quantum Electron.* 12: 1618.

187 Leuthold, J., Koos, C., and Freude, W. (2010). Nonlinear silicon photonics. *Nat. Photon.* 4: 535.

188 Boyraz, O. and Jalali, B. (2004). Demonstration of a silicon Raman laser. *Opt. Express* 12: 5269.

189 Rong, H., Jones, R., Liu, A., Cohen, O., Hak, D., Fang, A., and Paniccia, M. (2005). A continuous-wave Raman silicon laser. *Nature* 433: 725.

190 Takahashi, Y., Inui, Y., Chihara, M., Asano, T., Terawaki, R., and Noda, S. (2013). A micrometre-scale Raman silicon laser with a microwatt threshold. *Nature* 498: 470.

191 Rong, H., Xu, S., Cohen, O., Raday, O., Lee, M., Sih, V., and Paniccia, M. (2008). A cascaded silicon Raman laser. *Nat. Photon.* 2: 170.

192 Raghunathan, V., Borlaug, D., Rice, R.R., and Jalali, B. (2007). Demonstration of a mid-infrared silicon Raman amplifier. *Opt. Express* 15: 14355.

193 Liu, X., Osgood Jr, R.M., Vlasov, Y.A., and Green, W.M.J. (2010). Mid-infrared optical parametric amplifier using silicon nanophotonic waveguides, *Nat. Photon.* 4, 557 (2010).

194 Chavez Boggio, J.M., Zlatanovic, S., Gholami, F., Aparicio, J.M., Moro, S., Balch, K., Alic, N., and Radic, S. (2010). Short wavelength infrared frequency conversion in ultra-compact fiber device. *Nat. Photon.* 4: 561.

195 Griffith, A.G., Yu, M., Okawachi, Y., Cardenas, J., Mohanty, A., Gaeta, A.L., and Lipson, M. (2016). Coherent mid-infrared frequency combs in silicon-microresonators in the presence of Raman effects. *Opt. Express* 24: 13044.

196 Sabella, A., Piper, J.A., and Mildren, R.P. (2014). Diamond Raman laser with continuously tunable output from 3.38 to 3.80 μm. *Opt. Lett.* 39: 4037.

197 Latawiec, P., Venkataraman, V., Burek, M.J., Hausmann, B.J.M., Bulu, I., and Lončar, M. (2015). On-chip diamond Raman laser. *Optica* 2: 924.

198 Kippenberg, T.J., Spillane, S.M., Armani, D.K., and Vahala, K.J. (2004). Ultralow-threshold microcavity Raman laser on a microelectronic chip. *Opt. Lett.* 29: 1224.

199 Grudinin, I.S. and Maleki, L. (2007). Ultralow-threshold Raman lasing with CaF_2 resonators. *Opt. Lett.* 32: 166.

200 Vanier, F., Rochette, M., Godbout, N., and Peter, Y.-A. (2013). Raman lasing in As_2S_3 high-Q whispering gallery mode resonators. *Opt. Lett.* 38: 4966.

201 Kuyanov, K.E., Momose, T., and Vilesov, A.F. (2004). Solid hydrogen Raman shifter for the mid-infrared range 4.4–8 μm. *Appl. Opt.* 43: 6023.

202 Gladyshev, A.V., Kosolapov, A.F., Khudyakov, M.M., Yatsenko, Yu.P., Kolyadin, A.N., Krylov, A.A., Pryamikov, A.D., Biriukov, A.S., Likhachev, M.E., Bufetov, I.A., and Dianov, E.M. (2017). 4.4-μm Raman laser based on hollow-core silica fibre. *Quantum Electron.* 47: 491.

6

Supercontinuum and Frequency Comb Sources

6.1 Supercontinuum Sources

Broadband and spatially coherent sources in the mid-IR, especially in the molecular fingerprint region, 3–20 μm, attract enormous interest due to the plethora of applications they are enabling, including spectroscopy, multispecies trace-gas detection, hyperspectral imaging, and nano-IR imaging, just to name a few. Supercontinuum (SC) generation occurs when relatively narrowband input pulses undergo extreme nonlinear spectral broadening to yield a broadband, spectrally continuous "white light" output. SC sources typically use a single pump laser with nano-, pico-, or femtosecond pulse durations and take advantage of nonlinear effects in a wide variety of media including solids, liquids, and gases, but typically in waveguiding structures. In the mid-IR, SC sources can provide spectral coverage spanning up to several octaves and reach as far as >20 μm in wavelength. Unlike conventional broadband blackbody sources such as globars, SC laser sources are spatially coherent; they combine the brightness of lasers with extremely wide spectral coverage of blackbody radiation. Hence, SC radiation can be focused down to a diffraction-limited spot size and, due to the dramatically increased radiance (expressed in watts per steradian per square meter), they are very beneficial for such applications as remote sensing, time-resolved spectroscopy, the study of metamaterials, and nano-IR spectroscopy.

Alfano and Shapiro were the first to generate an SC, which was in the visible range and extended from 0.4 to 0.7 μm [1]. They used 0.53-μm picosecond pulses propagated through borosilicate BK-7 glass, where short-scale self-focused filaments were formed with optical intensities as high as 10^{13} W/cm^2. The authors described the effect in terms of a nondegenerate four-wave parametric process due to the third-order susceptibility $\chi^{(3)}$ in the glass. The first mid-IR SC, spanning 3–14 μm, was obtained by Corkum et al. in a 6-cm-long bulk GaAs crystal that was irradiated by picosecond CO_2 laser pulses (λ = 9.3 μm)

Laser-based Mid-infrared Sources and Applications, First Edition. Konstantin L. Vodopyanov.

with the focused intensity of 10^{11} W/cm^2 [2]. The spectral broadening mechanism originated from several nonlinear optical processes including self-phase modulation (SPM), parametric four-wave mixing, high-order harmonic generation, and stimulated Raman scattering.

Since SC generation is a nonlinear effect, beam confinement in waveguides and optical fibers facilitates nonlinear processes; thus, much lower threshold peak powers, as compared to bulk materials, are required. In addition, waveguides allow modification of group velocity dispersion (GVD), a condition that is critical for efficient SC generation.

Largely speaking, the following physical processes, mostly $\chi^{(3)}$ effects, contribute to SC generation [3–8]:

- Self-phase and cross-phase modulation, a result of intensity-dependent refractive index $n(I) = n_0 + n_2 I$, where n_0 is the linear, and n_2 is the nonlinear index of refraction, related to the third-order nonlinear optical susceptibility.
- Parametric four-wave mixing – the same as modulation instability if described in time domain.
- Raman scattering (which can be described as a time-delayed resonant nonlinear response involving lattice vibrations) that results in the occurrence of redshifted spectral components.
- Soliton generation and soliton fission, feasible in the anomalous dispersion regime.
- Soliton self-frequency shift (which can be viewed as intra-pulse Raman scattering in the anomalous dispersion regime) leading to the amplification of the lower-frequency part of the soliton spectrum at the expense of the higher-frequency part.
- Dispersive wave (Cherenkov wave) emission, a process in which an optical pulse propagating in a medium with Kerr nonlinearity, anomalous dispersion, and high-order GVD components, is scattered through a resonant-like process to a shifted frequency; this process requires wave-vector matching between the dispersive radiation and the excitation pulse.
- Formation of an optical intensity shock and wavebreaking, which happens in the normal dispersion regime.

The contribution of each mechanism can be very different, depending on the pulse duration, material GVD, length of the medium, and pump wavelength. With femtosecond pulses, the spectral broadening can be dominated by SPM. In the anomalous dispersion regime (that is, when the group delay increases with the wavelength), the combination of SPM and dispersion can lead to complicated soliton dynamics, including splitting of higher-order solitons into multiple fundamental solitons (soliton fission). With picosecond or nanosecond pump pulses, the two dominant mechanisms can be Raman scattering and four-wave mixing.

Photonic crystal fibers (PCF), tapered fibers, and waveguides are particularly attractive candidates for SC generation due to enhanced effective nonlinearity because of a very small effective modal area (a few square micrometers or less). In such structures, one can also engineer, through the design of the waveguide (WG) profile, the value and the sign of the GVD. With optical fibers, SC generation is achieved with fiber lengths that range from a few millimeters to several meters, depending on the pulse duration and the peak power [3–5].

Despite their wide application in telecommunications, the use of silica glass fibers for SC generation in the mid-IR is limited by strong material absorption above 2.4 μm. Several non-silica glass fibers, having much deeper mid-IR transparency, have been used in the last decade, including tellurite, fluoride, and chalcogenide glasses, as discussed below.

6.1.1 SC from Lead-silicate Glass Fibers

Omenetto et al. obtained a smooth SC, extending from 0.35 to 3 μm, using a highly nonlinear PCF – a special optical fiber with a built-in microstructure – made of lead-silicate Schott SF6 glass [9]. The reported value for lead-silicate glass, $n_2 = 2.2 \cdot 10^{-19}\ \mathrm{m^2/W}$, is almost an order of magnitude higher than that of fused silica glass. The pump source was at $\lambda = 1.55$ μm with a pulse duration of 110 fs and a repetition rate of 80 MHz. The PCF had a 2.6-μm core diameter (inset to Figure 6.1) and blueshifted zero-dispersion wavelength (ZDW) at 1.3 μm, so that the pump pulses were in the *anomalous* dispersion region. The

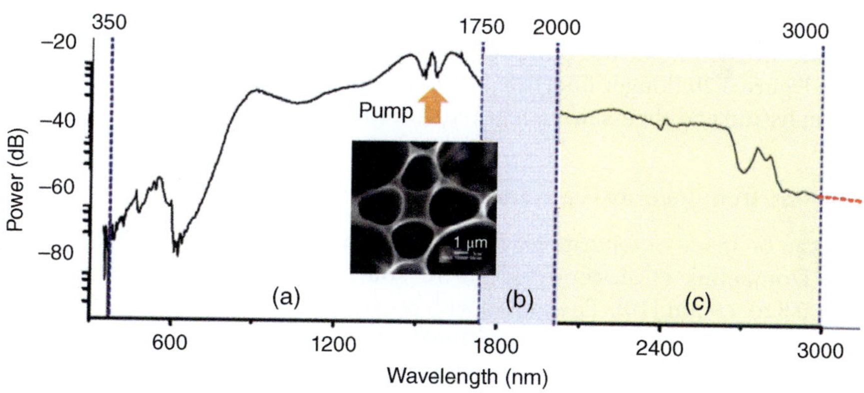

Figure 6.1 Supercontinuum spectrum resulting from the propagation of 1.55-μm, 110-fs pulses in a 5.7-mm-long lead-silicate PCF. The spectral region (a) was covered by an optical spectrum analyzer, region (c) by an MCT detector and a monochromator, and region (b) was not detected in the experiments. The inset shows the scanning electron microscope image of the tip of the sample. *Source:* reproduced from figure 2 of [9], with permission of OSA, The Optical Society.

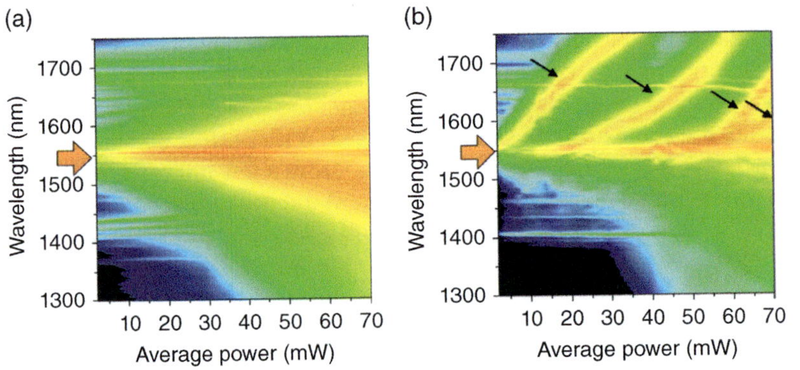

Figure 6.2 Measured spectral behavior near the pump wavelength as a function of the average pump power in the fiber for sample length of (a) 5.7 mm and (b) 70 cm. *Source:* reproduced from figure 3 of [9], with permission of OSA, The Optical Society.

average power of the SC pulses detected at the output of the 5.7-mm-long fiber was 70 mW. The corresponding SC spectrum is shown in Figure 6.1. The authors identify two regimes of nonlinear pulse transformation in this fiber: (i) when the fiber length is much shorter than the dispersion length $L_D \sim \tau^2/\beta_2$ (here, τ is pulse duration and $\beta_2 = d^2k/d\omega^2$ is the GVD), estimated to be 40 cm at $\lambda = 1.55$ μm, the soliton propagation is not important, and a symmetric SC spectrum arises from almost pure SPM, and (ii) when longer fibers are used and the SC is formed by the breakup of multiple Raman-shifted solitons. Figure 6.2 shows the measured spectral behavior as a function of the average pump power in the fiber for the two fiber lengths: 5.7 mm and 70 cm. Figure 6.2a shows a smooth broadening of the pump pulse, typical of a SPM-dominated process, while Figure 6.2b (longer fiber) clearly shows the spectral signature of multiple soliton fission and their subsequent redshift (indicated by the arrows).

6.1.2 SC from Tellurite Glass Fibers

PCF can be made of tellurite glass – another material suitable for SC generation. Domachuk et al. reported an SC source extending into the mid-IR from 0.8 to 4.9 μm [10]. The pulses with central wavelength 1.55 μm, duration 110 fs, average power 250 mW (150 mW at the input face of the fiber), and repetition rate 80 MHz were focused using a 0.5 NA aspheric lens into an 8-mm-long tellurite PCF. The fiber microstructure was a "wagon wheel" design shown in Figure 6.3. The microstructure is surrounded by a fiber that has a 120-μm outside diameter. Six 120-nm-wide and 16-μm long filaments support a 2.5-μm diameter core. The effective area of the fiber mode is 1.7 μm^2. The fiber design was such that the ZDW was at 1.38 μm; hence, 1.55-μm pump pulses were in the anomalous dispersion regime. The combination of fiber

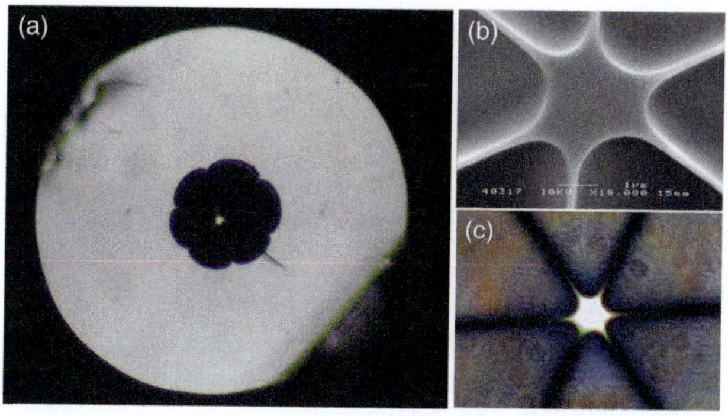

Figure 6.3 Cross-sectional profile of the tellurite PCF in the optical microscopy (a, c) and electron microscopy (b). Scale bar in (b) is 1 μm. *Source:* reproduced from figure 1 of [10], with permission of OSA, The Optical Society.

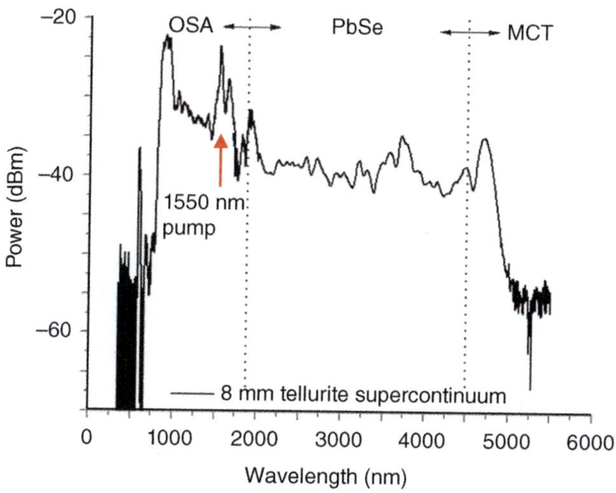

Figure 6.4 Spectrum of the SC generated by the 8 mm-long tellurite glass PCF. The wavelength regions analyzed by the OSA and the monochromator with lead selenide (PbSe) or mercury cadmium telluride (MCT) detectors are indicated. *Source:* reproduced from figure 3 of [10], with permission of OSA, The Optical Society.

microstructure and high nonlinear coefficient of the bulk tellurite glass ($n_2 = 2.5\cdot10^{-19}$ m^2/W) provided enhanced WG nonlinearity responsible for SC output after only 8 mm of pulse propagation. Shorter lengths are expected to result in the SC generation process that is mainly driven by SPM that typically preserves coherence of the pump and results in the smooth spectrum. Also, the use of very short fiber mitigates the relatively high loss of the tellurite glass. Figure 6.4 shows the SC spectrum, which extends from 0.79 to 4.87 μm at −25 dB level. The SC output power was measured to be 90 mW [10].

6.1.3 SC from ZBLAN Fibers

The family of glasses known as ZBLAN has a composition ZrF_4-BaF_2-LaF_3-AlF_3-NaF.[1] It is the most stable fluoride glass known, and is commonly used for making mid-IR optical fibers. ZBLAN has a broad optical transmission window extending from 0.3 to 7 μm, low refractive index (n = 1.5), and small dispersion. Its attenuation can be made as low as 3 dB/km at 2.6 μm. The nonlinear refractive index of ZBLAN is $n_2 = 2.1 \cdot 10^{-20}$ m^2/W.

Hagen et al. [11] performed SC generation spanning 1.8–3.4 μm using cascaded Raman self-shifting in two fibers connected in series: a silica fiber followed by a ZBLAN fiber. High-power pump pulses from an erbium-doped fiber laser (λ = 1.55 μm, 1.5 μJ, 200 kHz, 900 fs) were first coupled (Figure 6.5) into the silica fiber (Corning SMF-28, L = 21 cm) to establish an initial redshift to the wavelengths that are greater than the ZDW of the ZBLAN fiber (1.63 μm). Next, the light was coupled into the ZBLAN fiber (KDD Fiberlabs 05C-09, L = 91 cm) for further redshifting, with the silica and fluoride fibers coupled together either by lenses or via butt-coupling. A 1.8–3.4-μm continuum was generated with the spectrum shown in Figure 6.6. At an average power of the pump laser of 300 mW, 195 mW was coupled into the silica fiber, 37 mW was coupled into the fluoride fiber, and 5 mW was emitted at 1.8–3.4 μm after a germanium long-pass filter.

Using a two-stage (silica plus ZBLAN fiber) approach, Xia et al. produced mid-IR SC using *nanosecond* pulsed excitation [12]. The pump source was a seeded multistage erbium-doped fiber amplifier (λ = 1.553 μm) producing 2-ns pulses at a repetition rate of 5 kHz, peak power of 4 kW, and an average power of 40 mW. The first frequency conversion stage was a 1-to-2-m-long silica single-mode fiber (SMF). With pumping the SMF in the anomalous dispersion region, the phase-matched modulation instability triggered the breakup of the nanosecond pulses into femtosecond pulses. In the second stage, consisting of a 2-to-8-m-long ZBLAN fiber, the primary SC was broadened by fiber nonlinearities – predominantly SPM and stimulated Raman scattering. The generated SC spanned 0.8–4.5 μm (Figure 6.7a) with an average power of

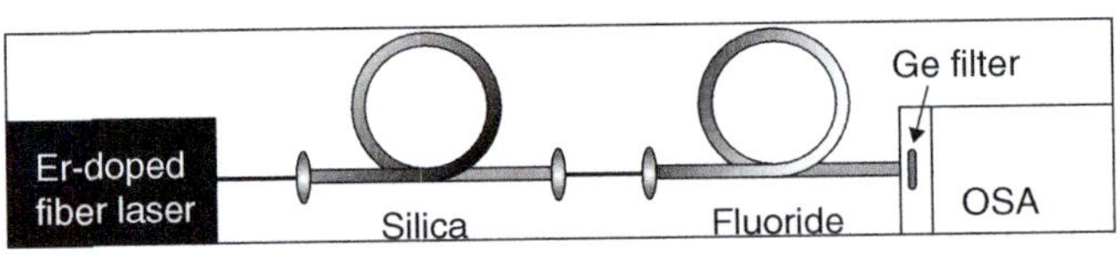

Figure 6.5 Schematic of the cascaded fiber system used to generate infrared SC. The germanium (Ge) filter is used as a long-pass filter in an optical spectrum analyzer (OSA). *Source:* reproduced from figure 1 of [11], with permission of IEEE.

1 http://en.wikipedia.org/wiki/ZBLAN

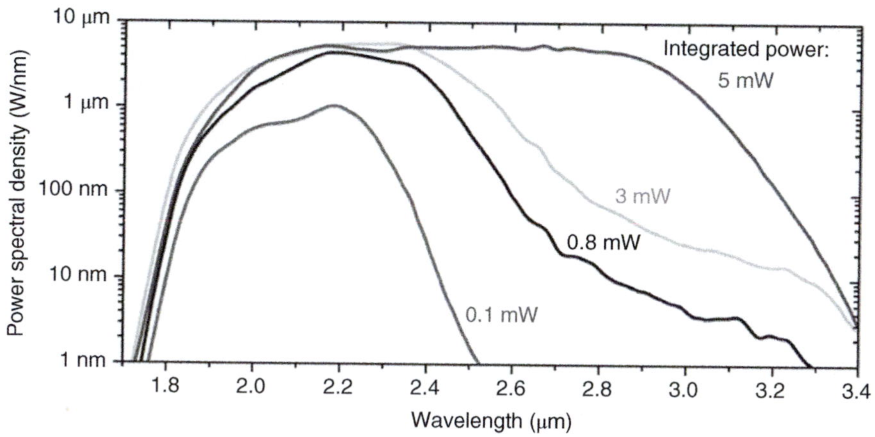

Figure 6.6 Spectral emission recorded at the output of the fluoride fiber. The average output power is shown for each curve plot. *Source:* reproduced from figure 2 of [11], with permission of IEEE.

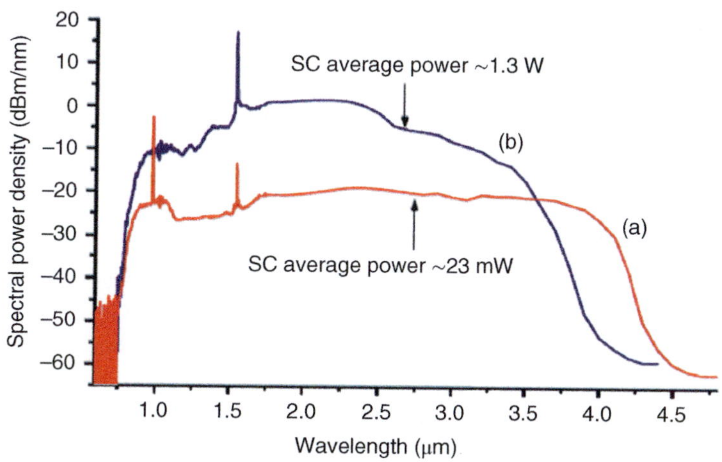

Figure 6.7 (a) SC Spectrum obtained from 1 m of a silica SMF followed by 8 m of ZBLAN. The feature at 980 nm is due to the undepleted pump of the EDFA. (b) SC spectrum from 3 m of SMF followed by 13 m of ZBLAN, obtained with higher pump power. *Source:* reproduced from figure 2 of [13], with permission of OSA, The Optical Society.

23 mW and pump-to-SC power conversion efficiency exceeding 50%. With an increased pump power, the same team scaled the average SC power to 1.3 W, with the spectral span of 0.8–4 μm (Figure 6.7b) [13]. Moreover, by scaling the pulse repetition rate to 3.3 MHz, the team reached an average SC power of 10.5 W [14].

Using *picosecond* pump pulses near 2 μm, Liu et al. achieved an SC spectrum extending from 1.9 to 3.8 μm with 21.8-W average power, from a single-mode ZBLAN fiber [15]. The pump source was a thulium-doped fiber master oscillator power amplifier system with the pulse width 24 ps and repetition rate 94 MHz. First, the seed laser spectrum was broadened to beyond 2.4 μm in a thulium-doped fiber amplifier delivering 42-W average power. This initial broadening in the amplifier was caused by the breakup of picosecond pulses into femtosecond solitons, modulation instability, stimulated Raman scattering, and SPM. Then, the output spectrum was further broadened to beyond 3.8 μm in a single-mode ZBLAN fiber. The ZBLAN spectrum-broadening mechanisms included a variety of nonlinear effects such as soliton self-frequency shift, stimulated Raman scattering, and four-wave mixing. Overall, the optical-to-optical conversion efficiency to the broadband SC, with respect to the 793-nm diode laser pump for the Tm-fiber system, was 17% [15]. Similarly, with a 2-μm picosecond pump, Yang et al. obtained SC generation in a single-mode ZBLAN fiber with 13-W average power and a spectrum extending from 1.9 to 4.3 μm [16].

By reducing the fiber length to 2 cm and using high-power femtosecond pumping (λ = 1.45 μm, pulse width 180 fs, energy per pulse 9 μJ, peak power 50 MW), SC spectrum obtained from a ZBLAN fiber was extended to 6.3 μm. The initial spectral broadening along the fiber was mainly caused by SPM; further broadening was produced by Raman scattering and four-wave mixing [17].

6.1.4 SC from Chalcogenide Glass Fibers

Chalcogenide glasses – amorphous semiconductors based on heavy elements, such as sulfur (S), selenium (Se), and tellurium (Te), bonded to As, Ge, Sb, Ga, Si, or P – are exceptionally well suited for mid-IR SC generation because of their (i) good long-wavelength IR transparency due to the low (300–450 cm^{-1}) phonon energy and (ii) high optical $\chi^{(3)}$ nonlinearity. Typically, sulfides transmit to ~11 μm, selenides to ~15 μm, and tellurides to beyond 20 μm in the IR [18–22]. For an optical WG, the strength of the nonlinear response is characterized by the nonlinear parameter $\gamma = n_2\omega_0/cA_{eff}$, where n_2 is a nonlinear refractive index, A_{eff} is the effective core area of the propagating mode, c is the speed of light, and ω_0 is the central angular frequency [3].

Strong confinement of the beam in PCF, associated with the air holes surrounding the fiber core, allows a dramatic increase in the nonlinear parameter γ of the fiber device. In addition, it allows engineering of total dispersion (e.g. from normal to anomalous). By using a suspended-core microstructured As_2S_3 optical fiber, El-Amraoui et al. attained SC between 1 and 2.6 μm using a 1.56-μm fs pump [23]. The As_2S optical fiber was 68 cm long with a core diameter of 2.6 μm surrounded by three holes (Figure 6.8a). By varying the input

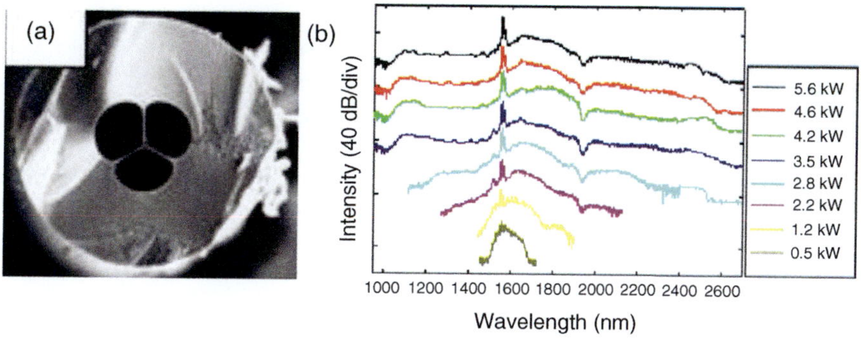

Figure 6.8 (a) Geometrical profile of a suspended core As_2S optical fiber. The external fiber diameter is between 120 and 160 μm. (b) Experimental SC spectra recorded for 1557 nm fs pumping at different peak powers. *Source:* reproduced from figure 9 of [23], with permission of OSA, The Optical Society.

peak power of the 1.56-μm pump between 0.5 and 5.6 kW (average power 3.4–38 mW), the authors observed continuous broadening of the initial spectrum, which extended at the maximum pump from 1 to 2.6 μm (Figure 6.8b). For the core diameter of 2.6 μm, the estimated ZDW was at $\lambda = 2.2$ μm, which is strongly shifted, due to the influence of geometry, toward shorter wavelengths. (For comparison, bulk As_2S_3 glass has ZDW at $\lambda \sim 5$ μm.)

Fiber tapering (a process of stretching an optical fiber while it is heated) provides a good alternative to PCF to dramatically increase the nonlinear parameter γ, due to a reduced cross section, and to engineer the dispersion [24, 25]. By using pulses from an erbium-fiber mode-locked laser (250-fs duration, 38.6 MHz repetition frequency, and 17 mW of maximum average power), Hudson et al. produced an octave-spanning SC in an As_2S_3 tapered fiber. The authors made a tapered region with a diameter of 1.3 μm and a length of 50 mm. The corresponding mode area A_{eff} was 0.8 μm^2, the nonlinear parameter $\gamma = 12.4\ m^{-1}W^{-1}$ at 1.55 μm, and ZDW ~1.4 μm. As the pump power was incrementally increased, the spectrum broadened to several hundred nanometers with relatively low (<1 kW) peak powers (Figure 6.9). The mechanisms responsible for SC generation include four-wave parametric process and high-order soliton fission, while Raman soliton self-frequency shift is responsible for extending the bandwidth on the long-wavelength side [26].

Using 1-ps pulses at 1.55 μm with 3.5-kW peak power, Shabahang et al. [27] obtained SC from 0.85 to 2.35 μm in a composite $As_2Se_{1.5}S_{1.5}/As_2S_3$ tapered fiber. The fiber was tapered without removing the polymer jacket that was an integral part of the fiber structure. Owing to the large index contrast, the field was strongly confined to the 480-nm diameter chalcogenide core, while a polymer jacket provided mechanical stability.

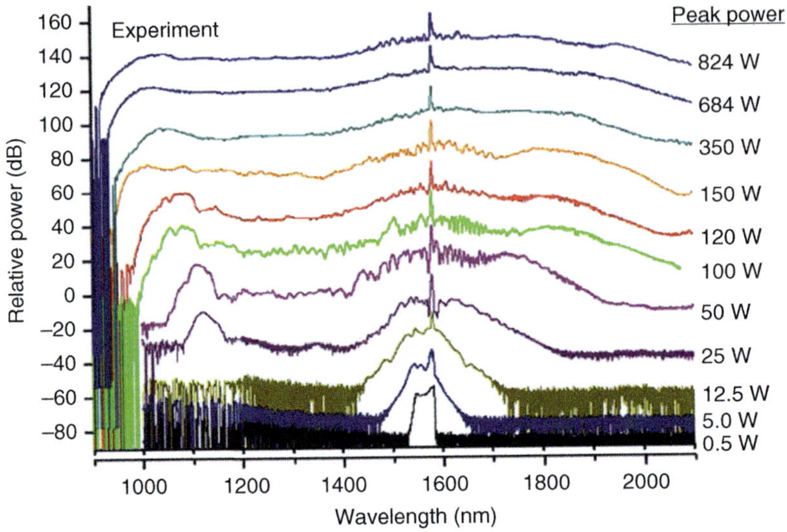

Figure 6.9 Output SC spectra from the As_2S_3 tapered fiber as a function of the peak pump power. *Source:* reproduced from figure 4 of [26], with permission of OSA, The Optical Society.

With pump pulses at longer wavelengths one can expect deeper penetration into the mid-IR. Sanghera et al. used λ = 2.5-μm, 100-fs pulses with 100-pJ pulse energy and 1-kW peak power to obtain mid-IR SC spanning 2–3.4 μm using small-core chalcogenide fibers with 1-m length [28]. Mouawad et al. reported SC generation from 0.6 to 4.1 μm (at −40 dB level) in a 20-mm-long suspended core As_2S_3 fiber, pumped at λ = 2.5 μm with pulse duration of 200 fs and peak power of 4.9 kW [29]. The central core of the fiber was 3.4 μm in diameter and it was separated from the clad by three surrounding holes. The fiber exhibited ZDW close to 2.4 μm, i.e. the fiber was pumped in a slightly anomalous dispersion regime. The SC process was characterized by a strong SPM followed by soliton fission. Subsequent propagation in the fiber was associated with soliton dynamics and dispersive wave generation. According to the authors, the fission occurs randomly, due to the role of modulation instability. This leads to a significant degradation of SC coherence properties [29].

Marandi et al. achieved SC generation with an instantaneous span of 2.2–5 μm in a tapered As_2S_3 fiber [30]. Pump pulses with λ = 3.1 μm and sub-100-fs pulse duration were produced as a subharmonic of a 1.56-μm erbium-fiber pump laser in a degenerate PPLN-based optical parametric oscillator (OPO) (Figure 6.10a). The authors used in-situ tapering, in which case the SC spectrum was measured in real time during the tapering process, and the tapering was stopped when the spectrum reached its maximum width.

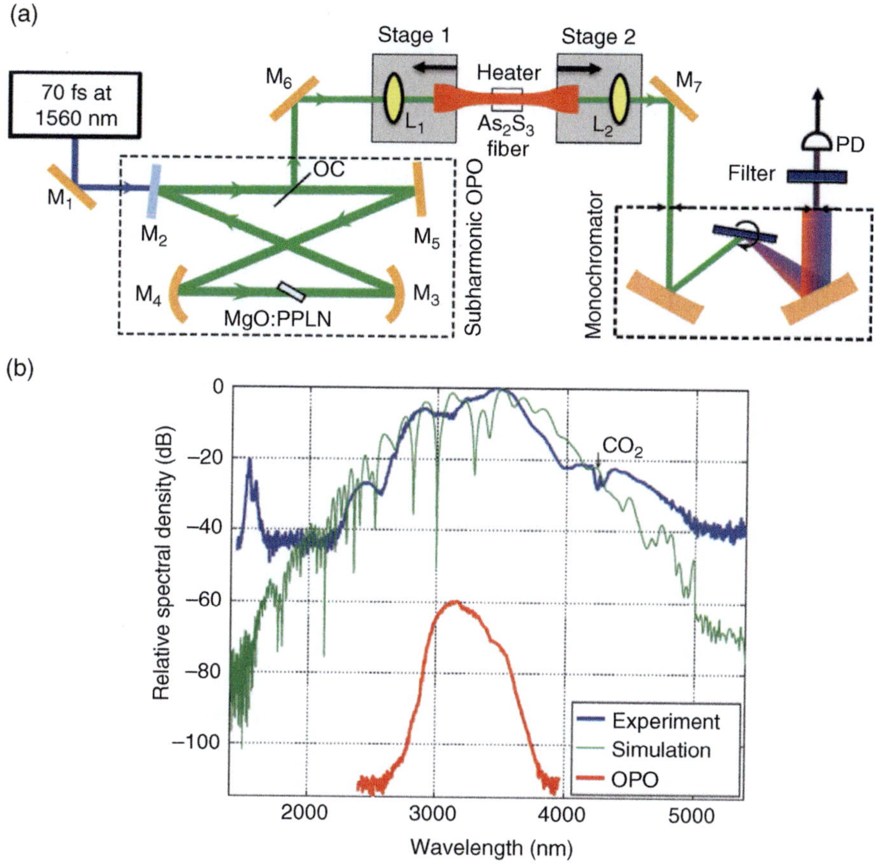

Figure 6.10 (a) Setup for SC generation in a tapered As_2S_3 fiber using 3.1-μm pump from a subharmonic OPO. M2–M5, OPO mirrors; OC, OPO output coupler. (b) Spectrum of the SC output compared with a simulated spectrum. The input pump spectrum is shown at the bottom. *Source:* reproduced from figure 5 of [30], with permission of OSA, The Optical Society.

The length of the tapered region was 18 mm and the diameter of the thinnest portion of the fiber was estimated to be 2.3 μm, resulting in ZDW close to 3.1 μm. The output spectrum, together with the pump spectrum, is shown in Figure 6.10b. The SC spectrum extended from 2.2 to 5 μm (at −40 dB) and was ~3 times broader (in optical frequency units) than that of the 3.1-μm pump. The average power at the fiber output was measured to be 15 mW. According to simulations, the spectral broadening was dominated by soliton fission and suggested a high degree of SC coherence [30].

A similar in-situ tapering technique was used to demonstrate a 1-to-3.7 μm SC in a tapered As_2S_3 2.1-mm-long fiber that was pumped by sub-100-fs pulses that originated directly from a mode-locked Tm-doped fiber laser with 75 MHz repetition rate, center wavelength of 2.04 μm, and 23 mW average power (pulse energy 300 pJ). To shift the GVD of the As_2S_3 fiber to the anomalous dispersion regime, the fiber was tapered to 1.95-μm diameter [31]. Also, with a 10-cm-long tapered As_2Se_3 fiber having a diameter of 1.6 μm, Al-Kadry et al. produced an SC that spanned the range 1.1–4.4 μm (at −30 dB). The pump was a mode-locked Tm-doped fiber laser at λ = 1.94 μm with pulse duration of 800 fs [32].

Yu and coauthors achieved SC in a small-core step-index chalcogenide fiber with a core made of Ge-As-Se and the cladding made of Ge-As-S glass. Both glasses show broad transparency and low losses in the mid-IR to beyond 9 μm. The fiber had an elliptical core of 4.1 × 4.6 μm and an outer diameter of 190 μm, and there were two ZDWs: at 3.2 μm and at 6 μm. By pumping an 11-cm-long fiber with λ = 4-μm pulses from a femtosecond optical parametric amplifier (OPA) system at 330-fs duration and 3-kW peak power, the team generated an SC spanning from 1.8 to 10 μm [33]. By using a suspended core $As_{38}Se_{62}$ fiber that was 18-cm long, had a core diameter of 4.5 μm, a ZDW of 3.5 μm, and combined with the same OPA pumping source tuned to λ = 4.4 μm, an SC spanning 1.7–7.5 μm with an average output power of 15.6 mW (peak power 5.2 kW) was obtained. The SC generation was initiated by SPM followed by soliton dynamics and soliton self-frequency shift [34].

By keeping the fiber geometry simple and robust, but at the expense of a higher-power pump source, Hudson et al. demonstrated broadband SC operation in a standard As_2S_3 step-index fiber. The fiber had a 9-μm diameter core and 170-μm cladding, and the pump source was an optical parametric chirped pulse amplifier (OPCPA) system operating at 3.1 μm with pulse energy of up to 18 μJ and pulse duration of 67 fs. The SC spectral bandwidth of 1.6–5.9 μm (at −20 dB level) was achieved at the peak pump power of 520 kW [35]. The SC average power was 8 mW. The pump wavelength of 3.1 μm was in the normal dispersion range of the fiber, and the spectral broadening was mainly driven by SPM, with Raman shifting extending the long wavelength edge.

Robichaud et al. reported a cascaded mid-IR SC source spanning 3–8 μm and based on a low-loss commercial As_2Se_3 step-index fiber (ZDW 8.9 μm) that was pumped by another SC source with a span of 3–4.2 μm. The latter was based on a seeded (λ = 2.8 μm) erbium-doped ZrF_4 fiber pumped at 980 nm, which served simultaneously as an amplifier and SC generator [36]. A maximum average output power of 1.5 mW was obtained at 3–8 μm at a repetition rate of 2 kHz. The SC output was single mode for the wavelengths >5 μm. Since the As_2Se_3 fiber exhibited normal dispersion up to ~8.9 μm, the spectrum broadening was mainly due to SPM.

By launching intense (>3 MW peak power) 100-fs pulses at 1 kHz repetition rate with a central wavelength of either 4.5 or 6.3 μm into an 85-mm-long high numerical-aperture step-index chalcogenide glass optical fiber, Petersen et al. generated mid-IR SC extending beyond 13 μm [37]. The fiber had a

16-µm-diameter $As_{40}Se_{60}$ core surrounded by a $Ge_{10}As_{23.4}Se_{66.6}$ cladding. Because of the large core, the fiber was effectively multimode with the calculated ZDW of 5.83 µm. When pumping in the normal dispersion regime at 4.5 µm with the peak power up to 3.3 MW (average power 350 µW, 350 nJ/pulse), the pulses initially undergo strong SPM, possibly leading to wave-breaking, owing to self-steepening and third-order dispersion, which causes a significant part of the light to be blueshifted. The redshifting part eventually crosses the ZDW, at which point soliton dynamics, in particular Raman-induced soliton self-frequency shifting, dominates further spectral broadening (Figure 6.11a and b). When pumping in the anomalous dispersion regime at 6.3 µm, just above the ZDW of 5.83 µm, with the peak power up to 7.15 MW (average power 760 µW, 760 nJ/pulse), the pump pulse transforms into a higher-order soliton that rapidly breaks up into multiple fundamental solitons through soliton fission and radiates dispersive waves at a wavelength that is phase-matched to the solitons in the normal dispersion regime (Figure 6.11c and d).

By combining a recently developed ultrafast fiber laser based on holmium $^5I_6-^5I_7$ transition at $\lambda = 2.9$ µm (pulse duration 230 fs, peak power 4.2 kW, repetition rate 42 MHz) with a polymer-protected all-chalcogenide (As_2Se_3 core and As_2S_3 cladding) tapered fiber, where the original As_2Se_3 core diameter of 14 µm was reduced in the tapered "microwire" section to 3 µm, Hudson et al. demonstrated an SC spanning 1.8–9.5 µm (at −20 dB) with an average power of >30 mW [38]. The microwire diameter has been optimized to minimize chromatic dispersion at 2.87 µm ($\beta_2 = 0.29\,ps^2/m$), leading to normal dispersion at and around the pump laser wavelength. The SC in this all-normal dispersion regime was driven, according to the authors, by two sets of processes: coherent SPM and optical wave-breaking, and incoherent stimulated Raman scattering and four-wave mixing. Figure 6.12 shows the layout of the SC source, as well as the spectrum as a function of the peak power.

Cheng et al. designed a step-index fiber with near-zero flattened dispersion based on As_2Se_3 (core) and $AsSe_2$ (cladding) glasses fabricated by the rod-in-tube drawing technique. The pump source was based on difference frequency generation (DFG) and was tunable from 2.4 to 11 µm (pulse duration 170 fs, repetition rate 1 kHz). At the pump wavelength of 9.8 µm and 3-mW average power, the authors demonstrated SC spectrum spanning 2–15.1 µm in a 3 cm-long fiber. The spectrum broadening took place in the region of anomalous dispersion and was dominated, according to the authors, by the fission of the higher-order solitons [39]. By using a low-loss double-cladding telluride glass fiber pumped by the output of an OPA at $\lambda = 7$ µm (pulse width 150 fs, repetition rate 1 kHz) and by operating in the normal dispersion regime, Zhao et al. produced a 2–16 µm mid-infrared SC. The fiber was fabricated using an extrusion method and had a $(Ge_{10}Te_{43})_{90}$-AgI_{10} core with a diameter of 20 µm and two claddings: the first was $(Ge_{10}Te_{40})_{90}$-AgI_{10} and the second $Ge_{10}Sb_{10}Se_{80}$ [40].

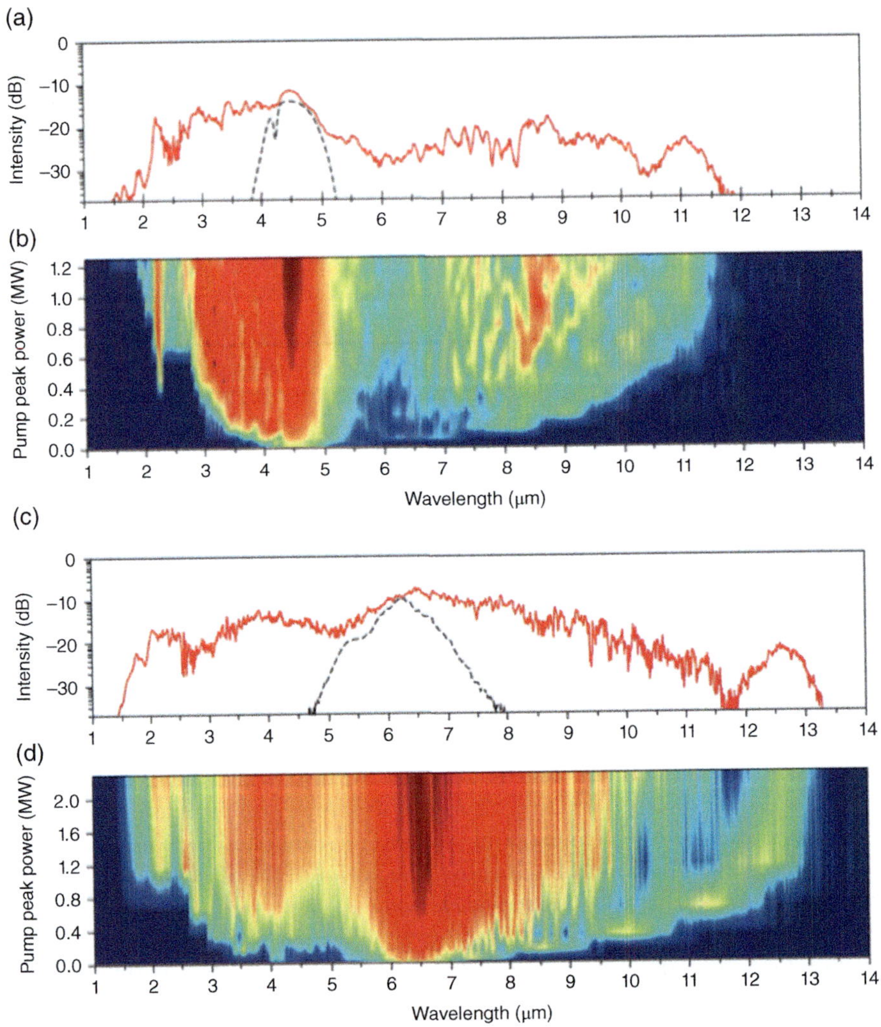

Figure 6.11 Experimental SC results with 4.5 and 6.3 μm pump [37]. (a) Input pump spectrum at 4.5 μm (dashed line) and spectral profile at maximum pump power (solid line). (b) Spectral evolution with increasing pump peak power at 4.5 μm, showing a gradual redshift of distinct soliton peaks above the ZDW and a combination of SPM and dispersive waves below the pump wavelength. (c) Input pump spectrum at 6.3 μm (dashed line) and spectral profile at maximum pump power (solid line), showing a broad, flat supercontinuum (1.64–11.38 μm at −20 dB) followed by a strong spectral peak extending the spectrum all the way to 13.3 μm. (d) Spectral evolution with increasing pump power, showing the gradual redshift of a distinct spectral peak at the long-wavelength edge and the corresponding formation and blueshift of dispersive waves. *Source:* reproduced from figures 4 and 5 of [37], with permission of Springer Nature.

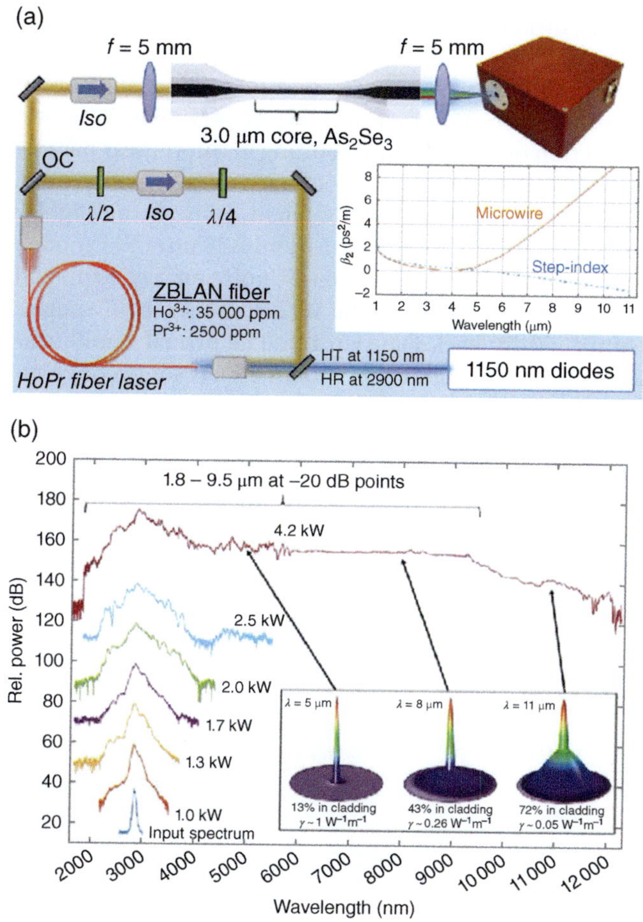

Figure 6.12 (a) Layout of the 1.8–9.5-μm supercontinuum generator [38]. The 2.9-μm holmium laser output passes through an isolator (>40 dB suppression) and is focused into the core of the tapered fiber. The output is collected using a ZnSe lens and focused into a spectrometer. The inset shows the calculated dispersion of both the untapered step-index fiber and the microwire (tapered) section. (b) Spectral broadening as a function of the peak power (vertically offset for clarity). The inset shows the mode profile in the tapered section (3 μm core diameter), as well as the nonlinear parameter γ at various wavelengths. *Source:* reproduced from figures 1 and 4 of [38], with permission of OSA, The Optical Society.

6.1.5 SC from Waveguides

Thanks to their small core size (typically <1 μm), nonlinear waveguides enable high peak power densities at modest average powers and hence show great promise for the realization of compact SC generators. Highly nonlinear materials such as chalcogenide glasses or silicon are extremely suitable for SC generation. Similar to fibers, dispersion engineering is an essential condition to achieve efficient SC generation.

Lamont et al. reported a mid-IR SC produced in a dispersion-engineered highly nonlinear As_2S_3 planar WG pumped at 1.55 μm [41]. The WG was 60 mm long with a cross section of 2 by 0.87 μm resulting in a nonlinear parameter $\gamma = 10\,m^{-1}W^{-1}$ ($\gamma = \omega n_2/cA_{eff}$, where n_2 is a nonlinear refractive index, A_{eff} is the area of the propagating mode, c is the speed of light, and ω is the angular frequency) and anomalous dispersion between 1.51 and 2.17 μm (TM mode). Laser pulses at $\lambda = 1.55$ μm with the peak power 68 W, pulse energy 60 pJ, and the average power 0.6 mW were launched into the WG, resulting in SC spectra with a span of 1.1–2.5 μm (−60 dB level, Figure 6.13). The primary nonlinear process driving the SC generation was four-wave mixing with some contribution of Raman shift and soliton fission.

Xie et al. [42] generated SC spanning 1.2–3.6 μm by using an As_2S_3-silica "double-nanospike" WG pumped by mode-locked Cr:ZnS laser pulses at $\lambda = 2.35$ μm (average power 360 mW, pulse repetition rate 90 MHz, pulse energy 4 nJ, pulse duration 100 fs). The waveguides used in the experiment had a hybrid As_2S_3-silica structure: a 3.2-μm-diameter chalcogenide wire was created inside a silica capillary fiber. In order to improve the coupling efficiency into such high numerical aperture waveguides, inverse nanotapers were created at both ends

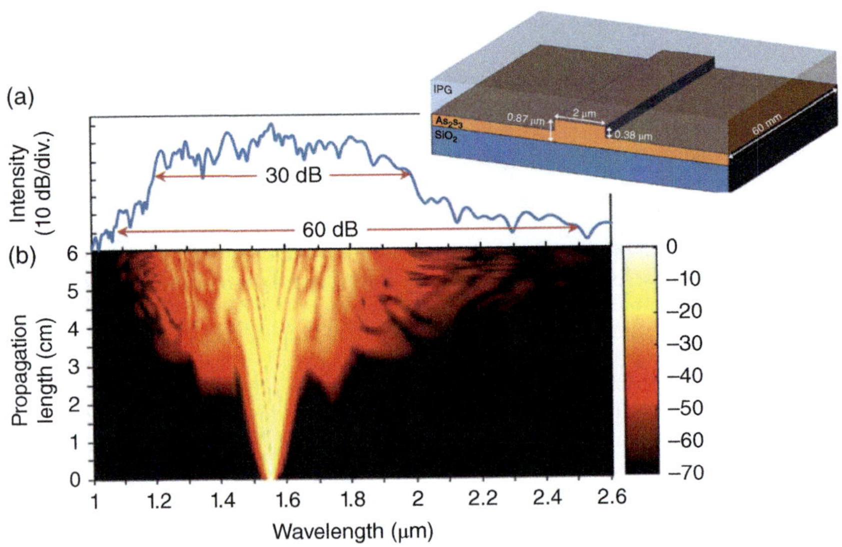

Figure 6.13 (a) Measured SC spectrum for an input pulse with 68 W peak power [41]. (b) Simulated spectral evolution along the waveguide for the same input pulse. Inset shows the waveguide structure (IPG, inorganic polymer glass). *Source:* reproduced from figure 6 of [41], with permission of OSA, The Optical Society.

of the WG (hence "double-nanospike"). The combination of the optimized GVD and high nonlinearity provided by the As_2S_3-silica hybrid WG enabled a 100-pJ-level pump pulse energy threshold for octave-spanning spectral broadening. Numerical simulations, according to the authors, show that the generated SC is highly coherent over the entire wavelength span, which is important for the future realization of broadband frequency combs. For direct comparison, the authors measured an SC generated from a commercially available single-mode As_2S_3 fiber (IRFlex IRF-S-5, effective mode area 28 μm^2) pumped by the same Cr:ZnS laser. Even with six times higher pulse energy (1.23 nJ), the SC bandwidth was only 1.7–3 μm, half as wide in frequency units as compared to the case of the nanospike WG [42].

A broadband SC spanning from 1.8 to >7.5 μm was produced by pumping a rib WG made of GeAsSe chalcogenide glass, with 320-fs pulses from an OPA at λ = 4 μm (Figure 6.14) [43]. The WG was 2.5 × 4 μm in cross section, 1-cm long, and was designed to have anomalous dispersion for the wavelengths between 3.06 and 5.8 μm, such that the 4-μm pump was approximately in the middle of this range. Given short pulse duration and short length of the WG, the SC process was dominated, according to the authors, by the soliton fission process. The total output SC power was 20 mW at the input power of 160 mW (3.3 kW peak inside the WG).

Silicon has attracted great interest as a platform for both linear and nonlinear integrated photonics for at least two decades. While its primary applications have been in the telecom window near 1.5 μm, the capability of exploiting its full transparency window to 8 μm in the mid-IR is highly attractive. Silicon offers high third-order nonlinearity that is similar to that of As_2S_3 chalcogenide glass and an additional important advantage of compatibility

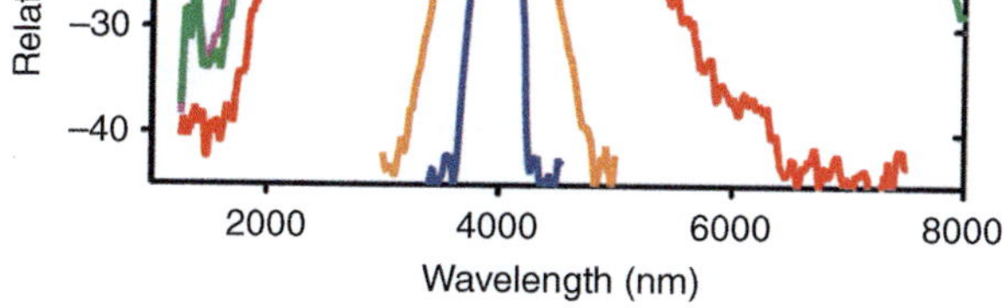

Figure 6.14 Output SC spectra obtained from a GeAsSe chalcogenide glass waveguide pumped with 320-fs pulses at λ = 4 μm at different peak powers. *Source:* reproduced from figure 3 of [43], with permission of Wiley.

with complementary metal-oxide–semiconductor (CMOS) technology. Lau et al. reported SC generation on a silicon chip, with the spectrum extending to >3.6 μm [44]. The silicon-on-insulator (SOI) WG had a cross section of 320 × 1210 nm, a length of 2 cm, and was engineered to exhibit zero GVD wavelength on each side of the 2.5-μm pump. Figure 6.15a shows the GVD of the WG for the fundamental transverse electric (TE) mode. The WG was pumped with 300-fs pulses at a repetition rate of 80 MHz and an average power of 3 mW (125 W peak). According to the authors, the SC was characterized by soliton fission and dispersive radiation across the two zero GVD wavelengths. As seen from Figure 6.15b, the spectral span at λ = 2.5-μm pump was from 1.5 to 3.67 μm (although with some gaps in the spectrum).

By utilizing dispersion-engineered silicon-on-sapphire (SOS) nanowires, Singh et al. demonstrated an octave-spanning SC [45]. The SOS WG used in experiments had a cross section of 2400 nm by 480 nm, and was engineered to exhibit zero GVD wavelengths at 3.3 and 7.1 μm. The WG was pumped into the TE mode of the nanowire with 320-fs pulses at λ = 3.7 μm from an OPA at a repetition rate of 20 MHz and an average power of 12 mW. The widest continuous spectrum achieved at the −30 dB signal level was from 1.9 to 5.5 μm. According to the authors, the SC bandwidth was mainly limited by the four-photon absorption (4PA), causing high loss as well as narrowing of the spectrum (the maximum pump intensity reached 100 GW/cm^2 in this experiment). Overall, these results establish SOS as a promising new platform for integrated nonlinear photonics in the mid-IR.

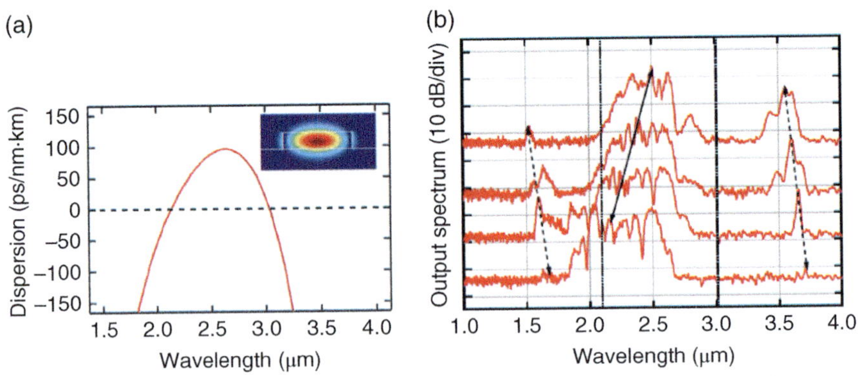

Figure 6.15 (a) Group-velocity dispersion curve for the fundamental TE mode of an SOI waveguide. The waveguide cross section and the mode profile are shown in the inset. (b) Experimentally measured output spectra as the pump was tuned from 2.165 to 2.5 μm. Dashed lines show the resulting shift of dispersive waves generated near 1.5 and 3.6 μm. Zero-GVD wavelengths (2.1 and 3.0 μm) are indicated by vertical lines. *Source:* reproduced from figure 2 of [44], with permission of OSA, The Optical Society.

Typically, SC generation has been performed in waveguides using third-order, $\chi^{(3)}$, nonlinearities. Phillips et al. demonstrated SC generation in a WG based on periodically poled lithium niobate (PPLN), a material with high $\chi^{(2)}$ nonlinearity [46]. The pump was a Tm-doped fiber oscillator–amplifier system at λ = 1.94 μm, with 97-fs pulse duration, 72 MHz repetition rate, and 9-nJ pulse energy (average power 650 mW). The pulses propagated through an 18.5-mm-long PPLN WG with a quasi-phase-matched period of 22.1 μm. The measured SC spectrum spanned from 1.35 to 2.8 μm at −40 dB level. According to the authors, the spectral broadening occurred from cascaded phase shifts associated with phase-mismatched second harmonic generation of the pump (a cascaded $\chi^{(2)}$ nonlinearity leading in effect to a $\chi^{(3)}$ nonlinearity). In addition, SC was enhanced by stimulated Raman scattering and SPM from the intrinsic $\chi^{(3)}$ nonlinearity. Furthermore, the authors have shown that the SC source preserves temporal coherence of the pump and can be used for self-referencing of the Tm fiber oscillator via *f*–2*f* and 2*f*–3*f* interferometry [46].

6.1.6 SC from Bulk Crystals

Broadband mid-IR supercontinua can be obtained by pumping *bulk* materials, although at the expense of high peak powers and low repetition rates. Silva et al. produced a 0.45–4.5-μm SC via filamentation in a 2-mm-thick YAG plate. The YAG sample was pumped with 85-fs pulses at 3.1-μm wavelength, repetition rate 160 kHz, pulse energy 6.9 μJ, and a beam spot size of 50 μm [47]. (See Table 6.1 for more details.) Yu et al. achieved SC spanning 2.5–7.5 μm by pumping a 5-mm-long GeAsS sample with 20-MW-peak-power 150-fs pulses at λ = 5.3 μm in the anomalous dispersion regime [49]. Four-wave mixing and SPM were, according to the authors, the dominant mechanisms for SC generation. Also, the authors indicated that the self-focusing effect could have contributed to SC generation [49]. Liao et al. achieved SC covering 0.2–8.0 μm in an 18-mm-thick piece of ZBLAN fluoride glass through beam filamentation. The researchers used 1.13-MW-peak-power 180-fs pulses at 1.6 μm (normal dispersion regime), with a repetition rate of 1 kHz. The pump-to-SC conversion efficiency was 67% [50].

Multi-octave mid-IR SC generation has been investigated in bulk ZnS and ZnSe, in the regime where the high peak power (~100 MW) of the incoming femtosecond pulses exceeded the critical power of self-focusing (~0.5 MW), thus leading to beam filamentation playing a significant role in the enhancement of the spectral broadening [51–54]. Liang et al. achieved a 0.5–4.5 μm SC span (this span includes second and third harmonics of the pump) in ZnS with the focused λ = 2.1-μm pulses from an OPCPA, having 5-μJ pulse energy, 27-fs duration, and 1-kHz repetition rate [51]. Suminas et al. produced an SC spanning 0.6–4.2 μm in ZnSe using pulses at λ = 2.4 μm (3-μJ pulse energy, 100-fs pulse duration, 1-kHz repetition rate) [52]. Mouawad et al. used

Table 6.1 Summary of supercontinuum (SC) laser sources.

SC medium	Pump laser source	SC span (μm)	Ave. SC power	Ref.
From fibers				
Lead-silicate glass PCF, L = 5.7 mm	λ = 1.55 μm, τ = 110 fs, 80 MHz, 230 mW, 26 kW peak	0.35–3	70 mW	[9]
Tellurite glass suspended core fiber, L = 8 mm	1.55 μm, 110 fs, 80 MHz, 250 mW,28 kW peak	0.79–4.87	90 mW	[10]
Tellurite glass suspended core fiber, L = 3 cm	2.4 μm, 90 fs, 80 MHz, 290 mW, 6.8 kW peak	1.9–3.4	49 mW	[48]
Silica fiber, L = 21 cm; ZBLAN fiber, L = 91 cm	1.55 μm, 900 fs, 1.5 μJ/pulse, 200 kHz, 300 mW, 1.7 MW peak	1.8–3.4	5 mW	[11]
Silica fiber, L = 2 m; ZBLAN fiber, L = 7 m	1.542 μm, 1 ns, 3.3 MHz, 20.2 W, 6 kW peak	0.8–4	10.5 W	[14]
ZBLAN fiber, few meters	2 μm, 24 ps, 94 MHz, 42 W,19 kW peak	1.9–3.8	21.8 W	[15]
ZBLAN fiber, L = 8.4 m	2 μm, 27 ps, 29 MHz, 62 W, 79 kW peak	1.9–4.3	13 W	[16]
ZBLAN fiber, L = 20 mm	1.45 μm, 180 fs, 1 kHz, 50 MW peak	0.4–6.3	~10 mW	[17]
Ag_2S_3 tapered fiber, L = 5 cm	1.55 μm, 250 fs, 39 MHz, 3 mW, 77 pJ, 308 W peak	0.97–2	~3 mW	[26]
Ag_2S_3 tapered fiber, L = 10 cm	1.94 μm, 800 fs, 30 MHz, 15 mW, 625 W peak	1.1–4.4	~10 mW	[32]
Ag_2S_3 tapered fiber, L = 2.1 mm	2.04 μm, <100 fs, 75 MHz, 15 mW, 2 kW peak	1–3.7	~15 mW	[31]
Ag_2S_3 tapered fiber, L = 18 mm	3.1 μm, 100 fs, 100 MHz, 47 mW, 4.7 kW peak	2.2–5	15 mW	[30]
Ag_2S_3 suspended core fiber, L = 20 mm	2.5 μm, 200 fs, 80 MHz, 80 mW, 5 kW peak	0.6–4.1	72 mW	[29]
$Ge_{12}As_{24}Se_{64}$ step-index fiber, L = 11 cm	4 μm, 330 fs, 21 MHz, 1.3 mW, 3 kW peak	1.8–10	~1 mW	[33]
$As_{38}Se_{62}$ suspended core fiber, L = 18 cm	4.4 μm, 330 fs, 21 MHz, 100 mW, 5.2 kW peak	1.7–7.5	15.6 mW	[34]
As_2S_3 step-index fiber, L = 7.2 cm	3.1 μm, 67 fs, 160 kHz, 18 μJ, 520 kW peak	1.6–5.9	8 mW	[35]
As_2S_3 step-index fiber, L = 3.5 m	3–4.2 μm SC, 400 ps, 2 kHz, 27 mW, 34 kW peak	3–8	1.5 mW	[36]

Table 6.1 (Continued)

SC medium	Pump laser source	SC span (μm)	Ave. SC power	Ref.
$As_{40}Se_{60}$ step-index fiber, L = 8.5 cm	6.3 μm, 100 fs, 1 kHz, 0.76 mW, 7 MW peak	1.4–13.3	0.24 mW	[37]
As_2Se_3/As_2S_3 tapered fiber, L = 5 cm	2.9 μm, 230 fs, 42 MHz, 140 mW, 4.2 kW peak	1.8–9.5	30 mW	[38]
$As_2Se_3/AsSe_2$ step-index fiber, L = 3 cm	9.8 μm, 170 fs, 1 kHz, 3 mW, 2.9 MW coupled peak power	2–15.1	μW level	[39]
Ge–Te–AgI double cladding fiber, L = 14 cm	7 μm, 150 fs, 1 kHz, 11.5 mW, 77 MW peak	2–16	—	[40]
From waveguides				
As_2S_3 waveguide, L = 6 cm	1.55 μm, 610 fs, 10 MHz, 1.2 mW, 68 W peak	1.1–2.5	0.6 mW	[41]
As_2S_3-silica nanospike waveguide, L = 5 mm	2.35 μm, ~100 fs, 90 MHz, 200 pJ, 2 kW peak	1.2–3.6	—	[42]
GeAsSe waveguide, L = 10 mm	4 μm, 320 fs, 21 MHz, 160 mW, 3.3 kW peak	1.8–7.5	20 mW	[43]
Si waveguide, L = 20 mm	2.5 μm, 300 fs, 80 MHz, 3 mW, 125 W peak	1.51–3.67	0.36 mW	[44]
Si-on-sapphire (SOS) nanowire, L = 16 mm	3.7 μm, 320 fs, 20 MHz, 12 mW, 1.8 kW peak	1.9–5.5	—	[45]
PPLN waveguide, L = 18.5 mm	1.94 μm, 97 fs, 72 MHz, 650 mW, 93 kW	1.35–2.8	—	[46]
From bulk crystals				
Bulk YAG, L = 2 mm	3.1 μm, 85 fs, 160 kHz, 6.9 μJ, 76 MW peak	0.45–4.5	—	[47]
Bulk GeAsS, L = 5 mm	5.3 μm, 150 fs, 3 μJ, 20 MW peak	2.5–7.5	—	[49]
Bulk ZBLAN, L = 18 mm	1.6 μm, 180 fs, 1 kHz, 0.2 μJ, 1.1 MW peak	0.2–8	—	[50]
Bulk ZnS, L = 8 mm	2.1 μm, 5 μJ, 27 fs, 1 kHz, 185 MW peak	0.5–4.5	—	[51]
Bulk ZnSe, L = 5 mm	2.4 μm, 3 μJ, 100 fs, 1 kHz, 30 MW peak	0.6–4.2	—	[52]
Bulk ZnSe, L = 8 mm	5 μm, 10 μJ, 67 fs, 1 kHz, 150 MW peak	0.5–11	—	[53]

(*Continued*)

Table 6.1 (Continued)

SC medium	Pump laser source	SC span (μm)	Ave. SC power	Ref.
Cr:ZnSe amplifier gain element, $L = 30$ mm	2.35 μm, 10 μJ, 50 fs, 1 kHz, ~200 MW peak	1.8–4.5	—	[54]
Bulk GaAs, $L = 10$ mm	5 μm, 3 μJ, 100 fs, 30 MW peak, focused intensity 6×10^{10} W/cm^2	3.5–7	—	[55]
Bulk GaAs, $L = 6$ cm	9.3 μm, 600 μJ, 2.5–8 ps, 75–240 MW peak, focused intensity 10^{11} W/cm^2	3–14	—	[2]
Bulk GaAs, $L = 6.7$ cm	10.6 μm, 6 mJ, 3 ps, 2 GW peak,focused intensity 10^{10} W/cm^2	2–20	—	[56]
By other methods				
Argon gas, $p = 4.5$ bar, $L = 80$ cm	3.9 μm, 6.5 mJ, 80 fs, 100 GW peak,focused intensity 10^{14} W/cm^2	0.35–5	—	[57]

high-peak-power (10-μJ pulse energy, 1-kHz repetition rate) fs pulses at a longer wavelength ($\lambda = 5$ μm), that is near ZDW (≈4.8 μm) of ZnSe to produce an extremely broad SC spanning 0.5–11 μm in an 8-mm-long bulk ZnSe crystal (Figure 6.16). According to the authors, the underlying physical mechanisms of SC generation were multi-filamentation accompanied by electronic and Raman-induced SPM, self-steepening, and ionization of the medium [53].

Vasilyev et al. demonstrated a compact mid-IR SC generator (10-μJ pulse energy, 1-kHz repetition rate), by taking advantage of the self-broadening effect in a laser amplifier [54]. The device used a single-pass amplifier based on a long (30 mm) polycrystalline Cr-doped ZnSe gain element (pumped by 2-mJ, 100-ns pulses at 1.6 μm from a Q-switched Er:YAG laser) seeded by 20-nJ pulses from a Cr:ZnS oscillator (center $\lambda \approx 2.38$ μm, pulse duration 50 fs). Figure 6.17 shows the amplifier output SC spectrum spanning 1.8–4.5 μm. The spectral broadening mechanism was attributed to SPM facilitated by self-focusing in the ZnSe amplifier medium. The polycrystalline Cr:ZnSe host was specially processed in this work for the large average grain size (500–1000 μm), in order to reduce near-IR and visible emissions, e.g. at the second, third, and fourth harmonic wavelengths that occur due to the random-phase matching process – to reduce the role of detrimental multiphoton ionization in ZnSe [54].

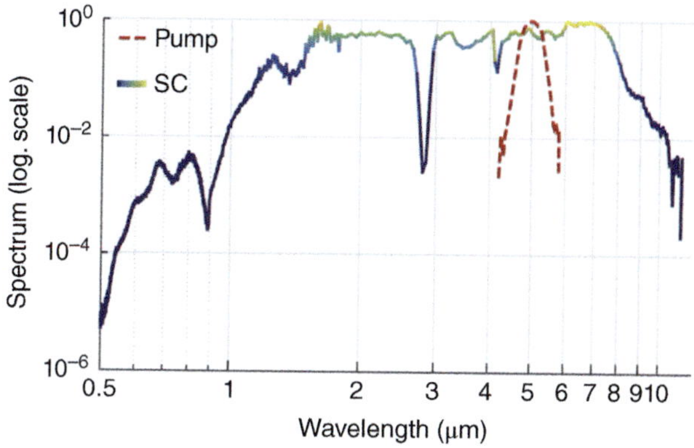

Figure 6.16 SC spectrum obtained in a bulk ZnSe crystal pumped by 100-MW peak power pulses at $\lambda = 5$ μm. The input pump spectrum is shown as a dashed red line. *Source:* reproduced from figure 2 of [53], with permission of Elsevier.

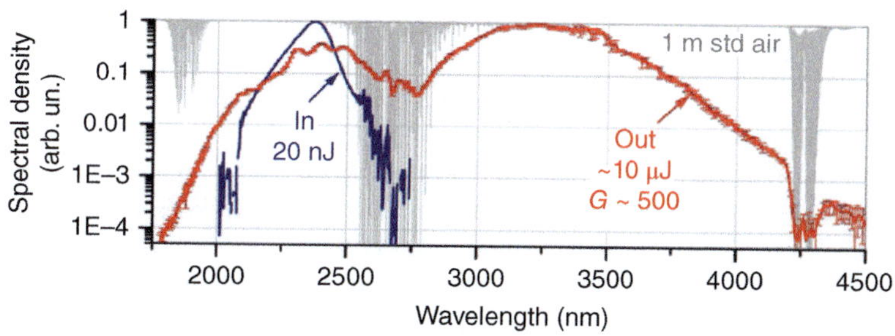

Figure 6.17 SC spectrum, produced in a single-pass high-gain ($G = 500$) Cr:ZnSe amplifier at 1 kHz repetition rate (red curve), along with the spectrum of the input pulses centered at 2.38 μm (blue curve). Shown in gray is the transmission of 1-m standard air. *Source:* reproduced from figure 10 of [54], with permission of OSA, The Optical Society.

With high-peak-power 5-μm femtosecond pump, Ashihara et al. demonstrated a SC spanning 3.5–7 μm (Figure 6.18) in bulk GaAs crystal, taking advantage of its high third-order Kerr nonlinearity ($n_2 \approx 10^{-17}$ m^2/W) [55]. The $\lambda = 5$ μm pulses (corresponding to the normal dispersion regime in GaAs) with 3-μJ energy and 100-fs duration were focused into a 10-mm-long GaAs crystal (beam diameter 240 μm, propagation direction <110>), such that the peak intensity at the entrance surface was 60 GW/cm^2. At half maximum (−3 dB) level, the authors observed the increase of the pump spectral bandwidth by more than four times: from 220 to 910 cm^{-1}. They attributed the main mechanism for broadening to SPM in GaAs. The role of cascaded $\chi^{(2)}$ process on SPM was estimated to be small, ~7% of that for the intrinsic Kerr nonlinearity.

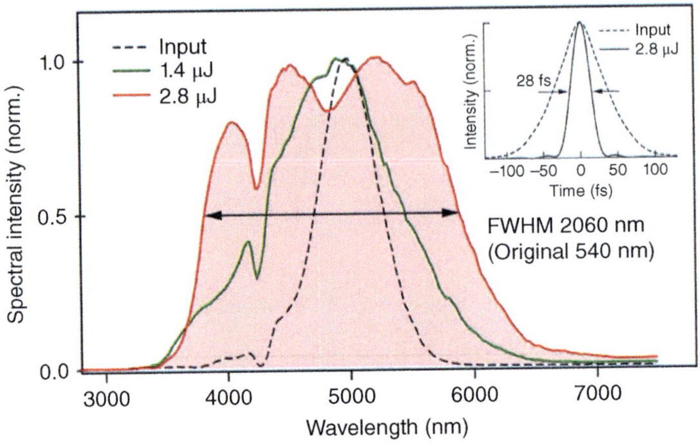

Figure 6.18 SC spectrum from bulk GaAs pumped at $\lambda = 5\,\mu m$. Input spectrum (dashed) and output spectra (solid) for the pump pulse energies of 1.4 and 2.8 μJ, respectively. Inset shows calculated intensity time profiles of the transform-limited pulses corresponding to the measured spectra. *Source:* reproduced from figure 1 of [55], with permission of OSA, The Optical Society.

Pigeon et al. reported on the generation of an SC from 2 to 20 μm (although with some gaps) in a 67-mm-long bulk GaAs crystal pumped by a train of 3-ps CO_2 laser pulses at $\lambda = 10.6\,\mu m$ (corresponding to the anomalous dispersion regime in GaAs), with 2-GW peak power and 10-GW/cm^2 incoming peak intensity [56]. Temporal measurements performed by the authors revealed splitting of the original pulse to sub-picosecond pulses, with harmonic generation, four-wave mixing, and stimulated Raman scattering being the dominant mechanisms for SC generation.

6.1.7 Other SC Sources

Kartashev et al. reported on a mid-IR SC generated by filamentation in argon gas. Experiments were performed with an OPCPA system delivering pulses centered at 3.9 μm with up to 12-mJ energy at 80-fs duration (peak power > 130 GW), at a repetition rate of 20 Hz. At 4.5-bar pressure of the argon gas, the authors observed a white-light beam surrounded by a ring conical emission. The measured spectrum spanned continuously from 0.35 to 5 μm. The filament in argon was measured to be 80 cm in length and 200–300 μm in diameter with a filamentation threshold pressure of around 3.5 bars. The intensity in the filament was estimated to be 45 TW/cm^2 [57].

The main results for mid-IR SC generation are summarized in Table 6.1. Also, a very good overview of experimental results on mid-IR SC generation can be found in [49].

6.2 Frequency Comb Sources

Frequency combs are typically produced by highly stabilized femtosecond lasers and are characterized by a broad spectrum consisting of evenly spaced, phase-coherent narrow spectral lines with the spectral width that can be as narrow as a few millihertz [58]. The output of a mode-locked laser can be regarded as a frequency comb, provided that the laser repetition rate, f_{rep}, and the comb offset from zero – the carrier-envelope offset (CEO) frequency, f_{CEO} – are both stabilized to a high degree of accuracy. (In other words, it is not only that the mode spacing remains constant, but that the absolute position of each mode is fixed as well.) In the time domain, the CEO is directly related to the exact position of the electric field underneath the pulse envelope. This enables generation of optical waveforms with deterministic electric field profile [59]. Extending the operating range of broadband frequency combs to the mid-IR is critical for a wide range of applications where temporal coherence and reproducibility of the optical field from pulse to pulse play a significant role. These include X-ray production via high harmonic generation [60], attosecond physics [61], laser-driven particle acceleration [62], dual-comb laser spectroscopy [63, 64], and molecular fingerprinting [65].

At the heart of the distinction between a broadband frequency comb and a broadband SC source is the shot-to-shot reproducibility of the optical waveform. For example, if a frequency comb is created by spectral broadening in a nonlinear (NL) fiber or a WG, one requirement is that the pulses should remain phase-stable after the nonlinear process of spectral evolution. The phase-stability should not be sensitive to the inevitable shot-to-shot fluctuations of the drive laser [66]. For example, SPM can be considered to be a deterministic coherence-preserving process [67], while SC generation due to modulation instability, which is seeded by the random fluctuations of the vacuum field, does not preserve coherence. (One test for phase coherence is whether neighboring pulses show stable interference.) In fact, the SC coherence depends strongly on the duration and wavelength of the input pulse, the length of the WG, and other experimental conditions [5, 66].

In this chapter we will consider several techniques for producing frequency combs including those based on:

- mode-locked lasers
- frequency conversion using second-order nonlinearities
- NL fibers and waveguides
- microresonators
- quantum cascade and interband cascade lasers.

An outstanding review by Schliesser et al. [68] surveys different techniques for producing frequency combs in the mid-IR region, as well as comb applications.

6.2.1 Direct Comb Sources from Mode-locked Lasers

Thulium-doped silica fiber lasers operate near $\lambda \approx 2\,\mu m$ and mode-locking in these lasers has been achieved through a variety of techniques (see Chapter 3). Amplifiers based on Tm-doped fibers can boost the power of these lasers to several watts. Lee et al. have demonstrated a fully stabilized frequency comb at $\lambda \approx 1.95\,\mu m$ using a fast intracavity graphene electro-optic modulator (EOM) [69, 70]. A Tm-fiber oscillator (Figure 6.19) was pumped at 1564 nm, with a total power 1.3 W. (Additionally up to 10 mW of the pump was provided by a second single-frequency diode laser, which allowed fast pump modulation.) Stable mode-locked operation with 100 MHz repetition rate, 20 mW of average output power, was achieved by a saturable-absorber mirror. Next, the oscillator output was amplified to 200 mW to generate a 1.1–2.4-μm SC in a highly NL fiber. The CEO frequency, f_{CEO}, was detected by an f–$2f$ interferometer at 1100 nm, involving a PPLN frequency doubling crystal. The f_{CEO} beat was phase-locked to a stable RF synthesizer via feedback to an intracavity graphene EOM with 5% insertion loss, 2% modulation depth, and 600-kHz bandwidth (Figure 6.19). Alternatively, f_{CEO} could be locked by fast pump power modulation. Using a slice of the generated SC around 1.55 μm, the laser comb was also

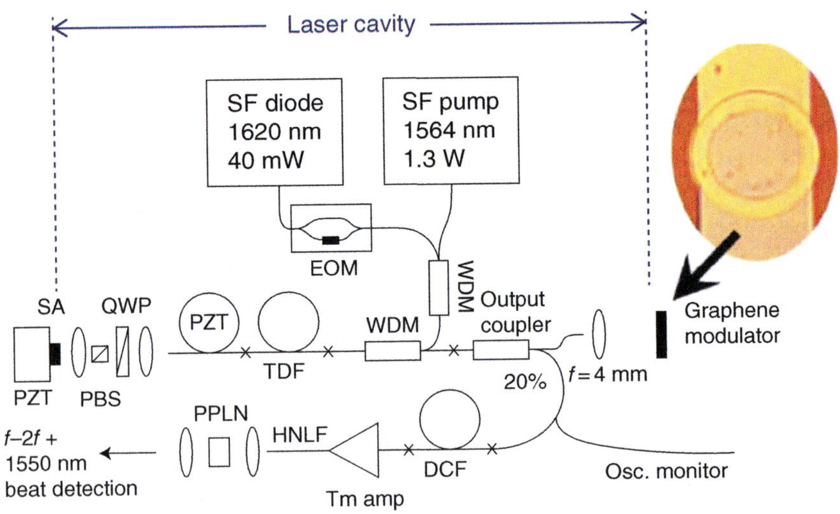

Figure 6.19 Fully stabilized frequency comb based on $\lambda = 2$-μm mode-locked thulium-doped silica fiber laser with a graphene modulator. WDM, wavelength division multiplexer; TDF, thulium-doped fiber; DCF, dispersion compensating fiber; PZT, piezoelectric actuator for repetition rate stabilization; QWP, quarter-wave plate; PBS, polarization beam splitter; SA, saturable-absorber mirror; HNLF, highly nonlinear fiber for SC generation; PPLN, periodically poled lithium niobate. The inset is an optical microscope image of the modulator (diameter 280 μm). *Source:* reproduced from figure 1 of [69], with permission of OSA, The Optical Society.

phase-locked to a narrow-linewidth (3 kHz) optical reference laser near 1.5 μm. This was done by detecting a heterodyne beat between one of the comb teeth and the reference laser, and by phase-locking it by feedback to the lead zirconate titanate (PZT) piezoelectric actuator cavity-length control. Generally, state-of-the-art 2-μm Tm-fiber frequency comb sources now produce sub-70-fs pulses with a bandwidth of 100 nm ($260\,cm^{-1}$) and an average power of 400 mW for comb-mode spacing 100 MHz, or as high as 2.5 W for comb-mode spacing 400 MHz [71, 72].

A promising class of solid-state mode-locked lasers suitable for producing broadband frequency combs is represented by the family of lasers based on transition metal ions such as Cr^{2+} or Fe^{2+} doped into group II–VI polycrystals operating correspondingly in the 2–3 and 4–5 μm regions (see Chapter 2). For example, Cr^{2+}:ZnSe and Cr^{2+}:ZnS lasers with center wavelength of 2.3–2.5 μm have broad emission bands exceeding 1000 nm and are regarded as mid-IR analogs of the Ti:sapphire laser, in terms of extremely broad emission band and high stimulated emission cross section [73]. In addition, these lasers can be conveniently pumped by robust erbium or thulium fiber lasers. A spectral span of 1.9–2.7 μm (bandwidth >$1500\,cm^{-1}$) has been demonstrated from a Kerr-lens mode-locked Cr:ZnS oscillator, with pulse duration down to few optical cycles [54]. In a recent paper, Vasilyev et al. reported an octave-wide (1.6–3.2 μm at −40 dB) span of an Cr:ZnS oscillator emitting few-cycle pulses with 4-W average power and 78-MHz repetition rate [74]. A unique feature of this system is that it enables the intrinsic common-path nonlinear interferometry: the CEO frequency, f_{CEO}, was detected by observing $3f$–$4f$ beat signal in the visible range (~633 nm) with a signal-to-noise ratio of 40 dB. The visible signal originated directly from the laser resonator, with the frequency beats being the result of interfering the broadband third and fourth harmonics of the mid-IR spectrum, both produced via random phase-matching in the Cr:ZnS polycrystal – a $\chi^{(2)}$ material. A fully referenced Cr:ZnS optical frequency comb has been reported in [75].

There is also ongoing research into ultrafast Er^{3+}-doped fluoride glass fiber lasers operating at longer (~3 μm) wavelengths that can potentially serve as frequency combs [76] (see Chapter 3).

6.2.2 Combs Produced by Spectral Broadening in NL Fibers and Waveguides

Generation of a broadband SC in a fiber pumped by a phase-coherent laser does not guarantee obtaining a phase-coherent frequency comb owing to the excess noise produced during SC generation. To ensure that the coherence is preserved, it is important to use a short (a few cm) piece of NL fiber in combination with short (~100 fs or less) laser pulses. Using a highly nonlinear fiber (HNLF) consisting of a small-effective-area (~14 μm^2) dispersion-flattened germanium-doped silica fiber, Washburn et al. [77, 78] demonstrated a phase-locked frequency comb spanning from 1100 to 2200 nm. The pump was a mode-locked erbium-doped fiber laser at 1.55 μm with an average power of 100 mW and 70-fs pulse duration. The SC output was a frequency comb with both spacing and CEO frequency that was set by the pump laser. The coherence of the comb has been proven by detecting frequency beats using the f-to-$2f$ heterodyne technique.

Marandi et al. have studied coherence properties of the 2.2–5 μm SC that was produced in an 18-mm-long tapered As_2S_3 fiber by pumping it at $\lambda \approx 3.1$ μm (a subharmonic of a femtosecond 1.56-μm Er-fiber laser; see also Section 6.1.4) [30]. To verify that the broad mid-IR SC output preserves the frequency comb structure of the laser, the frequency-doubled SC was interfered with a continuous-wave (CW) narrow-linewidth (3 kHz) laser at 1564 nm and the beat frequency was measured. Another beat frequency signal was obtained by interfering the mode-locked Er-fiber laser and the CW laser. The two beat frequencies were the same and tracked each other during tuning of the CEO frequency of the Er-fiber laser. This verified that the coherence of the 1.56-μm pump was preserved through the two frequency downconversion processes: (i) subharmonic generation in a subharmonic OPO and (ii) subsequent spectral broadening in the tapered fiber.

Generation of a phase-coherent broadband 1.8–3.8 μm SC was reported in an As_2S_3 WG embedded in silica [79]. The nonlinear WG consisted of a tapered silica capillary filled with As_2S_3. To create this structure, melted chalcogenide glass was pushed into the capillary by pressurized argon. The device was only 2 mm long with the first 0.3 mm being a tapered region. The untapered region had an inner diameter of 1 μm (Figure 6.20a) and the two ZDWs were at 1.35 and 2.5 μm. The pump pulses from a 2-μm Tm-doped fiber laser (pulse duration 65 fs, repetition rate 100 MHz) were launched into the WG from the tapered side, such that the incident light was transformed adiabatically into the fundamental mode of the As_2S_3 WG with a coupling efficiency of ~12%. Soliton fission and dispersive wave generation along the uniform section resulted in spectral broadening out to almost 4 μm (Figure 6.20b) for launched energies of only 18 pJ (the total pump average power used was 7 mW). In another experiment, coherence properties of the SC produced in this “nanospike” WG were tested. The pump was a fully stabilized 2-μm Tm-fiber frequency comb. A portion of the nanospike SC near λ = 3.3 μm was combined with the corresponding portion of the output of an OP-GaAs OPO that was itself coherent with the shared pump laser. A strong interference (a radiofrequency beat note) between the SC and the OPO output showed that SC generation in the WG was coherence-conserving [80].

Kuyken et al. used an SOI WG to generate a phase-coherent broadband frequency comb. As in the previous example, the comb generator required low

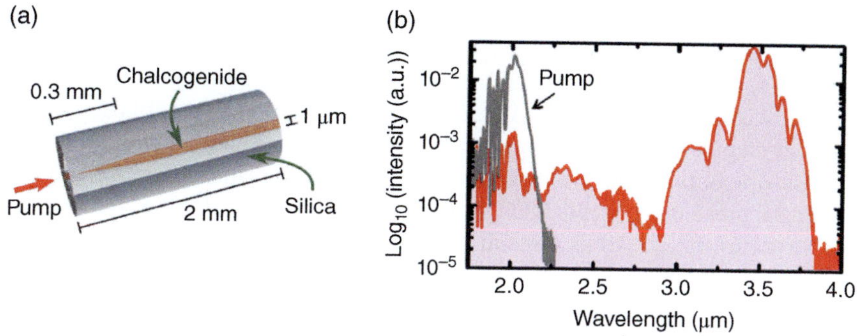

Figure 6.20 (a) Chalcogenide nanospike embedded in silica; light enters from the tapered side on the left. (b) Broadband spectrum of the supercontinuum produced after propagating along the chalcogenide-silica waveguide (sample length 1.7 mm). *Source:* reproduced from figure 6 of [79], with permission of OSA, The Optical Society.

pump power; it had a 30-dB bandwidth spanning from 1.54 to 3.2 μm [81]. A 1-cm-long air-clad photonic wire grown on top of a buried oxide layer had a rectangular cross section of 1600 × 390 nm. The pump source was tuned to 2.29 μm, close to the ZDW of the WG (2.18 μm), and delivered 35 mW of average power at 100 MHz repetition rate and 70-fs pulse duration. The broad spectrum was obtained with a coupled pulse energy of as low as 16 pJ. Coherence of the SC was confirmed by beat note measurements with a set of narrow linewidth CW lasers. The authors argue that the use of short pulses favored dispersive wave generation and SPM that maintain the coherence in the pump.

6.2.3 Combs Produced by Difference Frequency Generation

DFG is a widespread method of producing frequency combs in the mid-IR. Thanks to the coherent nature of the three-photon process, coherence of a near-IR pump comb is directly transferred to the mid-IR. For example, DFG can be performed by mixing outputs of a mode-locked and a CW laser. In this way, by mixing a 1.56-μm comb with a stable CEO frequency with the output of a tunable (1030–1070 nm) CW laser in a PPLN crystal, Maddaloni et al. created a new comb with central wavelength ranging from 2.9 to 3.5 μm with an instantaneous wavelength span of 180 nm and an average power of 5 μW [82].

Galli et al. presented a highly coherent mid-IR comb produced through an *intracavity* DFG process [83]. The output of a mode-locked Ti:sapphire laser was first spectrally broadened to 500–1100 nm by a PCF. The 1040-nm portion of this spectrum was amplified in an Yb-doped fiber and then mixed with the intracavity radiation inside the mode-locked Ti:sapphire laser, using a MgO:PPLN nonlinear crystal. A mid-IR comb was generated with an average power of 0.5 mW and 2-kHz tooth linewidth. The generated spectrum spanned 27 nm, with a center wavelength tunable from 4.2 to 5 μm. Also, the high repetition rate (1 GHz) ensured high spectral brightness (1 μW per comb tooth) [83].

High efficiency can be achieved in a DFG process that uses two phase-coherent amplified pulse trains that originate from the same oscillator [84–89]. In such a process, one of the pulse trains is frequency red- or blueshifted in an NL fiber. The mid-IR pulse train in this DFG scenario is expected to exhibit a "harmonic" comb structure ($f_{CEO} = 0$) as a result of the cancelation of CEO frequencies by subtracting frequencies of the two near-IR combs.

Erny et al. [84] generated mid-IR frequency combs by performing DFG between the outputs of two near-IR sources coherently linked to the same femtosecond mode-locked fiber oscillator. The oscillator (Figure 6.21a) seeded two separate amplifier stages, each delivering 65-fs pulses centered at ~1.58 μm with an average power of 250 mW and a repetition rate of 82 MHz. The output of the second amplifier was coupled into a highly nonlinear fiber (HNLF, core diameter of 3.7 μm, ZDW ~1.52 μm). The HNLF output was tunable near $\lambda \approx 1$ μm and was compressed with a prism pair to sub-40-fs duration. By nonlinear mixing of 1.58-μm pulses (average power 170 mW) with near-infrared pulses tunable between 1.05 and 1.18 μm (average power 11.5 mW) in a 2-mm-long PPLN crystal, the authors achieved femtosecond pulses tunable in the 3.2–4.8 μm range (Figure 6.21b) with an average power of 1.1 mW.

Using the same strategy of mixing the outputs of two near-IR fiber amplifiers (Yb-doped and Er-doped) seeded by the same oscillator, several groups have reported high-power (>100 mW) mid-IR frequency combs with the spectrum coverage 3–4.4 [85], 2.9–3.6 [86], and 2.3–3.6 μm [87]. Using a 3-mm-long MgO:PPLN crystal, Cruz et al. produced mid-IR femtosecond frequency combs with the average power 530 mW (3 μW/mode at 100-MHz mode

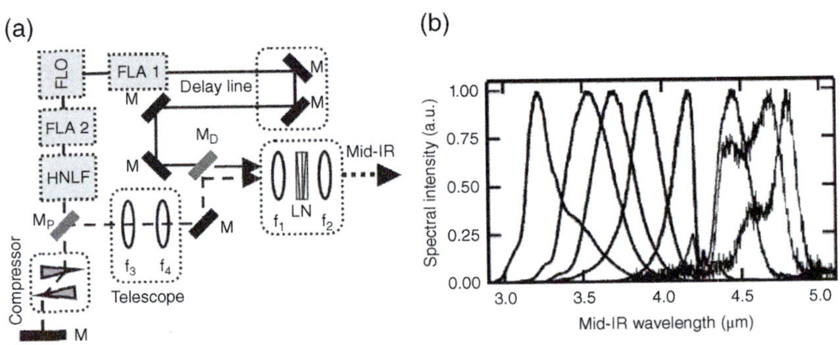

Figure 6.21 (a) Experimental DFG setup of [84]. FLO, fiber-laser oscillator; FLA1 and FLA2, fiber-laser amplifiers; HNLF, highly nonlinear fiber; LN, MgO-doped PPLN nonlinear crystal. (b) Normalized mid-IR DFG spectra. *Source:* reproduced from figures 1 and 2 of [84], with permission of OSA, The Optical Society.

spacing) with a spectral covering of 2.8–3.5 μm [88]. Also, Maser et al. demonstrated a *dual-comb* DFG system suitable for mid-IR molecular spectroscopy. The center wavelength was tunable between 2.6 and 5.2 μm with an instantaneous bandwidth of 33–230 cm^{-1} [89].

Through DFG in a 1-mm-long orientation-patterned gallium phosphide (OP-GaP) with multiple grating periods, Lee et al. generated frequency combs with the center wavelength varying from 6 to 11 μm, with 350–400 cm^{-1} instantaneous frequency span and up to 60 mW average power [90]. The DFG "pump" and "signal" waves were correspondingly the outputs of erbium and thulium fiber amplifiers seeded by the same erbium-fiber mode-locked oscillator, whose repetition rate was tightly stabilized to ~1 Hz by optical referencing to a narrow-linewidth diode laser.

DFG frequency combs with long-wavelength (>10 μm) coverage were attained with GaSe as a frequency mixer, thanks to its broad transparency range (0.6–20 μm) combined with high optical nonlinearity [91–94]. An extremely broad tuning of 4–17 μm, which covers most of the infrared "fingerprint" molecular vibration region, was reported by Keilmann and Amarie [94]. The mixing occurred in a 1-mm-long angle-tuned GaSe crystal. As a pump, the authors used amplified femtosecond pulses at 1.55 μm and SC pulses in the 1.7–2.3 μm region produced in a nonlinear optical fiber (Figure 6.22a). The near-IR SC was spectrally tuned by changing the chirp of the driving 1.55-μm pulses, and this tuned the spectrum of the DFG pulses. The resulting broadly tunable mid-IR harmonic frequency comb had an instantaneous spectral width of approximately 700 cm^{-1} (at −20 dB level, Figure 6.22b) and an average power of 1 mW [94].

Another way for generating longwave infrared (LWIR, 5–20 μm) frequency combs is through intra-pulse DFG (optical rectification, see Section 5.4.2).

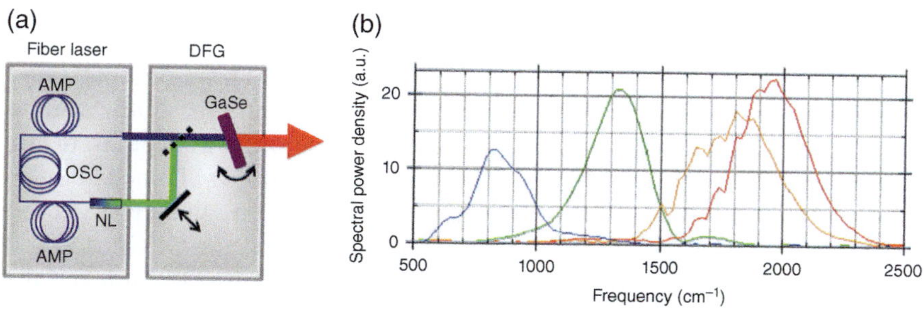

Figure 6.22 (a) Schematics of the long-wavelength mid-IR comb source [94]. The fiber laser system contains a common oscillator (OSC), two amplifiers (AMP), and a nonlinear optical fiber (NL) producing a redshifted near-IR continuum. The DFG unit combines both free-space output beams at zero pulse delay in a GaSe crystal which generates the mid-IR beam. (b) Corresponding DFG spectra. *Source:* reproduced from figures 1 and 2 of [94], with permission of Springer.

Timmers et al. presented a robust method for generating super-octave spanning LWIR frequency combs through intra-pulse DFG driven by spectrally broadened and compressed few-cycle pulses with 10.6-fs duration from an erbium fiber system [65]. Because LWIR modes occur as a difference between the n-th and m-th pump modes (m and n are integers), the CEO frequency of the pump is subtracted out. This provides an offset-free (harmonic) LWIR comb consisting of exact harmonics of the laser repetition frequency, $\nu_k = kf_{rep}$, where k is an integer. For frequency self-mixing, the authors used a 1-mm-thick OP-GaP crystal. With the pump power of 350 mW, they produced a frequency comb spanning 4–12 μm with 0.25 mW of the average power [65].

The main results on DFG combs are summarized in Table 6.2.

6.2.4 OPO-based Combs

An OPO needs only one pump laser, and typically provides much higher conversion efficiency and much higher average power, as compared to DFG. This comes at the expense of a more delicate setup that includes a stabilized synchronously pumped ("sync-pumped") resonant cavity. While the OPO automatically inherits the repetition rate of the pump laser, a fixed CEO frequency of the pump does not automatically guarantee that the CEO frequencies of the OPO "signal" and "idler" waves are fixed. The photon energy conservation law just connects the three CEO frequencies in the following way:

$$f_{CEO}(p) = f_{CEO}(s) + f_{CEO}(i), \tag{6.1}$$

where "p," "s," and "i" stand for pump, signal, and idler waves. The corresponding angular frequencies ω_p, ω_s, and ω_i are related (again, through the photon energy conservation law) as $\omega_p = \omega_s + \omega_i$. Thus, a separate CEO frequency control is required for the OPO waves.

Sun et al. produced a frequency comb ranging from the violet to the mid-infrared (0.4–2.4 μm) by phase locking a pump laser and a sync-pumped OPO to a common reference [118]. The mode-locked Ti:sapphire 800-nm pump laser (1.3 W average output power, 50-fs pulse duration) was repetition-rate locked to an external 200-MHz clock and part of its power was launched into a 30-cm PCF to generate an SC to stabilize the laser CEO frequency via f-to-$2f$ interferometry. The main laser output was used to pump the OPO, based on MgO-doped PPLN with the signal output tunable from 1.2 to 1.37 μm and the idler from 2.4 to 1.9 μm. By using the OPO combinational outputs at frequencies $2\omega_s$ (red) and $\omega_p + \omega_i$ (yellow) and measuring beat notes with the common SC reference, the OPO comb was CEO frequency locked through the use of a piezoelectric transducer (PZT) mounted on an OPO cavity end mirror.

Adler and coauthors reported on a *high-power* (~1.5 W) fully stabilized OPO mid-IR comb [95, 119]. The linear-cavity OPO was synchronously pumped by a

Table 6.2 Summary of mid-IR frequency combs.

Method	Pump source	Instantaneous comb span	Ave. power	Mode spacing	Ref.
Direct mode-locked laser combs					
Mode-locked Tm fiber laser	CW Er-fiber	1.9–2.0 μm	2.5 W	418 MHz	[72]
Kerr-lens mode-locked Cr:ZnS laser	CW Er-fiber	1.79–2.86 μm	3.25 W	80 MHz	[75]
Kerr-lens mode-locked Cr:ZnS laser	CW Er-fiber	1.6–3.2 μm	4.1 W	78 MHz	[74]
Combs based on SC generation					
Nanospike As_2S_3 waveguide in silica, L = 1.7 mm	2 μm, 65 fs, 7 mW, 1 kW peak	1.8–3.7 μm	1.8 mW	100 MHz	[79]
Si waveguide, L = 10 mm	2.3 μm, 70 fs, 35 mW, 230 W peak	1.54–3.2 μm	1.6 mW	100 MHz	[81]
DFG combs					
DFG, PPLN	CW 1.03–1.07 μm (0.7 W),fs 1.56-μm (0.7 W)	180 nm, tune 2.9–3.5 μm	5 μW	100 MHz	[82]
Resonant DFG, PPLN	fs 0.8-μm (30 W intracavity),fs 1-μm SC	27 nm, tune 4.2–5 μm	0.5 mW	1 GHz	[83]
DFG, PPLN	fs 1.05–1.18 μm SC (11 mW),fs 1.58 μm (170 mW)	500 nm, tune 3.2–4.8 μm	1.1 mW	82 MHz	[84]
DFG, PPLN	fs 1-μm + fs 1.5-μm Raman (2.4 W total)	170 nm, tune 3–4.4 μm	128 mW	100 MHz	[85]

(*Continued*)

Table 6.2 (Continued)

Method	Pump source	Instantaneous comb span	Ave. power	Mode spacing	Ref.
DFG, PPLN	fs 1.55 μm (450 mW), fs 1.05 μm SC (1.2 W)	2.9–3.6 μm	120 mW	250 MHz	[86]
DFG, PPLN	fs 1.04 μm (1.4 W), fs 1.55 μm (400 mW)	750 nm, tune 2.3–3.6 μm	150 mW (3.1 μm)	100 MHz	[87]
DFG, PPLN	fs 1.06 μm (4 W), fs 1.55 μm (140 mW)	700 nm, tune 2.8–3.5 μm	530 mW (3 μm)	100 MHz	[88]
DFG, PPLN dual comb	fs 1-μm SC (735 mW), fs 1.3–1.5-μm SC (81 mW)	33–230 cm^{-1}, tune 2.6–5.2 μm	20 mW (3.5 μm)	100 MHz	[89]
DFG, OP-GaP	fs 1.55 μm (1.6 W), fs 1.8–2.1 μm (0.6 W)	35–400 cm^{-1}, tune 6–11 μm	60 mW (8 μm)	93 MHz	[90]
DFG, GaSe	fs 1.55 μm (250 mW), fs 1.75–1.95 μm SC	200 cm^{-1}, tune 5–12 μm	160 μW (6.3 μm)	100 MHz	[91]
DFG, GaSe	fs 1.55 μm (550 mW), fs 1.76–1.93 μm SC(250 mW)	~300 cm^{-1}, tune 8–14 μm	4 mW (7.8 μm)	250 MHz	[92]
DFG, GaSe	fs 1.055 μm (1.9 W), fs 1.18–1.63 μm SC (20 mW)	~300 cm^{-1}, tune 3–10 μm	1.5 mW (4.7 μm)	151 MHz	[93]
DFG, GaSe	fs 1.5 μm (360 mW), fs 1.7–2.3 μm SC (160 mW)	~700 cm^{-1}, tune 4–17 μm	1 mW	40 MHz	[94]
Self-mixing, OP-GaP	10.6 fs, 1.56 μm, 350 mW	4–12 μm	0.25 mW	100 MHz	[65]
OPO combs					
PPLN OPO	fs Yb-fiber, 1.07 μm, 10 W	300 nm, tune 2.8–4.8 μm	1.5 W	136 MHz	[95]
PPKTP OPO	20-fs, Ti:Sapph.,0.8 μm, 1.4 W	200 nm, tune 0.4–3.2 μm	~mW	100 MHz	[96]

PPLN OPO	Chirped 3-ps, Yb-fiber, 1.05 μm, 2.2 W	200 nm, @ 3.6 μm	144 mW	94 MHz	[97]
PPLN OPO	fs, Yb-fiber, 1.04 μm, 5 W	300 nm, tune 2.25–4 μm	160 mW	51 MHz	[98]
PPLN OPO	fs, Yb-fiber, 1.04 μm, 2 W	350 nm, tune 2.7–4.7 μm	250 mW	90 MHz	[99, 100]
PPLN OPO	fs, Er-fiber, 1.56 μm, 580 mW	100–200 nm, tune 2.25–2.6 and 4.1–4.9 μm	20–60 mW	250 MHz	[101]
$AgGaSe_2$ OPO	fs, Tm-fiber, 1.95 μm, 2.5 W	500–700 nm, tune 8.4–9.5 μm	100 mW	110 MHz	[101]
Subharmonic OPO combs					
PPLN OPO	fs Er-fiber, 1.56 μm,300 mW	2.5–3.8 μm	60 mW	100 MHz	[102]
OP-GaP OPO	fs Er-fiber, 1.56 μm,300 mW	2.35–4.8 μm	29 mW	100 MHz	[103]
OP-GaAs OPO	fs Tm-fiber, 1.93 μm, 330 mW	2.6–7.5 μm	75 mW	115 MHz	[104]
OP-GaAs OPO	fs Cr:ZnS, 2.35 μm,650 mW	2.85–8.4 μm	110 mW	80 MHz	[105]
OP-GaAs OPO	fs Cr:ZnS, 2.35 μm, 6 W	3–8 μm	500 mW	900 MHz	[106]
OP-GaP OPO	fs Cr:ZnS, 2.35 μm, 1.2 W	3–12.5 μm	31 mW	79 MHz	[107]
Microresonator combs					
SiO_2 toroid	1.56 μm, 2.5 W	1–2.17 μm	—	850 GHz	[108]
SiN microring	1.56 μm, 2 W	1.17–2.35 μm	—	226 GHz	[109]
Si waveguide resonator	CW OPO, 2.59 μm, 1.2 W, 100 kHz linewidth	2.1–3.5 μm	—	127 GHz	[110]
Si waveguide resonator	CW OPO, 2.6 μm, 180 mW	2.46–4.28 μm	—	127 GHz	[111]
MgF_2 microresonator	CW OPO, 2.45 μm, 600 mW	2.35–2.55 μm	—	107 GHz	[112]

(*Continued*)

Table 6.2 (Continued)

Method	Pump source	Instantaneous comb span	Ave. power	Mode spacing	Ref.
MgF_2 disk	QCL, 4.4 μm, 80 mW	4.16–4.76 μm	—	14.3 GHz	[113]
MgF_2 disk	QCL, 4.5 μm, 55 mW	3.7–5.5 μm	—	2100 GHz	[114]
QCL combs					
InGaAs/InAlAs	Electric current	6.8–7.3 μm	Few milliwatts	7.5 GHz	[115]
InGaAs/InAlAs	Electric current	110 cm^{-1} @ 8 μm	880 mW	11.2 GHz	[116]
ICL combs					
InAs/GaInSb/AlSb	Electric current	35 cm^{-1} @ 3.6 μm	~6 mW	9.7 GHz	[117]

10-W femtosecond Yb-fiber laser centered at 1.07 μm and employed a 7-mm PPLN crystal with a variable poling period. The OPO idler wave was tunable from 2.8 to 4.8 μm with an instantaneous bandwidth of ~300 nm. The OPO comb stabilization was done by exploiting (as in the previous example) parasitic mixing signals in the visible ($\omega_p + \omega_s$) and near-IR ($\omega_\mathrm{p} + \omega_\mathrm{i}$) produced inside the OPO cavity, that were superimposed with a broad SC generated by a portion of the pump laser output in a PCF. Since the SC has f_{CEO} corresponding to that of the pump laser, the observed heterodyne beats corresponded to $f_{CEO}(s)$ and $f_{CEO}(i)$. Typically, $f_{CEO}(s)$ was locked to a stable RF reference via stabilizing the OPO cavity length with a piezo-actuated end mirror. On the other hand, $f_{CEO}(i)$ was stabilized via the relation (6.1) by feeding the error signal back to the fiber laser pump diode power, which acts on $f_{CEO}(p)$. The maximum idler power of ~1.5 W was obtained at 3.2 μm center wavelength, whereas the highest photon conversion efficiency (51%) was observed at 3.6 μm. The typical OPO idler spectra are shown in Figure 6.23.

Iwakuni et al. produced a long-wavelength phase-stabilized frequency comb in a synchronously pumped, singly resonant OPO, based on an angular phase-matched $AgGaSe_2$ crystal [120]. The pump laser was a mode-locked Tm:fiber laser at 1.95 μm, with a repetition rate of 110 MHz, and a maximum output power of 2.5 W. The OPO was continuously tuned, with the center wavelength range of 8.4–9.5 μm (instantaneous bandwidth 500–700 nm) and a maximum average idler power of 100 mW at $\lambda \approx 8.5$ μm. Both the repetition rate and the CEO frequency of the idler wave were phase-locked to microwave signals referenced to a Cs clock. The CEO of the idler wave was achieved by stabilizing the heterodyne optical beat of the sum frequency of the pump and idler wave ($\omega_p + \omega_i$), parasitically produced in the OPO cavity, and spectrally broadened pump wave (ω_p) using a highly nonlinear fiber.

Results for other frequency comb sources based on *singly resonant* OPOs pumped by Ti:sapphire [96], Yb-fiber [97–100], and Er-fiber lasers [101] are summarized in Table 6.2. The reader can find a very good review paper on frequency combs based on femtosecond OPOs in [121].

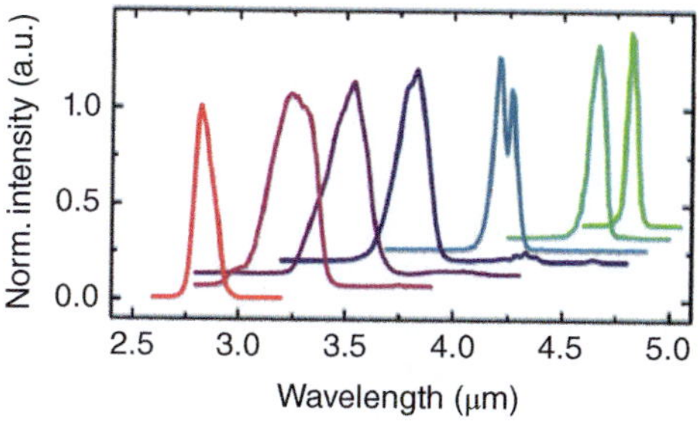

Figure 6.23 Typical PPLN-based OPO idler comb spectra of [95], normalized and vertically offset. *Source:* reproduced from figure 3 of [95] with permission of OSA, The Optical Society.

6.2.5 Combs Based on Optical Subharmonic Generation

A new method was implemented, suitable for generating extremely broadband mid-IR coherent outputs based on a degenerate (subharmonic) sync-pumped OPO [102, 122, 123]. A schematic of such an OPO is shown in Figure 6.24. The bottom panel displays, from top to bottom, the evolution of the spectrum when the OPO is driven from a nondegenerate to a partly degenerate, and finally to a degenerate (subharmonic) regime. In the last case, the co-polarized signal and the idler modes become indistinguishable, which results in some interesting properties.

For example, the subharmonic OPO source rigorously both down-converts (from the center frequency ω_p to the center frequency $\omega_p/2$) and dramatically augments the spectrum of a pump frequency comb provided by a mode-locked pump laser. The key advantages of the degenerate OPO approach are: (i) low pump threshold (typically ~10 mW) owing to the double (signal and idler) resonance, (ii) potential for high conversion efficiency (>50%) due to a non-dissipative nature of three-wave processes and recycling of all the generated photons, (iii) broad gain bandwidth at the OPO degeneracy, and (iv) phase locking to the pump laser (a subharmonic OPO can be considered as an ideal coherent frequency divider). These advantages come at the expense of the interferometric sensitivity of the cavity length alignment, owing to the double resonance. However, the cavity length can be actively stabilized, e.g. via the "dither-and-lock" method [102] involving a piezo actuator mounted on one of the OPO mirrors.

In simple terms, when the co-polarized signal (s) and the idler (i) waves become indistinguishable at degeneracy, $f_{CEO}(s) = f_{CEO}(i) = f_{CEO}(\mathrm{OPO})$ and Eq. (6.1) reduces to the two deterministic solutions [122]:

$$f_{CEO}(\mathrm{OPO}) = \frac{f_{CEO}(p)}{2} \tag{6.2}$$

$$f_{CEO}(\mathrm{OPO}) = \frac{f_{CEO}(p)}{2} + \frac{f_{rep}}{2} \tag{6.3}$$

The two solutions toggle when one toggles between adjacent cavity lengths (spaced by approximately pump wavelength, in terms of the cavity roundtrip length) that fulfill the doubly resonant condition [122]. In a similar way, a deterministic phase relation between any mode of the OPO and that of the pump is established through the usual relation for a three-wave nonlinear conversion process (for simplicity, the mode indices were omitted):

$$\varphi_p = \varphi_s + \varphi_i + \frac{\pi}{2}. \tag{6.4}$$

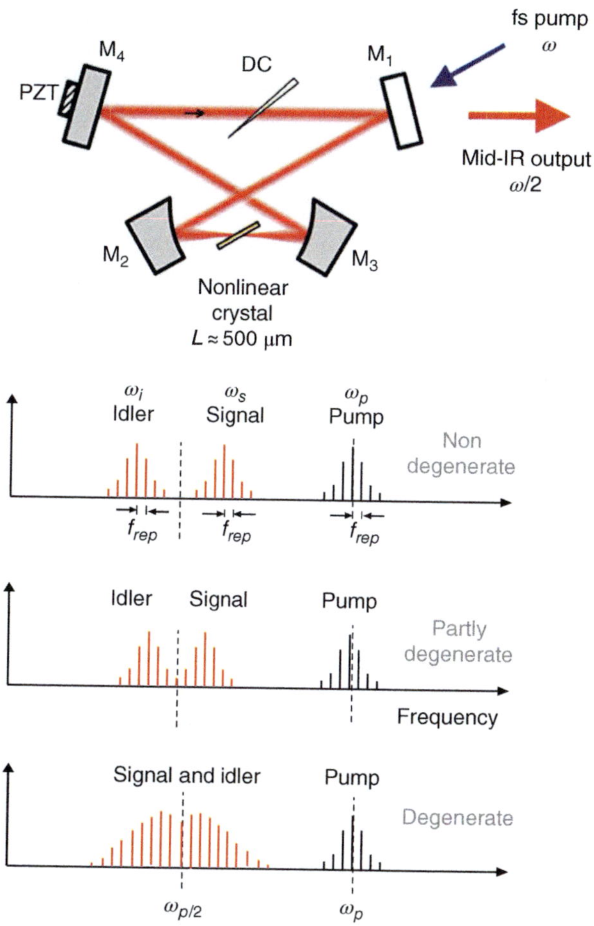

Figure 6.24 (Top) Schematic of a ring-cavity sync-pumped subharmonic OPO. M1, dielectric mirror for in-coupling of the pump; M2 and M3, concave, gold-coated mirrors; M4, flat, gold-coated mirror; DC, dispersion compensating wedge made of a suitable dielectric; PZT, piezo actuator for fine-tuning the cavity length. (Bottom) Conceptual sketch showing evolution of the spectrum when the OPO is driven (from top to bottom) from a nondegenerate to partly degenerate to degenerate (subharmonic) regime.

Since $\varphi_s = \varphi_i\ (=\varphi_{OPO})$, any mode of the OPO becomes phase-locked to that of the pump through the relation:

$$\varphi_p = 2\varphi_{OPO} + \frac{\pi}{2}. \tag{6.5}$$

A variety of ultrafast lasers with central wavelengths varying from 1.5 to 2.5 μm, in combination with different $\chi^{(2)}$ nonlinear materials, such as PPLN,

OP-GaAs, and OP-GaP, were used to achieve broadband mid-IR combs by subharmonic generation in sync-pumped OPOs [102, 103, 124, 125]. The first system of this kind producing a broadband mid-IR output was based on a PPLN crystal pumped by a mode-locked 1.56-μm Er-fiber laser (repetition frequency 100 MHz, pulse duration 70 fs, average power 300 mW). Low intracavity dispersion and large parametric gain bandwidth at degeneracy, associated with the use of a short (0.2–0.5 mm) PPLN crystal, combined with extensive cross-mixing of comb components, resulted in an instantaneous mid-IR bandwidth of 2.5–3.8 μm, with up to 60 mW of the average power [102]. This bandwidth was extended to an octave-wide (2.35–4.8 μm) with the use of the same pump laser and an OP-GaP as an OPO gain element [103]. (As compared to PPLN, OP-GaP has deeper IR transparency and smaller mid-IR GVD, which resulted in the achievement of a larger bandwidth.) The use of orientation-patterned GaAs (OP-GaAs) crystal as a nonlinear gain medium, combined with a mode-locked Tm-fiber laser frequency comb as a pump (center wavelength 1.93 μm, pulse duration 90 fs, repetition rate of 115 MHz, average power 330 mW) enabled a highly coherent comb, spanning 2.6–7.5 μm [104]. The pump threshold was as low as 7 mW and the output average power reached 73 mW at the full pump power. Figure 6.25 depicts the frequency comb spectrum measured with an FTIR spectrometer.

Ultrafast Cr^{2+}:ZnSe and Cr^{2+}:ZnS lasers operating near 2.4 μm (see Chapter 2) are extremely well suited for producing subharmonic combs in OP-GaAs and OP-GaP crystals, thanks to their small GVD near half the frequency of the pump (at ~4.8 μm) [105, 106, 126]. An instantaneous spectral span of 2.85–8.40 μm was achieved in a subharmonic OP-GaAs OPO pumped at 2.35-μm central wavelength with 62-fs pulse duration, 800-mW average power, and 79-MHz repetition rate. The OPO threshold was 8 mW and the maximum average output power reached 110 mW [105]. When the OPO was pumped at high (0.9 GHz) repetition frequency with 6-W average power and 77-fs pulse duration, the emitted average power reached 500 mW with a spectral span of 3–8 μm [106]. Even broader spectral span, exceeding two octaves (3–12.5 μm), was achieved with the use of the OP-GaP crystal (pump at 2.35 μm, repetition rate 79 MHz, average power 1.2 W, pulse duration 62 fs) [107]. So far, only free-running broadband combs have been achieved with subharmonic 2.35-μm pumped OPOs, since the f_{CEO} of the pump combs was not stabilized.

Coherence properties of subharmonic OPOs were studied in detail in [72, 103, 104, 127, 128]. Sub-Hz relative linewidths of the comb teeth with respect to the pump laser were experimentally demonstrated, as well as the absence of excess phase noise in the process of frequency division. Thus, subharmonic generators serve as ideal coherent frequency dividers for generating low-noise mid-IR combs.

The main results of the subharmonic OPO combs are summarized in Table 6.2.

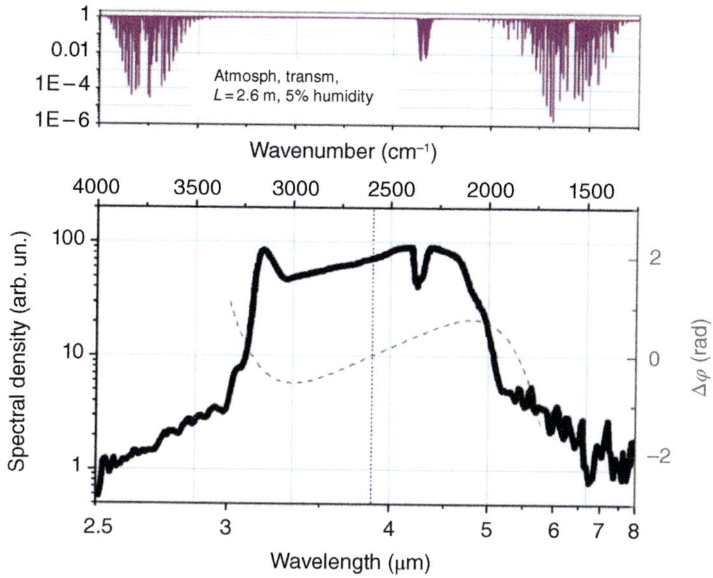

Figure 6.25 Frequency comb spectrum from a Tm-fiber laser pumped (1.93 μm) subharmonic OPO based on OP-GaAs crystal. The spectral span is 2.6–7.5 μm (−20 dB level). The vertical dotted line is the OPO degeneracy (3.86 μm). The dashed line is the calculated extra phase accumulated per round trip due to group velocity dispersion. The top panel shows the simulated transmission through 2.6 m (cavity round-trip length) of purged (5% humidity) air. *Source:* reproduced from figure 5 of [104], with permission of OSA, The Optical Society.

6.2.6 Microresonator-based Kerr Combs

Another new principle for optical frequency comb generation, "Kerr combs" [129, 130], uses parametric frequency conversion based on the third-order optical nonlinearity $\chi^{(3)}$ in whispering gallery mode resonators with high resonance quality (Q) factors. In such resonators, represented by microdisks, microspheres, microtoroids, and microrings, the light is confined by the total internal reflection around the perimeter of an air–dielectric or clad–dielectric interface, and extremely high Q-factors (>10^8) can be achieved. An optical microresonator is pumped with a narrow-linewidth CW laser; if the laser frequency is tuned to one of the cavity modes, the resonantly enhanced intensity reaches GW/cm^2 levels with modest (milliwatts to watts) optical pump powers. This gives rise to the four-wave mixing (parametric frequency conversion) also referred to as optical hyper-parametric oscillation [131, 132] that originates from the Kerr effect – an intensity-dependent refractive index, $n = n_0 + n_2 I$. The parametric frequency conversion annihilates two pump photons with angular frequency ω_p and creates a new pair of photons – a frequency upshifted signal ω_s, and a frequency downshifted idler ω_i in the following process:

$$2\omega_p = \omega_s + \omega_i \tag{6.6}$$

Or, in a pump nondegenerate case:

$$\omega_{p1} + \omega_{p2} = \omega_s + \omega_i \tag{6.7}$$

This frequency conversion mechanism can lead to a vast cascading of evenly spaced oscillating frequencies, as illustrated in Figure 6.26. If the signal and idler frequencies coincide with optical microresonator modes, the parametric process is enhanced, resulting in efficient spectral broadening. However, for this to happen, the dispersion of the microresonator needs to be properly adjusted in order to satisfy the phase-matching condition. As compared to frequency combs based on mode-locked lasers, microresonator-based combs have much larger line spacing – typically in the range 10–1000 GHz.

Microresonator-based frequency comb sources pumped by CW lasers have been reported at telecom wavelengths in a number of platforms, including

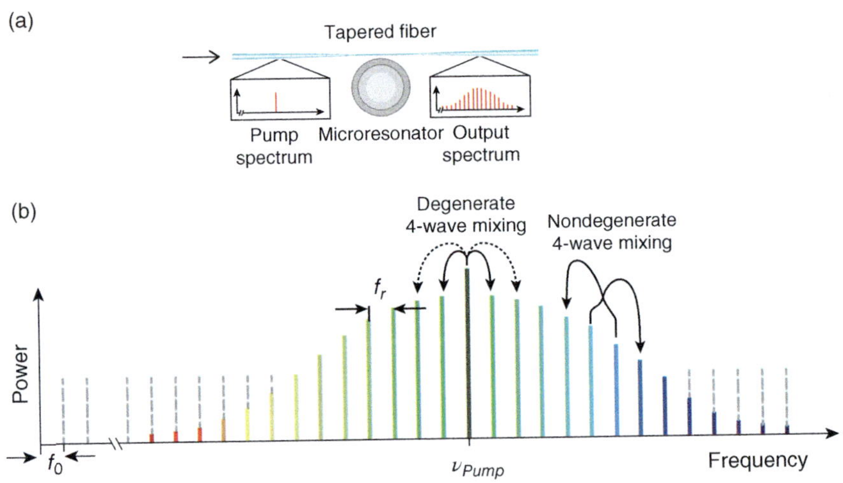

Figure 6.26 Frequency comb generation using microresonators. (a) An optical microresonator is pumped with a CW laser. The high intensity in the resonator (~GW/cm^2) gives rise to a parametric frequency conversion. (b) Optical frequency comb is characterized by the repetition rate (f_r) and the carrier envelope frequency (f_o). The comb is generated by a combination of degenerate and nondegenerate four-wave mixing. *Source:* adapted from figure 2 of [130], with permission of the American Association for the Advancement of Science.

silica [129], high-index-doped silica glass [133], crystalline calcium fluoride [134, 135], silicon nitride [136], aluminum nitride [137], and diamond [138].

Del'Haye et al. [108] reported the generation of an octave-spanning optical frequency comb reaching mid-IR in a CW laser pumped toroidal fused silica microresonator. The output of the pump laser (1560-nm diode laser amplified to 2.5 W) was coupled to a microresonator mode using a tapered optical fiber. The fused silica microresonator had a radius of 40 μm, and a cross-section radius of 2.8 μm (Figure 6.27a), mode spacing of 850 GHz, and optical quality factor $Q = 2 \times 10^8$. The generated comb spectrum (Figure 6.27b) covered the whole range from 1 to 2.17 μm. The ability to generate octave-spanning combs was attributed by the authors to the optimized (via adjusting the cross section) resonator GVD.

Generation of optical frequency combs from a CMOS-compatible integrated silicon nitride (Si_3N_4) microresonator was demonstrated in [109, 139]. Both the microresonator and the coupling WG were fabricated monolithically in a single silicon nitride layer using electron-beam lithography and subsequently clad with silica. The 112- or 200-μm-radius microring resonator (Figure 6.28a) had a free-spectral range of 204 or 226 GHz, and a quality factor $Q > 10^5$. The cross-sectional

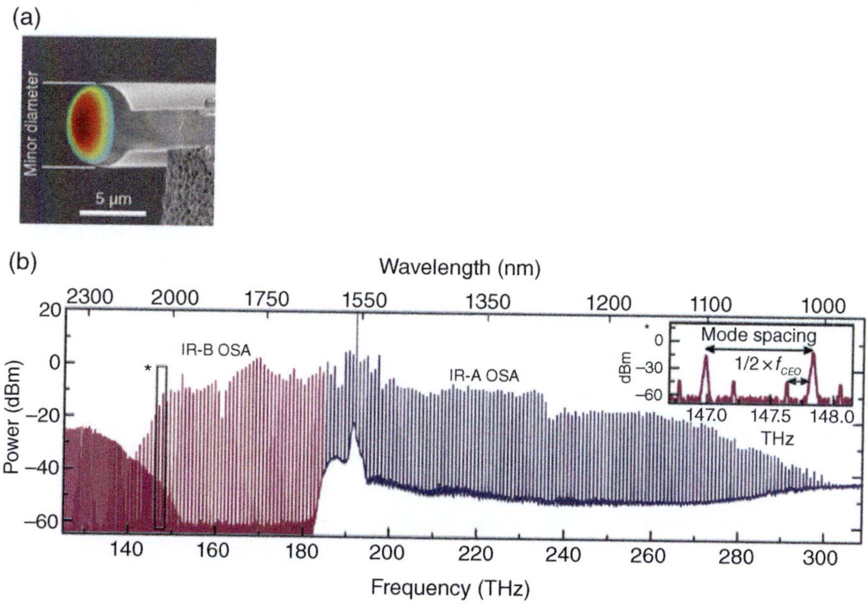

Figure 6.27 (a) Scanning electron microscope image of a cleaved fused silica microresonator (also shown is a simulated mode profile). (b) Frequency comb, generated in a 40 μm-radius silica microresonator with the modal spacing of 850 GHz. The feature on the low-frequency side (<140 THz) is a spectrometer artifact due to second-order diffraction. *Source:* reproduced from figure 1 of [108], with permission of APS.

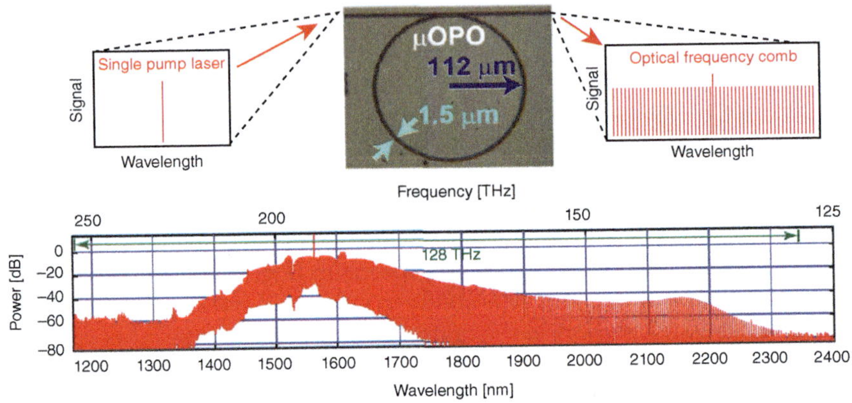

Figure 6.28 Schematic of a CMOS-compatible silicon nitride microresonator and experimentally measured frequency comb spectrum. *Source:* adapted from figure 2 of [139], with permission of OSA, The Optical Society.

dimensions of both the coupling WG and the ring (750 nm in height and 1500 nm in width) were designed to produce anomalous GVD at the pump wavelength of 1550 nm. Up to 2 W of power from an amplified tunable diode laser were injected into the coupling WG. As the pump laser was tuned into resonance, cascaded four-wave mixing generated new frequencies with the comb span from 1.17 to 2.35 µm (Figure 6.28b). Also, it has been verified in this work that the mode equidistance is better than 3×10^{-15} relative to the center comb frequency [139].

With a microring resonator made of a Si_3N_4 WG with silica cladding and 2.6-µm pump (an OPO output), Luke et al. demonstrated a comb with a free spectral range of 99 GHz and a wavelength span of 2.3–3.5 µm (at −60 dB level). The ring resonator had the radius 230 µm, the WG cross section 950 by 2700 nm, and the gap between the ring and the coupling WG of 860 nm (Figure 6.29) [140]. The pump power was 500 mW with the threshold for the comb generation of 80 mW. RF noise measurements of the comb with a fast photodiode did not show a reduction in RF noise when the spectral broadening occurred, indicating that the comb has not transitioned to a phase-locked state. Thus, further improvements to achieve comb coherence will be required.

Griffith et al. demonstrated a broadband frequency comb based on a silicon microresonator pumped at 2.59 µm [110]. The resonator had a SOI 1.4-µm-wide WG core (Figure 6.30a) and cavity-mode spacing of 127 GHz. To enable broadband comb generation, the WG geometry was engineered to have anomalous GVD from 2.2 to 3 µm. In this spectral region, the dominant loss in silicon comes from the three-photon absorption. To mitigate this nonlinear loss, the authors embedded the silicon microresonator in a reverse-biased positive-intrinsic-negative (PIN)-doped junction to sweep out carriers generated from

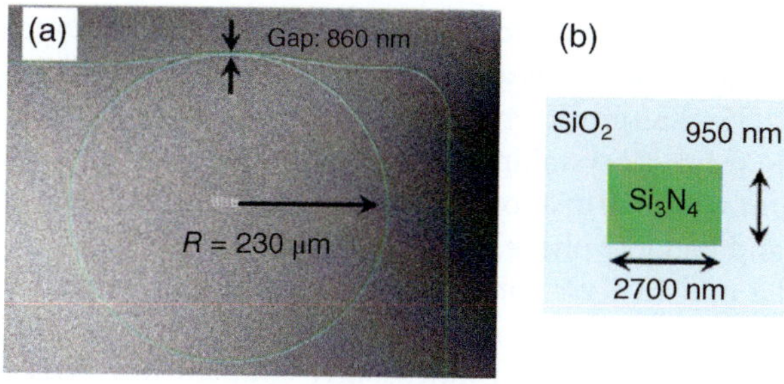

Figure 6.29 (a) Scanning electron microscope image of a 2.6-μm-pumped Si_3N_4 ring resonator. (b) Schematic of the resonator cross section. *Source:* adapted from figure 2 of [140], with permission of OSA, The Optical Society.

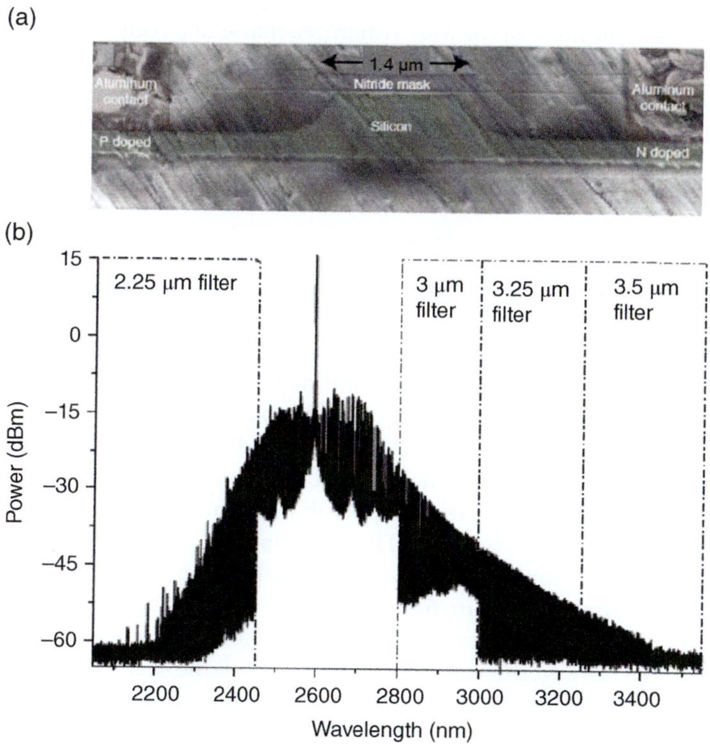

Figure 6.30 (a) Cross-sectional SEM image of a silicon waveguide. (b) Spectrum of a frequency comb from a silicon microresonator pumped at 2.59 μm. A series of optical bandpass filters were used to measure the spectrum. *Source:* adapted from figures 1 and 3 from [110], with permission of Springer Nature.

three-photon absorption. With this CMOS-compatible platform, a frequency comb span of 2.1–3.5 μm (Figure 6.30b) was achieved.

Wang et al. produced a 2.35–2.55 μm Kerr frequency comb using an ultra-high Q magnesium fluoride (MgF_2) crystalline resonator pumped by the output of a CW OPO at 2.45 μm with 600 mW of average power. The comb had a mode spacing of 107 GHz and had low phase noise [112].

Lecaplain et al. reported a mid-IR Kerr comb based on a high-Q crystalline MgF_2 resonator that was pumped by a CW quantum cascade laser (QCL) at 4.4 μm [113]. The light from the QCL was evanescently coupled to a MgF_2 resonator via a chalcogenide tapered fiber. For a pump power of 80 mW the generated mid-IR combs spanned from 4.1 to 4.7 μm with a mode spacing of 14.3 GHz, comprising more than 700 comb lines.

Similarly, Savchenkov et al. demonstrated mid-IR Kerr combs in high-finesse ($Q = 2 \times 10^8$) CaF_2 and MgF_2 whispering-gallery mode resonators pumped with a CW distributed-feedback QCL at 4.5 μm. Kerr combs were generated starting from the coupled pump power of only 2 mW. The largest Kerr comb stretch, from 3.7 to 5.5 μm, was attained in the MgF_2 microresonator at a coupled pump power of 55 mW [114]. However, the comb line spacing in this work was large, 2.1 THz – a multiple of the free spectral range of the resonator (19.3 GHz), and further study of coherence properties is required.

6.2.7 Combs from Quantum Cascade Lasers

Quantum cascade lasers (QCLs) are direct and compact mid-IR sources that use electrical injection to provide optical gain (see Chapter 4). Hugi et al. reported the first mid-IR frequency comb generator based on a QCL [115]. The authors demonstrated that under certain drive currents, the modes of a CW, free-running, broadband QCL could be passively phase-locked. Measurements of coherence properties [115] indicated that the phase relations between the comb lines are similar to that of a frequency-modulated (FM) laser, and the locking mechanism responsible for the formation of the frequency comb is four-wave mixing. This mechanism is facilitated by strong third-order optical nonlinearity and a low GVD of a broadband QCL structure. The comb spanned from 6.8 to 7.3 μm (−20 dB level) with the mode spacing of 7.5 GHz. The inter-mode beat linewidth was measured to be less than 250 kHz [115]. In the subsequent work, the same team has demonstrated that the equidistance of the comb lines is preserved to a fractional accuracy of 7.5×10^{-16} [141].

Most recently, Lu et al. demonstrated a frequency comb source at $\lambda \approx 8$ μm, based on a dispersion-compensated QCL structure, with a high-power output of 880 mW at room temperature [116]. The spectral coverage of the QCL comb device was 110 cm^{-1} with the mode spacing 11.2 GHz, which corresponds to

~290 lasing modes. The comb device exhibited a narrow intermodal beatnote linewidth of 50.5 Hz and a maximum wall-plug efficiency of 6.5%. This efficiency was a factor of 6 increase, as compared to previous demonstrations. Furthermore, narrow intermodal beatnote linewidths (<305 Hz) were observed over a wide range of pump currents (25% of the total range).

6.2.8 Combs from Interband Cascade Lasers

Interband cascade lasers (ICLs) – another class of electrically pumped semiconductor lasers (see Chapter 4) – operate at 3–6 μm and are ideal for many low-power spectroscopic applications since they consume much less drive power than QCLs. This is especially advantageous in battery-powered sensing systems, or for applications that require small size and weight. ICLs can be best operated in the important 3–4 μm band, a band where a large fraction of the absorption features associated with C—H bonds are clustered, and where QCL performance degrades.

Bagheri et al. demonstrated an electrically driven frequency comb source based on an ICL that spans more than 1 THz (35 cm^{-1}, 52 nm) of bandwidth near $\lambda \approx 3.6$ μm. This was achieved by passively mode-locking an ICL, where gain and saturable absorber sections were monolithically integrated on the same chip [117]. Figure 6.31a shows the schematic of a laser cavity with gain and saturable absorber sections, where the saturable absorber imposes pulsed operation by providing an intensity-dependent loss mechanism. Also shown (Figure 6.31c and d) is an FTIR spectrum of the mode-locked ICL operating at $I = 325$ mA current and $T = 15\,°C$ temperature, showing 35 cm^{-1} bandwidth and >120 modes spaced by 9.7 GHz. The intensity autocorrelation measurements performed by the authors have shown that the mode-locked ICL emitted pulses of ≈750 fs width.

The main results for frequency comb laser sources are presented in Table 6.2.

6.3 Summary

Producing "white light" in the mid-IR via SC generation can be achieved in various media, with a continuous wavelength span of up to several octaves and a longwave limit beyond 10 μm. With HNLFs and waveguides, SC generation can be accomplished with modest pump peak powers, which opens up new possibilities for creating a compact light source for the needs of mid-IR spectroscopy and sensing. Phase-locked frequency combs in the mid-IR – a time-coherent version of the SC – offer immense potential for high-precision, high-resolution, and high-sensitivity molecular spectroscopy. The most promising techniques for producing combs include those based on direct comb generation with

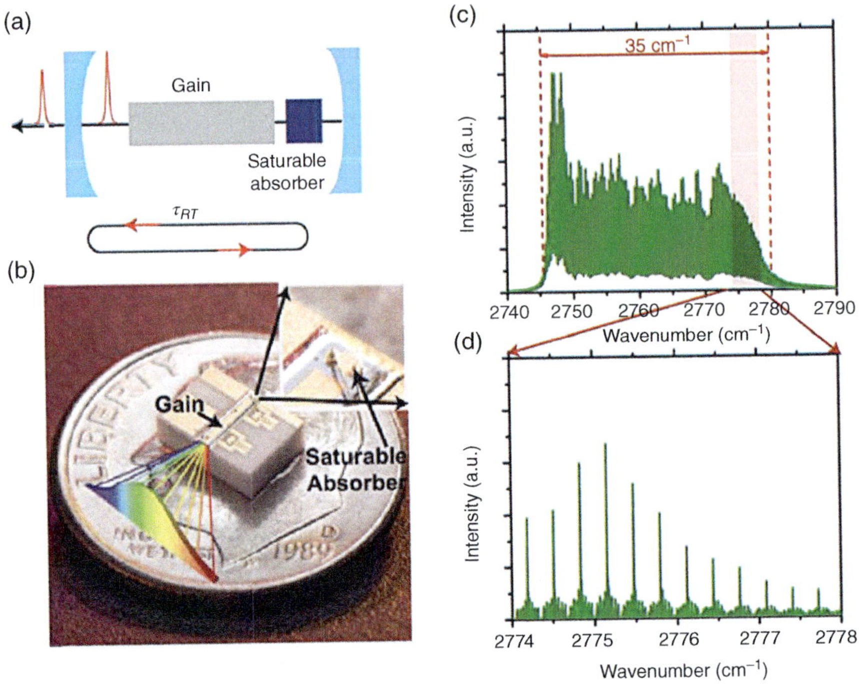

Figure 6.31 Frequency comb source based on passively mode-locked interband cascade laser (ICL). (a) Schematic of the laser cavity with gain and saturable absorber sections, where the saturable absorber imposes pulsed operation via intensity-dependent loss. (b) An ICL mode-locked laser with gain (front) and saturable absorber (back) sections. (c) FTIR spectrum of the ICL operating at $I = 325$ mA and $T = 15$ °C, with 35 cm^{-1} bandwidth and >120 modes. (d) A portion of the spectrum measured by a high-resolution (100 MHz) FTIR. *Source:* adapted from figures 1 and 2 of [117], with permission of Springer Nature.

mode-locked fiber- and solid-state lasers, parametric devices (using $\chi^{(2)}$ crystals), and hyper-parametric devices (based on $\chi^{(3)}$ in fibers, waveguides, and microresonators). The newly developed electrically pumped QCL and ICL combs show potential for efficient and compact comb sources.

References

1 Alfano, R.R. and Shapiro, S.L. (1970). Emission in the region 4000 to 7000 Å via four-photon coupling in glass. *Phys. Rev. Lett.* 24: 584.
2 Corkum, P.B., Ho, P.P., Alfano, R.R., and Manassah, J.T. (1985). Generation of infrared supercontinuum covering 3–14 μm in dielectrics and semiconductors. *Opt. Lett.* 10: 624.
3 Agrawal, G.P. (2013). *Nonlinear Fiber Optics*, 5e. Oxford: Academic Press.

4 Dudley, J.M., Genty, G., and Coen, S. (2006). Supercontinuum generation in photonic crystal fiber. *Rev. Mod. Phys.* 78: 1135.
5 Dudley, J.M. and Taylor, J.R. (eds.) (2010). *Supercontinuum Generation in Optical Fibers.* Cambridge: Cambridge University Press.
6 Islam, M.N., Sucha, G., Bar-Joseph, I., Wegener, M., Gordon, J.P., and Chemla, D.S. (1989). Femtosecond distributed soliton spectrum in fibers. *J. Opt. Soc. Am. B* 6: 1149.
7 R. Paschotta, Encyclopedia of Laser Physics and Technology, http://www.rp-photonics.com, accessed 21 November 2018.
8 Rothenberg, J.E. and Grischkowsky, D. (1989). Observation of the formation of an optical intensity shock and wave breaking in the nonlinear propagation of pulses in optical fibers. *Phys. Rev. Lett.* 62: 531.
9 Omenetto, F.G., Wolchover, N.A., Wehner, M.R., Ross, M., Efimov, A., Taylor, A.J., Kumar, V.V.R.K., George, A.K., Knight, J.C., Joly, N.Y., and Russell, P.St.J. (2006). Spectrally smooth supercontinuum from 350 nm to 3 μm in sub-centimeter lengths of soft-glass photonic crystal fibers. *Opt. Express* 14: 4928.
10 Domachuk, P., Wolchover, N.A., Cronin-Golomb, M., Wang, A., George, A.K., Cordeiro, C.M.B., Knight, J.C., and Omenetto, F.G. (2008). Over 4000 nm bandwidth of mid-IR supercontinuum generation in sub-centimeter segments of highly nonlinear tellurite PCFs. *Opt. Express* 16: 7161.
11 Hagen, C.L., Walewski, J.W., and Sanders, S.T. (2006). Generation of a continuum extending to the midinfrared by pumping ZBLAN fiber with an ultrafast 1550-nm source. *IEEE Photon. Technol. Lett.* 18: 91.
12 Xia, C., Kumar, M., Kulkarni, O.P., Islam, M.N., Terry, F.L. Jr., Freeman, M.J., Poulain, M., and Mazé, G. (2006). Mid-infrared supercontinuum generation to 4.5 μm in ZBLAN fluoride fibers by nanosecond diode pumping. *Opt. Lett.* 31: 2553.
13 Xia, C., Kumar, M., Cheng, M.-Y., Hegde, R.S., Islam, M.N., Galvanauskas, A., Winful, H.G., Terry Jr, F.L., Freeman, M.J., Poulain, M., and Mazé, G. (2007). Power scalable mid-infrared supercontinuum generation in ZBLAN fluoride fibers with up to 1.3 watts time-averaged power. *Opt. Express* 15: 865.
14 Xia, C., Xu, Z., Islam, M.N., Terry, F.L. Jr., Freeman, M.J., Zakel, A., and Mauricio, J. (2009). 10.5 W time-averaged power mid-IR supercontinuum generation extending beyond 4 μm with direct pulse pattern modulation. *IEEE J. Sel. Topics Quantum Electron.* 15: 422.
15 Liu, K., Liu, J., Shi, H., Tan, F., and Wang, P. (2014). High power mid-infrared supercontinuum generation in a single-mode ZBLAN fiber with up to 21.8 W average output power. *Opt. Express* 22: 24384.
16 Yang, W., Zhang, B., Xue, G., Yin, K., and Hou, J. (2014). Thirteen watt all-fiber mid-infrared supercontinuum generation in a single mode ZBLAN fiber pumped by a 2 μm MOPA system. *Opt. Lett.* 39: 1849.
17 Qin, G., Yan, X., Kito, C., Liao, M., Chaudhari, C., Suzuki, T., and Ohishia, Y. (2009). Ultrabroadband supercontinuum generation from ultraviolet to 6.28 μm in a fluoride fiber. *Appl. Phys. Lett.* 96: 161103.

18 Eggleton, B.J., Luther-Davies, B., and Richardson, K. (2011). Chalcogenide photonics. *Nat. Photonics* 5: 141.

19 Snopatin, G.E., Shiryaev, V.S., Plotnichenko, V.G., Dianov, E.M., and Churbanov, M.F. (2009). High-purity chalcogenide glasses for fiber optics. *Inorg. Mater.* 45: 14391460.

20 Shiryaev, V.S. and Churbanov, M.F. (2013). Trends and prospects for development of chalcogenide fibers for mid-infrared transmission. *J. Non-Cryst. Solids* 377: 225.

21 Hu, J., Menyuk, C.R., Shaw, L.B., Sanghera, J.S., and Aggarwal, I.D. (2010). Computational study of 3–5 μm source created by using supercontinuum generation in As_2S_3 chalcogenide fibers with a pump at 2 μm. *Opt. Lett.* 35: 2907.

22 Shabahang, S., Tao, G., Marquez, M.P., Hu, H., Ensley, T.R., Delfyett, P.J., and Abouraddy, A.F. (2014). Nonlinear characterization of robust multimaterial chalcogenide nanotapers for infrared supercontinuum generation. *J. Opt. Soc. Am. B* 31: 450.

23 El-Amraoui, M., Gadret, G., Jules, J.C., Fatome, J., Fortier, C., Désévédavy, F., Skripatchev, I., Messaddeq, Y., Troles, J., Brilland, L., Gao, W., Suzuki, T., Ohishi, Y., and Smektala, F. (2010). Microstructured chalcogenide optical fibers from As_2S_3 glass: towards new IR broadband sources. *Opt. Express* 18: 26655.

24 Yeom, D., Mägi, E.C., Lamont, M.R.E., Roelens, M.A.F., Fu, L., and Eggleton, B.J. (2008). Low-threshold supercontinuum generation in highly nonlinear chalcogenide nanowires. *Opt. Lett.* 33: 660.

25 Mägi, E.C., Fu, L.B., Nguyen, H.C., Lamont, M.R., Yeom, D.I., and Eggleton, B.J. (2007). Enhanced Kerr nonlinearity in sub-wavelength diameter As_2Se_3 chalcogenide fiber tapers. *Opt. Express* 15: 10324.

26 Hudson, D.D., Dekker, S.A., Mägi, E.C., Judge, A.C., Jackson, S.D., Li, E., Sanghera, J.S., Shaw, L.B., Aggarwal, I.D., and Eggleton, B.J. (2011). Octave spanning supercontinuum in an As_2S_3 taper using ultralow pump pulse energy. *Opt. Lett.* 36: 1122.

27 Shabahang, S., Marquez, M.P., Tao, G., Piracha, M.U., Nguyen, D., Delfyett, P.J., and Abouraddy, A.F. (2012). Octave-spanning infrared supercontinuum generation in robust chalcogenide nanotapers using picosecond pulses. *Opt. Lett.* 37: 4639.

28 Sanghera, J.S., Shaw, L.B., and Aggarwal, I.D. (2009). Chalcogenide glass-fiber-based mid-IR sources and applications. *IEEE J. Sel. Topics Quantum Electron.* 15: 114.

29 Mouawad, O., Picot-Clemente, J., Amrani, F., Strutynski, C., Fatome, J., Kibler, B., Dsvdavy, F., Gadret, G., Jules, J.-C., Deng, D., Ohishi, Y., and Smektala, F. (2014). Multioctave midinfrared supercontinuum generation in suspended-core chalcogenide fibers. *Opt. Lett.* 39: 2684.

30 Marandi, A., Rudy, C.W., Plotnichenko, V.G., Dianov, E.M., Vodopyanov, K.L., and Byer, R.L. (2012). Mid-infrared supercontinuum generation in tapered chalcogenide fiber for producing octave-spanning frequency comb around 3 μm. *Opt. Express* 20: 24218.

31 Rudy, C.W., Marandi, A., Vodopyanov, K.L., and Byer, R.L. (2013). Octave-spanning supercontinuum generation in in situ tapered As_2S_3 fiber pumped by a thulium-doped fiber laser. *Opt. Lett.* 38: 2865.

32 Al-Kadry, A., Amraoui, M.E., Messaddeq, Y., and Rochette, M. (2014). Two octaves mid-infrared supercontinuum generation in As_2Se_3 microwires. *Opt. Express* 22: 31131.

33 Yu, Y., Zhang, B., Gai, X., Zhai, C., Qi, S., Guo, W., Yang, Z., Wang, R., Choi, D.-Y., Madden, S., and Luther-Davies, B. (2015). 1.8-10 μm mid-infrared supercontinuum generated in a step-index chalcogenide fiber using low peak pump power. *Opt. Lett.* 40: 1081.

34 Møller, U., Yu, Y., Kubat, I., Petersen, C.R., Gai, X., Brilland, L., Méchin, D., Caillaud, C., Troles, J., Luther-Davies, B., and Bang, O. (2015). Multi-milliwatt mid-infrared supercontinuum generation in a suspended core chalcogenide fiber. *Opt. Express* 23: 3282.

35 Hudson, D.D., Baudisch, M., Werdehausen, D., Eggleton, B.J., and Biegert, J. (2014). 1.9 octave supercontinuum generation in a As_2S_3 step-index fiber driven by mid-IR OPCPA. *Opt. Lett.* 39: 5752.

36 Robichaud, L.-R., Fortin, V., Gauthier, J.-C., Châtigny, S., Couillard, J.-F., Delarosbil, J.-L., Vallée, R., and Bernier, M. (2016). Compact 3-8 μm supercontinuum generation in a low-loss As_2Se_3 step-index fiber. *Opt. Lett.* 41: 4605.

37 Petersen, C.R., Møller, U., Kubat, I., Zhou, B., Dupont, S., Ramsay, J., Benson, T., Sujecki, S., Abdel-Moneim, N., Tang, Z., Furniss, D., Seddon, A., and Bang, O. (2014). Mid-infrared supercontinuum covering the 1.4–13.3 μm molecular fingerprint region using ultra-high NA chalcogenide step-index fibre. *Nat. Photonics* 8: 830.

38 Hudson, D.D., Antipov, S., Li, L., Alamgir, I., Hu, T., El Amraoui, M., Messaddeq, Y., Rochette, M., Jackson, S.D., and Fuerbach, A. (2017). Toward all-fiber supercontinuum spanning the mid-infrared. *Optica* 4: 1163.

39 Cheng, T., Nagasaka, K., Tuan, T.H., Xue, X., Matsumoto, M., Tezuka, H., Suzuki, T., and Ohishi, Y. (2016). Mid-infrared supercontinuum generation spanning 2.0 to 15.1 μm in a chalcogenide step-index fiber. *Opt. Lett.* 41: 2117.

40 Zhao, Z., Wu, B., Wang, X., Pan, Z., Liu, Z., Zhang, P., Shen, X., Nie, Q., Dai, S., and Wang, R. (2017). Mid-infrared supercontinuum covering 2.0–16 μm in a low-loss telluride single-mode fiber. *Laser Photonics Rev.* 11: 1700005.

41 Lamont, M.R.E., Luther-Davies, B., Choi, D.-Y., Madden, S., and Eggleton, B.J. (2008). Supercontinuum generation in dispersion engineered highly nonlinear (γ = 10 /W/m) As_2S_3 chalcogenide planar waveguide. *Opt. Lett.* 16: 14938.

42 Xie, S., Tolstik, N., Travers, J.C., Sorokin, E., Caillaud, C., Troles, J., Russell, St.P.J., and Sorokina, I.T. (2016). Coherent octave-spanning mid-infrared supercontinuum generated in As_2S_3-silica double-nanospike waveguide pumped by femtosecond Cr:ZnS laser. *Opt. Express* 24: 12406.

43 Yu, Y., Gai, X., Ma, P., Choi, D.Y., Yang, Z.Y., Wang, R.P., Debbarma, S., Madden, S.J., and Luther-Davies, B. (2014). A broadband, quasi-continuous, mid-infrared supercontinuum generated in a chalcogenide glass waveguide. *Laser Photonics Rev.* 8: 792.

44 Lau, R.K.W., Lamont, M.R.E., Griffith, A.G., Okawachi, Y., Lipson, M., and Gaeta, A.L. (2014). Octave-spanning mid-infrared supercontinuum generation in silicon nanowaveguides. *Opt. Lett.* 39: 4518.

45 Singh, N., Hudson, D.D., Yu, Y., Grillet, C., Jackson, S.D., Casas-Bedoya, A., Read, A., Atanackovic, P., Duvall, S.G., Palomba, S., Luther-Davies, B., Madden, S., Moss, D.J., and Eggleton, B.J. (2015). Midinfrared supercontinuum generation from 2 to 6 μm in a silicon nanowire. *Optica* 2: 797.

46 Phillips, C.R., Langrock, C., Pelc, J.S., Fejer, M.M., Jiang, J., Fermann, M.E., and Hartl, I. (2011). Supercontinuum generation in quasi-phase-matched $LiNbO_3$ waveguide pumped by a Tm-doped fiber laser system. *Opt. Lett.* 36: 3912.

47 Silva, F., Austin, D.R., Thai, A., Baudisch, M., Hemmer, M., Faccio, D., Couairon, A., and Biegert, J. (2012). Multi-octave supercontinuum generation from mid-infrared filamentation in a bulk crystal. *Nat. Commun.* 3: 807.

48 Belal, M., Xu, L., Horak, P., Shen, L., Feng, X., Ettabib, M., Richardson, D.J., Petropoulos, P., and Price, J.H.V. (2015). Mid-infrared supercontinuum generation in suspended core tellurite microstructured optical fibers. *Opt. Lett.* 40: 2237.

49 Yu, Y., Gai, X., Wang, T., Ma, P., Wang, R., Yang, Z., Choi, D.-Y., Madden, S., and Luther-Davies, B. (2013). Mid-infrared supercontinuum generation in chalcogenides. *Opt. Mater. Express* 3: 1075.

50 Liao, M., Gao, W., Cheng, T., Xue, X., Duan, Z., Deng, D., Kawashima, H., Suzuki, T., and Ohishi, Y. (2013). Five-octave-spanning supercontinuum generation in fluoride glass. *Appl. Phys. Express* 6: 032503.

51 Liang, H., Krogen, P., Grynko, R., Novak, O., Chang, C.-L., Stein, G.J., Weerawarne, D., Shim, B., Kärtner, F.X., and Hong, K.-H. (2015). Three-octave spanning supercontinuum generation and sub-two-cycle self-compression of mid-infrared filaments in dielectrics. *Opt. Lett.* 40: 1069.

52 Suminas, R., Tamošauskas, G., Valiulis, G., Jukna, V., Couairon, A., and Dubietis, A. (2017). Multi-octave spanning nonlinear interactions induced by femtosecond filamentation in polycrystalline ZnSe. *Appl. Phys. Lett.* 110: 241106.

53 Mouawad, O., Béjot, P., Billard, F. Mathey, P., Kibler, B., Désévédavy, F., Gadret, G., Jules, J.-C., Faucher, O., and Smektala, F. (2016). Filament-induced visible-to-mid-IR supercontinuum in a ZnSe crystal: towards multioctave supercontinuum absorption spectroscopy. *Opt. Mater.* 60: 355.

54 Vasilyev, S., Moskalev, I., Mirov, M., Smolski, V., Mirov, S., and Gapontsev, V. (2017). Ultrafast middle-IR lasers and amplifiers based on polycrystalline Cr:ZnS and Cr:ZnSe. *Opt. Mater. Express* 7: 2636.

55 Ashihara, S. and Kawahara, Y. (2009). Spectral broadening of mid-infrared femtosecond pulses in GaAs. *Opt. Lett.* 34: 3839.

56 Pigeon, J.J., Tochitsky, S.Ya., Gong, C., and Joshi, C. (2014). Supercontinuum generation from 2 to 20 μm in GaAs pumped by picosecond CO_2 laser pulses. *Opt. Lett.* 39: 3246.

57 Kartashov, D., Ališauskas, S., Pugžlys, A., Voronin, A., Zheltikov, A., Petrarca, M., Béjot, P., Kasparian, J., Wolf, J.-P., and Baltuška, A. (2012). White light generation over three octaves by femtosecond filament at 3.9 μm in argon. *Opt. Lett.* 37: 3456.

58 Hänsch, T.W. (2006). Nobel lecture: Passion for precision. *Rev. Mod. Phys.* 78: 1297.

59 Cundiff, S.T. and Ye, J. (2003). Colloquium: Femtosecond optical frequency combs. *Rev. Mod. Phys.* 75: 325.

60 Popmintchev, T., Chen, M.-C., Arpin, P., Murnane, M.M., and Kapteyn, H.C. (2010). The attosecond nonlinear optics of bright coherent X-ray generation. *Nat. Photonics* 4: 822.

61 Corkum, P.B. and Krausz, F. (2007). Attosecond science. *Nat. Phys.* 3: 381.

62 Sears, C.M.S., Colby, E., England, R.J., Ischebeck, R., McGuinness, C., Nelson, J., Noble, R., Siemann, R.H., Spencer, J., Walz, D., Plettner, T., and Byer, R.L. (2008). Phase stable net acceleration of electrons from a two-stage optical accelerator. *Phys. Rev. Spec. Topics Accelerators Beams* 11: 101301.

63 Coddington, I., Newbury, N., and Swann, W. (2016). Dual-comb spectroscopy. *Optica* 3: 414.

64 Muraviev, A.V., Smolski, V.O., Loparo, Z.E., and Vodopyanov, K.L. (2018). Massively parallel sensing of trace molecules and their isotopologues with broadband subharmonic mid-infrared frequency combs. *Nat. Photonics* 12: 209.

65 Timmers, H., Kowligy, A., Lind, A., Cruz, F.C., Nader, N., Silfies, M., Ycas, G., Allison, T.K., Schunemann, P.G., Papp, S.B., and Diddams, S.A. (2018). Molecular fingerprinting with bright, broadband infrared frequency combs. *Optica* 5: 727.

66 Dudley, J.M. and Coen, S. (2002). Coherence properties of supercontinuum spectra generated in photonic crystals and tapered optical fibers. *Opt. Lett.* 27: 1180.

67 Bellini, M. and Hänsch, T.W. (2000). Phase-locked white-light continuum pulses: toward a universal optical frequency-comb synthesizer. *Opt. Lett.* 25: 1049.

68 Schliesser, A., Picqué, N., and Hänsch, T.W. (2012). Mid-infrared frequency combs. *Nat. Photonics* 6: 440.

69 Lee, C.-C., Mohr, C., Bethge, J., Suzuki, S., Fermann, M.E., Hartl, I., and Schibli, T.R. (2012). Frequency comb stabilization with bandwidth beyond the limit of gain lifetime by an intracavity graphene electro-optic modulator. *Opt. Lett.* 37: 3084.

70 Bethge, J., Jiang, J., Mohr, C., Fermann, M., and Hartl, I. (2012). Optically referenced Tm-fiber-laser frequency comb, Technical Digest. In: *Advanced Solid-State Photonics*. Washington, DC: Optical Society of America, Paper AT5A.3.

71 Lee, K.F., Jiang, J., Mohr, C., Bethge, J., Fermann, M.E., Leindecker, N., Vodopyanov, K.L., Schunemann, P.G., and Hartl, I. (2013). Carrier envelope offset frequency of a doubly resonant, nondegenerate, mid-infrared GaAs optical parametric oscillator. *Opt. Lett.* 38: 1191.

72 Lee, K.F., Mohr, C., Jiang, J., Schunemann, P.G., Vodopyanov, K.L., and Fermann, M.E. (2015). Midinfrared frequency comb from self-stable degenerate GaAs optical parametric oscillator. *Opt. Express* 23: 26596.

73 Mirov, S., Fedorov, V., Martyshkin, D., Moskalev, I., Mirov, M., and Vasilyev, S. (2015). Progress in mid-IR lasers based on Cr and Fe doped II-VI Chalcogenides. *IEEE J. Sel. Topics Quantum Electron.* 21: 1601719.

74 Vasilyev, S., Moskalev, I., Smolski, V., Peppers, J., Mirov, M., Fedorov, V., Martyshkin, D., Mirov, S., and Gapontsev, V. (2019). Octave-spanning Cr:ZnS femtosecond laser with intrinsic nonlinear interferometry. *Optica* 6: 126.

75 Vasilyev, S., Smolski,V., Peppers, J., Moskalev, I., Mirov, M., Barnakov, Y., Mirov, S., and Gapontsev, V. (2019). Middle-IR frequency comb based on Cr:ZnS laser. *Opt. Express.* 27: 35079.

76 Duval, S., Bernier, M., Fortin, V., Genest, J., Piché, M., and Vallée, R. (2015). Femtosecond fiber lasers reach the midinfrared. *Optica* 2: 623.

77 Washburn, B.R., Diddams, S.A., Newbury, N.R., Nicholson, J.W., Yan, M.F., and Jørgensen, C.G. (2004). Phase-locked, erbium-fiber-laser-based frequency comb in the near infrared. *Opt. Lett.* 29: 250.

78 Nicholson, J.W., Yan, M.F., Wisk, P., Fleming, J., DiMarcello, F., Monberg, E., Yablon, A., Jørgensen, C., and Veng, T. (2003). All-fiber, octave-spanning supercontinuum. *Opt. Lett.* 28: 643.

79 Granzow, N., Schmidt, M.A., Chang, W., Wang, L., Coulombier, Q., Troles, J., Toupin, P., Hartl, I., Lee, K.F., Fermann, M.E., Wondraczek, L., and Russell, P.St.J. (2013). Mid-infrared supercontinuum generation in As_2S_3-silica nano-spike step-index waveguide. *Opt. Express* 21: 10969.

80 Lee, K.F., Granzow, N., Schmidt, M.A., Chang, W., Wang, L., Coulombier, Q., Troles, J., Leindecker, N., Vodopyanov, K.L., Schunemann, P.G., Fermann, M.E., Russell, P.St.J., and Hartl, I. (2014). Midinfrared frequency combs from coherent supercontinuum in chalcogenide and optical parametric oscillation. *Opt. Lett.* 39: 2056.

81 Kuyken, B., Ideguchi, T., Holzner, S., Yan, M., Hänsch, T.W., Van Campenhout, J., Verheyen, P., Coen, S., Leo, F., Baets, R., Roelkens, G., and Picqué, N. (2015). An octave-spanning mid-infrared frequency comb generated in a silicon nanophotonic wire waveguide. *Nat. Commun.* 6: 6310.

82 Maddaloni, P., Malara, P., Gagliardi, G., and De Natale, P. (2006). Mid-infrared fibre-based optical comb. *New J. Phys.* 8: 262.

83 Galli, I., Cappelli, F., Cancio, P., Giusfredi, G., Mazzotti, D., Bartalini, S., and De Natale, P. (2013). High-coherence mid-infrared frequency comb. *Opt. Express* 21: 28877.

84 Erny, C., Moutzouris, K., Biegert, J., Kühlke, D., Adler, F., Leitenstorfer, A., and Keller, U. (2007). Mid-infrared difference-frequency generation of ultrashort pulses tunable between 3.2 and 4.8 μm from a compact fiber source. *Opt. Lett.* 32: 1138.

85 Neely, T.W., Johnson, T.A., and Diddams, S.A. (2011). Mid-infrared optical combs from a compact amplified Er-doped fiber oscillator. *Opt. Lett.* 36: 4020.

86 Zhu, F., Hundertmark, H., Kolomenskii, A.A., Strohaber, J., Holzwarth, R., and Schuessler, H.A. (2013). High-power midinfrared frequency comb source based on a femtosecond Er:fiber oscillator. *Opt. Lett.* 38: 2360.

87 Meek, S.A., Poisson, A., Guelachvili, G., Hänsch, T.W., and Picqué, N. (2014). Fourier transform spectroscopy around 3 μm with a broad difference frequency comb. *Appl. Phys. B* 114: 573.

88 Cruz, F.C., Maser, D.L., Johnson, T., Ycas, G., Klose, A., Giorgetta, F.R., Coddington, I., and Diddams, S.A. (2015). Mid-infrared optical frequency combs based on difference frequency generation for molecular spectroscopy. *Opt. Express* 23: 26814.

89 Maser, D.L., Ycas, G., Depetri, W.I., Cruz, F.C., and Diddams, S.A. (2017). Coherent frequency combs for spectroscopy across the 3–5 μm region. *Appl. Phys. B* 123: 142.

90 Lee, K.F., Hensley, C.J., Schunemann, P.G., and Fermann, M.E. (2017). Midinfrared frequency comb by difference frequency of erbium and thulium fiber lasers in orientation-patterned gallium phosphide. *Opt. Express* 25: 17411.

91 Gambetta, A., Ramponi, R., and Marangoni, M. (2008). Mid-infrared optical combs from a compact amplified Er-doped fiber oscillator. *Opt. Lett.* 33: 2671.

92 Gambetta, A., Coluccelli, N., Cassinerio, M., Gatti, D., Laporta, P., Galzerano, G., and Marangoni, M. (2013). Milliwatt-level frequency combs in the 8–14 μm range via difference frequency generation from an Er:fiber oscillator. *Opt. Lett.* 38: 1155.

93 Ruehl, A., Gambetta, A., Hartl, I., Fermann, M.E., Eikema, K.S.E., and Marangoni, M. (2012). Widely-tunable mid-infrared frequency comb source based on difference frequency generation. *Opt. Lett.* 37: 2232.

94 Keilmann, F. and Amarie, S. (2012). Mid-infrared frequency comb spanning an octave based on an Er fiber laser and difference-frequency generation. *J. Infrared Milli. Terahz Waves* 33: 479.

95 Adler, F., Cossel, K.C., Thorpe, M.J., Hartl, I., Fermann, M.E., and Ye, J. (2009). Phase-stabilized, 1.5 W frequency comb at 2.8–4.8 μm. *Opt. Lett.* 34: 1330.

96 McCracken, R.A., Sun, J., Leburn, C.G., and Reid, D.T. (2012). Broadband phase coherence between an ultrafast laser and an OPO using lock-to-zero CEO stabilization. *Opt. Express* 20: 16269.

97 Zhang, Z., Sun, J., Gardiner, T., and Reid, D.T. (2011). Broadband conversion in an Yb:KYW-pumped ultrafast optical parametric oscillator with a long nonlinear crystal. *Opt. Express* 19: 17127.

98 Gu, C., Hu, M., Zhang, L., Fan, J., Song, Y., Wang, C., and Reid, D.T. (2013). High average power, widely tunable femtosecond laser source from red to mid-infrared based on an Yb-fiber-laser-pumped optical parametric oscillator. *Opt. Lett.* 38: 1820.

99 Jin, Y., Cristescu, S.M., Harren, F.J.M., and Mandon, J. (2014). Two-crystal mid-infrared optical parametric oscillator for absorption and dispersion dual-comb spectroscopy. *Opt. Lett.* 39: 3270.

100 Jin, Y., Cristescu, S.M., Harren, F.J.M., and Mandon, J. (2015). Femtosecond optical parametric oscillators toward real-time dual-comb spectroscopy. *Appl. Phys. B* 119: 65.

101 Coluccelli, N., Fonnum, H., Haakestad, M., Gambetta, A., Gatti, D., Marangoni, M., Laporta, P., and Galzerano, G. (2012). 250-MHz synchronously pumped optical parametric oscillator at 2.25–2.6 μm and 4.1–4.9 μm. *Opt. Express* 20: 22042.

102 Leindecker, N., Marandi, A., Byer, R.L., and Vodopyanov, K.L. (2011). Broadband degenerate OPO for mid-infrared frequency comb generation. *Opt. Express* 19: 6296.
103 Ru, Q., Loparo, Z.E., Zhang, X., Crystal, S., Vasu, S., Schunemann, P.G., and Vodopyanov, K.L. (2017). Self-referenced octave-wide subharmonic GaP optical parametric oscillator centered at 3 μm and pumped by an Er-fiber laser. *Opt. Lett.* 42: 4756.
104 Smolski, V.O., Yang, H., Gorelov, S.D., Schunemann, P.G., and Vodopyanov, K.L. (2016). Coherence properties of a 2.6–7.5 μm frequency comb produced as a subharmonic of a Tm-fiber laser. *Opt. Lett.* 41: 1388.
105 Ru, Q., Zhong, K., Lee, N.P., Loparo, Z.E., Schunemann, P.G., Vasilyev, S., Mirov, S.B., and Vodopyanov, K.L. (2017). Instantaneous spectral span of 2.85–8.40 μm achieved in a Cr:ZnS laser pumped subharmonic GaAs OPO, Technical Digest (Online). In: *Conference on Lasers and Electro-Optics*. Washington, DC: Optical Society of America, Paper SM4M.3.
106 Smolski, V., Vasilyev, S., Moskalev, I., Mirov, M., Ru, Q., Muraviev, A., Schunemann, P., Mirov, S., Gapontsev, V., and Vodopyanov, K. (2018). Half-watt average power femtosecond source spanning 3–8 μm based on subharmonic generation in GaAs. *App. Phys. B* 124: 101.
107 Ru, Q., Schunemann, P.G., Vasilyev, S., Mirov, S.B., and Vodopyanov, K.L. (2019). A 2.35-μm pumped subharmonic OPO reaches the spectral width of two octaves in the mid-IR, Technical Digest (Online). In: *Conference on Lasers and Electro-Optics*. Washington, DC: Optical Society of America, Paper SF1H.1.
108 Del'Haye, P., Herr, T., Gavartin, E., Gorodetsky, M.L., Holzwarth, R., and Kippenberg, T.J. (2011). Octave spanning tunable frequency comb from a microresonator. *Phys. Rev. Lett.* 107: 063901.
109 Okawachi, Y., Saha, K., Levy, J.S., Wen, Y.H., Lipson, M., and Gaeta, A.L. (2011). Octave-spanning frequency comb generation in a silicon nitride chip. *Opt. Lett.* 36: 3398.
110 Griffith, A.G., Lau, R.K.W., Cardenas, J., Okawachi, Y., Mohanty, A., Fain, R., Lee, Y.H.D., Yu, M., Phare, C.T., Poitras, C.B., Gaeta, A.L., and Lipson, M. (2015). Silicon-chip mid-infrared frequency comb generation. *Nat. Commun.* 6: 6299.
111 Griffith, A.G., Yu, M., Okawachi, Y., Cardenas, J., Mohanty, A., Gaeta, A.L., and Lipson, M. (2016). Coherent mid-infrared frequency combs in silicon-microresonators in the presence of Raman effects. *Opt. Express* 24: 13044.
112 Wang, C.Y., Herr, T., Del'Haye, P., Schliesser, A., Hofer, J., Holzwarth, R., Hänsch, T.W., Picqué, N., and Kippenberg, T.J. (2013). Mid-infrared optical frequency combs at 2.5 μm based on crystalline microresonators. *Nat. Commun.* 4: 1345.
113 Lecaplain, C., Javerzac-Galy, C., Lucas, E., Jost, J.D., and Kippenberg, T.J. (2015). Quantum cascade laser Kerr frequency comb. ArXiv:1506.00626
114 Savchenkov, A.A., Ilchenko, V.S., Di Teodoro, F., Belden, P.M., Lotshaw, W.T., Matsko, A.B., and Maleki, L. (2015). Generation of Kerr combs centered at 4.5 μm in crystalline microresonators pumped with quantum-cascade lasers. *Opt. Lett.* 40: 3468.

115 Hugi, A., Villares, G., Blaser, S., Liu, H.C., and Faist, J. (2012). Midinfrared frequency comb based on a quantum cascade laser. *Nature* 492: 229.

116 Lu, Q., Wu, D., Slivken, S., and Razeghi, M. (2017). High efficiency quantum cascade laser frequency comb. *Sci. Rep.* 7: 43806.

117 Bagheri, M., Frez, C., Sterczewski, L.A., Gruidin, I., Fradet, M., Vurgaftman, I., Canedy, C.L., Bewley, W.W., Merritt, C.D., Kim, C.S., Kim, M., and Meyer, J.R. (2018). Passively mode-locked interband cascade optical frequency combs. *Sci. Rep.* 8: 3322.

118 Sun, J.H., Gale, B.J.S., and Reid, D.T. (2007). Composite frequency comb spanning 0.4–2.4 μm from a phase controlled femtosecond Ti:sapphire laser and synchronously pumped optical parametric oscillator. *Opt. Lett.* 32: 1414.

119 Adler, F., Maslowski, P., Foltynowicz, A., Cossel, K.C., Briles, T.C., Hartl, I., and Ye, J. (2010). Mid-infrared Fourier transform spectroscopy with a broadband frequency comb. *Opt. Express* 18: 21861.

120 Iwakuni, K., Porat, G., Bui, T.Q., Bjork, B.J., Schoun, S.B., Heckl, O.H., Fermann, M.E., and Ye, J. (2018). Phase-stabilized 100 mW frequency comb near 10 μm. *Appl. Phys. B* 124: 128.

121 Kobayashi, Y., Torizuka, K., Marandi, A., Byer, R.L., McCracken, R.A., Zhang, Z., and Reid, D.T. (2015). Femtosecond optical parametric oscillator frequency combs. *J. Opt.* 17: 94010.

122 Wong, S.T., Vodopyanov, K.L., and Byer, R.L. (2010). Self-phase-locked divide-by-2 optical parametric oscillator as a broadband frequency comb source. *J. Opt. Soc. Am. B* 27: 876.

123 Vodopyanov, K.L., Wong, S.T., and Byer, R.L. (2013). Infrared frequency comb methods, arrangements and applications. U.S. patent 8,384,990 (26 February 2013).

124 Leindecker, N., Marandi, A., Byer, R.L., Vodopyanov, K.L., Jiang, J., Hartl, I., Fermann, M., and Schunemann, P.G. (2012). Octave-spanning ultrafast OPO with 2.6–6.1 μm instantaneous bandwidth pumped by femtosecond Tm-fiber laser. *Opt. Express* 20: 7047.

125 Vodopyanov, K.L., Sorokin, E., Sorokina, I.T., and Schunemann, P.G. (2011). Mid-IR frequency comb source spanning 4.4–5.4 μm based on subharmonic GaAs OPO. *Opt. Lett.* 36: 2275.

126 Smolski, V.O., Vasilyev, S., Schunemann, P.G., Mirov, S.B., and Vodopyanov, K.L. (2015). Cr:ZnS laser-pumped subharmonic GaAs optical parametric oscillator with the spectrum spanning 3.6–5.6 μm. *Opt. Lett.* 40: 2906.

127 Marandi, A., Leindecker, N., Pervak, V., Byer, R.L., and Vodopyanov, K.L. (2012). Coherence properties of a broadband femtosecond mid-IR optical parametric oscillator operating at degeneracy. *Opt. Express* 20: 7255.

128 Wan, C., Li, P., Ruehl, A., and Hartl, I. (2018). Coherent frequency division with a degenerate synchronously pumped optical parametric oscillator. *Opt. Lett.* 43: 1059.

129 Del'Haye, P., Schliesser, A., Arcizet, O., Wilke, T., Holzwarth, R., and Kippenberg, T.J. (2007). Optical frequency comb generation from a monolithic microresonator. *Nature* 450: 1214.

130 Kippenberg, T.J., Holzwarth, R., and Diddams, S.A. (2011). Microresonator-based optical frequency combs. *Science* 332: 555.

131 Kippenberg, T.J., Spillane, S.M., and Vahala, K.J. (2004). Kerr-nonlinearity optical parametric oscillation in an ultrahigh-Q toroid microcavity. *Phys. Rev. Lett.* 93: 083904.

132 Savchenkov, A.A., Matsko, A.B., Strekalov, D., Mohageg, M., Ilchenko, V.S., and Maleki, L. (2004). Low threshold optical oscillations in a whispering gallery mode CaF_2 resonator. *Phys. Rev. Lett.* 93: 243905.

133 Ferrera, M., Razzari, L., Duchesne, D., Morandotti, R., Yang, Z., Liscidini, M., Sipe, J.E., Chu, S., Little, B.E., and Moss, D.J. (2008). Low-power continuous-wave nonlinear optics in doped silica glass integrated waveguide structures. *Nat. Photonics* 2: 737.

134 Savchenkov, A.A., Matsko, A.B., Ilchenko, V.S., Solomatine, I., Seidel, D., and Maleki, L. (2008). Tunable optical frequency comb with a crystalline whispering gallery mode resonator. *Phys. Rev. Lett.* 101: 093902.

135 Grudinin, I.S., Yu, N., and Maleki, L. (2009). Generation of optical frequency combs with a CaF_2 resonator. *Opt. Lett.* 34: 878.

136 Levy, J.S., Gondarenko, A., Foster, M.A., Turner-Foster, A.C., Gaeta, A.L., and Lipson, M. (2010). CMOS-compatible multiple-wavelength oscillator for on-chip optical interconnects. *Nat. Photonics* 4: 37.

137 Jung, H., Xiong, C., Fong, K.Y., Zhang, X.F., and Tang, H.X. (2013). Optical frequency comb generation from aluminum nitride microring resonator. *Opt. Lett.* 38: 2810.

138 Hausmann, B.J.M., Bulu, I., Venkataraman, V., Deotare, P., and Lončar, M. (2014). Diamond nonlinear photonics. *Nat. Photonics* 8: 369.

139 Foster, M.A., Levy, J.S., Kuzucu, O., Saha, K., Lipson, M., and Gaeta, A.L. (2011). Silicon-based monolithic optical frequency comb source. *Opt. Express* 19: 14233.

140 Luke, K., Okawachi, Y., Lamont, M.R.E., Gaeta, A.L., and Lipson, M. (2015). Broadband mid-infrared frequency comb generation in a Si_3N_4 microresonator. *Opt. Lett.* 40: 4823.

141 Villares, G., Hugi, A., Blaser, S., and Faist, J. (2014). Dual-comb spectroscopy based on quantum-cascade-laser frequency combs. *Nat. Commun.* 5: 5192.

7

Mid-IR Applications

The development of laser-based sources in the mid-infrared opened up unprecedented possibilities in many fields including spectroscopy and trace-gas detection, atmospheric science, greenhouse gas and pollution monitoring, homeland security, hyperspectral imaging, infrared countermeasures, free-space optical communications, biomedical diagnostics, surgery and neurosurgery, industrial process control, the study of combustion dynamics, organic material processing, and investigation of metamaterials, among others. This chapter reviews the most important applications of the mid-IR.

7.1 Spectroscopic Sensing and Imaging

Vibrational spectra provide fingerprints of molecular structures, and as such they are used extensively in science and technology. Coherent laser sources in the mid-IR, especially in the 3–20 μm spectral range, have long been recognized as important tools for both fundamental and applied spectroscopy and sensing. The vast majority of gaseous chemical substances exhibit fundamental vibrational absorption bands in that region, and the absorption of light by these bands provides a nearly universal means for their detection. The main advantage of optical techniques is the nonintrusive *in situ* detection capability for trace gases and their isotopologues (molecules that contain isotopes). As an illustration, the fundamental absorption spectra, associated with transitions between rotational–vibrational states, for 11 small molecules are shown in Figure 7.1.

The mid-IR region contains two important windows (3–5 and 8–14 μm) in which the Earth's atmosphere is relatively transparent. When the strongest rotational–vibrational ("ro-vibrational") molecular absorptions in the fingerprint region are targeted, one can detect small traces of vapors with sensitivities

Laser-based Mid-infrared Sources and Applications, First Edition. Konstantin L. Vodopyanov.

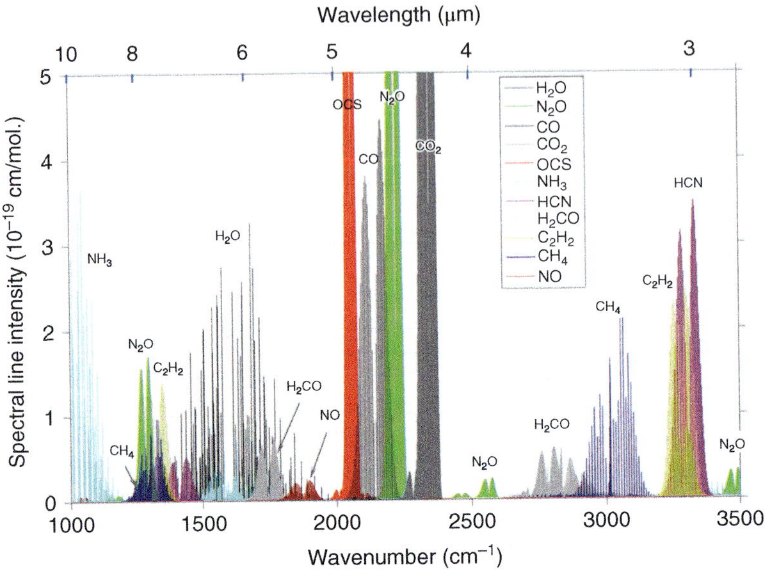

Figure 7.1 Fundamental absorption features (plotted as spectral line intensities) of 11 molecular gases. The plot is based on data from the HITRAN database [1]. *Source:* reproduced with permission of Elsevier.

of parts-per-billion (ppb) to parts-per-trillion (ppt), in a variety of atmospheric, security, and industrial applications.

There is a plethora of literature on spectroscopic applications in the mid-IR published to date. An outstanding review by Tittel et al. [2] provides an insight into different spectroscopic techniques using a variety of mid-IR laser sources. In this chapter, we will focus on the most recent developments with emphasis on the new techniques and advanced laser sources.

7.1.1 QCLs for Spectroscopy and Trace-gas Analysis

Quantum cascade lasers (QCLs) are almost ideal sources for trace-gas monitoring, thanks to their small size and high spectral purity, when operated in a single-longitudinal-mode (SLM) regime. Recent progress in the development of QCLs has led to the availability of robust mid-IR spectroscopic light sources capable of high output power and room temperature operation [3–7]. QCLs cover the mid-IR region from 3.5 to 19 μm, where most molecules have their fundamental absorption bands. For accurate spectroscopic analysis, in most cases both single transverse and single longitudinal mode operation for QCLs are required. The latter can be achieved with distributed feedback (DFB) QCLs [8, 9] or by using an external-cavity configuration [10]. QCLs can be used to

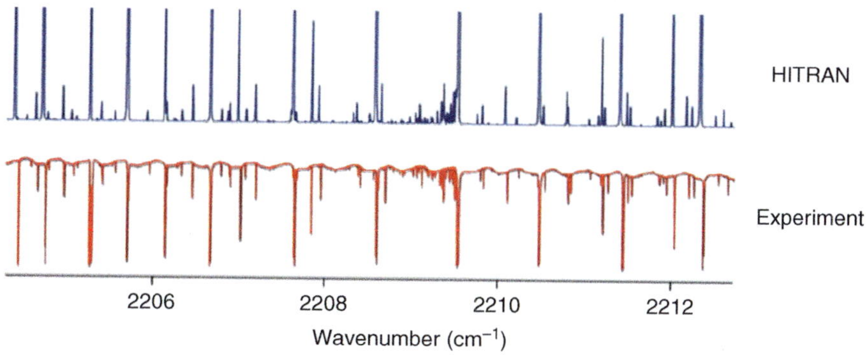

Figure 7.2 Portion of the absorption spectrum of nitrous oxide (N_2O) at 10-torr pressure obtained with a single scan of the EC-QCL. Shown are HITRAN simulation (top) and experimental absorption spectrum (bottom, inverted). *Source:* reproduced from figure 4 of [13], with permission of Elsevier.

carry out absorption spectroscopy with different techniques including multi-pass spectroscopy, wavelength- and frequency-modulation spectroscopy, cavity-enhanced spectroscopy, cavity ring-down spectroscopy (CRDS), intracavity spectroscopy, magnetic field rotation spectroscopy, photoacoustic spectroscopy (PAS), and photothermal spectroscopy [2, 11].

QCL-based trace-gas sensors that rely upon direct optical absorption in the mid-IR have wide applications for practical detection of trace gases in the atmosphere. Nelson et al. [12] reported measurements of nitric oxide (NO) at $\lambda = 5.26\,\mu m$ in outside air with a detection limit of less than 1 ppb using a thermoelectrically cooled QCL with a DFB grating that operated in pulsed (duration ~10 ns) frequency-sweep mode with an "instantaneous" full-width half-maximum (FWHM) linewidth down to $0.02\,cm^{-1}$. With a 210-m path-length multiple-pass absorption cell filled with NO gas at 50-torr pressure, a detection precision of $0.12\,ppb\,Hz^{-1/2}$ was achieved using a liquid-nitrogen-cooled detector.

Daylight Solutions demonstrated a high-resolution QCL spectrometer based on an external-cavity QCL (EC-QCL) that contains a diffraction grating and operates in a continuous-wave (CW) single-frequency regime. The instrument can be continuously tuned, without mode hops, over $>100\,cm^{-1}$ of the optical frequency, with a linewidth that can be as narrow as $0.001\,cm^{-1}$. A closed-loop servo with feedback was used to optimize the cavity length to select and support the desired single mode at every grating angle. Figure 7.2 shows a portion of the Doppler-resolved spectrum of N_2O gas at 10-torr (13-mbar) pressure obtained with such an instrument [13]. Wysocki et al. performed high-resolution background-free magnetic field rotation (Faraday rotation) spectroscopy of NO molecules using room-temperature CW mode-hop-free EC-QCL with a single-mode tuning range of 155 cm^{-1} near λ = 5.3 μm (maximum power of 11 mW). The authors also used another mode-hop-free EC-QCL with a single-mode tuning range of 7.77–9.05 μm (span 182 cm^{-1}, maximum power 50 mW), to measure high-resolution N_2O spectrum via direct absorption spectroscopy [14]. Chao et al. performed wavelength-modulation-spectroscopy (WMS) for real-time *in situ* NO detection in combustion gases at exhaust temperature of 600 K [15]. The laser used for these experiments was a commercial EC-QCL (Daylight Solutions) that could be tuned mode-hop-free from 1895 to 1951 cm^{-1} (5.1256–5.2770 μm) with a nominal spectral linewidth <0.0015 cm^{-1}. This tuning region can cover 20 pairs of NO absorption transitions in the R-branch of its fundamental vibrational band.

PAS is based, similar to the optical absorption spectroscopy, on resonant optical absorption process; however, it relies on a different physical principle [16]. When light at a specific wavelength is absorbed by the gas sample, the excited molecules subsequently relax to the ground state by means of non-radiative processes. This produces localized heating in the gas, which in turn results in the production of a sound wave. The key advantages of this technique are that (i) no optical detector is required – the resulting sound waves can be detected by an inexpensive microphone, and (ii) it is a background-free technique in the sense that the photoacoustic signal is present only in the presence of absorption and is proportional to the absorption coefficient of a measured sample. A recently developed novel approach to photoacoustic detection utilizes a quartz tuning fork that can be regarded as a resonant acoustic detector [11, 17, 18]. The basic idea of quartz-enhanced photoacoustic spectroscopy (QEPAS) is to accumulate the acoustic energy in a high-Q resonant system – a quartz tuning fork. The latter is used, for example, in electronic clocks as a frequency standard (its resonant frequency is ~32.77 kHz). If the incident light intensity is modulated at the same frequency, the photoacoustic signal from the quartz tuning fork can be dramatically enhanced. QEPAS is now widely used for detection of various chemical species, such as NH_3, H_2O, CO_2, N_2O, CO, CH_2O, and NO molecules. The best results in terms of minimum detectable gas concentration have been obtained by adding an acoustic micro-resonator tube (Figure 7.3) to further enhance the acoustic signal. In such a way, ppb-level concentrations can be measured for several small molecules [19, 20]. For example, Dong et al. achieved a sensitivity of detecting NO of 5 ppb at one-second averaging time at an optimal gas pressure of 210 torr by targeting the 1900.08 cm^{-1} (5.26-μm) absorption line of NO. The authors utilized an EC-QCL combined with QEPAS system with an added acoustic micro-resonator [21].

The CRDS method is based on measurement of the decay time of the intensity of light that is trapped in a low-loss optical cavity [22, 23]. The laser light is injected into an optical cavity formed by a pair of highly reflective (R > 99.9%) mirrors. (In the case of a narrow-band CW laser, the laser frequency should match one of the cavity resonances.) The light that is trapped inside the cavity reflects back and forth between the two mirrors, with a small fraction

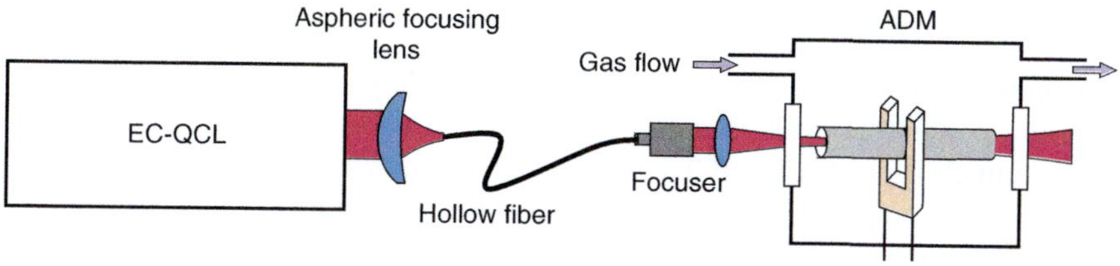

Figure 7.3 Quartz-enhanced photoacoustic spectroscopy (QEPAS) system. The incoming photons excite the target molecule at a resonant wavelength. Collisional de-excitation converts the absorbed energy into pressure waves, which are detected by a quartz tuning fork. EC-QCL, external-cavity QCL; ADM, absorption detection module. *Source:* reproduced from figure 11 of [19], with permission of OSA, The Optical Society.

transmitting through each mirror. The resultant leaking light is monitored as a function of time and this allows the decay time of the cavity to be determined. The apparatus is converted to an extremely sensitive absorption spectrometer by placing an absorbing medium (e.g. molecular gas) between the two mirrors and recording the frequency-dependent decay time of the cavity. Measurement of the ring-down time allows the absolute single-pass transmission coefficient of the cavity to be determined with high accuracy. The absorption intensities are obtained by subtracting the baseline absorption of the cavity, which is measured when the laser wavelength is off molecular resonance.

CRDS methods are widely used with QCLs. Kosterev et al. [24] demonstrated a spectroscopic gas sensor for NO based on the cavity ring-down technique by accessing absorption lines at 1921.599 and 1921.601 cm^{-1}. A CW DFB QCL operating near 5.2 μm was used as a tunable single-frequency light source. Measurements of ppb NO concentrations in nitrogen with a 0.7-ppb standard error for a data collection time of eight seconds have been performed.

Using CRDS, Galli et al. demonstrated mid-IR detection of carbon dioxide containing ^{14}C isotope ("radiocarbon dioxide," $^{14}CO_2$) at 4.5-μm wavelength with an unprecedented parts-per-quadrillion (10^{-15}) sensitivity level [25]. The spectroscopic apparatus consisted of a high-finesse cavity with a ring-down time constant of 17.5 μs (effective interaction path of 5.2 km). The system employed a narrow-linewidth (9-kHz) frequency-stabilized QCL with a tuning range 2208–2212 cm^{-1} and an output power 100 mW. The sample gas was kept at 12 mbar pressure and 170 K temperature. The authors used a newly developed saturated-absorption cavity ring-down spectroscopic technique, where, at the beginning of the ring-down process, the intracavity intensity is so high that it saturates the absorption of the sample gas, while at the end of the ring-down process, the decay rate is determined by the linear gas absorption. This allowed the authors to decouple the linear gas absorption decay rate from that determined by the passive cavity losses. In such a way the authors reached a sensitivity of measurement of $^{14}CO_2$ concentration of 5×10^{-15} with respect to the main isotopologue [25].

7.1.2 Spectroscopy with ICLs

Interband cascade lasers (ICLs) are currently widely used in spectroscopy in the 3–5 μm spectral band. Because the threshold drive powers of ICLs are far lower than those of QCLs, ICLs are viewed as the laser of choice for mid-IR laser spectroscopy applications that do not require high output power but need to be hand-portable and, ideally, battery operated.

Horstjann et al. [26] used a CW mid-IR DFB ICL at $\lambda = 3.53\,\mu m$ ($2832.5\,cm^{-1}$) to detect formaldehyde (H_2CO) using QEPAS. The laser was operated at liquid-nitrogen ($T = 77\,K$) temperatures and provided single-mode (linewidth <20 MHz) output power of up to 12 mW. The authors demonstrated the feasibility of detecting formaldehyde in ambient air with a detection limit of 0.6 parts per million by volume (ppmv) for a 10-second acquisition time. Wysocki et al. utilized a pair of ICLs mounted in the same Dewar for the simultaneous detection of two atmospheric trace gases, formaldehyde (H_2CO) and ethane (C_2H_6), in the wavelength range of 3.56 and 3.33 μm correspondingly [27]. The researchers used the WMS and a 100-m Herriott multipass cell. With a one-second integration time, the minimum detection limits of 3.5 ppb for H_2CO and 150 ppt for C_2H_6 were demonstrated.

Tunable ICLs operating at room temperature are most desirable for use in an outside laboratory, for example, for process monitoring in industry, where unattended operation is essential. Lundqvist et al. have demonstrated sensing of formaldehyde using a room-temperature DFB ICL operating around 3.493 μm with a detection limit better than 1 ppm × m [28].

7.1.3 Spectroscopy with DFG and OPO Sources

Mid-IR sources based on difference frequency generation (DFG) are widely used in spectroscopy. Typically, CW DFG-based systems use well-developed optical telecommunication lasers as pump sources, combined with robust optical fiber elements. They operate at room temperature and offer certain advantages (as compared to direct lasers sources like QCLs and ICLs) for high-precision spectroscopic measurements, since they offer inherent sub-MHz linewidths and do not need active feedback. Richter et al. used a mid-IR system near $\lambda = 3.5\,\mu m$ based on DFG in periodically poled $LiNbO_3$, with near-IR diode and fiber lasers at 1083 and 1562 nm, respectively, serving as pump sources [29]. With the WMS employing an astigmatic Herriott multipass gas absorption cell, the compact DFG system allowed detection of formaldehyde with a minimal detectable concentration of 74 parts-per-trillion by volume for one minute of averaging time. Excellent reviews on spectroscopic applications of CW mid-IR DFG sources can be found in [30, 31].

The versatility of optical parametric oscillators (OPOs) as a coherent light source in trace-gas detection has been demonstrated by numerous groups. For example, spectroscopic sources based on CW OPOs that can be tunable all the way from 2.5 to 4.7 μm with narrow (1 MHz) linewidth, high power (~W), and smooth, mode-hop-free tuning up to 60 GHz, suitable for PAS and CRDS, have been developed [32–34]. With the CRDS technique, von Basum et al. demonstrated detection of ethane in the 3-μm wavelength region, with a detectable concentration of 0.5 ppt at a three-minute integration time [35].

Todd et al. reported a mid-IR CRDS system that used a pulsed widely tunable long-wavelength (6–10 μm) OPO combined with a low-loss CRDS cavity with the ring-down decay time of 20 μs, equivalent to a 6 km path length. Vapor-phase mid-IR spectra of common explosives (TNT, TATP, RDX, PETN, and Tetryl) were acquired with ppb concentration levels [36].

7.1.4 Broadband Spectroscopy with Frequency Combs

Optical frequency combs (OFCs), initially conceived for precision spectroscopy of the hydrogen atom in the UV, are now commonly used for probing molecular ro-vibrational transitions throughout the whole mid-IR molecular fingerprint region [37]. Spectroscopic applications of broadband OFCs include several scenarios:

- A system where a frequency comb directly interrogates an absorbing sample after which the spectrum is dispersed in two dimensions and sensed with detector arrays [38–40].
- Michelson interferometer-based Fourier-transform spectroscopy [41–46] to enhance detection sensitivity; this technique may use a multipass cell [41] or an external high-finesse optical cavity [45, 47], or it may perform intracavity spectroscopy inside the comb source [43].
- Dual-comb spectroscopy (DCS). This is a form of high-speed Fourier-transform spectroscopy involving no moving parts and capable of combining vast spectral coverage with very high resolution [48–65]. The technique is capable of simultaneously probing molecular absorption on thousands of narrowly spaced wavelengths using a single high-speed photodetector.

By using an OPO broadband comb (span ~300 cm^{-1}) that operated in the 2.7–4.8-μm spectral range, combined with a multipass gas cell and a Fourier transform spectrometer (FTS), Adler et al. demonstrated rapid trace-gas detection with 0.0056-cm^{-1} resolution. The authors achieved ppb detection limits in 30 seconds of integration time for several important molecules including methane, ethane, isoprene, and nitrous oxide [41]. Figure 7.4 provides an example of detecting the Q-branch spectrum of methane with such a system.

Based on a comb source centered at 8.5 μm created by difference frequency mixing in an orientation-patterned GaP (OP-GaP) crystal, and in combination with cavity-enhanced direct frequency comb spectroscopy, Changala et al. obtained the first high-resolution ro-vibrational spectrum of the fullerene, C60, molecule [51].

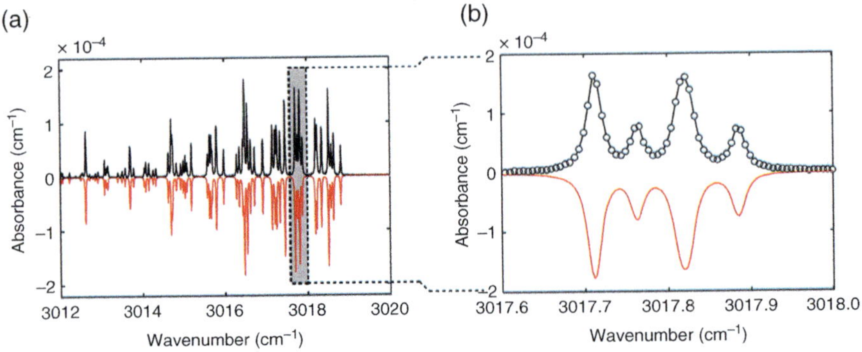

Figure 7.4 (a) Methane Q-branch spectrum (N_2 buffer gas at 100 torr) obtained by the use of a multipass cell and an OPO-based frequency-comb Fourier-transform spectrometer with the resolution of 0.0056 cm^{-1} (168 MHz). (b) A magnified view of the spectrum; open circles are measured data points. A HITRAN-based theoretical spectrum is plotted negative for clarity. *Source:* reproduced from figure 3 of [41], with permission of OSA, The Optical Society.

In DCS, which is the most advanced of the comb techniques, a second frequency comb, with a small offset of mode spacing (e.g. pulse repetition frequency), effectively plays the role of the time-delayed second arm in the Fourier-transform spectrometer. DCS requires a high degree of coherence between the two combs, and most reports that simultaneously take full advantage of the dual-comb technique (broad spectral coverage, comb-tooth resolved spectra, and rapid scans) have till recently privileged the near-IR domain [52–54]. The true promise of DCS for sensitive molecular detection lies in the mid-IR and proof-of-principle demonstrations of molecular detection have been carried out in different spectral regions of the mid-IR. These include: the 10-μm wavelength region (instantaneous spectral span 250 cm^{-1}, spectral resolution 2 cm^{-1}) [49], the 2.4-μm region (span 200 cm^{-1}/resolution 2 cm^{-1}) [55], the 3-μm region (150 cm^{-1}/0.2 cm^{-1} [56]; 250 cm^{-1}/0.8 cm^{-1} [57]; 350 cm^{-1}/0.2 cm^{-1} [58]; 250 cm^{-1}/0.07 cm^{-1} [59]; and 160 cm^{-1}/0.5 cm^{-1} [60]), and the 6.5-to-7.5-μm region (span 85 cm^{-1}/resolution 0.3 cm^{-1}) [61]. A very high spectral resolution has been demonstrated with a DFG-based dual comb in the 3.4-μm region, although with reduced simultaneous spectral coverage (spectral span 30 cm^{-1}/resolution 10 kHz $\approx 3 \times 10^{-7}$ cm^{-1}) [62], and in the 7-μm region with the use of QCL-based dual frequency comb (spectral span 16 cm^{-1}/ resolution 0.003 cm^{-1}) [63]. A broadband (2.6–5.2 μm) DCS was reported in [64] using dual combs created via DFG. The above span was covered in several steps, by adjusting the proper poling period of the frequency-mixing crystal for each step.

Most recently, using a pair of broad-bandwidth subharmonic OP-GaAs OPO combs (Figure 7.5) with a high degree of mutual coherence, the acquisition of

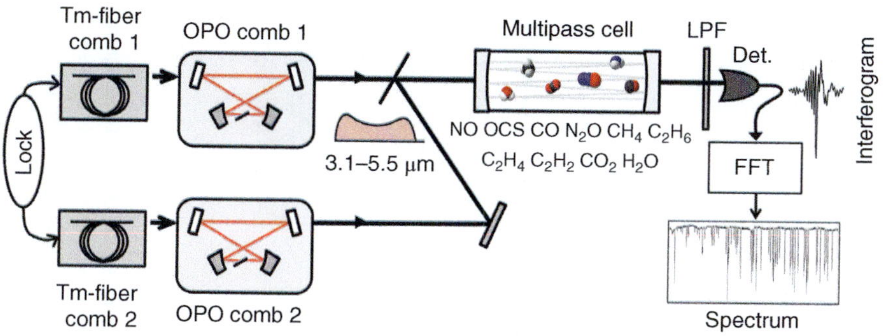

Figure 7.5 Schematic of the 3.1–5.5 μm DCS setup. A pair of phase-locked Tm-fiber laser combs pumps a pair of subharmonic OPOs. Their beams are combined, passed through a 76-m multipass gas cell, detected with an InSb detector, digitized, and Fourier transformed to retrieve the optical spectrum. LPF, long-pass (>2.5 μm) filter. *Source:* reproduced from figure 2a of [65], with permission of Springer Nature.

350 000 comb-tooth resolved data points over the entire (no gaps) range of 3.1–5.5 μm (spectral span >1300 cm^{-1}, spectral resolution <0.0038 cm^{-1}) has been achieved [65]. With a measurement time varying from 7 ms to 1000 s, the authors demonstrated parallel detection of 22 molecular species (including isotopologues) in a mixture of gases and demonstrated the full potential of the dual-comb technique in the mid-IR (spectral coverage, spectral resolution, speed, and the absolute frequency referencing, e.g. to atomic clock). Molecular spectra obtained with a 76-m multipass are shown in Figure 7.6. Also, an exceptional degree of coherence between the two subharmonic OPOs (residual linewidth 25 mHz, mutual coherence time 40 seconds) allowed resolving frequency comb modes with a finesse of 4000.

A paper by Vainio and Halonen gives a very good review of different methods for producing frequency combs based on nonlinear optics and gives examples of their use for molecular spectroscopy [66]. Another excellent paper, by Cossel et al. [67], provides an overview of broadband spectroscopy methods and applications, such as atmospheric measurements, chemical kinetics, breath analysis, astrochemistry, fundamental laboratory spectroscopy, and industrial applications.

7.1.5 Hyperspectral Imaging

The goal of hyperspectral imaging is to add spectral information for each pixel of the 2D image of a scene, with the purpose of finding hidden objects, identifying materials, detecting trace gases, etc. Stothard et al. demonstrated a scanning system for active, real-time hyperspectral imaging in the mid-IR [68]. The authors used a combination of a CW tunable OPO, an electromechanical

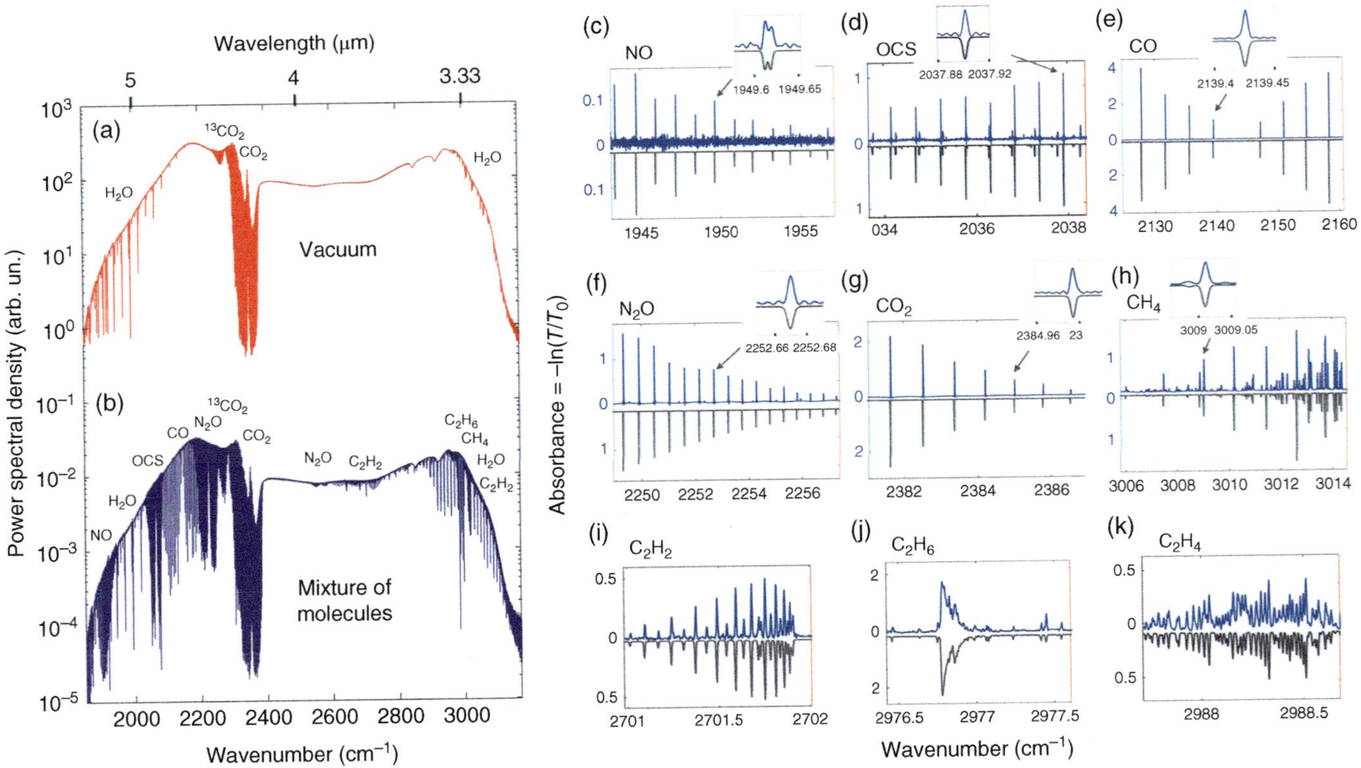

Figure 7.6 Dual-comb spectra of a mixture of gases. (a) Optical spectrum (log scale) retrieved from a single coherently averaged interferogram (100 000 averages), when the multipass ($L = 76$ m) gas cell was evacuated. Absorption dips originate from atmospheric gases outside the cell. (b) Optical spectrum with the gas cell filled with a mixture of gases (OCS, N_2O, NO, CO, CH_4, C_2H_6, C_2H_4, C_2H_2, CO_2, and H_2O) in N_2 buffer gas at 3 mbar total pressure. The two curves are vertically offset for clarity. (c–k) Absorbance spectra for nine molecules (out of 22 detected) obtained by normalizing the "sample" spectrum to that of the "vacuum." Also shown are theoretical (HITRAN) spectra (inverted for clarity). *Source:* reproduced from figure 3 of [65], with permission of Springer Nature.

scanning imager, and a single detector. The elegant design of a pump-enhanced CW OPO, which used periodically poled RTA (PP-RTA) crystal, and where the pump and the signal wave were both resonant, required only <500 mW of pump power from a 1064-nm Nd:YVO_4 laser to achieve the OPO threshold. With 1 W of pump power, the OPO provided adequate mid-IR idler-wave power (~50 mW) for 2D scanning.

The geometry of the scanner was such that the radiation from the OPO was directed to the scene under surveillance via the polygon scanner with 10 facets and a tilting mirror. The backscattered radiation (that can be approximated as Lambertian reflection into the 2π solid angle) returning from the scene was collected via the same tilting mirror and the polygon scanner, and was then focused by the collection lens onto a mid-IR detector with the sampling rate for digitization of 200 kHz. Due to the wide spectral coverage (3.18–3.5 μm) and high spectral resolution (<1 GHz), the system was capable of being selectively tuned into narrow (~5 GHz) absorption features of a wide variety of gaseous species in the atmosphere. As an example, the authors showed how a strong absorption peak exhibited by methane at 3057.7 cm^{-1} (3.27 μm) can be accessed and gas plumes can be imaged in concentrations as low as 30 ppm from the distance of several meters (Figure 7.7).

Dam et al. [69] demonstrated a compact imaging system where the incoming mid-IR light (coherent or incoherent) was upconverted in a single pass through a periodically poled lithium niobate (PPLN) nonlinear crystal, by mixing with a resonating 1064-nm Nd:YVO_4 laser beam, so that the mid-IR 2D image information was conserved in the process. The upconverted near-IR image near 800-nm wavelength, with quantum conversion efficiency 20%, was then detected by a charge-coupled device (CCD) camera. The system can be tuned to any wavelength band in the 2.85–5 μm region with the bandwidth,

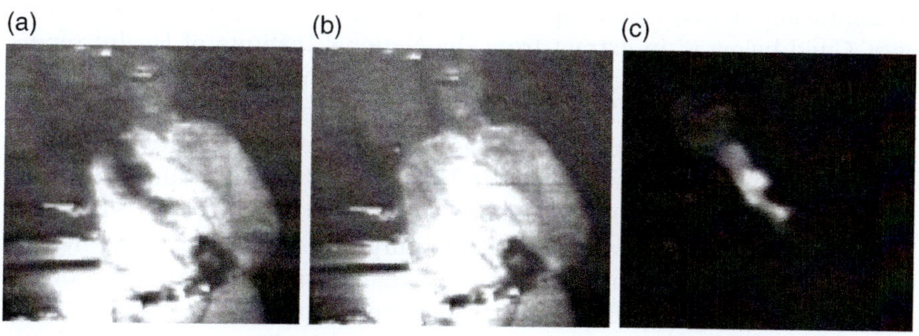

Figure 7.7 Hyperspectral imaging of methane gas. (a) Image of a methane plume acquired at the OPO wavelength in resonance with a strong methane line at 3057.7 cm^{-1} (3.27 μm). (b) Same when the OPO wavelength is tuned away from resonance. (c) The difference between images (a) and (b), which shows the methane plume. *Source:* reproduced from figure 6 of [68], with permission of OSA, The Optical Society.

determined by the phase matching, which varied from 5 nm at 2.9 μm to 25 nm at 3.8 μm. According to the authors, by combining image upconversion and modern low-noise electron-multiplying CCD cameras, one can obtain mid-IR images with a sensitivity of about one photon per pixel.

7.2 Medical Applications

7.2.1 Laser Tissue Interactions

The laser was envisioned as a surgical tool very soon after its first demonstration. Laser surgery provides a noncontact approach capable of cutting tissue with up to single-cell precision. Short-pulsed, mid-infrared lasers are promising tools for human surgical applications that complement other medical lasers, due to their ability to ablate tissue with relatively little damage to surrounding tissue and reduced photochemical side effects.

A residual tissue damage needs to be minimized, hence the most important laser parameters are wavelength, pulse duration, and pulse fluence (pulse energy per unit area). High optical absorption (and small penetration length) of the tissue, typical for the mid-IR wavelengths, is beneficial since it reduces the pulse fluence needed to superheat the tissue material and drive ablation; also, high optical absorption combined with short pulse duration help to reduce the residual tissue damage. By using laser sources operating in the mid-infrared, water-rich, lipid-rich, and protein-containing tissues can be ablated with minimal collateral damage.

7.2.1.1 Holmium and Thulium Surgical Lasers

Holmium and thulium lasers operating near 2 μm are most widely used in surgery. Cited below is an extract from the National Institute of Health report regarding the wide use of multi-watt pulsed Ho:YAG lasers (λ = 2.1 μm): "Ho:YAG lasers, like CO_2 lasers, offer precise cutting with minimal damage to adjacent tissue; however, unlike CO_2 lasers, they also offer fiber optic delivery (which is ideal for endoscopic use) and the ability to treat tissue in a liquid-filled environment (e.g. saline, blood). The initial specialty for which the Ho:YAG laser was used was arthroscopic surgery, especially discectomy (the surgical removal of herniated disc material that presses on a nerve root or the spinal cord). Today, it is effectively used in many surgical specialties, including general surgery, urology, laparoscopy, neurosurgery, lithotripsy, angioplasty, orthopedic surgery (e.g. meniscectomy, bone sculpting, plastic surgery, and some experimental surgery, such as cartilage shrinking to tighten loose joints), and dentistry. Because of its broad range of potential applications, it has been called the Swiss Army Knife of lasers" [70].

Also, in soft-tissue medicine, direct application of the Tm^{3+}-doped silicate fiber laser has been widespread, owing to the fact that its emission wavelength (1.94 μm) overlaps with the absorption wavelength of the combination vibration of the O—H bond in water. The high-power thulium fiber laser is capable of vaporization of the prostate and incision of ureteral and bladder tissues. This laser has several potential advantages over the Ho:YAG laser, including smaller size, more efficient operation, more precise incision of tissues, and operation in either the pulsed or the CW mode. Thulium fiber-based lasers of high average power and short pulse lengths are extremely beneficial for clinical use in rapid vaporization of the prostate and more precise incision of urethral/bladder neck strictures [71].

7.2.1.2 Er:YAG Lasers ($\lambda = 2.9\,\mu m$)

Wavelengths near 2.9 μm, which are close to the fundamental absorption peak of stretching vibrations in water, have significant benefits for laser surgery. In tissue, which is typically 70% water, the optical penetration depth near 2.9 μm is approximately 1 μm (water is a natural chromophore in human tissue). This means that the zone of thermal damage is minimized, especially when short pulses are used.

Figure 7.8 displays the absorption coefficient of liquid water over a large (six decades) span of frequencies, from the UV to THz. For clinical use, the major advantage of mid-IR 2.9-μm radiation over UV excimer laser radiation is that there are no mutagenic or carcinogenic effects of infrared radiation [72, 73].

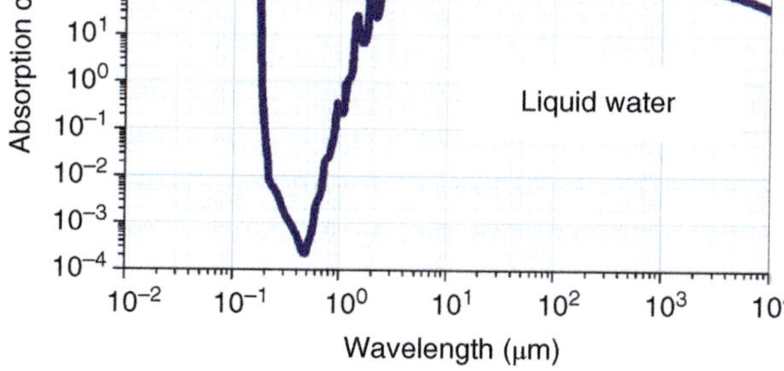

Figure 7.8 Absorption coefficient of liquid water: from UV to THz (http://en.wikipedia.org/wiki/File:Absorption_spectrum_of_liquid_water.png). The scale is logarithmic in both directions. *Source:* reproduced with permission of Wikipedia.

Kaufmann et al. have shown that pulsed 2.94-μm Er:YAG laser surgery allows the extremely precise etching of superficial skin lesions and also has a potential for skin resurfacing [74]. Franjic et al. reported on the highly efficient ablation of tooth enamel using 55-ps-long laser pulses at $\lambda \approx 2.95\,\mu m$ [75]. The 2.95-μm wavelength used in the experiment was both in resonance with the OH-stretch mode of water, and also near the absorption peak of hydroxyapatite crystals (hydroxyapatite is the predominant mineral of dental enamel). The reported fluence needed to achieve ablation was $0.75\,J/cm^2$, several times smaller than the typical ablation thresholds reported for longer laser pulses, with nanosecond and microsecond pulse durations.

Fiber lasers of the 3-μm range (see Chapter 3) have the potential to make a significant impact in laser surgery. Pierce et al. performed preliminary studies of the interaction between a CW erbium-doped fiber laser operating at around 3 μm and soft biological tissue [76]. The pulse-periodic nanosecond regime of such lasers would be particularly valuable for laser surgery. Also, solid-state pulse-periodic Cr^{2+}:ZnS/ZnSe lasers tuned to 2.9-μm wavelength (see Chapter 2) can be a good alternative to Er:YAG lasers for tissue ablation, thanks to their high efficiency and ability to operate at high repetition rates.

7.2.1.3 Importance of the Spectral Band of 6–7 μm

The ultimate goal of medical laser ablation is to remove a defined amount of material in an efficient manner with the least amount of collateral damage. Investigations were centered on two specific mid-IR wavelengths, 6.1 and 6.45 μm. These wavelengths coincide with, respectively, the amide I and amide II absorption bands of protein, and, in addition, with the bending mode of water at 6.1 μm [77]. At these wavelengths, energy is coupled into the protein matrix as well as the bound and unbound water within the tissue.

Edwards et al. reported that targeting laser radiation wavelength to the amide II band of proteins at 6.45 μm leads to the ablation of rat brain cortex tissue with minimal collateral damage while maintaining a substantial ablation rate [78]. The authors found that the collateral damage (coagulation necrosis) at $\lambda = 6.45\,\mu m$ was much smaller than at $\lambda = 2.94\,\mu m$, at the same dosage. They used a free-electron laser (FEL) in their experiments. FEL is a very flexible source of electromagnetic waves, in terms of tunability, from the UV to THz, although bulky and very expensive, since it involves an electron accelerator. FEL output was in the form of 5-μs-long macropulses consisting of a series of short, ~1 ps, pulses repeated at an electron beam repetition rate of 2.85 GHz; thus, there were ~15 000 micropulses per macropulse and each macropulse, with ~20 mJ energy, was emitted at a repetition rate of ~4 Hz.

Peavy et al. [79] made comparison of cortical bone ablations by using different FEL wavelengths, from $\lambda = 2.9$ to $9.2\,\mu m$. The main conclusion was that the use of wavelengths in the 6.1-μm amide I to 6.45-μm amide II region were the most efficient for cutting cortical bone, and produced less collateral thermal injury than cutting with a surgical bone saw.

Since FELs proved too costly and too complex for widespread surgical use, several alternative laser systems at 6–7 μm were developed and have demonstrated the ability to cleanly cut soft tissues with little collateral damage [80, 81]. Mackanos et al. [81] used a 2.8-μm Er:YSGG laser pumped-tunable (6–8 μm) ZGP OPO producing millijoule-level 100-ns pulses at 5 Hz repetition rate. Among several wavelengths studied, the best results were produced with the two wavelengths: $\lambda = 6.1\,\mu m$ (the range of high water and protein amide I band absorption), and $\lambda = 6.45\,\mu m$ (moderate water absorption and high protein amide II band absorption). When using porcine cornea, the average zone of thermal damage at $\lambda = 6.1\,\mu m$ was found to be 4.1 μm for the OPO and 5.4 μm for the FEL. At $\lambda = 6.45\,\mu m$ wavelength, the damaged zone was 7.2 μm for the OPO and about the same for the FEL. Thus, the OPO caused similar or less thermal damage when compared with the FEL while generating significantly deeper craters.

Edwards et al. developed a four-stage laser system that starts from a Nd:YLF laser, uses a combination of stimulated Raman scattering and difference frequency mixing, and generates 6.45-μm pulses with 3–5-ns pulse duration, 2-mJ pulse energy, and 1/2-Hz repetition rate [82]. The laser system was used to ablate rat brain tissue, where both a collateral damage and an ablation rate compare favorably with those previously observed with an FEL. Kozub et al. demonstrated a Raman-shifted alexandrite laser for soft tissue ablation (they used goat cornea, rat heart, rat skin, and rat kidney) in the 6–7-μm wavelength range [83]. In this setup, a tunable alexandrite laser pumped a two-stage Raman converter; the first-order Stokes shifted light (near 1 μm) was generated in a deuterium (D_2) convertor; this output then generated the second-order Stokes beam in a multipass hydrogen (H_2) convertor to yield mid-IR light. Tunable output, from 6 to 7 μm, was achieved by tuning the alexandrite laser from 771 to 785 nm. The mid-IR pulses had up to 9-mJ pulse energy, 10–20 ns pulse duration, and 10-Hz repetition rate.

7.2.2 Medical Breath Analysis

One emerging application of mid-IR lasers is medical exhaled breath analysis, based on spectroscopic methods of trace-gas detection. Exhaled breath contains trace quantities of some 1000 volatile organic compounds (VOCs). Abnormal concentrations for many of these molecules (biomarkers) have already been correlated to specific diseases and health conditions [84–93]. These include cardiovascular diseases, asthma, gastrointestinal issues, chronic obstructive pulmonary disease (COPD), diabetes, Helicobacter pylori infection, oxidative stress, and different types of cancers. Such correlations allow for fast and noninvasive methods of health screening for a wide variety of medical conditions. Spectroscopic methods can successfully compete in the clinical use with such existing methods as gas chromatography/mass spectroscopy (GC/MS), electrochemical methods, chemiluminescence, and electronic “noses” based on sensor arrays.

The biomarker concentrations in the exhaled breath vary dramatically. Aside from very high water (5%) and CO_2 (5%) concentrations, typically, the concentration levels are: 1–100 ppm (part-per-million in volume) for methane (CH_4), 1–5 ppm for carbon monoxide (CO), 0.3–2.5 ppm for acetone, 0.3–1.6 ppm for nitrous oxide (N_2O), 0.1–3 ppm for ammonia (NH_3), ~500 ppb (part-per-billion in volume) for methanol, 30–300 ppb for isoprene, 1–80 ppb for nitric oxide (NO), 1–200 ppb for 2,3-butanedione, and 2–20 ppb for ethane (C_2H_6) [88–90, 92, 93]. Although clinical breath analysis is currently in its infancy, it offers unique capabilities to the field of medicine, because sampling breath is noninvasive, inherently safe to the subject, and poses minimum risk to the person collecting the sample.

7.2.2.1 Ethane (C_2H_6)

Ethane is a volatile marker for lipid peroxidation in organisms causing oxidative stress and oxidative damage in the body. Using a tunable CO-overtone sideband laser tuned to a characteristic ethane absorption peak at 3000.3 cm^{-1}, von Basum et al. measured, via CRDS, ethane concentration in exhaled human breath [94, 95]. The researchers demonstrated that ethane traces down to 0.5 ppb with a time resolution of ~1 second were detectable. Similarly, Halmer et al. demonstrated breath measurements with a detection limit of 0.27 ppb/$Hz^{1/2}$ by using CRDS with a mid-IR light source based on bulk DFG [96]. Skeldon et al. used a technique based on a lead-salt tunable diode laser operating at 3.4 μm, combined with a multipass 170-m optical cell, to measure the concentration of ethane in exhaled breath with an accuracy of 0.1 ppb [97]. QCLs and ICLs are convenient mid-IR sources for ultrasensitive and highly selective trace-gas monitoring. These lasers overcome many of the difficulties encountered in high-resolution IR absorption with lead-salt tunable diode lasers: they have better stability, reproducible behavior, and operate without cryogens. Compact chemical sensors based on QCLs and ICLs were demonstrated to be effective in clinical studies that involved the detection of a large number of exhaled biomarkers (e.g. NO, CO, CO_2, NH_3, C_2H_6, and OCS) [12, 98, 99]. Parameswaran et al. used off-axis integrated cavity output spectroscopy (ICOS) with an ICL and demonstrated real-time measurements of breath ethane with a detection sensitivity of 0.48 ppb/$Hz^{1/2}$ (e.g. 0.48 ppb within one second integration time) [100].

7.2.2.2 NO

Measurements of exhaled nitric oxide constitute a new way to monitor the inflammatory status in respiratory disorders, such as asthma and other pulmonary conditions. Nitric oxide sensors based on off-axis ICOS and thermoelectrically cooled CW DFB QCLs operating at 5.45 μm (1835 cm^{-1}) were developed to determine NO in human exhaled breath at the sub-ppbv levels [99, 101–103]. A similar level of minimal detectable NO concentration was achieved in [104], where a thermoelectrically cooled CW QCL operating between 1847 and 1854 cm^{-1} was used in combination with WMS. Halmer et al. performed concentration measurements of NO for two nitrogen isotopes (^{14}N and ^{15}N) simultaneously, using a CO laser at $\lambda = 5$ μm and CW CRDS detection scheme [105]. The authors achieved 800-ppt sensitivity for ^{14}NO and 40 ppt for ^{15}NO in one-second averaging time. The detection of the ^{15}NO isotopologue allows use of this molecule as a tracer in biochemical processes.

7.2.2.3 NH_3

Exhaled ammonia (NH_3) is a potential noninvasive marker of liver and kidney function as well as peptic ulcer disease. A gas analyzer based on a pulsed, thermoelectrically cooled QCL operating near 970 cm^{-1} (10.3 μm) using CRDS was developed for the detection of exhaled NH_3 in human breath. A sensitivity of ~50 ppb with a 20-second time resolution was achieved [106].

7.2.2.4 CO

Exhaled carbon monoxide (CO) is a biomarker to assess such medical conditions as cardiovascular diseases, diabetes, nephritis, oxidative stress, and inflammation. Moeskops et al. demonstrated a thermoelectrically cooled DFB pulsed QCL operating between 2176 and 2183 cm^{-1} (~4.6 μm) that was used to measure CO in human breath [107]. The authors took advantage of the strong CO ro-vibrational transition at 2176.2835 cm^{-1} and achieved a detection limit of 175 ppb using a 0.2 second integration time.

7.2.2.5 OCS

Roller et al. reported on a tunable laser absorption spectroscopy system to detect ~30 ppb concentrations of carbonyl sulfide (OCS), which is considered a biomarker for failure of liver function, and for acute lung transplant rejection. The laser sensing system comprised a thermoelectrically cooled pulsed QCL operating at 4.86 μm (2054 cm^{-1}) and a 36-m-long optical multipass gas cell [108]. The selectivity of two stable isotopes in air, $^{12}C^{16}O^{32}S$ and $^{12}C^{16}O^{34}S$, was also demonstrated.

Shorter et al. presented a breath analyzer for detection of multiple VOCs based on tunable IR laser differential absorption spectroscopy. The instrument was equipped with two room-temperature-pulsed QCLs (Alpes Lasers) [109]. One QCL operated at 1900 cm^{-1} for NO and CO detection, with the second QCL at 2190 cm^{-1} for CO and N_2O monitoring. The researchers obtained detection precisions of 0.5 ppb for NO, 0.8 ppb for CO, and 0.8 ppb for N_2O with an instrument response time of ~1 second – a sensitivity level that is sufficient for real-world breath analysis.

7.2.2.6 Optical Frequency Comb Spectroscopy for Breath Analysis

While QCL- and ICL-based systems provide robust and highly sensitive detection of trace molecules, they typically operate in narrow spectral regions. OFCs offer enormous potential in the detection of molecules by combining their massive spectral coverage with high spectral resolution. Following the paper by Thorpe et al. [110], where broad-bandwidth frequency comb in the near-IR range (1.5–1.7 μm) combined with a high-finesse external optical cavity and an InGaAs detector array allowed detection of several biomarkers in human breath, including two stable isotopes of CO_2, CO, and NH_3, the same team later extended the frequency comb-based sensing technology to the mid-IR spectral region – in order to dramatically (by ~3 orders of magnitude) increase the detection sensitivity; an optical frequency comb-based Fourier transform spectrometer (FC-FTS), which operates in the spectral region of 2.7–4.8 μm (2100–3700 cm^{-1}) with an *instantaneous* spectral coverage of ~300 cm^{-1}, was introduced [41, 42, 47]. The frequency comb source was based on a high-power fiber-laser-pumped mid-IR OPO and was combined with an FTS using a fast-scanning Michelson interferometer. To minimize drifts of the spectrum during the measurement, the OPO comb was phase-stabilized with respect to the pump laser, whose repetition rate was in turn locked to a microwave reference for long-term stability. With a multipass gas cell or an external enhancement cavity, the FC-FTS system was capable of detecting ppb-level concentrations of various molecules with the spectral resolution of up to 0.0035 cm^{-1}, which is adequate for resolving Doppler-broadened fine spectral features at low gas pressures.

Overall, OFC spectroscopy has been widely used to detect different volatile compounds suitable for breath analysis. These include detection of hydrogen peroxide (H_2O_2), acetylene (C_2H_2), methane (CH_4), and nitrous oxide (N_2O) with cavity-enhanced frequency comb absorption spectroscopy using an OPO-based OFC combined with an FTS [47]; detection of methane, formaldehyde (CH_2O), ethylene (C_2H_4), and carbon monoxide (CO) using intracavity absorption method (by placing molecules under study inside the OPO resonator), combined with an FTS [43]; detection of methane using a difference frequency generator-based OFC and an optical spectrum analyzer [57]; and detection of methane, carbon monoxide, and NO using an OPO combined with a multipass cell and an FTS [45].

Novel spectroscopic approaches with broadband OFCs based on virtually imaged phased arrays (VIPAs) [39], Vernier spectroscopy [111], and DCS [56, 65, 112–114] show great promise for future breath analysis – in terms of their capability for performing fast and precise spectroscopic measurements.

An excellent review by Henderson et al. on laser spectroscopy for breath analysis and its prospective clinical implementation can be found in [115].

7.3 Nano-IR Imaging and Chemical Mapping

Infrared spectroscopy has been a benchmark technique in a broad range of sciences and industries to characterize and identify materials via vibrational resonances of chemical bonds on a macroscale. Because of the diffraction limit, traditional mid-IR techniques, such as Fourier-transform IR (FTIR) microscopy, cannot measure spectra on a scale that is below several wavelengths. The ability to identify material with nanoscale resolution under the tip of an atomic force microscope (AFM) has always been an ambitious goal in the AFM community. While AFM can measure mechanical, electrical, magnetic, thermal, and other properties of materials, it has lacked the robust ability to chemically characterize unknown materials. Keilmann and coworkers [116–119] introduced a new type of microscopy that provides simultaneous spectral identification and a spatial resolution that is far beyond the classical Abbe diffraction limit of one-half wavelength ($\lambda/2$), and also beyond the practical limit ($\sim\lambda/10$) of aperture-based scanning near-field optical microscopy (SNOM). The operation principle is based on the fact that the illuminated tip of an AFM can exhibit enhanced optical fields in its neighborhood. Such near fields are modified by the presence of a sample. As a consequence of this near-field interaction, the scattered light measured in the far field carries information on the sample's local optical properties. This near-field scattering is the basis of what has been named scattering-type scanning near-field optical microscopy (s-SNOM). In practice, the metallized tip of an AFM serves as a scatterer, and it is the radius of curvature at its apex (typically 20 nm or less) that determines both the mechanical and the optical resolution. Hence, there is no wavelength-related resolution limit, and a resolution of $\lambda/500$ was obtained in the mid-IR at $\lambda = 10\,\mu m$ [117]. The principle of s-SNOM is illustrated in Figure 7.9. Here, a focused IR beam illuminates the tip region of an AFM, where a sample is approached and scanned to produce a topographic image. By recording the backscattered light, an optical image is simultaneously generated at different wavelengths. To improve the signal-to-noise ratio, the tip oscillates at the cantilever's mechanical resonance frequency Ω (tapping mode) so that the near-field optical signal is modulated at Ω and its harmonics. This allows a synchronous lock-in detection. The IR laser wavelength has a clear effect on the IR image (Figure 7.9c and e). As the wavelength is tuned to a vibrational resonance of polystyrene embedded in polymethylmethacrylate (PMMA), the IR image is darker. This effect disappears at absorption-free wavelengths, giving clear evidence of near-field vibrational image contrast [116].

Dazzi and colleagues introduced a different principle of nano-chemical mapping with an AFM using the photoacoustic (photothermal) effect [120–123]. When the wavelength of a pulsed (nanosecond) laser is tuned to the absorption resonance, a fast heat deposition results in rapid thermal expansion of the absorbing region. This expansion pulse is then detected with the

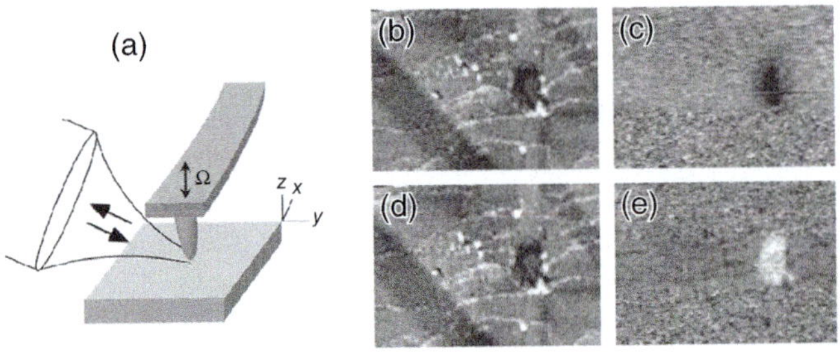

Figure 7.9 (a) Principle of s-SNOM. A focused laser beam illuminates the tip region of an AFM, where a sample is approached and scanned to produce a topographic image. By recording the backscattered light, an optical image is simultaneously generated at different laser wavelengths. The tip oscillates at the cantilever's mechanical resonance frequency Ω, so that the near-field optical signal is modulated at Ω and its harmonics, allowing a synchronous lock-in detection. (b and d) Topographic images and (c and e) corresponding infrared images of polystyrene (PS) embedded in polymethylmethacrylate (PMMA). The image (c) was taken at a wavelength $\lambda = 9.68\,\mu m$, and the image (e) at $\lambda = 10.17\,\mu m$. One can see a wavelength-dependent contrast that allows to identify chemical composition. The field of view in (b–e) is $3.5 \times 2.5\,\mu m$. *Source:* reproduced from figures 1 and 4 of [116], with permission of Springer Nature.

tip of an AFM through detecting the oscillations of the cantilever at its resonant modes. Thus, the system acts as a detector of extremely small motions induced by optical absorption. The detection scheme used here is analogous to PAS, except that it is the AFM tip and cantilever that are used to detect the thermal expansion, instead of a microphone in a gas cell. In life sciences, the AFM-IR technique provides a label-free method for mapping IR-absorbing species in biological materials. For example, Dazzi et al. performed chemical spectroscopy at the subcellular level and demonstrated chemical mapping of the distribution of viruses in the infected bacteria *Escherichia coli* [123]. By tuning the laser to a DNA absorption peak around $1080\,cm^{-1}$ ($9.26\,\mu m$), the authors were able to localize single viruses (bacteriophage T5) inside the bacteria. Figure 7.10 shows that while the topography performed by an AFM was not able to reveal the viruses, the AFM-IR was able to do so. As a tunable source in their AFM-IR experiments, the researchers used a pulsed mid-IR FEL "CLIO" located in the Centre Laser Infrarouge d'Orsay, France [121–123]. In 2010, Anasys Instruments, Inc., commercialized a compact version of the AFM-IR device that uses a broadly tunable benchtop IR source based on a nanosecond OPO [124].

Lu and Belkin reported a photothermal technique that allows obtaining AFM-IR spectra under low-average-power illumination from a pulsed tunable QCL (energy per pulse 4 nJ, duration 40 ns) [125]. To detect minute thermal expansion in a polymer deposited on a silicon substrate, the authors tuned the repetition frequency of the QCL to be in resonance with the mechanical

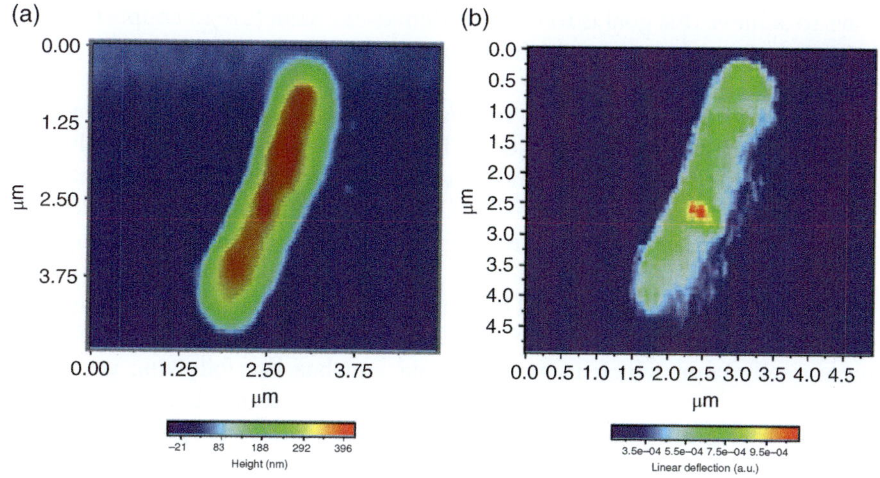

Figure 7.10 (a) Topography and (b) the corresponding AFM-IR chemical mapping of the infected *Escherichia coli* cell. Although the topography performed by an AFM cannot image the viruses, the AFM-IR was able to do so via infrared absorption contrast. *Source:* reproduced from figure 4 of [122], with permission of Elsevier.

oscillation frequency of the AFM cantilever, which is in the range 10–200 kHz. The enhancement of the AFM response, as compared to a single pulse excitation, is equal to Q, where $Q \approx 100$ is the quality factor of the cantilever oscillation. The achieved spatial resolution of 50 nm ($\lambda/170$) was determined by the thermal diffusion length in a sample during the laser pulse duration; it may be further improved using shorter QCL pulses [125]. This technique can be regarded as a nanoscale analog of the QEPAS [17].

7.4 Plasmonics in the Mid-IR

Plasmons – charge oscillations at the interface between a dielectric and a metal – were first studied in the visible regime, with increased interest now extending to the mid-IR [126]. Plasmons give rise to very strong local fields and can be guided along the interface in the form of a traveling wave, known as a surface plasmon polariton. The excitement generated by mid-IR plasmonics comes from the interplay between circuit (antenna) theory and wave optics. To date, research in mid-IR plasmonics is focused on emitters (in particular, related to QCLs), improved IR detectors, and chemical sensing surfaces. Another goal of mid-IR plasmonics is to increase the absorption in a given volume of material: in mid-IR detectors, smaller volumes provide lower noise, and higher material absorption results in a stronger output signal. One of the ways to achieve this goal is to use an antenna-like structure to couple incident light to surface plasmons, which are focused by the antenna to a small detection region.

Based on mid-IR plasmonics, Yu, Wang, and Capasso demonstrated a QCL with improved beam characteristics. This was achieved by monolithically integrating specially designed plasmonic structures on the cleaved substrate of a QCL, below the laser waveguide [127]. The plasmonic structure creates desirable far-field radiation by coupling laser emission into surface waves propagating on the substrate and by coherently scattering the energy of the waves into the free space. This is illustrated in Figure 7.11a, which shows a scanning electron microscope (SEM) image of a $\lambda = 9.9\,\mu m$ QCL with an integrated 1D plasmonic collimator, which is made of gold and has the following geometry: aperture size = $2 \times 25\,\mu m$, grating period = $8.9\,\mu m$, distance between aperture and the first grating groove = $7.3\,\mu m$, groove width = $0.8\,\mu m$, and groove depth = $1.5\,\mu m$. One can see from Figure 7.11c that the measured divergence of the beam is dramatically improved in the "fast" (vertical) direction when the plasmonic collimator is used.

The other avenue of plasmonics is spectroscopic chemical sensing with surface-enhanced IR absorption using metamaterials. Metamaterials are artificially constructed materials that consist of subwavelength elements whose spatially averaged response can be treated as that of a homogeneous medium with a characteristic effective permittivity and permeability. This permits the optical properties to be engineered, enabling, for example, near-unity

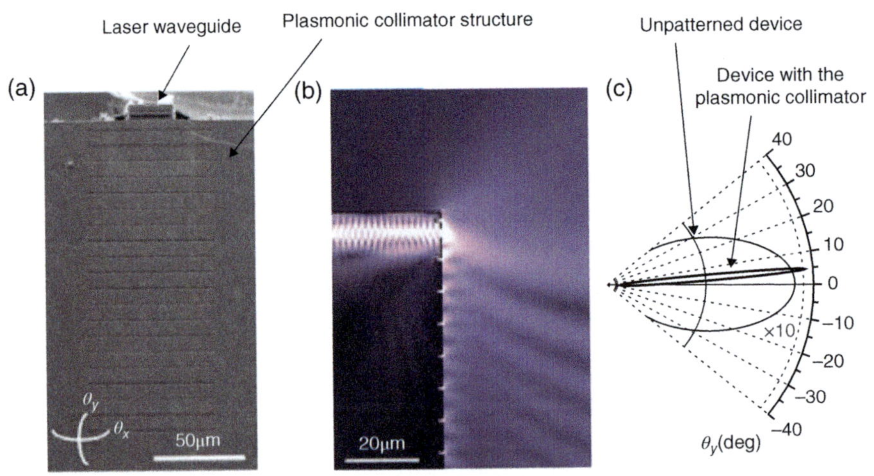

Figure 7.11 (a) SEM image of a $\lambda = 9.9\,\mu m$ QCL with an integrated 1D plasmonic collimator, which is made of gold. (b) Simulated electric-field distribution of the QCL. (c) Measured far-field intensity profiles in the vertical direction of the device with and without the plasmonic collimator. *Source:* reproduced from figure 2 of [127], with permission of Wiley.

absorbance at the surface [128, 129]. Chen et al. demonstrated a dual-band perfect absorber for plasmon-enhanced infrared spectroscopy using an asymmetric cross-shaped antenna made of gold nanoparticles that were placed on top of a thick gold film and separated from it by a magnesium fluoride (MgF_2) spacer layer [129]. The authors demonstrated, by mid-IR reflection spectroscopy, strong absorption enhancement for the two absorption bands: the C—H band at 2850 cm^{-1} (3.5 μm) and C=O band at 1700 cm^{-1} (5.9 μm) of a thin (4-nm-thick) polymer film. Even despite the fact that the two bands were well separated in wavelength, the authors demonstrated that both of them have shown a large field enhancement.

7.5 Infrared Countermeasures

Infrared countermeasure (IRCM) systems are used to defend the host aircraft from heat-seeking attacking missiles by detecting the approaching threat and disabling it through the use of directed laser energy (Figure 7.12). The laser intensity should be higher than that of the target (usually blackbody radiation from a jet engine) in the appropriate mid-IR range. When the mid-IR radiation produced by an IRCM system is seen by a missile seeker, it overwhelms the signal from the aircraft jet and confuses the missile.[1] The missile will begin to deviate from the target, rapidly breaking lock. Once an infrared seeker breaks

Figure 7.12 Infrared countermeasure (IRCM) system (circled in red color) on board of an airplane.

1 http://en.wikipedia.org/wiki/Infrared_countermeasure

lock (a typical field of view of 1–2°), it rarely reacquires the target. Typically, high-power broadband pulse-periodic ZGP OPO systems with 10-W average power at 3–5 μm (BAE systems) or high-power QCL systems with 10-W average power (the high-power beam is obtained by combining outputs from few QCLs) at 4–5-μm wavelength (Daylight Solutions, Pranalytica) are utilized. Northrop Grumman uses similar technologies (OPOs and QCLs) in their multiband IRCM laser systems.

7.6 Extreme Nonlinear Optics and Attosecond Science

When energetic few-cycle laser pulses interact with an atomic gas (He, Ne, Ar, etc.), with peak intensities $>10^{14}\,\mathrm{W/cm^2}$, one can generate high odd-number optical harmonics reaching the extreme UV and even X-ray spectral range. High-harmonic generation (HHG) in this case can no longer be explained by perturbative nonlinear optics, which predicts decrease of the harmonic power with harmonic number. Rather, the process is initiated by electron tunneling and completed by the recollision of a highly energetic electron with its associated ion. Three decades of research into high-order harmonic generation in gases has uncovered several practical routes for generating phase-matched coherent X-ray beams using tabletop femtosecond lasers [130, 131]. A bright supercontinuum that spans the entire spectrum from the UV to more than 1.6 keV with the harmonic number as high as >5001 can now be produced, allowing, in principle, the generation of pulses as short as 2.5 attoseconds. This opens up two major classes of applications: (i) creating a tabletop source of coherent X-rays that might be used for super-resolution imaging in medicine and nanotechnology; (ii) in fundamental science – the attosecond time durations of high-harmonic pulses can be applied to capture the fastest events in the natural world, for instance, for imaging molecular orbitals.

The highest energy for the HHG photon emitted is given by the microscopic single-atom cutoff rule:

$$h\nu_{X-ray} = I_p + 3.17U_p, \tag{7.1}$$

where I_p is the ionization potential of the gas atom and U_p is the ponderomotive energy of a free electron (the average energy of an electron driven by the oscillating laser field), which is proportional to the driving laser intensity and the wavelength squared. The quadratic wavelength dependence suggests a counterintuitive result that using a longer-wavelength driving laser should result in a higher cutoff photon energy (and thus more energetic X-rays). This prediction was verified by several experiments in different laboratories [131].

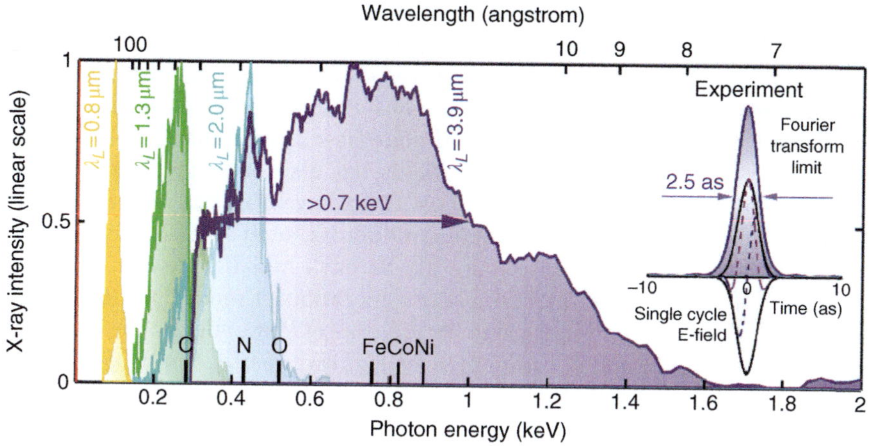

Figure 7.13 Experimental HHG spectra emitted under phase-matching conditions at different driving-laser wavelengths: 0.8, 1.3, 2, and 3.9 μm. One can see that for a longer wavelength driving laser, larger X-ray cutoff photon energy was achieved. The inset shows that the bandwidth of the coherent supercontinuum produced with $\lambda = 3.9$-μm laser could support a single-cycle X-ray pulse with 2.5-attosecond duration (Fourier-transform limit). The plot also shows at the bottom the inner-shell absorption edges of several materials (Fe, Co, Ni, etc.). *Source:* reproduced from figure 1(B) of [131], with permission of AAAS.

Figure 7.13 shows experimental HHG spectra emitted under phase-matching conditions as a function of driving-laser wavelength of 0.8, 1.3, 2, and 3.9 μm. One can see that for a longer-wavelength driving laser, larger X-ray cutoff photon energy was achieved. Popmintchev et al. used 80-fs, 10-mJ pulses at 20-Hz repetition rate at $\lambda = 3.9$ μm from an optical parametric chirped-pulse amplification (OPCPA) system. The phase-matched HHG emission was produced by focusing the optical beam into a 200-μm diameter in a 5-cm-long, high-pressure He gas-filled hollow waveguide, with the spectrum extending beyond 1.6 keV ($\lambda < 7.7$ Å). The broadband X-ray supercontinuum could support (if properly compressed) a single-cycle X-ray pulse with 2.5-attosecond duration (inset to Figure 7.13) [131].

Weisshaupt et al. have demonstrated the generation of hard X-ray photons, consisting of the characteristic X-ray lines of copper (Cu) at energies $K_\alpha \approx 8.0$ keV and $K_\beta \approx 8.9$ keV and of a broad Bremsstrahlung component that extended to ~100 keV, by the interaction between a Cu target and intense 15-mJ femtosecond mid-IR pulses at 3.9 μm, generated by OPCPA [132]. X-ray generation occurred in this case via inner-shell ionization of copper followed by a radiative transition of an outer-shell electron into the unoccupied inner shell. The flux of X-ray photons was about 25 times higher than in the case when 800-nm pulses were used. This enhancement of the X-ray flux is caused by the much more efficient electron acceleration at a longer wavelength, which results in higher kinetic energies of the electrons accelerated in the optical field. This makes high-energy, long-wavelength (>3 μm) lasers with few-cycle pulse duration and high repetition rate a highly desired driving source for the generation of coherent high-harmonic X-rays and isolated attosecond pulses in atomic gases (see Chapter 2 for the recent solid-state laser development).

HHG with strong electromagnetic fields has also been demonstrated in bulk crystalline materials. Chin and coauthors have observed extreme nonlinear optical phenomena in bulk semiconductors, such as ZnTe, ZnS, ZnSe, and GaAs, produced by intense (10^9–10^{11} W/cm^2) mid-IR pulses at λ = 3.5–3.9 μm [133]. These phenomena include generation of multiple mid-IR harmonics (up to the seventh harmonic in the case of ZnSe) below the bandgap edge of semiconductors, accompanied by significant broadening of mid-IR harmonic spectra, which the authors attribute to the onset of self-phase modulation (SPM) as well as cross-phase modulation (XPM) at these high intensities.

Ghimire et al. demonstrated nonperturbative (i.e. that cannot be described in the frame of nonlinear optics based on perturbation theory) HHG in a strongly driven wide-bandgap (3.2 eV) semiconductor ZnO using 3.2–3.7-μm (0.34–0.38-eV) few-cycle laser pulses having peak intensity 5×10^{12} W/cm^2 and field strength 60 MV/cm (corresponding to 0.6 V/Å, comparable to atomic field strength) [134]. The authors measured harmonics up to the 25th order (Figure 7.14), extending to the photon energy of 9.5 eV (λ = 130 nm) – more than 6 eV above the band edge. They also found that the high-energy cutoff scaled linearly with the driving field (in contrast to the atomic case where the scaling is linear to the intensity). The spectrum comprised odd, or both odd and even, harmonics, depending on crystallographic orientation. Vampa et al. suggested, based on their high-field HHG experiments in ZnO, that at similar field strengths and laser wavelengths, a recollision between an electron and its associated hole is the primary source of high harmonics [135].

Also, using few-cycle mid-IR laser excitation at the driving laser center wavelength of 10 μm (30 THz), with peak fields of up to 72 MV/cm, high harmonics have been generated in a 220-μm-thick semiconducting gallium selenide (GaSe) [136]. The authors observed the emission of phase-stable transients, covering the entire terahertz-to-visible spectral domain between λ = 3 mm and λ = 440 nm (0.1–675 THz). Emission between 0.1 and 10 THz originated from the optical rectification in GaSe, while the shorter wavelengths were generated via HHG (up to order n = 23) in the non-perturbative regime. The authors attributed the main mechanism of HHG to coherent interband polarization combined with dynamical Bloch oscillations [136].

Langer et al. studied the effects of strong optical field in a monolayer of tungsten diselenide (WSe_2). Electron–hole pairs were generated by a near-IR pulse and were subsequently accelerated by a mid-IR transient with center

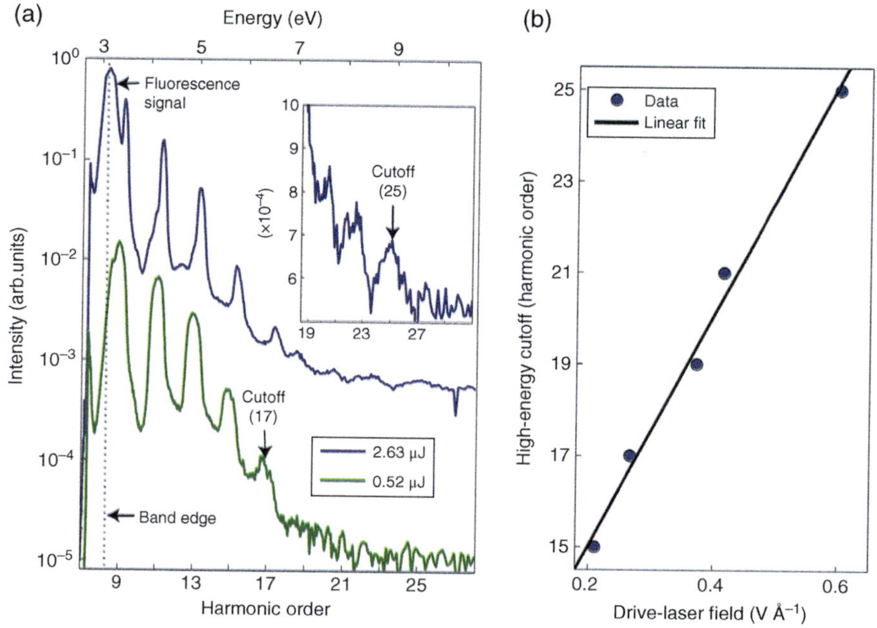

Figure 7.14 (a) High-order harmonic spectra from a 500-μm-thick ZnO crystal for two $\lambda = 3.25$-μm driving pulse energies of 0.52 and 2.63 μJ (in the latter case the peak intensity was ~5 TW/cm^2 and the field strength ~0.6 V/Å). The inset shows an expanded linear-scale view near the cutoff for the case of pulse energy 2.63 μJ. The dotted vertical line is the ZnO band edge, where the residual long-lived fluorescence signal is present. (b) Drive-laser field dependence of the cutoff photon energy showing that the latter scales linearly with the driving laser field. *Source:* reproduced from figure 1 of [134], with permission of Springer Nature.

wavelength 7.5 μm (40 THz), ~3-cycle pulse duration, and peak electric field (in air) of 18 MV/cm, followed by a recollision of the constituent electron and hole. Recolliding electrons and holes emitted their kinetic energy in high-order sidebands of the near-IR pump. Despite a low sample thickness, harmonic sidebands (spaced by mid-IR laser frequency) up to order $n = 12$ ($\lambda = 340$ nm, 880 THz) were observed [137].

7.7 Other Applications

7.7.1 Laser Wake-field Accelerators

Particle accelerators are used in a wide variety of fields, ranging from high-energy physics to medicine and biology. The accelerating fields in conventional radio frequency (RF) accelerators are limited to a few tens of

MeV per meter, owing to material breakdown at the walls of the structure. Thus, the production of energetic particle beams currently requires large-scale accelerators and expensive infrastructures. Replacing RF accelerator technology with compact laser-plasma accelerators promises an economical alternative to attaining super-relativistic electron energies [138–140]. Laser pulses, as they propagate through plasma, create a strong longitudinal accelerating field, proportional to the square root of plasma density; for example, one can get ~1 GeV/cm at a plasma density of $10^{18}\,cm^{-3}$. Trapped in such a wake, electrons pick up several GeV of kinetic energy from a compact (1–10 cm) accelerating stage. Recent demonstrations of such high-gradient electron acceleration, combined with good control over the properties of the accelerated beam, encourage researchers to consider plasma accelerators up to TeV energies that can be used for next-generation e^-e^+ colliders and beyond. Pogorelsky et al. [141] compared the performance of laser-plasma accelerators based on near-IR solid-state lasers (Ti:sapphire) and those based on long-wavelength (9–11 μm) mid-IR lasers and identified several advantages of selecting mid-IR lasers for these applications.

7.7.2 Laser Acceleration in Dielectric Structures

An alternative path to creating lab-size particle accelerators is laser acceleration in specially fabricated periodic dielectric microstructures, which directly couple an electron beam and an accelerating optical field [142, 143]. At optical frequencies, dielectric materials withstand high fields, roughly two orders of magnitude larger than metals. Together with the large optical field strength attainable with short laser pulses, dielectric laser accelerators may support acceleration gradients in the multi-GeV per meter range. Peralta et al. demonstrated particle acceleration in a periodic structure fabricated by bonding two fused-silica wafers, processed via standard optical lithography and reactive ion etching techniques, and powered by a 800-nm-wavelength mode-locked Ti:sapphire laser [142]. The authors observed a maximum acceleration gradient of ~190 MeV/m. However, dielectric structures based on silicon are considered to be the most promising for future laser accelerator applications, and accelerating gradients in excess of 200 MeV/m have already been demonstrated [144]. Ideally, for silicon, the accelerating laser needs to have low photon energy, less than half the bandgap of silicon (0.55 eV), in order to avoid detrimental linear as well as two-photon absorption. This might be possible with the use of mid-IR lasers with $\lambda > 2\,\mu m$, for example solid-state or fiber-based thulium laser systems.

7.7.3 Free-space Communications

Free-space communications is an important technique for bridging the so-called "last mile" in telecommunication networks, especially in urban areas.

QCLs are very suitable for such applications because (i) they have fast internal lifetimes that might allow for modulation frequencies of up to 5–10 GHz, (ii) their emission wavelength can be chosen to be in the atmospheric transmission windows, i.e. around 5 or at 8–10 μm, and (iii) these emission wavelengths are barely affected by atmospheric conditions such as rain and fog.

Blaser et al. showed a free-space optical data link between two buildings separated by 350 m using a Peltier-cooled QCL operating at $\lambda = 9.3\,\mu m$ and a room-temperature HgCdTe detector, with the highest transmitted modulation frequency 330 MHz. The signal was successfully transmitted even under foggy conditions with a visibility range below 100 m [145]. Martini et al. demonstrated a high-speed optical link over ~200 m of free space using a QCL with an emission wavelength of 8.1 μm; the laser was packaged and processed for high-frequency modulation. Complex digital data (comprising up to 650 multimedia satellite channels) in the frequency region from 900 MHz to 1.45 GHz were transmitted [146]. Corrigan et al. performed a multiwavelength test in the New York metropolitan area under realistic weather conditions and showed that a mid-IR QCL source at $\lambda = 8.1\,\mu m$ provides much better optical link stability as compared to near-IR wavelengths (1.3 and 1.5 μm) during fog formation [147].

7.7.4 Organic Material Processing

Thin polymer films are widely used in electronics, photonics, sensor technology, organic electronics and optoelectronics, and medicine. Resonant IR pulsed laser deposition (RIR-PLD) [148] is a promising new approach for vacuum-phase deposition of polymers and organic molecules. As a dry, clean, and environmentally friendly technique for producing organic films, it avoids many problems caused by solvents associated with commonly used spin- or dip coating methods. The method employs a pulsed mid-IR laser source with a wavelength tuned to one of the molecular vibrational bands of the organic molecule to be ablated. RIR-PLD has been successfully employed for deposition of a variety of technologically important polymers with applications ranging from micro-electro mechanical systems (MEMS), biological and chemical sensors to biocompatible medical-device coatings and biodegradable time-release drug-delivery encapsulation. In all these experiments, an FEL was used as an optical source for RIR-PLD, most often tuned to the C—H or O—H stretch vibrational bands around 3.4 and 2.9 μm, respectively. A number of key laser parameters that appear to be important for this technology include: laser pulse durations in the *ps* range, pulse energies around 1 μJ, and pulse repetition periods much shorter than the cooling time of the laser-heated layer (100 μs to 1 ms). An elegant tabletop optical source for polymer ablation based on two-stage pulsed optical parametric amplification, as an alternative to FEL, was developed for this purpose [149].

References

1 Gordon, I.E., Rothman, L.S., Hill, C., Kochanov, R.V., Tan, Y., Bernath, P.F., Birk, M., Boudon, V., Campargue, A., Chance, K.V., Drouin, B.J., Flaud, J.-M., Gamache, R.R., Hodges, J.T., Jacquemart, D., Perevalov, V.I., Perrin, A., Shine, K.P., Smith, M.-A.H., Tennyson, J., Toon, G.C., Tran, H., Tyuterev, V.G., Barbe, A., Császár, A.G., Devi, V.M., Furtenbacher, T., Harrison, J.J., Hartmann, J.-M., Jolly, A., Johnson, T.J., Karman, T., Kleiner, I., Kyuberis, A.A., Loos, J., Lyulin, O.M., Massie, S.T., Mikhailenko, S.N., Moazzen-Ahmadi, N., Müller, H.S.P., Naumenko, O.V., Nikitin, A.V., Polyansky, O.L., Rey, M., Rotger, M., Sharpe, S.W., Sung, K., Starikova, E., Tashkun, S.A., VanderAuwera, J., Wagner, G., Wilzewski, J., Wcisło, P., Yu, S., and Zak, E.J. (2017). The HITRAN 2016 molecular spectroscopic database. *J. Quant. Spectrosc. Radiat. Transf.* 203: 3–69.

2 Tittel, F.K., Richter, D., and Fried, A. (2003). Mid-infrared laser applications in spectroscopy. In: *Solid-State Mid-Infrared Laser Sources*, Topics in Applied Physics, vol. 89 (eds. I.T. Sorokina and K.L. Vodopyanov). Berlin: Springer.

3 Blaser, S., Yarekha, D., Hvozdara, L., Bonetti, Y., Muller, A., Giovannini, M., and Faist, J. (2005). Room temperature, continuous-wave, single-mode quantum-cascade lasers at $\lambda \approx 5.4$ μm. *Appl. Phys. Lett.* 86: 041109.

4 Evans, A., Yu, J.S., David, J., Doris, L., Mi, K., Slivken, S., and Razeghi, M. (2004). High-temperature, high-power, continuous-wave operation of buried heterostructure quantum-cascade lasers. *Appl. Phys. Lett.* 84: 314.

5 Yu, J.S., Slivken, S., Evans, A., Darvish, S.R., Darvish, R., Nguyen, J., and Razeghi, M. (2006). High-power $\lambda \sim 9.5$ μm quantum-cascade lasers operating above room temperature in continuous-wave mode. *Appl. Phys. Lett.* 88: 091113.

6 Diehl, L., Bour, D., Corzine, S., Zhu, J., Hofler, G., Loncar, M., Troccoli, M., and Capasso, F. (2006). High temperature continuous wave operation of strain-balanced quantum cascade lasers grown by metal organic vapor-phase epitaxy. *Appl. Phys. Lett.* 89: 081101.

7 Maulini, R., Mohan, A., Giovannini, M., Faist, J., and Gini, E. (2006). External cavity quantum-cascade lasers tunable from 8.2 to 10.4 um using a gain element with a heterogeneous cascade. *Appl. Phys. Lett.* 88: 201113.

8 Aellen, T., Blaser, S., Beck, M., Hofstetter, D., Faist, J., and Gini, E. (2003). Continuous-wave distributed-feedback quantum-cascade lasers on a Peltier cooler. *Appl. Phys. Lett.* 83: 1929.

9 Bakhirkin, Y., Kosterev, A., Curl, R., Tittel, F., Yarekha, D., Hvozdara, L., Giovannini, M., and Faist, J. (2005). Sub-ppbv nitric oxide concentration measurements using cw thermoelectrically cooled quantum cascade laser-based integrated cavity output spectroscopy. *Appl. Phys. B* 82: 149.

10 Wysocki, G., Curl, R.F., Tittel, F.K., Tittel, F.K., Tittel, K., Maulini, R., Bulliard, J.M., and Faist, J. (2005). Widely tunable mode-hop free external cavity quantum cascade laser for high resolution spectroscopic applications. *Appl. Phys. B* 81: 769.

11 Kosterev, A., Wysocki, G., Bakhirkin, Y., So, S., Lewicki, R., Fraser, M., Tittel, F., and Curl, R.F. (2008). Application of quantum cascade lasers to trace gas analysis. *Appl. Phys. B* 90: 165.

12 Nelson, D.D., Shorter, J.H., McManus, J.B., and Zahniser, M.S. (2002). Sub-part-per-billion detection of nitric oxide in air using a thermoelectrically cooled mid-infrared quantum cascade laser spectrometer. *Appl. Phys. B* 75: 343.

13 Curl, R.F., Capasso, F., Gmachl, C., Kosterev, A.A., McManus, B., Lewicki, R., Pusharsky, M., Wysocki, G., and Tittel, F.K. (2010). Quantum cascade lasers in chemical physics. *Chem. Phys. Lett.* 487: 1.

14 Wysocki, G., Lewicki, R., Curl, R.F., Tittel, F.K., Diehl, L., Capasso, F., Troccoli, M., Hofler, G., Bour, D., Corzine, S., Maulini, R., Giovannini, M., and Faist, J. (2008). Widely tunable mode-hop free external cavity quantum cascade lasers for high resolution spectroscopy and chemical sensing. *Appl. Phys. B* 92: 305.

15 Chao, X., Jeffries, J.B., and Hanson, R.K. (2012). Wavelength-modulation-spectroscopy for real-time, in situ NO detection in combustion gases with a 5.2 μm quantum-cascade laser. *Appl. Phys. B* 106: 987.

16 Zharov, V.P. and Letokhov, V.S. (1986). *Laser Optoacoustic Spectroscopy.* Springer Series in Optical Sciences, vol. 37. Berlin: Springer.

17 Kosterev, A.A., Bakhirkin, Y.A., Curl, R.F., and Tittel, F.K. (2002). Quartz-enhanced photoacoustic spectroscopy. *Opt. Lett.* 27: 1902.

18 Kosterev, A.A., Tittel, F.K., Serebryakov, D., Malinovsky, A., and Morozov, A. (2005). Applications of quartz tuning forks in spectroscopic gas sensing. *Rev. Sci. Instrum.* 76: 043105.

19 Lewicki, R., Wysocki, G., Kosterev, A.A., and Tittel, F.K. (2007). QEPAS based detection of broadband absorbing molecules using a widely tunable, cw quantum cascade laser at 8.4 μm. *Opt. Express* 15: 7357.

20 Patimisco, P., Scamarcio, G., Tittel, F.K., and Spagnolo, V. (2014). Quartz-enhanced photoacoustic spectroscopy: a review. *Sensors* 14: 6165.

21 Dong, L., Spagnolo, V., Lewicki, R., and Tittel, F.K. (2011). Ppb-level detection of nitric oxide using an external cavity quantum cascade laser based QEPAS sensor. *Opt. Express* 19: 24037.

22 Scherer, J.J., Paul, J.B., O'Keefe, A., and Saykally, R.J. (1997). Cavity ringdown laser absorption spectroscopy: history, development, and application to pulsed molecular beams. *Chem. Rev.* 97: 25.

23 Romanini, D., Kachanov, A.A., Sadeghi, N., and Stoeckel, F. (1997). CW cavity ring down spectroscopy. *Chem. Phys. Lett.* 264: 316.

24 Kosterev, A.A., Malinovsky, A.L., Tittel, F.K., Gmachl, C., Capasso, F., Sivco, D.L., Baillargeon, J.N., Hutchinson, A.L., and Cho, A.Y. (2001). Cavity ringdown spectroscopic detection of nitric oxide with a continuous-wave quantum-cascade laser. *Appl. Opt.* 40: 5522.

25 Galli, I., Bartalini, S., Ballerini, R., Barucci, M., Cancio, P., De Pas, M., Giusfredi, G., Mazzotti, D., Akikusa, N., and De Natale, P. (2016). Spectroscopic detection of radiocarbon dioxide at parts-per-quadrillion sensitivity. *Optica* 3: 385.

26 Horstjann, M., Bakhirkin, Y.A., Kosterev, A.A., Curl, R.F., Tittel, F.K., Wong, C.M., Hill, C.J., and Yang, R.Q. (2004). Formaldehyde sensor using interband cascade laser based quartz-enhanced photoacoustic spectroscopy. *Appl. Phys. B* 79: 799.

27 Wysocki, G., Bakhirkin, Y., So, S., Tittel, F.K., Hill, C.J., Yang, R.Q., and Fraser, M.P. (2007). Dual interband cascade laser based trace-gas sensor for environmental monitoring. *Appl. Opt.* 46: 8202.

28 Lundqvist, S., Kluczynski, P., Weih, R., von Edlinger, M., Nähle, L., Fischer, M., Bauer, A., Höfling, S., and Koeth, J. (2012). Sensing of formaldehyde using a distributed feedback interband cascade laser emitting around 3493 nm. *Appl. Opt.* 51: 6009.

29 Richter, D., Fried, A., Wert, B.P., Walega, J.G., and Tittel, F.K. (2002). Development of a tunable mid-IR difference frequency laser source for highly sensitive airborne trace gas detection. *Appl. Phys. B* 75: 281.

30 Chen, W., Cousin, J., Poullet, E., Burie, J., Boucher, D., Gao, X., Sigrist, M.W., and Tittel, F.K. (2007). Continuous-wave mid-infrared laser sources based on difference frequency generation. *C R Phys.* 8: 1129.

31 Richter, D., Fried, A., and Weibring, P. (2009). Difference frequency generation laser based spectrometers. *Laser Photon. Rev.* 3: 343.

32 van Herpen, M.M.J.W., Bisson, S.E., and Harren, F.J.M. (2003). Continuous-wave operation of a single-frequency optical parametric oscillator at 4–5 μm based on periodically poled $LiNbO_3$. *Opt. Lett.* 28: 2497.

33 Henderson, A. and Stafford, R. (2006). Low threshold, singly-resonant CW OPO pumped by an all-fiber pump source. *Opt. Express* 14: 767.

34 Ngai, A.K.Y., Persijn, S.T., von Basum, G., and Harren, F.J.M. (2006). Automatically tunable continuous-wave optical parametric oscillator for high-resolution spectroscopy and sensitive trace-gas detection. *Appl. Phys. B* 85: 173.

35 von Basum, G., Halmer, D., Hering, P., Mürtz, M., Schiller, S., Müller, F., Popp, A., and Kühnemann, F. (2004). Parts per trillion sensitivity for ethane in air with an optical parametric oscillator cavity leak-out spectrometer. *Opt. Lett.* 29: 797.

36 Todd, M.W., Provencal, R.A., Owano, T.G., Paldus, B.A., Kachanov, A., Vodopyanov, K.L., Hunter, M., Coy, S.L., Steinfeld, J.I., and Arnold, J.T. (2002). Application of mid-infrared cavity-ringdown spectroscopy to trace explosives vapor detection using a broadly tunable (6–8 μm) optical parametric oscillator. *Appl. Phys. B* 75: 367.

37 Schliesser, A., Picque, N., and Hansch, T.W. (2012). Mid-infrared frequency combs. *Nat. Photonics* 6: 440.

38 Diddams, S.A., Hollberg, L., and Mbele, V. (2007). Molecular fingerprinting with the resolved modes of a femtosecond laser frequency comb. *Nature* 445: 627.

39 Nugent-Glandorf, L., Neely, T., Adler, F., Fleisher, A.J., Cossel, K.C., Bjork, B., Dinneen, T., Ye, J., and Diddams, S.A. (2012). Mid-infrared virtually imaged phased array spectrometer for rapid and broadband trace gas detection. *Opt. Lett.* 37: 3285.

40 Fleisher, A.J., Bjork, B.J., Bui, T.Q., Cossel, K.C., Okumura, M., and Ye, J. (2015). Mid-infrared time-resolved frequency comb spectroscopy of transient free radicals. *J. Phys. Chem. Lett.* 5: 2241.

41 Adler, F., Maslowski, P., Foltynowicz, A., Cossel, K.C., Briles, T.C., Hartl, I., and Ye, J. (2010). Mid-infrared Fourier transform spectroscopy with a broadband frequency comb. *Opt. Express* 18: 21861.

42 Foltynowicz, A., Maslowski, P., Ban, T., Adler, F., Cossel, K.C., Briles, T.C., and Ye, J. (2011). Optical frequency comb spectroscopy. *Faraday Discuss.* 150: 23.

43 Haakestad, M.W., Lamour, T.P., Leindecker, N., Marandi, A., and Vodopyanov, K.L. (2013). Intracavity trace molecular detection with a broadband mid-IR frequency comb source. *J. Opt. Soc. Am. B* 30: 631.

44 Meek, S.A., Poisson, A., Guelachvili, G., Hänsch, T.W., and Picqué, N. (2014). Fourier transform spectroscopy around 3 μm with a broad difference frequency comb. *Appl. Phys. B* 114 (4): 573.

45 Khodabakhsh, A., Ramaiah-Badarla, V., Rutkowski, L., Johansson, A.C., Lee, K.F., Jiang, J., Mohr, C., Fermann, M.E., and Foltynowicz, A. (2016). Fourier transform and Vernier spectroscopy using an optical frequency comb at 3–5.4 μm. *Opt. Lett.* 41: 2541.

46 Maslowski, P., Lee, K.F., Johansson, A.C., Khodabakhsh, A., Kowzan, G., Rutkowski, L., Mills, A.A., Mohr, C., Jiang, J., Fermann, M.E., and Foltynowicz, A. (2016). Surpassing the path-limited resolution of Fourier-transform spectrometry with frequency combs. *Phys. Rev. A* 93: 021802(R).

47 Foltynowicz, A., Maslowski, P., Fleisher, A.J., Bjork, B.J., and Ye, J. (2013). Cavity-enhanced optical frequency comb spectroscopy in the mid-infrared application to trace detection of hydrogen peroxide. *Appl. Phys. B* 110: 163.

48 Keilmann, F., Gohle, C., and Holzwarth, R. (2004). Time-domain mid-infrared frequency-comb spectrometer. *Opt. Lett.* 29: 1542.

49 Schliesser, A., Brehm, M., Keilmann, F., and van der Weide, D.W. (2005). Frequency-comb infrared spectrometer for rapid, remote chemical sensing. *Opt. Express* 13: 9029.

50 Coddington, I., Newbury, N., and Swann, W. (2016). Dual-comb spectroscopy. *Optica* 3: 414.

51 Changala, P.B., Weichman, M.L., Lee, K.F., Fermann, M.E., and Ye, J. (2019). Rovibrational quantum state resolution of the C60 fullerene. *Science* 363: 49.

52 Coddington, I., Swann, W.C., and Newbury, N.R. (2008). Coherent multiheterodyne spectroscopy using stabilized optical frequency combs. *Phys. Rev. Lett.* 100: 013902.

53 Coddington, I., Swann, W.C., and Newbury, N.R. (2010). Time-domain spectroscopy of molecular free-induction decay in the infrared. *Opt. Lett.* 35: 1395–1397.

54 Zolot, A.M., Giorgetta, F.R., Baumann, E., Nicholson, J.W., Swann, W.C., Coddington, I., and Newbury, N.R. (2012). Direct-comb molecular spectroscopy with accurate, resolved comb teeth over 43 THz. *Opt. Lett.* 37: 638.

55 Bernhardt, B., Sorokin, E., Jacquet, P., Thon, R., Becker, T., Sorokina, I.T., Picqué, N., and Hänsch, T.W. (2010). Mid-infrared dual-comb spectroscopy with 2.4 μm Cr^{2+}:ZnSe femtosecond lasers. *Appl. Phys. B* 100: 3.

56 Zhang, Z., Gardiner, T., and Reid, D.T. (2013). Mid-infrared dual-comb spectroscopy with an optical parametric oscillator. *Opt. Lett.* 38: 3148.

57 Cruz, F.C., Maser, D.L., Johnson, T., Ycas, G., Klose, A., Giorgetta, F.R., Coddington, I., and Diddams, S.A. (2015). Mid-infrared optical frequency combs based on difference frequency generation for molecular spectroscopy. *Opt. Express* 23: 26814.

58 Jin, Y.W., Cristescu, S.M., Harren, F.J.M., and Mandon, J. (2015). Femtosecond optical parametric oscillators toward real-time dual-comb spectroscopy. *Appl. Phys. B* 119: 65.

59 Zhu, F., Bicer, A., Askar, R., Bounds, J., Kolomenskii, A., Kelessides, V., Amani, M., and Schuessler, H. (2015). Mid-infrared dual frequency comb spectroscopy based on fiber lasers for the detection of methane in ambient air. *Laser Phys. Lett.* 12: 095701.

60 Kara, O., Zhang, Z., Gardiner, T., and Reid, D.T. (2017). Dual-comb mid-infrared spectroscopy with free-running oscillators and absolute optical calibration from a radio-frequency reference. *Opt. Express* 25: 16072.

61 Kara, O., Maidment, L., Gardiner, T., Schunemann, P.G., and Reid, D.T. (2017). Dual-comb spectroscopy in the spectral fingerprint region using OPGaP optical parametric oscillators. *Opt. Express* 25: 32713.

62 Baumann, E., Giorgetta, F.R., Swann, W.C., Zolot, A.M., Coddington, I., and Newbury, N.R. (2011). Spectroscopy of the methane ν_3 band with an accurate midinfrared coherent dual-comb spectrometer. *Phys. Rev. A* 84: 062513.

63 Villares, G., Hugi, A., Blaser, S., and Faist, J. (2014). Dual-comb spectroscopy based on quantum-cascade-laser frequency combs. *Nat. Commun.* 5: 5192.

64 Maser, D.L., Ycas, G., Depetri, W.I., Cruz, F.C., and Diddams, S.A. (2017). Coherent frequency combs for spectroscopy across the 3–5 μm region. *Appl. Phys. B* 123: 142.

65 Muraviev, A., Smolski, V.O., Loparo, Z., and Vodopyanov, K.L. (2018). Massively parallel sensing of trace molecules and their isotopologues with broadband subharmonic mid-infrared frequency combs. *Nat. Photonics* 12: 209.

66 Vainio, M. and Halonen, L. (2016). Mid-infrared optical parametric oscillators and frequency combs for molecular spectroscopy. *Phys. Chem. Chem. Phys.* 18: 4266.

67 Cossel, K.C., Waxman, E.M., Finneran, I.A., Blake, G.A., Ye, J., and Newbury, N.R. (2017). Gas-phase broadband spectroscopy using active sources: progress, status, and applications. *J. Opt. Soc. Am. B* 34: 104.

68 Stothard, D.J.M., Dunn, M.H., and Rae, C.F. (2004). Hyperspectral imaging of gases with a continuous-wave pump-enhanced optical parametric oscillator. *Opt. Express* 12: 947.

69 Dam, J.S., Tidemand-Lichtenberg, P., and Pedersen, C. (2012). Room-temperature mid-infrared single-photon spectral imaging. *Nat. Photoics* 6: 788.

70 National Institute of Health Report: Health Devices 24(3), 92–122 (1995). https://www.ncbi.nlm.nih.gov/pubmed/7782226?report=abstract, accessed 21 September 2017

71 Fried, N.M. and Murray, K.E. (2005). High-power thulium fiber laser ablation of urinary tissues at 1.94 μm. *J. Endourol.* 19: 25.

72 Walsh, J.T. Jr., Flotte, T.J., and Deutsch, T.F. (1989). Er:YAG laser ablation of tissue: effect of pulse duration and tissue type on thermal damage. *Lasers Surg. Med.* 9: 314.

73 Vogel, A. and Venugopalan, V. (2003). Mechanisms of pulsed laser ablation of biological tissues. *Chem. Rev.* 103: 577.

74 Kaufmann, R. and Hibst, R. (1996). Pulsed erbium:YAG laser ablation in cutaneous surgery. *Lasers Surg. Med.* 19: 324.

75 Franjic, K., Cowan, M.L., Kraemer, D., and Miller, R.J.D. (2009). Laser selective cutting of biological tissues by impulsive heat deposition through ultrafast vibrational excitations. *Opt. Express* 17: 22937.

76 Pierce, M.C., Jackson, S.D., Dickinson, M.R., King, T.A., and Sloan, P. (2000). Laser-tissue interaction with a continuous wave 3-μm fibre laser: preliminary studies with soft tissue. *Lasers Surg. Med.* 26: 491.

77 Venugopalan, V., Nishioka, N.S., and Mikic, B.B. (1996). Thermodynamic response of soft biological tissues to pulsed infrared-laser irradiation. *Biophys. J.* 71: 3530.

78 Edwards, G., Logan, R., Copeland, M., Reinisch, L., Davidson, J., Johnson, B., Maciunas, R., Mendenhall, M., Ossoff, R., Tribble, J., Werkhaven, J., and O'Day, D. (1994). Tissue ablation by a free-electron laser tuned to the amide II band. *Nature* 371: 416.

79 Peavy, G.M., Reinisch, L., Payne, J.T., and Venugopalan, V. (1999). Comparison of cortical bone ablations by using infrared laser wavelengths 2.9 to 9.2 μm. *Lasers Surg. Med.* 26: 421.

80 Edwards, G.S. (2009). Mechanisms for soft-tissue ablation and the development of alternative medical lasers based on investigations with mid-infrared free-electron lasers. *Laser Photon. Rev.* 3: 545.

81 Mackanos, M.A., Simanovskii, D.M., Contag, C.H., Kozub, J.A., and Jansen, E.D. (2012). Comparing an optical parametric oscillator (OPO) as a viable alternative for mid-infrared tissue ablation with a free electron laser (FEL). *Lasers Med. Sci.* 27: 1213.

82 Edwards, G.S., Pearlstein, R.D., Copeland, M.L., Hutson, M.S., Latone, K., Spiro, A., and Pasmanik, G. (2007). 6450 nm wavelength tissue ablation using a nanosecond laser based on difference frequency mixing and stimulated Raman scattering. *Opt. Lett.* 32: 1426.

83 Kozub, J., Ivanov, B., Jayasinghe, A., Prasad, R., Shen, J., Klosner, M., Heller, D., Mendenhall, M., Piston, D.W., Joos, K., and Hutson, M.S. (2011). Raman-shifted alexandrite laser for soft tissue ablation in the 6- to 7-μm wavelength range. *Biomed. Opt. Express* 2: 1275.
84 Phillips, M. (1992). Breath tests in medicine. *Sci. Am.* 267: 74.
85 Miekisch, W., Schubert, J.K., and Noeldge-Schomburg, G.F.E. (2004). Diagnostic potential of breath analysis--focus on volatile organic compounds. *Clin. Chim. Acta* 347: 25.
86 Risby, T.H. and Solga, S.F. (2006). Current status of clinical breath analysis. *Appl. Phys. B* 85: 421.
87 Wang, C. and Sahay, P. (2009). Breath analysis using laser spectroscopic techniques: breath biomarkers, spectral fingerprints, and detection limits. *Sensors* 9: 8230.
88 Schwarz, K., Pizzini, A., Arendacká, B., Zerlauth, K., Filipiak, W., Schmid, A., Dzien, A., Neuner, S., Lechleitner, M., Scholl-Bürgi, S., Miekisch, W., Schubert, J., Unterkofler, K., Witkovský, V., Gastl, G., and Amann, A. (2009). Breath acetone—aspects of normal physiology related to age and gender as determined in a PTR-MS study. *J. Breath Res.* 3: 027003.
89 Shorter, J.H., Nelson, D.D., McManus, J.B., Zahniser, M.S., Sama, S., and Milton, D.K. (2011). Clinical study of multiple breath biomarkers of asthma and COPD (NO, CO_2, CO and N_2O by infrared laser spectroscopy). *J. Breath Res.* 5: 037108.
90 Cikach, F.S. Jr. and Dweik, R.A. (2012). Cardiovascular biomarkers in exhaled breath. *Prog. Cardiovasc. Dis.* 55: 34.
91 de Lacy Costello, B.P.J., Ledochowski, M., and Ratcliffe, N.M. (2013). The importance of methane breath testing: a review. *J. Breath Res.* 7: 024001.
92 Mochalski, P., King, J., Klieber, M., Unterkofler, K., Hinterhuber, H., Baumann, M., and Amann, A. (2013). Blood and breath levels of selected volatile organic compounds in healthy volunteers. *Analyst* 138: 2134.
93 Kim, S., Young, C., Vidakovic, B., Gabram-Mendola, S.G.A., Bayer, C.W., and Mizaikoff, B. (2010). Potential and challenges for mid-infrared sensors in breath diagnostics. *IEEE Sensors J.* 10: 145.
94 Dahnke, H., Kleine, D., Hering, P., and Mürtz, M. (2001). Real-time monitoring of ethane in human breath using mid-infrared cavity leak-out spectroscopy. *Appl. Phys. B* 72: 971.
95 von Basum, G., Dahnke, H., Halmer, D., Hering, P., and Mürtz, M. (2003). Online recording of ethane traces in human breath via infrared laser spectroscopy. *J. Appl. Physiol.* 95: 2583.
96 Halmer, D., Thelen, S., Hering, P., and Mürtz, M. (2006). Online monitoring of ethane traces in exhaled breath with a difference frequency generation spectrometer. *Appl. Phys. B.* 85: 437.
97 Skeldon, K.D., McMillan, L.C., Wyse, C.A., Monk, S.D., Gibson, G., Patterson, C., France, T., Longbottom, C., and Padgett, M.J. (2006). Application of laser spectroscopy for measurement of exhaled ethane in patients with lung cancer. *Respir. Med.* 100: 300.

98 McCurdy, M.R., Bakhirkin, Y., Wysocki, G., Lewicki, R., and Tittel, F.K. (2007). Recent advances of laser spectroscopy- based techniques for applications in breath analysis. *J. Breath Res.* 1: 014001.

99 Risby, T.H. and Tittel, F.K. (2010). Current status of midinfrared quantum and interband cascade lasers for clinical breath analysis. *Opt. Eng.* 49: 111123.

100 Parameswaran, K.R., Rosen, D.I., Allen, M.G., Ganz, A.M., and Risby, T.H. (2009). Off-axis integrated cavity output spectroscopy with a mid-infrared interband cascade laser for real-time breath ethane measurements. *Appl. Opt.* 48: B73.

101 McCurdy, M., Bakhirkin, Y.A., and Tittel, F.K. (2006). Quantum cascade laser-based integrated cavity output spectroscopy of exhaled nitric oxide. *Appl. Phys. B* 85: 445.

102 Bakhirkin, Y.A., Kosterev, A.A., Curl, R.F., Tittel, F.K., Yarekha, D.A., Hvozdara, L., Giovannini, M., and Faist, J. (2006). Sub-ppbv nitric oxide concentration measurements using CW room-temperature quantum cascade laser based integrated cavity spectroscopy. *Appl. Phys. B* 82: 149.

103 McCurdy, M.R., Bakhirkin, Y., Wysocki, G., and Tittel, F.K. (2007). Performance of an exhaled nitric oxide and carbon dioxide sensor using quantum cascade laser based integrated cavity output spectroscopy. *J. Biomed. Opt.* 12: 34034.

104 Moeskops, B.W., Cristescu, S.M., and Harren, F.J. (2006). Sub-part-per-billion monitoring of nitric oxide by use of wavelength modulation spectroscopy in combination with a thermoelectrically cooled, continuous-wave quantum cascade laser. *Opt. Lett.* 31: 823.

105 Halmer, D., von Basum, G., Horstjann, M., Hering, P., and Muertz, M. (2005). Time resolved simultaneous detection of ^{14}NO and ^{15}NO via mid-IR cavity leak-out spectroscopy. *Isot. Environ. Health Stud.* 41: 303.

106 Manne, J., Sukhorukov, O., Jager, W., and Tulip, J. (2006). Pulsed quantum cascade laser-based cavity ring-down spectroscopy for ammonia detection in breath. *Appl. Opt.* 45: 9230.

107 Moeskops, B., Naus, H., Cristescu, S., and Harren, F. (2006). Quantum cascade laser-based carbon monoxide detection on a second time scale from human breath. *Appl. Phys. B* 82: 649.

108 Roller, C., Kosterev, A.A., Tittel, F.K., Uehara, K., Gmachl, C., and Sivco, D.L. (2003). Carbonyl sulfide detection with a thermoelectrically cooled mid-infrared quantum cascade laser. *Opt. Lett.* 28: 2052.

109 Shorter, J.H., Nelson, D.D., McManus, J.B., Zahniser, M.S., and Milton, D.K. (2010). Multicomponent breath analysis with infrared absorption using room-temperature quantum cascade lasers. *IEEE Sensors J.* 10: 76.

110 Thorpe, M.J., Balslev-Clausen, D., Kirchner, M.S., and Ye, J. (2008). Cavity enhanced optical frequency comb spectroscopy: application to human breath analysis. *Opt. Express* 16: 2387.

111 Khodabakhsh, A., Rutkowski, L., Morville, J., and Foltynowicz, A. (2017). Mid-infrared continuous-filtering Vernier spectroscopy using a doubly resonant optical parametric oscillator. *Appl. Phys. B* 123: 120.

112 Jin, Y.W., Cristescu, S.M., Harren, F.J.M., and Mandon, J. (2014). Two-crystal mid-infrared optical parametric oscillator for absorption and dispersion dual-comb spectroscopy. *Opt. Lett.* 39: 3270.

113 Timmers, H., Kowligy, A., Lind, A., Cruz, F.C., Nader, N., Silfies, M., Ycas, G., Allison, T.K., Schunemann, P.G., Papp, S.B., and Diddams, S.A. (2018). Molecular fingerprinting with bright, broadband infrared frequency combs. *Optica* 5: 727.

114 Ycas, G., Giorgetta, F.R., Baumann, E., Coddington, I., Herman, D., Diddams, S.A., and Newbury, N.R. (2018). High-coherence mid-infrared dual-comb spectroscopy spanning 2.6 to 5.2 μm. *Nat. Photonics* 12: 202.

115 Henderson, B., Khodabakhsh, A., Metsälä, M., Ventrillard, I., Schmidt, F.M., Romanini, D., Ritchie, G.A.D., Hekkert, S.t.L., Briot, R., Risby, T., Marczin, N., Harren, F.J.M., and Cristescu, S.M. (2018). Laser spectroscopy for breath analysis: towards clinical implementation. *Appl. Phys. B* 124 (161).

116 Knoll, B. and Keilmann, F. (1999). Near-field probing of vibrational absorption for chemical microscopy. *Nature* 399: 134.

117 Keilmann, F. and Hillenbrand, R. (2004). Near-field microscopy by elastic light scattering from a tip. *Philos. Trans. R. Soc. Lon. A* 362: 787.

118 Taubner, T., Hillenbrand, R., and Keilmann, F. (2004). Nanoscale polymer recognition by spectral signature in scattering infrared near-field microscopy. *Appl. Phys. Lett.* 85: 22.

119 Taubner, T., Keilmann, F., and Hillenbrand, R. (2005). Nanoscale-resolved subsurface imaging by scattering-type near-field optical microscopy. *Opt. Express* 13: 8893.

120 Dazzi, A., Prazeres, R., Glotin, F., and Ortega, J.M. (2005). Local infrared microspectroscopy with subwavelength spatial resolution with an atomic force microscope tip used as a photothermal sensor. *Opt. Lett.* 30: 2388.

121 Dazzi, A., Prazeres, R., Glotin, F., and Ortega, J.M. (2007). Analysis of nano-chemical mapping performed by an AFM-based (AFMIR) acousto-optic technique. *Ultramicroscopy* 107: 1194.

122 Dazzi, A., Prazeres, R., Glotin, F., Ortega, J.-M., Alsawaftah, M., and De Frutos, M. (2008). Chemical mapping of the distribution of viruses into infected bacteria with a photothermal method. *Ultramicroscopy* 108: 635.

123 Dazzi, A., Prater, C.B., Hu, Q., Chase, D.B., Rabolt, J.F., and Marcott, C. (2012). AFM–IR: combining atomic force microscopy and infrared spectroscopy for nanoscale chemical characterization. *Appl. Spectrosc.* 66: 1365.

124 Prater, C., Kjoller, K., Cook, D., Shetty, R., Meyers, G., Reinhardt, C., Felts, J., King, W., Vodopyanov, K.L., and Dazzi, A. (2010). Nanoscale infrared spectroscopy of materials by atomic force microscopy. *Microsc. Anal.* 24: 5.

125 Lu, F. and Belkin, M.A. (2011). Infrared absorption nano-spectroscopy using sample photoexpansion induced by tunable quantum cascade lasers. *Opt. Express* 19: 19942.

126 Stanley, R. (2012). Plasmonics in the mid-infrared. *Nat. Photonics* 6: 409–411.

127 Yu, N., Wang, Q., and Capasso, F. (2012). Beam engineering of quantum cascade lasers. *Laser Photon. Rev.* 6: 1.

128 Adato, R., Yanik, A.A., and Altug, H. (2011). On chip plasmonic monopole nano-antennas and circuits. *Nano Lett.* 11: 5219.

129 Chen, K., Adato, R., and Altug, H. (2012). Dual-band perfect absorber for multispectral plasmon-enhanced infrared spectroscopy. *ACS Nano* 6: 7998.

130 Popmintchev, T., Chen, M.-C., Arpin, P., Murnane, M.M., and Kapteyn, H.C. (2010). The attosecond nonlinear optics of bright coherent X-ray generation. *Nat. Photonics* 4: 822.

131 Popmintchev, T., Chen, M.-C., Popmintchev, D., Arpin, P., Brown, S., Ališauskas, S., Andriukaitis, G., Balčiunas, T., Mücke, O.D., Pugzlys, A., Baltuška, A., Shim, B., Schrauth, S.E., Gaeta, A., Hernández-García, C., Plaja, L., Becker, A., Jaron-Becker, A., Murnane, M.M., and Kapteyn, H.C. (2012). Bright coherent ultrahigh harmonics in the keV X-ray regime from mid-infrared femtosecond lasers. *Science* 336: 1287.

132 Weisshaupt, J., Juvé, V., Holtz, M., Ku, S., Woerner, M., Elsaesser, T., Ališauskas, S., Pugžlys, A., and Baltuška, A. (2014). High-brightness table-top hard X-ray source driven by sub-100-femtosecond mid-infrared pulses. *Nat. Photonics* 8: 927.

133 Chin, A.H., Calderón, O.G., and Kono, J. (2001). Extreme midinfrared nonlinear optics in semiconductors. *Phys. Rev. Lett.* 86: 3292.

134 Ghimire, S., DiChiara, A.D., Sistrunk, E., Agostini, P., DiMauro, L.F., and Reis, D.A. (2011). Observation of high-order harmonic generation in a bulk crystal. *Nat. Phys.* 7: 138.

135 Vampa, G., Hammond, T.J., Thiré, N., Schmidt, B.E., Légaré, F., McDonald, C.R., Brabec, T., and Corkum, P.B. (2015). Linking high harmonics from gases and solids. *Nature* 522: 462.

136 Schubert, O., Hohenleutner, M., Langer, F., Urbanek, B., Lange, C., Huttner, U., Golde, D., Meier, T., Kira, M., Koch, S.W., and Huber, R. (2014). Sub-cycle control of terahertz high-harmonic generation by dynamical Bloch oscillations. *Nat. Photonics* 8: 119.

137 Langer, F., Schmid, C.P., Schlauderer, S., Gmitra, M., Fabian, J., Nagler, P., Schüller, C., Korn, T., Hawkins, P.G., Steiner, J.T., Huttner, U., Koch, S.W., Kira, M., and Huber, R. (2018). Lightwave valleytronics in a monolayer of tungsten diselenide. *Nature* 557: 76.

138 Tajima, T. and Dawson, J.M. (1979). Laser electron accelerator. *Phys. Rev. Lett.* 43: 267.

139 Mangles, S.P.D., Murphy, C.D., Najmudin, Z., Thomas, A.G.R., Collier, J.L., Dangor, A.E., Divall, E.J., Foster, P.S., Gallacher, J.G., Hooker, C.J., Jaroszynski, D.A., Langley, A.J., Mori, W.B., Norreys, P.A., Tsung, F.S., Viskup, R., Walton, B.R., and Krushelnick, K. (2004). Monoenergetic beams of relativistic electrons from intense laser-plasma interactions. *Nature* 431: 535.

140 Faure, J., Glinec, Y., Pukhov, A., Kiselev, S., Gordienko, S., Lefebvre, E., Rousseau, J.-P., Burgy, F., and Malka, V. (2004). A laser–plasma accelerator producing monoenergetic electron beams. *Nature* 431: 541.

141 Pogorelsky, I.V., Polyanskiy, M.N., and Kimura, W.D. (2016). Mid-infrared lasers for energy frontier plasma accelerators. *Phys. Rev. Accel. Beams* 19: 091001.

142 Peralta, E.A., Soong, K., England, R.J., Colby, E.R., Wu, Z., Montazeri, B., McGuinness, C., McNeur, J., Leedle, K.J., Walz, D., Sozer, E.B., Cowan, B., Schwartz, B., Travish, G., and Byer, R.L. (2013). Demonstration of electron acceleration in a laser-driven dielectric microstructure. *Nature* 503: 91.

143 Breuer, J. and Hommelhoff, P. (2013). Laser-based acceleration of nonrelativistic electrons at a dielectric structure. *Phys. Rev. Lett.* 111: 134803.

144 Leedle, K.J., Pease, R.F., Byer, R.L., and Harris, J.S. (2015). Laser acceleration and deflection of 96.3 keV electrons with a silicon dielectric structure. *Optica* 2: 158.

145 Blaser, S., Hofstetter, D., Beck, M., and Faist, J. (2001). Free-space optical data link using Peltier-cooled quantum cascade laser. *IEEE Electron. Lett.* 37: 778.

146 Martini, R., Bethea, C., Capasso, F., Gmachl, C., Paiella, R., Whittaker, E.A., Hwang, H.Y., Sivco, D.L., Baillargeon, J.N., and Cho, A.Y. (2002). Free-space optical transmission of multimedia satellite data streams using mid-infrared quantum cascade lasers. *IEEE Electron. Lett.* 38: 181.

147 Corrigan, P., Martini, R., Whittaker, E.A., and Bethea, C. (2009). Quantum cascade lasers and the Kruse model in free space optical communication. *Opt. Express* 17: 4355.

148 Bubb, D.M., Horwitz, J.S., McGill, R.A., Chrisey, D.B., Papantonakis, M.R., Haglund, R.F., Jr., and Toftmann, B. (2001). Resonant infrared pulsed-laser deposition of a sorbent chemoselective polymer. *Appl. Phys. Lett.* 79: 2847.

149 Kolev, V.Z., Duering, M.W., Luther-Davies, B., and Rode, A.V. (2006). Compact high-power optical source for resonant infrared pulsed laser ablation and deposition of polymer materials. *Opt. Express* 14: 12302.

Index

a

ablation
  biological tissue 13, 16, 260–261
  dental enamel 260
  polymer 23, 275
absorption
  atmospheric 1–2, 11, 13, 34
  Cr^{2+}-dopant 19–20
  diode-laser-pump 11
  Er^{3+}-dopant 49
  excited state 44, 48, 50
  Fe^{2+}-dopant 19–20
  four-photon (4PA) 206
  free carrier 66, 96
  human tissue 23–24, 164, 258–261
  'in-well' 96
  long-wave edge 112, 138
  molecular 4–5, 79, 132, 235, 265–266
  multiphonon 116, 127, 130, 170, 172
  nonlinear 26
  residual near-IR tail 136, 141
  ro-vibrational 72, 247–257, 262–264
  saturable 56
  spectrum 4–5, 11, 13, 248–249, 254–256, 259
  surface-enhanced 267–269
  three-photon (3PA) 171–172, 232, 234
  two-photon (2PA) 116–120, 140, 163, 171–172
  water 14, 48, 51, 259
absorption spectroscopy 5, 65, 70, 72, 85, 87, 123, 132, 189, 247–267, 269
acetylene *see* C_2H_2
active region, laser 66–67, 74–78, 82–85, 90, 92, 95, 98–99
$AgGaS_2$ (AGS) 35, 115, 117, 119, 130–131, 134–135, 145, 154–155, 158, 161, 166
$AgGaSe_2$ (AGSe) 115, 117, 128–130, 135, 145, 154–156, 158–160, 163, 223, 225
AlGaInAsSb barrier 67–69
AlGaIn/AsSb material system 95
angle tuning 135–137, 142, 149
anomalous dispersion 59, 169, 190–192, 194, 198, 200–201, 204–205, 207, 212
anti-Stokes 171
arsenic selenide glass *see* As_2Se_3
arsenic sulfide glass *see* As_2S_3

Laser-based Mid-infrared Sources and Applications, First Edition. Konstantin L. Vodopyanov.

As_2S_3 chalcogenide glass 53
 Raman medium 53–54, 57, 172
 step-index fiber 53, 200, 208–209
 suspended core fiber 196–198
 tapered fiber 197–200, 208–209, 216
 waveguide 204–205, 216, 221
As_2Se_3 chalcogenide glass 53
 Raman medium 53, 172
 step-index fiber 200, 209
 tapered fiber 200–201, 203, 209
atmospheric transparency window 1–2, 11, 13, 34, 85–86
attosecond 270–273
Auger recombination 66, 68, 72, 96

b

$BaGa_4Se_7$ (BGSe) 135, 142, 145
band alignment (semiconductors) 89
bandwidth
 laser gain 19, 21, 27, 56, 83
 phase-matching 122, 135, 226, 228
 spectral output 156, 166, 172, 200, 205–206, 211, 215, 217, 219, 225, 235–236, 254–257, 264
 X-ray supercontinuum 270–273
beam filamentation 207, 210, 212
beam profile 28, 78, 138, 140, 152
birefringent 12
 nonlinear crystal 110–121
Bloch oscillation 272
bound-to-continuum QCL 81, 83, 85, 88–89
Bragg grating
 chirped 30
 fiber 50–54, 59
 laterally coupled 73
Bragg reflector 95–97
broadening
 homogeneous 18–19, 24
 inhomogeneous in QCLs 83
 type-I OPO phase matching 167
buried
 channel waveguide 26
 gain region, QCL 76
 grating, QCL 82
 heterostructure QCL 84
 ridge waveguide, QCL 77–78, 88

c

carbon dioxide *see* CO_2
carbon monoxide *see* CO
cascaded emission of photons
 ICL 87, 89–91
 QCL 73–74
cascaded lasing
 Er^{3+}-fiber 50, 57
 Ho^{3+}-fiber 47–48, 57
cascaded $\chi^{(2)}$ nonlinearity 207, 211
cascaded OPO 135, 146
cascaded pumping, injection lasers 68–69, 72
cascaded Raman lasing 53, 169–172, 194
cascaded Raman self-shifting 194
cascaded SC generation 200
cascaded Tm-Ho solid state laser 11
cascading frequencies, microresonators 230, 232
cavity-enhanced frequency comb spectroscopy 253, 264
cavity finesse 25, 143, 234, 251, 253, 255, 264
cavity ring-down spectroscopy (CRDS) 251
$CdGeAs_2$ (CGA) 115, 117, 130–131
CdS 7, 18
CdSe
 laser host 7, 18, 20, 22–23, 30, 32, 35
 nonlinear crystal 115, 117, 161, 166
$CdSiP_2$ (CSP) 115–117, 121, 140–141, 160–161, 164, 168, 174
C_2H_2 248, 255–256, 264
C_2H_6 252, 255–256, 262
CH_4 248, 255–256, 262, 264
chalcogenide 7, 18, 53, 95, 142, 172, 196–199, 205, 216–217
 chromium-doped *see* Cr^{2+}-doped chalcogenides

fiber 198, 200
iron-doped *see* Fe^{2+}-doped chalcogenides
charge oscillation 267
CH_2O 121, 250, 252, 264
CO 5, 132, 248, 250, 255–256, 262–264
CO_2 2, 34, 199, 248, 250–251, 255–256, 262, 264
coherence length 118
compressive strain 68
conduction band 66, 73–74
offset 77, 85
confinement
carrier 77
hole 68, 72
quantum 91
conversion efficiency
frequency combs 225–226
nonlinear optical 112, 121–122, 126, 128, 130, 132, 133, 135, 136, 140–142, 148–154, 162, 166, 257
optical-to-optical 9, 12, 23–24, 27, 31, 35
Raman 169, 172, 174
semiconductor lasers 68, 93
supercontinuum 195–196, 207
Cr^{2+}-doped chalcogenides 18
CdMnTe 23
CdS 7, 18, 30
CdSe 7, 18, 20, 22–23, 30, 32, 33
ZnS 18–20, 23, 27–36, 210–211, 215, 228, 260
ZnSe 18–36, 210–211, 215, 228, 260
cross relaxation, 'two-for-one' process 8–9, 15, 44–46, 61

d

damage threshold 116, 119–120, 135, 141, 171
detection limit, gas sensors 249, 252–253, 262
diamond
microresonator 172
submount 76
diode pumping 7, 47–49
dispersion-engineered 204
dispersion length 192
dispersive wave generation 198, 201–202, 216–217
distributed Bragg reflector (DBR) 95
distributed feedback (DFB) 70, 72–73, 79, 87
double-clad fiber 49, 51, 54–55, 201
double-heterostructure (DH) laser 66
double-pass
laser amplifier 12
OPG 141, 166
OPO cavity 136–137, 142, 150
double QW laser 67–69, 71
double-stack active region, QCL 85
doubly resonant OPO (DROPO) 125, 128, 144–145, 226–229
dual-wavelength pumping, fiber laser 51–52, 61
Dy^{3+} fiber laser 44–45, 52

e

effective nonlinear coefficient 111–114
electron injector 68, 74, 90
emission spectra 19–20, 45, 87, 92
energy-level diagram 3–4, 7–11, 14–15, 43–48, 66, 74
energy transfer 46–49, 55
erbium *see* Er^{3+}
Er^{3+}, Cr^{3+}: YSGG laser 18
Er^{3+}-doped 35–36
emission cross section 45
energy-level diagram 14–15, 18, 44, 46
fluoride fiber laser 51, 53, 59
fluorozirconate (ZrF_4) laser 52, 200
GGG laser 15–16
GSGG laser 14–15
YAG laser 13–18, 30, 31–34, 35, 259–260
YSGG laser 13–18, 31, 261
ZBLAN fiber laser 49–50, 52, 54, 59–60
zirconium fluoride glass 59

Er^{3+}, Pr^{3+}: ZBLAN fiber laser 49
ethane *see* C_2H_6
excited state absorption (ESA) 44, 48, 50
excited state lifetime 19–20
external cavity
  injection laser 69, 72,
  QCL 82–87, 248–249, 251
  VECSEL 94

f

Fabry–Pérot laser cavity 53, 79, 84
Fe^{2+}-doped chalcogenides 18
  CdMnTe 19–20
  CdSe 32, 35
  CdZnTe 20
  ZnSe 18–21, 30, 31–36
  ZnTe 32, 35
ferroelectric oxide 116, 119, 133, 151
fiber amplifier 45, 55, 194, 196, 218–219
first Stokes emission 53, 169, 171
fluorescence lifetime 9–10
fluoride glass fiber 49–54, 59–60, 200
formaldehyde *see* CH_2O
Fourier transform IR (FTIR) spectrometry 253–255, 264–265
four-level laser 3, 8, 9
four-wave mixing (FWM) 171, 190, 196, 201, 204, 207, 212, 230, 232, 234
free carrier absorption 66, 96
free spectral range, microresonator 232, 234
frequency down conversion 109–110
fundamental ro-vibrational absorption 72, 247–257, 262–264

g

GaAlSbAs/InGaSbAs/GaAlSbAs, double heterostructure 66
GaAs/AlGaAs QCL 75
GaInAs/AlInAs/InP QCL structure 77–78, 88
GaInAs/AlInAs, QCL 76
GaInAsSb/ AlGaAsSb
  QW laser 67–69
  semiconductor disk laser 69
GaInSb/AlGaAsSb laser 68–69
GaInSb/InAs heterojunction 89
gain-switched laser 24–25, 32–33
gallium arsenide (GaAs) 189, 210–212, 272
  orientation patterned (OP-GaAs) 118–124, 127–129, 136–140, 145–146, 149–150, 153, 155–156, 158, 174, 216, 223, 228, 229, 254, 272
gallium phosphide (GaP)
  orientation patterned (OP-GaP) 115, 118–121, 160, 164–165, 174, 219–220, 222–223, 228, 253
gallium selenide (GaSe) 115, 117, 119, 130–131, 154–155, 157–159, 161, 166–167, 219, 222, 272
GaSb/InAs heterojunction 89
gas sensing 71, 132
germanate-based Raman gain cavity 53
grating-coupled QCL 82–83
group velocity dispersion (GVD) 163–164, 190, 206, 229

h

heavily strained 67
  type-I QW 68
Herriott multipass gas cell 252–255, 264
heterojunction diode laser 65–66
HF laser 33
High-harmonic generation (HHG) 213, 270–273
HITRAN database 2, 248–249, 254, 256
Ho^{3+}-doped
  emission cross section 11, 45
  energy-level diagram 11, 44
  fiber laser 47–48
  YAG laser 11–13, 24, 33, 129, 148–149, 157, 258–259
  YLF laser 12–13, 30, 151–153, 168

YSGG laser 35
ZBLAN fiber laser 48
hole confinement 68, 72
homogeneous broadening 18–19, 24
Ho^{3+}, Pr^{3+} codoped fiber laser 55
Ho^{3+}, Tm^{3+} codoped YLF laser 148
hydride vapor phase epitaxy (HVPE) 120, 127

i

idler wave, OPO 113
InAs/GaSb QCL 75, 85–87
indium arsenide (InAs) 56, 58–59, 65
InGaAs/AlInAs/InP 77–78, 88
InGaAs/InAlAs 75, 85, 89, 224
inhomogeneous broadening 83
InP 75
cladding layer 76
intersubband transition 75
intrapulse DFG 119, 157, 272
iron doped lasers *see* Fe^{2+}-doped

k

Kerr-lens mode locking (KLM) 27
$KTiOAsO_4$ (KTA) 116, 131–132, 150, 162, 168
periodically poled (PPKTA) 117
$KTiOPO_4$ (KTP) 116, 135, 146–148, 150, 160–161, 168
periodically poled (PPKTP) 117, 146–147, 174

l

laser spectroscopy
cavity ring-down (CRDS) 251
frequency comb 253–256, 264
photoacoustic (PAS) 87, 180, 467, 478, 528
photothermal 249, 265–266
laser tissue interaction 54, 164, 258–261
laterally coupled distributed feedback (LC-DFB) 69, 72–73, 92
lattice-matched 66, 74, 76, 96, 120
lifetime
fluorescence 9–10
lower laser level 14
upper laser level 8, 13, 19
lithium niobate ($LiNbO_3$) 112, 115, 146, 150
periodically poled (PPLN) 115, 117, 118, 120–122, 146, 151, 154, 162, 252
lithium tantalate ($LiTaO_3$)
periodically poled (PPLT) 117, 127, 151
Littrow configuration 23–24, 49–50, 72, 82, 84

m

master oscillator–power amplifier (MOPA) 23, 142, 153
metal–organic chemical vapor deposition (MOCVD) 77
methane *see* CH_4
MgO:$LiNbO_3$ 154
mode-locked fiber lasers 56–60, 214–215
molecular spectroscopy 219, 255
momentum conservation 109
multipass gas cell 252–255, 264
multiphonon absorption 116, 127, 130, 170, 172
multiple quantum well (MQW) 78, 96–97
multiwavelength QCL array 81–82

n

narrow-linewidth
DFB laser 70, 72, 79
DFG 121, 132
OPO 127, 143–146
QCL 251
nested-cavity OPO 144–145
noncollinear phase matching 163, 167–168

noncritical phase matching 116, 162
nonlinear optical
  coefficient 112, 117–118
  effective coefficient 112
  figure of merit (FOM) 112
nonlinear parameter γ, waveguides 204
nonlinear polarization evolution (rotation) 56, 59–60
nonlinear refractive index 194, 196, 204
nonlinear susceptibility, 2nd order 109–112
nonlinear susceptibility, 3rd order 189–190

o

optical parametric
  gain 114–115
  quantum noise 133
optical rectification (OR) 119, 157, 272

p

periodic poling 110–112
phase-matching 109–111
  noncollinear 163, 167–168
  noncritical 116, 162
  quasi-(QPM) 109–112
photoacoustic spectroscopy (PAS) 132, 249–251
photon energy conservation 109, 220
photonic crystal fiber (PCF) 155, 191
photothermal IR spectroscopy 249, 265–266
potassium titanyl phosphate *see* $KTiOPO_4$
pump absorption spectrum
  Ho:YAG 11
  Ho:YLF 13

q

Q-switching 141
  acousto-optic 12, 148
  electro-optic 17
quantum defect 96, 98
quantum efficiency 8, 15, 31, 96, 128, 162
quantum well (QW) 66–67
quasi-CW 13, 17, 53–54
quasi-phase matching (QPM) 109–112

r

Raman
  cascaded 170
  converter using solid para-hydrogen 173
  intra-pulse scattering of solitons 154
  self-frequency shift 59, 155–156
  soliton pulses 60
rare-earth ions 7–10
$RbTiOAsO_4$ (RTA) 116
  periodically poled (PP-RTA) 117, 150, 257
ridge-waveguide laser 68–73, 76–81, 88, 91–93

s

second Stokes wave 33, 54, 171
self-focusing 207, 210
side pumping, Er:YAG laser 17
signal wave, OPO 113
silicon (Si) waveguide 206, 216–217, 232
single longitudinal mode (SLM)
  ICL 91–93
  injection laser 69–70
  quantum cascade lasers (QCL) 82–83, 88
soliton 190
  fiber laser 56, 58–60
  fission 192, 197–199, 201, 204–206, 216
  higher-order 201
  self-frequency shifted 155–156, 192, 196–197, 200–202, 204

spinning-ring amplifier 23–24, 28
stimulated emission cross section 10, 11, 19, 32, 43–45
Stokes
limit 48–50, 123
wave output 24, 33, 52–54, 155, 169–173, 261
synchronous pumping, OPO 113, 161–164

t

tellurite glass 53, 192–193
Tm^{3+}-doped,
emission cross section 11, 45
energy-level diagram 8–9, 44, 46
fiber master oscillator power amplifier 196, 219
mode-locked fiber laser 56, 214
silica fiber laser 27, 45–47, 53, 96, 150, 258–259
YAG laser 8–11, 24
YALO laser 21, 24
YAP laser 22
YLF laser 11–12
YSGG laser 9
Tm^{3+}, Ho^{3+}-codoped silica fiber laser 55
trace-gas detection 252–253, 261
transmission spectrum, atmosphere 1–2, 11, 13, 35, 85–86
type-I QW diode laser 66, 68–69, 71–72
type-II QW ICL 89–90, 93

u

upper-level lifetime 8, 13, 19
uranium-doped calcium fluoride (U^{3+}: CaF_2) 3–4

v

vertical-external-cavity surface-emitting laser (VECSEL). 94
vibronically broadened transition 18–20
vibronic laser 18–20

w

water, liquid phase absorption 14, 48, 51, 259
waveguide (WG)
As_2S_3 embedded in silica 216
nanospike 205
nonlinearity 193, 204
planar 204
PPLN 207
rib 205
silicon-on-insulator (SOI) 206, 216–217, 232
silicon nitride (Si_3N_4) 232
silicon-on-sapphire (SOS) 206
wavelength tuning, via
acousto-optic modulator (AOM) 84
cavity length 98, 144–145
chirp of the driving pulses 219
coupled-in power 155
diffraction grating 21–24, 82–83, 85–87
intracavity prism 35
multiwavelength QCL array 81
quasi-phase-matched (QPM) period 125, 134
phase-matching angle 113, 135–137, 166–167
pump wavelength 112–113, 123, 127, 138–140, 152–154, 173
temperature 80, 92, 98, 139–140

x

X-ray supercontinuum 270–273

z

ZBLAN 194
fiber lasers 48–50, 52, 54, 59–60
$ZnGeP_2$ (ZGP) 106, 121, 153, 155, 160, 163–173
ZnS 207, 272
chromium-doped *see* Cr^{2+}-doped chalcogenides

ZnSe 115, 118, 120, 138–139, 141–142, 165, 173, 202, 207–210, 272
chromium-doped *see* Cr^{2+}-doped chalcogenides
iron doped *see* Fe^{2+}-doped chalcogenides
ZnTe 272
iron doped *see* Fe^{2+}-doped chalcogenides

The Universe is Indifferent

The Universe is Indifferent

Theology, Philosophy and Mad Men

Edited by
Ann W. Duncan
and
Jacob L. Goodson

The Lutterworth Press

I hate to break it to you, but there is no big lie. There is no system.
The universe is indifferent.
Don Draper, "The Hobo Code", Season 1, Episode 8

The Lutterworth Press
P.O. Box 60
Cambridge
CB1 2NT
United Kingdom

www.lutterworth.com
publishing@lutterworth.com

ISBN: 978 0 7188 9518 1

British Library Cataloguing in Publication Data
A record is available from the British Library

First published by The Lutterworth Press, 2017

Published by arrangement
with Cascade Books

CONTENTS

Acknowledgments | xi
Introduction | xiii
—Ann W. Duncan and Jacob L. Goodson

PART 1: BUSINESS ETHICS

Chapter 1
"It's the Real Thing": Identity and Sincerity in *Mad Men* | 3
—Howard Pickett

Chapter 2
The Business of Creativity: From SCDP to the Modern Creative Enterprise | 28
—Jennifer Phillips

Chapter 3
Mad Manners: Courtesy, Conflict, and Social Change | 52
—Sarah Conrad Sours

Chapter 4
"All the Research Points to the Fact that Mothers Feel Guilty": Maternal Desire and the Social Construction of Motherhood in *Mad Men* | 78
—Ann W. Duncan

Chapter 5
Supporting this World's Ballerinas: Learning from Mad Men's Female Workers | 101
—Kristen Deede Johnson

Chapter 6
"If I don't go in that office every day, who am I?": Culture, Identity, and Work in *Mad Men* | 123
—David Matzko McCarthy

PART 2: WHO IS DON DRAPER?

Chapter 7
Counterculture Beatrices? Don Meets Dante | 145
—Gabriel Haley

Chapter 8
"Moving Forward" as Return: The Redemptive Journey of Don Draper | 169
—Jackson Lashier

Chapter 9
"You Are Okay": Don*sein*'s Despair and Our Road to Recovery | 192
—Seth Vannatta

Chapter 10
Don Draper, Double Consciousness, and the Invisibility of Blackness | 215
—Nsenga K. Burton

Chapter 11
The Erotic Reduction of Don Draper: Iconicity, Idolatry, and Madness | 237
—Carole L. Baker

PART 3: POLITICS AND SOCIAL THEORY

Chapter 12
Zou Bisou Bisou*: Feminist Philosophy and Sexual Ethics in *Mad Men | 265
—Jacob L. Goodson

Chapter 13
***Exitus et Reditus* in Marriage: *Mad Men* vs. Hollywood Remarriage Comedies** | 292
—Brandon L. Morgan and Jonathan Tran

Chapter 14
Uneasy Bedfellows: On Pete and Trudy's Marriage | 308
—Matthew Emile Vaughan and Christopher J. Ashley

Chapter 15
Mad Men, Bad Parents: Representations of Parenting in *Mad Men* | 329
—Susan E. Frekko

Chapter 16
"I Can't Believe That's the Way God Is": Sexism, Sin, and Clericalism in Peggy's Pre-Vatican II Catholicism | 345
—Heidi Schlumpf

Chapter 17
"We Don't Know What's Really Going On": *Mad Men* as a Bellwether of the Politics to Come | 368
—Jared D. Larson

Contributors | 391
Subject Index | 393
Author Index | 399
Episode Index | 403

ACKNOWLEDGMENTS

The editors of this volume wish to thank Goucher College and Southwestern College for their support of this project. Specifically, we thank the Goucher College Aitchison Faculty Development Fund for summer research support and Southwestern College for providing research assistance through its Honors program. Thank you to Lindsey Graber for her research and editorial support throughout the project. We also owe thanks to Chris Spinks, our editor at Wipf and Stock, who provided the initial encouragement to undertake this project and constant support throughout its development. Finally, we wish to thank our respective spouses—Daniel and Angela—who were our faithful companions on innumerable evenings watching the stories of *Mad Men* unfold and who have been our first and most constant conversation partners about the series.

INTRODUCTION

—ANN W. DUNCAN
and JACOB L. GOODSON

The AMC television series, *Mad Men*, takes its viewers on an emotional, psychological, and sociological roller coaster ride in the setting of New York City. The years depicted span March 1960 through either the autumn 1970 or winter 1971—depending upon one's interpretation of the Coca-Cola advertisement, an ad that first aired on radio on February 12, 1971. Created by Matthew Weiner, *Mad Men* develops and follows the lives of Donald Draper/Dick Whitman, Peggy Olson, Joan Holloway/Harris, Roger Sterling, Pete Campbell, Megan Calvet/Draper, Betty Draper/Francis, and Sally Draper as they negotiate careers, love and marriage, and dramatic social and political change. This collection helps readers navigate the exciting roller coaster ride that is *Mad Men*, through exploring unknown depths of its characters and storylines. After enjoying the ride of watching all seven seasons of *Mad Men*, we hope that you take pleasure in the ride of reading *'The Universe Is Indifferent': Theology, Philosophy, and* Mad Men.

"THE UNIVERSE IS INDIFFERENT"?

In Season 1, Don receives a bonus check in the amount of $2,500 and brings it to his mistress's apartment with the hopes of inviting her to join him on a trip to Paris. He finds her relaxing with a group of marijuana-smoking friends markedly different from his colleagues on Madison Avenue. These friends engage with Don about the vices of his work and conclude, "You make the lie. You invent want. You're for them, not us." Don responds with an equal amount of clarity: "I hate to break it to you, but there is no big lie.

There is no system. The universe is indifferent." His mistress rises and comes to Don, and he again invites her to Paris. She claims that she cannot go with him; he signs his bonus check, puts it in her brassiere, and tells her to buy a car. Don turns to leave, but one of her friends gives Don a warning: "There are cops. You can't go out there." Don responds without hesitation, "No, *you* can't."

These beatniks remain critical toward power and society but not nihilistic about potential change within the world. Don claims that their criticism has no object: how does one "bring the man" down when "the man" does not exist? Their protests remain directed toward a ghost, a phantom. For them, Don *is* the face of this ghost—this phantom. But that becomes too easy of a target. Don cannot be the face of "the man" because there is no "man" for them to bring down and critique. The claim, "the universe is indifferent," is not only an existentialist claim—because it means that we must create meaning for ourselves—but also a nihilistic claim in the sense that nothingness has as much power and sway as the meaning we make.

According to the German philosopher, Martin Heidegger, we can distinguish between two forms of nihilism: (a) "Incomplete nihilism does indeed replace the former values with others, but it still posits the latter always in the old position of authority that is, as it were, gratuitously maintained as the ideal realm of the suprasensory"; (b) "Complete nihilism, however, must in addition do away even with the place of value itself, with the suprasensory as a realm, and accordingly must posit and revalue values differently."[1] Which version of nihilism fits the world of *Mad Men* better remains up to the judgment of the viewer, but we certainly hope that this collection informs the question in helpful and interesting ways.

CHAPTER PREVIEWS

This book contains seventeen chapters divided into three sections. While the subject matters of the chapters have a decidedly socio-historical focus, the authors use basic topics as starting points for philosophical, religious, and theological reflections. *Mad Men* reveals deep truths concerning the social trends of the 1960s and early 1970s in American life. Because of this, *Mad Men* deserves a significant amount of reflection from philosophical, religious, and theological perspectives. Some of the chapters go beyond mere reflection and make deeper inquiries into what these trends say about American cultural habits, the business world within Western capitalism,

1. Heidegger, "The Word of Nietzsche," 69.

and the rapid social changes (gender, race, and sexuality) during this period. In what follows, we provide brief previews of each chapter.

Part 1 is ordered from general-to-particular-topics, with its bookends as two different perspectives on identity and work. In "'It's the Real Thing': Identity and Sincerity in Mad Men," Howard Pickett addresses both ethical and unethical business practices found in *Mad Men*. Pickett claims that the series, itself, represents what he calls a narrative ethic: a storyline, plot twists, and character development. All of these elements allow businesses in the real world to assess their own ethical approaches. Throughout *Mad Men*, we find a close connection between advertising, deception, and lying. Surprisingly, Don Draper refuses the temptation to lie in relation to his client, Lucky Strikes. The ambiguity of Don's character comes into play in the sense that while he is a liar (he consistently lies to those closest to him: Betty, Megan, and his children), *Mad Men* raises philosophical questions concerning authenticity, sincerity, and truth—especially in relation to Don's identity. Additionally, Pickett observes that *Mad Men* connects sex to selling, religion to retail. These connections play out in multiple advertisements produced throughout the series. Pickett claims that we need to maintain distinctions between *public* and *private* sincerity in order to establish healthy and moral business practices within the real world.

Jennifer Phillips begins her chapter by distinguishing between artists, businesspersons, and creative workers. In "The Business of Creativity: From SCDP to the Modern Creative Enterprise," Phillips describes the ways in which creative workers find inspiration. Places of inspiration come from the internal, through the production of ideas, or the external, for sake of awards and profits. Phillips argues that it is more virtuous to find inspiration inwardly, through the production of ideas. Creative workers need freedom and flexibility in order to express themselves. Business people need to manage their creative workers by allowing them flexibility with hours, lots of paid vacation, and promotions relating to the common good. Phillips concludes by claiming that our social responsibility concerns promoting advertising for the sake of the common good—which involves avoiding advertisements and business practices that harms or objectifies people.

Sarah Conrad Sours highlights the theme of courtesy found throughout *Mad Men*. In her chapter, "Mad Manners: Courtesy, Conflict, and Social Change," Sours shows how courtesy takes on different purposes; for instance, Peggy's acts of courtesy are connected to her genuine moral character whereas Pete's acts of courtesy come across as insincere and manipulative. Courtesy plays a role in power dynamics, perpetuating the socially constructed authorities. Wealthy white men carry the most power, which becomes evident in their relationships with women and people of color. White men have the freedom to make sexual comments and gestures toward the women in the office, asking crass personal questions—questions the women would never dream of asking the men. The few African-American characters in this series always act with courtesy but rarely receive much courtesy in return. The two African-American secretaries know all of the names of the employees and refer to them with their proper titles—courteous acts that are never reciprocated. The more powerful party bends the rules of social courtesy, while the weaker parties adhere to courtesy at all costs. Courtesy becomes a way for characters to navigate social relationships in their professional and personal lives.

In "'All the Research Points to the Fact that Mothers Feel Guilty': Maternal Desire and the Social Construction of Motherhood in *Mad Men*," Ann W. Duncan explores the confluence of maternal desire, societal paradigms of motherhood and vocation in the lives of Betty Draper/Harris, Peggy Olson and Joan Holloway/Harris. As a societal construction mediated by the often-conflicting maternal desire, career ambition and quest for personal fulfillment, motherhood finds expression in dramatically different ways in the characters of *Mad Men*. Through attention to theories of motherhood and feminism through the twentieth century, Duncan's chapter examines these three women as a lens through which to understand the complexities of motherhood in the contemporary age. Such an examination reveals that these character's stories are both deeply conditioned by their historical context and directly relevant to contemporary discourse about motherhood and feminism.

In "'Supporting this World's Ballerinas': Learning from *Mad Men*'s Female Workers," Kristen Deede Johnson focuses on the question of women's ability to pursue their dreams—both professionally and personally—through the lens of a theology of the home. Throughout *Mad Men*, women struggle to balance their personal and professional lives. Megan achieves her professional dream as an actress, but her marriage crumbles into divorce. Betty achieves her dream of marriage, children, and a home in the suburbs but gives up her professional dream as a model. Peggy achieves her professional dream of being a copywriter but foregoes motherhood by giving up her own child up for adoption. Joan succeeds within the firm, but her marriage also ends in divorce and she then regrets how much she relies on her own mother for childcare for her son. What is the response, perhaps the solution, to these tragic conflicts between professional careers and home economics? Somewhat counter-intuitively, Johnson claims that a Christian theology of the home contributes to a positive perception of housework. If housework becomes significant enough to acquire God's attention, as detailed in the Bible, then it deserves our human attention. Humans have certain limitations and the repetitive tasks of care-giving and domestic work serve as moments of "holy leisure."

In "'If I Don't Go In that Office Everyday, then Who am I?': Culture, Identity, and Work in *Mad Men*," David Matzko McCarthy argues that the shift from Dick Whitman to Don Draper comes strictly through the process of Don defining his identity in terms of he who is at work. McCarthy claims that the only "constancy" Draper has in his life concerns his role at work while several other relationships—family members, friends, mistresses, and even his sibling—fade away. While Don works, he operates as an artist: not the Romantic sense of artist, but in terms of being patient and concerned with getting it "right." He allows time for ideas to form and forces bad ideas onto neither clients nor colleagues. Don also operates as a craftsman. A craftsman differs from an artist because a craftsman identifies with a particular community. Don finds that his community depends upon him, but he relies on them as well. Throughout *Mad Men*, Don searches for authentic community, cultural unity, and an overall sense of belonging.

McCarthy's chapter provides an ideal transition from the section "Business Ethics" to the section "Who Is Don Draper?" Originally, we planned for Part 2 to contain studies of multiple characters within *Mad Men*; as it happens in the storylines of *Mad Men*, however, it all comes back to Don! Once we recognized that we had five chapters—counting neither McCarthy's nor Goodson's chapter as part of the five—reflecting upon Don's character, we did what Don would expect us to do: we gave him his own section. Part 2 contains five chapters: the first two chapters are explicitly theological in the sense that they employ religious or theological scholarship to understand Don's character and how Don gets perceived by others; the next three chapters are more philosophical in the sense that they employ arguments and sources from the canon of Western philosophy.

In "Countercultural Beatrices?: Don Meets Dante," Gabriel Haley uses Dante's *Inferno* as a lens through which to view Don's actions and moral progress. First, Haley discusses the role of Beatrice in Dante's account; Beatrice symbolizes love as a physical presence and a spiritual guide. According to Haley, Megan represents Beatrice for Don in *Mad Men* by taking on an identity of the other and surprising Don with both countercultural and counterintuitive actions. Secondly, Haley explains the notion of how properly ordered desire leads to moral progress within Dante's writing. Though Megan represents Beatrice, Haley argues that Don displays a disordered desire for Megan—which inhibits the moral progress of Don's character. Haley suggests, however, that the lighthearted tone at the end of the series reflects the potential for Don's moral progress: Don's willingness to be part of the "counterculture" and to confess his past mistakes, prepares him for healthier relationships in the future.

In "'Moving Forward' as Return: The Redemptive Journey of Don Draper," Jackson Lashier addresses the duplicity of Don Draper's life: false self vs. true self. Relying on categories from Thomas Merton's theology, Lashier connects living in the false self to being in exile. The Israelites experienced exile as a result of their failure to live faithfully into their covenant with God. As a displacement from one's homeland, the Israelites experienced exile as being displaced from the promise land; exile, however, also refers to not being who you were created to be. Draper lives in a place of "false self," which leads to a feeling of emptiness like a never-ending meaningless carousel ride. Lashier argues that the way out of exile, ironically, involves death. This death is not literal death but a metaphorical death—death to the "false self." Draper needs to die to his "false self" in order to start over. Through multiple storylines, we gain a glimpse of Draper dying to his unrelenting desire to start over and then begin to embrace his "true self"—an authentic self.

Engaging the existentialist philosophies of Søren Kierkegaard and Martin Heidegger, Seth Vannatta provides the strongest conceptual description for what we mean by the title of the book: "*'The Universe Is Indifferent.'*" In his chapter, "'You Are Okay': *Don*sein's Despair and our Road to Recovery," Vannatta treats Don's character on the terms of Heidegger's nihilistic existentialism vs. Kierkegaard's Christian existentialism. Kierkegaard calls for passionate action whereas Heidegger calls for authenticity. In order for Don to find authenticity, according to Kierkegaard, he must reject the world and begin to embrace his inner self. Through a Heideggerian lens, Vannatta demonstrates that Don can achieve authenticity only through modifying his anonymous modes of existence. Don must recover himself by resisting anxiety and despair. Vannatta suggests that anxiety, in fact, awakens Don from his slumber and causes him to feel uncanny and not-at-home-in-the-world. For Kierkegaard, the road to recovery involves a road that turns away from despair and turns towards faith. For Heidegger, self-recovery remains grounded in Don's actions and his anticipation of the future.

Nsenga K. Burton, in her chapter "Don Draper, Double Consciousness and the Invisibility of Blackness?" employs W. E. B. DuBois's notion of "double consciousness" to explore the multiplicity of Don's character as well as the representation of racial, class and religious minorities throughout the series. In so doing, Burton unsettles a frequent critique of the show—that it fails to address race in any substantial way. Burton argues that the series addresses race alongside many markers of identity and social change tackled throughout the seven seasons by exploring Don's complicated relationship to his own whiteness. His hidden identity and early experience of poverty mark him as an "other" or an outsider in ways similar to other characters marked by their race, gender, ethnicity or religion. In this way, Burton argues, Don is coded as a black man and much of his existential struggle can be explained through this lens. The "myth of whiteness" evident in Don's character parallels the myth of the non-racial narrative of the series.

In "The Erotic Reduction of Don Draper: Iconicity, Idolatry, and Madness," Carole Baker explains and utilizes Jean-Luc Marion's phenomenology of the erotic. Marion's philosophy offers a way to reflect upon Don's character, within both personal and professional contexts, in terms of "icon" and "idol." Baker interprets the title, *Mad Men*, as suggestive of a philosophy of "madness"—revealing the connections between idolatry and madness. We, as viewers, recognize the fictional nature of Draper's character even as we recognize the very real aspects of his character. The realism within this paradoxical man invites the viewer to moral judgment and reflection. Through application of elements of the erotic reduction to *Mad Men*'s leading man, Baker explores the paradoxical nature of the phenomenology of love.

Part 3 addresses topics relating to politics and social theory: marriage, parenting, political identity, and the role of religion in individual and public life. Part 3 begins with Jacob L. Goodson's "*Zou Bisou Bisou*: Feminist Theory and Sexual Ethics in *Mad Men*" as a transition from Don's character (Part 2) to the social impact of feminism on sexuality within marriage. Goodson navigates the sexual aspects of Don's marriages to both Betty and Megan. Goodson makes a case for viewing *Mad Men* through the philosophical lenses of Immanuel Kant's deontological sexual ethics, Simone de Beauvoir's existentialist-feminist theory of sex and sexuality, and Catherine MacKinnon's radical feminist sexual ethics. Doing so helps to reveal the empowerment experienced by Betty and Megan through the sexual aspects of their marriage to Don.

While Goodson merely suggests the significance of ordinary life for understanding marriage, Brandon Morgan and Jonathan Tran take the next step and establish the full significance of the connections between marriage and ordinary life. In their chapter, "*Exitus et Reditus* in Marriage: *Mad Men* vs. Hollywood Remarriage Comedies," Morgan and Tran look to the work of Stanley Cavell to make sense of the significance of ordinary life for sustaining marriage. As their sub-title suggests, they compare and contrast *Mad Men* with the genre of "Hollywood remarriage comedies"—a genre carefully studied by Cavell.

In their chapter, "Uneasy Bedfellows: On Pete and Trudy's Marriage," Matthew Emile Vaughan and Christopher J. Ashley guide the reader through Pete and Trudy's rocky relationship from the first through the seventh seasons of *Mad Men*. Pete Campbell comes from old money and feels entitled to his position and status. Although Trudy's character is less developed than Pete's, we know that she represents new money and she takes on the role of Pete's ethical surrogate throughout the series. Pete's desire for leadership and power carries over into his marriage with Trudy. Pete wants to be in charge, but Trudy wants to an equal partnership. Pete and Trudy's marriage encompasses both hardship and success. For viewers who consider themselves as part of the millennial generation, Pete and Trudy's marriage represents the ways of their grandparents' marriage. As a result of seeing the hardships of their grandparents' marriage, millennials now tend to wait until their late twenties or early thirties to get married and have kids.

Power and sex are two key themes that saturate *Mad Men*. This leaves children on the margins throughout the series. In her chapter, "Mad Men, Bad Parents: Representations of Parenting in Mad Men," Susan Frekko addresses the themes of parenting found in Betty's and Don's parenting styles. Frekko emphasizes both the adult-centric and neoliberal contexts in which Sally and Bobby are raised. Responsibility and self-regulation become the two highest values of neoliberal parenting. Betty fails to show responsibility and self-regulation through her addiction to cigarettes and vain concerns about her appearance. While Don tends to show some restraint when it comes to punishing the children, he has his own addictions to alcohol and sex—as well as his habitual tendencies to deceive and lie. Frekko argues that viewers see these failings and are invited to develop feelings of superiority to the characters on *Mad Men*. Gender equality, racial equality, and bad parenting are not yet behind us; society must continue to work for a positive change in all three of these areas.

In "'I Can't Believe that's the Way God Is': Peggy's Pre-Vatican II Catholicism," Heidi Schlumpf begins with Season 2 of *Mad Men*—in the midst of the Cuban Missile Crisis and the threat of the end of the world. Characters find their escape in multiple ways, but Peggy—in particular—goes to church. She finds herself in a pre-Vatican II Roman Catholic Church—which displays a traditional view of women, an emphasis on the sins of sex and sexuality, and a strict hierarchical structure. Peggy rebels against the Roman Catholic view of womanhood by taking birth control and having sex outside of marriage. At her job, Peggy encounters the secular culture and quickly realizes that both the secular view of womanhood and the Christian view of womanhood remain necessarily connected to female sexuality. Peggy obtains an illusion of respect from Gill, the young Catholic priest, when Gill asks for her advice on his sermons and asks for her help with a promotional flyer. Gill, however, refuses to fully support her decisions concerning the flyer. Eventually, Peggy stands up for herself and seemingly walks away from the Roman Catholic Church.

Jared D. Larson's chapter, "'We Don't Really Know What's Going On': *Mad Men* as a Bellwether of the Politics to Come," brings the collection to a strong conclusion. As the sub-title suggests, Larson connects the political problems of the 1960s to the current (2000s) political problems of the United States of America. Arguing that we never learn from our mistakes, Larson shows how American politics continues to struggle with issues of race, gender, capitalism, nationalism, and political identity. In the first two seasons of *Mad Men*, the threat of the Cuban Missile attack and nuclear war with the Soviets propelled many citizens and politicians into fear. This fear was paired with a polarization between America and foreign countries. The presidential campaign between Nixon and Kennedy established a new connection between Presidential Campaigns and advertising agencies. *Mad Men* also shows the level of privilege held by some people in society: in the 1960s, police officers favor white, wealthy, men; in the 2000s, this favoritism remains. Lastly, Larson connects politics and big business during the 1960s, the nascent years of the now omnipotent military-industrial complex. Not much has changed in our current government: America remains run by fear with strict polarization, certain privileges, and the end goal of making money. One possible solution to these problems, offered by Larson, concerns how more television series—in the model of *Mad Men*—ought to be produced allowing its audiences to reflect more deeply upon the past, present, and future.

This reflection upon the past, present, and future, hopefully manifests throughout this entire collection. While multiple volumes could be authored to address the internal plot twists and character development alone, our understandings of the show itself are only deepened by connections to happenings in our past and present reality. Whether in relation to gender, race, business ethics, personal morality, or marriage, our reactions to the seemingly outdated social mores and paradigms represented in this series reveal both how far we have come and how deeply entrenched our own existential struggles, moral quandaries and deep seated anxieties lie. This reflectivity makes *Mad Men* a classic for this generation and one that viewers watch again and again. We hope this collection will only enhance the joy, amusement, disgust, and reflection that experience entails.

BIBLIOGRAPHY

Martin Heidegger. "The Word of Nietzsche." In *The Question Concerning Technology and Other Essays*, translated by William Lovitt, 69. New York: Harper, 1982.

Part 1

BUSINESS ETHICS

Chapter 1

"IT'S THE REAL THING"

Identity and Sincerity in *Mad Men*

—HOWARD PICKETT

"EVERY GREAT AD TELLS A STORY"—NARRATIVE ETHICS AND *MAD MEN*

"Don't fool yourself. This is some very dirty business." Madison Avenue executive Roger Sterling's admonition to his younger colleague Pete Campbell comes on the heels of a decision to ask office manager Joan Holloway to prostitute herself in order to win a coveted account with British automaker Jaguar.[1] Although Sterling's words ostensibly condemn Campbell's plan, the ad exec's actions tell a different story. Immediately prior to his comment, Sterling has acquiesced (along with the other partners in the room) to the "dirty business" Campbell proposes. Furthermore, it is Sterling who prevents any further protest or change of plans, lodging his (apparent) complaint as he stands up to leave Campbell and partner Lane Pryce behind to make the arrangements with Joan. In light of these actions, Sterling's comment seems less like a condemnation and more like a cynical acceptance of the amoral, even immoral, nature of corporate survival. "This *is* some very dirty business," and, he implies, there is no avoiding dirty

1. "The Other Woman," S5/E10.

hands. "Don't fool yourself," he says; business opposes (even *precludes*) the other-oriented "moral point of view."[2]

To anyone familiar with Roger Sterling's perspective, "business ethics" may seem a dubious notion, if not a contradiction in terms. To anyone familiar with the AMC television series, an essay on business ethics in *Mad Men* may seem doubly dubious. Best-known for the worst tendencies of corporate America (e.g., the sexism, dishonesty, and greed on display in this episode among most others), Matthew Weiner's series seems, at first glance, incapable of cultivating ethical business practices. Upon closer investigation, however, *Mad Men*'s serial narrative may be uncommonly well suited to business ethics. The series is, admittedly, no treatise on advertising ethics; what it offers, though, may be more valuable than a treatise. It serially dramatizes the thoroughly human (i.e., the emotionally complex, temporal) human beings at the heart of business and ethics alike. With its attention to these thoroughly "realistic" (read: "messy") human subjects, this essay highlights questions the series raises *as a series*, questions about: (1) the discernibility and value of truth and sincerity; (2) the mutable, even malleable quality of desire, personal identity and moral character; and (3) the complicated intersections between office and world: i.e., the knotty relationships among the personal, the professional, and the political. In short, the essay traces the moral ramifications of the radically temporal (if not existential) view of the self at the heart of Weiner's series. In doing so, the essay also models an alternative strategy for business ethics.

As a quick glance at a number of standard textbooks and course syllabi reveals, business ethics typically focuses on a narrow set of practical decisions for managers.[3] It also typically assumes a narrow scope of moral attention and a narrow view of truth, personal identity, and autonomy. What *Mad Men* offers, in contrast, is a "critique by expansion," inserting into business ethics oft-neglected questions and perspectives. What *Mad Men* simultaneously offers is a "narrative ethics" approach to advertising and business. Its critique by expansion is due, not only to *what* the series presents, but also to *how* it presents it—namely, *as* a series. Broadly speaking, "narrative ethics" refers to any intersection between narrative and ethics, whether: (1) the use of ethics to illuminate narrative (e.g., considerations of moral responsibilities among authors, readers, and characters)[4] or (2) the use of narrative to illuminate ethics (e.g., the view that moral judgments depend on an agent's

2. Baier, *Moral Point of View*; for the suspicion about the unethical side of business and advertising especially, see Godin, *Marketers are Liars*.

3. For a history of the field, see Bowie, "Business Ethics."

4. Newton, *Narrative Ethics*.

role within a particular life-narrative).[5] Because what Weiner and team offer is not just a powerful story, but a powerful story about the power of stories, narrative ethics seems especially apt to *Mad Men*.

To be sure, with its reliance on case studies, business ethics has long been attuned to the indispensability of narrative and narrative ethics. The point of this essay, then, is not that *Mad Men* introduces narrative where once there was none. Rather, *Mad Men* emphasizes and extends narrative's place in practical thought. Unlike the typical business ethics casebook, *Mad Men* presents a detailed story that complicates the relationship between public and private, past and present, even appearance and reality. Most significantly (if ironically), Weiner's fictitious series replaces the unified, rational fiction of a self at the center of business ethics with "the real thing"—i.e., "mad men" and women better represented by narrative than theory.[6]

"WHERE THE TRUTH LIES"—TRUTH IN ADVERTISING

Unsurprisingly, a television series about the inner-workings of an ad agency has a great deal to say (and show) about the principal issue in advertising ethics: deception.[7] Tellingly, the series introduces main character Donald Draper in the midst of a moral dilemma. Creative director at a Madison Avenue ad agency, Don is looking for a new way to sell cigarettes now that the Federal Trade Commission (FTC) is cracking down on lies about the safety of smoking."[8] If Don's fictional agency is anything like the real agency described in Samm Sinclair Baker's *The Permissible Lie* (1968), then the culture "[i]nside the agency . . . is hardly conducive to truth telling"; instead, "the usual thinking in forming a campaign is first what can we say, true or not, that will sell the product best," and, only second, how can we avoid "censure by the FTC."[9] Faced with a similar office climate, Don initially tosses the

5. MacIntyre, *After Virtue*.

6. The hyper-rational, unified agent stands at the heart of much business ethics and economics alike. The close reading that follows takes its narrative ethics cues less from MacIntyre than from Nussbaum, *Love's Knowledge*.

7. "Most criticisms of advertising focus on the deceptive aspects of modern advertising." Velasquez, *Business Ethics*, 297. "Deceptive advertising" is a technical term of art with a distinct legal meaning. "A deceptive advertisement is one that involves representation, omission, or practice likely to mislead a reasonable consumer." Furthermore, to suffer legal consequences, the misleading advertisement must also have "material" (i.e. actual, detrimental) effects on the "consumer's purchasing decisions." *Sage Brief Guide*, 137.

8. "Smoke Gets in Your Eyes," S1/E1.

9. Baker, *Permissible Lie*, 16.

health report on smoking into his office wastebasket, while his boss, Roger Sterling, accuses the health advocates of their own deceptive "manipulation of the media." With this opening dilemma, the series highlights both the deception and the harm unethical advertising can do.

In lieu of simplistic moralizing, however, the show offers something more than a series-long indictment of the ad industry. By providing its main character the moral integrity lacked by many of his coworkers (e.g., unctuous Pete Campbell), *Mad Men* complicates our view of human character and the world of advertising. As we come to see, Don, whatever his faults, refuses to lie in his ads. Yet, having ruled out lying in the strict sense (and having nearly lost the cigarette account as a result), Don adopts another sales technique, one with questionable moral dimensions of its own. Don will, as he says elsewhere, "change the conversation."[10] In short, Don sells products by making true, yet distracting claims about them—in this case, that Lucky Strikes are "toasted."

Granted, one might still condemn Don and his ad—e.g., for selling a dangerous (albeit legal) product. By doing business that degrades and destroys lives, Don and Lucky Strike violate what Buddhist thinkers label "right livelihood."[11] If he has violated a moral duty here, however, it is not the duty of truth-telling. Despite his infamous opposition to lying (including to a murderer on your doorstep inquiring into the whereabouts of your loved one), Immanuel Kant (1724–1804) insists that one (Don included presumably) has a duty to avoid telling untruths; he does not have a corresponding duty to tell the whole truth. One has a duty to truthfulness, not candor.[12]

Nonetheless, one might wonder if Don and the managers from Lucky Strike have a duty to tell, if not the *whole* truth, then, at least, key aspects of it. Does Don's ad, despite the accuracy of its claims about Lucky Strikes, do wrong by omitting central claims—e.g., about cigarettes' fatal health consequences? In their desire to preserve customer autonomy, some critics of modern advertising would say yes. Drawing on the so-called "golden rule," David Holley insists, "As a practical guide, a salesperson might consider, 'What would I want to know, if I were considering buying this product?'"[13] If

10. "Love Among the Ruins," S3/E2. Peggy Olson later repeats the line during a campaign for Heinz Beans: "As I always say, if you don't like what they're saying, change the conversation" ("To Have and to Hold," S6/E3).

11. Saddhatissa, *Buddhist Ethics*, 107.

12. Mahon, "Kant on Lies," 114–15. The whole truth (i.e., everything that is true) never can be stated and, therefore, need not be revealed. If ought implies can, then *cannot* implies *need not*. Kant, "Supposed Right to Lie," 612.

13. Holley, "Moral Evaluation," 467. Like the golden rule, Holley's rule corrects

adopted, Holley's golden rule would presumably remind Don that customers want to know that a product is likely to kill them. Customer autonomy, in short, requires adequate information. In that case, the Lucky Strike ad seemingly deserves condemnation.

More precisely, if the salesperson alone is tasked with providing that adequate information, then the Lucky Strike ad violates the stipulation that "No advertisements should mislead by inaccuracy, ambiguity, exaggeration, *omission*, or otherwise."[14] However, *if* the salesperson alone must provide that information, then the job of the salesperson has become exceedingly different from—and more demanding than—what it had been before. This level of responsibility would require sellers to know every major detail about the market, and to inform buyers of those details, including the lower prices of competitors's products at other stores in town—or even online.[15] Yet, we need not assume that the salesperson alone is tasked with the responsibility of preserving customer autonomy. The key question, therefore, is not whether or not Don's ad omits valuable information (whether it says, "It's toasted—*and deadly*"). The key question is whether that omission is "likely to mislead consumers acting reasonably under the circumstances" (in the words of the FTC's Deception Policy Statement).[16] As the episode's opening scene makes clear, information about the health risks of smoking is readily available to the American public from other sources; as Don's waiter observes, "I love smoking [. . .] My wife hates it. *Reader's Digest* says it will kill you."

Nonetheless, Don knows advertising has its questionable moral aspects, acknowledged most memorably, perhaps, in an exchange he imagines with his deceased father.[17] As Don imagines that conversation (more accurately, hallucinates it), Don's father (and so, perhaps, Don himself) thinks advertising is: worthless ("You're a bum"); effeminate ("Look at your hands. They're as soft as a woman's"); unproductive ("What do you do? What do you make?"); and generally dishonest ("You grow bullshit"). As another character says to Don, "You make the lie," a not uncommon view of advertising, in fact.[18] According to one recent survey, "Seventy-four percent of

---

self-interest by asking agents to reverse positions with the other.

14. See the British Codes of Advertising and Sales Promotion, emphasis added. Qtd. in Spence and Van Heekeren, *Advertising Ethics*, 45.

15. If one were required to tell customers everything, ads would no longer be ads, but *Consumer Reports* articles.

16. Spence and Van Heekeren, *Advertising Ethics*, 45. Released in 1983, the statement, if not the sentiment, postdates Don's work on Lucky Strikes.

17. "Seven Twenty Three," S3/E7.

18. "The Hobo Code," S1/E8.

American consumers either 'strongly' or 'somewhat strongly' believe that 'most advertisements deliberately stretch the truth about the products they advertise.'"[19]

A primetime cable series, *Mad Men* is, needless to say, interested in more than a moral evaluation of the ad industry. Madison Avenue advertisers (the "mad men" of the show's title) occasion reflection on the moral complexity and inscrutability of humankind, not just ad-men. Don is, as viewers know, particularly complex and inscrutable. His integrity in the ad notwithstanding, Don is, in fact, a liar—most often covering over his marital infidelity. Resisting simplistic moralizing once again, *Mad Men* presents neither devil nor angel, but a complex, messy, actually existing human.[20] Furthermore, by giving Don more integrity at work than at home, the show maps a surprising fissure between the private and the public and raises intriguing questions in the process. Does Don's integrity in the boardroom, which he lacks in the bedroom, demonstrate America's obsession with work? Does it figure a modern desire to find identity in the office rather than the home?

By highlighting Don's inconsistent truthfulness, the show also raises questions about the consistency of human (moral) character. In light of Don's split moral identity (something more easily represented by narrative than theory), the series echoes a growing chorus of voices challenging the view that the cultivation of virtue is the centerpiece of ethics. By exposing the inconsistency and contextuality of character, *Mad Men* echoes the "situationist" critique of "virtue ethics."[21] As the label suggests, "situationists" hold that a situational stimulus has substantial effects on our moral decisions, in fact, more so than does character.[22] In a well-known (admittedly ironic) experiment, seminary students tasked with giving a talk on the Good Samaritan parable (the best-known New Testament story about helping neighbors in need) were more likely to stop and help someone in

19. Spence and Van Heekeren, *Advertising Ethics*, 44.

20. With its attention to self-deception, human finitude, and the ever-unfolding temporal (i.e., "serial") quality of human life, *Mad Men* echoes the existential thought of Soren Kierkegaard, Martin Heidegger, and Jean-Paul Sartre.

21. "Virtue ethics" is a normative ethical theory that emphasizes the moral character of agents over or, at least, in addition to their duties and the consequences of their isolated actions.

22. A psychological (not a moral) theory, situationism contends that situations *do* have this impact, not that they *should*. For information on the "fundamental attribution" of intention and character to behavior, see Ross, "Intuitive Psychologist"; and Gilbert and Malone, "Correspondence Bias." For a discussion of these psychological studies and business ethics, see Solomon, "Victims of Circumstances"; and Harman, "No Character or Personality."

distress (actually a confederate in the study) if they thought they had ample time before the talk (a seemingly trivial factor) than if they thought they were running late (63 percent compared with 10 percent).[23] Like *Mad Men*, situationism reminds us of the messy, unstable complexities of flesh-and-blood human beings. Insofar as they aim to be practical, business and advertising ethics ought to keep this messiness in mind.

Mad Men also raises questions about truth by underscoring the finitude of human perception and our vulnerability to deception. Not only is character inconsistent; moral judgment is, too. For that matter, not only are the characters on the show deceived (e.g., by their cheating partners); we, too, are deceived, especially by means of the series' many surprising twists. Thanks to its serial quality, *Mad Men* not only *states* (as philosophical essays might) but also *performs* this vulnerability to deception. The series opener, to take a prominent case, concludes with the surprising revelation that Don is married with kids. Given Don's involvement with Midge (with whom he spends the night in the first half of the episode) and Rachel (with whom he has a flirtatious dinner date), not to mention given the absence of his wedding ring (the "Band of Gold" about which Don Cherry sings in the opening scene), a first-time viewer might rightly wonder where Don is, let alone who Betty Draper is, when Don enters her suburban home and bedroom in the final minutes of the first episode.[24] Only in retrospect (so crucial in *Mad Men*) do we learn that the ad-man, who refuses to lie in the profession known for lying, lies to those closest to him and also to the audience.

To be fair, like virtue ethics, the show is, in the end, less interested in what we do than in who we are. In short, it is less interested in the lies we tell than in the lies we are. Don is lying, not only about his fidelity, but also about his identity. As we learn by the end of Season 1, Don is not Don, but Dick; at least, he used to be Dick. Like Ragged Dick (the title character from Horatio Alger's story of another up-and-coming New Yorker), Dick Whitman (alias Donald Draper) rose from rags to, if not riches, then middle-class respectability. Like Ragged Dick, *Mad Men*'s main character also embodies the "American Dream." Like another fictional antecedent, Jimmy Gatz (F. Scott Fitzgerald's Jay Gatsby), Dick Whitman embodies that dream precisely because of his creative, even deceptive, techniques of self-fashioning.[25] It is his morally ambivalent talent for storytelling that makes

23. For an introduction to situationism, see Upton, "Situationism Debate."

24. As Ann Duncan will show, in Chapter 4, this scene also communicates much about Betty Draper's façade.

25. Through similar main characters, both *The Great Gatsby* and *Mad Men* imply that the American Dream's obsession with transformational upward mobility stands in tension with the value of character consistency. Like Gatsby, Don depends on his

him successful. As anyone familiar with the show knows, Don Draper is a master ad-man; however, as anyone familiar with the show also knows, Don is his own best ad. For good and for bad, he is a man who can tell a "story" (i.e., narrative or fib).

Don is, of course, not the only one whose identity is not what it seems. Don may be an actor, but so is his second wife Megan (quite literally in her case). Her roles as French twins on a television soap opera parallel Don's own multiple identities.[26] Salvatore Romano, Don's closeted gay colleague, has also been disguised for seasons, lying *about* (if not also *to*) himself.[27] After learning of Sal's affair with a male bellhop, Don, an expert at identity manipulation himself, offers Sal the same tagline he offers London Fog raincoats: "Limit your exposure."[28] Notably, Peggy Olson also has her share of secrets, including the baby she abandons at the end of Season 1.[29] Together, these secrets drive home the inaccessibility of others' insides (quite literally in Peggy's case). After all, if Peggy can be pregnant for nine months without anyone's knowledge (including her own), what other surprises lurk beneath the surface?

Through both the characters it represents and the way it represents them (namely, *serially*), *Mad Men* unsettles our usual confidence in the transparency or legibility of character. As the series would have us remember: People are not always what they seem.[30] That observation has ramifications not only for anthropology and epistemology, but also for ethics. How valuable could sincerity (i.e., "the congruence between avowal and actual feeling," between outsides and insides) be in a world in which it cannot be confirmed?[31]

How valuable could sincerity be in a world in which one cannot confirm it even in oneself? The show recognizes better than most the formidable

---

outward appearance in his act of "self-fashioning," insinuated by his new name, "Draper" (i.e., a cloth-merchant and haberdasher). See Greenblatt.

26. Cf. "All the world's a stage and all the men and women merely players." William Shakespeare, *As You Like It*, II.vii.

27. Only in retrospect is Sal's expressed interest in women known to have been a lie.

28. "Out of Town," S3/E1.

29. "The Wheel," S1/E13.

30. Not only is there a Dick Whitman we never knew; there is also an earlier Don we never knew. An initially simple dichotomy between Dick and Don turns into a complex of characters, including multiple Dons. In the years following the war in Korea, Don worked as a used car salesman, the archetypal marketplace liar. See John Steinbeck, *Grapes of Wrath* (Chapter 12).

31. Trilling, *Sincerity and Authenticity*, 2. One and the same realization (namely, that people are opaque and deceptive) motivates both exhortations to, and skepticism about, sincerity.

human capacity for self-deception.[32] In Peggy's Season 2 flashback, Don's advice, whether real (the result of Don's actual hospital visit) or imagined (the result of Peggy's medicated state), is to block out traumatic events from her mind in order to "move forward."[33] Referring to Peggy's baby, his own past, and even this visit to the hospital ward, Don (again, whether real or imagined) tells Peggy: "This never happened. It will shock you how much it never happened."[34]

At its most provocative, then, the show insinuates that the truth about one's identity may be mutable. Confronted by Betty's discovery of his past life, Don insists that he is not Dick, but Don. When Betty asks, "What's your name?" Don replies, with neither irony nor evasion, "Don Draper."[35] While he admits that it "used to be Dick Whitman," from his own point of view (not just ours or Betty's), Don is Don. His conversion from Dick to Don has been so complete that it makes as much sense to call him Dick as it makes to call St. Paul by his pre-conversion name, Saul of Tarsus. From the looks of it, Don believes in the mutability of the self, if not also the truth. Given our temporal existence, humans possess the power of habituation, i.e., the power to transform ourselves through temporary pretension and habit into something we are not or, at least, *were* not.[36] Unsurprisingly, moral judgments about such pretension (a word with pejorative connotations) and inauthenticity (an equally pejorative term) are usually negative.

More surprisingly, though, those judgments are sometimes positive. To take what may be the most surprising case, Immanuel Kant (again, known for his demanding prohibition against lying, including to the murderer on the doorstep) elsewhere defends pretension and habituation. Having acknowledged that the "more civilized human beings are, the more they are actors," Kant adds:

32. Though paradoxical, lying to oneself is not just possible, but common. For an early discussion, see Butler, "Self-deceit."

33. This obsession with forward movement, characteristic of both Don and Peggy, is also characteristic of the race from the post-war years to the present that serves as the backdrop of the series. In the words of Henry Ford, another icon of the American Dream: "History is more or less bunk [. . .] the only history that is worth a tinker's damn is the history that we make today" (Chicago Tribune 1916).

34. "The New Girl," S2/E5.

35. "The Gypsy and the Hobo," S3/E11.

36. Arguably, the truth about Don has not changed, but what people call the "truth" is rooted in a gross oversimplification of human existence. In effect, what seem to be Don's struggles with sincerity often result from the fact that the human (due to its temporal existence and its complex, sometimes contradictory, psychology) is much more complex than exhortations to sincerity allow.

> And it is also very good that this happens in the world. For when human beings play these roles, eventually the virtues, whose illusion than have merely affected for a considerable length of time, will gradually really be aroused and merge into the disposition.[37]

In this discussion of "permissible moral illusion" (his term), Kant recommends nothing less than "virtuous theatricality" or, in another critic's words, "hypocrisy upward" (from the Greek *hypocrites*, "stage actor").[38] Kant realizes that, despite the demands of ideal moral theory, actual human beings sometimes need to act their way to virtue.[39] Given both human imperfection and temporality, most of us need to "fake it"—at least, in order to "make it." From another angle, given the imperfection of others (e.g., their bigotry), we should be allowed to pretend to be something we are not. Given American anti-Semitism, comedian Jimmy Barrett (née Brownstein) seems justified in using a stage-name, not his own. Furthermore, theatrical hypocrisy upward is the American way. As wife and manager Bobbie Barrett puts the point, "This is America. Pick a job and then become the person who does it."[40]

To complicate matters, though, the effectiveness of habituation and human theatricality (i.e., imitation, self-fashioning)—e.g., the success of Don's conversion from the awkward, inarticulate poor boy of the Midwest to the smooth-talking ad-man of Manhattan—is questionable. The complexity of human identity shows itself, not only in allowing transformation over time, but also in allowing slippage back and forth between earlier and later identities. When she confronts her husband with his secret past, Betty is taken aback by Don's silence and awkwardness.[41] As Don clumsily drops his cigarette on the floor, Betty looks on in dismay; her husband is a man who never fumbles—or so she thought. Yet, to our eyes, the dropped cigarette is reminiscent of Dick's awkwardness—in particular, his dropped cigarette lighter in Korea, which led to the first Don Draper's death and Dick's

37. Kant, *Anthropology*, 263.

38. Booth, *Company*, 254.

39. In Booth's words, "Many of the virtues that we most honor are originally gained by practices that our enemies might call faking." Booth, *Company*, 253.

40. "The New Girl," S2/E5. By adopting a theatrical persona, Jimmy (like Don) ambiguously embodies both the possibility and the duplicity of the American Dream. Of course, the show acknowledges the crucial role theatricality or social imitation plays in all human lives, regardless of socio-political context or oppression. To Pete's question—"What does one do? [. . .] What am I supposed to do?"—i.e., after the death of a loved one—Don replies, "What people do. Go home" ("Flight 1," S2/E2).

41. "The Gypsy and the Hobo," S3/E11.

rebirth as Don.[42] By confronting Don with his past, Betty has revived Dick Whitman, as much to her horror as his. When Sally and Bobby disguise themselves for Halloween as gypsy and hobo later in the same episode, we are reminded of Don's disadvantage and rootlessness, but also his theatricality. In the final scene, a neighbor, having successfully identified the kids' Halloween costumes, looks to Don and asks: "And who are you supposed to be?"

Additionally, the series raises pressing questions not only about truth's hiddenness and malleability, but also about its value. As a minor character remarks, "Your mistake is that you're assuming that, because something is true, that it is good."[43] Given our imperfections—again, given the pain, even injustice that might result from full disclosure—how desirable is truth about oneself (i.e., "sincerity")? Confronted by Betty about his hidden identity, Don asks: "What difference does it make?" What if the present self is better than the past self? For that matter, what if the "fake" self is better than the "real" self? In light of Don's past traumatic and abusive childhood or Sal Romano's vulnerability as a gay man in mid-century America, is sincerity always a virtue to cultivate? Or might it be a luxury prized by the privileged, by those who risk less through self-disclosure and by those who wish to keep the Dons and Sals of the world out of power and prominence? Given the disunified character of the self (cf. people's complex, contradictory feelings), is it even possible to be sincere? What would it mean for Don to be sincere—how would he externalize both Don and Dick?

Like others before him (including Kant), Weiner is ambivalent about sincerity and truthfulness. When Pete Campbell exposes Don as Dick (i.e., as a liar, a fraud, and a criminal), boss Bert Cooper surprises both Don and Pete by responding, "Who cares?"[44] Bert's indifference to Don's past—and to sincerity, more pointedly—is borne by three admissions: 1) the ubiquity of human (and especially American) wrong-doing ("This country was built and run by men with worse stories"); 2) the socially constructed quality of character ("A man is whatever room he is in, and right now Donald Draper is in this room"); and 3) the priority of sales over sincerity ("I assure you, there is more profit in forgetting this [. . .] I'd put your energy into bringing in accounts").

Lest we grow too comfortable with hypocrisy upward, however, the show reminds us of the darker side of "hypocrisy downward." As we

42. "Nixon vs. Kennedy," S1/E12.

43. "Far Away Places," S5/E5. Regardless of its source (in fact, a young Timothy Leary, the LSD advocate), the statement bears further consideration—or so the series implies.

44. "Nixon vs. Kennedy," S1/E12.

discover (again, only in retrospect), Don was not the only one disguised in Bert's office. Faced with the opportunity to score an account with hotel magnate Conrad Hilton, Bert uses what he knows about his past to blackmail Don into signing a contract with Sterling Cooper.[45] There was, indeed, "more *profit* in forgetting" Don's little secret, at least until later. Once again, *Mad Men*'s seriality exposes the impermanence and imperfection of our moral judgments; Bert's apparent magnanimity about the problems with sincerity was arguably just insincere self-interest. Ironically, it may take a fictional narrative (about the sometimes fictional narratives of others) to keep us honest about the lives affected by our moral endorsement of truth and sincerity.

As the series would have us know, the truth may "set you free," but being free is not always simple or painless.[46] Due to his commitment to truthfulness in the advertising pitch—more exactly, due to his excessive sincerity, his imprudent candor—Don eventually loses his job in advertising. Following an uncharacteristically dishonest pitch to Hershey (a sentimental story about Don, his loving father, and their trips to the corner candy store), Don decides to come clean. At the end of his pitch, Don concludes, "That's the story we're going to tell"—i.e., that's the *lie* we're going to tell. Committed as he is to the integrity of storytelling, Don eventually confesses his own true story.[47] As a boy, Dick received his Hershey bar from, not a loving father, but an indifferent prostitute, who rewarded him for stealing change from the johns who frequented his stepmom's whorehouse. For him, the chocolate bar is reminiscent of, not love, but pretense and isolation. Eating his candy bar "alone, with great ceremony, feeling like a normal kid [. . .]," Don (or rather Dick) experienced "the only sweet thing in my life." That "sweet thing" has now somehow called him back to himself. The authenticity the bar embodies ("The wrapper looked like what was inside") has challenged him to be sincere, to make outsides and insides, as well as past and present, align. Thanks to the nostalgia it invokes, the bar has transformed Don back into Dick. However, it has, in the process, also called him back to the problems he harbors with advertising. Echoing longstanding complaints about the superfluity of advertising, Don tells the Hershey execs, "If I had my way, you would never advertise. You shouldn't have someone like me telling that boy what a Hershey bar is. He already knows."[48]

45. "Seven Twenty Three," S3/E7.

46. John 8:32.

47. "The Quality of Mercy," S6/E12.

48. "Some economists also charge that advertising is wasteful and inefficient. In their view, advertising is largely inefficient and nonproductive activity." Boatwright, *Ethics and Conduct*, 261.

With that comment, Don has become too candid—perhaps, too honest—for advertising. With that, he has also rejected the foundation on which the trade depends—namely, that people need to be told what to want. Having returned to his identity as Dick, he no longer belongs among his colleagues; knowing it, they send him packing.[49] Ironically, fellow partner Jim Cutler observes that Don's disturbing disclosure is the "kind of theater that makes our work so different." The sentimental lie (about the hair-tussling dad and perfect suburban life) sounds like the truth; the horrific truth (that Don watched his abusive father die and was raised by a stepmom who hated him) couldn't be anything but a theatrical lie. Truth and falsehood have become all mixed up, as hard to disentangle as the show's ambiguous tagline: "*Mad Men:* Where the truth lies [i.e., resides or misrepresents]."

"IT'S THE REAL THING"—ADVERTISING DESIRE

Continuing its critique (of business ethics) by expansion, and its inquiry into the complexities of human identity, the series questions identity's relation, not only to truth, but also to desire. If it seems no coincidence that Don Draper is both master liar and master ad-man, it should seem no coincidence that he is also a master of (if not mastered by) desire, as his infidelities attest.[50] Underscoring the close connections between advertising and desire, Don remarks in the first episode, "Advertising is based on one thing: Happiness." Happiness is, as he says, "a billboard on the side of the road that screams with reassurance that whatever you're doing, it's okay. You are okay."

In the world of *Mad Men*, however, the desire at work in a good ad campaign is not just the desire for happiness (or freedom from fear), but sexual desire, more narrowly. Again, it seems no coincidence that Don is both ad-man and adulterer, a master in the art of seducing buyers and lovers alike. Informed by his own insatiability, Don understands the "itch" of the "new" that drives the ad market.[51] As the son of a prostitute and the stepson of a madam, Don (or, better, Dick) can hardly remember a time when sex

49. Technically, Don, a firm partner, is not fired, but put on indefinite leave.
50. Arguably, Don is not so much master *of* desire, as mastered *by* it.
51. "The Wheel," S1/E13.

and sales were not bound together.[52] He knows more than most about, in the words of his Hershey pitch, "the currency of affection."[53]

That complex interaction between selling and seduction comes most clearly to the fore in the campaign for Jaguar.[54] Don's pitch to the British carmakers offers the closest the series gets to a treatise on desire. Although the campaign makes the usual, predictable comparisons between cars and women, Don's commentary on desire reaches beyond the usual clichés. As he says, "deep beauty" creates desire, not because it is beautiful or titillating, but because it is unattainable, transcendent (i.e., lying outside our reach). Against the modern belief "that function is all that matters"—i.e., against the modern obsession with utility—"we have a natural longing for this other thing." Don knows from personal experience that the itch will never cease. As the show's narrativity and seriality reveal (i.e., through Don's insatiability), the worldly pleasures we want are unsatisfying stand-ins for something beyond us. In that respect, both Don and the show also realize that desire is always, in philosopher Emmanuel Levinas's words, "an unquenchable desire for the Infinite."[55]

Disappointingly, perhaps, Don uses what he knows about insatiable desire—initially at least—to make the next sale: "Jaguar: At last, something beautiful you can truly own." This association between commerce and (sexual) desire is further emphasized (and condemned) by cross-cutting Don's visit to the Jaguar boardroom with Joan's visit to a Jaguar executive's hotel bedroom. Again, as the cross-cutting visually intimates, advertisers are involved in something akin to prostitution.

To the complaint that sexual ads are indecent, corrupting, or even stereotyping (i.e., constraining a group's liberty within society and even self-perception through simplistic portrayal), one might also complain that, by

52. Don is certainly not the only one to equate selling and seduction. Japanese woodblock artist Hokusai's "The Dream of the Fisherman's Wife," which portrays two octopi ravishing a naked woman and which hangs in the agency's top office, reminds Bert Cooper of "our business" ("Out of Town," S3/E1). To write a successful ad, Bert explains, one must successfully orchestrate the ecstasy of the other. Consequently, there is something both sexual and theatrical about advertising. The point is reiterated near series end when Peggy makes her dramatic entrance into McCann Erickson, complete with male strut, cigarette, sunglasses, and erotic Hokusai print (once belonging to her late boss) in hand ("Lost Horizon," S7/E12).

53. "The Quality of Mercy," S6/E12.

54. "The Other Woman," S5/E11.

55. When he writes of an "unquenchable Desire for the Infinite," Levinas signals the fact that our desires aim at something that necessarily transcends and eludes our grasp. Levinas, *Totality and Infinity*, 150. For a discussion of Levinas's understanding of desire, see Ferreira, "Misfortune."

virtue of their manipulative aspects, ads (especially sexual ads) violate personal autonomy. According to one critic, autonomy-threatening "behaviour control [. . .] occurs when a person causes another person to act for reasons which the other person could not accept as good or justifiable reasons for the action."[56] If we assume that sex ads are causally efficacious (and if they aren't, one wonders why companies spend so much on them)—*and* if we assume that very few people would welcome sexual persuasion—then such ads do violate autonomy. (How many people wish to be persuaded to buy a car because of their sexual frustration? At the very least, how likely is it that *everyone* in a commercial's mass media audience welcomes such persuasion?) Given this demanding view, the only permissible ads might be strictly informational, rational appeals characteristic of the classroom, perhaps, but not the TV.

The same complaint—i.e., that ads override autonomy—is sometimes also made about ads insofar as they create new desires. Advertising works, not just because humans have desires, but also because ads can create new desires. As Don says, one of the "most important idea[s] in advertising is '*new!*'" It "*creates* an itch."[57] While Don may refuse to convey false information to trick consumers, his is not the business of sharing information for the satisfaction of pre-existing desires. Like most ads today, Don's ads have little, if anything, to do with product information. (One study found that over half of all television ads contained no relevant product information for consumers.[58]) Like most memorable ads today, the ads at Sterling Cooper Draper Pryce offer, not information, but a vision of the good life (i.e., in lieu of empirical commentary about what *is*, normative commentary about what *ought to be*.) In Don's words, "Every great ad tells a story," but the "story" in question is less a "lie" than it is a "myth" (from the Greek for "story").

Granted, one might complain that marketers disrespect personal autonomy and identity by stimulating desires agents do not already have. Don is to be distrusted not only because, "You make the lie," but also because, in the same character's words, "You invent want."[59] In 1958 (a year before the series begins), John Kenneth Galbraith lodged a similar complaint about the "dependence effect," the paradoxical, even perverse, nature of market production and advertising.[60] Although markets and manufacturers ostensibly exist in order to satisfy desires, "[p]roduction only fills a void that it

56. Crisp, "Persuasive Advertising," 453.
57. "The Wheel," S1/E13, emphasis added.
58. Velasquez, *Business Ethics*, 290–91.
59. "The Hobo Code," S1/E8.
60. Galbraith, *Affluent Society*, 126.

has itself created."[61] As a result, capitalist production chases its own tail; desire will never be satisfied so long as the techniques used to satisfy it only stimulate it further.

Unfortunately, what Galbraith's complaint or, at least, less sophisticated versions of it fail to acknowledge is the good that comes from stimulating new desires. Certainly one might argue that productive markets raise living standards. In a different vein, however, one might argue that productive markets facilitate new identities and new conceptions of the good. *Mad Men* implies that there may be no specific desire—and, so, no thick personal identity—prior to its production (including through advertising).[62] Don implies that certain central aspects of (modern) identity are themselves the creation of the advertising world, for instance, when he tells unmarried Rachel Menken that the "reason you haven't felt [true love for someone] is because it doesn't exist. What you call love was invented by guys like me to sell nylons."[63]

What Don's comment indicates is that advertising (like similar forms of storytelling) may be necessary for us to have recognizable identities at all. As *Mad Men* repeatedly demonstrates, a master ad-man like Don is as moving a storyteller as any dramatist or poet. In Theodore Levitt's words, "while the ad man and designer seek only to convert the audience to their commercial custom, Michelangelo sought to convert its soul. Which is the greater blasphemy?"[64] *Mad Men* suggests, to the contrary, that the modern advertiser might provide something remarkably akin to the normative vision of self and world found in religious traditions. Like Galbraith's advertising, religion also has a "dependence effect" on the human.[65]

This view of advertising and desire further challenges our usual confidence in the virtue of sincerity. If the previous section asked: Is it necessarily good to externalize a thoroughly imperfect self in an imperfect world?; then this section asks: Are we even able to externalize the inconsistent, insubstantial self? If the self is a bundle of inconsistent, fickle responses to situations,

61. Galbraith, *Affluent Society*, 127.

62. There may be a basic desire for happiness, for reassurance, even for sex, but the thin, aimless identity such vague impulses connote would hardly constitute a *particular* human identity.

63. "Smoke Gets in Your Eyes," S1/E1.

64. Levitt, "Advertising and its Adversaries," 421. Connecting the normative, the artistic, and the sexual, Levitt writes: "Like advertising, poetry's purpose is to influence an audience; to affect its perceptions and sensibilities [. . .] Poetry's intent is to convince and *seduce*," emphasis added.

65. German theologian Friedrich Schleiermacher is the most famous thinker to define religion as the "feeling of absolute dependence." Schleiermacher, *Christian Faith*, proposition 4.

then how would one ever be self-congruent? If the self is insubstantial (e.g., if, in Sartrean terms, "existence precedes essence"), then how could one be true to oneself? In both cases (the inconsistent self and the insubstantial self), there is no *there* there; at least, there is no externalizable self there.

While the associations between sexual and consumerist desire are relatively easy to come by (here and elsewhere), *Mad Men* makes a bolder, if more ambivalent, claim by equating consumerist desire with religious communion. Most obvious are the religious echoes in Peggy Olson's popsicle ad.[66] Picking up Sal Romano's offhand remark about the way his mother used to break the twin pops and give them to the kids "like Jesus at the Last Supper," Peggy highlights the popsicle's place in a central family "ritual." In the words of her tagline for the product, "Take it. Break it. Share it. Love it," something Peggy knows could also be said of the consecrated bread of communion: "While they were eating, Jesus took bread, and when he had given thanks, he broke it and gave it to his disciples, saying, 'Take and eat; this is my body'" (Matt 26:26).

Given her upbringing by an Irish Catholic mother, Peggy also knows, as she says in the same episode, that "the Catholic Church knows how to sell things." Read one way, the line implies that religion, like advertising, is in the self-serving business of convincing people to want what they neither want nor need. In effect, the association criticizes religion, evacuating it of its importance and power. However, the association might just as easily raise advertising up. Read another way, Peggy's comment implies that advertising is the new religion, offering people a meaningful vision of the good.

Although his imagination usually gravitates towards familial desire (cf. his ads for the Kodak Carousel, the Hershey bar, Mohawk Airlines), Don turns in his final ads towards the religious. His ad for the Royal Hawaiian Hotel combines the morbid and the mystical.[67] A businessman's suit sprawls on the edge of the waterfront above the words: "Hawaii. The jumping off point." (Is it an ad for a resort or for spiritual release, or, for that matter, for suicide?) However, the series' final ad is, perhaps, most suggestive of Don's spiritual quest.[68] With everything gone—job, wife, home, clothes, kids—Don finds himself (like the businessman of the Royal Hawaiian ad) overlooking the coast, stripped of everything he owns, everything he is. Having made his telephone confession of sin to Peggy—Peggy: "What

66. "The Mountain King," S2/E13. On Peggy's Catholicism, see also Heidi Schlumpf's chapter 16.

67. "The Doorway," S6/E1.

68. Granted, Peggy's popsicle ad admits there is often no hard and fast line between the familial and the religious; however, the religious aspects of Don's main ads are all but eclipsed by domestic aspects.

did you ever do that was so bad?"—Don attends group therapy, where he tearfully embraces another man, whose confessions of being unloved and unloving resonate with Don's own story.[69]

In the series' final scene, Don's ambiguous final vision relates communion and the commercial (quite literally, in the latter case). While his meditation instructor speaks of "the new day," of "new hope," of "the lives we've led" and "the lives we've yet to lead," of "new ideas," and (what Don has wanted most of all) a "new you," Don chants with eyes closed until a smile eventually steals across his face. In light of the show's many twists, viewers rightly wonder: What has finally brought Don peace? Ironically, what appears on the immediate heels of Don's smile is the 1971 hilltop Coke ad (aka "I'd like to teach the world to sing")—an ad that has been called, including by series creator Matthew Weiner, the "best ad ever made."[70]

Whatever else it may be, that ad is also religious—one of the most religious ads ever made. The "apple trees and honey bees and snow-white turtle doves," not to mention the hilltop vision of the good, are reminiscent of the biblical garden (Gen 2:16) and Moses's view of the promised land of milk and honey (Deut 26:9), mentioned in the title of the previous episode.[71] Whether or not they resemble the singing members of the "new Jerusalem" (Rev 14:3), the singers in the ad figure a harmonious "heavenly" place. If so, then, the final scene suggests that the ad-man is less profiteer than prophet.

Far from other-worldly, though, the ad combines the religious with the familial and the political as well. Certainly Don's preceding confession and meditation provide a more-or-less religious context in which to read the ad. However, what the ad imagines is a longing for the domestic tranquility young Dick never quite had ("I'd like to buy the world a home and furnish it with love"). Yet, because that home is intended for the whole "world," the ad transcends (even while it incorporates) the domestic sphere, moving to the socio-political dimension underscored by the (ostensibly) global cast of the commercial. In effect, the song imagines, not just a happy home, but also a happy world of diversity and peace (i.e., "harmony"), free of war and poverty and homelessness. More than that, it also imagines a world beyond this world or, at least, one more perfect than our current world. In short, the ad simultaneously evokes the domestic, the political, and the paradisiacal, even the eschatological. In this culminating scene, the show seems intent on reminding us that, for good or bad, ads are myths, promoting, not just sodas, but soteriologies.

69. "Person to Person," S7/E14.

70. Weiner said so just a few days after the series finale. Egner, "Coke Ad."

71. "The Milk and Honey Route," S7/E13.

As a culmination to the series, the ad draws together the series' various strands, and in doing so draws together various strands of this analysis as well. It not only offers meditations on identity, desire, and truth; it does so in a fittingly ambiguous way. The hilltop ad was part of a larger Coke campaign, "It's the Real Thing," a campaign line that recurs in the central part of the song. In light of *Mad Men*'s series-long meditation on truth and reality, one might understandably wonder just how to make sense of an ad for "the real thing"? Is it real—that is, is it truthful or sincere? In a certain sense, it *is* "the real thing." The footage we see at the end of this historical tour through America's recent past is the same footage many of us actually did see on television in the early 1970s.[72] Furthermore, it advertises real, authentic Coca-Cola, not some generic cola, and does so through a real ad, not some *Mad Men* campaign made to look old. As a result, the television has become (to borrow Don's phrase for the Kodak slide-projector the Carousel) a "time machine."[73]

In another sense, though, the ad is anything but "the real thing." It is, after all, an ad and one within a fictional television series, at that. Is it "the real thing" or is it yet another inauthentic, borderline deceptive, seduction? Worse, is it an inauthentic appeal to authenticity for the purpose of selling something filled with artificial flavors? Has Don finally gotten real or has he simply found a new way to sell a product? Given the twists of the series, has Don finally found release or will he, instead, return to morally suspect Madison Avenue—this time, to perpetuate the next great health crisis among millions of Americans, not cancer from tobacco, but Type II diabetes from sugary soft drinks? The ambiguity of the ad is characteristic of the series as a whole and of Don, in particular. Assuming he realizes anything in the end, it seems likely that he realizes something about who he is and ought to be. As actor Jon Hamm sees it: "The next day, he wakes up in this beautiful place, and has this serene moment of understanding, and realizes who he is. And who he is, is an advertising man."[74] What might it mean, though, to be an advertising man—to be, in all sincerity, a person marked by insincerity, a person whose calling is to channel the desires (which is to say, the *identities*) of others?[75]

72. Some may applaud Coca-Cola for receiving no money for the show's use of the ad. (The Beatles, in contrast, received $250,000 for "Tomorrow Never Knows" in Season 5.) Steinberg, "'Mad Men Finale." However, neither did Coke pay the show for what was, in effect, the most well-placed, well-integrated product placement spot in television history.

73. "The Wheel," S1/E13.

74. Itzkoff, "Jon Hamm Talks."

75. See Miller, *Death of a Salesman* (Act II). Main character, Willy Loman (also a

Again, it seems no coincidence that the man with the least identity here is also the man with the greatest genius for advertising. Negatively, one could say Don Draper has no character (moral or otherwise). Positively, he is, in Ralph Waldo Emerson's terms, the "transparent eyeball," with which another ad-man compares Peggy.[76] Like Emerson, Don might admit: "I become a transparent eyeball; I am nothing; I see all; the currents of the Universal Being circulate through me."[77] More accurately, whereas Emerson channels the powers of nature in solitude, Don channels the bustling masses of the city streets. Consequently, Don Draper (or, better, Dick Whitman) echoes Emerson's contemporary, poet Walt Whitman (whose "O Captain, My Captain" is selected for Bert's eulogy).[78] Like the earlier Whitman, *Mad Men*'s main character is the quintessential Manhattanite, obsessed by the "blab of the pave" and the "City of Orgies."[79] Whatever its complicated assessment of Don's character, the show (itself a story not unlike the stories Don tells) reminds us that the ability to imagine and tell others' stories has value. However much it may offend the modern sensibility, the show challenges our obsession with self-congruence (i.e., sincerity, authenticity). In effect, the series asks: Given human imperfection, insubstantiality, even temporality or seriality, might things go better if more of us were incongruent or, at least, paradoxically honest about our incongruence?

"THE GYPSY AND THE HOBO"—A CONCLUSION ON THE POVERTY OF BUSINESS ETHICS

In addition to the narrow view of truth and identity often assumed in business and advertising ethics, *Mad Men*'s critique by expansion also opens up the narrow scope of the field's moral attention. The show's attention to both home and office (not to mention the blurring of the boundaries between the two) raises questions about the relationship between our personal and professional lives. Of course, it raises questions about the appropriateness of mixing the two (e.g., in the case of intra-office affairs). More interestingly, though, it also raises questions about the inappropriateness of *not*

salesman in a show about the American Dream), is similarly insubstantial: "You don't understand: Willy was a salesman. And for a salesman, there's no rock bottom to the life."

76. "Christmas Waltz," S5/E10.

77. Emerson, "Nature," 52.

78. "Waterloo," S7/E7.

79. Whitman, *Leaves of Grass*. "The blab of the pave" appears in "Song of Myself" (Section 8); "City of Orgies" is the title of a short poem within *Leaves of Grass*.

mixing the two. Both Don and family lack integrity or unity due to an overpronounced rift between what goes on with Don at work and what goes on with Betty and the kids at home.

Opening up a business ethics textbook after an evening with *Mad Men,* a reader may be struck by how much business—and how little home—is included in it. Ironically, Weiner populates his world with multidimensional, albeit fictional, characters, while business ethics characteristically populates its world with flat, albeit actual, people. In the former, we cut back and forth from boardroom to bedroom, from office to kitchen; in the latter, the manager's responsibilities rarely extend to anyone outside the work environment. Although one might argue that such responsibilities fall outside the realm of business ethics, what *Mad Men* reveals is that Don's work impacts his family, perhaps above all, and vice versa. Furthermore, although one might also argue that familial concerns are self-centered—and, therefore, not in keeping with the "moral point of view"—*Mad Men* reminds us that concern for others at home is, nevertheless, a concern for *others.* Notwithstanding the popularity of "the stakeholder theory of management" (i.e., the view that businesses should consider everyone they affect), business ethics textbooks still pay scant attention to the way managerial responsibilities impact family responsibilities.[80]

The show also offers a complicating look at how professional decisions are often entangled with one's personal circumstances. Indeed, the show reveals that even ostensibly ethical decisions may, by the light of personal circumstances, seem morally ambiguous, if not worse. The Jaguar ad, for instance, appeals to Don in the end because of his own frustrations over Megan's burgeoning career and independence.[81] Don's full-page *New York Times* tirade against unscrupulous tobacco companies, and the ad agencies that represent them, thinly veils his grief over former lover Midge's heroin addiction—as well as his own sour grapes over losing the Lucky Strike account.[82] As that latter case reveals, an apparently upright, uncalculated statement of principle turns out to be, by another light, a personally motivated, thoroughly calculated full-page ad that serves the self-interest of Sterling Cooper Draper Pryce.[83]

80. Freeman, *Stakeholder Theory.*

81. "The Other Woman," S5/E11. For more on Megan's career trajectory, see Kristen Deede Johnson's chapter 5.

82. "Blowing Smoke," S4/E12.

83. The line between personal and professional, familial and commercial is blurry in this world of advertising insofar as advertisers channel and create market desires in private individuals and families.

Mad Men also sometimes expands the scope of moral attention beyond the office and the home and into the larger macro-economic, social world in which business operates—most obviously, to the racial and sexual politics of the day. As most viewers and critics have observed, that socio-political attention typically orbits around issues of sex, religion, and race. In that sense, what the show shares in common with standard business and advertising ethics is too little attention to political economy and the questions of economic justice that always accompany, if in an unacknowledged way, any ethical considerations about business. Although often treated more as a literary trope than as a complex moral and social reality, poverty does figure into the series. Notably, through its many flashbacks, the series implies that Don's childhood poverty is the driving force behind his ambition and, so, his success in the world of big business advertising. In effect, if it comes as little surprise that the master ad-man is a master liar and seducer, the series wants us to believe that it should also come as no surprise that the ad-man grew up impoverished. A walking, talking embodiment of the economy itself, Don, despite his glitz and glamour, is founded on and fueled by the poverty and social exclusion hidden within. As the series intimates, Don's (or, rather, Dick's) poverty leaves an indelible impression capable of explaining both his insatiable desire (i.e., his hunger and emptiness) and also (more disturbingly) his immorality.

At its worst moments, then, the series perpetuates two longstanding, poverty-related myths: (1) the romanticized "hobo" (e.g., the gentleman of the rails, who inspires young Dick's search for a better life and a new identity); and (2) the vilified "libertine" (e.g., Dick's dad, whose infidelity prefigures Don's adulterous tendencies and mistreatment of his family).[84] At its best moments, however, the series intimates a self-consciousness about these portrayals of poverty. Tellingly, Sally and Bobby wear Halloween costumes ("gypsy" and "hobo," respectively) that underscore both the theatricality of identity and the need for a more complex representation of poverty in the American imagination and in business ethics, specifically. At its most insightful moments, then, the show acknowledges the poverty of its (and our) usual treatment of poverty.

More to the point, at its most insightful moments, the series—not least of all, by being a series (i.e., a temporally unfolding narrative)—reminds us that the true subject matter of ethics is not some miraculously static and autonomous fictional self conjured up for the convenience of business ethics textbooks and classrooms. On the contrary, the subject matter of ethics is or, at least, should be (in the words of the Coke ad) "the real thing"—namely,

84. "The Hobo Code," S1/E8.

the fragmented, confused, thoroughly embedded (i.e., socially, historically, politically dependent and contextualized) creature whose character is both actually and (often) acceptably unstable. In other words, what we study when we do ethics responsibly is a self whose sincerity is variously undetectable, unachievable, and undesirable, and whose character and constitutive desires are so mutable and malleable that the notion of *a* self (i.e., a single enduring character or persona) becomes suspect, at best. Ultimately, what *Mad Men* offers business ethics is a challenge—specifically, to make moral judgments about actual, flesh-and-blood human beings, i.e., the "mad men" and women whose temporality and societal constitution better suit serial narrative than ideal theory.

BIBLIOGRAPHY

Alger, Horatio. *Ragged Dick, of Street Life in New York with the Boot-blacks.* England: Dodo, 2008.

Baier, Kurt. *The Moral Point of View: A Rational Basis of Ethics.* Ithaca, NY: Cornell University Press, 1958.

Baker, Samm Sinclair. *The Permissible Lie.* New York: World Publishing, 1968.

Boatwright, John R. *Ethics and the Conduct of Business.* 2nd ed. Upper Saddle River, NJ: Prentice Hall, 1997.

Booth, Wayne C. *The Company We Keep: An Ethics of Fiction.* Berkeley, CA: University of California Press, 1988.

Bowie, Norman E. "Business Ethics." In *New Directions in Ethics: The Challenge of Applied Ethics,* edited by Joseph P. DeMarco and Richard M. Fox, 158–72. New York: Routledge & Kegan Paul, 1986.

Butler, Joseph. "Sermon X. Upon Self-deceit." *Fifteen Sermons Preached at the Rolls Chapel.* New York: Robert Carter, 1846.

Crisp, Roger. "Persuasive Advertising, Autonomy, and the Creation of Desire." In *Ethical Issues in Business: A Philosophical Approach,* edited by Thomas Donaldson and Patricia H. Werhane, 503–10. 6th ed. Upper Saddle River, NJ: Prentice Hall, 1999.

Egner, Jeremy. "Matthew Weiner on the Coke Ad and the Meaning of 'Mad Men.'" *Arts Beat* (May 20, 2015).

Emerson, Ralph Waldo. "Nature." In *The Essential Transcendentalists,* edited by Richard G. Geldard, 52–103. New York: Penguin, 2005.

Ferreira, M. Jamie. "The Misfortune of the Happy: Levinas and the Ethical Dimensions of Desire." *Journal of Religious Ethics* 34 (2006).

Fitzgerald, F. Scott. *The Great Gatsby.* New York: Scribner, 1953.

Freeman, R. Edward. *Stakeholder Theory: The State of the Art.* Cambridge: Cambridge University Press, 2010.

Galbraith, John Kenneth. *The Affluent Society.* Boston: Houghton-Mifflin, 1958.

Gilbert, Daniel, and Patrick S. Malone. "The Correspondence Bias." *Psychological Bulletin* 117 (1995).

Godin, Seth. *All Marketers are Liars: The Power of Telling Authentic Stories in a Low-trust World.* London: Penguin, 2005.

Greenblatt, Stephen. *Renaissance Self-Fashioning.* Chicago: University of Chicago Press, 1980.

Harman, Gilbert. "No Character or Personality." *Business Ethics Quarterly* 13 (2003).

Holley, David M. "A Moral Evaluation of Sales Practices." *Ethical Theory and Business,* edited by Tom Beauchamp and Norman Bowie, 462–72. 4th ed. Englewood Cliffs, NJ: Prentice Hall, 1993.

Itzkoff, Dave. "Jon Hamm Talks About the 'Mad Men' Series Finale." *ArtsBeat: The Culture at Large* (May 18, 2015).

Kant, Immanuel. "Anthropology from a Pragmatic Point of View." In *Anthropology, History, and Education,* edited by Gunter Zoller and Robert B. Louden, 227–449. Cambridge: Cambridge University Press, 2007.

———. "On a Supposed Right to Lie from Philanthropy." In *Practical Philosophy,* edited by Mary J. Gregor, 605–15. Cambridge: Cambridge University Press, 1999.

Levinas, Emmanuel. *Totality and Infinity: An Essay on Exteriority.* Translated by Alphonso Lingis. Pittsburgh, PA: Duquesne University Press, 1969.

Levitt, Theodore. "Advertising and its Adversaries." In *Business Ethics: Corporate Values and Society*, edited by Milton Snoeyenbos, Robert Almeder, and James Humber, 421–24. Buffalo, NY: Prometheus, 1983.
MacIntyre, Alasdair. *After Virtue: A Study in Moral Theory.* Notre Dame, IN: Notre Dame Press, 1981.
Mahon, James. "Kant on Lies, Candour, and Reticence." *Kantian Review* 7 (2003).
Miller, Arthur. *Death of a Salesman: Certain Private Conversations and a Requiem.* New York: Viking, 1949.
Newton, Adam Zachary. *Narrative Ethics.* Cambridge, MA: Harvard University Press, 1995.
Nussbaum, Martha. *Love's Knowledge: Essays on Philosophy and Literature.* New York: Oxford University Press, 1990.
Ross, Lee. "The Intuitive Psychologist and His Shortcomings: Distortions in the Attribution Process." *Advances in Experimental Social Psychology* 10 (1977).
Saddhatissa, Hammalawa. *Buddhist Ethics.* Boston, MA: Wisdom, 1997.
Sage Brief Guide to Marketing Ethics. Los Angeles: Sage, 2012.
Schleiermacher, Friedrich. *The Christian Faith.* Edited by H.R. Mackintosh and J.S. Stewart. Philadelphia: Fortress, 1976.
Shakespeare, William. *As You Like It.* Ed. Alan Brissenden. Oxford: Oxford University Press, 1998.
Solomon, Robert C. "Victims of Circumstances? A Defense of Virtue Ethics." *Business Ethics Quarterly* 13 (2003).
Spence, Edward, and Brett Van Heekeren. *Advertising Ethics.* Upper Saddle River, NJ: Prentice Hall, 2005.
Steinbeck, John. *The Grapes of Wrath.* New York: Viking, 1939.
Steinberg, Brian. "'Mad Men Finale Revises Landmark Coke Ad." *Variety* (May 18, 2015).
Trilling, Lionel. *Sincerity and Authenticity.* London: Oxford University Press, 1972.
Upton, Candance L. "Virtue Ethics and Moral Psychology: The Situationism Debate." *The Journal of Ethics* 13 (2009).
Velasquez, Manuel. *Business Ethics: Concepts and Cases*, second edition. Englewood Cliffs, NJ: Prentice Hall, 1988.
Whitman, Walt. *Leaves of Grass.* New York: New York University Press, 1965.

Chapter 2

THE BUSINESS OF CREATIVITY

From SCDP to the Modern Creative Enterprise

—JENNIFER PHILLIPS

Mad Men animates an important transition in twentieth-century American economic life: the rise of what Richard Florida names the "creative class," whose creative processes and outputs are their commodity.[1] This group forms a particular subset of the knowledge society heralded by Peter Drucker in 1959,[2] and its increased influence presents new opportunities for ethical management and work. Creative workers assert the humanness of work in a new and forceful manner. According to Drucker, "Knowledge is always embodied in a person; carried by a person; created, augmented, or improved by a person; applied by a person; taught and passed on by a person; used or misused by a person. The shift to the knowledge society therefore puts the person in the center."[3] This idea belies traditional assumptions about the fungibility of workers and invites a more humane paradigm.

In addition, knowledge work invites more diverse economic participation. Becasue it is free of the physical constraints that burden manual work, it is accessible by those who lack the brawn of the prototypical industrial

1. Florida, *Rise of the Creative Class*, 5.
2. Wartzman, "What Peter Drucker Knew About 2020."
3. Drucker, *Essential Drucker*, 287.

worker—in particular, many women, people with physical disabilities, and the elderly. The *creative* economy builds on this: because innovation generates success, diverse perspecitves and experiences find new purchase.[4] In the modern competitive environment, where success demands innovation and creativity, an enterprise that promotes and embodies diversity is better poised to thrive than a closed or exclusive one. This shift opens the door to meaningful economic participation beyond the realm of white, male, heterosexual privilege.

However, the creative economy is far from perfect. The exploitation of workers persists. The ideal of a diverse *and equitable* economy is unfulfilled. The impact of new ideas can take years to assess. Social responsibility is muddled. The alliance of creativity with business capital amplifies creative workers' influence, making social concern more important than ever.[5] These challenges are all the more significant, Florida argues, because thinkers and creators have "been living in a world of our own concerns—selfishly pursuing narrow goals with little regard for others or for broader social issues."[6]

How should creative workers and their employers respond to these challenges? I will suggest answers, by way of examining *Mad Men's* midcentury, Madison Avenue characters and its fictional advertising agency Sterling Cooper Draper Pryce (SCDP).[7] I will begin by contrasting romantic notions of autonomous or mystical creativity with for-profit creativity, to clarify our idea of the creative worker. I will then address the challenge of managing creative workers not only in advertising, but also across a broad spectrum of industries where creativity has become a commodity. Finally, I will prioritize three ethical considerations for creative workers and their employers: employee well-being, diversity and inclusion, and social responsibility.

Florida's work is influential for this project because his use of the phrase *creative class* illuminates a modern macroeconomic phenomenon. In addition, he recommends strategies for building an economy that promotes broad creative thinking and ongoing innovation.[8] While those contributions are important, they need microeconomic complements that assess creative work *in practice* and suggest discrete and useful improvements for workers and employers. For example, Florida's assumption that creative workers are more autonomous than other workers is logical from a macroeconomic

4. Florida, *Rise of the Creative Class*, 35.

5. Ibid., 12.

6. Ibid., xi.

7. In this chapter, I will use the acronym "SCDP" to represent the agency's various iterations, beginning with Sterling Cooper Advertising Agency and culminating with the sale of Sterling Cooper & Partners to McCann Erikson.

8. Florida, *Rise of the Creative Class, Revisited*, Kindle loc. 5788–999.

perspective, but it should be unpacked and tested at ground level. This chapter will employ the *Mad Men* series as a tool for such efforts: juxtaposing individual stories with macro-level trends in the creative economy. This approach insists on accountability for the individual decisions that creative workers and their employers make daily, and it can better incorporate the experiences of marginalized groups. By employing it, I invite creative workers and their employers to participate in their industries' ethical discourse, as well as the ethics of economics writ large.

Mad Men offers a window into the concrete experiences of workers and employers who comprise the creative class. It depicts the earliest days of managed creativity—a phenomenon that is now considered essential not only for advertising, but also across industries and sectors. The series is fictional; however, this aspect allows for a kind of ethical voyeurism that exposes the real economy. In addition, plot itself (as a device requiring characters to *do something*) affirms the efficacy of creative workers and their employers. It asserts individuals as active participants—working, consuming, hiring, and firing—that generate the economy, sustain or change its ethos, and deal with its externalities. Such an intimate portrait of a company, its owners, and its employees is a welcome addition to the theoretical discourse of ethics and economic life.

FOR-PROFIT CREATIVITY: ART OR BUSINESS?

Throughout the series, the characters of *Mad Men* embody various forms of creativity. There is advertising creative, of course, which is represented by many characters, including the series protagonist, mid-century adman Don Draper. The show also portrays illustration (Midge Daniels), writing (Ken Cosgrove, Paul Kinsey), acting (Megan Draper), and photography (Joyce Ramsay, Stan Rizzo, Pima Ryan)—all set amid the countercultural backdrop of beatnik New York and the radical political energies of the 1960s. Of these, advertising creative is depicted as a particularly agonizing process, marked by late nights, heavy drinking, frustrating pitches, and ephemeral satisfaction, at best. It is uniquely vulnerable to the advent of technology. On the show, an IBM 360 mainframe computer[9] literally and symbolically displaces the creative team. These tensions beg the question: what kind of creativity is advertising, exactly?

Elliot Paul and Scott Kaufman trace the mystical understanding of creativity to Plato's *Ion* and *Phaedrus*, in which the ancient philosopher

9. Ovans, "That Mad Men Computer."

articulates a creative state of "madness" generated by the Greek Muses.[10] They ask whether we must move beyond consciousness in order to be truly creative, and they offer the muse, the Eureka moment, and mind-altering drugs for consideration.[11] Each of these examples is fully recognizable within the *Mad Men* landscape—moments of blind inspiration leading to the pitch that shocks with its out-of-nowhere genius ("Lucky Strike: It's toasted."),[12] in an office where bar carts seem to outnumber typewriters and, in later seasons, pot smoke wafts from the creative lounge.

Nonetheless, *Mad Men* is set in an ad agency, where creative work aims to sell things and, ultimately, to enrich the partners. It is commercial and exists to create profit, rather than something intrinsically new or inspired. Because of this, advertising creative is subject to business needs and best-practice developments, both of which treat creative work as means to an end. *Mad Men* takes place during the 1960s, a time when real-world advertisers began to embrace data analysis and market research. We witness this transition in the series in the forms of the IBM 360 mainframe and researchers Greta Guttman and Faye Miller. *Advertising Age* credits this evolution to two important changes: new data available from the US Census that enabled nuanced "lifestyle" targeting, and the advent of a recession that necessitated cautious, data-driven marketing.[13] The tension between creative epiphany and hard numbers persists to this day, with current practice incorporating the best of both worlds. Harvard Business Review gives a precise example of this dynamic by showcasing Google Creative Lab:

> Executives demand brilliant, creative thinking and insights at the core of any campaign. But they don't expect perfect translation of these insights into marketing content and execution right out of the gate. Instead, they deliberately champion post-launch testing, expecting that the creative content will be refined and perfected based on a data-driven analysis of live feedback from the marketplace.[14]

Matthew Kieran offers a concept that further distinguishes for-profit creative: virtue. While he doesn't name advertising specifically, he makes

10. Paul and Kaufman, *Philosophy of Creativity*, 3.

11. Ibid., 10.

12. The series premiere is tethered emotionally by Don's creative anxiety, which increases steadily as an important client meeting nears without an idea. Don solves the puzzle in the nick of time with the slogan, "Lucky Strike: It's toasted." "Smoke Gets In Your Eyes," S1/E1.

13. "1960s Creativity and Breaking the Rules."

14. Joshi, "Use 'Both-Brain' Marketing."

a distinction between creativity that is intrinsically motivated (by an idea, an aesthetic, etc.) and creativity that is extrinsically motivated (by awards, by profit, etc.). The former, he argues, is virtuous, while the latter is not. Creativity requires a degree of freedom for the artist to follow the idea down whatever unpopular or unprofitable path it may lead. Kieran writes,

> It takes honesty to evaluate the nature and value of what one is doing properly; it takes courage to be prepared to fail; it takes humility and open-mindedness to recognize when one has gone wrong; and it takes perseverance and fortitude to continue to work at something for its own sake or in seeking to do justice to an idea.[15]

Creative virtue is "corrupted," he argues, when the work is motivated by something outside itself.[16]

This tension permeates the *Mad Men* series. There are moments of pure joy when its characters land upon the perfect pitch (e.g., Kodak Carousel Slide Projector), but there are also darker moments. For example, the characters are berated as sellouts: cue Season 1 beatnik asking Don, "How do you sleep at night?" ("In a bed made of money," Don answered.)[17] Stifled creative aspirations cause escalating angst in characters like Stan Rizzo and Ken Cosgrove. Recall an agonized Stan nervously showing personal work to photographer Pima Ryan, so distracted by his own self-doubt that he could not recognize her seduction as the hustle it was. Or a just-fired Ken planning "the life not lived" as a writer in Vermont. Rather than pursue his dream, he follows revenge and greed to a corporate marketing job with an SCDP client. The show's protagonist ends the series on a personal journey that takes him far afield of advertising, finally shedding his signature suits for a more relaxed wardrobe complete with bare feet. Before Don leaves, one of his final professional acts is to conduct Peggy Olson's performance review. In a scene that reveals both characters' struggle with the tension inherent in for-profit creativity, Don poses a question: "What do you see for the future?" He rejects Peggy's initial answers ("to be the first woman creative director at this agency . . . to land something huge . . . to create a catchphrase") and prods her to answer more profoundly:

> Don: What else?
>
> Peggy: I don't know.
>
> Don: Yes you do.

15. Kieran, "Creativity as a Virtue of Character," 132–33.
16. Ibid., 133.
17. "Babylon," S1/E1.

> (Pause.)
>
> Peggy: Create something of lasting value.
>
> Don: In advertising?"
>
> (Don laughs openly at the idea.)
>
> Peggy: This is supposed to be about my job, not the meaning of life.
>
> Don: So you think those things are unrelated?
>
> (Peggy stands to leave.)
>
> Peggy: I didn't know you'd be in a mood. Why don't you just write down all of your dreams, so I can shit on them?[18]

This scene, like the entire series, reflects an unmistakable ambiguity about for-profit creative work. Both an emerging talent (Peggy) and an established leader (Don) are uncertain regarding their profession's deeper meaning. They sense that, as creative workers, they inhabit a kind of liminal state, where they exist as neither artists nor businesspeople. This generates a feeling of disquiet for both characters, which in turn creates tension between them.

The ambiguity expressed by Peggy and Don invites a closer look at what it means to be a creative worker. It compels viewers to understand for-profit creatives as a unique subset of professionals, whose relation to the wider world is simultaneously influential and remote. Such a complicated characterization allows important questions to surface, such as: How can employers manage creative workers? What are the ethics of managing creativity? And, are creative workers ethically responsible for the impact of their work? The remainder of this chapter will explore these questions by examining the series, connecting it to contemporary enterprise, and suggesting priorities for creative workers and their employers.

MANAGING CREATIVITY

Even though creativity-for-profit may be a shadow of virtuous, autonomous, or some other creative ideal, it remains the lifeblood of the creative industry—and managing it well separates companies that thrive from companies that wither. The future of each firm will hinge, for better or worse, on its ability to generate creative thinking and creative products or services. Certainly, a good deal of agency work is more technical than creative, such

18. "The Forecast," S7/E10.

as the infrastructure supporting account teams; the implementation of campaigns; and budgeting, data processing, and statistical analysis. However, even these should have a creative aspect; in fact, they *must* for an agency to succeed. Further, an agency's core services, typically brand strategy with art and design concepts to support it, are *explicitly* creative. At the end of the day, an agency's directors must inspire and employ the creative energies of their staff, so the firm can actually *do* the work it was hired to do and win or retain the clients that will keep it in the game.

This stress—the pressure to be creative—is as real for modern agencies as it was for SCDP in the *Mad Men* series. For example, Ogilvy & Mather founder David Ogilvy offers this sound byte: "Unless your advertising contains a big idea, it will pass like a ship in the night."[19] The Martin Agency claims, "Our strength is our ability to help brands find and tap into new energy."[20] A third agency, McKinney, positions itself as follows:

> In an era when people have as much power as brands, we need creativity more than ever. Not just to get people's attention, but also to change what they feel, think and do. This is influence: the power to cause change by connecting your brand to what people really want. In the world of marketing today, influence is the payoff, creativity is the currency, and we are the mint.[21]

The McKinney statement's careless treatment of power and influence deserves critical analysis, which I will revisit later in this chapter. However, it is relevant here because, like the other examples, it illustrates the centrality of creative energy to agency business models.

Following this logic, creative energy is extremely lucrative raw material. If, as Drucker posits, "The essence of management is to make knowledge productive,"[22] then the essence of agency management is to make creativity productive. In order to be effective, agency leadership must harness that energy and direct it, as efficiently as possible, to the bottom line—a reality that introduces a host of questions, such as: Is it possible to contrive a creative environment? Can you train employees to be creative? If creativity has a mystical aspect, can it be controlled? And so forth. Fifty years ago, business creativity was not revered, as it is today, and modern buzzwords like "innovation" and "disruption" were meaningless. However, creativity *was* standard fare in the advertising industry, and *Mad Men* is representative of that value.

19. Ogilvy.com, "Ogilvy & Mather: Our History."
20. Martinagency.com, "Martin Agency: What We Do."
21. Mckinney.com, "Why McKinney."
22. Drucker, *Essential Drucker*, 313.

Advertising managers in the *Mad Men* era employed a hierarchical mode of management, which they inherited from their forebears in manufacturing, with all its problematic power structures and ethical baggage. Creative employees were given firm deadlines and told to get it done or get out. Alcohol at work was pervasive. The substance may have loosened creative inhibitions, but it also intensified agencies' privileged, authoritarian cultures. At their worst, businesses condoned or explicitly expected sexual currency—a theme that is prominent from the first episode of *Mad Men*, when Roger Sterling instructs his then-employee Don, to Don's obvious distaste, to "charm" Rachel Menken into hiring the agency. It culminates with Joan's eventual promotion to partner, bought at the price of sexual favors to secure the Jaguar account.[23]

Since the 1960s, the management of creative output has become a robust and broadly relevant topic not only for advertising agencies, but also for business writ large. Managing creativity is a fundamental twenty-first-century business skill, widely considered to be both an art and a science that impacts organizations of all stripes. The topic is pervasive in the current business press, with coverage addressing creativity throughout every imaginable business process, from the artistic to the mundane. A cottage industry of "innovation consulting" and "disruption consulting," headlined by both established consulting firms like McKinsey and edgy design agencies like IDEO, has sprung up to meet this need for organizations around the world, across industries, and across sectors. Organizations have money to save and investment potential to maximize, and creativity is an essential tool.

Because applications of creative thinking have expanded, contemporary management techniques are more thoroughly developed. Ample advice can be found on everything from office design,[24] to vacation policies,[25] to wellness programs,[26] to offsite meetings,[27] each of which is justified or dismantled on the basis of creative productivity. Best practices for managing creativity—diversity, employee engagement, committed leadership, and flexible management—are widely accepted.

Notwithstanding this zeitgeist, however, the successful leadership of creative enterprise remains an elusive accomplishment. As a team of

23. For further discussion of Joan's road to becoming partner, see Howard Pickett's Chapter 1, Ann Duncan's Chapter 4, and Kristen Deede Johnson's Chapter 5.

24. For example, Stillman, "10 Office Design Tips."

25. For example, Thomas, "Vacation Policy."

26. For example, Vanderkam, "Do Corporate Wellness Programs Really Boost Productivity?"

27. For example, Dahle, "Can This Off-Site Be Saved?"

Harvard Business Review contributors explains, "At the core of leading innovation lies a fundamental tension, or paradox, inherent in the leader's role: leaders need to *unleash* individuals' talents, yet also *harness* all those diverse talents to yield a useful and cohesive result."[28] This recommendation directly echoes Drucker's assessment, which underscores the interdependency of the thinker and the manager:

> The intellectual's world, unless counterbalanced by the manager, becomes one in which everybody "does his own thing" but nobody achieves anything. The manager's world, unless counterbalanced by the intellectual, becomes bureaucracy and the stultifying greyness of the "organization man." But if the two balance each other, there can be creativity and order, fulfillment and mission.[29]

ETHICAL CONSIDERATIONS

The development of philosophies and techniques for the management of creativity introduces notable positive developments, most significantly the evolution of employee-centered policies and enhanced diversity. However, these improvements are far from complete—certainly not in *Mad Men*, nor in the context of contemporary creative work. Also problematic, creative workers are dogged by patterns of stress and overwork. Finally, it is unclear who bears responsibility for the *content of* creative work. Workers? Employers? Consumers? It is necessary to examine creative management in light of these ethical considerations, so we can advocate for continued incremental improvement on behalf of both workers and the common good.

Employee Well-being

One set of positive outcomes has to do with employee-centered policies and the discourse around them. As managers seek to inspire and support creativity, best practice has evolved to consider the humanity of workers. Earnest concern has eclipsed outmoded bureaucratic approaches, and *caring* has emerged as a twenty-first-century management virtue.[30] Managers are assessed not only on their intelligence and ability to deliver hard results, but also on their success in retaining talent, their team-building skills, and

28. Hill et al., "Inescapable Paradox."
29. Drucker, *Essential Drucker*, 292.
30. Florida, *Rise of the Creative Class*, 130–32.

their EQ, or emotional intelligence. Perks like paid vacation, onsite fitness centers, catered meals, nap rooms, and game rooms are commonplace in creative firms.[31] For example, creative tech giant Google's benefits are legendary, exemplifying managerial caring and serving as a point of comparison for creative employers of all stripes. In the company's own words, "Our hope is that . . . you become a better person by working here"; "You're valuable to us, and our benefits and perks are there to show it"; and "Hey, we're family."[32]

In addition to exhaustive benefits, modern workplaces have undergone important changes in work styles and schedules.[33] Flexible scheduling, where employees set their own hours, and telecommuting, where employees work from locations of their choosing, afford workers significant freedoms that were unknown in 1960s agencies like SCDP. *Mad Men* protagonists, with the possible exception of SCDP's established partners, simply . . . work. They work nearly all the time, and they are tethered to either their desks or their clients. This particular contrast between the culture of the *Mad Men* agencies and the culture of modern creative organizations is striking and perhaps foreshadowed by Peggy's final scenes, in which she has no office but asserts her identity as a creative worker. She is simultaneously independent (her confidence comes from her own accomplishments, not her agency's) and determined to navigate advertising's corporate landscape. Her romantic denouement with creative colleague Stan Rizzo is a similarly prescient nod to work-life integration, that now-familiar drumbeat in contemporary business publications like *Harvard Business Review*,[34] *Fortune*,[35] *Forbes*,[36] and *Fast Company*.[37]

In theory, these developments should alleviate the burden on employees. In reality, modern work falls far short of the dream. Florida writes, "Those in the Working Class and the Service Class are primarily paid to execute according to plan, while those in the Creative Class are primarily paid to create and have considerably more autonomy and flexibility than the other two classes to do so."[38] While of course creative workers maintain

31. Ibid., 127–28. Excerpt includes benefits statistics for creative technology firms.

32. Google.com, "Google Careers: Benefits."

33. Florida, *Rise of the Creative Class*, 120–22. Excerpt includes statistics on creative workers' schedules and hours.

34. Friedman, "What Successful Work and Life Integration Looks Like."

35. Vanderkam, "Work-Life Balance Is Dead."

36. Schawbel, "Work Life Integration."

37. Douglas, "Why Work-Life Integration Trumps Work-Life Balance."

38. Florida, *Rise of the Creative Class*, 8.

relative autonomy and flexibility (a vital point for discussions of inequality), those in service of for-profit enterprise are by definition required to "execute according to plan." They remain employees who, while given certain latitude to encourage creative output, are nonetheless accountable to an agenda that is not their own. Rich benefits, managerial caring, and employee well-being ultimately serve the interests of employers, stockholders and investors, though they can be advantageous for workers. They typically last only as long as they are perceived to advance profit, and they are vulnerable in times of financial crisis.[39]

In addition, creative work environments are persistently stressful. This is an unavoidable aspect of creative work, according to Florida: "Stress increases because the Creative Economy is predicated on change and speed. . . . The result could be called a 'caring sweatshop,' but there's not really any contradiction here. It's just the reality of a workplace in the Creative Economy."[40] By accepting this outcome as "just the reality" in which modern creative workers must operate, Florida's assessment falls short. We should instead question weak stakeholder analyses that do not recognize workers as intrinsically valuable constituents. We should withhold approval of sweatshops, no matter how caring. We should be critical of a Creative Economy that rewards change and speed for their own sake, rather than for their impact on the common good. In other words, we should think *more creatively* about our economy, its purpose, and our own roles within it.

Workplace Diversity

Workplace diversity provides a second opportunity for positive outcomes. As with employee well-being, progress in this area is simultaneously noteworthy and lagging. Diversity can be enhanced in spaces that self-identify as creative because new ideas require openness to difference, fresh modes of thinking, and multi-dimensional perspectives. Again, *Mad Men*'s SCDP is representative. Series creator Matthew Weiner explains, "I think it [differentness] gives you unique skills for advertising, acknowledging that you're an outsider, and using that outsider status to look at how people function and what they want."[41] The show embraces the opportunity to tell a story about increasing diversity, most notably in the development of its female and Jewish characters, secondarily in its plots around closeted homosexuality and the racial integration of SCDP's secretarial pool.

39. Ibid., 128.

40. Ibid., 132.

41. Lednicer, "Q&A: 'Mad Men' Creator Matthew Weiner."

Both Peggy and Joan are central protagonists of the *Mad Men* series, charting feminist paths in creative industry with remarkable success. Joan moves from office manager to partner at SCDP, and she ends the series as entrepreneur-owner of her own agency. We meet Peggy in the secretarial pool, but she soon advances to copywriter on her creative merits. She ends the series aspiring to a creative directorship—if not at McCann Erickson, with its brutally misogynistic culture, then at her next place of employment, wherever that may be.

However, notwithstanding Peggy and Joan's successes, both characters inhabit the sexist realities of a typical 1960s workplace. They represent the first tenacious generation of businesswomen whose reach for the brass ring not only was set to a chorus of jeering and outright harassment, but also hinged too precariously on their choices to use (or not) their own sexual power. Two moments stand out particularly in this regard: Joan's decision to trade sexual favors for an equity stake in SCDP, and Peggy's symbolic decision to hang Bert Cooper's sexually suggestive octopus painting in her new office after worrying, "They won't take me seriously," and, "You know that I need to make men feel at ease." ("Who told you that?" retorted Roger, both naively and wisely.)[42] However loudly we cheer Peggy and Joan's accomplishments, we nevertheless cringe at the context in which the two women achieve them.

Also navigating sexual difference and power, both Salvatore Romano and Bob Benson walk a nimble line between closeted and open homosexuality. They reveal their sexual orientation to their colleagues—Bob to Joan purposefully and Sal to Don accidentally—in moments that, while not undermining them immediately, do reveal a hefty dose of uncertainty. Bob's professional trajectory is reoriented completely after his affair with a GM executive opens the door for a future with the car company. However, he fears for his security in that environment and therefore offers a marriage of convenience to Joan. Sal hides behind an established marriage of convenience, but this fails to spare him at SCDP, where his career unravels after he rebuffs an important client's advances. Sal's romantic refusal of Lee Garner, Jr. seals his fate as decisively as Joan's liaison guarantees hers. None of these characters can unravel their sexuality from their professional lives, and both Joan and Sal are expected to relinquish sexual consent to the firm. The complicated varieties of sexual harassment are portrayed deftly in the juxtaposition of these plotlines.

Mad Men's Jewish characters are also well developed and complicated. They alternately thrive and suffer against a backdrop of anti-Semitic

42. "Lost Horizon," S7/E12.

comments and assumptions. This story begins in Season 1 with Rachel Menken. Menken is a client, rather than an employee, a status that would normally earn her a degree of deference from her ad agency. As a Jewish person and a woman, however, she has to prove both her own business savvy and her store's potential beyond a Jewish market. She has to convince the admen that her business merits their time.

Inside the agency, art mimics life in the firm's hiring of Michael Ginsberg. According to Weiner, the hire represents a nonfictional aspect of 1960s advertising, a period during which agencies began mining Jewish humor for fresh creative edge. This development was a reflection of the humor in popular entertainment, which propelled advertisers to purposefully assimilate Jewish humor into their modus operandi.[43] Nonetheless, as Weiner explains in an interview with *The Washington Post*, this happened within a context where the *differentness* of Jewish colleagues was front and center: "These were white agencies, populated by white people, using all of the typical philosophies that are used to exclude people. Which is, 'I'm not comfortable around them; they're selling to a minority,' and the irony being, of course, that a lot of the entertainment that the ads are sitting in is being written by Jewish comedy writers."[44]

In contrast, there is significantly less to say regarding *Mad Men*'s portrayal of African-Americans or other nonwhite characters. Whether or not Weiner was propelled by historical accuracy, as he claims,[45] it remains that he did not choose to highlight African-American stories.[46] On the one hand, the series captures major moments of the civil rights movement, and SCDP breaks new ground in the New York advertising community by integrating its secretarial pool. However, the show utilizes these developments as time markers and to illustrate white characters' attitudes, not to tell the stories of African-American characters.[47] Weiner has criticized both advertising's "shameful past" and its current lack of racial diversity, using these failures to explain his own creative decisions.[48] But, as Alex Madison writes for *Salon*, "it's baffling that Weiner would fail to bring even one black character into central focus."[49] In this light, the wonderful moment in which Dawn

43. Lednicer, "Q&A: 'Mad Men' Creator Matthew Weiner."

44. Ibid.

45. Colby, "Mad Men and Black America."

46. See also, Nsenga Burton's Chapter 10, which explores how racial themes emerge in the analysis of white characters, most notably Don Draper.

47. Madison, "Mad Men Failed Its Black Women."

48. McDermott, "'Mad Men' Creator."

49. Madison, "Mad Men Failed Its Black Women."

Chambers and Shirley make fun of their colleagues' tendency to confuse them is as damning for *Mad Men*'s writers as it is for the women's fictional employers.[50]

In addition to portraying specific groups' experiences in 1960s corporate culture, these characters, taken together, tell us something important about workplace diversity—then and now. Specifically, they depict the uncomfortable reality that, while diversity finds a theoretical home in creative enterprise, meaningful inclusion is unfulfilled in practice. This is an important distinction to understand, because "mission accomplished" illusions obstruct further progress. A false sense of completion masks the struggle against unfair privilege, and, in doing so, it entrenches an unjust status quo. *Mad Men* depicts this error and the harm it can cause by juxtaposing marginalized employees, who suffer exclusion and harassment, with managers who either don't care about them or can't see the problem due to outsized pride for SCDP's nascent diversity. We witness SCDP's ultimate failure at inclusion in *Mad Men*'s final season, put gracefully to Roger by Shirley when she resigns from the agency. As she leaves, she tells a surprised Roger, "Advertising is not a comfortable place for everyone."[51]

A half-century later, Shirley's observation resonates and is validated by the industry's diversity statistics. In 2009, the National Association for the Advancement of Colored People's "Madison Avenue Project" released dismal data on racial diversity in advertising. The study cites both unequal pay ($.80 to the dollar for black college graduates) and problematic hiring practices (16 percent of large firms failed to employ *any* black managers).[52] Gender statistics also reflect diversity failures: In 2013, only 11.5 percent of ad agency creative directors were female, and women represent only 24 percent of the leadership teams on *Ad Age*'s 2014 Agency A-List.[53]

For creative enterprise more broadly, Florida provides a well meaning but flawed example of the "mission accomplished" illusion. He makes a connection between aesthetically creative movements and future enterprise in the regions that spawned them, specifically examining the rise of creative technology firms.[54] For example, he suggests these lineages: the Seattle of Jimi Hendrix and Nirvana gave birth to Microsoft and Amazon; Dell was inspired by the Austin of Willie Nelson; and San Francisco's Haight-Ashbury

50. Drumming, "On 'Mad Men.'"

51. "Lost Horizon," S7/E12.

52. National Association for the Advancement of Colored People, "NAACP Calls on Top 25 Advertisers."

53. Andjelic, "Lack of Gender Diversity in Agencies Is an Organizational Failure."

54. Florida, *The Rise of the Creative Class*, 202–3.

culture made Silicon Valley possible.[55] "All of these places," he writes, "were open, diverse, and culturally creative first. *Then* they became technologically creative."[56]

Mired in diversity failures, today's Silicon Valley is particularly illustrative. According to Florida in 2002, "What set Silicon Valley apart was not just Stanford University or the warm climate. It was that the place was open to and supportive of the creative, the different, and the downright weird. The Valley was able to integrate those who were offbeat, not ostracize them or discourage them."[57] However, it appears that "downright weird" and "offbeat" are assets largely embodied by white men. Ellen Pao's 2014 lawsuit against the venture capital firm Kleiner Perkins Caufield & Byers brought Silicon Valley's sexism into the light, but her story is one example among many. In spite of a few publicly recognizable female leaders, all-male boards of directors are too common, and recent statistics report that women's share of programming positions has dropped from a third to a quarter—a low participation rate that nonetheless dwarfs the share of new ventures and venture capital partnerships held by women.[58] And, although *The Economist* reports that more than half of Silicon Valley's firms are founded by immigrants,[59] major tech firms have a long road ahead to achieve racial diversity. Recent equal opportunity disclosures report that a mere 1.8 percent of employees at Facebook, Google, and Twitter are African American.[60] The free expression and musical exploration of 1960s San Francisco may have spawned a new form of creative enterprise in the Valley, but it could not guarantee that everyone's contributions would be equally welcome or rewarded.

Social Responsibility

In addition to employee well being and diversity, we must also attend to the ethics of creative *content*. Whether we focus our gaze specifically on advertisers or widen it to include creative enterprise more broadly, we must think critically not only about the management of creative input, but also about the value of creative output—for the firm, for society, and for the world. Too

55. Ibid., 206–7.
56. Ibid., 207.
57. Ibid., 206.
58. Ryder, "Valley of the Dudes."
59. Ibid.
60. Harkinson, "Combined Black Workforces."

often, businesses engage in the pursuit of something new, entertaining, or profitable without pausing to assess the externalities of the thing created.

Mad Men has moments of genius in portraying the tension between advertisers' creative purpose and the common good, but cynicism permeates the characters' resolution of that tension. The best example is surely Don's letter repudiating Lucky Strike and the tobacco industry. The letter is a satisfying tirade against advertising without conscience, against peddling goods to consumers that will only harm them. However, it is also a publicity stunt—a desperate rebranding after the loss of an anchor client. We see more of this tension with Peggy, who chafes against crafting sexist messages, but nonetheless does the work and even excels at it. Lastly, this tension bubbles aggressively when agency characters interact with bohemians or beatniks. For example, Peggy's relationship with future domestic partner Abe Drexler begins with the following tense conversation:

> Abe: I'm sure they're perfectly nice for racists, you know, and obviously your company has an investment in looking the other way.
>
> Peggy: Well, it's a complicated idea, but in advertising, we don't really judge people.[61]

It is worrisome that contemporary agencies may reflect a similar agnosticism regarding clients' products, missions, and ethics. They have become increasingly effective at selling and shaping public opinion—skills propelled forcefully by modern advances in technology and data, decades of best practice, and increasing financial power. Let's reconsider the McKinney statement discussed earlier:

> In an era when people have as much power as brands, we need creativity more than ever. Not just to get people's attention, but also to change what they feel, think and do. This is influence: the power to cause change by connecting your brand to what people really want. In the world of marketing today, influence is the payoff, creativity is the currency, and we are the mint.[62]

McKinney boldly—and accurately—describes the power of advertising to exert influence. The advertiser's mission is to impact public engagement with its clients' brands and to translate that impact into action: "to change what they feel, think, and do."[63] What change, other than increased

61. "The Beautiful Girls," S4/E9.

62. Mckinney.com, "Why McKinney."

63. Ibid.

revenue and net profit, does McKinney have in mind? Does the concentration of power among marketers and their corporate clients benefit anyone other than shareholders? What happens when marketers get more adept at generating this power, and their interests conflict with the broader concerns of society, humankind, or the natural world? How is their power checked?

McKinney's statement raises myriad questions about the nature of the advertiser's mission—questions that, one fears, get paltry attention in the trenches. Too many advertisers—and creative workers more generally—turn a blind eye to their work's outcome and their employers' and clients' ethics. This may be due to naked greed, a feeling of entrapment by golden handcuffs, or an unexamined assumption that what is good for business is good for society (and vice versa). Peggy seems to embody the last of these in the above confrontation with Abe, during which she goes on to muse that her clients' unethical behavior would be "bad for business." She naively suggests that advertisers shouldn't judge their clients, but should instead "help them out of these situations" and "try and stop this."[64] One presumes that she will accomplish this with—what else?—more advertising.

More advertising serves to smooth the waters at best, and to reward and entrench injustice at worst. What hope do consumers and the general public have that unjust and unethical business practices will cease? More effective—and, no doubt, more courageous—responses from advertisers would include critical examination of their work's merit and a clear refusal to enable unjust practices. This should be happening at every level of creative enterprise, in advertising and beyond: creative workers should raise questions when their employers fall short at social responsibility; managers should do the same and encourage critical thinking to support it; and executives should lead businesses with high standards for their own missions and practices. Creative enterprise will never fulfill its promise until every single person engaged in it refuses to spend energy, time, or creativity on businesses that chase profit at any expense.

CONCLUSION

In *The Affluent Society*, John Kenneth Galbraith criticizes private wealth that impoverishes the commons. One of the book's most widely cited passages reads like a scene from *Mad Men:*

> The family which takes its mauve and cerise, air-conditioned, power-steered and power-braked automobile out for a tour

64. "The Beautiful Girls," S4/E9.

> passes through cities that are badly paved, made hideous by litter, blighted buildings, billboards and posts for wires that should long since have been put underground. They pass on into a countryside that has been rendered largely invisible by commercial art. (The goods which the latter advertise have an absolute priority in our value system. Such aesthetic considerations as a view of the countryside accordingly come second. On such matters we are consistent.) They picnic on exquisitely packaged food from a portable icebox by a polluted stream and go on to spend the night at a park which is a menace to public health and morals. Just before dozing off on an air mattress, beneath a nylon tent, amid the stench of decaying refuse, they may reflect vaguely on the curious unevenness of their blessings. Is this, indeed, the American genius?[65]

In fact, Thomas Frank points out that a Season 2 scene echoes this passage.[66] The Drapers have an idyllic picnic that concludes with casual littering: Don tosses an empty beer can into the park, and Betty shakes trash off the picnic blanket onto the ground.[67] Frank interprets this scene as homage to Galbraith's critique—one of several instances in which *Mad Men* alludes to "the literature of consumer society and suburban anomie."[68] The show goes too far, Frank argues, to skewer post-war admen as establishment baddies who contradict modern progressive values, all the while generating an enjoyable air of nostalgia. It goes so far, in fact, that viewers are distracted from real problems. He concludes: "Those were corporate sins we could actually comprehend, and that we found downright soothing to contemplate. . . . Forget credit-default swaps or the misanthropic billionaires of Greenwich."[69]

To avoid this error, we must connect the ethical subject matter of *Mad Men* to our own time and understand how it remains relevant. By animating the rise of the creative class during and beyond the postwar era, Florida provides a valuable construct for doing so. He rightly calls on creative workers to accept responsibility for the power they wield:

> We thus find ourselves in the puzzling situation of having the dominant class in America—whose members occupy the power centers of industry, media and government, as well as the arts

65. Galbraith, *Affluent Society*, 187–88.
66. Frank, "Ad Absurdum."
67. "The Gold Violin," S2/E7.
68. Frank, "Ad Absurdum."
69. Ibid.

> and popular culture—virtually unaware of its own existence and thus unable to consciously influence the course of the society it largely leads. . . . It's time for the Creative Class to grow up and take responsibility.[70]

Unfortunately, as forceful as this call to action sounds, it comes early in his tome and is not accompanied by a constructive vision for those responsibilities. This shortcoming is as disappointing as Florida's economic and social history is compelling. Thankfully, however, he revisits it in a later edition, prescribing a fully formed *creative society* that moves beyond consumerism and toward a richer understanding of well-being.[71] He concludes the revised edition of *The Rise of the Creative Class* with a passionate call for a new "creative compact" built on six tenets for the new economy (listed here verbatim):

1. Invest in developing the full human potential and creative capabilities of every single human being.
2. Make openness and diversity and inclusion a central part [*sic*] of the economic agenda.
3. Build an education system that spurs, not squelches, creativity.
4. Build a social safety net for the creative economy.
5. Strengthen cities; promote density, clustering, and concentration.
6. [Move] from dumb growth to true prosperity.[72]

This is a welcome and robust macroeconomic vision that needs a microeconomic complement. What changes should *individual* creative workers and their employers make in order to support and fully participate in Florida's creative compact?

First, both workers and employers should deepen their understanding of employee well-being—refusing to be distracted by amenities like pool tables and onsite gyms. We must both criticize the business model that Florida names "the caring sweatshop" and give it a constructive complement: disruptive business models that assess the value of "new" things thoughtfully and emphasize the *quality* of creative output as much as or more than the quantity. For an example of creative business modeling, consider Patagonia's move to what some call an "anti-growth strategy," complete with a

70. Florida, *Rise of the Creative Class*, xi–xii.
71. Florida, *Rise of the Creative Class, Revisited*, Kindle loc. 208.
72. Ibid., Kindle loc. 5788–999.

"Worn Wear" campaign that matches well-made products with purchasers who desire to consume thoughtfully.[73] Though Patagonia is not strictly in the business of creativity, and though the jury is out on the net environmental impact of this effort,[74] the model itself is highly creative and potentially disruptive. By alleviating the pressure of constant growth, companies like Patagonia give their employees the breathing room necessary for sustainable creativity and whole lives. They teach us that employee well-being can be more than an elusive ideal.[75] It won't be easy, however. As Patagonia's example reveals, it may require a comprehensive overhaul of a company's business model.

In addition to employee well-being, creative firms must engage more deeply on diversity. This requires a commitment that is both top-down (diversity hiring for influential positions) and bottom-up (aggressive diversity recruitment at entry level). Further, because diversity is incomplete without genuine inclusion, a complementary strategy should assimilate both new and existing employees into an organizational culture that welcomes meaningful participation from *all.* Step one for most firms will be prioritizing diversity in the budget, which may mean increased recruiting expenses; new positions for accountability, such as Chief Diversity Officer; or an investment in diversity assessment and consulting. Large firms have an important role to play here, because they have the financial capacity for meaningful investment, and their success will generate more diverse hiring pipelines for their industries—thereby positively affecting diversity among smaller players.

Third, creative workers and creative firms must hold themselves accountable for their output in new ways. Simple profitability metrics will not suffice; triple-bottom line and other stakeholder metrics are requried. The business case for this change may not be obvious, and many companies—particularly those that are publicly traded—will be slow to pursue these measures in any meaningful way. However, others will make strides. Two indicators of this are the rise of social business models and the growing popularity of the B-Corporation designation, by which the nonprofit B Lab certifies businesses based on "social and environmental performance, accountability, and transparency."[76] These movements are further supported by research on the millennial generation's desire for meaningful work,[77] a

73. McKinnon, "Patagonia's Anti-Growth Strategy."
74. Ibid.
75. Schulte, "Company That Profits."
76. Bcorporation.net, "What Are B-Corps?"
77. Smith and Aaker, "Millennial Searchers."

trend one hopes will strengthen the business case for social responsibility and eventually impact even the most entrenched organizations.

Finally, we must not allow profitability to co-opt the raw potential of human creativity. This potential becomes thematic in *Mad Men*: the show's characters manage it, chase it, exploit it, corrupt it, exalt it, and even ridicule it. As a viewer, I have rooted for unbridled creativity as fervently as I have rooted for Joan and Peggy to overthrow the patriarchy. This is because taming the creative spirit for profit is yet another exertion of traditional hegemony. We champion the freedom of artists not only for aesthetic reasons, but also because creativity is essential to our hope for a better world. Justice is impossible without creative solutions, and peace demands a mode of thinking that sees beyond the world we inhabit and the structures we inherit.

Creativity that is bold enough to do, say, or create something truly new is the lifeblood of progressive change. It is Martin Luther King Jr.'s creative theological treatment of violence, which simultaneously indicts racial injustice, poverty, and the very profitable war machine.[78] It is Abraham's Lincoln's courage to dismantle the profit apparatus of slave labor by ordering, "all persons held as slaves within said designated States, and parts of States, are, and henceforward shall be free."[79] It is the genius of so many authors, artists, and songwriters who evoke the injustice of a profitable status quo or sketch a vision of peace. These examples should inspire creative workers to exert creative freedom and accept responsibility for it. Such resolve will enliven and reorient the creative economy in its next iteration, for the sake of the common good.

BIBLIOGRAPHY

"1960s Creativity and Breaking the Rules." *Advertising Age*, March 28, 2005. Online: http://adage.com/article/75-years-of-ideas/1960s-creativity-breaking-rules/102704/.

Andjelic, Ana. "Lack of Gender Diversity in Agencies Is an Organizational Failure." *The Guardian*, December 11, 2014. Online: http://www.theguardian.com/media-network/womens-blog/2014/dec/11/gender-diversity-agencies-women-advertising.

Bcorporation.net. "What Are B-Corps?" Online: https://www.bcorporation.net/what-are-b-corps.

Colby, Tanner. "Mad Men and Black America." *Slate*, March 14, 2012. Online: http://www.slate.com/articles/arts/culturebox/features/2012/mad_men_and_race_the_

78. King Jr., "Remaining Awake Through a Great Revolution." See also *Where Do We Go From Here*.

79. Lincoln, "Emancipation Proclamation."

series_handling_of_race_has_been_painfully_accurate_/mad_men_and_race_the_series_handling_of_race_has_been_painfully_accurate_.html.

Dahle, Cheryl. "Can This Off-Site Be Saved?" *Fast Company*, September 30, 2001. Online: http://www.fastcompany.com/43727/can-site-be-saved.

Douglas, Dean. "Why Work-Life Integration Trumps Work-Life Balance." *Fast Company*, May 7, 2014. Online: http://www.fastcompany.com/3030120/bottom-line/why-work-life-integration-trumps-work-life-balance.

Drucker, Peter. *The Essential Drucker: Selections from the Management Works of Peter F. Drucker*. New York: HarperBusiness, 2001.

Drumming, Neil. "On 'Mad Men,' Dawn and Shirley Are Stealing the Show." *Salon*, April 21, 2014. Online: http://www.salon.com/2014/04/21/mad_mens_dawn_and_shirley_are_stealing_the_show/.

Florida, Richard. *The Rise of the Creative Class and How It's Transforming Work, Leisure, Community and Everyday Life*. First ed. New York: Basic, 2002.

———. *The Rise of the Creative Class, Revisited*. Kindle ed. New York: Basic, 2012.

Frank, Thomas. "Ad Absurdum and the Conquest of Cool: Canned Flattery for Corporate America." *Salon*, December 22, 2013. Online: http://www.salon.com/2013/12/22/ad_absurdum_and_the_conquest_of_cool_canned_flattery_for_corporate_america/.

Friedman, Stew. "What Successful Work and Life Integration Looks Like." *Harvard Business Review*, October 7, 2015. Online: https://hbr.org/2014/10/what-successful-work-and-life-integration-looks-like.

Galbraith, John Kenneth. *The Affluent Society*. 40th anniv. ed. United States: John Kenneth Galbraith, 1998.

Harkinson, Josh. "The Combined Black Workforces of Google, Facebook, and Twitter Could Fit on a Single Jumbo Jet." *Mother Jones*, July 2, 2015. Online: http://m.motherjones.com/mojo/2015/07/black-workers-google-facebook-twitter-silicon-valley-diversity.

Hill, Linda, Greg Brandeau, Emily Truelove, and Kent Lineback. "The Inescapable Paradox of Managing Creativity." *Harvard Business Review*, December 12, 2014. Online: https://hbr.org/2014/12/the-inescapable-paradox-of-managing-creativity.

Joshi, Aditya. "Use 'Both-Brain' Marketing to Balance Creativity and Analytics," August 13, 2014. Online: https://hbr.org/2014/08/use-both-brain-marketing-to-balance-creativity-and-analytics/.

Kieran, Matthew. "Creativity as a Virtue of Character." In *The Philosophy of Creativity*, edited by Elliot Samuel Paul and Scott Barry Kaufman, 125–44. New York: Oxford University Press, 2014.

King, Jr., Martin Luther. "Remaining Awake Through a Great Revolution." In *A Testament of Hope: The Essential Writings and Speeches of Martin Luther King, Jr.*, edited by James M. Washington, 268–78. New York: HarperCollins, 1986.

Lednicer, Lisa. "Q&A: 'Mad Men' Creator Matthew Weiner Talks 'Other-Ness' and Jewish Identity on Eve of Finale." *The Washington Post*, May 14, 2015. Online: http://www.washingtonpost.com/blogs/style-blog/wp/2015/05/14/matthew-weiner-on-jewish-identity-in-mad-men/.

Lincoln, Abraham. "The Emancipation Proclamation." *National Archives*, January 1, 1863. Online: http://www.archives.gov/exhibits/featured_documents/emancipation_proclamation/transcript.html.

Madison, Alex. "Mad Men Failed Its Black Women—and Squandered Those Opportunities to Tell a Richer Story." *Salon*, May 11, 2015. Online: http://www.salon.com/2015/05/11/mad_men_failed_its_black_women_—%C2%A0and_squandered_those_opportunities_to_tell_a_richer_story/.

Martinagency.com. "Martin Agency: What We Do." Accessed July 10, 2015. http://www.martinagency.com/about.

McDermott, John. "'Mad Men' Creator: Advertising Has a PR Problem." *Advertising Age*, September 30, 2013. http://adage.com/article/special-report-mad-men/mad-men-creator-talks-diversity-advertising/244462/.

Mckinney.com. "Why McKinney." Online: http://mckinney.com/us/why-mckinney.

McKinnon, J.B. "Patagonia's Anti-Growth Strategy." *The New Yorker*, May 21, 2015. Online: http://www.newyorker.com/business/currency/patagonias-anti-growth-strategy.

National Association for the Advancement of Colored People. "NAACP Calls on Top 25 Advertisers to Hold Advertising Agencies Accountable; Issues Letter to Procter & Gamble CEO to Address Dramatic Racial Discrimination in U.S. Advertising Industry." Online: http://www.naacp.org/press/entry/naacp-calls-on-top-25-advertisers-to-hold-advertising-agencies-accountable—-issues-first-letter-to-procter—gamble-ceo-to-address-dramatic-racial-discrimination-in-us-advertising-industry/.

Ogilvy.com. "Ogilvy & Mather: Our History." Online: http://www.ogilvy.com/About/Our-History/Ogilvy_Mather.aspx.

Ovans, Andrea. "That Mad Men Computer, Explained by HBR in 1969." *Harvard Business Review*, May 15, 2014. Online: https://hbr.org/2014/05/that-mad-men-computer-explained-by-hbr-in-1969/.

Paul, Elliot Samuel, and Scott Barry Kaufman. *The Philosophy of Creativity: New Essays*. Edited by Elliot Samuel Paul and Scott Barry Kaufman. New York: Oxford University Press, 2014.

Ryder, Brett. "Valley of the Dudes: Tech Firms Can Banish Sexism Without Sacrificing the Culture That Made Them Successful." *The Economist*, April 5, 2015. Online: http://www.economist.com/news/business/21647611-tech-firms-can-banish-sexism-without-sacrificing-culture-made-them-successful-valley.

Schawbel, Dan. "Work Life Integration: The New Norm." *Forbes*, January 21, 2014. Online: http://www.forbes.com/sites/danschawbel/2014/01/21/work-life-integration-the-new-norm/.

Schulte, Bridget. "A Company That Profits as It Pampers Its Workers." *The Washington Post*, October 25, 2014. Online: http://www.washingtonpost.com/business/a-company-that-profits-as-it-pampers-workers/2014/10/22/d3321b34–4818-11e4-b72e-d60a9229cc10_story.html.

Smith, Emily Esfahani, and Jennifer L. Aaker. "Millennial Searchers." *The New York Times*, November 30, 2013. Online: http://www.nytimes.com/2013/12/01/opinion/sunday/millennial-searchers.html?_r=0.

Stillman, Jessica. "10 Office Design Tips to Foster Creativity." *Inc.*, October 23, 2012. Online: http://www.inc.com/ss/jessica-stillman/10-office-design-tips-foster-creativity.

Thomas, Maura. "Vacation Policy in Corporate America Is Broken." *Harvard Business Review*, June 26, 2015. Online: https://hbr.org/2015/06/vacation-policy-in-corporate-america-is-broken.

Vanderkam, Laura. "Do Corporate Wellness Programs Really Boost Productivity." *Fast Company*, July 24, 2014. Online: http://www.fastcompany.com/3033411/do-corporate-wellness-programs-really-boost-productivity.

———. "Work-Life Balance Is Dead—Here's Why That Might Be a Good Thing." *Fortune*, March 6, 2015. Online: http://fortune.com/2015/03/06/work-life-integration/.

Wartzman, Rick. "What Peter Drucker Knew About 2020." *Harvard Business Review*, October 16, 2014. Online: https://hbr.org/2014/10/what-peter-drucker-knew-about-2020/.

Chapter 3

MAD MANNERS

Courtesy, Conflict, and Social Change

—SARAH CONRAD SOURS

A strange little scene occurs late in the first season of *Mad Men*. One would be hard-pressed to justify isolating any one strange little scene in a series full of them, especially to justify the claim that one's pet scene reveals what the series is "all about." The series is about many things, and there are many strange little scenes to tip us viewers off to the writers' interests and to engage and encourage our own. Our most compelling interests are always ourselves, of course, and the strangeness of strange scenes provides a thin barrier to our seeing ourselves in them or through them. Like good science fiction and speculative fiction, good period fiction is as much about the creator's period as about the period depicted; we must meet ourselves as strangers in order better to understand ourselves.

The strange scene I mean takes place during the overnight festivities accompanying the 1960 presidential election.[1] Two supporting characters enact a parody for the rest of the office, a one-scene play that will show the gathered crowd its own fundamental conflict—what Dr. Miller will later describe as the fundamental conflict of "what I want versus what's expected

1. "Nixon vs. Kennedy," S1/E12.

of me."[2] Ken Cosgrove and Allison the secretary[3] enact the scene. She walks by, and he initiates a chase: "You better run." She accepts (that is, she starts giggling and running). The men in the room know Ken's purpose before she does. They start guessing colors. Only when Ken "catches" her does he reveal what game they have been playing: he will forcibly discover the color of her underwear, literally uncovering it and displaying it to the public eye. When he has done so ("Who had blue?"), he pulls her up, offers his arm, and says, "Can I walk you home?" making an elaborate show of the ending of a pleasant date. This parody of rape followed by a parody of courtesy plays to laughs and cheers and inspires at least one copycat couple. (Immediately afterward, a man takes down another giggling secretary in the background.) Peggy Olson quietly makes her exit, but no one else seems to understand the full meaning of the little play. It serves, for them, as mere entertainment.

This little one-scene play represents what the entire series will be for us viewers: a play that shows us who we are, but by parodying us to ourselves so well that we risk being satisfied with the entertainment of it. The series has signaled from the first episode that courtesy will be a key lens through which to view the characters: small gestures of courtesy and discourtesy will reveal their personalities, motivations, and moral frameworks. More importantly, courtesy will be a lens through which to examine the immense change being wrought both in the time period depicted and, possibly, in our own. By the time Ken comically offers Allison his arm, we understand that this series will not allow us to content ourselves with comfortably simplistic analyses.

COURTESY AND SOCIAL CONTROL

A simplistic definition of courtesy might place it on the "what's expected of us" side of that fundamental conflict Dr. Miller names. Courtesy is, minimally speaking, a set of social behaviors broadly considered normative in a given society. Other people in society expect certain behaviors of us when we interact with them, and they sufficiently communicate those expectations such that fulfilling them is possible. It is expected that I offer my girlfriend my arm as I walk her home; it is expected that I smile up at my boyfriend as he walks me home. If these expectations are not met, one or both parties will notice. This basic understanding of courtesy will get us through the first

2. "Christmas Comes But Once a Year," S4/E2.

3. Secretaries don't require last names, at least not until they manage to become engaged to one of the partners. Unlike "Miss Calvet" ("Tomorrowland," S4/E13), Allison never manages to do that, so she must remain without a last name.

few episodes of the series well enough, but as the series progresses, understanding its cultural critique requires a fuller description of what courtesy is and how it functions. The series does not itself offer such a description, even in bits and pieces, but it does evidence a sophisticated understanding of the cultural function of courtesy. Thus, it enables the viewer to see that courtesy is not simply on the "what's expected of me" side of the conflict; it is instead an integral part of the fruitfulness of the conflict itself. Courtesy may depend on expectations, but it is not simply a list of expectations that one may adhere to or transgress. It is instead a system of communicating, renegotiating, and enforcing desirable social behaviors.

One can distinguish this system from other systems of communicating, renegotiating, and enforcing behavior, even when those systems overlap. Although the first communicators and enforcers of courtesy are parents, courtesy is distinguishable from familial culture in that the parents' wishes are subject to norms beyond themselves. The parent teaches the child *society's* expectations, apart from and in addition to his own.[4] Although religiously-framed moral values may undergird both the content and the goal of courtesy rituals, courtesy is distinguishable from religion in that it can and does exist in religiously plural societies. Different courtesy rules may apply between religious strangers than between religious friends, but the existence of rituals that do not presume religious agreement, or that negotiate potential hazards of cross-religious interactions, suggests that courtesy has a different function than religion.[5] Moreover, courtesy norms are often framed without specific reference to the ultimate reality addressed by religious systems, again suggesting that its functions are separable from the functions of religion.[6] Although courtesy norms, like legal ones, are systematic, promulgated, changeable, and enforced, courtesy is distinguishable from law in that its norms are promulgated, enumerated, adapted, and enforced without reference to the coercive power of the state. Discourtesy is not criminal and does not incur civil liability, and responsibility for its correction is diffused throughout society rather than concentrated in any

4. This is most obvious when a parent distinguishes between public and private behavior: "When we're at home, you can do such-and-such, but when we go out to eat, you must thus-and-so."

5. There is, admittedly, a certain Euro-centrism to this claim. It presumes the particularly Western notion of religion as an entity distinguishable from culture. Confucianism, if it is a religion at all, makes no appreciable distinction between religion and courtesy.

6. Again, this claim is easier to sustain in a predominately Western discourse. The substitution of "ultimate reality" for "a divine figure" is intended to make room for various forms of monism, non-theism, and humanism as religious entities, but such an inclusion is not self-evidently justified.

governing body. Courtesy, then, is a form of social control that extends beyond the family unit, that is intelligible without reference to ultimate reality, and that functions without recourse to state-sanctioned violence.

As a form of social control, courtesy has two primary elements: the grammatical and the political. These two factors are distinguishable but inseparable: the gestures of courtesy are a kind of language shared by the participants, the goal of which is the management of power differentials. Actions have meaning apart from the verbal utterances that may accompany them, a meaning that may reinterpret the verbal utterances themselves. Gestures must be capable of making statements, asking questions, correcting misunderstandings, offering persuasion, and signaling assent. That shared grammar of behavior allows the communication of relative status and power within a particular social body, of compliance and violation, and of the imposition of sanctions. If actions can have meaning, they must also have structures of intelligibility that allow for communication and secondary reflection on their meaning in ways that may be equally non-verbal. We might call these structures the rituals of courtesy—complex interactions that are often minutely scripted such that all parties involved can be trained in their performance and meaning.

These communications facilitate relatively peaceable social interaction. To be "in society"—that is, to be with at least one other person—is to be in a relationship at least partly structured by status and power differences. Where power differences exist, so too does danger. Personal insults can lead to interpersonal violence,[7] or they may poison future relationships,[8] so courtesy norms guide individuals in not offering insults, whether intentionally or unintentionally. Courtesy cannot manage power differentials, however, without attending to them rather minutely. One needs to know one's status and power *relative to* another person, so that one can show appropriate deference, exercise appropriate authority, use appropriate forms of address, and so on. Courtesy norms, then, allow one to judge one's position in and relationship to other members of the social body so that positional and relational differences may be observed. Compliance with those norms both acknowledges *and perpetuates* the status and power differentials being negotiated. By last-naming those whom I am expected to last-name and by answering to my first name when those who are authorized to first-name me do so and by insisting that those unauthorized to first-name me use

7. Witness Lane Pryce's and Pete Campbell's fisticuffs in "Signal 30," S5/E5.

8. So, Roger Sterling can't do client hospitality late in the series because of his prior discourtesies toward Burt Peterson. ("New Business," S7/E9.)

my last name, I submit to and enforce the distribution of status that drives naming conventions.

To describe courtesy in such a way is already to acknowledge that courtesy is a moral undertaking, or at least an activity that admits of moral inflection. Any of the elements of courtesy described above may be described in moral terms. The preference for peaceable rather than violent interactions may be described as a moral preference rather than a self-interested one; the obligation to preserve social order may be described as a moral obligation rather than a pragmatic one; submission to socially constituted authority may be described as a just recognition of morally constituted authority. As a careful study in the minutia of social courtesies, *Mad Men* is also at least potentially a careful study in the ethics of being in society.

COURTESY AS MORALITY

How courtesy relates to morality is a difficult question. Many characters in the series are portrayed operating under the conventional assumption that courtesy is transparent to substantive morality. Courteous people are good people, and rude people are not. Roger Sterling's secretary, Caroline, seems genuinely grieved by the death of Roger's mother—whom she liked because "she was always so polite to me, when she could hear me."[9] Mrs. Sterling must have been good, however little Caroline knew her, because she was courteous to Caroline. Courteous people are morally good, and morally good people are courteous. Conversely, failures of courtesy indicate moral failures. Rebecca Pryce—comparing American to British life with a pair of fellow expatriates—says, "I don't find Britain any more moral than here nowadays. The boys look like girls, and the girls . . . they don't seem to be concerned with manners, do they?" She equates manners with morals, and both can be judged by dress—that is, by how an individual has decked his body out in symbols of submission to or rejection of cultural expectations.[10] Betty Draper, wishing to deliver a stinging rebuke to squatters who didn't help her locate a runaway, accuses them of the worst thing she can think of—bad manners: "You have bad manners. You deserve to live in the street of this pigsty, and I hope you get tetanus or crabs or whatever else is crawling around here."[11] Bad manners not only merit disapproval, they

9. "The Doorway," S6/E1.

10. "Signal 30," S5/E5. For an example of how such social pressures function without recourse to governmental control, see Graybill and Arthur, "Social Control," 25–26.

11. "The Doorway," S6/E1.

merit divine or karmic retribution in the form of the worst things she can think of—disease and a filthy home.

These are unsophisticated, unreflective judgments about trivial matters. But courtesy is treated as a guide to character in more substantive matters as well. When Joan and Greg Harris give a dinner party in their home for his boss, a colleague, and their wives, his boss's wife clearly feels confident evaluating Greg's fitness to be a surgeon (which entails more than his competence *as* a surgeon) through *his* wife's ability to host a dinner party. "The fact that Greg can get a woman like you" means that his career prospects must be good.[12] Joan's creative confidence in entertaining evidences her good character, and her good character evidences his. Greg understands this dynamic as well as Joan, even to the extent that they fight over the seating arrangements for the dinner. Greg wants to cede the head of the table to his boss, knowing the man will expect such deference, while Joan wants to seat people according to Emily Post's guide, knowing that their wives will expect such evidence of correct etiquette. Both the deference and the etiquette signal fitness for membership in the society of surgeons and their wives; they signal a willingness to conform to the expectations of the group to which the Harrises wish to belong. Good manners portend good professional and personal relationships.[13]

Similarly, Don Draper unknowingly woos an important client with his easy manners. Discovering the country club's bar untended and another guest looking for a drink, he jumps behind the bar, makes a drink for himself and the guest, and engages him in a brief conversation. His small kindness—offering a drink, offering conversation, offering sympathy and a funny story—makes an outsized impression on the stranger, later revealed to be hotel magnate Conrad Hilton.[14] The anonymity of the exchange elicits Connie's positive evaluation: Don's casual courtesy is extended to the other man as a fellow man, not as a potential client. Overly accustomed, we are perhaps meant to infer, to a certain deference because of his name, he appreciates Don's courtesy all the more because it was offered freely, without reference to Connie's status. Of course, Connie misreads his own motivation. We have only to imagine Hollis, the black elevator operator, or Betty, the proper housewife, in Don's place to understand why.[15] Only a gentle-

12. "My Old Kentucky Home," S3/E3.

13. Creatively confident hostess that she is, Joan comes up with a solution to this argument, on which more below.

14. "My Old Kentucky Home," S3/E3.

15. Roger Sterling has primed us to think of men like Hollis, by performing in blackface in the preceding scene, and the comparison with Betty will suggest itself a few scenes later when she meets her own mysteriously significant stranger.

man of a certain race and class can share a drink with Connie, man to man. A "negro" or a "girl" doing the same would merely be fulfilling an expected service, one he would scarcely notice. Don's careless confidence signals membership, just as Joan's good housekeeping does. Don, like Connie, was not born to wealth, but now Don, like Connie, belongs in formalwear at a country club and can thus be trusted with Connie's business. Whatever Connie's self-understanding, the near-identity of manners and morals remains: the small kindness of ducking behind the bar to make himself and a stranger a drink is taken as evidence of something more substantive, something worth using to guide business decisions.

Shared courtesy norms communicate belonging, just as courtesy missteps can signal not-belonging. Not-belonging is expressed in subtle ways: through ignorance of another's language of courtesy, through blocked or repulsed offers of courtesy, and through refusal to extend courtesies to perceived outsiders.[16] Though Peggy Olson's date attempts to be courteous (accepting a meal he didn't order), he misfires at first because Peggy is "the kind of girl that doesn't put up with things."[17] She doesn't initially interpret his reluctance to correct the order as an expression of courtesy. Presumably intended as a kindness both to the waiter, by not troubling him, and to his date, by not causing a potentially embarrassing scene, Stevie's offer of courtesy is misunderstood because he offers it in a social language Peggy does not share. Peggy, accustomed to issuing directives by this point in the series, does not see the trouble to the waiter, and would not be embarrassed. Don knows enough to bring food to a family sitting shiva, but not enough to know what to do with his shoes or the definition of a minyan or why he can't be counted in one. His offer to help make up the minyan is blocked, but not even directly—his status as an outsider is disclosed from one insider to the other: "He can't. He's not Jewish."[18] Shirley and Dawn can know themselves to be outsiders—if they were confused on that point—because few at SC&P trouble to distinguish between the two black secretaries, confusing their names often enough that it becomes a shared joke between the two.[19] Learning names is an offer of courtesy no one bothers to extend the women—it

16. Although one could reference the economic language of blocked exchanges, the language here is from improvisatory theater. I am indebted to Samuel Wells's discussion on offering, accepting, and blocking, in *Improvisation*, 103–14.

17. "Severance," S7/E8.

18. Ibid. Even after these small proofs of his ignorance, he interrupts Barbara in order to pretend to knowledge he doesn't entirely have—"I've lived in New York a long time"—blocking his hostess's gentle offers of cross-cultural courtesy, her willingness to serve as cultural translator for him.

19. "A Day's Work," S7/E2.

is blocked by social structures that influence perception and that explicitly name "negroes" as outsiders.

The cumulative effect of such blocked courtesies can be permanent hostility. Though we viewers are not witness to Ken Cosgrove's relationship with other execs at McCann Erickson, we see the result. Ferguson Donnelly accuses him of ethnocentric arrogance: unlike Cosgrove, other people "don't walk around the office like their shit doesn't stink, and then go out and tell the world we're, what, a bunch of black Irish thugs?"[20] That it is phrased as a question tells us what we need to know: there need have been no direct confrontations, no blatant insults (of the sort to which Peggy and Joan will be subjected at McCann), no overt shows of contempt. Their mutual dislike was likely cemented by a series of miscues, blocked courtesies willfully misinterpreted as petty discourtesies, originating, perhaps, in cultural distance and spurred on by intellectual dissimilarity. Cosgrove, in his turn, knows he is not one of them: "I never fit in. I'm not Irish, I'm not Catholic, I can read." Their private lives—the literary pursuits Cosgrove carefully hides from his colleagues and the narrowness and chauvinism McCann execs try to gloss over with charm—cannot mesh, and even though they are kept rigorously private, the difference generates public strife. Ken and Ferg have nothing to share apart from work, no social or religious ties in common, nothing to bridge the gap between the civility owed relative strangers and the kindnesses exchanged between genuine intimates. Their attempts at connection have gone awry, and nothing remains but to assume the worst of each other.

Courtesy can have this function because courtesy norms both presume and enact a kind of agreement on moral norms. Only shared judgments about good and bad, right and wrong, things worth pursuing and things worth avoiding sustain shared social rituals. Only a shared understanding of how claims to status are sorted out at a dinner table can provide guidance in arranging the seats (or in selecting the right manual for providing such guidance). Only a shared understanding of the moral status of the nuclear family and its relationship to the state can provide a justification for preferring the liturgy of the Draper's family meal to the squatter's scavenged goulash. Only a shared understanding of who ought to serve and who ought to be served and in what manner can provide a foil against which making someone a drink can be understood as exceptional and therefore indicative of anything in particular. By acting out the rituals correlative to these shared moral judgments, and by objecting to and correcting any transgression of them, group members confirm their membership in the group and signal their compliance with its moral norms. The acknowledged and implicit

20. "Severance," S7/E8.

goods of a society are in a mutually reinforcing relationship with the rituals of social interactions.

Self-control is the primary moral value inscribed in courtesy expectations of the period of *Mad Men*. Where good morals and good manners coincide, their mutual expression bespeaks self-possession and restraint. Even good things must be expressed and enjoyed "tastefully" (that is, moderately). Roger, confronted with Don's disgust at Roger's behavior with his new wife, quotes his mother's dictum, "it's a mistake to be conspicuously happy."[21] But Roger has misapplied the dictum. He thinks his happiness with Jane has prompted anger and jealousy; he interprets his mother's advice as a counsel to self-interested prudence. He should rather have understood that his unguarded displays of affection toward his second wife transgressed the boundaries of self-control. They were tasteless and vulgar precisely because they were so exuberant. Don does not allow Roger's interpretation to stand: "No one thinks you're happy. They think you're foolish." Roger's exuberant affection becomes a social faux pas that opens him up to the private criticism of his peers, and shows him to be a man without wisdom. Don's words prove apt, as, quite predictably, Roger's open infatuation will soon turn to patent contempt. Roger ridicules Jane's ignorance and youth to Don ("Which one is Mussolini?"), and later tells her to "shut up" when she asks the time.[22] The man who cannot channel his emotions into socially acceptable forms cannot be trusted to guard those emotions when they endanger his relationships. He cannot be trusted to feel the right emotions at all.

Social expectations express this moral value but they also cultivate it. Peggy exemplifies moral formation through good manners.[23] Peggy seems not only to offer sincere expressions of courtesy, but also to cultivate her own self-mastery through adherence to courtesy norms. As such, she is implicitly contrasted with Pete Campbell—whose courtesies are often pro forma or transparently insincere and who is therefore never quite his own master. He is often shown as inept, someone whose attempts at courtesy are too ham-handed or ill-informed to be successful.[24] Most of his failures of

21. "My Old Kentucky Home," S3/E3.

22. "A Little Kiss," S5/E1.

23. The language of formation is different than the language of control used earlier. Both are necessary fully to understand the role of courtesy. One must describe the processes of social control in order to see how courtesy norms differ from laws and religious rituals. But the language of formation is necessary to see how courtesy norms are integrated into more explicit religious or philosophical commitments such that they cease to be merely external controls.

24. See, for example, his gift-giving missteps with Japanese executives from Honda ("The Chrysanthemum and the Sword," S4/E5), or his clumsy attempt to school Lane Pryce in the etiquette of engagement congratulations ("Tomorrowland," S4/E13). One

courtesy, though, are shown as failures of self-control. He is too selfish to master his negative emotions—which often leak out as peevish comments toward secretaries, his wife, and other subordinates.[25] He is excessively attentive to what is due his status, thus showing his insecurity with that status. A characteristic moment comes after the merger between SCDP and CGC: at their first joint conference meeting, to which he arrives late, he finds one chair too few at the table.[26] He asserts his authority over the recording secretary, asking her (not impolitely) to find him a seat; she instead offers him hers, which he accepts, prompting Ted Chaough, in turn, to offer his seat to the displaced woman. Though Roger Sterling diffuses the effect of Ted's gallantry by calling attention to it, the gesture has already shown Pete's status-consciousness to be clumsy and selfish. Similarly, when the firm rehires Freddy Rumsen after he achieves sobriety, Pete's concern with his own position blinds him to the fact that everyone else in the room intentionally avoids mention of Freddy's former drunkenness.[27] He intends to embarrass Freddy by bringing it up, but Sterling interrupts him (a superior can do that) to wrestle the conversation back to where it will preserve Freddy's dignity. Less concerned with others' status and feelings than his own, and too childish to exercise control over those feelings, he often causes embarrassment with his tactlessness.

Though Peggy, too, occasionally misspeaks in ways that create embarrassment, she more frequently exhibits courtesy toward those with whom she interacts. Many examples of Peggy's courtesy toward others are played for comic effect. Early in Season 3, she finds herself trapped in an incomprehensible conversation with Roger Sterling, who is emoting because his daughter has just threatened to disinvite him from her wedding.[28] True to form, his self-centeredness prevents him from engaging what she says, but she is enough master of herself to play her assigned role in the conversation. Similarly, much later in the series, she becomes trapped in a telephone conversation with Ted Chaough's pastor while trying to reach Ted.[29] Though on a matter of some urgency, she allows herself to be distracted by his in-

---

is "not supposed to congratulate the bride," true enough, but one is also not supposed to call attention to others' breaches of etiquette.

25. See, e.g., "Out of Town," S3/E1, after learning of Cosgrove's and his joint promotion; or "The Grown-Ups," S3/E12, where he takes out his bad mood on his secretary.

26. "Man with a Plan," S6/E7.

27. "Christmas Comes But Once a Year," S4/E2.

28. "Love Among the Ruins," S3/E2.

29. "The Doorway," S6/E1. This conversation immediately follows the exchange between Caroline and Roger Sterling, above, about his mother's courteousness. Thus the audience is primed to be attentive to Peggy's.

creasingly detailed questions about her family and religious upbringing. She cannot quite bring herself to refuse his questions or insult him, despite her frustration. (She does allow herself to challenge his handwriting speed.) She carefully couches her suggestion that he is a bad message-taker in respectful language. She even returns his closing salutations (presumably "Peace be with you") with the correct response ("And also with you"), despite their religious differences and her disinclination to be involved in a religious community by this point in the series. He has made an offer of good will, and she has accepted and reciprocated, albeit with a pardonable eye-roll.

These moments might be mistaken for proof of Peggy's unreflective conventionality. We viewers might view her as trapped in these conversations because she is too "nice," rather than because of her self-conscious formation of her own character through intentional acts of kindness. Her interaction with Pete Campbell after the announcement of his wife's pregnancy holds the key to interpreting these scenes.[30] It fits the same pattern: awkward or unpleasant conversation in which she nonetheless participates with verbal marks of respect and kindness toward the other person. This time, however, the viewer sees that her actions are carefully chosen and undertaken at some cost to herself. The conversation is more than awkward: it is positively painful, dredging up her profound sense of loss at having (secretly) borne Pete's child and given it away. She becomes aware of the expected child when Clara, a secretary, hands her a card while asking for a contribution to a gift for the Campbells. She masters herself enough to say, innocently but not *too* innocently, "Trudy's pregnant?" but hands the card wordlessly back and leaves the room. She appears overmastered by her emotion, and the viewer perhaps expects she will not participate in the congratulatory moment. (Clara has signaled what we should feel by pursing her lips in annoyance; why does Peggy refuse? Without a word? How rude!)

Her next move is a surprise: she has not left to wallow in or even to hide her feelings—she has gone to congratulate Pete personally. Her words are as warm as they are correct, and though the viewer understands the full emotional import of the moment, an in-universe observer would see nothing untoward. The act is one of quiet but astonishing bravery. Given the easy path to meeting social expectations—a signature on a card requires little and betrays nothing—Peggy instead faces down emotional pain and performs a *more* substantive act than that which was required.[31] We realize that

30. "The Rejected," S4/E4.

31. Lest we think that she is callous rather than courteous, the next scene confirms that it has indeed been painful for Peggy: she retreats to her office and quietly bangs her head on her desk. Her cool demeanor does not spring from indifference or disregard but from self-possession.

her small acts of courtesy have been neither unreflective nor (merely) conventional. She has chosen acts of kindness and respect even when she could have chosen otherwise without offense. She has met social expectations not out of timidity or childishness but out of a sense of moral obligation, and therefore she does more than merely or grudgingly meet them. Because she has done what she ought, over and over, she is able to do what she ought even when she feels otherwise.[32] Her self-mastery goes all the way down.

Yet, this very scene suggests an interpretive problem with courtesy: does the self-control it cultivates represent a genuine virtue or a kind of deception? If Peggy does one thing while feeling another, if her calm courtesies mask turbulent emotions, if her behavior meets expectations while denying desire, is it not possible that courtesy precludes rather than cultivates integrity, exactly to the extent that it divides the internal from the external self? Late in the series, Megan Calvet's father, Émile, raises this exact issue about another character, and thereby raises a question the attentive viewer has been wondering all along: are Don Draper's efforts to please sincere? "*Il a l'air un peu faux*," Émile opines—there is something a bit false about his "air," his manner.[33] The possibility of pretense does not bother him: he has just said of his daughter that she "pretends to find interesting what I find interesting because she loves me." To offer such pretense evidences affection, not untrustworthiness. To smooth over a potentially awkward situation (hosting the new bride's parents after a rather sudden engagement and marriage) through expressions of more warmth than one feels is an overture of friendship, not deceit. What bothers Émile is that his wife responds to those overtures with more warmth than she should feel; Don lights Marie Calvet's cigarette and remembers her favorite drink, and she takes more pleasure in his attention than she ought. Still, he correctly adduces that Don's charming manners are not self-interpreting. No act of courtesy has only one possible meaning, and the possibility that the person performing the act intends a double meaning—that is, intends to deceive some or all of the recipients and observers of the act—is a troubling one. The series suggests, indeed, that courtesy is (or at least may be) a lie, but an artful one, a sophisticated and powerful one.

32. Thomas Aquinas aptly describes the process of self-mastery through habituation: repeated actions form an inclination to the same action. If the repeated actions are consciously chosen, the habit, once formed, makes the performance of the action more natural, even in the presence of difficulty. *Summa Theologica* I-II.51–52, I-II.56.4.

33. "At the Codfish Ball," S5/E7.

COURTESY AS DECEPTION

The least artful and least successful forms of deceptive courtesy are those that call attention to themselves as courtesy. Self-professing courtesies are those most transparently insincere. The more one calls attention to one's own courtesy, the less likely it is that one actually intends to be courteous at all. Don Draper masks his reluctance to answer questions about himself with various excuses, but the pretense of self-effacing courtesy is a favorite. So, he deflects a reporter's questions about his childhood with, "We were taught that it's not polite to talk about yourself."[34] This recalls the scene early in the series where he deflected Roger Sterling's innocuous question about whether or not he had had a nanny growing up: "I can't tell you about my childhood; it would ruin the first half of my novel."[35] He is clearly jesting, and the remark suits the humorous badinage Sterling seems to prefer. Betty Draper, though, is unsatisfied (presumably having heard dozens of such remarks), and she resumes the subject in the car on the way home. Required not just to deflect a question but to deflect a question about his pattern of question-deflecting, he appeals to courtesy: "Maybe it's just manners but I was raised to see it as a sin of pride to go on like that about yourself." The dramatic irony here is no mere stylistic flourish: it calls into question the sincerity of all such deflections. Yes, the viewer knows more about Don's identity than the characters to whom Don speaks, but even without such questions, we know enough to distrust people who draw attention to their acts of courtesy.

Whether or not the ostensibly courteous person names her actions as courtesies, courtesies are often used to disguise frank self-assertion. Roger Sterling murmurs plausible-sounding condolences over the phone, presumably to the wife of someone from whom he'd hoped to drum up business—all the while flipping through his rolodex for another hopeful client.[36] When that fails, the principal members of the firm attend a competitor-colleague's funeral, scoping for the decedent's disaffected clients.[37] The rituals of mourning prove an apt cover for their self-interested pursuits. Similarly, offers of social friendship are often thinly-disguised attempts to secure status or curry favor. Pete Campbell's offers are often the most thinly disguised of anyone's: witness his clumsy attempts to glad-hand Lane Pryce

34. "Public Relations," S4/E1.
35. "Ladies Room," S1/E2.
36. "Hands and Knees," S4/E10.
37. "Chinese Wall," S4/E11.

after receiving news of his promotion,[38] his smarmy thanks to a Korean War veteran interviewing Don,[39] or his transparent glee at having scored the Drapers as dinner guests.[40] He comes to be so well-known for this tactic that Bert Cooper can openly mock one such display late in the series: "Crocodile tears. How quaint!"[41] This is the trouble with ostensible courtesy that is transparently self-assertive: no one is fooled. We are all Bethany, the young woman who recognizes and calls out Don Draper's attempt to manage his way up to her apartment (and presumably into her bed) after their first date. When he says, "Let me walk you up," (such a polite thing for a date to do!), she responds, "I know that trick."[42] We all do.

When courtesy succeeds, it can function not only to cover self-assertion but actually to collapse any difference between generosity and self-asserting power. Magnanimity displays security in one's own status; thus small gestures of kindness, which may seem to be gifts or offers of friendship, are at the same time proclamations of superiority.[43] Ted Chaough shows himself to be more secure than Pete Campbell in the scene where there were too few seats at the conference table (see discussion above). Magnanimously offering his seat to the secretary whose seat Pete usurped, Ted confidently perches on a console along the back wall of the conference room. He loses nothing by giving away his seat. His status is secure, and his generosity only further cements it. Unlike Pete, he does not need to fight for a place at the table—he belongs there no matter where he sits. Earlier in the series, Roger Sterling has taken Freddy Rumsen for a night on the town as a send-off—essentially firing him for drinking excessively (even by their standards).[44] After their night of drinking and gambling, Roger makes a show of offering Freddy the first cab home. Again, this costs Roger nothing—another cab

38. "Out of Town," S3/E1.

39. "Public Relations," S4/E1.

40. "Signal 30," S5/E5.

41. "The Quality of Mercy," S6/E12. Pete has just pretended to be "reluctant" to take over one of Ken Cosgrove's accounts.

42. "Public Relations," S4/E1. No one is fooled—including the person attempting to be charming. By the time Don reaches his nadir, he doesn't even bother with the cover of charm. Summoned to his apartment in the middle of the night, Diana-the-waitress appears to expect a little enticement. Don instead abruptly cuts short their conversation: "It's three in the morning. You know why you're here. Do you want a drink or not?" ("New Business," S7/E9).

43. Aristotle defines magnanimity as the superior person's sense that he really is superior. Generosity belongs to the magnanimous person precisely because it shows his greatness; the greater person gives, the weaker receives. Aristotle, *Nicomachean Ethics*, 4.3.

44. "The New Girl," S2/E5.

arrives straightaway—and in no way changes the essential reality that Roger does the firing and Freddy has been fired. Their statuses fixed, this magnanimous act (like their ostensible kindness in taking Freddy out on the town to begin with) is scarcely generosity at all. The signal characteristic of such acts is that they do not imperil the actor in any way; indeed, they often smooth the way for even greater exercises of raw power. It is not that the act of courtesy disguises an act of self-assertion—it *is* an act of self-assertion. It is a proclamation of status, power, and even dominance, for all that it presents itself in the garb of friendship.

If, in courtesy, an act of kindness can be identical with a display of dominance, so too can an act of command be identical with a display of deference. Courtesy is a tactic for managing others, one whose relative deceptiveness often depends on the status differential at play. Faye Miller has alerted us to this dynamic by telling us (and Don) the fable of The North Wind and the Sun, the point of which is that "kindness, gentleness, and persuasion win where force fails."[45] Kindness and gentleness are not goods in themselves but are strategies to acquire what one wants from others. Courtesy thus manipulates in its very appearance of renouncing coercive power. Lane Pryce's father, Robert, manages Lane's mistress, Toni, out of the room so that he can beat his son into submission in private. Toni has no idea that she is about to be managed out of Lane's life as completely as she has been managed out of the room: Robert has spoken to her with nothing but gentleness.[46] Joan Harris reproves the self-asserting John Hooker for not understanding how to use courtesy to manage people out the door: when Burt Peterson is fired, he makes a spectacular show of anger and defiance, trashing his office and insulting all his coworkers before leaving. "If you had spoken to me," she chastises Hooker, in an unfailingly sweet voice, "I would have been waiting with his coat and his rolodex."[47] Master of people that she is, she would have managed the situation to the benefit of the company and for the preservation of the fired man's dignity without ever having to use threats or coercion to do so.

Lane Pryce is the master of these sorts of manipulative shows of deference. In order to convince Pete Campbell to accept the hiring of a rival, he produces an apparently sincere, glasses-off, manly-look-in-the-eye pseudo-apology before blatantly appealing to Pete's vanity: "I apologize. It was wrong of me not to consult you. Roger Sterling is a child, and frankly

45. "The Summer Man," S4/E8.
46. "Hands and Knees," S4/E10.
47. "Out of Town," S3/E1.

we can't have you pulling the cart all by yourself."[48] Similarly, discovering a picture of a beautiful girl in a found wallet, he masks his salacious interest in the girl as deferential courtesies toward her when he telephones her about returning the wallet. When the girl's boyfriend later comes to collect the wallet, the man demonstrates just how effectively Pryce's courtesies had disguised his interests: "my girl said you were real polite."[49] So convinced is this man of (and by) Pryce's courtesy that he insists on giving Pryce a cash reward for the wallet, not even missing the picture Pryce has already stolen. "You're a real gentleman," he says, but "this" (the cash reward) "is the way we do things."

The supreme example of Pryce's ability to manipulate others through apparent deference comes when he persuades Joan to sleep with the Jaguar executive in order to win the account.[50] Desperate for a Christmas bonus to settle back taxes, he persuades her to prostitute herself out by pretending to object to her doing it all. Pete Campbell has already clumsily made the same attempt, but Pryce will succeed where he failed. His pretense of objection and affection and *delicacy* (unlike Campbell, he can manage the conversation such that neither of them need to use the word "prostitution") not only persuade her to do the deed, but to do it for a financial reward that suits his interest. She will sleep with the Jaguar exec in exchange for a partnership rather than an immediate cash settlement, preserving (so he had hoped) his Christmas bonus.

That this ploy succeeds so brilliantly—just as the rest of Pryce's attempts to control his life fail—suggests that we might look to his character for the beginnings of an answer to the question with which this section began: whether courtesy is essentially deceptive. Deception may or may not be at the heart of courtesy, but when courtesy consists of nothing *but* deception (charm, manipulation, strategy, control), it cannot sustain the pursuit of genuine moral goods. Like Don Draper, whose easy charm eventually gives way to petulant self-gratification, and unlike Peggy Olson, whose self-mastery confers genuine moral strength, Pryce has no resources on which to draw when his financial difficulties overwhelm him. His courtesies have masked his lack of self-control, all the more so because they have enabled him to be so successful at controlling others.

48. "Waldorf Stories," S4/E6.
49. "A Little Kiss," S5/E1.
50. "The Other Woman," S5/E10.

COURTESY AND ITS ABROGATION

This is finely wrought. The writers' attention both to courtesy in the time period depicted and to how courtesy works in any time period is excellent. What makes the series exceptional, though, is how it is equally attentive to the meaning and role of courtesy's abrogation: that is, of intentional acts of discourtesy. Just as acts of courtesy may be multivalent and equivocal, discourtesies may function and may be interpreted in a variety of ways.

The very first episode alerts us to the threat discourtesy (or the perception of discourtesy) poses to those in tenuous social positions. Rachel Mencken notes that Don can flout convention (asking an overly personal, implicitly insulting question) with impunity, while her gender constrains her to observe conversational conventions more carefully: "If I weren't a woman, I could ask you the same question."[51] Men can survive open discourtesies toward women, especially women who have transgressed social boundaries. Freddy Rumsen can slap Peggy's bottom with a file folder and call her "sweetheart;"[52] male account executives can make sexual jokes at their female colleagues ("You worried that L'eggs are gonna spread all over the world?");[53] and junior copywriter Joey can draw and publicly post a salacious cartoon of Joan.[54] Even when the discourtesy extends to threats of violence (witness Joey's "walking around like you want to get raped"[55] comment in the same episode), men understand themselves to be insulated from any social costs to these discourtesies and veiled threats because other men will reliably laugh off their misbehavior. (Don gives the typical "Boys will be boys" dismissal—in those actual words—in response to Joan's complaint about Joey's "ungentlemanly" behavior.[56])

Women cannot afford such discourtesies. They cannot even afford to call men out for their misbehavior in any but the most oblique way. Similarly, "negroes" cannot afford any but the most formal and deferential behavior. Thus, Hollis pointedly refuses Pete Campbell's disingenuous offer of informality: in response to Pete's, "It's just us, it's just Hollis and . . ." Hollis knows better than to respond anything other than, "Mr. Campbell."[57] Hollis no doubt remembers better than "Mr. Campbell" how "Sonny, from

51. "Smoke Gets in Your Eyes," S1/E1.
52. "Maidenform," S2/E6.
53. "Severance," S7/E8.
54. "The Summer Man," S4/E8.
55. Ibid.
56. Ibid.
57. "The Fog," S3/E5.

the elevator, and some janitor" were scapegoated after Peggy reports a theft early in the series.[58] Those with the lowest status can least afford any kind of misstep, because their positions are in constant peril.

Roger Sterling, on the other hand, can act with the confidence that his petty discourtesies will be read as a sort of charming insouciance. He can walk in late to a meeting where an underling will be fired, just as he can (later) take malicious delight in re-firing that same underling.[59] He can insult a waitress ("Hey, Mildred Pierce!") with no cost save the tepid protests of his companions ("I'm sorry about my friend, he's—" "Witty?").[60] He can use a crass sexual analogy to describe a business relationship, all the while apologizing to the recording secretary (whose "delicacy" as a woman he is presumably offending) yet continuing the analogy, too entranced by his own cleverness to desist.[61] Transgressive humor, particularly humor at others' expense, is only acceptable coming from those with power.

Whether the relevant power disparity is social or professional, acknowledged or implicit, innate or constructed, courtesy missteps imperil the weak(er) more than the strong(er). Conversely, just as the powerful can afford generosity without threatening their position, the powerful can transgress cultural expectations without threatening their cultural status. Indeed, the crassest and most deliberate offenses may function as dominance displays. Thus, a transgression may be calculated to claim and proclaim status; to flout common courtesies is to flaunt one's social power. Conrad Hilton can call Don Draper at any hour, can usurp Don's position in his own office, and can make any number of outrageous demands; he is both absolutely (because of his immense wealth) and situationally (because of the service provider's dependence on the good will of the client) more powerful. In the context of such a power differential, Connie's "Did I wake you" is as disingenuous as Don's "No, no, I'm up."[62] Connie is saying, rather, "Do you acknowledge my right to wake you whenever I choose?" and Don replies, as he must, in the affirmative. It is Don's responsibility not only not to be offended by Connie's discourtesy, but to preempt any suggestion of guilt Connie might feel for having exercised his power in this way.

If Connie is a needy, demanding infant (as Betty observes[63]), Lee Garner, Jr., is a schoolyard bully. Not only is he, like Connie, a wealthy client,

58. "Nixon vs. Kennedy," S1/E12.

59. "Out of Town," S3/E1; "Man with a Plan," S6/E7.

60. "Severance," S7/E8.

61. "A Day's Work," S7/E2.

62. "Souvenir," S3/E8.

63. "Wee Small Hours," S3/E9.

but with the formation of SCDP he is also The Client—the client on whom the entire firm depends, the one whose wishes must be met at all costs. He makes demands that transgress the boundaries of courtesy—insisting, e.g., that Roger Sterling don a Santa suit after he has politely declined—simply to prove he can; the more power he has to effect his will, the more urgently his will wills to be unchecked. He portrays his own pleasure-seeking as the innocent hedonism of a child: "Reminds me of when I was a kid. Remember that? You ask for something, and you'd get it, and it made you happy."[64] But this is patently false: the joy of getting his way comes not from the thing he has gotten nor even the getting of it but the offense he has given in the process. The offense is necessary to prove his power. If courtesy is that which protects the weaker from the brute power of the stronger, then discourtesy is a declaration by the stronger that they will not be constrained. The frank exercise of raw power regardless of social expectations displays a belief that the powerful will not suffer for having flouted convention. The weak(er) must constrain themselves by observing courtesy norms, because flouting them costs too much. The strong(er) pays no cost at all—or, no appreciable cost.

There is another kind of violation of courtesy—one that performs neither the weakness of the weak nor the strength of the strong but something entirely other. Some apparent discourtesies are actually strategic negotiations, carefully calculated risks that are taken not to *display* social power but to *acquire* it. These strategic violations may, for example, communicate an invitation to greater intimacy. Joan Harris signals her and her husband's desire for a warmer relationship with his superiors with the phrase, "We don't stand on ceremony."[65] The phrase is, itself, a kind of ceremony: the protestation against formal courtesies *is* the formal invitation to more familiar courtesies. The proposed relationship would still be guided by courtesy expectations, the set of expectations appropriate to intimates replacing those appropriate to acquaintances. The warmth of the Holloway-Harris party is intercut with scenes from the Sterlings's Derby party to highlight both the success of Joan's strategy and the appeal of the intimacy won thereby. The Sterlings cannot secure those sorts of friendships; people who do not particularly like each other have nothing but ceremony to stand on. Joan's risky move has proved profitable, but it was nonetheless a risk: her warmth may have been misread as proof of ignorance, or judged as tasteless informality, or rebuffed as a premature overture of intimacy.

64. "Christmas Comes But Once a Year," S4/E2.

65. "My Old Kentucky Home," S3/E3.

In an environment that demands aggressiveness and ambition, a certain willingness to dare disapproval is actually an asset. Still, these dominance displays are fraught, and Don Draper deploys them masterfully. When Honda Motorcycles sets up a highly structured competition among rival ad firms to win their business, Don strategically goads their rival into breaking the rules of the competition and then withdraws, brusquely accusing the Honda execs of unfairness.[66] His rudeness is calculated to impress, and it does. When he gets word that his partners are trying to force him out of the firm by picking up a tobacco company as a client, he arrives uninvited to a meeting with them.[67] His forceful persona once again turns a losing situation into a relative win, and the risk proves well-calculated. Certainly these scenes function as both narrative device and character study: Don is the principle agent of his own narrative, or at least he imagines himself to be. More important for our purposes, though, is the attention given to the politics of these calculated offenses. Don is not rude to Honda execs because he is more powerful than they—quite the opposite. He is in such a weak position, relative to them, relative to his competitors, that his only opportunity for advancement is this tactical discourtesy. He negotiates through rudeness, enacted through the language of courtesy. By enacting an offense typical of a stronger person, a weaker person can secure the power or position he lacks.

COURTESY AND SOCIETAL CHANGE

These moments of calculated offense are among the most important of the series. In these moments of strategic discourtesy, we witness the mechanism of social change at work. If strategic discourtesies may signal an individual's renegotiation of his own position in a given society, they may also signal a renegotiation of *the givens of society* themselves.

It is no accident that the characters in whom we see this phenomenon most clearly are those at some significant social disadvantage at the beginning of the decade: white women struggling against "traditional" mindsets that limit their employment, and black women struggling to find employment at all. Tactically appropriating courtesy and discourtesy to their own comparative advantage, these characters effect a renegotiation not just of their place in the social order but of the social order itself. Strategic transgressions of courtesy may be an opening proposal, as it were, for

66. "The Chrysanthemum and the Sword," S4/E5.

67. "The Runaways," S7/E5.

new patterns of interaction in light of societal changes. Carla the maid,[68] presumably in light of the slowly changing role of African-Americans in American society, uses calmly commanding language to forestall attempts to intimidate and insult her. When Gene Hofstadt, Betty's father, attempts to dominate and insult her, she quietly but forcefully says, "There'll be no more of that," to which Gene acquiesces.[69] While ever at the mercy of the whims of her employer—as her unjust dismissal from the Draper-Francis household will later show—growing cultural sympathy for the Civil Rights Movement allows her to exercise some authority over a relatively weak member of her employer's family. As social boundaries change, courtesy norms can be renegotiated, and strategic transgressions become the means by which those rules are negotiated. The social order can only change so much, however, and her attempt to enact a similar renegotiation with Betty a year or so later ("It was a mistake—there's no need for that kind of talk") fails.[70] Betty rejects her renegotiation and fires her without a reference.

Courtesy norms must also be renegotiated when moral boundaries change. If courtesy enacts and reinforces shared moral judgments, courtesy also becomes the technique through which the meaning of new moral arrangements is negotiated. Sometimes those renegotiations are unsuccessful: witness Peggy's failed attempt to signal the moral equivalence of marriage and co-habitation through the ritual of an "announcement dinner." Her mother refuses to accept Peggy's proposed renegotiation, and in defiance of all conventions of proper guest behavior, Peggy's mother demands her cake back when she discovers she has brought it to dinner at her daughter's apartment in error: "I'm not giving you a cake to celebrate youse living in sin."[71] Her words and her actions are an intentional affront, but the affront is necessary to describe the moral gulf that has opened between Peggy and her mother. They disagree on the moral status of Peggy's new venture (although Peggy's demeanor when Abe first asks her suggests that she, too, finds cohabitation less satisfying than marriage), and that disagreement cannot be smoothed over by a shared ritual of celebration. Such a shared ritual would be a miscommunication: the manners would not cohere with the morals they are intended to enact.[72]

68. Again, women in serving roles do not require last names.

69. "My Old Kentucky Home," S3/E3.

70. "Tomorrowland," S4/E13.

71. "At the Codfish Ball," S5/E7.

72. It would be easier to dismiss Mrs. Olson's words as merely cruel or judgmental if she had not earlier been shown in a more favorable light. When Peggy unexpectedly delivers a baby and suffers a mental breakdown because of it, she is shown supporting rather than condemning Peggy ("The New Girl," S2/E5). Although she sometimes

Sometimes renegotiations are too inchoate to effect any change, but merely articulate the recognition that change must occur. Henry Francis proposes to Betty, before her marriage to Don has ended. He recognizes that the situation does not match traditional courting rituals, yet his first wish, after having expressed his love for Betty and his desire to marry her, is for, precisely, a traditional courting activity: "I wish I could take you to the movies right now, that some theater was playing your favorite movie."[73] He struggles to redefine such conventions as courting when his very courting transgresses convention.

Madison Avenue, the new moral universe created by the Mad Men themselves, is a toxic and intoxicating mix of avarice, sexuality, intoxication, and constant terror.[74] In this new moral universe, confusion inevitably arises, and courtesy norms seem in a constant state of flux, if they function at all. The rituals of something like courtesy are still there, though. The trappings of class and education are what differentiate an acceptable LSD party[75] from a mere drug den.[76] Adultery is expected, but there are rules, for the sake of both the wife and the mistress. Open flirtation is acceptable in the new suburbia but Pete commits a real violation by sleeping with an actual neighbor rather than someone from the city.[77] If a woman comes to a married man's apartment in the city, she should know better than to "linger in the hallway."[78] Don's girlfriend does not want him to mention his wife at her apartment because it "makes me feel cruel."[79] Allison the secretary betrays her adherence to other moral conventions than the ones operative on Madison Avenue by appearing to expect a relationship after having had sex with her boss. Don must tutor her in the moral boundaries operative at the office: he speaks brusquely and matter-of-factly to her after she clearly

---

speaks sharply to Peggy, her supportiveness is marked enough to prompt resentment from Peggy's sister: "She does whatever she feels like, with no regard at all. You're too easy on her, you know that" ("Three Sundays," S2/E4).

73. "The Grown-ups," S3/E12.

74. There are a number of memoirs of the period, but one that has enjoyed a revival due to the success of the series is Jerry Della Femina's *From Those Wonderful Folks Who Gave You Pearl Harbor: Front-Line Dispatches from the Advertising War*. Della Femina's is remarkable not only for how it describes but how it displays the moral world he inhabited. It was, to put it mildly, less than admirable.

75. "Far Away Places," S5/E5.

76. "The Doorway," S6/E1.

77. "Collaborators," S6/E2.

78. "Collaborators," S6/E2.

79. "Ladies Room," S1/E2.

signals a desire for shared intimacy at work.[80] Megan, on the other hand, displays her understanding of these new moral conventions. After she and Don have sex on his office couch, she declines his pro-forma offer of a meal and continues calling him Mr. Draper.[81] She operates comfortably with the new conventions.

Peggy, however, proves the most adept at managing new moral boundaries with skill. Her transgressive career—success in a "man's" job in a "man's" world—opens her up to criticism and jealousy, and changes in the broader society force her to recognize that the Bay Ridge manners she came to the City with will not suffice.[82] She has a learning curve in this respect, to be sure. In the first season, Peggy is frustrated at how her conventional behavior is mocked rather than respected: "I follow the rules, and people hate me. . . . And other people, people who are *not good*, get to walk around doing whatever they want."[83] At some points, she attempts to mimic Don's directness and finds instead that it opens her to criticism from her coworkers[84] and does not particularly succeed with clients.[85] She continues addressing the other Mad Men by their last names, even after she has become one of them. Indeed, it is Joan who first begins to insist that the other "girls" use her last name—they had been continuing to call her "Peggy," as befitted her as a secretary.[86]

When Bobby Barrett advises her to treat Don as an equal, she does begin to call him "Don."[87] She is emboldened, no doubt, by the fact that he has required her help bailing him out of jail and covering up his affair with Bobby Barrett: she is not as absolutely powerless as she has been. That boldness continues to grow throughout the second and third seasons. She grows increasingly comfortable adapting to courtesy norms appropriate to her position in the company (even when her gender threatens to trump her job title in others' eyes)—asking for Freddy Rumsen's office after he is fired, refusing to get Roger Sterling coffee during their midnight sack of the Sterling Cooper offices.[88] She does not always succeed—once, for example, she receives a sharp dressing-down from Don for attempting to worm her way

80. "Christmas Comes but Once a Year," S4/E2.
81. "Chinese Wall," S4/E11.
82. "I'm from Bay Ridge! We have manners," ("Ladies Room," S1/E2).
83. "Nixon vs. Kennedy," S1/E12.
84. "Collaborators," S6/E2.
85. "Far Away Places," S5/E5.
86. "For Those Who Think Young," S2/E1.
87. "The New Girl," S2/E5.
88. "The Mountain King," S2/E12; "Shut the Door. Have a Seat," S3/E13.

onto an account.[89] But those failures primarily serve to show that she has understood and responded to the change in her position. She begins to act like an ad man, although with far more self-possession than the Mad Men around her. Her prior conventionality has not trapped her in a conventional position; instead, it has given her the wherewithal to manage subtle courtesy cues to her own benefit.

Peggy's mastery of these cultural forms of negotiation gives her tremendous power in a period of moral and social change. She is clever enough to come up with creative solutions to novel situations—more creative than Don's standby, the risky dominance-display-cum-power-play. She wittily deflects a lesbian come-on with enough humor and kindness that the rejected suitor remains a friend.[90] (This is the essential role of courtesy, after all: to manage potential social conflicts.) She had already learned to do so with male colleagues, but she adapts these courtesies to new moral situations in a way others cannot.[91] She challenges Stan's pretense at unconventional morality by herself violating convention in the most direct and challenging way possible: offering to work in complete nudity with him.[92] He proves unable to do so, and she has freed herself from his barrage of insults. Freedom is, in fact, the defining characteristic of her personality. Marshall McLuhan suggested that his generation's preoccupation with courtesy manuals resulted from a childish need for security among those who were fundamentally not free.[93] Peggy's social creativity shows us that she is genuinely free; manners are, to her, less a set of rules to follow than a set of tools by which to live well.

An exceptionally well-crafted series, *Mad Men*—like all good period fiction—says as much about the present as it does about the period it depicts. We, too, live in a period of immense social and moral change; we, too, struggle to articulate the requirements of civility and courtesy in the face of that change; we, too, find ourselves unexpectedly called upon to

89. "Seven Twenty Three," S3/E7.

90. "The Rejected," S4/E4.

91. Indeed, the series implicitly compares her with two other characters who find themselves deflecting similar same-sex advances and with far less success: Megan rebuffs the wife of her director ("The Better Half," S6/E9) somewhat more forcefully, and Pete repulses Bob Benson's oblique come-on by calling such a relationship "disgusting" ("Favors," S6/E11).

92. "Waldorf Stories," S4/E6.

93. "The socially immature cling aggressively to the books of Emily Post with the same baleful discomfort as the mentally exempt latch onto *Reader's Digest*," Marshall McLuhan, *Mechanical Bride*, 51. This creativity is also a prime characteristic of Joan Holloway-Harris's character: caught between slavish adherence to Emily Post and her husband's boss's potential expectations, she is able to come up with a creative solution to her dinner party problem.

respond to rituals of courtesy or discourtesy, offers of intimacy or enmity, displays of dominance or deference, often while being too shocked at the strangeness of the moment to be able to respond to it as we might after due deliberation. What are the new courtesies surrounding cell phone use: is it my obligation to strive not to overhear a private conversation being held in public, or is it the cell phone user's obligation to hold such conversations in private? Which is more important when one sees a co-worker who appears pregnant: avoiding potential offense or offering affirmation and support? Are offers of assistance for persons with handicapping conditions belittling or kind? What is the correct terminology for unmarried romantic partners? When should sexual orientation be indicated on a famous person's Wikipedia entry? Which personal questions count as invitations to friendship and which count as micro-aggressions? If we are to learn anything from *Mad Men*'s careful study, it must surely be that such questions will not be answered by some self-proclaimed expert producing a guide to which the socially timid must minutely adhere. They will be answered organically and democratically, in the mundane embodied wisdom of ordinary folk struggling to live well.

BIBLIOGRAPHY

Aquinas, Thomas. *The Summa Theologica of St. Thomas Aquinas*. Translated by Fathers of the Dominican Province. Allen, TX: Christian Classics, 1948.

Aristotle. *Nicomachean Ethics*. Translated by Terence Irwin. Indianapolis: Hackett, 1999.

Della Femina, Jerry. *From Those Wonderful Folks Who Gave You Pearl Harbor: Front-Line Dispatches from the Advertising War*. New York: Simon & Schuster, 2010.

Graybill, Beth, and Linda B. Arthur. "The Social Control of Women's Bodies in Two Mennonite Communities." In *Religion, Dress, and the Body*, edited by Linda B. Arthur, 9–30. New York: Berg, 1999.

McLuhan, Marshall. *The Mechanical Bride: Folklore of Industrial Man*. Berkeley: Gingko, 2001.

Wells, Samuel. *Improvisation: The Drama of Christian Ethics*. Grand Rapids: Brazos, 2004.

Chapter 4

"ALL THE RESEARCH POINTS TO THE FACT THAT MOTHERS FEEL GUILTY"

Maternal Desire and the Social Construction of Motherhood in *Mad Men*

—ANN W. DUNCAN

"All the research points to the fact that mothers feel guilty." So says Peggy Olson in one of the most personal pitches of her career—a campaign for Burger Chef.[1] Coming after extensive research interviewing tired and resigned mothers pulling out of Burger Chef with meals for their families, Peggy uses this psychological reality as the basis for her campaign. In this single pitch and the conversations that surround it, the show reveals the pervasive discontent underlying the experience of even the most ideal family unit while also highlighting Peggy's discontent in not conforming to those deeply entrenched paradigms of womanhood and maternal desire. In a brainstorming session about the pitch with her mentor Don Draper, Peggy voices this tension:

> Peggy: What the hell do I know about being a Mom? I just turned thirty.

1. "The Strategy," S7/E6.

> Don: Shit, when?
>
> Peggy: A couple of weeks ago. It doesn't matter. I kept it secret as long as I could. And now I'm one of those women lying about her age. I hate them.
>
> Don: I worry about a lot of things but I don't worry about you.
>
> Peggy: What do you have to worry about?
>
> Don: That I never did anything and that I don't have anyone.
>
> Peggy: I was in Ohio, Michigan, Pennsylvania. I looked in the window of so many station wagons. What did I do wrong?

Like the mother interviewed in an opening scene of the episode—frantic, guilty, brusque and harried—they represent a paradigm of womanhood that Peggy sees as separate from herself. Yet, the rapidity with which she switches from this self-loathing to creative inspiration shows that it is in the latter that her true passion lies. She snaps out of her depressing reverie and smiles, saying "What if there was a place where you could go where there was no TV and you could break bread and whoever you were sitting with was family? . . . That's it." The scene closes with a satisfied Don and Peggy dancing to Sinatra's "I Did It My Way," secure again in the choices each had made to pursue career above family.[2]

Media critics and audiences have focused extensively on the gender dynamics at play in *Mad Men* and the various expressions of feminism, sexism, treatment of sexuality and femininity that play out in the plotlines and character development of the series.[3] Yet, these themes do not only manifest in the blatant sexism of the workplace or the carefully prescribed gender roles in the home. They manifest in the most revealing and complex ways within the female characters themselves. Peggy does not smugly note the guiltiness and unhappiness of those mothers who conform to the societal paradigm of motherhood as an affirmation of her own choices. Instead, it sends her into a spiral of self-doubt, depression, and second-guessing. For Peggy and other female characters, choices in relationships and career belie challenges faced by women in these changing times. Through the various plotlines of the series, we see the often exciting, sometimes devastating life events of these women as they negotiate social constructions of gender and family and the effects of social change on their self-image, career opportunities, and gender identities. These negotiations effectively demonstrate, and sometimes complicate, the many feminist discourses of the time.

2. Ibid.

3. An exhaustive list of such pieces is beyond the space limitations of this chapter. An example of this type of criticism can be seen in Coontz, "Why 'Mad Men.'"

Even as second-wave feminism moved to separate women's identity from their biological functions, one cannot understand the role of sex and gender in *Mad Men* without attention to motherhood. As Nancy Chodorow wrote in 1978, "Women's mothering is one of the few universal and enduring elements of the sexual division of labor."[4] In a time of carefully proscribed gender roles and dramatic social change, motherhood provides an illuminating lens through which to examine the evolution of feminism in mid-twentieth-century America generally and in the *Mad Men* series more specifically. As a societal construction mediated by the often-conflicting maternal desire, career ambition and quest for personal fulfillment, motherhood finds expression in dramatically different ways in the characters of *Mad Men*.

This chapter will focus on three of the most prominent characters—Betty, Peggy, and Joan—and their own experiences of motherhood. Though starkly different in their maternal choices and styles as well as personalities and career paths, these women each demonstrate the growing complexity and inextricability of maternal, personal and professional concerns in the wake of increased but incomplete gender equality. Through attention to theories of motherhood and feminism through the twentieth century, this chapter will examine these three women as a lens through which to understand the complexities of motherhood in the contemporary age. Such an examination will reveal that these character's stories are both deeply conditioned by their historical context and directly relevant to contemporary discourse about motherhood and feminism.

BETTY AS THE SECOND SEX

Simone de Beauvoir's *The Second Sex* (1952) set the stage for second-wave feminism's emphasis on transforming women's roles in the family and in society at large. Moving beyond a mere acceptance of greater legal rights within traditional family structures, the *Second Sex* illustrates the second-wave feminist perspective of rejecting all stereotypical or traditional roles for women in favor of those previously unavailable. Though de Beauvoir stopped short of arguing that motherhood was incompatible with women's liberation, she indicated that a nearly unavoidable contradiction exists between motherhood and personal fulfillment. Indeed, according to de Beauvoir, the very biological make-up of a woman predestines her to subservience. Biologically, she explained, the woman's body works to protect the egg and its development above the less evolutionarily beneficial personal

4. Chodorow, *Reproduction of Mothering*, 3.

desires and aspirations of the woman. Her life, then, is a gradual unfolding of a process in which she plays only a passive and supportive role.

De Beauvoir asserts that this disjuncture between a woman's soul and body ultimately dooms her to imbalance and frustration. Biological determinism has meant women are "doomed . . . to domestic work" and a lack of access to meaningful work due to "her enslavement to the generative function."[5] Society supports this enslavement through the perpetuation of characterizations of women and women's reproductive functions as impure, unsavory, and dangerous. Furthermore, the mother-child relationship itself represents both fulfillment and imprisonment.[6] This is because "the advantage man enjoys which makes itself felt from his childhood, is that his vocation as a human being in no way runs counter to his destiny as a male."[7] Though a woman can benefit from the satisfaction of her biological destiny, such fulfillment comes at the expense of other destinies along with her peace, success, and wholeness. Women who have been raised in this culture, groomed for motherhood, rely completely on it for their sense of self-worth, fulfillment, and purpose—even if that fulfillment will always be incomplete. This culturally bred maternal desire can be seen no more clearly than in Trudy Campbell. In the midst of infertility struggles, she advocates adoption or any means necessary to secure a child "because otherwise what's all this for?"[8]

In the character of Betty Draper/Francis, we see the paradigm against which second-wave feminism fought and the ways in which this set of life circumstances acts as a barrier to emotional maturity and personal fulfillment and, ultimately, serves as a death sentence. Strictly limited by her gender even as she enjoyed all the luxuries of privilege accorded her race and class, we see a subtle but steady evolution of character throughout the seven seasons. More than any other player, Betty demonstrates the limitations brought by a patriarchal society but also the negative repercussions of those limitations in her as a wife, a mother, as well as in terms of her general well-being and happiness. Though seemingly perfect on the outside and conceivably having all she could ever need, Betty again and again demonstrates frustration, restlessness, and even bitterness about her life's path and the complicity of those around her.

In life circumstances, Betty's path is completely defined by her gender and roles as wife and mother. We are first introduced to Betty as mother

5. De Beauvoir, *Second Sex*, 117.

6. Ibid., 518.

7. Ibid., 682.

8. "The New Girl," S2/E5.

and wife at the end of the pilot episode. Sleepy and beautiful, lying waiting in their marital bed, Betty notes Don's late arrival home but also seductively calls him to bed. Later, wistfully watching him caressing the sleeping head of their child, the camera pans out on an ideal couple in a beautiful suburban home.[9] In subsequent episodes, we learn a bit more of Betty's history: her mother deeply influenced Betty's notions of femininity,[10] and she once had a modeling career that ended abruptly upon her marriage and pregnancy.[11] When Don surreptitiously secures Betty a modeling job (while allowing her to believe she earned it on her own merit), her reaction upon being let go was enough to earn the show an award from the Art Director's Guild. In "Shoot," Betty is let go from the modeling job rather abruptly but her pride keeps her from revealing this to Don. She breaks the news over a carefully prepared dinner:

> Don: How'd it go today?
>
> Betty: Good! Fun! They were talking about a whole string of other possibilities for me. But honestly, I don't think I want to work anymore. I don't like you coming home to some whipped together mess of whatever's left in the fridge. And, frankly, I don't like Manhattan on my own. It's harsh.
>
> Don: It can be.
>
> Betty: And what am I going to do? Run around the city with my book, like some teenager, making a fool out of myself?
>
> Don: If that's what you want. It's my job to give you what you want.
>
> Betty: You do. Look at all this. I don't know what I was thinking.
>
> Don, taking Betty's hand: Birdy, you know I don't care about making my dinner or taking in my shirts. You have a job. You are mother to those two little people and you are better at it than anyone else in the world. . . . Or, at least, in the top five hundred.
>
> Betty laughs.
>
> Don: I would have given anything to have had a mother like you. Beautiful, and kind, and filled with love like you're an angel.
>
> Betty: You're sweet.
>
> Don, eating: And this does not taste whipped up.

9. "Smoke Gets in Your Eyes," S1/E1. In Chapter 1, Howard Pickett notes the significance of this scene from the perspective of Don's character development.

10. "Red in the Face," S1/E7.

11. "Shoot," S1/E9.

Betty, smiling: No, it's good, isn't it.

The scene ends with Betty seemingly fully content with this new decision. In the morning, this glow continues as she dotes on her children, excited about the rather mundane plans of "going to the community center to watch them fill the pool." In the next scene, however, we see her get up from her seat where she had been quietly smoking, go outside with her son's BB gun and shoot her neighbor's pigeons as Bobby Vinton's "My Special Angel," plays louder and louder in the background and the credits roll.[12] From Don's "Birdy" and "angel" to pigeon assassin, Betty exhibits what has been called "fierce lioness behavior " in this "greatest act of . . . motherhood and frustration."[13]

After this act of impulsive aggression, Betty expresses her lack of identity with a kind of listless defeat, complicated by the crumbling of her birth family and the family she created through marriage. Visiting her ailing father in Season 2, Betty comes face to face with the memory of her idealized mother. Gazing at her portrait or fretting over the trivialities of her father's girlfriend or the change in her childhood home, Betty rebuffs Don's attempt to get her to eat by saying, "Stop it Don, nobody's watching."[14] Her own marriage is crumbling and the mother she once idolized as the pinnacle of femininity and motherhood, has been nearly forgotten soon after her death, and her own remaining parent is fading quickly into dementia.

The episode ends with an unexpected heart to heart with a neighbor friend. When Betty reveals that Don has moved out, the friend responds that she found it easier without her husband around. Betty pauses and then says, "Sometimes I feel like I'll float away if Don's not holding me down."[15] As her father loses his grip on reality and Don steps out of her life, her sense of grounding and identity have been deeply unsettled. After an early attempt to seek relief through psychiatric care, Betty finds some relief through discussion with a counselor hired to treat Sally in Season 4. That it takes a child psychiatrist to support and hear Betty belies the immaturity and base emotion at the core of her troubles. In a less than subtle move, the writers show the ways in which Betty's emotional intelligence has been stunted by the protective coating of a privileged life and the easy and early transition from childhood to a married life of ease.

The automation of her life choices seems to have muted Betty's maternal desire and created a mothering style that is minimalist, often harsh

12. "Shoot," S1/E9.

13. Lipp, "'Shoot' Wins ADG."

14. "Inheritance," S2/E10.

15. Ibid.

and distant. This disengagement can be seen in her response to the news of her last pregnancy. Don and Betty are both stone-faced when Betty reveals this pregnancy that she will carry to term because the doctor discouraged abortion due to her life situation.[16] In labor, Betty is to be shaved, given an enema and heavily medicated. We see her horror as she yells out, "I'm just a housewife, why are you doing this to me?"[17] Coming in and out of a twilight sleep, we see the physical manifestation of Betty's conflicted feelings about the birth, dreams that remind us of her still child-like sensibilities and real distress as she, at times, wakes to realize what her body is currently going through.

As a mother, Betty freely relies on nannies to provide the majority of care for the children. Her interactions with them are, for the most part, harsh and questionable at best from a modern perspective.[18] When she catches Sally smoking, Betty's anger stems from a concern about fire safety alone and she responds by locking Sally in a closet.[19] In a scene meant to shock modern ears by its outrageousness and yet telling of Betty's complicated relationship to children, sexuality, and her marriage, Betty jokes with Henry about raping Sally's visiting friend:

> Betty: She's just in the next room. Why don't you go in there and rape her? I'll hold her arms down.
>
> Henry: Betty, what the hell?
>
> Betty: You said you wanted to spice things up. Will it ruin it if I'm there? You know what? If you want to be alone with her I'll put on my housecoat and take Sally for a ride. You can stick a rag in her mouth and you won't wake the boys.
>
> Henry: Alright. Alright, Betty.
>
> Betty, smiling: My goodness, you're blushing.[20]

The viewer feels shock and horror but also recognizes Betty's inner turmoil. Now an overweight housewife, Betty recognizes the distance between herself and the youthful beauty and allure of her daughter's friends. She searches for a way to harness that allure to reinvigorate her own lackluster relationship. She seeks to find excitement in her otherwise mundane life. We

16. "Meditations in an Emergency," S2/E13.

17. "The Fog," S3/E5.

18. For more on the questionable parenting of Betty and others, see Susan Frekko's chapter 15.

19. "The Mountain King," S2/E12.

20. "The Doorway," S6/E1.

also see her tendency to dehumanize the women in lower socio-economic or cultural positions than her—whether a nanny or her daughter's friends.

This transfer of internal discord to relations with children comes to a head in the episode "Field Trip." This episode presents one of the most striking statements on the contradictions of motherhood and the divide between Betty's children's views of her, her maternal desire and the reality of their conflicted relationship. After hearing her friend Francine praise the benefits of being a working mother, Betty agrees to chaperone Bobby's field trip in an impulsive attempt to reaffirm her life choices and commitment as a mother. When Bobby gives away Betty's sandwich in exchange for gumdrops, Betty is incensed. Bobby responds, incredulously, "I didn't know you were going to eat."[21] Betty responds coldly, "There were two sandwiches," and then forces him to eat the gumdrops while she dons her dark sunglasses, turns away and placidly smokes a cigarette. Betty's coldness and Bobby's incredulity reflect a story from showrunner Matt Weiner's own childhood. As he describes in an interview with *Fresh Air's* Terry Gross, he once traded his own mother's sandwich for gumdrops because "I'd never seen her eat, so it was kind of shocking to me that she had brought lunch for her." He reflects that the act "was irreversible and something that I would wake up and be crying about, you know, for years, you know, just like, oh, why did I do that?"[22]

For all her parenting fails, Betty surprises viewers with moments of maternal tenderness and evident affection. These moments suggest that her distance comes less from her own lack of maternal desire and more from projections of her own existential dissatisfaction. These moments of tenderness come most often during times when she is transferring gendered socialization to Sally—such as her discussion with Sally about first kisses,[23] and her treatment of Sally when she has her first period.[24] This tenderness finds expression again in the final season when Betty becomes a figure of greater maturity, hopeful change and begins to experience a level of redemption. After numerous instances of beginning to speak her mind in political discussions with her husband and guests and frustration at her husbands' reprimands, a fire is lit inside Betty.[25] She returns to school with a kind of determination and light in her eyes not seen before. A chance encounter with Don provides one of the most genuinely loving and sweet moments between Don and Betty as she tells him of his plans and, reminiscent of

21. "Field Trip," S7/E3.
22. "'Mad Men' Creator," *Fresh Air.*
23. "Souvenir," S3/E8.
24. "Commissions and Fees," S5/E11.
25. "The Runaways," S7/E5.

their talk at the dinner table after the end to Betty's short revival of her modeling career, tenderly supports her in the new venture.[26] And yet, Betty will barely make it to school before the fall that will ultimately lead to her dire diagnosis.

Her rapid and tragic decline shocks viewers as well as the other characters and quickly cuts short any dreams of education and career. She begins the series an immature but seemingly perfect suburban housewife with a lovely house and car, beautiful dresses, healthy children and a handsome husband who provides all the money the family could want. Her growing sense that this is not enough and that she is in some sense trapped by these realities leads to her treatment of her children in a way that reveals the frustration and bitterness slowly growing over time. She has been, in the words of de Beauvoir, "enslaved to the generative function."[27] She feels the dissatisfaction of vocational dreams cut short, a life not as perfect as it seems from the outside and a sense of entrapment by her children, her marriage, her class, and the very material possessions that are otherwise a mark of her happiness.

Betty's role as mother makes her ending less tragic. Through her mother's decline, Sally emerges as a mature young woman who begins to mother her younger brothers and seems aware enough of her parents' various weaknesses and failings to resolve to herself not to repeat these patterns in her own life. We have hope for Sally and have the sense that her lack of strong parenting has only made her stronger and wiser as she faces her own adulthood. Betty recognizes this generational shift as well. In the note she writes to Sally, meant to be read only after her death, Betty writes, "I always worried about you because you marched to the beat of you own drum, but now I know that's good. I know your life will be an adventure."[28] For a woman who had to find adventure through pigeon shooting, a clandestine affair, and the suggestion that her husband rape her daughter's friend, this is hopeful indeed. Yet, in this hope we see the dramatic contrast between generations. For *Mad Men*, the feminine mystique is fatal—to the spirit, the soul and, in this case, even the body. The bonds of the biological and the cultural trappings of maternal paradigms can lead to a life that has every appearance of comfort, happiness, and ease but that is underlain with dissatisfaction, frustration, and unrealized potential.

26. "Don't Make Plans for Me," S7/E12.

27. De Beauvoir, *Second Sex*, 117.

28. "The Milk and Honey Route," S7/E13.

PEGGY AND THE OPPRESSION OF THE FEMININE MYSTIQUE

If Betty Draper/Francis represents the dangers of socially determined maternal paradigms and the tragedy of a spark lit too late, Peggy Olson represents a competing paradigm, still rife with frustrations. While Betty represents the frustration inherent in a woman's existence when biologically determined, Peggy represents an attempt to break free and pursue personal vocational goals. Simone de Beauvoir's landmark work in feminism provides an indictment of the Betty Drapers of the world or, at the least, the society that makes their life's path the path of least resistance.[29] With such a deeply embedded social construction of motherhood as the reality in 1960s America, many women rejected motherhood completely as a means of avoiding this trap and finding personal fulfillment. In many ways, Peggy Olson represents such a woman, particularly in her earlier career. Yet, as her story progresses and as the opening vignette suggests, her unconventional path leaves her with a new sense of dissatisfaction—one that she has trouble attributing to her own desires or societal pressure.

This tension has been explored by Betty Friedan as she describes this liminal state as a reality for many women and a problem incumbent upon women and society as a whole to address. In *The Feminine Mystique* (1963), another landmark book in second-wave feminism, Friedan wrote that women must find a balance between the reality of motherhood and the desire to mother, and the desire to seek personal fulfillment. These dual impulses, Friedan asserted, come from the particular problem of womanhood, a "problem that has no name." Describing the situation of the suburban woman in America, Friedan writes, "As she made the beds, shopped for groceries, matched slipcover material, ate peanut butter sandwiches with her children, chauffeured Cub Scouts and Brownies, lay beside her husband at night—she was afraid to ask even of herself the silent question—'Is this all?'"[30] We see in the tragic figure of Betty Draper, the potential futility of this questioning for those living the maternal paradigm and we have to wonder at the choice of "Betty" as the name for this character.

The American mother faced these existential dilemmas alongside societal demands that she fulfill maternal and wifely functions. As she found her way through this labyrinth, society characterized her as "a frustrated, repressed, disturbed, martyred, never satisfied unhappy woman. A demanding, nagging, shrewish wife. A rejecting, overprotecting, dominating

29. For a defense of Betty's character, from de Beauvoir's existentialist-feminist perspective, see Jacob Goodson's chapter 12.

30. Friedan, *Feminine Mystique*, 15.

mother."[31] Such characterizations along with warnings about the effects on children of mothers who work outside of the home or otherwise pursue non-domestic interests pushed women to remain a happy, smiling, very active and nurturing mother and wife. By upholding the "feminine mystique," a woman lives according to the limited role society has placed before her. She is giving in, not fighting and not progressing. She argued that, for religious women, breaking away from this model was even more difficult because the feminine mystique "is enshrined in the canons of their religion, in the assumptions of their own and their husbands' childhoods, and in their church's dogmatic definitions of marriage and motherhood."[32]

The danger of perpetuating the feminine mystique, Friedan wrote, is not just the effect on the woman herself. Instead, she writes,

> If we continue to produce millions of young mothers who stop their growth and education short of identity, without a strong core of human values to pass on to their children, we are committing, quite simply, genocide, starting with the mass burial of American women and ending with the progressive dehumanization of their sons and daughters.[33]

Were women to reject this mystique and act freely, the possibilities would be endless; such self-limitation would cease.

The storyline of Peggy Olson shows a different path that responds to the same societal pressures and ideals faced by Betty Draper/Francis. Subject to the same sexism and assumptions about a woman's place and proper path, Peggy quietly bucks the status quo throughout the seven seasons as she strives for career advancement and the respect of her male colleagues. That this is a path of rejecting family and motherhood in favor of career (and the fact that this is an either-or choice) only crystalizes as Peggy becomes pregnant. The audience and Peggy do not realize the weight gain is due to pregnancy until late in the pregnancy.[34] She makes this realization at virtually the same time she achieves a promotion. The choice is stark and clear.

While Peggy's emotional distancing mirrors Betty's detached experience of childbirth in the next season, Peggy's choices are more cut and dried. Whereas Betty, as a married housewife of means, possesses all the societal structures and culturally acceptable life circumstances within which to welcome a child, Peggy has none. Betty has no choice but to accept

31. Ibid., 189.
32. Ibid., 351.
33. Ibid., 364.
34. "The Wheel," S1/E13.

motherhood. Peggy has a choice and rejects this version of motherhood. The viewer does not learn of the circumstances of the birth until Season 2. In a flashback, we see Peggy refuse to hold her baby once she gives birth. We see Don—the closest Peggy has to a father figure—sitting by her bedside after she has given birth and encouraging her to move on, saying to her "it will shock you how much it never happened."[35]

Though Don's comments here could be read as cold compartmentalization, the autobiographical element should not be dismissed. In another flashback at the beginning Season 3, we see Don's mother—a prostitute who suffered many unintended pregnancies with complications ending, eventually, in her death. In this episode, Don flashes back to his mother who has just given birth to a stillborn child whose father accuses her by saying, "You killed another!"[36] We learn that Don was given the name "Dick" because of his mother's threat to castrate his father if he got her pregnant again. In Don's memory, pregnancy is a cause of anger—an invitation to pain and a cause of death. This wall, once erected, changes the course of one's life in irreparable and negative ways. While this association does not play out in his own cookie-cutter suburban family life, it does in relation to Peggy, the character who arguably most mirrors his own creative personality and own ambitions.

This absolute discomfort with the idea of motherhood and emotional brick wall that Peggy erects to protect herself comes through in several subsequent episodes. The young priest that she befriends tries to encourage her to talk about the baby,[37] and he recommends reconciling this life even with God.[38] When Peggy finally tells Pete that she had a baby and he is the father, the audience holds its breath. Though familiar with Pete's questionable character, we hope Peggy will find some outlet to process the emotions, sadness and fear of these events by sharing this information and claiming its truth. However, Pete expresses only disgust and disbelief; he says only, "Why did you tell me?!"[39] She neither has motivation to share the information nor the promise of any support in her decision. There are no societal structures for motherhood in this context and it is clear that her vocational aspirations would end should her choice be different. The biological remains a persistent limiting reality in her life but she is unable to claim any middle ground that maintains both motherhood and a career.

35. "The New Girl," S2/E5.
36. "Out of Town," S3/E1.
37. "A Night to Remember," S2/E8.
38. "Meditations in an Emergency," S2/E13.
39. Ibid.

Peggy's relationship to motherhood and traditional feminine roles evolve as the show moves onward. In Season 4, we see Peggy try on Faye's wedding rings when she thinks no one is looking.[40] Yet, early in the next season, she shows extreme discomfort in holding Joan's new baby;[41] throughout the next three seasons, we see her struggle to find and maintain any kind of love interest while pursuing her career. The bitterness, harshness, and long hours brought on by her devotion to work give her an air of desperation, loneliness but also coldness that make such relationships both logistically and emotionally impossible. She suffers from breaking free of the feminine mystique in a society not yet structured to support such a choice. She has gone from Betty's extreme to another, and neither brings happiness or fulfillment.

Peggy's story finally brightens in the last season. Her continued career success satisfies her aspirations even as she continues to aspire to more. Yet, she also begins to realize that she wants and needs more to make her life one that includes happiness and fulfillment outside of the world of advertising. We see glimmers of this in her brainstorming for the Burger Chef commercial. In Season 7, Episode 6, Peggy talks with Don about motherhood and family in a way that, on the surface, is focused on an effective ad campaign but, on a deeper level, seems to speak to her innermost desires. The wistfulness and hope with which she describes this life give the audience a glimmer of the future Peggy hopes to, in some small way, experience. This continues as Peggy's maternal desire emerges, focused on a neighbor boy. Peggy surprises herself in how deeply she grieves his upcoming departure.[42] When giving the Burger Chef narrative in the sales pitch, it becomes clear that the dreaminess with which she describes the paradigm of the family meal is not just for the benefit of the client. She wants this for herself—and not simply because this is what society prescribes for her.

In the final episodes of the series, Peggy finally finds some personal satisfaction to mirror the vocational satisfaction she has worked so hard to achieve. As she and Stan finally bring their relationship to fruition, the audience is hopeful. Not only do they seem to balance each other's personalities in a way that might bring Peggy some peace in an otherwise uptight disposition but we see that she can freely share with Stan all her past and her hopes for the future in a way that starkly contrasts with Peggy's revelation to Pete about the unwanted baby. These revelations come as the agency shoots a commercial involving child actors and the office is filled with these children

40. "The Rejected," S4/E4.
41. "A Little Kiss," S5/E1.
42. "The Astronauts," S7/E7.

and their mothers. As each of the characters project their own issues with parents and parenthood onto their interactions with the children and each other, so too does Peggy. When Peggy and Stan argue about the mother of a child who inadvertently injured herself with a stapler, Peggy rebuffs Stan for arguing that this mother should never have had children. She then blurts out the defensive non sequitur, "I don't hate kids."[43] An interaction with another mother later in the scene leaves Peggy feeling unfit to be a mother herself and tongue tied and confused in determining how to respond. Yet, as her relationship with Stan develops in the final episodes, we see that a family might be a real possibility for her. The audience hopes that this relationship with someone who both understands her work but also balances her headstrong and driving focus might give her fulfillment of these very human desires.

In comparison to Betty, Peggy's path suggests a feminist sensibility. She rejects the paradigms latent not only in society but also within the Catholic Church in which she grew up.[44] She works tirelessly to follow a career path that is rocky and, at times, impassable, as she achieves success in a field almost exclusively populated by men. She rejects motherhood even when its reality hits her square in the face and she preferences her own career above any romantic relationship even as it leaves her lonely and romantically frustrated. This conflicted relationship with romance emerges most heartbreakingly and humorously in the Valentine's Day flower mix-up when Peggy mistakes her secretary's flowers for her own. Her embarrassment combines with her anger at receiving the flowers and then disappointment in not.[45]

Yet, Peggy's feminism is much more personal than it is political. In a 2013 interview with *Entertainment Weekly*, actress Elizabeth Moss refused to characterize Peggy as a feminist: "She doesn't care about politics unless it relates to her job. As the show goes on, the generation passes her by. She's not going to be a hippie. She's a professional woman."[46] Her choices demonstrate this new feminist sensibility even as she is narrowly and singly focused on her own path. Her complicated life demonstrates what can be lost when the focus shifts from the maternal to the vocational. The balance that Betty Friedan suggests between femininity (and, by association, maternity) and individual autonomy remains largely elusive to Peggy until the end of the series. The viewer has some hope for Peggy and her future with Stan and in her career, even as the viewer mourns for so many years spent in

43. "The King Ordered It!," S7/E11.

44. For more on Peggy's Catholicism, see Heidi Schlumpf's chapter 16.

45. "A Day's Work," S7/E2.

46. Bahr, "'Mad Men' and Feminism."

frustration and loneliness. Our last vision of Peggy involves her typing away at her typewriter with her newly declared lover massaging the tension from her shoulders.[47]

JOAN—RIDING THE THIRD WAVE

In Betty and Peggy, we see the deep ramifications of a patriarchal and restrictive society on women of the 1960s and 1970s. Forced to choose between family and career, we see that neither path is sufficient for a fulfilling and balanced life and within both choices lie significant challenges that lead to frustration and limitation. We see Betty's sad end as she begins to claim some agency for her life but is cut short by a terminal disease and resigned to be satisfied in hope for her daughter's brighter future. We suffer through Peggy's constant internal struggle of stress, frustration and anger as she fights tooth and nail to rise up the career later and the deep loneliness and unhappiness that results. Even as she achieves some balance later on, we grieve the time spent frustrated and alone. As two types of extremes in the dichotomy that Simone de Beauvoir draws between the biological and the feminist, they demonstrate that neither fully satisfies. Turning now to the third of the main female characters in *Mad Men,* Joan Holloway/Harris, we see yet another complex paradigm of womanhood. Through this complexity, however, Joan's story ultimately reflects an attempt to balance these two extremes and embrace both career and motherhood as life-giving and fulfilling aspects of female existence.

This middle ground has been theorized by third-wave feminist scholars such as Adrienne Rich who questions the direct connection between biological function and oppression in women's lives. As Rich explains in *Of Woman Born* (1976), patriarchy—not motherhood itself—causes this oppression. Indeed, she describes her own experience in originally rejecting motherhood for its restrictiveness and then being moved deeply by an encounter with a young mother with a two-week-old infant. To Rich, this woman appeared enraptured by the "pure pleasure of having this new creature, immaculate, perfect."[48] Yet, Rich indicates that this woman's life had become more complicated by this birth and that "she is living even now in the rhythms of other lives."[49] This duality of experience—the patriarchal connection between motherhood and female identity or fulfillment and the physical, emotional, and psychological benefits and life-changing

47. "The Real Thing," S7/E14.
48. Rich, *Of Woman Born,* 33.
49. Ibid.

experiences a woman experiences through motherhood—leads the feminist both to desire and reject such a vocation. For Rich, it is not the biological restrictions of motherhood that oppress women but what patriarchy has done to motherhood—its restrictiveness, society's expectations, and compounding family dynamics.

The modern feminist must not reject motherhood but must re-imagine it, reclaim it, and use it as a vehicle for liberation:

> Patriarchal thought has limited female biology to its own narrow specifications. The feminist vision has recoiled from female biology for these reasons; it will, I believe, come to view our physicality as a resource, rather than a destiny. In order to live a fully human life we require not only *control* of our bodies (though control is a prerequisite); we must touch the unity and resonance of our physicality, our bond with the natural order, the corporeal ground of our intelligence.[50]

In her analysis, Rich echoes the sociological work of Mary Frances Berry and the theological work of Rosemary Reuther, Elisabeth Schussler-Fiorenza, and Mary Daly—all of whom re-examine social, biblical, and theological traditions to reveal more fully inclusive roles for women. According to these theorists and theologians, motherhood was more restrictive because of the patriarchal system in which it developed, and from which it could be liberated.

This difficult middle ground that avoids the extremes of patriarchy on the one hand and rejection of motherhood on the other was particularly difficult during the *Mad Men* era. Though contemporary parental leave policies and other accommodations to make working motherhood possible leave much to be desire, they are leaps and bounds beyond what was available in 1960s and 1970s corporate America. And this is not to even consider societal attitudes towards the possibility of achieving both maternal and career success. Yet, something like dual success gets achieved through one of the most intriguing and changing characters in *Mad Men*.

Joan's outward appearance gains her significant attention from men. As a result, she must constantly negotiate and sometimes leverage this often disingenuous or lecherous attention to achieve career success as she climbs the corporate ladder. Joan's love life ebbs and flows through a number of troubled relationships, a failed marriage and the inextricable ties to Roger Sterling who secretly fathers her son. In Season 4, she approaches motherhood as a married woman by visiting her doctor to see if she is able to

50. Ibid., 40.

conceive and revealing past abortions.[51] By the end of the season, Joan is visibly pregnant after a chance encounter with Roger, her on-and-off lover.[52] She speaks wistfully with her husband overseas about the pregnancy as any first-time parents might. Acknowledging the truth about the pregnancy is out of the question and Joan carries on a façade that temporarily maintains her marriage as well as Roger's. Yet, by Season 5 when Joan now has a baby, the struggle begins to fit this baby into her already busy life. Her mother, invaluable in her willingness to help Joan care for the child, also vocally objects to Joan's attempt to be a working mother.[53] Despite these tensions, her mother remains a constant support throughout the series. A later episode ends with a vision of Joan, the baby and Joan's mother lying together in bed after Joan kicks the husband out of the house and, the viewer assumes, her life.[54]

As her career progresses, Joan faces numerous choices in both her romantic and professional career. The audience applauds Joan as she rises from secretary to full partner in the firm and yet cringes deeply as Joan is forced to sleep with a business associate to achieve the promotion. Don comes too late to stop this, and the audience sees that this incident will forever remain a stain on Joan's career and affect her relationship with her male partners. Her tough outer shell has been strengthened but, as always, her head is held high and she continues onward. In Season 7, we see her new work responsibilities and assertion of authority. As the agency is subsumed under McCann Erickson, Joan finds that the upward momentum that propelled her rise in Sterling Cooper Draper Price holds little weight in this new setting; she becomes subject to new subjugations, sexual innuendo, and threat. When she finally puts her foot down, she does not get what she wants. The audience is devastated for her and yet she rises again. As the series ends, we see her beaming and busy as she begins her own company—including her two last names on the masthead. Joan is finally succeeding on her own terms without reliance on another soul.[55]

Joan's romantic life never reaches fulfillment but the audience can live with this outcome. We have seen Joan's mistreatment by Roger who clearly loves her deeply but also devalues her loyalty and presumes her constancy. We see Joan offered marriage by her gay friend and her brave refusal and articulation of her dream of future love rather than a marriage of convenience

51. "The Good News," S4/E3.
52. "Tomorrowland," S4/E13.
53. "A Little Kiss," S5/E1.
54. "Tea Leaves," S5/E3.
55. "The Real Thing," S7/E14.

and societal acceptance.[56] Her romantic liaison in Season 7 promises to bring her fulfillment at last, but she refuses to have to choose between her son and love.[57]

Her life is nothing but messy, and her climactic moments make the audience cringe and anger. Yet, Joan arguably achieves more balance and fulfillment than any other woman in the show. She has a healthy son with whom she is involved and loves dearly. She has made her own career path and ends with an ambitious new venture that promises to be devoid of the problems that made that path so difficult thus far. The audience has a sense that she will likely find love but that this love will come when Joan is firmly on her own two feet and, therefore, will be on her terms. It will be a relationship without the compromises implied in every relationship she has had until now.

In an interview with *Entertainment Weekly*, Christina Hendricks denied the feminist label for Joan, largely due to her methods for getting to the top. Hendricks also noted the gradual way in which Joan's intentions shifted over time: "The whole goal was to play the cards right, get a husband, and move out to the country. And what a surprise to find out that the husband and the child are not as fulfilling as the work." Joan denies her gay friend's marriage proposal and offer of security by saying, "I want love and I would rather die hoping that would happen than to make some arrangement."[58]

Joan's story, in many ways, achieves the goals of what Shulamith Firestone called a "Feminist Revolution." In her 1970 book, *The Dialectics of Sex: The Case for Feminist Revolution,* Firestone demanded "the freeing of women from the tyranny of their reproductive biology,"[59] "full self-determination, including economic independence,"[60] full integration of women and children into larger society, and sexual freedom for women.[61] While the society of *Mad Men* as the show ends falls far short of this ideal, Joan's life does not. She lives a life of sexual liberation, she has worked towards her own economic independence and she has fulfilled and maintained her maternal desire outside of a traditional family structure. Though she has sacrificed in ways unimaginable, the viewer feels hopeful that she can, in fact, have it all.

56. "Waterloo," S7/E7.
57. "The Real Thing," S7/E14.
58. "The Strategy," S7/E6.
59. Firestone, *Dialectics of Sex*, 233.
60. Ibid., 234.
61. Ibid., 236.

LESSONS FROM THE MOTHERS OF *MAD MEN*

Much ink has been spilt over the role of feminism in *Mad Men.* Though the series pays homage to other social issues of the time such as race and class disparities, political upheavals, countercultural movements and the like, the centrality of the changing narrative of feminism underlies the life choices, attitudes, plotlines and dialogue of every female character. In ways both subtle and blatant, the writers demonstrate a keen concern with these changing gender dynamics just as they capitalize on the now outdated paradigms of gender and sexuality of a bygone era. The resonance of the show today, however, does not lie only in the sense of nostalgia or even horror at these much different times. The gender issues at play, particularly as they relate to motherhood and the ways in which women balance their biological and vocational aspirations speak to contemporary debates and concerns. These issues have not been resolved as time has progressed. Rather, as increasing numbers of women have chosen a path different from Betty and Peggy, choosing instead to embrace both career and family life, the conversation among feminists has grown more complex. In this way, Joan can be seen as an early manifestation of the "third-wave feminism" that would develop in subsequent decades. Refusing to downplay her beauty and sexuality, she embraces romantic relationships and even motherhood but also desires and works toward personal vocational goals. The path is of her own making and not without great difficulty and sacrifice, but it allows her to maintain both sides of her identity—satisfying her maternal desire without enslaving herself to it.

Yet, the term "third-wave feminism" comes with its own methodological problems. As Lisa Jervis observed, "what was at first a handy-dandy way to refer to feminism's history, present, and future potential with a single metaphor has become shorthand that invites intellectual laziness, an escape hatch from the hard work of distinguishing between core beliefs and a cultural moment."[62] Perhaps the difficulty of definition lies at the heart of what contemporary feminism is all about. Though Astrid Henry's discussion of third-wave feminism noted its divergence from earlier forms of feminism, the difference was in its resistance to essentializing.[63] Instead of being a concrete guideline for living a fulfilled life, feminism was "a way for [women]

62. Lisa Jervis, "Goodbye to Feminism's Generational Divide," in Berger, *We Don't Need Another Wave*, 14. This essay was adapted from an address to the 2004 National Women's Studies Association and was published in *Ms.* Magazine in Winter 2004. Jervis the founding editor and publisher of *Bitch: Feminist Response to Pop Culture* and an editor at large of *LiP: Informed Revolt.*

63. Henry, *Not my Mother's Sister*, 1.

to be an individual and break free of society's many rules about women's proper place."[64] While only a segment of the generation at question actually labels itself feminist, this tendency to buck regulation points to a trend reaching beyond the realm of feminism. Without a clear sense of what the proper place, goals or life journey of a woman might be, women are experiencing greater freedom to choose but also the stress, tension and in-fighting that can come from the weight tied to each of those choices.

We must also note that race and privilege play a large part in the narratives presented in *Mad Men*. This is particularly true in relation to feminist narratives of career and motherhood. As bell hooks argues, motherhood was never seen as an obstacle to liberation for black women who want more time with family, not liberation from the home.[65] The black women who appear in *Mad Men* as nannies—here doing the mothering for white women as their own children are cared for by another—or as secretaries, are not fully developed characters in which these themes might manifest.[66] The particular dilemmas of Betty, Peggy, and Joan are struggles of privilege and their choices are informed by that position at every step.

So how did Americans move from this opening up of feminism's prescriptiveness and wealth of opportunity to the "Mommy Wars"? Why has this cultural conflict emerged in a time when women have more opportunity, more rights and more choices than in any preceding generation? As Betty Friedan wrote eighteen years after the *Feminine Mystique*, generational differences play a large part:

> The grandmothers, who had no choices—no pill, no IUD or diaphragm, no professions open—made the best of their necessities. Most took what life offered them, a few rebelled, or secretly burned. The mothers swung between the drastic choice defined by the feminine mystique . . . and it was a no-win choice. . . . Those few who combined motherhood and professions were "exceptional," and had, indeed to be superwomen. Or they had to settle for second best in career, and/or were oppressed by terrible conflicts and guilt in their relationship with husband and children.[67]

According to Friedan, the issue of family is the remaining complication in women's struggle to find personal peace in the American world. With rights in hand, birth control options available, and careers in motion, women have

64. Henry, *Not My Mother's Sister*, 1–2.

65. hooks, *Feminist Theory*, 133.

66. They are not even, as Sarah Sours notes in chapter 3, given last names.

67. Friedan, *Second Stage*, 89.

yet to determine how to combine successfully these aspects of their lives and are conditioned by feminists past not to reject either.

Articulating this tension another way, Judith Warner writes of a "Mommy Mystique," pulling from Friedan's phrase, that has infused modern society.[68] This "mystique" amounts to the following dilemma: stay-at-home mothers who society says have made all the right choices and who enjoy nice homes and financial security can experience boredom and a lack of fulfillment, while working mothers can become stressed beyond coping and guilt-ridden over what they are supposedly not giving their families and jobs. This "mommy mystique," Warner writes,

> Tells us that we are the luckiest women in the world—the freest, with the most choices, the broadest horizons, the best luck, and the most wealth. It says we have the knowledge and know-how to make "informed decisions" that will guarantee the successful course of our children's lives. It tells us that if we choose badly our children will fall prey to countless dangers. . . . To admit that we cannot do everything ourselves, that indeed we need help—and help on a large, systematic scale—is tantamount to admitting personal failure. . . . We are consumed with doing for our children in mind and soul and body—and the result is we are so depleted that we have little of ourselves left for ourselves. And whatever anger we might otherwise feel—at society, at our husbands, at the experts that led us to this pass—is directed, also, just at ourselves. Or at the one permissible target: other mothers.[69]

Indeed, though Friedan's ideal of a woman with infinite choices and the ability to manage both motherhood and career has largely been realized, the existential dilemmas women face remain the same.[70] This is true, Warner argues, because women have embraced the new opportunities, while simultaneously reclaiming the ideal of the earlier time—in which a mother can provide all to her children and her husband.

The first half of Season 7 ends with Peggy confronting her own maternal desire. When she learns that the young boy that she has befriended in her building is leaving, he hugs her and she hugs him back with tears in her eyes. Her tears are partly brought about by her stress at the approaching final pitch for Burger Chef. At that meeting, we see her and Don exchange a slow-motion glance and start their pitch. After beginning her remarks with

68. Warner, *Perfect Madness*, 13.

69. Ibid., 32–33.

70. Ibid., 40.

reference to the moon landing of the night before, Peggy uses the communal ritual of watching that momentous event to signal a desire for connection and ritual lacking in the present world. She, like others, was transported to another world through that event. She continues,

> Tonight, I'll go back to New York. And I'll go back to my apartment and find a ten-year-old boy parked in front of my TV, eating dinner. Now I don't need to charge you for a research report that tells you that most television sets are no more than six feet away from the dining room table. And I don't need to tell you that that table is your battlefield and your prize. It's the home that your customers really live in. The TV is always on, the news wins every night. And you're starving, and not just for dinner. . . . What if there was another table, where everyone gets what they want, when they want it. It's bright, it's clean and there's not laundry and no telephone and no TV and we can have the connection that we're hungry for. There may be chaos around but there's family supper at Burger Chef.[71]

For the women of *Mad Men,* negotiating the social constructions of motherhood, femininity and the "feminine mystique" becomes their battlefield and finding some level of fulfillment through it all, their prize. Matthew Weiner has been transparent about the ways in which his show directly addresses the feminist voices of the time—in particular Betty Friedan.[72] While so much of the series represents a life so foreign to our own contemporary experience, the struggles of women in balancing motherhood, vocational aspirations, and societal expectations mirror so many struggles continuing today. This makes the treatment of motherhood yet one more reason for the remarkable success of the show, the relatability of the often-outrageous characters and the timelessness of its plotlines. On the subject of motherhood, *Mad Men* presents case studies of the evolution of feminist thought and experience as well as narratives that explore the timeless existential struggle many women feel as they mitigate the complexities, joys and heartbreaks of maternal desire.

71. "Waterloo," S7/E7.

72. Poniewozik, "Time Machine."

BIBLIOGRAPHY

Bahr, Linsey. "'Mad Men' and Feminism: A Place for Peggy, Joan, and Megan in the Movement." *Entertainment Weekly,* April 9, 2013. Online: http//www.ew.com/article/2013/04/09/mad-men-and-feminism.

Beger, Melody, ed. *We Don't Need Another Wave: Dispatches from the Next Generation Feminists*. Emeryville, CA: Seal, 2006.

Chodorow, Nancy. *The Reproduction of Mothering: Psychoanalysis and the Sociology of Gender.* Berkeley: University of California Press, 1978.

Coontz, Stephanie. "Why 'Mad Men' is TV's Most Feminist Show." *The Washington Post,* October 10, 2010. Online: http://www.washingtonpost.com/wp-dyn/content/article/2010/10/08/AR2010100802662.html?sid=ST2010101103521.

De Beauvoir, Simone. *The Second Sex*. Translated and edited by H. M. Parshley. New York: Vintage, 1989.

Firestone, Shulamith. *The Dialectics of Sex: The Case for Feminist Revolution.* New York: Morrow Quill Paperbacks, 1970.

Friedan, Betty. *The Feminine Mystique.* New York: W. W. Norton, 1963.

———. *The Second Stage.* New York: Summit, 1981.

Henry, Astrid. *Not My Mother's Sister: Generational Conflict and Third-Wave Feminism.* Bloomington: Indiana University Press, 2004.

Hooks, Bell. *Feminist Theory from Martin to Center,* 2nd ed. Cambridge, MA: South End Press, 1984.

Lipp, Roberta. "'Shoot' Wins ADG, Matt Weiner's Visions, and Birds." *Basket of Kisses,* February 17, 2008. Online: https://madmenmad.wordpress.com/2008/02/17/shoot-wins-adg-matt-weiners-visions-and-birds/.

"'Mad Men' Creator on Don Draper's Losses and the End of the Road." *Fresh Air,* National Public Radio, aired May 7, 2015. Online: http://www.npr.org/2015/05/07/404904172/mad-men-creator-on-don-drapers-losses-and-the-end-of-the-road.

Poniewozik, James. "Time Machine: How Mad Men Rode the Carousel of the Past into Television History," *Time Magazine*. Online: http://time.com/mad-men-history/.

Warner, Judith. *Perfect Madness: Motherhood in the Age of Anxiety.* New York: Riverhead, 2005.

Chapter 5

SUPPORTING THIS WORLD'S BALLERINAS

Learning from *Mad Men*'s Female Workers

—KRISTEN DEEDE JOHNSON

> Not every little girl gets to do what they want. The world could not support that many ballerinas.
>
> —Marie Calvet to her daughter, Megan Draper[1]

As much as *Mad Men* reveals the inner workings of the Madison Avenue advertising industry, the mysteries and malaise of Donald Draper, and the cultural changes that occur between 1960 and 1970, it is equally about the dramatic shifts in the experiences and dreams of the female worker. As one viewer writes on *The Huffington Post*, "The title of AMC's hit series may be 'Mad Men', but in many ways the show is about its women and the evolution of their revolution."[2] As the series takes its viewers on a decade-long journey, it reveals the ongoing employment and sexual revolution that had profound effects on women and their workplace dreams.

At the same time as *Mad Men* portrays the initial impact of these cultural revolutions on women in the workplace, it also highlights that in many

1. "The Phantom," S5/E13.
2. Diller, "Mad Men, Madder Women."

ways these revolutions have not yielded the dramatic changes we might have expected to be in place by today. In contemporary discourse, Sheryl Sandberg urges us to acknowledge that the employment revolution that brought so many women into the work force has stalled—particularly when it comes to women in leadership positions in the workplace.[3] Anne-Marie Slaughter refers to this same phenomenon as "the Great Stall."[4]

On one level, the experiences faced by the female workers of *Mad Men* reflect a number of dynamics that continue to impact female workers today. When, for example, we learn that Joan has implemented a rule that there is no crying in the office, including the break room, we realize that women today are still wondering if they should bring their emotions into the workplace.[5] As we watch Peggy learn to negotiate, we are reminded of the extensive literature documenting that women today are at a significant disadvantage because they do not negotiate as effectively as men.[6] When we see Peggy excluded from the informal socializing and networking with clients that her male colleagues naturally undertake, we think of the recent research demonstrating how important informal time with colleagues is to an employee's ability to build consensus and establish power.[7] As we compare how Peggy and Joan find their voices and relate to male colleagues as they move up the career ladder, we are aware that the conversation around whether women need to act like men, be treated like men, or can carve out space for their femininity in the workplace continues to this day.[8]

In her important contributions to the contemporary conversation around women and work, Slaughter tries to get to the root of these ongoing tensions. As a society, Slaughter argues, we have identified the competitive world of breadwinning as important. We have sought to support involvement in that world through our policies, laws, and structures. The social revolution of the 1960s and 1970s rightly sought liberation from "the gray suits, perfect wives, and nonstop martinis of *Mad Men*."[9] But the liberation of women came at the expense of valuing the world of caregiving. This devaluation of caregiving is, Slaughter argues, the underlying problem beneath

3. Sandberg, *Lean In*, 7.

4. Slaughter, *Unfinished Business*, 81.

5. Sandberg, *Lean In*, 90–91.

6. See Babcock and Laschever, *Women Don't Ask;* Babcock and Laschever, *Ask For It.*

7. See Heath, Flynn, and Holt, "Women, Find Your Voice," 4.

8. See, for example, Wittenberg-Cox, "To Hold Women Back."

9. Slaughter, *Unfinished Business*, 87–88.

the ongoing tensions that women and men experience as they try to balance breadwinning and caregiving. She writes:

> Fifty years ago middle-class women stayed home, cared for their families, and were manifestly unequal to their breadwinning husbands. To make them equal, we liberated women to be breadwinners too and fought for equality in the workplace. But along the way, we left caregiving behind, valuing it less and less as a meaningful and important human endeavor.[10]

In short, as a culture we rightly celebrate that women can pursue their workplace dreams in ways that were unimaginable in previous eras, but we are still in need of conceptual and structural support for the activities that were traditionally associated with the home. We need to imagine creative ways to support the world of caregiving alongside the world of competition.

This imaginative process needs to include questioning the very assumption that the worlds of caregiving and competition must stand in tension with one another. In our current cultural framework, the public world of work is considered the opposing counterpart to the private world of home. The logical outcome of this public-private divide between work and home is that the real contributions to society happen in the working world. This leaves us without the conceptual space to acknowledge the importance of non-paid functions and roles in our society, from caring for elderly relatives and dependent children to preparing meals and cleaning our living spaces to volunteering in community organizations. "Care," as Slaughter notes, "has many faces."[11] And caregiving, in its many forms, contributes just as much to the public good as competition.

Perhaps ironically, then, after all these decades of encouraging women to pursue their workplace dreams, what we most need to grapple with to make the realization of those dreams possible is why matters of the home matter—for all of us, male and female, and for society at large. If this perspective is right, we need to muster helpful resources to articulate the significance of caregiving and support its contribution to society. In a society as diverse and pluralistic as America, we are unlikely to find collective language and shared concepts to undergird policy recommendations that might be put forward as helpful resources to address the contemporary tensions between the market economy and the household economy. But we can mine our own traditions and probe our own commitments for resources that can help further the public conversation around the significance of supporting the

10. Ibid., 79.

11. Ibid., 98.

work of caregiving.[12] What might a Christian theology of home contribute to the conversation as we continue to look for ways to acknowledge the importance of our caregiving dreams alongside our workplace dreams? Using the arguments of David Matzko McCarthy, Kathleen Norris, and Margaret Kim Peterson, I conclude the chapter with an exploration of a theology of the home to help us consider ways to articulate the significance of home and caregiving alongside the significance of work and competition.

EMPLOYMENT REVOLUTION

Beginning before the time period that *Mad Men* depicts, World War II served as a catalyst for a season of significant change for the female worker. As many men left the country to fight in the war, many women—including white married women—entered the workforce and took on more skilled and higher-paying positions than they had been allowed to hold before. Although at the war's end women were supposed to return to their roles in the home, their years of successfully holding positions during the war proved that women could perform in jobs that had previously been restricted to men. This cultural lesson was not immediately heeded, however, as a significant cultural movement emerged to reinforce that women's proper place was in the home rather than in paid labor.[13] This cultural norm in the post-wars years that encouraged women to seek their entire fulfillment through marriage, housekeeping, and child-rearing is what Betty Friedan—one of the pioneering leaders of the modern women's movement—came to call "the feminine mystique."[14] Friedan argued that many women in the home were deeply dissatisfied and unhappy while trying to embody this role, suffering from what she called "the problem that has no name."

This discontent with prevailing societal assumptions around housekeeping and child rearing in the decades following World War II seems to be presented on *Mad Men* by suburban housewife Betty Draper. College-educated, Betty works as a model first in Italy and then in New York, where she meets her future husband, Don Draper. By the time we meet her, she is married and living in the suburbs with two children, having given up her career when she got engaged, according to the prevailing cultural norm. She is also deeply unhappy, for reasons that are mysterious to her, her husband, and her doctor, who suggests she see a psychiatrist. Eventually she shares with her therapist that she was raised by her mother to be beautiful so she

12. See Johnson, *Theology, Political Theory, and Pluralism*, 233–49.

13. Fox and Hesse-Biber, *Women at Work*, 24–27.

14. Friedan, *Feminine Mystique*.

could find a husband, but she also expresses deep dissatisfaction with how that has played out in her life. "But then what," she asks, "just sit and smoke and let it go until you're in a box?"[15] These comments are offered in the midst of an episode in which she decides to go back to work as a model, using a contact of her husband Don's to land a Coca-Cola ad. Unfortunately Don's contact was simply using Betty to try to hire Don, so she loses the contract—and her sense of excitement and anticipation of fulfillment—when Don refuses to take the job.

For the remainder of the show, we see the discontent (identified by Friedan) playing out in Betty's life as she seeks more of a voice and role within society than she sensed she had as a homemaker. This seems to be a large part of her attraction to her eventual second husband Henry Francis, who works in New York politics, as she imagines that life with him would be more meaningful even as she continues playing the roles of wife and mother. In the show's final season, she expresses disapproval that her friend Francine—her former suburban housewife neighbor—has gone to work three days a week.[16] Shortly after, she shares her opinion about political events in a social setting (a major cultural taboo for women at that time), gets reprimanded by her husband, and vents frustration about not being allowed to think and communicate her own thoughts. She also gives viewers an initial glimpse into what she ponders doing outside of the home.[17] By the show's end, she has chosen to enroll in a masters-level program in psychology, a degree she continues to pursue even after being diagnosed with a terminal form of lung cancer. Betty and her friend Francine reflect a growing trend in the 1960s—married women entering the workforce after their children were in school.[18] They are just one part of a larger movement within the decades after World War II known as women's "employment revolution."[19]

To better understand this employment revolution, we need to pause to look at the longer history of women in the workforce. Behind the employment revolution depicted in *Mad Men* lies an economic arrangement that was largely dependent upon women dedicating themselves to making sure those in their homes were fed, clothed, and clean. Historically, this had required a considerable expenditure of time and labor. Within the agrarian economy of the eighteenth and nineteenth centuries, the home sphere was essential for survival, requiring full-time attention. As the economy became

15. "Shoot," S1/E9.

16. "Field Trip," S7/E3.

17. "The Runaways," S7/E5.

18. Fox and Hesse-Biber, *Women at Work*, 27–28.

19. Thistle, *From Marriage to Market*, 35, citing Frank Mott, *Employment Revolution*.

increasingly industrialized, factory work that initially depended upon young, single women and then upon new waves of immigrants eventually became the purview of men. Meanwhile, the realm of home still required full-time attention. Typical homes in the early twentieth century lacked running water and electricity, which meant no indoor access to fresh water, plumbing, or refrigeration. Water had to be carried from nearby sources, chamber pots had to be emptied, coal or wood had to be collected, and while basic provisions could now be purchased, almost all food still had to be cooked from scratch.[20]

Even by the 1940s, running water was not available in about a third of American homes, while two-fifths of homes did not have flush toilets. Beginning in the 1950s, however, access to utilities and basic household appliances grew dramatically. The household inventions of earlier decades were now mass-produced; combined with increasing postwar incomes, the majority of households across racial and economic divides had running water, indoor plumbing, stoves, refrigerators, and access to washing machines. The hours involved in meal preparation and care of clothing were significantly reduced, while time and energy spent on things such as tending fires for cooking and heating and caring for animals and gardens were almost entirely eliminated.

In short, these developments opened the way for the employment revolution. These same developments also made it possible for women like Betty, who were at home, to experience the malaise and ennui of Friedan's "problem that has no name." In Betty's ongoing struggle with her role in the home and her voice in society, we see reflected an ambiguity about the importance of the home, the traditional functions required in the home, and the caregiving roles associated with the home that marked this time of cultural and economic transition. As economic realities no longer required the home to receive full-time attention, a dominant cultural and political consensus about the home's significance ceased to exist.

SEXUAL REVOLUTION

If Betty's storyline shows us one way in which the employment revolution intersected with increasing cultural ambiguity around the home, Joan Holloway's storyline offers a different but complementary perspective. As the show opens, we learn that Joan's dreams for the workplace primarily revolve around meeting a future husband who will enable her to move to the

20. Fox and Hesse-Biber, *Women at Work*, 17–21; Thistle, *From Marriage to Market*, 23–26.

country and stop working. By the third season, Joan's dream seems to have come true. The reality that the viewer sees, however, becomes much more complex. Joan gives her notice to the agency, only to learn from her husband late one night that he is not going to succeed as a surgeon so she will need to get another job. After learning this news, she turns out the light: the viewer senses that the light has gone out, not just in her bedroom, but also on her dream.[21] By the time the show ends, Joan has been in and out of marriage, had a child, returned to the workplace, and finally given up a lover in order to pursue her dream of launching her own business. Joan's character invites us further into the complexities that shifting economic, technological, and cultural norms introduced into traditional hopes and dreams for marriage, child-rearing, and the home. Her experience shows us a woman finding more fulfillment in work than at home, feeling more success in the field of competition than the realm of caregiving. As she represents the sizeable demographic that begin working with young children at home, she also conveys the cultural reality that no external support was provided for those in the workplace who needed childcare. Without her mother's ability to move to New York permanently and care for her children, viewers remain uncertain how she would have made it.

Prior to entering into her roles as wife and mother, Joan represented another significant revolution of the day: the sexual revolution. Appropriately, it was a book written by a former copywriter for an advertising firm, Helen Gurley Brown, that launched the sexual revolution. Matthew Weiner modeled Joan's character on the real-life Brown.[22] In her 1962 book, *Sex and the Single Girl,* Brown shared her own premarital sexual encounters and encouraged other women to follow her lead. It was the first book of its kind, reflecting a dramatic shift in public discourse related to unmarried women and sex. Prior to the sexual revolution, the dominant cultural mentality was that a woman's virginity was to be protected at all costs. Regardless of actual behavior, books, movies, magazines, and television shows largely reinforced this message. While met with considerable controversy, Brown's book—along with her role in transforming the magazine *Cosmopolitan* into a self-help guide for sexually engaged women—helped shift this public message.[23]

From the first episode of the series, Joan makes it clear, without apology, that she has been sexually involved with people at the ad agency. Also in the first episode, she sends brand-new secretary Peggy to the Midtown Medical Center to see a doctor who can prescribe the birth control pill.

21. "Guy Walks into an Advertising Agency," S3/ E6.

22. See Tuttle, "Christina Hendricks."

23. Allyn, *Make Love Not War,* 10–12, 14, 22.

Indeed, the birth control pill was invented in 1957 and the Food and Drug Administration (FDA) licensed it in 1960—the same year as Season 1 of *Mad Men*. This was the first oral contraceptive and proved revolutionary in its impact. It enabled its users to separate the use of birth control from the act of sexual intercourse, while simultaneously enabling them to separate sex from pregnancy. With Peggy and Joan's use of the pill, we see the ways this new form of birth control influenced the lives of single women. But this oral contraceptive also impacted wives and mothers, who could now more easily control the number of children they had. As the size of their families grew smaller, and women had more years before them in which they were not directly responsible for raising children, they found themselves with yet another reason to question the necessity of a life spent at home.

THE SILENT REVOLUTION

Mad Men invites us into a time of profound cultural transformation, with special significance for the female worker. We have seen how the show reflects the employment revolution and the sexual revolution—as its characters and their storylines demonstrate the increasing cultural momentum pulling women into the paid work force. Susan Thistle notes that alongside this "pull" into the workforce, as an expanding economy offered women new opportunities while changing cultural norms made it more acceptable for them to take them, we have to acknowledge that many women experienced this as a "push" out of the home. The support that used to be provided to women to dedicate their time to the economies of their households lessened and, on a legal and public policy level, was eventually eradicated.[24] This "silent revolution," while harder to convey within a television show, is an essential part of the picture of women, work, and home in *Mad Men*.

One aspect of the silent revolution portrayed on *Mad Men* concerns the show's depiction of marriage and divorce. On a pragmatic level, marriage had once served to solidify an economic arrangement in which women provided the domestic labor that made home life and eating possible for men. As the need for women to engage in all-consuming domestic chores lessened, not only did women's economic lives have the potential to shift, but men could become less dependent on women. It was now possible for marriage to become less about economic necessity and more about emotional intimacy and love. This cultural shift had taken place by the 1960s, and it also seems to have been a contributing factor to the rise of the divorce rate, which doubled between 1964 and 1975. Marriages based on the meeting of

24. Thistle, *From Marriage to Market*, 37.

emotional and sexual needs turned out to be more fragile than those based on mutual economic need.[25] We see this turn reflected in the character of Roger Sterling, who leaves his first wife to try to find emotional and sexual fulfillment with a much younger second wife; in his character's storyline we also see the fragility of this basis for marriage, as his second marriage ends fairly quickly and rather unsurprisingly. By the time the show ends, in 1970, the first no-fault divorce law had been passed, reflecting the new cultural assumption that women could, and should, find their own financial source of income by entering the workforce.

Another aspect of the silent revolution concerns sexual harassment in the workplace. The term "sexual harassment" did not even exist during the time period portrayed in *Mad Men*. It was first used in 1974 in a claim filed with the Equal Employment Opportunity Commission on behalf of a female worker at Cornell University. In 1979, legal scholar Catherine MacKinnon articulated a legal framework for sexual harassment—arguing that it is a form of sex discrimination that should be covered by Title VII of the 1964 Civil Rights Act. The Equal Employment Opportunity Commission, which had been established in 1964 to help implement the Civil Rights Act and mediate disputes, changed its guidelines a year later. From this point onwards, Title VII, which prohibits firms from discriminating on the basis of sex, includes the prohibition of sexual harassment.[26] *Mad Men* depicts how overt and acceptable sexual harassment of female workers was prior to the legal and cultural shifts. Of the many instances of sexual harassment in the show, the one that perhaps most galvanized viewers was that of Joan as she was asked to consider sleeping with one of the agency's clients.[27]

Without the splash of the sexual revolution or the visibility of the employment revolution, Thistle argues that the "silent revolution" entailed the dismantling of the last legal buttresses supporting the traditional gender division of labor. In an earlier time of economic and cultural transition, when faced with the initial implications of the Industrial Revolution, the predominant cultural and legal response was to protect the domestic economy. Battles were waged so that men in factories would be paid a family wage, enough to allow their wives to remain at home tending to domestic

25. Ibid., 46–47.

26. Giele and Stebbins, *Women and Equality*, 66–67, 139–40.

27. "The Other Woman," S5/E11. Weiner decided to put that in the show because more than one hundred women shared stories like that with him as they reflected on their time in the labor force in the 1960s. The aspect of the story that did not mirror real experience was that she was able to negotiate so much—becoming a partner—out of the arrangement; usually women acquiesced to such work-place "requests" for far less (see Itzkoff, "Weiner Reflects").

chores. Single mothers received monetary stipends so they could remain at home. Protective laws were passed so that women could not be exploited in the paid labor force and be pulled too far away from their domestic and maternal roles.

In the period following World War II, the cultural momentum goes in the other direction. Protective laws are repealed, having been deemed antiquated and repressive. New laws are passed seeking to enforce equal opportunities, equal treatment, and equal wages for women in the workforce. Divorce laws and welfare reform shift to reflect the cultural assumption that women will support themselves and their families through their paid employment. No longer was the domestic realm protected and supported by the legal realm.[28]

As economic arrangements shifted, leaving the household economy less essential than it had once been, our culture struggled to articulate the role and importance of the home and caregiving. Within the silent revolution of legal decision-making, this manifested itself in a quiet dismantling of every public policy in which support had been provided for women to attend to matters traditionally associated with the home. In the decades since, we have not replaced this dismantled policy that depended upon a gender division of labor in which women attended to the home, with any new policies. We have not yet envisioned different or innovative public ways to support the activities of caregiving and the home within an altered economic and cultural landscape.

REVOLUTIONS AND DREAMS

Megan Draper serves as the best character on *Mad Men* to shift our focus from the historical setting of the 1960s to questions women face today. On the show, Megan represents the first generation of women who has the

28. Thistle, *From Marriage to Market*, 48–49. One holdout in our social policy from the earlier era was a welfare program that provided support to impoverished women to enable them to stay home with their children, based on the dominant family and economic model in place when the law was first implemented in 1935. This program (Aid to Families with Dependent Children) became harder for the public to support as increasing numbers of married women and mothers entered the workforce. In 1988 the Family Support Act (FSA) was passed based on a welfare-to-work model. The most significant change to welfare law in the previous fifty years, this new model had its goal that all recipients enter the workforce, regardless of whether they were married or had children. Provisions for childcare were included in the bill for those who were parents. This changed law reflected the working cultural assumption that women, wives and mothers included, should be a part of the paid workforce. See Giele and Stebbins, *Women and Equality*, 92–94, 193–94.

possibility of pursuing dreams in the workplace. Joan and Betty's workplace dreams were secondary to the dreams they for held marriage and family. Peggy never imagined she could be more than a secretary and relishes the unexpected opportunity to become a copywriter and then a copy chief. By the end, she dreams big, too—at least, compared to where she started.[29] Megan becomes the only female character we meet on *Mad Men* who makes big decisions—moving first to New York and then to California—primarily to pursue dreams for *her* career.

The generational shift that Megan represents seems evident in her (older) husband's response when she articulates her desire to pursue her workplace dream. He is clearly baffled, even as he tells her that he does not want to get in the way of her pursuing her dream. He processes Megan's decision with his colleague Roger in a telling exchange:

> Roger: What happened?
>
> Don: She's following her dream.
>
> Roger: That's admirable. I sure as hell didn't get to choose what I wanted to do. My father told me.
>
> Don: I was raised in the 30s, my dream was indoor plumbing. [Silence] Why shouldn't she do what she wants. I don't want her to end up like Betty, or her mother.[30]

Reflecting on this exchange and Megan's desire to pursue her dreams, Weiner says, "We take it for granted that you can choose what you want to do. That's all part of a new generation, and very soon there's going to be a generation doing whatever it wants . . ."[31] Weiner intentionally created the character of Megan to represent a new generation—describing her as extroverted, joyful, as well as confident and strong with her own opinions.[32] This is part of why Don loves her, but the reality of how that plays out in their marriage conveys the struggles that transpired as women who have their own voice and seek their own workplace dreams also try to pursue their dreams for marriage and the home. Megan's departure from the ad agency to pursue acting spells the beginning of the end of their marriage. The independence that initially draws him to her becomes the independence that drives him away.

In the final scene of the final episode of the season in which all of this plays out, a song plays in the background. Weiner has said that the lyrics to

29. "The Forecast," S7/E10.

30. "Lady Lazarus," S5/E8.

31. Sepinwall, "'Mad Men' Creator."

32. Yuan, "*Mad Men's* Matthew Weiner."

that song tell what this entire season is about.[33] In a scene loaded with symbolism, Don sits alone—having walked away from the set on which Megan films her first ad, an acting job that Don procured for her. Out of love for her, he helps her workplace dream come true. The viewer senses, however, that the realization of this dream comes at the expense of their marriage. The song in the background, "You Only Live Twice," has telling lyrics: "This dream is for you, so pay the price. Make one dream come true, you only live twice." One dream had come true but at the price of another.

Yet, the viewer remains deeply aware that Don knows that forcing Megan to stay at home—as his first wife Betty did—would equally have led to the demise of their marriage. And there we find the tension! On a micro-level, the storyline exhibits the complexity that one husband experiences as he tries to enter into a new cultural moment in which a man is supposed to support his wife's workplace dreams. Indeed, he *wants* to support her dreams. But, as they navigate this pursuit of dreams, the reality for both of them is much harder than they had imagined. On a macro-level, it reflects the tensions created by the employment, sexual, and silent revolutions as they impact American notions of work and home.

What are our options in the face of these enduring tensions? Should we simply heed the advice of Megan's own mother, who tells her daughter, "not every little girl gets to do what they want. The world could not support that many ballerinas"?[34] Or are there other options?

TOWARDS A THEOLOGY OF THE HOME

Megan Draper represents a new generation of women who dared to dream that they could have it all. Megan faced obstacles from her mother, who told her the world could not possibly hold the dreams of all the little girls wanting to be ballerinas. She faced obstacles in her marriage, from a husband who wanted to support her independence and dreams, but in the end could not handle the move away from a traditional marriage. She faced obstacles in the workplace, in an industry that still operated under norms that would today be defined as sexual harassment. Fifty years later, can the world support any more ballerinas? Can the world support any more executive directors of ballet companies while also supporting them in their caregiving commitments, responsibilities, and desires? Can *Mad Men* simultaneously make us grateful that little girls can pursue their dreams in ways and under conditions that once seemed culturally impossible, while also reminding

33. Sepinwall, "'Mad Men' Creator."
34. "The Phantom," S5/E13.

us that the tensions navigated by the show's beloved characters continue to need resolution? Can it inspire us not only to continue to pursue our dreams, but to creatively seek cultural solutions to the problems that perpetually keep many from realizing their dreams? Can it even help us to reshape the dreaming process, so that alongside the time, attention, and support we dedicate to our workplace dreams we recognize the importance of giving time, attention, and support to our caregiving dreams?

To engage some of these questions, we turn now to consider what a Christian theology of the home might contribute to the ongoing public conversation around these issues of work, home, competition, and caregiving. To adequately probe the Christian tradition for ways to conceptualize work and home would require a lifetime of research along with a wellspring of careful discernment to wade through the good, the bad, the beautiful, and the ugly offered through the centuries. Minimally, we can say that all throughout the biblical narrative run themes of home, homemaking, and caregiving. We can even understand the whole biblical story as, in the words of Margaret Kim Peterson, "a story that moves from home to home."[35] God created a home for humanity in which they could dwell with God and care for each other and creation; humanity lost their home when they were expelled from the garden which God intended to be their home; God makes a covenant with Abraham, promising Abraham's descendants Israel a promised land in which they can make their homes; Jesus tells his disciples that they will join him in his Father's home, full of many rooms, and he speaks of the kingdom of God as a great banquet feast taking place in a full house; and God promises that in the age to come God will dwell with God's people in this renewed creation.[36]

Alongside this rich significance given to the idea of home within the story of Scripture, we find God involved in the activities of homemaking and caregiving. In addition to creating and providing homes for God's people, God provides food, drink, and clothing to God's people. Biblical imagery throughout the scriptures affirms that God willingly and regularly connects Godself to the caregiving of the people God loves. Citing Psalm 104, Peterson notes that "the psalmist's portrayal is of God as a great housekeeper, pitching a tent, clothing himself with light and the earth with water as with garments, ordering boundaries, making homes for creatures, giving them food, sustaining all life, creating and re-creating through the Spirit."[37]

35. Peterson, *Keeping House,* 16.

36. Gen 1–2; Gen 3; Gen 12; John 14 and Luke 14; Rev 21. See Peterson, *Keeping House,* 16.

37. Peterson, *Keeping House,*13.

If "housekeeping" is significant enough for God to attend to, surely it is significant enough to deserve our sustained human attention.

Indeed, according to the teachings of Jesus, activities related to housekeeping and caregiving will provide evidence of his followers' righteousness and serve as criteria for the final judgment. In the gospel of Matthew, Jesus shares that the righteous who will get to inherit God's kingdom consist of those who feed the hungry, provide drink to the thirsty, welcome the stranger, provide clothing to the naked, take care of the sick, and visit the imprisoned.[38] While this passage is typically associated with serving the impoverished people in our midst, Peterson pushes us to consider that this passage also has meaning for the work that occurs in our homes, for "housework is all about feeding and clothing and sheltering people who, in the absence of that daily work, would otherwise be hungry and ill-clad and ill-housed."[39] Daily housework matters, enormously.

Kathleen Norris makes this same point, from a different perspective, when she relays the story of attending a Roman Catholic mass for the first time as an adult—after years of being away from any kind of church. She was delighted to discover that in the middle of the ceremony the priest was doing the dishes! She writes:

> I found it remarkable, and still find it remarkable—that in that big, fancy church, after all of the dress-up and the formalities of the wedding mass, homage was being paid to the lowly truth that we humans beings must wash the dishes after we eat and drink. The chalice, which had held the very blood of Christ, was no exception.[40]

This cleaning up of the chalice by the priest serves as an occasion for Norris to reconsider what we have termed menial work. "Menial," she discovers, derives from a Latin word connected to dwelling in a household. Menial labor, jobs that were traditionally done privately in the home by women and servants, are usually the lowest-paying jobs in our culture. Norris points out that one of the strangest things about American culture is that we pay garbage collectors more than we pay those who care for our children in day-care centers.[41]

The devaluation of this work is part of a deeper rejection of the significance of the ordinary, daily, repetitive components of life. She argues that, "our culture's ideal self, especially the accomplished, professional self, rises

38. Matt 25:34–40.

39. Peterson, *Keeping House*, 3.

40. Norris, *Quotidian Mysteries*, 3.

41. Ibid.

above necessity, the humble, everyday, ordinary tasks that are best left to unskilled labor."[42] And we tell ourselves "comfortable lies" that allow us to believe that the daily chores we do to care for our homes and ourselves don't matter.[43] But these quotidian tasks do matter. These daily, repetitive tasks of caregiving can ground us in this world without grinding us down. They can remind us that most of our stories unfold over the course of ordinary life.[44] They can serve as invitations to play and holy leisure, reminiscent of the way many children love to play with water, bowls, and cups.[45] They can help us experience the daily, repetitive "round of sunrise and sunset" as "a symbol of hope, or of God's majesty, . . . or an expression of God's love for us and for all creation," rather than merely as a reminder that another day has passed in which we didn't have enough time to do what we think needs to get done.[46]

Both Norris and Peterson invite us to view the daily rhythms of housework as an antidote to the relentless pace of the cultural moment in which we currently live. Slaughter joins them in recognizing that the tasks people are trying to complete each day are impossible to achieve within a twenty-four hour period, causing men and women to seek to chronically live beyond their limits and capacities. She points to recent best-selling books about stressed-out workers, stories of workers all across the socioeconomic spectrum working twelve- to sixteen-hour days and experiencing exhaustion and anxiety attacks, and an epidemic of stress that has begun to be identified by public health experts as evidence of the problem.[47]

In the time use studies we discussed earlier, we saw that the average American wife today is working seventeen more hours per week than the average American wife in 1965, when paid work and domestic work are both included. Sandberg discusses this in terms of "the new normal" that has developed for the workplace and for the home. Many Americans are putting in more hours in the workplace every week while also embracing the use of technology that brings the workplace into the home. This impacts workers of all ages and stages of life. Many sacrifice sleep, relationships, social lives, community involvement, and family time to meet these increased workplace demands. Simultaneously, expectations for mothers spending time focused on their children have increased. According to the research cited by Sandberg, shifting cultural expectations have led mothers

42. Ibid.
43. Ibid., 40.
44. Ibid., 76–77.
45. Ibid., 27.
46. Ibid., 16–17.
47. Slaughter, *Unfinished Business*, 51.

to dedicate ever increasing amounts of time to directly nurturing their children, so that employed women today are spending the same amount of time engaging with their children as full-time stay-at-home mothers did forty years ago.[48] In the face of these impossible demands coming from work and home, Sandberg counsels that, "long-term success at work often depends on *not* trying to meet every demand placed on us. The best way to make room for both life and career is to make choices deliberately—to set limits and stick to them."[49]

The cultural pressure that many are feeling today at work and at home can be countered by a healthy appreciation of our God-given limits. We were created to need regular food, rest, and relationships. Our human embodiment is a gift affirmed in both creation and the incarnation, and this includes the embodied realities that we need to take time to nurture our bodies every single day through sleep and eating. To take time for this daily, repetitive caregiving work, in our own lives and in the lives of others, can serve as a kind of daily litany. This "litany of everyday life" can bring a sense of meaning, structure, order, and significance to the many different tasks that together make up the caregiving dimension of life.[50] At the same time, inhabiting these tasks as a daily litany can help us resist the cultural pressure to have perfect housekeeping and child-raising as our goals. Peterson points out the irony that in our materialistic society, we so often neglect or resent attending to basic material needs. Instead of valuing providing basic food, clothing, and rest to those in our homes and strangers we might welcome, we accept commercially established standards of housekeeping that are impossible to attain and add stress to our lives.[51]

Or we seek to attend to these matters of housekeeping in the most efficient manner possible, consistently seeking to do the most amount of work in the least possible amount of time (without margins for those frequent moments when things don't go according to our efficient plan, which add its own stress). This approach unquestioningly accepts the cultural value that efficiency is to be prized above all else. Furthermore, it implicitly suggests that the work of housekeeping itself is not valuable.[52] This perspective is

48. Stay-at-home mothers today spend seventeen hours per week on primary child care, compared to stay-at-home mothers in 1975 who spent eleven hours per week attending to primary child care. We see a similar rise in the hours when we consider employed mothers: eleven hours per week today, compared to six hours per week in 1975. See Sandberg, *Lean In*, 134; see also 131–37.

49. Sandberg, *Lean In*, 126.

50. Peterson, *Keeping House*, 19.

51. Ibid., 148–49.

52. Ibid., 156.

connected to the often unquestioned assumption of the market economy that unless something is connected to a monetary value, it is not worthwhile. As David Matzko McCarthy writes, "housework is not productive in the dominant economy . . . to the degree that value is determined by the market, housekeeping is nearly meaningless."[53] Putting it more strongly, Wendell Berry argues, "the triumph of the market economy is the fall of community . . . and all the things that only community life can engender and protect: the care of the old, the care and education of children, family life, neighborly work, the handing down of memory."[54]

When we consistently devalue the work that belongs to the household economy, we are allowing the logic of the market economy to take over. When we regularly try to do our housekeeping in the least possible amount of time, we are allowing the market to position the work of caregiving as nothing but a competing interest to a market economy that, as McCarthy writes, "makes voracious demands upon our time."[55] Surely there is more substance to the work of caregiving than simply being a competing interest. Surely there is more significance to the household economy than is acknowledged by the mentality that we are to get the work of the home done in the cracks and margins leftover after we've given everything else to the work that pays.

McCarthy's theological reflections on the home point to the same cultural ambiguity around the "status of the household" that we have been exploring throughout this essay. He argues that while the household economy was once a "basic social and economic institution," the economic function of the home has been reduced to "consumption" of "household items, automobiles, and the never-ending products children and adults come to want and need."[56] The modern home is now considered part of the "private realm," with the concomitant assumption that to contribute to the public good one has to leave the home to enter the working world.[57] The home is now considered a private refuge from the public world of competition. The dream of the suburban home encapsulates this, according to McCarthy. By design, a house in the suburbs—"a self-sufficient home, inhabited by affable kin and graced with plenty of yard to provide a buffer between neighbors"[58]—is separated from surrounding houses and the wider so-

53. McCarthy, *Sex and Love*, 106.

54. Berry, *Sex, Economy, Freedom and Community*, 133.

55. McCarthy, *Sex and Love*, 107.

56. Ibid., 20.

57. Ibid., 77.

58. Ibid., 80.

cial world. These homes intentionally isolate families from the demands that neighbors might make of them if in closer proximity. They accept the assumption that the home serves a private role, with an intentionally inward-focus, in contradistinction from the public world of work.

Accepting this private, inward-looking conception of the home, most in our culture have assumed that for equity to be reached between the sexes, both women and men need to have "equal access to that world of competition and work and meaningful activities outside the home."[59] McCarthy argues that this working assumption is problematic because it is rooted in a "shallow conception of social life" that allows contemporary notions of economics and politics to define work outside of the home as truly "public" and "social," while minimizing the potential of the household to make a public contribution to the social good.[60] In an ironic twist, McCarthy argues that modern households marked by a "traditional" gender division of labor *and* modern households in which all adults are pursuing careers outside the home accept the same problematic dichotomy between work and home.

McCarthy asks us to reimagine our notions of the home in relation to the public good. Rather than pursing the ideal of a closed family unit, which pays for help if needed but otherwise operates independently of other families, households, and communal needs, he offers the notion of an "open household" that is marked by interdependence, porous boundaries, wider networks of exchange and cooperation, and reciprocity. This is a vision of neighborhood involvement and exchange that is admittedly messy and risky. On a fundamental level, it moves the household from an inward to an outward orientation: "The open household is a story of discovery and a training ground for coming outside of oneself and 'moving forward' in love."[61] It recognizes that the household can be the source of "social transformation" based in the conviction that "family plays a vital part in neighborhood and community life, making for a rich and venturesome social world."[62] In short, it implicitly rejects the notions that work is the public, social realm while home is the private realm, and that work must exist in a competitive relationship with home.[63]

This "open household" vision also celebrates the quotidian realities of family life. As McCarthy writes, "the 'we' of the household is more quotidian than romantic or emotional. . . . They pass through unpredictable stages and

59. Ibid., 81.

60. Ibid., 183.

61. Ibid., 164.

62. Ibid., 172, 110.

63. Ibid., 174.

events: rearing infants and arguing with teenagers, sponsoring neighborhood clean up days and coaching football, scraping by with little income, enjoying times of plenty, cultivating friendships, sharing trouble. . ."[64] McCarthy worries that just as romantic love has not proved to be a strong basis for marriage, so having a family rooted in a private vision of receiving and giving love is not a strong enough basis for a household. This internal orientation needs to be replaced by an outward orientation, for "love and affection attain their depth through common work and through a common vocation."[65] Or, to put it more concretely, "'relationships' are represented by dinner for two, face-to-face, enjoying intimate moments and long conversations. The household, by contrast, orients a couple to matters of raising children, leaky faucets, and the politics of the neighborhood association."[66] When families enter into these common tasks, they find themselves strengthened as families even as they strengthen the larger community.

McCarthy's reflections help us to see that the world of caregiving is deeply connected to the public good. Affirming the importance of caregiving entails, in part, rejecting that work and home correspond to the opposing spheres of public and private. As Slaughter writes, "at least since the Industrial Revolution, we have split work and family into two different spheres . . . the relationship between work and family become one of profound tension, each tugging at the other."[67] Her own journey has involved coming to reject the inevitability of this tension as she's realized that the world of caregiving contributes as much to society as the world of competition.[68] This realization lies behind her call for our society to become "exceptional once again, not only for the speed of our computers and the power of our armies, but for the strength of our communities and the quality of our care."[69] To accomplish this, she believes we must invest not only in the world of competition but in the establishment of an infrastructure of care, to creatively implement a network of arrangements and institutions that can support women and men as they pursue their workplace dreams and their dreams for relationships and community.

If *Mad Men* gives us a glimpse of the significant cultural and economic change female workers have experienced in recent decades, contemporary voices like those of Anne-Marie Slaughter remind us that we have yet to

64. Ibid., 164.

65. Ibid., 155.

66. Ibid., 154.

67. Slaughter, *Unfinished Business*, 255.

68. Ibid., 249–53.

69. Ibid., 247.

find satisfactory cultural resolutions to the lingering tensions produced by this change. At one point in time, our economic and cultural arrangements depended upon a robust household economy supported largely by women. As those economic and cultural arrangements have shifted, we have been left unsure of the economic and cultural significance of matters of the household. In our contemporary imagination, we tend to pit work and home against each other. We have developed conceptual and economic categories to capture the significance of what transpires in the world of paid work, while we have not yet been able to name the significance of what happens in the home within these new economic and cultural arrangements. This leaves matters of the home and the work of caregiving unvalued and unsupported, on economic, political and cultural levels, even as these areas continue to require considerable time and attention.

This final section has represented one effort to explore why caregiving matters. Within a Christian theological framework, we can find resources to articulate why the home is significant and why daily acts of caregiving and housekeeping matter. We can also find reasons to acknowledge God-given limits, so that even as we seek to embrace the daily tasks of caring for self, home, and others, we can recognize the need to resist the dominant cultural formation that encourages a relentless pace and unattainable expectations related to work, home, and family.

These explorations of a theology of home are not offered with the assumption that they would be accepted by the culture at large, but as one way to engage the lingering strains our society faces as it continues to navigate the ramifications of the rise of the market economy on the household economy. These explorations have pushed us to consider the need for a deeper questioning that an inherent tension exists between work and home. When "work" is considered public while the home is considered "private," we fail to recognize the significant contributions to the public good that the work of caregiving and housekeeping make. We perpetuate the idea that matters of the home are socially inconsequential, while things done in the competitive world of work matter. As we continue to think creatively and collectively about how to support women and men as they pursue their dreams, we need to re-imagine the relationship between work and home so that both are considered essential not only for individuals but for communities and society to flourish. We need to recognize that for many, it is not just their dreams of being ballerinas, but their ability to sustain, nurture, and care for themselves, their families, and the larger community that is at stake.

BIBLIOGRAPHY

Allyn, David. *Make Love Not War: The Sexual Revolution: An Unfettered History.* Boston: Little, Brown, and Company, 2000.

Babcock, Linda, and Sara Leschever. *Ask for It: How Women Can Use the Power of Negotiation to Get What They Really Want.* New York: Bantam, 2009.

———. *Women Don't Ask: The High Cost of Avoiding Negotiation and Positive Strategies for Change.* New York: Bantam, 2007.

Berry, Wendell. *Sex, Economy, Freedom & Community: Eight Essays.* New York: Pantheon, 1993.

Blake, Meredith. "Women of 'Mad Men' Make a Strong Impression Amid their Fight for Change." *Los Angeles Times.* Online: http://www.latimes.com/entertainment/tv/la-et-st-ca-mad-men-women-20150405-48-story.html.

Buckingham, Marcus. "Why Are Women Unhappier Than They Were 40 Years Ago?" *Bloomberg Business.* No pages. Online: http://www.bloomberg.com/bw/stories/2009-10-16/why-are-women-unhappier-than-they-were-40-years-ago-businessweek-business-news-stock-market-and-financial-advice.

Diller, Vivian. "Mad Men, Madder Women: Have Roles Really Changed in the Workplace?" *Huffington Post.* Online: http://www.huffingtonpost.com/vivian-diller-phd/mad-men-madder-women-have_b_720893.html.

Ely, Robin J., et al. "Rethink What You 'Know' About High-Achieving Women." *Harvard Business Review,* December 2014. Online: https://hbr.org/2014/12/rethink-what-you-know-about-high-achieving-women.

Fox, Mary Frank, and Sharlene Hesse-Biber. *Women at Work.* Mountain View, CA: Mayfield, 1984.

Friedan, Betty. *Beyond Gender: The New Politics of Work and Family,* edited by Brigid O'Farrell. Washington, DC: Woodrow Wilson Center Press, 1997).

———. *Feminine Mystique.* New York: Norton, 1963.

Giele, Janet Zollinger, and Leslie F. Stebbins. *Women and Equality in the Workplace: A Reference Handbook.* Santa Barbara: ABC-CLIO, 2003.

Heath, Kathryn, et al. "Women, Find Your Voice." *Harvard Business Review,* June 2014. Online: https://hbr.org/2014/06/women-find-your-voice/ar/1.

Itzkoff, Dave. "'Mad Men' Creator Matthew Weiner Reflects on the Season So Far." *New York Times,* June 10, 2012. Online: http://artsbeat.blogs.nytimes.com/2012/06/10/mad-men-creator-matthew-weiner-reflects-on-the-season-so-far/?ref=arts&_r=0.

Johnson, Kristen Deede. *Theology, Political Theory, and Pluralism: Beyond Tolerance and Difference.* Cambridge: Cambridge University Press, 2007.

McCarthy, David Matzko. *Sex and Love in the Home.* London: SCM, 2004.

Norris, Kathleen. *The Quotidian Mysteries: Laundry, Liturgy and "Women's Work."* New York: Paulist, 1998.

Peterson, Margaret Kim. *Keeping House: The Litany of Everyday Life.* San Francisco: John Wiley & Sons, 2007.

Pope John Paul II. *The Genius of Women.* Washington, DC: United States Catholic Conference, 1997.

Rosin, Hannah. "The End of Men." *The Atlantic* (2010) 56–72.

Sandberg, Sheryl, and Nell Scovell. *Lean In: Women, Work, and the Will to Lead.* New York: Alfred A. Knopf, 2013.

Sepinwall, Alan. "'Mad Men' Creator Matthew Weiner Talks Peggy, Joan, Sci-Fi and More about Season 5." *Hit Fix*. Online: http://www.hitfix.com/whats-alan-watching/mad-men-creator-matthew-weiner-talks-peggy-joan-sci-fi-and-more-about-season-5.

Slaughter, Anne-Marie. *Unfinished Business*. New York: Random, 2015.

———. "Why Women Still Can't Have It All." *The Atlantic* (2012) 84–102.

Thistle, Susan. *From Marriage to Market: The Transformation of Women's Lives and Work*. Berkeley: University of California Press, 2006.

Tuttle, Mike. "Christina Hendricks Dishes Dirt After Mad Men." Online: http://flavorwire.com/510877/its-about-class-matthew-weiner-and-mad-mens-cast-on-the-shows-final-episodes.

Wittenberg-Cox, Avivah. "To Hold Women Back, Keep Treating Them Like Men." *Harvard Business Review*. Online: https://hbr.org/2015/07/to-hold-women-back-keep-treating-them-like-men.

Yuan, Jada. "Mad Men's Matthew Weiner on How He Found the Perfect Megan Draper." No pages. Online: http://www.vulture.com/2012/05/mad-mens-matt-weiner-on-jessica-pare.html#.

Chapter 6

"IF I DON'T GO IN THAT OFFICE EVERY DAY, WHO AM I?"[1]

Culture, Identity, and Work in *Mad Men*

—DAVID MATZKO MCCARTHY

If we judge by the Coca-Cola ad at the closing cut of the series, the happy ending for *Mad Men*'s Don Draper comes through an ingenious idea—through a moment of creative inspiration. At the close of the series, Don sits in a lotus position at a cliff top. His route to this ocean-side retreat has been accidental and indirect, and he sits, with irritation and acceptance, among a modest group of seekers. They appear to be in peaceful meditation. Then, he is struck with a faint smile. Immediately, we cut to the iconic "Hilltop" ad—a host of young singers, representing the peoples of the world, who would like to "buy the world a Coke." The implication is that Don makes his way back to the advertising industry in order to define it with "the greatest commercial ever made."[2]

1. The question is asked to Don Draper by Freddy Rumsen when Freddy is put on leave because of drunkenness at work, "Six Month Leave," S2/E9.

2. See Snyder, "*Mad Men* Creator," who writes, "It turns out the 1971 Coke ad that aired during the *Mad Men* finale was supposed to be written by protagonist Don Draper . . . [S]how creator Matthew Weiner said that the ad the episode cuts to during the show's final minutes, a Coca-Cola "Hilltop" commercial," suggests Draper's return to McCann-Erickson. . . . 'In the abstract, I did think, why not end this show with the greatest

This conclusion to the series is full of irony. If we know Don, we cannot expect that he will be happy for long. The series ends at an inspired moment and a new start but, in time, Don will unravel. To add to the irony, happiness and harmony are visualized in terms of consumer and corporation. We all know that Coke does not actually confer harmony, but the Hilltop commercial makes harmony seem possible—at least until the end of its song . . . "It's the Real Thing." Likewise, we are encouraged to believe that Don's new start is the real thing. If we step back from the scene, we will regain our critical edge. But we accept for the moment that Don will have changed for good; he has achieved a final, series ending, eternal moment of the "happily ever after." Further, the Coke ad is not only a cultural artifact but sets a standard for the advertising industry. With this inspiration, Don shows—for the last great moment of the series—his creative genius. He is able to return to the ad-factory that is McCann Erickson on his own terms, as a conquering hero.

Mad Men, as the title suggests, is about advertising men at work. More precisely, it is set on the landscape of work and the workplace in a world of professional men. The struggle for women—represented primarily by Peggy and Joan—is how to succeed and find fulfillment in that world as servile outsiders, always as resident aliens. *Mad Men* is about the quest for personal fulfillment—the cultivation of one's identity, purpose, and place—on changing and often unforgiving social, economic, and interpersonal terrain. Throughout the series, McCann Erickson represents a corporate Goliath. It is an ad factory; people become cogs in the machine. For this reason, Don's cliff-top inspiration (the perfect harmony of the Coke ad) is a victory. Will Don be happy? Probably not. Yet, we do know that he will return to McCann as a free individual—as a creative genius who is not defined by industry and mass-production, but defines who he is in relationship to his own work. The victory of the Coca Cola ad is not that Don will enjoy lasting peace and happiness, but that he re-enters the working world (he faces the leviathan) with enough sense of himself to find his own way.

In effect, Don finds himself and is liberated by the work. Following *Man Men*'s themes of identity and fulfillment, this chapter will focus especially but not exclusively on Don as we consider the male worker and his workplace. My perspective is shaped within the tradition of Catholic social teaching.[3] In brief, Catholic thought understands the work and the worker

---

commercial ever made?'"

3. Catholic social teaching refers to a set of modern documents beginning with Pope Leo XIII's *Rerum novarum* in 1891. Two phrases provide an apt summary for our purposes: Paul VI, in *Populorum progressio* (1967), sets economic development in terms of "authentic human development," and John Paul II, in *Laborum exercens* (1981), explains that "work is 'for man' and not man 'for work.'" In other words, "the

in terms of the development of the whole person. In this consideration of the person, Catholic social teaching sees various spheres of life as interconnected; individual, interpersonal, economic, social, and political spheres of life are woven together through the development of persons in community. Work connects us with people in "ever widening circles," from a community of workers to communities and peoples who share, enjoy, and depend upon the products of our labors.[4]

The perspective of Catholic social teaching fits well with the cycle of grasping and fumbling for wholeness in *Mad Men*. The series revolves, not as a wheel, but like a rolling egg: Relationships, careers, advertising firms, and identities begin, take a conventional arc toward fulfillment, bend downward to dissolution, and begin again with a renewed hope for fulfillment. In the context of an ad agency, *Mad Men*—as a series—is able to weave personal narratives (e.g., family life), the ongoing life of a "community of work,"[5] and professional development together with social changes in cultural conceptions of family, men, women, and work itself. The main tensions related to work, represented in historical perspective, remain unresolved today: the tensions between work and home, workplace culture and the equality of women *as* women, individual creativity and corporate demands, and money-making and personal fulfillment along with corporate profit-making and social responsibility (e.g., cigarettes and cigarette ads). *Mad Men* succeeds as entertainment because relationships, loves, hopes, and personal development are placed in the center of work. Although the point might be obvious, it is important to emphasize: In order to secure an audience, *Mad*

---

basis for determining the value of human work is not primarily the kind of work being done but the fact that the one who is doing it is a person" (*Laborum exercens*, no. 6). Also see McCarthy, "Modern Economy," 129–40.

4. See John Paul II's understanding of work within the web of common life in *Centesimus annus*: "Mention has just been made of the fact that *people work with each other*, sharing in a "community of work" which embraces ever widening circles. A person who produces something other than for his own use generally does so in order that others may use it after they have paid a just price, mutually agreed upon through free bargaining. It is precisely the ability to foresee both the needs of others and the combinations of productive factors most adapted to satisfying those needs that constitutes another important source of wealth in modern society. Besides, many goods cannot be adequately produced through the work of an isolated individual; they require the cooperation of many people in working towards a common goal. Organizing such a productive effort, planning its duration in time, making sure that it corresponds in a positive way to the demands which it must satisfy, and taking the necessary risks—all this too is a source of wealth in today's society. In this way, the *role* of disciplined and creative *human work* and, as an essential part of that work, *initiative and entrepreneurial ability* becomes increasingly evident and decisive" (no. 32).

5. In *Centesimus annus* nos. 35 and 49, the phrase, a "community of work," is used by John Paul II to refer to both business and the family.

Men is ostensibly about a line of work but really seems to be about persons in community. Likewise, Catholic social thought holds that the essential subject of work is the person while such things as profit and success in the market are instruments, certainly good and necessary, but still only a means to the end of personal development and the good of common life.[6]

Work (both the process of production and the good of the product) is directed to the fulfillment of persons in community. This good end and purpose of work is accentuated but ultimately fragmented and overpowered in *Mad Men*. The personal and common goods of work are disconnected, not only by the various economic and cultural conditions that are displayed in *Man Men*, but also by the narrative frame of the series, that is, by its deployment of heroic individualism. The first section of the chapter introduces this narrative arc. It attends to the tropes "artist" and "artistic genius" as they are framed in the professional world of *Mad Men*. In this professional world, work and career are conceived as expressive of personal identity, character, and one's meaning-giving place within the world. However, the expressive nature of the work constitutes a fundamental tension for individuals (or expressing one's individuality) in relation to the community of work. In short, Don needs a working community, but cannot help but undermine it as well.

The balance of the chapter develops these tensions between work and common life. The second section will treat questions of personal identity within the culture of the workplace, where persons are formed and compete (as allies and adversaries). In American culture, the workplace is assumed to be a complement to the household. The worker supports a family. However, the workplace is also a competing institution, in conflict with the home. *Man Men* goes further and fashions a working community as a family (Bertram Cooper as grandfather, Roger Sterling as father and so on). The third section considers work in terms of essential and instrumental goods: how fame, glory, wealth, and honor function in relationship to the good of the work

6. Consider John Paul II's discussion of profit in *Centesimus annus*. "The Church acknowledges the legitimate *role of profit* as an indication that a business is functioning well. When a firm makes a profit, this means that productive factors have been properly employed and corresponding human needs have been duly satisfied. But profitability is not the only indicator of a firm's condition. It is possible for the financial accounts to be in order, and yet for the people—who make up the firm's most valuable asset—to be humiliated and their dignity offended. Besides being morally inadmissible, this will eventually have negative repercussions on the firm's economic efficiency. In fact, the purpose of a business firm is not simply to make a profit, but is to be found in its very existence as a *community of persons* who in various ways are endeavouring to satisfy their basic needs, and who form a particular group at the service of the whole of society. Profit is a regulator of the life of a business, but it is not the only one; *other human and moral factors* must also be considered which, in the long term, are at least equally important for the life of a business" (no. 35).

itself. The tension between essential and instrumental goods highlights the romantic ideal of Don as the artist who lives for the work.

Finally, the chapter concludes by considering work in relationship to the arc of the modern tragic hero who, in *Mad Men*, comes to a triumphant end. The form of Don's victory is entertaining and uplifting; the Hilltop Coke ad signifies Don's genius and his success over and within machine of McCann Erickson. But Don's quest and his victory are ultimately unsatisfying as a representation of the worker and the good of the work. As individual hero, he must find himself by setting himself apart from a community of work and its fulfillment. His triumph is only his own; good work is shared.

THE ARTIST

The relationships between work, identity, and community in *Mad Men* are set up by likening Don to an artist—to an artist in a romantic (nineteenth century) sense of the originality, sensitivity, and independence (that is, the artistic isolation) of the creative genius.[7] The artist's creative suffering and sensitivity set him apart. He (historically conceived as a he) is alone in the world.[8] We will have gone too far if we think of Don as an actual artist in this romantic sense. But Don's talent, grasping for love, independence, and need for the work develop similar tensions between the creative individual and economic production, between his empathetic, intuitive genius and the ordered (rationalized) workplace, between his insight into the human heart and the ordinary work of his unexceptional co-workers.[9] These tensions play a major role in the drama of work in *Mad Men*. Certainly there are other important tensions, such as a conflict with home and for women in finding a role among men. However, the conflicts of the creative genius set up an opposition, more often than not, between working men. These tensions put Don at both the center and periphery of the community of work. Further, as Don is framed as creative genius—at the center and the periphery—he takes the role of the hero. Don is the hero in at least two ways: His genius compensates for his constant failures and deeply engrained flaws, and we

7. See Shiner, *Invention of Art*, 99–115. Shiner emphasizes the role of institutions of fine art, galleries and museums, in the (eighteenth and nineteenth century) process of separating art from "utility," artisans and their products, and commonplace activities. Through the museum and gallery, fine art has a "place" to inhabit a transcendent sphere of individual originality, creativity, and inspiration.

8. Jones, *Machine in the Studio*, 1–59. Jones shows that the transcendent and unique characteristics of art correspond to a sense of the artist as a set apart and alone and of artistic activity as essentially secluded.

9. Ibid., 10–11.

(the audience) take his side when the meaning of work is at stake when he argues with account executives or clients about what constitutes good work.

The analogy of the artist is established straight away, as Don Draper is introduced and his character is developed in the first season of *Mad Men*. In the very first scene of the first episode, we find Don at a bar but alone and focused on his work, scribbling ideas on a napkin.[10] He is absorbed by the work. His concentration forms a contrast with scenes at the office that immediately follow (and crowd the whole first season). The office is cluttered with men guffawing, politicking, and scheming for importance. Don, in contrast, is alone in the bar, working while others play. He consults a busboy (an elderly black man) to learn his perspective on cigarettes. Don's interaction with the only black man in the room has an authentic feel. The man is supposed to be silent and unseen. His boss intimates as much if we were to miss the point. Don wants to hear what the man has to say; he wants to know what he thinks, why he smokes his brand of cigarettes. Don scribbles down the man's ideas, while others obviously take this busboy for granted. Afterward, Don scans the crowded bar, pondering the relationships between the patrons and their cigarettes. In the next scene Don is knocking on an apartment door, obviously late in the evening. At this point, he enters the artist's studio, where Midge Daniels is wide awake and working late into the night, alone and independent. Their tryst is about far more than sex. Don is sensitive, frustrated by a creative block, and wistful about his comfort and peace with Midge. In this studio and with the artist is where Don wants to stay forever (he wants to marry Midge). It is fitting that Midge is too independent, and Don heads to the office in the morning: exiled by the artist.

An adumbration of the studio is carried over to Madison Avenue. From Don's sensitive, vulnerable presence in Midge's studio, we go to the revolving doors, crowded elevators, and loud, chauvinistic jockeying of Pete, Paul, and Ken. We see Peggy Olsen introduced to the office and the secretarial pool, where women are servile and always open to objectification. In effect, we are introduced to the contrived (that is, ugly) side of the workplace. Then, Don's office is established as part refuge and part workshop. Alone he thinks, obsesses over ideas, and sleeps. He is interrupted, in their turn, by Roger, Peter, Salvador, and a marketing consultant, all coming in for the work. The contrast is clear: The office is the world, and Don's office is set apart. The office culture, in a sense, is "manufactured" and inauthentic—marked by constant posturing, male-bonding, misogyny, and symbolic chest thumping—marked, in short, by scheming for position

10. "Smoke Gets in Your Eyes," S1/E1.

without actually working. Don focuses on the work. And while in his office, others focus on the work as well.

Don's authenticity is seen in his willingness to suffer for good work. He does not force himself upon the muse. Likewise, his great temptation and downfall will be to betray the work. For example, he tries a strategic maneuver and takes the moral high ground with his "Why I'm Quitting Tobacco" ad in the *New York Times*. But his maneuver is rash and sardonic; as a result, the ad only serves to alienate him from his co-workers and clients (current and future) of SCDP.[11] At his best, Don has no formula or plan; he has no mechanism to produce ideas. The process is to agonize, endure, start over again and again, and wait for the idea to show itself. In effect, Don can be selfless in relationship to the work. For instance, in the episode, "Wee Small Hours" (S3/E9), Don meets with Peggy, Kurt, and Smitty to work on the Hilton account. When he criticizes one of their tag lines, Peggy tells him that the line was his. He replies, "That doesn't make it good. If it's bad, don't use it. . . . There is no deadline. I want to see work as you think of it. Give me more ideas to reject. I can't do this all by myself." Consider also a conversation between Don and Peggy in "The Suitcase" (S4/E7). The two have been frustrated and dissatisfied with their work on a Samsonite account. Their own lives are in disarray. Peggy would rather work on her birthday as her family and boyfriend gather at a restaurant to celebrate the day. Instead, Peggy and Don sit together in a diner. Peggy complains, "I can't tell the difference anymore between something that's good and something that's awful." Don replies, "Well, they're very close. But the best idea always wins and you know it when you see it. Keep banging your head against the wall, then it happens." The passive voice in this statement is important. Creativity cannot be managed.

Don's creativity sets him apart. He (along with Peggy and Michael Ginsberg) has something that others cannot quite grasp. Roger is mystified and refuses to understand how Don thinks. Bert sees something special in him: a talented, self-made individual who is worthy of Ayn Rand's *Altas Shrugged*. He says of Putnam, Powell & Lowe (the British owners of Sterling Cooper) that "they have shown a great deal of interest in you . . . they study and dissect your work, trying to decipher what is your particular American genius."[12] In the final season, when Don is forced to go on leave, his singular (and former) presence in the office is contrasted with his replacement, Lou Avery. Lou imposes his prosaic imagination and mechanical process upon the creative team. He conspires against Don with Jim Cutler, who is

11. "Blowing Smoke," S4/E12.

12. "Guy Walks Into an Advertising Agency," S1/E6.

pragmatic and unimaginative and has no patience for Don's singularity. Don represents the division between "creative" and business "accounts."[13] Although he tries some clever business schemes, he represents the contrast between office machinations and the good and creative side of the work itself.[14]

In the first season, as Don's creativity is being established, an important contrast is developed between Don and Peter. Peter is ambitious; he wants to be brilliant and to enjoy the swagger of the singular man. He has an uneasy relationship to his prominent family, alternating between using the family name to his advantage and wanting to make his own name on his own. In any case, he has noteworthy progenitors. He is critical of Don and suspects that he (Peter) can be just as creative. Sitting surreptitiously, after hours, in Don's office, he discovers that Don is really Dick Whitman. Peter feels snubbed when Don hires Duck Phillips as head of accounts, and he tries to use his knowledge of Don's false identity to blackmail him and then to destroy Don's career.[15] His scheme falls flat because Bert says, simply, "Who cares?"

The contrast and conflict between Don and Peter in the first episode (S1/E1) sets Don's creative arc for the rest of the series. First, he is completely devoid of ideas for a Lucky Strike campaign, especially in light of FDA rulings about the proven health risks of smoking. The agency's head of research, Dr. Greta Guttman, suggests that Sterling Cooper ads begin to tap the dangerous side of smoking and draw on our subconscious "death wish." Don throws her research into the trash. When Don has to face executives from Lucky Strike, he has nothing to say. He shuffles his papers. Peter takes

13. At the end of Season 3, Don wants to fight the selling of Sterling Cooper to McCann Erickson. He tells Bert Cooper, "I'm sick of being batted around like a ping pong ball. Who the hell is in charge? A bunch of accountants trying to make $1 into $1.10? I want to work. I want to build something of my own" ("Shut the Door. Have a Seat," S3/E13).

14. When Don tries to make business maneuvers, like his letter to the *New York Times* about refusing to represent tobacco companies, his plans turn out badly ("Blowing Smoke," S4/E12).

15. Upon hearing Peter's accusation, Bert Cooper responds philosophically, "Who cares? . . . The Japanese have a saying: a man is whatever room he is in, and right now Donald Draper is in this room" ("Nixon v. Kennedy," S1/E12). After the first season, the relationship between Peter and Don takes several corners, and they end up allies and friends. As their characters develop, Don and Peter come to depend upon one another. Don helps to bring Peter on as a founding member (and junior partner) of Sterling Cooper Draper Pryce ("Shut the Door. Have a Seat," S3/E13). In S4/E12, "Blowing Smoke," Don covers Pete's $50,000 responsibility to put of collateral for a bank loan. In Season 7, Peter works to reintegrate Don into the business of Sterling Cooper & Partners.

advantage of Don's apparent failure and introduces the "death wish" idea (obviously he had been in Don's office and has found the research in the trash). Lee Garner, Sr., the executive from Lucky Strike, is appalled. He gets up from the table and starts for the door. When all seems lost, an idea comes to Don. The idea cuts through the distractions of health risks and Freudian psychology. It's simple; it focuses on what is pleasing about a cigarette, "It's toasted." He explains, "Advertising is based on one thing: happiness. And you know what happiness is? Happiness is the smell of a new car. It's freedom from fear. It's a billboard on the side of the road that screams with reassurance that whatever you're doing it's okay."[16] Peter schemes; Don waits. Peter seeks approval; Don wants to get it right. Peter digs it out of the trash; Don finds it in a flash of insight.

Don continually betrays those he loves—grasps for and then throws away people and happiness—but his own flaws and failures become part of his noble artistic quest. He is a tortured soul who, amid his own failures, has insight into what people desire—not the death wish but reassurance, freedom from fear, and moments when we can forget our pains and anxieties. Time and again, Don fails to be true to his own heart. Better yet, his heart betrays him. For instance, while married to Megan, Don's infidelity includes a tryst with his ex-wife Betty. Lying in bed together, Betty muses with regret about Don's wife. "That poor girl. She doesn't know that loving you is the worst way to get to you."[17] Don ruins almost every relationship, even his relationship to "work." In the end, he finds a bit of redemption through artistic inspiration (on the hilltop). Don, at key moments of decision, does betray the work, but his failures become part of the fabric of his sensitivity and isolation. He can be likened to the suffering artist. His route to heroism is his art. Don, however, is not an actual artist.[18] Rather, a cultural image of the artist—his depth of inspiration, originality, his tortured soul, lone quest, and his conflicts with a community of work—presents viewers with elements of his character. If we want victory for Don (despite his obvious faults), it is his creative gifts and the triumphs of these gifts (despite his modest beginnings) that keep us on his side.

16. "Smoke Gets in Your Eyes," S1/E1.

17. "The Better Half," S6/E9.

18. When all is said, Don is the keeper of a craft. His personal life embodies a modern conflict between the artist and the craftsman, between the quest for uniqueness and passing on traditions of mastery. Don is a master craftsman in the sense described my Richard Sennett in his sociology of work, *The Craftsman*. But he is also the artist who sets himself apart from the fidelity of the workshop and craft. In this regard, Don's infidelity serves to emphasize his artistry.

THE WORKSHOP

Although I have likened Don's gifts to the artist, allow me to shift to images of a craft, the craftsmen, and their workshop. I am using both art and craft, not as distinct methods of production, but as cultural tropes.[19] The artist is the inspired individual. Craft will be developed further below, but at this juncture, it is enough to say that a craft is cultivated with in a community and place.[20] Its goal is a product that judged by shared standards of excellence (rather than by the originality of fine art, which points to the inspiration of the artist). By framing Don through the analogy of the lone artist, I am suggesting an inherent tension between Don and the workshop; indeed, there is an inherent tension within Don himself. In Season 6, when Don loses the Jaguar account, Peter yells, "You're Tarzan swinging from vine to vine." The image is spot on. In the same scene, Joan says with frustration, "Just once I would like to hear you use the word 'we.'"[21] The analogy of the craft provides an image of a community at work. It provides a contrast between the workshop and the factory (represented by McCann Erickson).[22] The craft—within a community of work—shapes a person's identity; the threat of McCann Erickson is that it has no place for community and individuality.[23]

When Freddy Rumsen asks, "If I don't go in that office every day, who am I?"[24] his meaning is not entirely clear. What is it "in that office" that

19. My use of the term craft follows Glenn Adamson in his *Thinking Through Craft*. Adamson holds that craft is "an approach, or habit of action. Craft only exists in motion. It is a way of doing things, not a classification of objects, institutions, or people. It is also multiple: an amalgamation of interrelated core principles, which are put into relation with one another through the overarching idea of 'craft'" (3–4). While Adamson locates his concerns in the "modern craft movement" (as opposed to modern art), I am using craft as a trope to organize key characteristics of a craft that are identified by Adamson: knowledge and skill are developed in within common practices of a workshop, its goal is a material product (rather than "inspiration"), and common work is woven together with a shared love for excellence in performance and product.

20. Rather than an individual making crafts at home and selling at flea markets, think about carpenters in a cabinet shop.

21. S6/E6, "For Immediate Release."

22. Berg, *Machinery Question*. Berg shows that the machine challenges the craft and factory challenges workshop, not because machine take the place of craftsmanship, but because the division of industrialized labor makes all labor, including the craft, an appendage to machinery (143). It seems to me that this problem is similar to contrast, in *Mad Men*, between the small ad agencies (SC, SCDP, SC&P) and McCann Erickson.

23. As Don says in "Waterloo" (S7/E7), "we created this agency to get away from them [McCann]." The episode when the remains of SC&P move over to McCann Erickson is called "Lost Horizons" (S7/E12).

24. "Six-Month Leave," S2/E9.

points to who he is? In the wider context of *Mad Men*—in the narrative arc of the main characters—it is arguable that the identity-building and identity-sustaining elements of "that office" are found in its community of work, and in particular, its community of creative work—what David Pye, in his classic *The Nature and Art of Workmanship*, calls the workmanship of risk.[25] The craft of advertising, at least as it is framed within Sterling Cooper (as well as Sterling Cooper Draper Pryce), requires personal investment and risk. When Peggy appears to be too attached to her work, it is a set of practices and performance of skill that she loves.[26] The work is attractive because it makes demands upon who she is and who she is becoming. In the historical context of *Mad Men*, this kind of personal development is an opportunity given to men.[27] In Season 5, when Peggy is working out issues of identity, she discusses with Dawn (Don's secretary) her struggles as a copywriter and asks, "Do you think I act like a man?" Dawn's response is, "I guess you have to, a little."[28] Peggy has to "a little" because cultivating one's identity through work is the prerogative of men within a community of men.

Workmanship of risk, as understood by David Pye, depends continually on the judgment, skill, and engagement of the craftspersons. Workmanship of "certainly," in contrast, takes risk out of production. Pye's example is the use of the printing press (as opposed to writing) where risk and uncertainty are controlled before production begins.[29] Workmanship of risk depends upon particular people and their skill; who they are and the skills that they embody are the means of production.

Workmanship of risk is an important theme in *Mad Men*. Consider the scenes in S1/E1 when Don throws Dr. Guttman's research in the trash and risks going into his meeting with Lucky Strike with no stock ideas. Think of the contrast between Peggy, Stan, and Michael's workroom and the "Lost Horizons" episode (S7/E12) where a young man in a white clinical jacket leads Don to the Miller Beer meeting. On one hand, the workroom at SCDP is full of color, variation, and a history of ideas displayed on bulletin boards; on the other hand, McCann's workroom is impersonal, full of men in nearly identical suits (all creative directors) with their boxed lunches (exactly the same), sitting around a table to listen to the director of a research firm who

25. Pye, *Nature and Art.*

26. Peggy's attachment to work becomes a point of contention with her boyfriend Abe Drexler in "Far Away Places," S5/E6.

27. See "Shoot," S1/E9, when Don tells Betty that her job (her identity) is taking care of the children. Today, men still have greater opportunities to define who they are in relationship to work. See Hochschild, *Second Shift.*

28. "Mystery Date," S5/E4.

29. Pye, *Nature and Art*, 5.

admittedly offers a pile of "facts and figures."[30] Contrast how Freddy recognizes Peggy's creativity when she is only a secretary[31] with how Peggy, after establishing herself as Copy Chief at SC&P, is thought to be a secretary and not given an office when SC&P has to move over to McCann Erickson. In effect, Sterling Cooper and its various incarnations represent a community work, a workshop where particular people make a difference and develop as people.[32]

Seasons 1 to 3 follow the cast of characters as they attempt to find their place at Sterling Cooper. In the process, they establish a working community.[33] Young men—Peter, Ken, Harry, and Paul—attempt to assert their dominance and to establish their importance. Toward the end of Season 3, Peter learns that Ken has been appointed vice president in charge of accounts.[34] But in the final episode, Roger and Don visit Peter at home to ask him to join them. They agree to make him a partner as they form a new company. Before he agrees, Peter wants Don to articulate his talents. When Roger starts to explain, Peter says, "No, I want to hear it from him [Don]."[35] Don is the master craftsman and the focal point of the community of work. Likewise, at the end of Season 3, Peggy feels like she has been taken for granted as she tries to clear a path for herself as a woman. Don admits that he has been insensitive, speaks of Peggy as an extension of himself, and makes cryptic remarks about her talents.[36] When the new group, who are

30. For the creative chaos of the workroom and the risk of the creative process see "Dark Shadows," (S5/E9) where Don is competing with Michael for the best ideas for the "Snow Ball" account. Also take a look at the boardroom at SCDP turned into a creative room at the beginning of "The Other Woman," (S5/E10).

31. "Babylon," S1/E6.

32. It should be added that "accounts" also begins to be understood as a matter of skill, embodied judgment and engagement. This point is made in terms of Roger's abilities in contrast to Lane and Peter in "Signal 30," S5/E5.

33. After this working community establishes itself—after it becomes independent and self-sustained, we witness a cycle of fragmentation (largely influenced by Don's own unraveling) and attempts to buttress and recapture the independence of the working community. By the end of season 7, the pathways of the community and Don diverge completely. The original working community is dismantled and people go their own way. Yet, Don—as the final episode implies—returns to McCann Erickson on his own terms.

34. "The Grown-Ups," S3/E12.

35. Don says, "You've been ahead on a lot of things: Aeronautics, teenagers, the Negro market. We need you to keep us looking forward. I do, anyway" ("Shut the Door. Have a Seat," S3/E13).

36. Don explains, "Do you know why I don't want to go to McCann? . . . Because there are people out there who buy things, people like you and me, and something happened, something terrible [presumably the assassination of JFK]. And the way that

forming a new company, gathers at the offices of Sterling Cooper to gather (i.e., steal) what they need, Joan arrives. Lane Pryce asks, "What is she doing here?" Before he gets an answer, Bert addresses Joan, "Tell them what to do." Immediately, she gives tasks and a plan. The message is clear: Whatever has transpired between members of this group, they are now starting new. They have become a community of work, where each has been growing into their role and taking a place.

From the creation of Sterling Cooper Draper Pryce at the end of Season 3 to its merger with Cutler Gleason Chaough in Season 6,[37] this community of work takes a downward path. At the beginning of Season 4, Don and Peggy argue, and Peggy tells Don, "We are here because of you and all we want to do is please you."[38] But the artist can't be tied to the journeymen in the workshop. The artist cannot be bond (faithful) to a community and its procedures. By the end of the season, Don is a cliché: an executive who marries his twenty-five-year-old secretary.[39] In between, there are ups and downs; both successes and failures create acrimony. The firm wins a Cleo Award and loses Lucky Strike. In Season 5, Peter becomes full of himself; Roger dictates his memoirs as he becomes irrelevant in the office. Don is largely absent and becomes envious of Michael Ginsberg's energy and creativity. Joan (with Peter's initiative and encouragement) prostitutes herself for the Jaguar account. Peggy leaves and joins CGC. Lane commits suicide. In "Commissions and Fees" (E5/12), Don is frustrated with the work. He complains to Roger, "I don't like what we're doing. . . . I'm tired of this piddly shit. . . . I don't want Jaguar; I want Chevy. I don't want Mohawk; I want American. I don't want Dunlop; I want Firestone." The common work and the methods of the craft are not good enough.

A sense of common work is reinvigorated in the middle of Season 6 when Don joins forces with Ted Chaough to secure an account with Chevy. With the merger of SCDP and CGC, Peggy returns, and it appears that the office, once again, will be focused on common endeavor. But a new hope for a good working community is short lived. Don is the cause of its fragmentation. He spirals downward. Ironically, he hits bottom and is put on leave when he reveals the truth of his past (that he was raised in a whorehouse). In a pitch to Hershey, he tells the potential clients that they do not need advertising: "If I had my way, you would never advertise. You shouldn't have

---

they saw themselves is gone. And nobody understands that. But you do."

37. "For Immediate Release," S6/E6.

38. "Public Relations," S4/E1.

39. Joan makes this intimation when talking on the phone to her husband, "Tomorrowland," S4/E13.

someone like me telling that boy [Don as a child] what a Hershey bar is; he already knows."[40] The creative genius has unraveled; he is not suited to lead a workshop. Lou—precisely because he is prosaic and uniform—takes Don's position. Peggy and the creative staff are frustrated. Michael Ginsberg rages against the newly installed computer, which fills a whole room and imposes a constant hum.

The hope for common work is invigorated again in the middle of Season 7. Jim and Lou scheme against Don, but Roger helps Don return to work. Don finds himself under Peggy's supervision, writing tag lines for a Burger Chef account. Amid Don's self-pity, Freddy enters the narrative to issue an imperative, "Do the work, Don."[41] The Burger Chef account gives us the last vision of our working community as we know it should be.[42] It sets up the following episode ("Waterloo," S7/E7) where Jim tries to terminate Don's contract, Bert dies after watching the moon landing, and Roger makes a deal with McCann Erickson in order to buy out Jim, to save Don's job, and to keep Roger, Don, Joan, Peter, and Peggy together. On the Burger Chef account, Don is working for Peggy, but Peter wants Don to make the pitch to Burger Chef (Don disagrees). Peggy is working hard and has doubts about their progress and her ideas. Don finally reassures her, "I am here to help you do whatever you want to do."[43]

The scenes that follow capture some of the best working moments of the entire series. Peggy is skeptical that Don really wants to help; finally, she says, "If you really want to help me, show me how you think." The first thing Don says is that, "You can't tell people what they want. It has to be what you want." This point retrieves the core of the work: Don's sensitivity as an artist. Later he adds, "Whenever I'm really unsure about an idea, first I abuse the people whose help I need. And then I take a nap." Peggy quickly responds, "Done." From there Don suggests that they start from the beginning—to see if we end up in the same place. Their conversation about the ad reaches into their own worries and loneliness. Peggy admits, "What the hell do I know about being a mom? I just turned thirty. . . . I kept it as secret as I could. Now I'm one of those women lying about her age. I hate them." Don says that he

40. "In Care of," S6/E13. In context: "Closest I got to feeling wanted was from a girl who made me go through her john's pockets while they screwed. If I collected more than a dollar, she'd buy me a Hershey bar. I would eat it alone in my room . . . feeling like a normal kid. . . . It was the only sweet thing in my life. Do you want to advertise that? If I had my way, you would never advertise. You shouldn't have someone like me telling that boy what a Hershey bar is. He already knows."

41. "The Monolith," S7/E4.

42. "The Strategy," S7/E6.

43. Ibid.

worries, "that I never did anything, and that I don't have anyone." Then the inspiration comes to Peggy, "What if there was a place where you could go where there was no TV and you could break bread and whoever you were sitting with was family? . . . *That's it.*"

In the final scene of the episode (S7/E6), Peter, Don, and Peggy have a business meeting (of sorts) over a meal at Burger Chef. We know each of them is alone: Peter's girlfriend and Don's wife, Megan, have flown back to Los Angeles. Don is an absent father. Peter is hoping to overcome his estrangement from his ex-wife and daughter. Peggy has no one. Her life is her work. Throughout *Mad Men*, work and the workplace have been set against family. The work of men and their successes are divided from women, motherhood, and the realm of home. These conflicts are familiar to us (the audience) and to modern people generally; they are part of our lives. As Peter sits down with Peggy and Don at Burger Chef, Peggy alludes to her pitch: It will be filmed at Burger Chef rather than at a home, and it will be about family rather than motherhood. Peter is critical; he scowls, "I hate even the word family. It's vague." Don takes Peggy's side. "She's doing it the way she wants to do it. You want it right or not?" Peter shakes his head, resigns himself to the idea, grins, and starts to enjoy their company. Don signals to him with a polite gesture that he has ketchup on his cheek; Peggy affectionately hands him a napkin. As the credits flash on the screen, we see that Pete, Don, and Peggy are a family.[44]

THE WORK ITSELF

Mad Men is cynical about relationships—personal and professional. Ulterior motives, misdirected desires, and personal dysfunction intrude in romance, marriage, friendship, and the workplace. *Mad Men* is critical of sexism in our culture and the antagonism set up between careers and parenthood—social and economic practices that exclude mothers and allow fathers to opt out of childrearing. As the series concludes, Don makes a lame overture to raise his children after Betty dies.[45] Joan leaves her beau, Richard, when he lays down an ultimatum; he will not accept that she is starting a new business. Peggy finds love, with Stan, at work. Work has always been home for her. Roger puts his family affairs in order, putting Joan and their son, Kevin, in his will. He marries Marie, someone suitable in age. He seems to accept his aging, finally happy in marriage because he no longer expects

44. "The Strategy," S7/E6.

45. Betty says that they need a mother figure and stability, neither of which Don can provide.

to make a mark at work.[46] Amid the cynicism about the life of work and its antagonism toward a full life, *Man Men* is romantic about creativity and the craft, about moments of happiness in discovering the good of the work itself. Accordingly, we know that Don will never be a good father and probably will never be a good husband or friend, but he's got the Hilltop ad in the end. At least for a transcendent moment, he is elevated above the troubles of the world.[47]

Given that *Mad Men* is about a line of work and that Don Draper is a tragic protagonist, he carries—throughout the narrative—the possibility of redemption and transcendence though the good of the work itself, that is, through work as an activity which brings out the human good and in particular Don's own dignity as a person.[48] Don lives through a tragic cycle—his brokenness and desire for wholeness are at the core of his sensitivity and creative genius as well as his pattern of self-destruction. Through it all, he courageously takes decisive action in relationship to the work. In order to return to work at SC&P, Don is given stipulations that require him to refrain from thinking or acting on his own. To break the stipulations will mean that he will lose his job, his partnership, and his shares in the company. No one expects him to accept or to be able to keep the new set of rules. But Don simply says, "OK."[49] Then he breaks the rules by crashing a meeting that Jim and Lou have set up with Phillip Morris.[50] Later, when he is trying to convince Ted to go to McCann after a buyout, he explains, "I really lost *this* [the work] last year, and I realized I would do anything to get back in. And I did anything. I wrote tags; I wrote coupons, things I haven't done since I first got my start. . . . You don't want to see what happens when it's really gone."[51]

46. The last we see of Roger, he is like Christopher Nolan's Bruce Wayne, who at the end of *Dark Knight Rises* has been freed from the labors of both Batman and Bruce Wayne (he wills it all away) and sits with the reformed Catwoman at a Paris café.

47. Caroline Jones, in her study, *Machine in the Studio*, notes the connection between the twentieth-century life, the work of the artist, and nineteenth century romanticism: "What allied the twentieth-century Americans to the nineteenth-century Romantics was a shared search for transcendence, and an overwhelming sense of isolation in an increasingly crowded, explosive, aggressive world" (8).

48. In a 1949 essay, "Tragedy and the Common Man," Arthur Miller holds that "the tragic feeling is evoked in us when we are in the presence of a character who is ready to lay down his life, if need be, to secure one thing-his sense of personal dignity." The essay is available at *New Times on the Web*, https://www.nytimes.com/books/00/11/12/specials/miller-common.html.

49. "Field Trip," S7/E3.

50. "The Runaways," S7/E5.

51. "Waterloo," S7/E7.

Don's dedication and love of the work is fundamental to his character and provides continuity throughout the *Mad Men* series. There are points when be betrays the work, but he bounces back, choosing to lose everything for it. Consider what Aristotle calls "external goods" in book one of his *Nicomachean Ethics*—pleasure, wealth, fame, and honor.[52] Don struggles with them all. But he gives away money freely.[53] He does not want fame.[54] He knows that pleasures are a diversion. Amid every affair, he hopes for true love and personal peace. The most tempting external good for Don (although not the most destructive) is honor. On the day when Don wins a Cleo Award, he is drunk and full of himself. As a result, he makes a fool of himself during his pitch to Life cereal.[55] His self-satisfaction creates a rift with Peggy, who notes to Don that the original ideas for the award winning commercial were hers. Peggy finally cries, "You never say, 'thank you.'"[56] It is fitting that honor is likely to attract Don because, of all the external goods, it is most closely attached to the excellences of good work.[57]

Don's good work and creative brilliance cannot be understood apart from his false life, his brokenness and failures as a person, along with his constant and unquenchable desire to be whole and true. Don's genius is his openness and ability to give himself to his work: He is sensitive to his own brokenness and desire for wholeness and empathetic when he encounters the same fragmentation and yearning in others. Advertising (at least in Don's world) is an attempt to discover the whole and true from within a contrived putting together of random pieces—from within a false world. When the true within the false is found, the result is the ad for the Kodak Carousel.[58] In his pitch to Kodak, a nostalgic yearning for unity is brought to the surface by Don's own family photos, which are found amid the effects of Don's fragmented life: his brother's suicide, Betty's growing disillusionment at home, and Don's refusal to join his family as they travel for Thanksgiving.

52. Aristotle, *Nicomachean Ethics.*

53. Don's detachment from money is signaled in season 1 when Bert gives him a bonus check for $2500 and advice about reading Ayn Rand. Don gives the entire check to Midge when Midge declines to run away with him.

54. In "Advertising Age," S4/E1, Don embarrasses the firm because of his reticence to talk about himself to *Advertising Age.*

55. "Waldorf Stories," S4/E6.

56. "The Suitcase," E4/S7. To this conflict with Peggy, we should add the conflict with Michael Ginsberg over the Snow Ball account in "Dark Shadows," S5/E9. Michael is working hard and producing creative ideas. Don is threatened and uses his authority to suppress Michael's ad ideas and to promote his own.

57. Aristotle, *Nicomachean Ethics,* Bk 1, Ch 5.

58. "The Wheel," S1/E13.

In the final scenes of the series, Don is the artist, luminous and alone. In the end, establishing a workshop, like "workmanship of risk" that animated Sterling Cooper and SCDP, is no longer a possibility. McCann Erickson has absorbed SC&P into its ad factory. Don's struggle to be a craftsman within a community of work is no longer in play. In the final episode, his best work comes to him when he is far away from the workplace. Don is meditating with others on retreat, but he is really alone with his pain and his fragmented life. He is a suffering artist who finds inspiration in a strike of lightening. Throughout the chapter, I have developed the tension between this figure of the artistic genius and the trope of the workshop—the community of work that is held together and matures in relationship to its craft. With the image of the craft, I have drawn upon key themes in Catholic social teaching, not only a focus on the community of work, but also a concern for the growth of the person and for the development of a product that serves the common good. In *Mad Men*, the drama of working men, *is* the conflict between these common goods of the work and the desires of individuals to be set apart, to stand alone.

THE HERO

Man Men is successful because it engages tensions between individualism and an orientation to common life, between self-directed pleasures and fidelity to others, between work that attains external goods only (i.e., fame, etc.) and work that is good in itself and forms persons in relationship to that good. These conflicts mark professional, skilled, and specialized work in the modern world.[59] These dramatic tensions are sustained throughout *Mad Men* and are basic to the development of friendships, the personal growth of main characters (e.g., Peter), and the hopes to build a community of work. But, in the end, the tensions are resolved in relationship to Don Draper as a tragic hero in a modern Romantic mode.[60] As tragic hero, he must be alone in the end. Indeed, Don ends the series with nothing. He triumphs over the impersonal, utilitarian machine (McCann Erickson). But the victory is only his own; he stands alone. With the inspiration of the Hilltop ad, he will return to McCann Erickson on his own terms—as the artist.

59. *Mad Men* give little treatment of the struggles and desires of the working class. By and large, working class people are background and props. This is not a criticism. It is not reasonable to expect that a TV series would be able to give wide treatment of work (and to keep character development in focus).

60. By Romantic, I am referring Don's common origins, his fight with an economic leviathan (the advertising corporation), and his artistic genius, which comes to the common man without education or training.

To this degree, the resolution of *Mad Men* is unsatisfying, heroic and inspiring but finally disappointing. Don's victory is provisional; there is no indication that he will be free from his cycle of attachment and infidelity to loved ones, colleagues, and the work itself. In deciding for the lone artist and setting aside the tensions of modern work, the conclusion of the series can be a reminder that the tensions might be our best hope for finding goodness and nobility in our work. In our world (and Don's) the conflicts are not likely to be resolved. We have to live with conflicts between self-directed desires and a community of work, between "standing out" and fidelity to a craft and the workshop, between competing with co-workers, professional envy, and becoming a better person. These are the struggle we live and work with, and *Mad Men*, despite its conclusion, has done us a service by putting these personal and social tensions of a community of work in the forefront.

BIBLIOGRAPHY

Adamson, Glenn. *Thinking Through Craft*. New York: Berg, 2007.

Aristotle. *Nicomachean Ethics*. Translated by Martin Ostwald. New York: Macmillan, 1962.

Berg, Maxine. *The Machinery Question and the Making of Political Economy*. New York: Cambridge University Press, 1980.

Hochschild, Arlie. *The Second Shift: Working Families and the Revolution at Home*. Rev. ed. New York: Penguin, 2012.

John Paul II. *Laborum exercens*. Vatican City: Libreria Editrice Vaticana, 1981.

Jones, Caroline A. *Machine in the Studio: Constructing the Postwar American Artist*. Chicago: University of Chicago Press, 1996.

Leo XIII. *Rerum novarum*. Vatican City: Libreria Editrice Vaticana, 1891.

McCarthy, David Matzko. "Modern Economy and the Social Order." In *The Heart of Catholic Social Teaching*, edited David Matzko McCarthy, 129–40. Grand Rapids: Brazos, 2009.

Miller, Arthur. "Tragedy and the Common Man." *New York Times*, February 27, 1949.

Paul VI. *Populorum progressio*. Vatican City: Libreria Editrice Vaticana, 1967.

Pye, David. *The Nature and Art of Workmanship*. New York: Cambridge University Press, 1968.

Sennett, Richard. *The Craftsman*. New Haven: Yale University Press, 2008.

Shiner, Larry. *The Invention of Art: A Cultural History*. Chicago: University of Chicago Press, 2001.

Snyder, Benjamin. "*Mad Men* Creator Says Don Draper Wrote that Iconic Coke Ad." *Fortune*, June 11, 2015. Online: http://fortune.com/2015/05/21/mad-men-coke-don-draper/.

Part 2

WHO IS DON DRAPER?

Chapter 7

COUNTERCULTURE BEATRICES?

Don Meets Dante

—GABRIEL HALEY

> Don (to Pete): The day you sign a client is the day you start losing them.
>
> [. . .]
>
> Roger (to Don): The day you sign a client is the day you start losing them.
>
> Don: You don't really believe that.[1]

For all his mystique, Don Draper has a type. Generally, he exhibits the kind of discontented behavior that the Sterling-Cooper mantra, repeated by both Don and Roger above, attributes to its clients. Don wants what he does not have, and this trait appears most commonly in his relationships with women. In his quest for the new, for the unknown, he tends to be attracted to women who are *other* than him.[2] Thus, to some extent, Betty represents for him the privileged, upper-class life that he lacked during his unfortunate childhood. For him, their marriage shores up a new identity,

1. "Long Weekend," S1/E10.

2. One notable exception is the waitress Diana in Season 7, who appears to be Don's female doppelganger.

distinct from the Dick Whitman of his low-class past. But, having attained what he at first did not have, Don remains true to type. Once he settles into the glamorous, wealthy lifestyle of a Madison Avenue advertising executive, he increasingly finds Betty's charms less seductive. The conventional, familiar nature of his new life disappoints. Furthermore, though Don's behavior takes center stage, viewers of *Mad Men* know that the trait isn't Don's alone. Betty herself struggles with the social conventions imposed upon her. The life of suburban domesticity proves to be, for both of them, its own ghetto. They look elsewhere.

Part of what makes Don a superb ad-man is his ability to recognize that a desire for *something else* motivates just about everybody around him. Don displays awareness of this tendency, this capacity for infinite desire, when he relays the corporate mantra to the Sterling-Cooper's newest account man, Pete. With this self-awareness, though, also comes self-critique. When Don questions Roger's use of the commonplace, he expresses a discontentment about discontentment itself ("You don't really believe that"). Don knows that seduction and the chase have mass appeal, but he also intuits that more appealing than the chase is the expectation of fulfillment, a hope that does not disappoint. This counter-recognition, likewise, helps him in his craft. Among his most momentous triumphs in the boardroom is his presentation to Jaguar, where his tag proclaims, "At last, something beautiful you can truly own."[3] Of course, viewers might remember that this pitch intersects with one of the seediest transactions in Sterling Cooper's history, as the partners (Don excluded) have agreed to prostitute Joan for a chance at Jaguar's business. Ah, there's the rub.

Early in the show's history, *Mad Men*'s creator Matthew Weiner explained: "*Mad Men* is about the conflicting desires in the American male and the people who pay the price for that, who are women."[4] The unsettling transaction with Jaguar in Season 5 indicates how persistent this theme remains throughout the series, but by the later seasons the show has also grown into exploring the depths of this idea. "The people who pay the price" come to include more than just women, and the "conflicting desires" are attributable to more than just the white American male. More intriguingly, these desires, while often harmful or devastating to others, nevertheless seem, in their ubiquity, essential to human nature—even, somehow, a good worth considering in itself.

This recognition, that human desires hold some redeeming quality, despite the repulsive outcomes they regularly bring, is central to the

3. "The Other Woman," S5/E11.
4. Quoted in Krouse, "Every Woman," 188.

characterization within *Mad Men*. The show understands people as desiring agents, and part of the way it demonstrates this assessment is by turning to two cultural icons: one expected in a series about the 1960s, and one less so. The first is the show's portrayal of the counterculture, the second its use of the medieval poet, Dante Alighieri. Both the counterculture and Dante's poetry posit a way to live that is otherwise than the norm—that is, they each give a form to the human desire for otherness. Clearly, the counterculture of the 1960s is of greater concern to the series as a whole, but we might better understand *Mad Men*'s depiction of the counterculture by considering Dante. It is through Dante that we can articulate how the show represents the counterculture in a way that is neither entirely naïve nor entirely cynical.

Mad Men portrays the counterculture of the sixties in a manner that reveals, like Dante's poetry, an especially nuanced understanding of human desire. In the show, the counterculture embodies a desire for another way of being, in contrast to the Madison Avenue mainstream which the show often paints as so corrupt. Yet the counterculture is not a utopian solution to all the social ills in *Mad Men*. It inevitably falls to some of the same corruptions that are embedded within the mainstream culture; over the course of the series, the counterculture even becomes part of the mainstream. We need not see this portrayal as a unilateral critique of countercultural desire. Part of *Mad Men*'s appeal seems to be its belief that desire itself, while destructive, demands some consideration as an essential feature of being human. This is an idea that has a long heritage in Christian thought, regularly appearing in Christian apologetics from St. Augustine to C. S. Lewis. Lewis, for instance, argues a case for the existence of God from the fact of human desire: "If I find in myself a desire which no experience in this world can satisfy, the most probable explanation is that I was made for another world."[5] Dante's poetry is among the most dramatic and influential versions of this theology of desire.

Throughout his writings, including his magnum opus, known today as the *Divine Comedy*, Dante explores the concept of desire as a key structural feature of his poetic theology. When set alongside *Mad Men*, Dante offers revealing insights into the show's treatment of desire, especially in its treatment of the counterculture and the way it elevates the good of desire over and against the corruptions of the mainstream, which tends to preserve power structures shown to be inherently corrupt. Since *Mad Men* never identifies the counterculture as a salutary movement in itself, not even an authentic diversion from the mainstream and its errant ways, its representation of the counterculture demands closer attention. *Mad Men* uses the counterculture

5. Lewis, *Mere Christianity*, 136–37.

in Dantean fashion simultaneously to promote and to critique an endless human capacity for desire.

BEAUTY AND THE BEAST

The poetry of Dante receives special attention at a pivotal moment in the series. After the first episode of Season 6 begins with a point-of-view shot from the ground looking up—as Don's neighbor, Dr. Arnold Rosen, says, "That's it, hang in there," amidst shocked bystanders—the scene cuts to a reading of the first lines of Dante's *Inferno* as the camera fixates on a midsection of a bikinied female body.[6] The scene's framing does not allow the viewer immediate recognition of the body, only a bit of navel gazing, while Don intones Dante's famous opening lines: "Midway in our life's journey, I went astray from the straight road and woke to find myself alone in a dark wood." Viewers soon discover that they are looking at Megan, and that she and Don are on vacation in Hawaii (Megan's first word is *mahalo*, Hawaiian for "thank you"). Potentially a vision of marital bliss, the image instead sends competing signals—especially in the early frames, which voyeuristically cut off Megan's head. Don's reading material itself—coming from Dante's description of hell—is already foreboding, but viewers soon learn that he obtained it from his latest mistress, Sylvia Rosen. Since Don and Sylvia's affair ends what had appeared throughout Season 5 to be a more promising marriage than Don's first, the mise-en-scene casts a dark shadow on what proves to be a very dark season.

Mad Men often sets out such literary Easter eggs to tell us something about its characters. We see Don dipping into Frank O'Hara's *Meditations in an Emergency* (which consequently caused a spike in the sales of O'Hara poetry).[7] We find Betty Draper reading Mary McCarthy's novel *The Group*.[8] Burt Cooper, naturally, has Ayn Rand's objectivist manifesto, *Atlas Shrugged*.[9] Sally, intriguingly, reads *Rosemary's Baby*.[10] Yet of all the books placed in the hands of its cast, Dante's *Inferno* is by far the oldest. *Inferno* is also the only book to begin a season, even to begin an episode.[11] In many

6. "The Doorway," S6/E1.

7. "For Those Who Think Young," S2/E1, Dubner, "Publishers."

8. "The Color Blue," S3/E10.

9. "The Hobo Code," S1/E8.

10. "The Crash," S6/E8.

11. O'Hara's *Meditations*, read aloud by Don in the concluding montage of "For Those Who Think Young," S2/E1, is given nearly the same priority, although critical commentary has attended to its appearance more than *Inferno*.

ways, the show's placing of Dante's *Inferno* does more than give viewers a peek into Don's psyche. The special priority given to Dante suggests that there are important thematic parallels in *Mad Men* that are best understood through some of Dante's most distinctive ideas. At the very least, the appearance of Dante's work at the beginning of Season 6 foreshadows the downward trajectory Don follows throughout the season. Even more, however, by attending to features within Dante's work as a whole, we can observe points that resonate with *Mad Men*—the first being the assumption that *desire must be embodied*. Both *Mad Men* and Dante share this perspective: human desire, though often vague and abstracted, will identify a physical "source" that motivates it.

Embodied desire is a distinctive feature of Dante's poetry, one that often surprises first-time readers. *Inferno*, the work quoted by *Mad Men*, is only the first part of a poetic trilogy recounting an otherworldly journey through hell, purgatory, and heaven (*Inferno*, *Purgatorio*, and *Paradiso* respectively)—which Dante wrote while in exile from his hometown of Florence during the early years of the fourteenth century. Throughout the trilogy, Dante makes his way through the uncharted terrain of the afterlife with the help of guides, most notably the poet Virgil and a woman named Beatrice. A vital character in all of Dante's works, Beatrice is perhaps the most remarkable feature of his understanding of human desire. She embodies Dante's desire, though she is not, in the end, what Dante desires. Through her, Dante charts embodied desire's relationship to epistemology; that is, he believes that physical desires tell us something about how humans come to know the true from the false, heaven from hell. Virgil, who leads Dante through hell and purgatory, also appears important in this regard—he is the premier author of the classical Roman past, the one Dante looks to as a creative mentor—yet Beatrice, Dante's guide in heaven, is a woman from Dante's own time, whom Dante knew. It is she whom Dante first acknowledges as the beginning of his poetic accomplishments. In fact, he goes further than that. She is, as he puts it, the beginning of a New Life.

Notice the religious implication of the language of "New Life." The first meeting of Dante and Beatrice, on an outside viewing, would appear quotidian, simply two children noticing one another. Dante first sees Beatrice at the age of nine. She is eight. They do not even speak. But Dante portrays the moment as transformational, a kind of conversion. While the two do end up speaking eventually, they never have a romantic relationship. Dante, in fact, marries another woman. Nevertheless, Beatrice makes an irrevocable mark on Dante's imagination and on his writing. Even after Beatrice's untimely death, Dante continues to write under her auspices. Throughout Dante's life, she remains a muse, representing not just poetic inspiration but the path to God. Though "muse" might not, in fact, be the right word here. The role of Beatrice in Dante's writings is more complex. In some ways, she is a muse-like abstraction, a symbol, though she is also a flesh-and-blood reality. The critic Charles Williams describes Beatrice's position in this way: "Beatrice is [Dante's] knowing. To say so is not to reduce her actuality nor her femininity. The reason for the insistence on her femininity is simple—it is what Dante insisted on."[12] One way to understand Dante's Beatrice figure is to think in terms of his medieval theology. While the inclusion of Beatrice makes Dante's *Divine Comedy* idiosyncratic, his vision of the afterlife is undoubtedly a medieval Christian one. While it contains numerous hermeneutical complexities, the work sustains a traditional, devotional impetus—as it points readers to a Trinitarian and incarnate vision of God. Beatrice fits within Dante's incarnational understanding of the world in the way she remains entirely physical while at the same time a spiritual guide.

Dante's entire *Comedy* thus works along the principle that spiritual experiences cannot be divorced from physical ones. Anthony Esolen, a recent translator of Dante's works, explains that hell itself is part of this assertion that "things have meaning." The doctrine of Hell makes sense to Dante because it is the natural outcome of a meaningful life: "One of the happy corollaries of this belief in the meaning of things is that life is tremendously dramatic, with eternal consequences. For, in his providence, God has allowed human beings free will, and the use of one man's free will, at one moment, can mean life or death for a city, or salvation or damnation for himself and even, ultimately, for countless others."[13] Both Dante's vision of hell and his figure of Beatrice participate in the incarnational principle that physical realities are connected to spiritual realities. This life matters.

To be clear, I do not wish to contend here that *Mad Men* shares the same theology as Dante; however, I do see parallels in the way *Mad Men* understands human experience. Despite multiple levels of remove, Dante's work can appeal to a wide audience today because its views affirm meaning in life without falling into a naïve form of idealism. Dante asserts simultaneously the value of human desire and its dangers. This balance is only manageable within an ethical framework and, as I will show later in this chapter, whatever their theologies, both Dante and *Mad Men* confirm the possibility of real moral progress. Dante's figure of Beatrice is one way to understand the intersection of physical desire and moral progress, and *Mad Men* similarly seems interested in the way particular women—especially counterculture women, like Megan and Midge (Don's first mistress in the

12. Williams, *Figure of Beatrice*, 232.

13. Alighieri, *Inferno*, trans. Anthony Esolen, xxii.

series)—offer Don the potential for moral progress. By including women as both desired and desiring agents, Dante and *Mad Men* share a belief that physical desire itself can be the starting point for one's moral development.

That *embodied desire can lead to moral progress* is thus a second observation we can glean from Dante. We see this point suggested in the lines of poetry used by *Mad Men.* In their lament for the loss of a "straight road," they emphasize a need for moral direction. The immediate question to follow would be: "Where is the straight road?" For the pilgrim Dante, the answer is not immediately apparent. Readers of the *Divine Comedy* are often startled, along with the work's protagonist, that the path to God involves the guidance of a pagan poet and a beautiful woman. Furthermore, this "straight road" entails a downward journey through hell before the upward journey through purgatory and heaven. This downward movement is echoed in *Mad Men*'s sixth season, as Don hits numerous professional and personal lows, including being dismissed from his job[14] and, at another level of degradation, scandalizing his daughter Sally.[15]

To an extent, the course of *Mad Men*'s final two seasons encourages a narrative of redemption, culminating with a moutaintop experience. Dante's lines that start Season 6 anticipate a moral darkness which pervades the season. Moreover, they resonate with a sequence of eschatological language that is embedded within the subsequent season. Afterlife imagery persists into Season 7, for instance, when Peter says, "Sometimes I think I died and I'm in some sort of . . . I don't know if it's heaven or hell or limbo."[16] Likewise, when McCann Erickson absorbs SC&P, Jim Hobart calls it "advertising heaven."[17] Such language even takes on a parodic edge when Peggy chastises Don's metaphysical approach to a job review: "This is supposed to be about my job, not the meaning of life."[18] If Season 6 looks like hell, the various eschatological references in Season 7 appear as deliberate variations of the afterlife metaphor. Indeed, the movement from the inferno to "advertising heaven" suggests an intentional attempt on the part of the writers to map out a narrative of redemption, but it is a self-conscious narrative. We cannot forget that despite Hobart's claim that working for McCann Erickson offers some kind of paradise, Don quickly feels its mundane realities, memorably

14. "In Care Of," S6/E13.
15. "Favors," S6/E11.
16. "A Day's Work," S7/E2.
17. "Time & Life," S7/E11.
18. "The Forecast," S7/E10.

depicted by the well-oiled, yet tedious, Miller presentation of another corporate executive.[19]

So, *Mad Men* never fully embraces the recurrent afterlife imagery. While the later seasons may faint in the direction of a hell-to-heaven journey, it would be foolhardy to read the plotlines as direct analogues to Dante's *Divine Comedy*. *Mad Men* is never pure allegory, and we should not try to turn it into one. Attempting to read Season 6 simply as an allegorical version of Dante's *Inferno*, for example, might encourage us to see Don's new mistress Sylvia as the Beatrice figure.[20] Sylvia owns the copy of Dante's *Inferno* that Don reads. She is never seen without her crucifix. Yet, she also appears in the episode as the first proof that Don's monogamy with Megan has been broken. If a Beatrice figure, she is an unusual one. Sylvia prays for Don's salvation, yet their extramarital affair nearly destroys Don's relationship with his daughter Sally. Whatever parallels may be made, *Mad Men*'s intersection with Dante must be understood fundamentally as a thematic intersection, not as a systematic pairing. We should not seek a one-to-one correspondence. Nevertheless, the comparison remains a valuable way to understand the function of desire within the series as a whole. Through Dante, *Mad Men*'s countercultural women, and the counterculture itself, may best reveal their function within the show.

More fitting to the Beatrice figure of Dante's moral imagination, then, is Megan's earlier role in Season 5. Before the dissolution and downward spiral of Season 6, Don and Megan's marriage offered Don a possible way forward, as Megan seemed to bring him actual happiness for a sustained period (more so than a regular viewer might suspect of Don). The possibility that Don might be able to live in a monogamous marriage motivates the entire fifth season, and Megan is represented, like Beatrice, as an outsider who might actually change Don's world. Through Megan's *joie de vivre*, Don glimpses an alternative to the ego-driven interactions of Madison Avenue. Although Megan enters the series gradually (she appears at first as one of the show's many ancillary characters), by the time she accompanies Don to California it is clear that her part in the series has become substantial. She embodies everything that Don believes he is missing. Don surprises Megan, and perhaps the viewers, by proposing to Megan out of the blue, having only decided to do so while on vacation.[21]

19. "Lost Horizon," S7/E12.

20. James Meek has warned against this critical tendency; see Meek, "Shock of the Pretty."

21. "Tomorrowland," S4/E13.

The very title of the proposal episode, "Tomorrowland," suggests the optimism that surrounds Megan's character. Megan's ability to save Don from the joyless, mutually-damaging elements of the New York City is central to her role in the series. Her difference from the world of advertising executives, and from the warzone of a marriage that Don built with Betty, is particularly emphasized during their whirlwind courtship. At a meal with the children, Megan's calm response to a spill of a milkshake contrasts with Don's immediate enervation ("Don't be upset. It's just a milkshake.").[22] In contrast to the bitterness of Don's usual Madison Avenue cohort, Megan brings levity and confirms the truism that there should be no crying over spilt milk. Don's petulant reaction appears, by comparison, childish, and in light of Megan's response he realizes this. It is Megan's buoyancy, and Don's self-recognition, that prompts the surprise proposal.

Megan's ability to shock Don out of his apparent ennui reflects the "love at first sight" trope used by Dante when he describes his first viewing of Beatrice. Dante recounts the memory of the event (which occurred, remember, when he was nine years old) in his first major work, *Vita Nuova*:

> At that moment, and what I say is true, the vital spirit, the one that dwells in the most secret chamber of the heart, began to tremble so violently that even the least pulses of my body were strangely affected; and trembling, it spoke these words: "Here is a god stronger than I, who shall come to rule over me."[23]

The feelings expressed here have only a partial connection to our current understandings of romantic love. Dante writes the *Vita Nuova* within a community of poets who recognized rhetorical conventions, and a passionate devotion to romantic love, which some today might find excessive or embarrassing, was a part of these conventions. Still, we can see that Dante's first sight of Beatrice gives him the impression that his "vital spirit" (we might say "heart") has met "a god stronger than I." What Dante sees in Beatrice is a distinct vision of otherness, proof of a power beyond the ego. While egoism makes a god of the self, Beatrice's appearance makes self-worship an impossibility.

Beatrice thus suggests neither the "angel" nor the "monster" made famous by Sandra Gilbert and Susan Gubar's foundational study of nineteenth-century novels.[24] For Gilbert and Gubar, these two literary types

---

22. "Tomorrowland," S4/E13.

23. Alighieri, *Vita Nuova*, 4.

24. Gilbert and Gubar, *The Madwoman in the Attic*, 20–44. Gilbert and Gubar cursorily include Beatrice in the tradition of angel figures, and while she may be a literary influence, she should not be placed the same category as the nineteenth-century "angels

pervade the characterization of women in the nineteenth century. The angel is pure, virginal, and submissive, while the monster is dangerous, corrupted, and uncontrollable. In the *Vita Nuova*, however, Beatrice, combines many of qualities associated with both the "angel" and the "monster": she is pure and virginal, but also uncontrollable, and therefore dangerous. By identifying Beatrice as a stronger god, Dante articulates the way her sudden appearance in his life made him reconsider any egoism, by which he would consequently reconsider his understanding of the greater good. Nor is Beatrice a static figure in Dante's writings; over the course of Dante's writings, her meaning extends and deepens. The early vision of Beatrice's captivating beauty is later supplemented by the *Divine Comedy*. Here, *Purgatorio* shows Beatrice commanding autonomy. By the time Dante writes of Beatrice in *Paradiso*, she stands within a celestial hierarchy, herself subordinate to Christianity's one God. What remains constant in Dante's Beatrice is the way she draws Dante outside of himself, causing him to reassess his life.[25] Beatrice poses, as gods usually do, a danger to the ego, in part because of her unknowability. Of all the characters in *Mad Men*, Megan best exhibits this power over Don, at least for the longest duration. Since Don's tendency to want what he does not have makes him particularly attracted to Beatrice figures, Beatrice figures, like Megan, remain appealingly elusive to him as the "stronger god."

We cannot ignore, however, that Megan's power over Don does not last. Don's relationship troubles usually start once the unknowable becomes the familiar. Where Betty Draper largely plays the role of the "angel of the house," at least insofar as she and Don are concerned to maintain conventional social proprieties during the majority of their marriage, the outsider Megan appeals to Don as a power that he lacks.[26] But both of these situations change. Betty stops heeding conventional expectations, and Don gets bored with Megan, just as he did with Betty. Some have taken this plot reiteration to be a flaw in Megan's characterization as a whole.[27] However, in order to understand Megan's place in the show, we must note that Megan is a part of the show's treatment of the counterculture as a whole—as it weighs both the legitimacy and the illegitimacy of the counterculture as a social critique

in the house" that are the subject of their study.

25 The literary theorist Elaine Scarry has articulated how the decentering of the self caused by the recognition of beauty results in a revaluation of one's priorities. Scarry, *On Beauty and Being Just*.

26. For further discussion of the power dynamics in both of Don's marriages, particularly in relation to sexuality, see Jacob Goodson's Chapter 12.

27. The return of Don's marital discontentment sparked, for example, complaints by Joyce Carol Oates which dismissed Megan as a completely disposable character. "Joyce Carol Oates Tweets Epic Dissatisfaction."

of the mainstream. In this light, the failure of Megan and Don's marriage reflects the failure of the counterculture to offer a legitimate alternative to the mainstream's corruptions. The dynamic can best be explained through a third point learned from Dante: *disordered desire*. In other words, for Dante a utopian ideal can only be identified through embodied desire (hence Beatrice). Moral progress does, in fact, occur when initiated by embodied desire, yet desire is disordered when it equates the embodiment with the utopian ideal. To understand this process in detail, we need to consider the show's characterization of its countercultural figures more closely.

COUNTERCULTURE'S COUNTERPOINT

Mad Men portrays Megan's otherness from Don through her freewheeling optimism, but also by visually associating her with the countercultural. She wears countercultural clothes (one outfit in particular sent Sharon Tate buzz through internet circles).[28] Her friends are an eclectic group of bohemian artists. When Megan invites her social set to the surprise birthday party that she arranges for Don, the visual contrast is striking.[29] Throughout the series, Don is, in appearance at least, the fifties mainstay. Until the finale, when he dresses in linen on a mountaintop, the counterculture has little aesthetic influence on him. Where other characters evolve into the colors, fashions, and facial hair of the late sixties—even Roger sports a handlebar mustache in Season 7!—Don remains steadfastly "The Man in the Gray Flannel Suit." He even voices his dislike of sixties-era trends, ridiculing the Volkswagen "Lemon" ad,[30] and its many imitators, and scorning the bohemian art scene as flashing "too much art."[31]

One early confrontation with the counterculture displays Don's general aversion to the counterculture's aesthetic pretentions:

> Beatnik: Toothpaste doesn't solve anything. Dacron sure as hell won't bring back those ten dead kids in Biloxi.
>
> Don: Neither will buying some Tokaj wine and leaning up against a wall in Grand Central pretending you're a vagrant.

28. Williams, "Does Sharon Tate."
29. "A Little Kiss," S5/E1.
30. "The Marriage of Figaro," S1/E3.
31. "Babylon," S1/E6.

> Beatnik: You know what it's like to watch all you ants go into your hive? I wipe my ass with the Wall Street Journal.[32]

The counterculture's critique of the mainstream, as might be expected, targets consumerism. More pointedly it connects a consumeristic malaise with the deaths resulting from a racial riot in Biloxi, Mississippi. In his rebuttal, Don does not defend enforced segregation, but neither does he take the offensive stance against social injustice. Instead, he scorns the counterculture's aesthetic technique of opting out of the mainstream. He posits that the counterculture's "artistic" response is just as ineffective, and even more absurd, than the mainstream's complicity.

Just as Don views the outward pretensions of the counterculture as preposterous and phony, viewers are often encouraged to think likewise. The beatnik's retort, to the twenty-first-century viewer, feels clichéd. In this way, *Mad Men*'s portrayal of the counterculture resonates with Joseph Heath's and Andrew Potter's negative assessment of countercultural movements in their study *Nation of Rebels*. Heath and Potter argue that the counterculture failed because it formed around a false premise, that of a mainstream "system" that needed to be resisted. Heath and Potter claim on the contrary, "There is no single, overarching system that integrates it all. The culture cannot be jammed because there is no such thing as 'the culture' or 'the system.' There is only a hodgepodge of social institutions, most tentatively thrown together, which distribute the benefits and burdens of social cooperation in ways that sometimes we recognize as just, but that are usually manifestly inequitable."[33] Building of the false premise of a coherent "system," the counterculture seeks to step outside of mainstream norms, yet end up simply asserting new desires that, when given enough popularity, themselves become mainstream. Which is to say, hippies will inevitably turn into yuppies.

The show's dismissal of countercultural pretensions dramatizes the trend *avant le lettre*. Economic transactions, power positioning, egotistical struggles—the counterculture does not transcend these structures. It only supplies its own wares, often in a socially detrimental way. We watched a version of this critique unfold as late as Season 7 when Margaret attempts to join an upstate commune. Margaret's experiment as a hippie plays out as merely an updated version of the previous generation's vices. Roger neglected Margaret, now Margaret neglects Ellery.[34] What was once done covertly is now done out in the open. Likewise, Irene Small has argued

32. "The Hobo Code," S1/E8.
33. Heath and Potter, *Nation of Rebels*, 8.
34. "The Monolith," S7/E4.

that Don's forays into the counterculture tend to reveal the counterculture's bankruptcy as a legitimate alternative to the mainstream. "The Hobo Code," according to Small, portrays the counterculture as "all play acting and histrionics, it [the counterculture] offers little alternative to the narratives spun by Madison Avenue advertisers uptown."[35] Don is perspicaciously aware of this tendency. Like Heath and Potter, Don thinks that the counterculture's delusion is that they can remove themselves from a mainstream system. In this regard, his pessimism takes on a metaphysical grandeur; "There is no big lie, there is no system. The universe is indifferent."[36]

If the counterculture misfires in practice, nevertheless the show suggests that the counterculture might at least be correct in its assessment of mainstream attitudes. As such, Don's claim that "the universe is indifferent" inevitably smacks of its own pretentions. Don often spouts such nihilistic utterances, but his actions suggest that he, in some way, believes otherwise. The counterculture imagines another way to be, and Don is, if anything, constantly interested in alternatives. Don is not a hippie, but neither is he a mainstream mensch in the way Bert Cooper tries to peg him. Cooper (mis)identifies Don through the lens of the Objectivist virtues proposed by Ayn Rand.[37] Don, he says, is a "productive and reasonable man, and in the end, completely self-interested."[38] Now, it may be true that Don is self-interested, but he does not rest in this characteristic as a laudable virtue. Don, unlike Bert, has read Jack Kerouac, and, unlike Bert, he lives in the existential rift exhibited in that book.[39] Don may be productive, but he can also be self-destructive, as his Hersheys pitch demonstrated.[40] Finally, it is never entirely clear that Don is reasonable. Don straddles the outward pretentions of Bert's mainstream culture while holding to alternative desires of the counterculture. Randian notions that consumer culture should satisfy human desires and that with enough material goods and a position of power one ought to be happy never appeal to the counterculture, nor to Don.

Consider Midge. She functions in the show as Don's first potential Beatrice figure in that she holds a position of autonomy outside of the purviews of Don's Madison Avenue/Westchester County world. Besides Megan, Midge is the other woman to whom we see Don propose marriage, however

35. Small, "Against Depth," 184.

36. "The Hobo Code," S1/E8.

37. White, "Egoless Egoists." White is inclined to see Don's inability to keep to a Randian virtue ethic as a flaw, though I argue otherwise.

38. "The Hobo Code," S1/E8.

39. "Lost Horizon," S7/E12.

40. "In Care Of," S6/E13.

impractically. Midge's autonomy, and wide-eyed perception of Don's nature, is evident in her reply: "You think I would make a good ex-wife?"[41] Unlike other autonomous women in the show, such as Peggy and Joan, Midge inhabits the otherworld of the counterculture.[42] She is thus the type of woman Don pursues and, by doing so, he seeks ways to align himself with an alternative reality. In fact, when Midge appears too much like a denizen of Don's world, as when he realizes that she owns a television, Don thinks less of her.[43]

Midge's position within the counterculture is thus a crucial part of her characterization. The show reveals Midge's social sphere only after viewers learn of her relationship to Don, and without this world she would be little more than a generic "mistress" character (more like the flight attendant of "Out of Town").[44] She, like Megan, mingles with the Greenwich Village subculture of beatniks and other Bohemian artists, and it is their social otherness that solidify them as more essential characters. As a part of the counterculture, they offer a simultaneous critique of the mainstream and a potential misdirection. It is a double-edged portrayal that is well illustrated in the episode "Babylon."

In this episode, as is often the case, *Mad Men* offers moments where Don's response stands in for the viewers'. Of course, part of his response is idiosyncratic: upon meeting Roy Hazelitt, one of Midge's beaus from within this subculture, Don, in a spirit of male competitiveness, flaunts his status. But when he scorns the counterculture's attempt to distinguish a more authentic form of art, he reflects a view encouraged by the show's direction. The following exchange, for instance, allows Don's retort to hold rhetorical weight:

> Roy: Not the soul-less bullshit kept alive by the middleclass with their season tickets to Dick Van Dyke and Mary Martin. I bet Don here can tell you first hand. Broadway is the birthplace of mediocrity.
>
> Don: Maybe it's born there but I think it may be conceived right here.[45]

Along with Midge's apparent agreement with Don's jibe (she chuckles), Don's remark reflects the show's criticism of the counterculture's artistic ambitions. Viewers inevitably cringe at the spoken word poetry that

41. "Smoke Gets in Your Eyes," S1/E1.
42. For more on Peggy's and Joan's autonomy see Rogers, "Mad Men / Mad Women."
43. "Ladies Room," S1/E2.
44. "Out of Town," S3/E1.
45. "Babylon," S1/E6.

begins "I dreamed of making love to Fidel Castro in a king size bed at the Waldorf-Astoria."

The dissonance between Roy's serious appeals for artistic authenticity and the juvenile pretensions of the artistic performances on display makes the counterculture appear farcical at first. Yet by the episode's conclusion, a more serious tone captures Don's, and the viewers', attention. A trio of folk musicians perform a rendition of a biblical Psalm of exile ("By the rivers of Babylon, we sat down and wept"), which transcends the moment in the club, becoming a non-diegetic accompaniment to the episode's concluding montage. Pairing this with a close up of Don's sudden shift toward attentiveness, the show's direction suggests that the artistic authenticity Roy proposed might have actually been found. At least, the song of a people exiled from their homeland adds a level of historical and cultural depth, in some way legitimizing Don's personal ennui.

Midge's role as a potential Beatrice figure becomes most apparent here. Having been the initial enticement to Don's journey to the underworld of Bohemian artists, she fades into the background as Don sees a truth beyond her. This does not diminish her importance; in fact, it enhances her personhood. Like Midge, the counterculture is not unilaterally dismissed. Rather, "Babylon" is an important early portrayal of the show's double assessment of the counterculture—identifying its inevitable complicity in the mainstream structures it claims to critique, as well as allowing a space for the authenticity of its critique. Without the guidance of Midge, Don would not have experienced the connection to a larger culture of exile, one that is simultaneously American, Jewish (as the appearance of Rachel Menken emphasizes), and (the show seems to imply) indicative of the human condition. Dante's view of the world thus provides us with a non-polarizing way to understand the mainstream/counterculture dichotomy. The counterculture may be wrong in presuming it can offer an alternative world, yet as "Babylon" implies, it is not wrong or misguided to desire a better world.

DISORDERED DESIRE

If early in the series Don's character is defined by a countercultural desire to live outside the system, despite his resistance to the pretentions of the counterculture, Season 5 continues this trend, but at greater length. His marriage to Megan initially promises a way out of the corruption associated with Don's work and lifestyle. However, Don conflates an ideal of escape with Megan herself. He thus denies Megan's personhood, fashioning her instead as a kind of archetype of transcendent happiness. So when Megan shows interest in this-worldly affairs by seeking a career in acting, Don resents her choice (recall the way Don walks off Megan's set in "The Phantom," S5/E13). Part of his instinct may be that, out of personal experience, he sees inherent corruption in business and he seeks to protect Megan from this corruption, but by attempting to isolate her from the world he inhabits, he neglects her reality within that world. Hanna Rosin has argued that *Mad Men* tends to treat its women characters as archetypes, and I am inclined to think that this is true, but only insofar as the show often places its viewers within Don's perception of the world.[46] While Midge and Megan do offer Don a way to perceive an alternate ethic apart from Madison Avenue's, Don begins to treat them as a separate reality altogether. By treating others as an abstracted good, Don falls to a temptation that Dante illustrates early in *Inferno*.

Dante learns early on that his initial guide Virgil was sent to help him find his way by the request of Beatrice. As a resident of *Paradiso*, Beatrice herself holds a background role in *Inferno*, but she remains a motivating attraction for Dante (she eventually appears in person in Dante's *Purgatorio*, at the top of Mt. Purgatory, and from then takes over Virgil's guide duties). Throughout the *Divine Comedy*, this desire for Beatrice is portrayed as appropriate because it is properly *ordered*, a disposition that is set in contrast to the various examples of *disordered* desire found in hell. Disordered desire is defined early in the poem as Dante meets a pair of lovers named Francesca and Paulo. Condemned for the sin of lust because of their illicit affair, these two lovers swirl around each other forever pulled by their insatiable desire for the other. Of the two, Francesca is the only one who speaks, and scholars have noted how her foregrounded position places her as a kind of negative image of Beatrice. Contrary to Beatrice, who directs desire properly, Francesca speaks of the potential misdirection caused by desire.

When Francesca tells Dante of how she came to be included among those punished for the sin of lust, she poetically expresses Love's undeniable power.

> Love, quick to kindle in the gentle heart,
> Seized this man with the fair from taken from me.
> The way of it afflicts me still.
>
> Love, which absolves no one beloved from loving,
> Seized me so strongly with his charm that,
> As you see, it has not left me yet.
> Love brought us to one death.[47]

46. Rosin, "I Wish 'Mad Men.'"
47. Alighieri, *Inferno*, trans., Hollander and Hollander, 5.100–6.

Francesca speaks of a mutual attraction as a violent shifting of one's attention that results in death. By using the word "Love," she employs the same word Dante uses throughout the *Divine Comedy* to define God. Yet, for Francesca, Love instigates transgression rather than salvation. She and Paulo become involved in their extramarital affair because Love seizes them "so strongly." The language also echoes the amatory poetry Dante wrote earlier in his career, so by viewing the plight of Francesca and Paulo, Dante begins to understand the dangers that accompany any narrative that provokes desire. In Dante, love, desire, and lust can form what one recent critic calls an "indistinguishable discourse."[48]

While it may be difficult to distinguish between the Love Francesca utters and the Love that Beatrice leads to, the two women undoubtedly end up in very different places because of the way they order their understanding of love. Francesca and Paulo identify Love with the person that incited their desire, thereby undermining each other's personhood and separating them from God. Beatrice, however, remains distinct from Love itself. She thus avoids becoming altogether an abstraction or a transcendent mystery. A similar distinction occurs in *Mad Men*. Don's actions end up reflecting Francesca's idealization of the other more than they reflect Dante's salutary vision of Beatrice. In the case of Midge, Don conveniently discards her when she stops looking like a promising escape. And his and Megan's marriage suffers as a result.

Season 5 thus explores at length the movement from the beginnings of a hopeful relationship to the beginnings of its dissolution. As I observed previously, Megan offers Don the promise of a way of living that is distinct from what he has known. So, in getting close to Megan, Don distances himself from work. His colleagues consequently notice his detachment. When Don's work suffers, Bert accuses him of being on "love leave," forgetting to attend to his job's duties.[49] This decline in productivity, however, changes when Megan leaves the office to pursue acting. Don starts to assert himself in work again, and viewers watch a revival of Don's egoism concomitant with a reenergized ad-man. The subsequent moral decline is visually marked by the eerie image of an empty elevator shaft, whose depths anticipate the *Inferno* reference at the beginning of the following season.[50]

We may be tempted to see these plotlines as a critique of the business world, to read Megan's farewell descent in an elevator as a choice to join the mainstream's moral abyss. Yet attending to Dante's structures of desire

48. Lombardi, *Wings of the Doves*, 17.

49. "Far Away Places," S5/E6.

50. "Lady Lazarus," S5/E8.

helps us see otherwise; we need not assume that the show is faulting Megan's pursuit of acting or Don's reengagement with his own career. Instead, Don exhibits disordered desire, assuming that Megan herself was the escape, the alternate world, he desired, rather than seeing her personhood. She may have been Don's guide beyond the corrupt ethic he inhabited, but as a person she remains embedded in the world. Don's misdirected perception, which sought to idealize Megan beyond personhood, twists him back to where he was already. In his first professional exertions after Megan's departure, he steamrolls the young copywriter Michael Ginsberg's Sno Ball campaign—underhandedly asserting his own campaign in order to reposition his status within the company. When Ginsberg calls him out, Don belittles him, saying, "I don't think of you at all." And, in yet another of the show's infernal allusions, the image and tag Don settles on is that of a cartoon Satan pronouncing, "This could change everything."[51]

And change does happen: Don reverts back to a dissolute ethic. The way he maneuvers himself back into his previous job reveals the extent to which he is willing to ignore moral questions in order to assert his professional status. Don even presents an impassioned speech before Dow Chemical executives in which he suggests that support for Napalm is patriotic.[52] Yet, even this self-interested moment is not purely Machiavellian; it holds its own bit of authenticity. Tellingly, Don once again builds his argument by making a virtue of discontentment. With a 50 percent market share in nearly all of their products, the Dow Chemical executives claim that they are happy with their current marking strategies. Don's rebuttal is to offer himself as someone who will never be satisfied; he will never settle. He commandingly argues that 50 percent should not make one happy, only everything can do that.

In this, Dante would see Don's point. The *Divine Comedy* is built on the idea that moral progress only comes from the movement first incited by desire. However, desire requires direction and proper ordering. The way forward might take unexpected turns, but there is a way forward. Does *Mad Men* agree in this regard? Does Don make any moral progress? At the beginning of Season 3, Don opines to the contrary, "I keep going to a lot of places and keep ending up somewhere I've already been."[53] However, by the conclusion of the series Don does indeed find a way to direct desire toward moral reform. In the finale, the counterculture once again plays a part in

51. "Dark Shadows," S5/E9.

52. "Commissions and Fees," S5/E12.

53. "Out of Town," S3/E1.

identifying the mainstream culture's need for reform. But, once again, the counterculture is not itself the locus of that moral reform.

PERFECT HARMONY

The show's final image of Don meditating on a mountaintop alone would have been enough to provoke debate about the final episode of *Mad Men*, but the way this image fades to a Coca-Cola commercial sparked a flood of competing internet commentary. If Don regularly finds himself in a position of renewal (consider his "baptism" in "The Mountain King," S2/E12),[54] only to revert back to his old ways, we might assume that Don is either a hopeless case or that, as he says, "The universe is indifferent," there is no real moral progress. Don persistently acts on the assumption that actual change is possible, but in the end it is possible that the show may be suggesting otherwise.

To determine if there is any chance of moral progress supported by the show, we must consider what counts as moral progress. As the moral philosopher Alasdair MacIntyre and others have argued, morality is only made possible when there is an assumed set of virtues that can be practiced.[55] We should determine whether the show assumes any set of virtues and then we can say whether Don achieves these. To conclude this essay, I propose that *Mad Men* includes an identifiable set of virtues that, upon examination, reveal Don's moral progress by the end of the series. The virtues I cover are not intended to be comprehensive, but they are all virtues that come to a head in the final episode—which, significantly, places Don at an Esolen-like retreat (a final location, which confirms the series' regular use of the counterculture as a means of moral critique but not as an end in itself).

Some critical examinations of *Mad Men*'s virtues have sought to locate virtues from within modern systems—in the Objectivism of Ayn Rand, for example—or to argue a fundamentally existentialist system—where the virtues are contingent upon the occasion, detached from a traditional set of virtues.[56] Such attempts seem to side with the ethical systems promoted by individual characters (Bert in the case of Objectivism, Don for existentialism). By considering characters as well as the counterculture vis-à-vis the mainstream, I have attempted to shift the emphasis onto larger cultural structures portrayed within the show so that the identification of virtues may be a less singular endeavor. As a protagonist, Don holds our attention,

54. See the next chapter in this volume.

55. MacIntyre, *After Virtue.*

56. White, "Egoless Egoists"; Elia, "Don Draper."

but we cannot forget the he is embedded within a world. Don overlaps two competing cultures, and this dialectical relationship reveals something about both the mainstream and the counterculture.

It is safe to assume that, for most viewers, Don's actions are morally sound when he behaves in a way that is (at least overall) beneficial to the underprivileged. The premier episode shows Don stepping outside of the mainstream's standard practice by conversing with his African-American waiter, even rebuffing the restaurant's owner when he suggests the irregularity of the interaction. In the same episode, however, Don shows his complicity in the mainstream's prejudice when he storms out of a meeting with Rachel Menken, saying, "I'm not going to let a woman talk to me like that!"[57] Despite contradictory behaviors, Don's more heightened awareness of the marginal members of society, an awareness associated with the counterculture, makes him a more sympathetic character than most of his male coworkers. A modern ethical system where values are existentially and contingently defined only goes so far in determining this moral judgment. The existentialist condition, to have to create one's own meaning, may be a difficult place to be, and in this Don succeeds at times and fails at others. But persisting in a state of self-creation does not appear to be the final aim of the show. In one final-season interview, Weiner suggests dissatisfaction with existentialist philosophy, saying "You just stop having a confident attitude about the meaninglessness and disorder of the universe. Some of that is about having children, but I do think it's also about getting older."[58] Instead, *Mad Men* sets forth some identifiable, even traditional, virtues (like the fundamental need to recognize human equality) that distinguish moral behavior from immoral behavior.

Julia Turner points out that one of the show's central virtues is that good people do not abandon their children.[59] Of the virtues exhibited by the series, this is one of the most blatantly traditional; yet it is a virtue ignored in the show by both the counterculture and the mainstream, as Margaret's and Roger's interactions in "The Monolith" make clear. The series finale similarly signals a crossover phenomenon by echoing Don's "It will shock you how much it never happened" speech to Peggy, when Don tells Peggy that she can forget about her child, move on, and build a new life.[60] Stephanie, the countercultural figure who brings Don to the retreat, is pregnant and not wishing to be a mother—but she also has a certainty that she is

57. "Smoke Gets in Your Eyes," S1/E1.

58. Elder, "Matthew Weiner."

59. Turner, "'Mad Men' Stuck the Landing."

60. "The New Girl," S2/E5.

unable to forget her child. Don gives Stephanie the same advice he gave Peggy, "You can put this behind you. It'll get easier as you move forward." But Stephanie responds to Don's confidence that he "know[s] how people work" with a simple, powerful rejoinder: "Oh, Dick, I don't think you're right about that."[61] That we are meant to take Stephanie's wisdom over Don's is suggested in one of Weiner's post-series interviews.[62] Weiner conflates this direct retort with the response Rachel Menken gives to Don's claim that love was invented to sell nylons, where Rachel actually responds with the question, "Is that so?" ("Smoke Gets in Your Eyes," S1/E1). By marking these two moments as bookends of the series, Weiner undermines Don's confidence in favor of each woman's point of view. Don's own progress in this virtue is developed over the course of Season 7, as his relationship with Sally grows to previously unseen levels.

Stephanie's realistic observation that you cannot forget a child ends up triggering an emotional breakdown in Don as he seems to realize that he built his life on the false assumption that you can leave behind the past. The confessional moment that follows his breakdown indicates another of the virtues the show champions, that is, a self-scrutinizing sense of honesty. The countercultural retreat offers a setting where this virtue finds a natural home, as group therapy sessions are devoted to honest expression of how one feels. Yet, Don's self-scrutiny and confession of his past misdeeds occurs outside of these group therapy session. In a phone call to Peggy, Don, in an act more reminiscent of a religious confession, finally verbalizes the major misdeeds of his life: "I have kept none of my vows, I have scandalized my child, I took a man's name and have made nothing of it."[63] Peggy also had her own confessional moment a few episodes earlier, as she tells Stan of her decision to give her child up for adoption, an exchange whose honesty seems to establish the best argument for their relationship going well.[64] Don's and Peggy's confessions about their past reveal moral progression within the show's system of virtues. This progress perhaps even makes it possible for them to find some of the contentment they desire.

Finally, one dominant virtue promoted by *Mad Men* is a disciplined contentment found within life's imperfect situations. Once again, this virtue is traditional and Weiner, in an interview with Matt Zoller Seitz, associates it particularly with Rachel's secular version of Judaism.[65] The finale returns

61. "Person to Person," S7/E13.
62. *Matt Weiner Talks 'Mad Men' Final Season.*
63. "Person to Person," S7/E14.
64. "Time and Life," S7/E11.
65. *Matthew Weiner on the Final Season of "Mad Men."*

to the theme of contentment in an unsettled world by juxtaposing a shot of Don racing across the Bonneville Salt Flat with the concluding image of meditative bliss. While the end feels jarring to many viewers, Don's Esolen experience breaks free from the (admittedly enjoyable-to-watch) ennui that saturated the series from its inception. With the Coke commercial included, the silly, even joyful, conclusion implies a salutary contentment in the profession of selling mundane things. While the commercial drips of feigned authenticity, its artifice is more apparent than its authenticity when juxtaposed with Don's ridiculous grin. And this is okay. The lighthearted tone is a sign of Don's moral progress. His numerous journeys into the counterculture, involving finally his honest confession of his past and of his wrongdoing, has prepared him to have a better relationship to his mainstream work life. By no means is he one of the saints whom Dante would place in *Paradiso*, but in the final analysis Don's journey appears more purgatorial than infernal.

Here we can conclude by distinguishing contentment from a lack of restlessness, and we turn one more time to Dante for guidance. Dante's view of the human is heavily indebted to the writings of St. Augustine, who believed that humans lived in a persistently restless condition until finding rest in God. In the *Divine Comedy*, Dante dramatized this restlessness as an effect originating not within ourselves but from divine love which stretches throughout the universe. Those who identify God as the source of all fulfillment have hope of finding rest and can use their restlessness as motivation toward God. Those who seek satisfaction in other things, ideas, or people are damned to eternal hopelessness, since these cannot satisfy. The relevant point here is: The condition of restlessness, on its own, is neither good nor bad. It can result in either moral progress or moral decline. In fact, restlessness is needed to spark moral progress, and when it does, that restlessness is salutatory. Discontentment, though, is a negative version of this restlessness in that it scorns earthly goods for their inability to offer complete fulfillment. In light of this distinction, then, *Mad Men*'s conclusion suggests that, while Don may ever remain restless, he may have found a way to move beyond discontentment.

BIBLIOGRAPHY

Alighieri, Dante. *Inferno*. Translated by Anthony Esolen. Bilingual edition. New York: Modern Library, 2005.

———. *Inferno*. Translated by Robert Hollander and Jean Hollander. New York: Anchor, 2002.

———. *La Vita Nuova*. Translated by Mark Musa. Oxford: Oxford Paperbacks, 2008.

Dubner, Stephen. "Publishers: Get Your Books in Don Draper's Hands." *Freakonomics*, July 28, 2008. Online: http://freakonomics.com/2008/07/28/publishers-get-your-books-in-don-drapers-hands/.

Elder, Sean. "Matthew Weiner on the Last Season of 'Mad Men' and Don Draper's Terrible Year." *Newsweek*, April 3, 2015. Online: http://www.newsweek.com/2015/04/17/matthew-weiner-lastseason-mad-men-don-draper-terrible-year-319371.html.

Elia, John. "Don Draper, on How to Make Oneself (Whole Again)." In *"Mad Men" and Philosophy: Nothing Is as It Seems*, edited by James B. South, Rod Carveth, and William Irwin, 168–85. Hoboken, NJ: Wiley, 2010.

Gilbert, Sandra M., and Susan Gubar. *The Madwoman in the Attic*. New Haven: Yale University Press, 1984.

Heath, Joseph, and Andrew Potter. *Nation of Rebels: Why Counterculture Became Consumer Culture*. New York: HarperBusiness, 2004.

"Joyce Carol Oates Tweets Epic Dissatisfaction with 'Mad Men' Finale." *Salon.com*, May 18, 2015. Online: http://www.salon.com/2015/05/18/joyce_carol_oates_tweets_epic_dissatisfaction_with_mad_men_finale/.

Krouse, Tonya. "Every Woman Is a Jackie or a Marilyn: The Problematics of Nostalgia." In *Analyzing "Mad Men": Critical Essays on the Television Series*, edited by Scott F. Stoddart, 186–204. Jefferson, NC: McFarland, 2011.

Lewis, C. S. *Mere Christianity*. New York: Harper Collins, 2001.

Lombardi, Elena. *Wings of the Doves: Love and Desire in Dante and Medieval Culture*. Montreal: McGill-Queen's University Press, 2012.

MacIntyre, Alasdair. *After Virtue: A Study in Moral Theory, Third Edition*. 3rd ed. Notre Dame: University of Notre Dame Press, 2007.

Matthew Weiner on the Final Season of "Mad Men." YouTube, 2015. Online: https://www.youtube.com/watch?v=oW-tlMf8AZE.

Matt Weiner Talks "Mad Men" Final Season. YouTube, 2015. Online: https://www.youtube.com/watch?v=h1oLCH9quE4.

Meek, James. "The Shock of the Pretty." *London Review of Books* (April 2015) 29–34.

Rogers, Sara. "Mad Men / Mad Women: Autonomous Images of Women." In *Analyzing "Mad Men": Critical Essays on the Television Series*, edited by Scott F. Stoddart, 155–65. Jefferson, NC: McFarland, 2011.

Rosin, Hanna. "I Wish 'Mad Men' Had Handled Its Themes about Women in the Workplace with a Little More Subtlety." *Slate*, May 18, 2015. Online: http://www.slate.com/articles/arts/tv_club/features/2015/mad_men_season_7_part_2/episode_7/mad_men_season_7_reviewed_i_wish_mad_men_had_handled_its_themes_about_women.html.

Scarry, Elaine. *On Beauty and Being Just*. Princeton: Princeton University Press, 2001.

Small, Irene. "Against Depth: Looking at Surface through the Kodak Carousel." In *Mad Men, Mad World: Sex, Politics, Style, and the 1960s*, edited by Lauren M. E. Goodlad, Lilya Kaganovsky, and Robert A. Rushing, 181–91. Durham: Duke University Press, 2013.

Turner, Julia. "'Mad Men' Stuck the Landing." *Slate*, May 19, 2015. Online: http://www.slate.com/articles/arts/tv_club/features/2015/mad_men_season_7_part_2/wrapup/mad_men_season_7_reviewed_the_finale_stuck_the_landing.html.

White, Robert. "Egoless Egoists: The Second-Hand Lives of Mad Men." In *"Mad Men" and Philosophy: Nothing Is as It Seems*, edited by James B. South, Rod Carveth, and William Irwin, 79–84. Hoboken, NJ: Wiley, 2010.

Williams, Alex. "Does Sharon Tate Hold a Clue to the 'Mad Men' Finale?" *The New York Times*, May 9, 2014. Online: http://www.nytimes.com/2014/05/11/fashion/does-sharon-tate-hold-a-clue-to-the-mad-men-finale.html.

Williams, Charles. *The Figure of Beatrice: A Study in Dante*. New York: Noonday, 1961.

Chapter 8

"MOVING FORWARD" AS RETURN

The Redemptive Journey of Don Draper

—JACKSON LASHIER

> Every one of us is shadowed by an illusory person: a false self.
> —Thomas Merton

MAD MEN AND THE RESOURCES OF MONASTIC THEOLOGY

In the pilot episode of *Mad Men,* Don Draper gives an inspired pitch to the owners of the cigarette company Lucky Strike. The year is 1960, and tobacco's link to lung cancer makes it illegal to advertise that smoking is not detrimental to one's health. While the despondent owners hope for a clever way to address the health risk, Draper advises them to ignore it: "We can say anything we want . . . 'Lucky Strikes: It's Toasted.'" The owners, of course, object that everybody's tobacco is toasted, to which Draper responds, "No. Everybody else's tobacco is poisonous. Lucky Strikes' [tobacco] is toasted."[1] His pitch, in other words, focuses on a positive image of tobacco in order to make the consumer forget the negative and contradictory reality that smoking kills.

1. "Smoke Gets in Your Eyes," S1/E1.

In this prescient pitch, Draper foreshadows the central struggle of his life, namely, the contradiction between image and reality. This struggle plays out over the show's narrative through Draper's dual and often contradictory lives: Dick Whitman or Don Draper, family man or cavorting playboy, loyal businessman or utilitarian executive, and the like. The struggle of his contradictory lives is so pervasive, in fact, that the question "Who is Don Draper," uttered at numerous points and in various ways over the show's seven seasons, becomes the question that drives the series.

Draper's struggle between image and reality has been the centerpiece of previous analyses of the show.[2] Notably absent from the secondary literature, however, has been a sustained theological interpretation of Draper's struggle.[3] The character Draper's expressed lack of religious convictions has likely discouraged such an interpretation. Nevertheless, the preponderance of theological images running throughout the seven seasons of *Mad Men* warrants a theological approach, regardless of the stated religious preferences of the character. More to the point, a theological interpretation may be able to illumine both the nature of Draper's struggle and his subsequent development in ways not open to other lines of interpretation. To take a recent example, Jeremy Varon, writing from a psychological perspective, insists that the question "'Who is Don Draper?' remains grounded in the question of whether Don Draper—whoever he is—is a good man."[4] While certainly a valid perspective, this psychological approach struggles to account for Draper's somewhat erratic behavior, how he can appear a changed man in one episode and return to his cavorting ways in the very next. The answer to the question from this perspective necessarily depends on what point in the narrative one asks it.[5] By contrast, a theological interpretation avoids the reduction inherent in equating identity with action and, because of the nature of the narrative of Christian Scripture, opens up categories of interpretation such as growth, change, and redemption.[6] In particular,

---

2. Of many such available essays, one notes in particular the fine essay by Christopher Bigsby in his book *Viewing America*, 356–403 and those by Jim Hansen and Lilya Kaganovsky in *Mad Men, Mad World*, 145–60 and 238–56.

3. By "theological interpretation," I mean an interpretation oriented to the particular set of questions and answers arising from the narrative of Christian Scripture and the two millennia of writings and development of that foundational narrative known often by the shorthand "tradition."

4. Varon, "History Gets in Your Eyes," in *Mad Men, Mad World*, 257–78.

5. See, for example, Andrew Terjeson, "Is Don Draper a Good Man?" in *Mad Men and Philosophy*, 154–67.

6. Incidentally, a theological interpretation does not demand that the *character* Draper experience a renewal in the religious sense, i.e., that Draper literally becomes a Christian in the show. Indeed, I will nowhere claim that he does. Nevertheless, the

the Christian monastic tradition offers both the language and the categories for understanding Draper's struggle and subsequent development from a theological perspective.[7] A brief overview of the consistent narrative told by these writers will set the parameters for the theological analysis of Draper's storyline in the remainder of the essay.

The monastics speak of the archetypal human struggle between image and reality as the contrast between the "false self" and "true self."[8] This language develops the New Testament language of the contrast between the "old man" and "new man" used to describe human existence before and after Christ.[9] In Christ, say the monastics, a human being finds his or her true self, which turns out to be the self that God originally intended when he "created humankind in his image, in the image of God he created them; male and female he created them."[10] This means, among other things, that God creates human beings to dwell with him, a divine intention thwarted by the entry of sin into the world.[11] Nevertheless, the image of God is never completely lost, though wounded by sin and hidden by the false self. One of the defining characteristics of the false self, therefore, is a confusion of that self's true identity, the practical result of which is a sort of double life. "I do not understand my own actions," Paul writes, "For I do not do what I want, but I do the very thing I hate."[12]

Redemption comes only by *putting to death* the false self and embracing the true self. This significant turning point is precipitated by confession and initiated by baptism. But while redemption happens in a moment, the

theological categories I am pursuing remain instructive to Draper's progress as a character.

7. Christian monastics are those figures who have voluntarily withdrawn from society to live in cloistered communities, often taking vows of obedience, poverty, and chastity. The monastic tradition is quite varied, spanning a large geographical area, numerous historic ecclesial traditions, and nearly two millennia. Nevertheless, similar theological emphases present across the varied traditions justify referring to monastic theology as a singular entity.

8. The specific language of "false self" and "true self" comes from the twentieth century Trappist monk Thomas Merton, but other monastic figures refer to the same contrasting reality with different language.

9. "[P]ut off, concerning your former conduct, the old man which grows corrupt according to the deceitful lusts, and be renewed in the spirit of your mind, [and] put on the new man which was created according to God, in true righteousness and holiness." Eph 4:22–24, New King James Version.

10. Gen 1:27, New Revised Standard Version.

11. The famous story of Adam and Eve reveals that the ultimate curse of their unfaithfulness is expulsion from the Garden of Eden, a poetic way of narrating the loss of the divine presence in human history. See Gen 3:24.

12. Rom 7:15, New Revised Standard Version.

growth of the new Christian takes place over the course of a lifetime. Consequently, the life of discipleship in the monastic tradition takes the form of a journey away from the false self and its dualities toward the true self, which is characterized by inner peace.[13] Pilgrims enact this journey communally as they embrace their true selves only with the help of and in the company of others. But the journey is neither easy nor predestined to be finished. Indeed, according to the monastics, the deep wound of sin cannot be rooted out in a moment, meaning that the false self always lurks, eager to return. But those who press forward in the journey, those who truly put to death the false self, find the eternal peace of the true self.

With this brief overview in place, I am now prepared to embark on the theological analysis of Draper's struggle and development. I will proceed according to the steps of the redemptive story as told by the monastics, which, perhaps not coincidentally, follows the general narrative sequence of *Mad Men's* seven seasons. In the first section, I will address Draper's struggle between image and reality through the first two seasons, corresponding in monastic theology to the life of the false self prior to the moment of redemption. In section two, I will analyze the episodes toward the end of Season 2 where Draper experiences both a low point in his struggle and a subsequent redemptive moment that holds the promise of a new beginning. In section three, I will interpret the narrative arc of the series beginning with Season 3 through the middle of Season 7 as Draper's convoluted journey away from the dualities of the false self. Finally, in section four I will interpret the second half of Season 7 as bringing a denouement to Draper's struggle, a denouement admittedly open to numerous interpretations, as subsequent analysis of the series finale has borne out. Likewise, we cannot anticipate the ending of Draper's story from a theological perspective because the nature of sin may keep a person's true self from being realized. Accordingly, we might say that the question that drives the series is not "Who is Don Draper?" because this is actually clear from the outset. What remains to be seen is whether Draper's *true* self will emerge or whether the dualities of the false self will result in his eternal contradiction, which is the theological way of saying madness.

13. The monastic tradition is shot through with the journey motif, the most common being the image of one who ascends a mountain after the biblical figure of Moses who climbs Mount Sinai to meet with God. See, in particular, Gregory of Nyssa, *Life of Moses*, Ephrem, *Hymns on Paradise*, esp. Hymn 1, and Pseudo-Dionysius Areopagites, *The Mystical Theology*.

"BY THE WATERS OF BABYLON": THE EXILE OF DRAPER'S FALSE SELF

Over the course of the first two seasons of *Mad Men*, Don Draper is obsessed with maintaining an image of himself: a confident and successful advertising executive, who can boast of being a decorated war hero and having a beautiful wife and two adoring children. While a crack in this image appears early to the audience—his wife is unaware of his infidelity—the season progresses to reveal the depth of the double life he is living. For unbeknownst to his wife Betty, his boss Roger Sterling, or anyone who knows him, Don Draper is not his real name.

Born Dick Whitman, he is the bastard child of a prostitute and an alcoholic father.[14] Following his father's untimely death, his stepmother raises him in a brothel and never hides her disdain for his existence. Later in life, he enlists in the army and is stationed in Korea where he inadvertently causes the death of his commanding officer, one Lieutenant Donald Draper, by igniting an explosion that burns the Lieutenant beyond recognition. He switches dog tags with Lieutenant Draper and, upon returning home, allows all who knew him as Dick Whitman to believe he died while using his new name and identity to craft for himself a different and, he hopes, better life.[15]

The fear of being exposed, however, persists and leads him to duplicitous behavior in order to hide the truth of his past, all the while deepening his internal contradiction. One notable example occurs when a character named Adam Whitman appears claiming to be Draper's brother. Although Adam wants nothing more than to renew their relationship, Draper sends him away with five thousand dollars for his silence. The sincere embrace the brothers share at the episode's end suggests Draper once valued this relationship, yet he ultimately sacrifices it to maintain the image he has built up.[16] Adam Whitman later hangs himself, symbolically revealing the mounting casualties that result from Draper's deceitful efforts.[17]

Ironically, the image Draper so frantically tries to protect becomes increasingly meaningless to him. We have hints of this realization in Season 1, notably, the episode "Babylon,"[18] to which I will return momentarily, as well as that season's closing shot which features Draper sitting alone, visibly

14. These biographical details are slowly revealed through a series of flashbacks in various episodes spanning the seven seasons.

15. "Nixon vs. Kennedy," S1/E12.

16. "5.G.," S1/E5.

17. "Indian Summer," S1/E11.

18. "Babylon," S1/E6.

disturbed that he has missed Thanksgiving with his family.[19] But Draper's inner turmoil becomes palpable in Season 2, coinciding with his affair with Bobbi Barrett. Although Draper has had many affairs to this point, his relationship and behavior with Barrett breaks an implied code of ethics established in the first season. For example, Draper has always kept his philandering outside the office, notably when he respectfully dismisses Peggy Olson's advances in the pilot episode, telling her later: "I have to keep rules about work."[20] By contrast, he seems out of control with Barrett as if the affair happens to him, and he is powerless in stopping it. Barrett initiates their first tryst despite his persistent denials, and later she forces herself on him in his office.[21] More to the point, Barrett, unlike his other paramours, is married. Barrett's husband Jimmy gives voice to Draper's code on this count when he tells him, "You want to step out, you go to a whore. You don't screw another man's wife."[22] Draper's facial expression indicates his tacit agreement. His codes are breaking down, indicative of his loosening grip on his carefully crafted image.

The middle episodes of Season 2 are punctuated with several close ups on Draper's face—the long, forlorn looks reveal his growing sense of meaninglessness. Nowhere is this clearer than in the episode, "Maidenform," which occurs over the 1962 Memorial Day weekend.[23] Early in the episode, Draper's image as the war hero emerges as he stands with other veterans to be honored at a luncheon. His evident discomfort with the praise recalls the secret of his rather dishonorable actions in the war. Later, he meets up with Barrett who claims that his infidelity makes them the same. This thought repulses Draper, and he leaves her tied to the bed, ironically confirming her statement. At the episode's end, while he shaves in the bathroom, his daughter Sally looks up at him as her hero, paralleling the episode's opening shot at the luncheon. Draper smiles at her but pauses when he looks at his reflection in the mirror and, remembering Barrett's words, realizes that he is far from the man his daughter thinks he is. Draper sends her away, and the camera slowly pans out to capture him sitting adjacent to his reflection in a hallway mirror—two Drapers, two lives; one a hero, one a bastard. The

19. "The Wheel," S1/E13. The pitch for Kodak's product Carousel, which featured Draper showing a series of pictures of his wife and kids, keeps him from actually being with his family on the holiday. I will revisit this rather sad image below.

20. "The Suitcase," S4/E7. Peggy starts out as Draper's secretary but slowly progresses in the company to become a copywriter. The platonic nature of their relationship elicits a deeper, theological meaning, which I will develop in section four below.

21. "Three Sundays," S2/E4.

22. "The Gold Violin," S2/E7.

23. "Maidenform," S2/E6.

remarkable shot symbolizes both the deep contradiction of Draper's life and his profound desperation at the full realization of it.[24]

From a theological perspective, drawing particularly on the monastic thought outlined in the introduction, "Don Draper" is this character's false self. This false image, despite its promise of a better life, has produced for the character only contradiction and meaninglessness precisely because it denies reality, the true self this character was created to be.[25] Thomas Merton's description of the false self aptly captures Draper's desperation in "Maidenform":

> I wind experiences around myself and cover myself with pleasures and glory like bandages in order to make myself perceptible to myself and to the world, as if I were an invisible body that could only become visible when something visible covered its surface. But there is no substance under the things with which I am clothed. I am hollow, and my structure of pleasures and ambitions has no foundation. I am objectified in them. But they are all destined by their very contingency to be destroyed. And when they are gone there will be nothing left of me but my own nakedness and emptiness and hollowness, to tell me that I am my own mistake.[26]

Like the brand new Cadillac he buys in the episode following "Maidenform," the false self promises so much, its allure so great.[27] But it's just an image and, in the end, images lack substance. What's underneath turns out to be hollow and rotten. And when Betty learns of Draper's affair with Barrett, she vomits all over their brand new Cadillac.[28]

Importantly, such analysis of Draper's struggle precludes a facile identification of the false self with poor behavior. While sin certainly characterizes the false self, the wound goes much deeper and, therefore, is better captured in the sense of disequilibrium that marks a double life. Draper's

24. For a fascinating, non-theological interpretation of "Maidenform" related to Draper's struggle, see Kaganovsky, "'Maindenform': Masculinity as Masquerade" in *Mad Men, Mad World*, 238–56.

25. By this interpretation, I do not mean to identify 'Dick Whitman' as the character's true self. Rather, his efforts to live into the image of 'Don Draper' have prevented him from discovering his true self; thus, the character's true self remains unknown at this point in the show's narrative.

26. Merton, *New Seeds of Contemplation*, 34.

27. "The Gold Violin," S2/E7. Draper's purchase seems a deliberate attempt to deny his sense of unhappiness and recalls his own words from the pilot pitch: "Happiness is the smell of a new car." "Smoke Gets in Your Eyes," S1/E1.

28. "The Gold Violin," S2/E7.

poor behavior is not the root cause, but the damaging effect of this deeper issue. Incidentally, this interpretation explains why Draper, although firmly in his false self during the first two seasons, can at times manifest admirable behavior, as he does, for example, in his loyalty to Mohawk Airlines or at various points in his relationship with Peggy.[29] Draper's struggle goes beyond behavior; he lacks substance. Since assuming the "Don Draper" image, he has been objectified as a war hero, an advertising executive, a family man. And none of it is true. "You're garbage and you know it," Jimmy sneers.[30] Draper knows it; he's always known it.

The problem of the false self, then, is one of contradiction. The false self contradicts the theological reality of a person as an image of the creator God. The false self, in turn, suffocates that reality and leaves the person empty and hollow, with no real substance to call his or her own. Merton writes, "To say I was born in sin is to say . . . I was born in a mask. I came into existence under a sign of contradiction, being someone that I was never intended to be and therefore a denial of what I am supposed to be."[31] Likewise, Draper's existence is a contradiction and its weight threatens his collapse. When he stares into the mirror in "Maidenform," perhaps he is asking himself the same question the audience wonders: "Who is Don Draper?" He fears he is nothing; in Merton's words, "if I never become what I am meant to be, but always remain what I am not, I shall spend eternity contradicting myself by being at once something and nothing . . ."[32] God meant people to be, Merton assumes, true selves created in the image of God and, therefore, created to dwell in his presence. By contrast, the false self cannot abide in the presence of God and, despite the divine intention, will remain alone.

The Bible portrays this estrangement from God as exile.[33] My theological conclusions at this point are, thus, foreshadowed by the stunning theological imagery in Season 1's episode "Babylon" which, although coming earlier in the show's narrative, forms a fitting conclusion to this section.[34] The episode centers on an advertising campaign for the Israeli Ministry of

29. "Flight 1," S2/E2.

30. "The Gold Violin," S2/E7.

31. Merton, *New Seeds of Contemplation*, 33.

32. Ibid., 32.

33. The history of the nation of Israel, God's "chosen nation" according to the Hebrew Scriptures, historically ends in exile. They are forcibly taken from the land of Canaan and made to dwell in the foreign lands of Assyria and Babylon apart from their Temple, where the presence of God was thought to dwell. Nevertheless, "exile" can also take a metaphorical sense in Scripture as separation from God, which we saw above with the Adam and Eve story.

34. "Babylon," S1/E6.

Tourism. In the initial meeting, Draper receives a copy of Leon Uris's popular 1958 novel *Exodus* as evidence that Americans will want to go to Israel. The title, which flashes across the screen no less than six times through the course of the episode, refers to the biblical story of the release of the enslaved Hebrews from Egypt.[35] In conjunction with the metaphorical meaning of exile as separation from God, 'exodus' can take the broader, metaphorical sense of a return to home, a return to the person one is supposed to be.

But for Draper and the rest of *Mad Men's* characters, exodus eludes them. They are trapped in loveless marriages, like Roger, whose affair with Joan Holloway is revealed in this episode. They are 1960s women who exist for nothing more than the amusement of men, like the secretaries who try on Belle Jolie lipstick in front of a two-way mirror. And they are 1960s men, like Draper himself, defined by images they can never match. These characters are in exile, moving through this world alone, separated from God and one another. The episode ends with a montage of close ups on these lonely characters over which a bohemian artist sings the haunting refrain, "By the waters of Babylon / We lay down and wept for thee Zion / We remember, we remember, we remember thee Zion."[36]

"THEN YOU CAN CHANGE": THE BAPTISM OF DRAPER

In the closing episodes of Season 2, Don Draper's carefully crafted image of a competent ad executive/family man, which I have labeled with the theological language of the false self, quickly unravels. As noted above, Betty has learned of Draper's affair with Barrett and kicks him out, telling him not to return.[37] The historical setting of these episodes—the Cuban Missile Crisis—frame this loss as a sort of death.[38] While all the characters worry about whether the world will survive, Draper faces the end of the image that has so long defined him. Nevertheless, *Mad Men* takes one of its many surprising narrative turns at this point, a turn that offers Draper not the eternal contradiction of the false self, but the hope of a new beginning. Notably, as

35. The biblical story of the exodus climaxes in the parting of the Red Sea, which Salvatore Romano alludes to in a brainstorming session. Incidentally, the Uris novel itself is about the founding of the modern state of Israel in 1948.

36. The words come from Ps 137, an exilic Psalm in which the Israelite exiles in Babylon cry out for their distant homeland of Jerusalem.

37. "A Night to Remember," S2/E8.

38. Bigsby has helpfully observed that in *Mad Men*, "history is played out in the background of personal dramas which seem detached from that history but which echo its dislocations, tensions, and transformations." Bigsby, *Viewing America*, 385.

with Season 1's "Babylon," the series uses poignant theological imagery to express this unexpected development in Draper's journey.

First, though, Draper is offered the false hope of reinventing himself by starting over with a new image, false because it would keep him at the level of the false self. In the episode "The Jet Set," Draper flies to California for a conference but, perhaps suggestive of the breakdown in his image, he completely shirks his work responsibilities and instead pursues an attractive and mysterious young woman named Joy—the name is surely significant—and her nomadic friends.[39] No one in her group is employed or rooted in any stability; they simply move about as they feel led. They live life with a hedonistic philosophy that attempts to defy death. "To not being carried out in a box," they toast at dinner. After a tryst, Joy asks Draper to join them, which gives him a chance to start over, to leave his crumbling image and to begin again.

Everything known about Draper to this point suggests he'll go with them. Indeed, the nomads' philosophy matches his own. "You're born alone and you die alone and this world just drops a bunch of rules on top of you to make you forget those facts but I never forget," he said in the pilot episode, "I'm living like there's no tomorrow because there isn't one."[40] More importantly, he has started over before, particularly when he left behind his life as Dick Whitman to make a new life as Don Draper. Indeed, the nomads' dinner toast recalls the scene when Draper leaves his family by allowing them to think it was Dick Whitman in the coffin delivered to them. The stranger on the train tells him, "You've got your whole life ahead of you. Forget that boy in the box."[41] But as Draper and Joy are in the pool discussing plans for a nomadic and hedonistic future, a man named "Christian" arrives with a young girl in his arms and a boy at his side, children he shares with one of the nomads. Draper and the young boy exchange a look that clearly disturbs him. Perhaps the children remind him of his own childhood where he too was the last priority of self-centered caretakers. Perhaps they convict him about his abandonment of his own little boy and girl. Whatever the reason, Draper does not go with the nomads.

Instead, he visits Anna Draper—the widow of the late, and real, Lieutenant Donald Draper.[42] Anna, we learn through a flashback, had tracked Draper down prior to his Sterling Cooper days.[43] Caught in his lie, Draper

39. "The Jet Set," S2/E11.

40. "Smoke Gets in Your Eyes," S1/E1.

41. "Nixon vs. Kennedy," S1/E12.

42. "The Mountain King," S2/E12.

43. "The Gold Violin," S2/E7.

had no choice but to tell her the truth about his identity, which is the first and only time to this point in the show, he had revealed this truth to anyone. Contrary to his greatest fears, however, this moment of authenticity did not ruin everything for him, but freed him. And as a result of this freedom, he develops with Anna an authentic relationship, one not based on sex or business or convenience. When with her, he seems at peace because he is with someone who knows his true identity. Poignantly, she never calls him "Don," the name of his image. And when he visits her in California at the end of Season 2, he is lighter and happier, despite the turmoil of his life. Without hair gel and dressed in casual clothes, he looks more like the innocent version of Draper seen in flashbacks.

Reunited with his one, authentic friend in his darkest moment, Draper offers a confession, which, in monastic theology as in the wider Christian tradition, initiates the healing process. "I ruined everything," Draper confesses, "My family, my wife, my kids. My brother came to find me. I told him to go away."[44] The confession reads like a summary of the first two seasons and, significantly, focuses on the many relationships destroyed by his selfishness. He confesses, thus, not simply his infidelity but the deeper problem, namely, the contradiction of his life. The maintenance of his false image has resulted in hurtful behavior and broken relationships, Adam Whitman being the central example. And Draper himself has also been a casualty of this manner of living, for the image of "Don Draper" has choked out any true life he might have experienced. "I have been watching my life," he continues, "It's right there. I keep scratching at it, trying to get into it. I can't."[45] In this profound statement, one hears the cry of the false self, desperate for something true or, perhaps, that of the exile longing to come home. This extraordinary scene marks the first time Draper has been so vulnerable because to this point he had lacked the authentic relationship that makes such vulnerability possible.

Here, we meet an important aspect of monastic theology—namely, the centrality of human community. While it may be easy to overlook this communal emphasis in the monastics because they are known for craving solitude and withdrawing from the world, solitary monastics become the exception in history; the great rules of the monastic tradition all affirm the need for monks to live in community.[46] The few monastics who pursue

44. "The Mountain King," S2/E12.

45. Ibid. These words recall the episode "The Wheel," S1/E13, where, as I noted above, Draper views a series of pictures of his family in a pitch for Kodak's Carousel. In this scene, he literally watches his life from outside it and, as a result, misses being with them on Thanksgiving.

46. The desert monastics of the fourth century, some of whom we will encounter

solitude do so not to live in isolation, but precisely so they can learn to love others and live in human community more deeply. Merton writes, "I must look for my identity, somehow, not only in God but in other men. I will never be able to find myself if I isolate myself from the rest of mankind as if I were a different kind of being."[47] Indeed, isolation marks the false self for it contradicts the very purpose for which humans were created. "So God created humankind in his image, in the image of God he created them; *male and female he created them*."[48] The significance of this foundational text is not simply that humans are created in the image of God, but that human communities—not individuals—are the primary bearer of that image. Humans, therefore, need one another: so their true selves can be actualized, and so they can experience the full presence of God. "The only justification for a life of deliberate solitude," Merton writes, "is the conviction that it will help you to love not only God but also other men."[49]

Confession, thus, negates the isolation of the sinner. I speak my sin to *another* who embraces me in forgiveness, thereby reconciling me to the human community from which my sin had separated me. In finding this true community, I find God. Anna, who initially did not turn away from Draper when he was exposed as a fraud, also refuses to turn away from him at this crucial point. Their authentic relationship is, in many ways, his absolution and his hope. Despite what he has always believed, people *can* change. "The only thing keeping you from being happy," she tells him, "is the belief that you are alone."[50]

The episode ends with a baptism. Draper takes his clothes off and wades into the ocean with the sun shining on his face. He lets the waves crash on him and is submerged in the water. George Jones's "Cup of Loneliness" plays over the scene, drawing attention to the deep theological imagery of the moment: "I say Christian pilgrim / so redeemed from sin / called out of darkness / a new life to begin . . . "[51] In the season's final episode, Draper

below, are the paradigmatic solitary monastic figures. Following the legalization of Christianity under Constantine, they intentionally withdrew from society in order to live sacrificially by themselves in the vast Egyptian desert. Nevertheless, it did not take long for these figures to realize their need to be among other monastics. This longing produced monastic communities and organizers, like Pachomius, who developed common ways of life for the monastics. This was the precursor to the more well known rules of later centuries, the most important being that of St. Benedict.

47. Merton, *New Seeds of Contemplation*, 51.

48. Gen 1:27, New Revised Standard Version, italics added.

49. Merton, *New Seeds of Contemplation*, 52.

50. "The Mountain King," S2/E12.

51. Ibid.

returns to his family and as he sits on a bed with his daughter and son who he had only recently considered abandoning, he writes his wife a letter of confession: "I think about you and how I behaved and my regret . . . *without you, I'll be alone forever*."[52] These stunning words imply that something has changed from Draper's contradictory existence that dominated the first two seasons. They imply the departure from his false self—and its dualities and isolation—and the commencement of a journey toward his true self and the experience of peace in authentic human relationship.

"MIDWAY IN OUR LIFE'S JOURNEY": DRAPER MOVES FORWARD

As Season 3 opens, the audience anticipates a different Don Draper. He has had his crisis moment, he responded with a confession and a symbolic baptism and he returns to his wife who, now pregnant, invites him back to the house. Surprisingly, however, Draper does not seem to have changed. In the season's first episode, he engages in a one-night stand with a stewardess, foreshadowing—as we will see—continued infidelity and isolation throughout the following seasons.[53] The dissonance between this behavior and Draper's redemptive moment at the end of Season 2 leads most commentators to reinterpret the latter as something other than a baptism. For example, Varon's essay, which I addressed in the introduction, focuses on the dominating image of an ocean tide coming in and going out as symbolic of repetition.[54] Such interpretation suggests that nothing new occurred at the end of Season 2 and Draper's contradictory lives, and poor behavior, will continue.[55]

Nevertheless, the resources of monastic theology suggest the possibility of a more hopeful interpretation, one that accounts for Draper's relapse into infidelity without denying the reality of a change in his character. For the monastics, a person who falls back into old patterns of sin following his or her baptism is far from an anomaly; in fact, the depth of the wound of sin makes such reversion common. The fourth-century monastic figure, known as Pseudo-Macarius, speaks to this when he writes:

52. "Meditations on an Emergency," S2/E13.

53. "Out of Town," S3/E1.

54. Varon, "History," 267.

55. This interpretation is unavoidable when one makes Draper's behavior, as Varon does, the sole determiner of his "goodness."

> Simple-minded and foolish persons, when grace begins to some degree to work in them, believe that they are simply freed from sin. But those who have discretion and prudence would never dare to deny that, even if they are gifted by divine grace, we are still tested by wicked and obscene thoughts. . . . Therefore, no one in his sane mind should dare to say: "Because I am in grace, I am thoroughly free from sin." But the two principles exercise their proper force upon the human mind.[56]

The two principles Pseudo-Macarius alludes to here are those of grace and sin. The "force," as it were, of grace, which has been imparted at baptism, battles with the force of sin, which has taken root in human existence and which continues to lure each individual human. Indeed, it would be fitting to use Pseudo-Macarius's image to interpret Draper's journey in Season 3 through Season 7. He experiences "two principles [exercising] their proper force upon [his] mind."

The problem in these middle seasons, from this perspective, is not that Draper has not experienced any kind of redemption, but rather that the redemption is not yet complete, the journey, as it were, not yet finished. The primary indication of this incompleteness is not simply his poor behavior but, even more profoundly, the endurance of the false self. Although Draper had confessed his hurtful behavior to Betty, he never revealed to her his true identity as Dick Whitman. Thus, he returns to her in the hopes of living in an authentic relationship, the kind he has with Anna, but without being fully known by her. In the terms of monastic theology, his false self has not fully died for it persists in Betty's mind. And very soon it captures Draper with his relapse into infidelity.[57] Draper finally confesses the full truth of his secret identity to Betty, but only after she discovers it for herself and confronts him.[58] The confession is somewhat meaningless at this point as it comes not from a place of authentic relationship but from the fear of being exposed. Neither does Betty react as Anna had. She kicks him out for good.

As Season 4 opens, Betty and Draper are divorced, and he is slipping deeper into isolation. He has chances to develop relationships, but goes instead to prostitutes, where he can get the quick satisfaction of a physical encounter without ever revealing anything about himself. In one episode, he goes to bed with one woman and wakes up with another, so drunk that

56. Pseudo-Macarius, Homily 17.5–6, in *Fifty Spiritual Homilies*, 137. Pseudo-Macarius is considered one of the desert monastics referenced above.

57. In addition to the one-night stand with the stewardess, Draper begins a longer affair with Sally's former teacher, Suzanne Farrell in "Souvenir," S3/E8.

58. "The Gypsy and the Hobo," S3/E11.

he does not remember the intervening hours.[59] But the point is clear, Draper has no intimacy in these relationships. The only relationship with the potential for authenticity he develops that season is with Dr. Faye Miller, a psychologist who consults with the firm. The relationship appears to work, but just as he seems poised to reveal himself to her, Draper asks his secretary Megan to marry him. When Draper calls Faye to tell her, she poignantly diagnoses his problem: "I hope she knows you only like the beginnings of things."[60]

We have already witnessed Draper's proclivity to start over—this effort initiated the deep contradiction of his life when he assumed the image "Don Draper." Faye's words now reveal that starting over has been a defense mechanism of sorts, a means of avoiding true relationships, of short-circuiting authentic human community. Two of the series more memorable moments demonstrate the false hope that lies in starting over. First, in "The New Girl," a flashback shows Draper with Peggy in the hospital after she delivered an unplanned and unwanted baby.[61] He tells her, "Get out of here and move forward. This never happened. It will shock you how much it never happened." Although Peggy goes on to build a successful career as a copywriter, memories of the son she gave up continue to plague her throughout the series; Peggy, in fact, never moves on. Second, in "Commissions and Fees," Draper has to fire Lane Pryce because he embezzled money.[62] In an effort to assuage Pryce's pain, Draper says, "The next thing will be better because it always is . . . I've started over a lot. This is the worst part." Pryce, however, commits suicide. Both events expose the lie that starting over provides the best way to move forward. It simply masks pain that emerges at a later point, often with destructive consequences. In Draper's case, starting over prevents him from progressing toward his true self in these seasons because it permits him to avoid an authentic relationship wherein his true self might emerge.

Monastic theology, thus, articulates Draper's difficulty in these middle seasons. While given the hope of new life with his symbolic baptism at the end of Season 2, the force of sin continues to work in his life by keeping him from true relationships. Through these middle seasons he remains isolated, from his wife Megan, from his kids, and from his friends. Most significantly, Anna dies of cancer, causing him to lament to Peggy, "Someone

59. "Waldorf Stories," S4/E6.

60. "Tomorrowland," S4/E13. This statement turns out to be both a summary of the show to that point as well as a prophecy. His marriage to Megan ends in divorce after Draper returns to his infidelity in Season 6.

61. "The New Girl," S2/E5.

62. "Commissions and Fees," S5/E12.

very important to me has died . . . the only person in the world who really knew me."[63] Importantly, this moment comes after Draper has been sick from drinking. As he says these deeply unhappy words to Peggy, his white shirt is covered in vomit, the symbolism reminding the viewer that his false self is still very much alive.

Monastic theology also helps us understand the only solution for Draper's deep unhappiness. As noted in the introduction, the monastics believed that the "putting off of the old man"[64] required not simply a change of one's actions, but, in accordance with the deep wound of sin, a death of the false self. Significantly, this is the hard road modeled by Christ himself, as Paul had observed:

> For if we have been united together in the likeness of [Christ's] death, certainly we also shall be in the likeness of His resurrection, knowing this, that our old man was crucified with Him, that the body of sin might be done away with, that we should no longer be slaves of sin. For he who has died has been freed from sin.[65]

The teaching that the false self needs to die is clearest in that strange collection of sayings by the Desert Fathers, who saw their renunciation of the comforts of society as an act of dying. To cite one of many such examples, a saying of Abba Moses goes as follows: "A brother questioned Abba Moses saying, 'I see something in front of me and I am not able to grasp it.'"[66] What the monastic laments here is the inauthenticity of his life, the same problem Draper complained of in his remarkably similar confession to Anna: "I have been watching my life. It's right there. I keep scratching at it, trying to get into it. I can't."[67] The monastic saying proceeds to reveal the solution: "The old man said to him, 'If you do not become dead like those who are in the tomb, you will not be able to grasp it.'"[68] By the middle of Season 7, Draper remains very much in the same place of inauthenticity and contradiction. He has yet to grasp the solution provided by the monastics, namely, that in accord with the example of Christ, the false self must die, for only in dying does one come to find true life.

63. "The Suitcase," S4/E7.

64. Eph 4:22.

65. Rom 6:5–7, New King James Version.

66. Ward, *Sayings of the Desert Fathers*, 140–41.

67. "The Mountain King," S2/E12.

68. Ward, *Sayings of the Desert Fathers*, 140–41.

"YOU CAN COME HOME": THE RETURN OF DRAPER

Having established how monastic theology provides a lens through which to understand Draper's struggle in the first two seasons and his slow progression through the following four and a half seasons, only the outcome of Draper's story remains to be analyzed. I will pursue the question of this outcome largely through an examination of the second part of the final season, that is, the last seven episodes.[69] Consistent with monastic theology, as well as the end of Season 2, the only mechanism by which Draper can realize his true self is an authentic relationship. Thus, the question of how Draper's story will resolve can be framed as follows: which of the two forces working on Draper since his symbolic baptism at the end of Season 2 will win out? Will he continue on the path of the false self, marked by an avoidance of reality and a series of shallow relationships, or will he find the peace of his true self by resisting his destructive tendency to start over and by embracing an authentic human relationship?

As episode eight, "Severance," opens, the year is now 1970, a decade removed from the show's beginning. Draper is a bachelor again, as his divorce to Megan is in its final stages, and he seems entrenched in his old ways. While in a diner, for example, he meets and is taken with a waitress named Diana Bauer who resembles many of Draper's previous paramours; Draper even asks her, "Don't I know you?" They share a brief and mostly sexual relationship, and after only a few meetings, she disappears without a word.

At work, he learns that their firm, which now works as a subsidiary of McCann Erickson, is being absorbed by the larger firm. The head of McCann Erickson, Jim Hobart, entices Draper to remain with the company by offering him the Coca-Cola account. But when Draper gets to McCann Erickson, he finds the work empty and unsatisfying. Everything he has worked for over the course of the show, in fact, appears absolutely meaningless to him. So he leaves. He packs no bags and says nothing to anyone; he simply walks out of a meeting and drives west in search of Diana. Draper is, once again, starting over, cutting off any authentic relationships he had with his coworkers and his children. "He does that," Roger later tells Hobart.[70]

Although he is pursuing Diana—"Di" for short, surely another significant name—this relationship is hardly authentic; indeed, she turns out to be the counterpart of Draper's false self. "She's a tornado," her ex-husband tells

69. The final seven episodes aired a year after the first seven episodes of Season 7 and together were called "Season Seven, Part Two: End of an Era."

70. "Lost Horizon," S7/E12.

him, "just leaving a trail of broken bodies behind her."[71] Perhaps Draper asked Diana if he knew her because she reminded him of himself. Draper does not find her but, instead of returning to New York, he continues west with seemingly no plan or purpose. At one point he picks up a hitchhiker who says, "I don't want to take you out of your way man." Draper replies, "It's not a problem."[72] The journey that Draper has been on, however convoluted to this point, seems now to have derailed altogether.

In the series finale "Person to Person," Draper learns from Sally that Betty has cancer, and he offers to return to take care of his children.[73] Both Betty and Sally reject the offer; they do not trust him or his ability to make good on his promises. Feeling rejected and hurt, Draper returns to California where he visits Stephanie, Anna Draper's niece, longing perhaps for some connection.[74] Stephanie, who was pregnant earlier in the season,[75] has given up her baby and is clearly wounded by the decision, a hurt that mirrors Draper's own state of isolation. They have both abandoned their children; they are both reeling with regret. Stephanie convinces Draper to go to a spiritual retreat with her to seek healing.[76] Although skeptical, Draper goes with her.

During a sharing session at the retreat, Stephanie reveals the pain she feels from leaving her child. One of the participants tells her the brutal truth that she will never get over this pain, causing Stephanie to run from the room crying. Draper follows her and, in a moment that resembles his interaction with Peggy in "The New Girl," tells her, "I just know how people work. You can put this behind you. It will get easier as you *move forward*." It's Draper's personal motto; effectively, he tells her to forget her baby and start over. But for the first time, someone is honest with him. "Oh Dick," she says, "I don't think you are right about that." Stephanie knows the woman at the retreat is right, that she will never recover from this pain, no matter how

71. "Lost Horizon," S7/E12.

72. Ibid.

73. "Person to Person," S7/E14. Betty's disease is first revealed in "The Milk and Honey Route," S7/E13. Draper's relationship with Sally had been strained ever since she caught him cheating on Megan in "Favors," S6/E11 and had lost all respect for him.

74. Stephanie is first introduced in "The Good News," S4/E3 and reappears at various points in the show. Significantly, Stephanie knows his true identity as Dick Whitman and remains his only connection to Anna Draper.

75. "The Runaways," S7/E5.

76. Presumably this is the famous Esalen Institute, although it is never mentioned by name. It is a communal living situation with Zen mediation exercises, seminars, and the like.

much she attempts to move forward. Stephanie leaves the next day without saying goodbye.

Upon learning of her departure, a distraught Draper expresses his frustration to a retreat worker: "People just come and go, and no one says goodbye?" Draper's extreme hurt is difficult to interpret—few relationships to this point have caused him any grief—unless one understands that in Stephanie he had found a connection to Anna, the only person with whom he had shared an authentic relationship. Perhaps he had pursued Stephanie in order to find that kind of relationship. But she had told him, "You're not my family" and had effectively abandoned him. Her departure, thus, only underscores his isolation. Don Draper is alone in the world and, with the impending death of his ex-wife and the recent rejection of his daughter, he finally realizes it. In his desperation, he calls Peggy.

Draper's friendship with Peggy is the most consistent relationship he has had throughout the series. To be sure the relationship had its tumultuous moments, but somehow they had always been there for each other. Draper championed Peggy in the early part of her copywriting career while Peggy had always believed the best about Draper and was the only person in the office who consistently stood up to him and told him the truth.[77] The honesty between them, not to mention the platonic nature of their relationship, makes Peggy the mirror image of Anna Draper by Season 7. This is clearly on display in the episode "The Strategy," which centers on a pitch for Burger Chef.[78] While Draper and Peggy had collaborated before, they had never worked together so well. Their chemistry leads to a poignant moment of honesty. Peggy shares her fears of turning thirty and not being married. "I worry about a lot of things," Draper tells her, "But I don't worry about you." Peggy replies, "What do you have to worry about?" Draper says, "That I never did anything and that *I don't have anyone.*"[79] The moment of honesty ends with them physically connecting, dancing together in an embrace. Of all the relationships Draper has had, this one endures.

The authenticity of their relationship discovered in "The Strategy" leads Draper to call Peggy in his crisis moment. And, following the parallel between Peggy and Anna, Draper offers a confession. "I messed everything up," he says, "I broke all my vows, I scandalized my child." Like his Season 2 confession, Draper repents of broken relationships. More importantly, he

77. Notably in "Man With a Plan," S6/E7.

78. "The Strategy," S7/E6.

79. The scene even takes on a quasi-Eucharistic quality. After they are honest with each other about their fears of loneliness, Peggy comes up with the Burger Chef pitch: "What if there was a place where you could go where there was no TV, and you could break bread and whoever you were sitting with was family?"

confesses to Peggy the existence of his false self, a confession he had never made to Megan and only made to Betty when she forced it by her discovery. "I'm not the man you think I am," he tells Peggy, "I took another man's name and made nothing of it." His willing confession of the secret he had so long tried to hide, the secret of his false self, suggests that he has finally come to the point where he is done with it, where he wants its death. And, like Anna, Peggy provides this safe place through her presence. She does not reject him but tries to convince him of the true way forward. "I know you get sick of things and you run," she says, "but *you can come home*." Draper's response to Peggy, however, does not parallel his response to Anna. He seems unmoved by her words and he simply hangs up; he appears almost suicidal. It seems that he has no ability to imagine life apart from his false self. Who will he be if he returns?

In this desperate state, Draper mindlessly attends a sharing group where he hears another man's story. The man feels invisible: "I work in an office, people walk right by me," he says, "I know they don't see me. I go home and I watch my wife and my kids; they don't look up when I sit down." Draper maintains the same faraway, disengaged look as the man speaks, but when he hears the words, "It's like no one cares that I'm gone," Draper looks up. This man's story is his own. Draper has hidden behind another name and another story his entire life. He has hidden behind his false self, not allowing people to see him. When aspects of that self would begin to erode, he simply picked up and started over in a new place, building a new image, but continuing the same false self. But in the finale, he fully sees the poverty of his attempts to start over and, it would seem, the poverty of his false self.

Draper finally allows his false self to die in this moment. As a result, he has no words of advice for this man, no encouragement to simply start over and move forward, as "Don Draper" had for Peggy, Pryce, Stephanie, and countless others over the course of the show. The only thing he can do is share in the man's pain which parallels his own. As the man weeps, Draper embraces him, letting his own tears fall. Thus, with the death of the false self, Draper's true self emerges and does so, significantly, in a moment of authentic human connection. Draper embraces this man to let him know he is not alone, perhaps simultaneously realizing that he himself has not been alone, that true relationships, in the form of his children and of Peggy, are waiting for him to come home and be his true self. In this moment, Draper decides not to start again or, worse, to end his life, but rather to return to these authentic relationships. This moment of renewal recalls Season 2's baptism moment as the very next scene shows Draper on a bluff, overlooking the ocean with its crashing waves. He does not go into the ocean again because he has experienced the baptism of new life already. But standing there, he *remembers* it. The next time we see him he sits in the lotus position of meditation, wearing the white garb of the new man. The leader expresses the theological significance of the moment: "The new day brings new hope. The lives we've led, the lives we get to lead. A new day, new ideas, a new you." The sun shines on Draper's face, and he smiles. The exile is returning home, a new man.

CONCLUSION

The theological interpretation of Draper's narrative arc using the resources of monastic theology has produced several fruitful results. First, it provides a more compelling diagnosis of his struggle in the first two seasons than is typically offered when simply emphasizing his behavior. Focusing instead on the sense of disequilibrium created by the contradictory nature of the false self (pictured poignantly in the character's false identity as "Don Draper") helps us understand both why he can still act admirably and, perhaps, why we as an audience are drawn to him despite such terrible behavior. Theologically, we affirm Draper's true self as created in the image of God and for community with others. He has been kept from realizing the true self precisely because he avoids human relationship through constantly reinventing his life. Accordingly, he finds his true self, first in Season 2 and then in the series finale, through embracing authentic human relationship. Second, the theological interpretation explains the dissonance created by the hopeful moments in his story (such as the baptism scene at the end of Season 2) and the destructive moments in his story (such as his serial infidelity). This juxtaposition does not make Draper a disjointed character or the show somehow overly repetitious, but is in keeping with the nature of his struggle as understood from the perspective of monastic theology. Third, the theological interpretation accounts for the rich theological imagery running throughout the show without forcing a literal religious experience on the character himself. The point has not been to suggest that Draper finds renewal in the literal Christian sense, but rather that the categories of monastic theology help us better understand the change and renewal Draper experiences.

The net result of this theological interpretation is that Draper's story ends on a hopeful, redemptive note. Having experienced a rebirth of some sort in Season 2, he remembers that hope at the end and begins living into it not by starting his life anew, but by returning home to the authentic relationships waiting for him. His return is not signified by a reunion with Peggy or with his children. Such a tidy wrap would not have been in keeping with the complexity and brilliance of the show. Rather, *Mad Men* closes with the famous 1971 Coca-Cola ad, "I'd Like to Buy the World a Coke." The hill-top setting and the hippy clothing of the varied individuals in that ad bear an unmistakable resemblance to the people Draper met at the retreat. The images suggest that Draper returns to New York, takes up the Coca-Cola account Hobart promised him, and writes that famous advertisement based on his experiences at the retreat.

While ending the series with an advertisement is in keeping with the theme of the show, the key for the purposes of this theological interpretation is more simply that he returns. And he returns not because he has the idea for the ad but because he now knows he can be his true self there, the self he rediscovered at the retreat. Peggy, in her parallels to Anna, offers this to him as does his children who need him more than ever with Betty's looming death. That he also returns to his role at McCann Erickson does not negate the altruism of this return but, rather, suggests that the advertising profession is a key part of who he really is for it involves both his passion and his skill set. But, significantly, his best work—the 1971 Coke ad is one of the most famous and influential ads ever created—comes not from the false self but from the true self which has finally emerged in its fullness. To quote the famous refrain, which repeats several times as *Mad Men* fades to black for good: "It's the real thing."

Mad Men has ended, but the real Don Draper—whatever he chooses to call himself—finally begins.

BIBLIOGRAPHY

Bigsby, Christopher. *Viewing America: Twenty-First-Century Television Drama.* Cambridge: Cambridge University Press, 2013.

Hansen, Jim. "Mod Men." In *Mad Men, Mad World: Sex, Politics, Style, and the 1960s*, edited by Lauren M. E. Goodlad, Lilya Kaganovsky, and Robert A. Rushing, 145–60. Durham: Duke University Press, 2013.

Kaganovsky, Lilya. "'Maindenform': Masculinity as Masquerade." In *Mad Men, Mad World: Sex, Politics, Style, and the 1960s*, edited by Lauren M. E. Goodlad, Lilya Kaganovsky, and Robert A. Rushing, 238–56. Durham: Duke University Press, 2013.

Merton, Thomas. *New Seeds of Contemplation.* Vol. 337. New York: New Directions, 1972.

Pseudo-Macarius. "Homily 17." In *Pseudo-Macarius: The Fifty Spiritual Homilies and the Great Letter*, edited and translated by George A. Maloney, 135–41. New York: Paulist, 1992.

Terjesen, Andrew. "Is Don Draper a Good Man?" In *Mad Men and Philosophy: Nothing is as it Seems*, edited by Rod Carveth, and James B. South, 154–67. Hoboken: John Wiley & Sons, 2010.

Varon, Jeremy. "History Gets in Your Eyes: *Mad Men,* Misrecognition, and the Masculine Mystique." In *Mad Men, Mad World: Sex, Politics, Style, and the 1960s,* edited by Lauren M. E. Goodlad, Lilya Kaganovsky, and Robert A. Rushing, 257–78. Durham: Duke University Press, 2013.

Ward, Benedicta, trans. *The Sayings of the Desert Fathers: The Alphabetical Collection.* Kalamazoo, MI: Cistercian, 1975.

Chapter 9

"YOU ARE OKAY"

Don*sein*'s Despair and Our Road to Recovery

—SETH VANNATTA

> Roger: Don, Do you believe in energy?
>
> Don: What? Do you mean like the thing that gives ya the get up and go?
>
> Roger: No like a human energy [. . .] a soul?
>
> Don: What do you want to hear?
>
> Roger: Jeez, I've been living the last twenty years like I'm on shore-leave. What the hell is that about?
>
> Don: Living, like you said.
>
> Roger: God, I wished I was going somewhere.[1]

In the pivotal exchange between Roger Sterling and Don Draper at the end of Season 1, Roger imagines the possibility of his own death. He has had a heart attack and questions his mode of existence, living like he is "on shore-leave," a navy man's phrase for thoughtless, care-free drinking and philandering. Don appears stoic and unmoved, but the event foreshadows

1. "Long Weekend," S1/E10.

a telling moment in the structures of Don's own existence—being towards death.

In *Being and Time,* Martin Heidegger discloses the essential structures of *Dasein*—that human entity that comports itself toward the Being of "existence." The anxious realization of the possibility of one's own death evokes the call of conscience, where an autochthonous voice asks, "How have I lived?" On the hospital bed looking weak and stripped of his slick three-piece suit and suave pocket square, Roger reminds us of Leo Tolstoy's Ivan Ilyich, who, on his death bed finally asked "I can do 'right,' but what is 'right?'"[2] These terrifying calls to oneself have many names—despair, dread, anxiety, guilt, and sin. For all of us, they aim to initiate a recovery of sorts, from inauthenticity to authenticity or from debilitating despair to passionate action. But since human existence is an open affair, Roger's question, "What the hell is that about?" or Ivan's "What is right?" can only be addressed through existing itself. An inquiry into Don Draper's world and the structures of his existence, as disclosed by both Søren Kierkegaard's and Martin Heidegger's philosophies, enables us to ask whether Draper's self-recuperative project embodies the being-toward death and recognition of his own most possibilities that Heidegger describes or illustrates the leap of faith Kierkegaard calls for.

Kierkegaard's existentialism and Heidegger's phenomenological ontology help us see the structures of Draper's existence and, in doing so, help us learn something about our own recuperative existential projects. Two of Kierkegaard's works significant to an analysis of Don Draper's existence are *The Present Age* and *The Sickness unto Death.* The former foreshadows many of Heidegger's insights in *Being and Time* regarding the anonymity of being registered in *das Man*, but also the modes of being *Dasein* exhibits in everydayness from "idle talk," "curiosity," and "ambiguity" to its states of "fallenness" and "thrownness." While Kierkegaard's call for passionate action parallels Heidegger's call for authenticity, Kierkegaard turns to the "inwardness of religion" to access authentic existence. Heidegger shows that authenticity can only emerge as a function of inauthenticity, which is a brute fact, not to be dismissed as unethical. *The Sickness unto Death* illustrates Kierkegaard's understanding of despair as sin. An exposition of despair and comparison to Heidegger's understanding of anxiety will serve to depict Don Draper, whose character is presented as ostensibly opaque, as an archetype of the human condition. Further, since I argue that his journey reflects our own, the exposition will enable us to measure Draper's path to self-recovery as either making the faithful Kierkegaardian leap or as

2 Tolstoy, *Death of Ivan Ilyich*, 52.

an authentic being living toward his "ownmost possibilities," more in line with Heidegger's description of *Dasein*. Kierkegaard thinks Don needs to turn inward, while Heidegger thinks Don can only modify his anonymous modes of existence to achieve authenticity.

The answer to the question as to which prescription Don follows, and by extension which we should consider following, is no simple one—not only because we are forced to discover Don's self over seven seasons in a show that needs to perpetuate its central character's mystery, but also because its revelation is shrouded in cycles of hope and despair, false bottoms of guilt, and failed attempts at living authentically. When it comes to Don Draper, Kierkegaard's and Heidegger's diagnoses are more accurate than either of their divergent prescriptions for his self-recovery. When it comes to our own, as we will see, reading Don's despair as representative of our own through the lens of two existentialisms is a necessary step in understanding both our inauthenticity and our potential for achieving authentic character.

ANONYMITY AND INAUTHENTICITY

Don Draper finds himself in need of recovery from the states of anonymity and inauthenticity. Kierkegaard contributes to an analysis of these by portraying the age in which he found himself. Kierkegaard's "present age" was mid-nineteenth century European Christendom. For our purposes, the present age is the age of post-Korean War, Madison Avenue advertising. It is an age of reflection, devoid of passion.[3] Kierkegaard contrasts his present age with eras of rebellion and revolution. In *Mad Men,* the civil rights and anti-war struggles of the mid to late 1960s serve as the show's backdrop and present the audience with a Gestalt effect—where seeing genuine revolutionary struggles puts the reflective repose of advertising in clearer view. This age of reflection recoils in the face of decisive action, content to comment on it and explain it away. In this way, Kierkegaard deems it unimaginable that a young man of his age would renounce the world in religious self-denial.[4]

Can we imagine any young man at the Draper's ad firm denying the world? The world of advertising inevitably embraces the world as it is given. Draper tells the executives at Lucky Strike, "whatever you are, you're OK."[5] Even when Roger tells Don that we live in troubling times, Draper responds,

3. Kierkegaard, *Present Age*, 33.
4. Ibid., 36.
5. "Smoke Gets in Your Eyes," S1/E1.

"We do? Who could not be happy with all this?"[6] The young men of *Mad Men*, such as Pete Campbell, long *for* the world. They enviously seek what others have—a corner office, a partnership, even the bourgeois nuclear family, reflected back to them by the advertising they create. They walk the path of Ivan Ilych, attempting to climb the corporate ladder, earn a higher salary, and snag a socially acceptable spouse. Mad men are "happy with all this."

Kierkegaard calls this age "reflective and passionless," empty of significance.[7] In such an age, the "inward reality of relationships" becomes condensed into a "reflective tension." Draper offers the hope of inwardness, (by selling Lucky Strikes or Jaguars), but what is in fact purchased is a loss of inwardness and an embrace of worldliness. In the age of advertising what does exist becomes masked by "a dialectical deceit," which through "a feat of dialectics," provides "a secret interpretation," which is precisely what does *not* exist in fact.[8] Consider Don's many pitches. He condenses the inward reality of relationships into this reflective tension in order to mask deceitfully what in fact exists. The mask is the reflective lure of desires, the embrace of the world as it is in order to become or acquire what it holds out for us. That which is masked is the desire of a company to profit off of our embrace of the mask and belief in the lie. The ad world is passionless because it lacks inwardness. It is all veneer, all spin, all surface. Witnessing what appears to be Roger's genuine, passionate love for Jane Siegel, his second wife, and his resultant conspicuous happiness, Don dismisses it reflectively: "No one thinks you're happy. They think you're foolish."[9]

Kierkegaard states that the "negative unifying principle" of the "reflective tension" (posited by the ad men) is envy. Pete Campbell's jealous desires represent the energetic force of envy, this "negative unifying principle" of the reflective age of advertising. Pete is enraged with selfish jealousy when he is co-promoted as head of accounts with Ken Cosgrove, as he selfishly wanted the job for himself.[10] As Kierkegaard tells us, envy results in selfishness. This selfishness prevents the individual from acting passionately and decisively out of the inwardness of character.[11] Selfishness prevents authenticity. The envy of reflection, embodied by Pete Campbell, becomes a "moral *ressentiment*."[12] When, in Season 2, an American Airlines flight crashes, Pete

6. "Ladies Room," S1/E1.
7. Kierkegaard, *Present Age*, 42.
8. Ibid., 42–43.
9. "My Old Kentucky Home," S3/E3.
10. "Out of Town," S3/E1.
11. Kierkegaard, *Present Age*, 47.
12. Ibid., 49.

Campbell, whose own father (we later find out) was on board and died in the crash, jokes that the flight had golfers on it, and that the ocean turned plaid as a result. This defeatist and cynical attitude offers up the suspicion that the world is in fact indifferent to us as individuals. The ad world is full of cynics. The plight of a Guy MacKendrick, whose foot is shredded by a John Deere riding lawn mower steered poorly by Lois Sadler in the office, is merely fodder for jokes and petty witticisms. Roger nonchalantly comments: "Somewhere in this business, this has happened before."[13] The cynicism of the ad world makes people morally lazy, cowardly, and indecisive. In the face of character manifest in action, the individual demonstrates *ressentiment* by making a joke of it, being insulted by it, or being content to dismiss it as really nothing at all.[14] When, in Season 4, Peggy Olson suggests hiring Harry Belafonte to be a company spokesman to show a racially progressive image, senior management displays its *ressentiment*: "Our job is to make men like Fillmore Auto," Don Draper says, "not make Fillmore Auto like Negroes."[15]

Kierkegaard asserts that *ressentiment* is the principle of a lack of character.[16] In order for a lack of character to level all distinctions and distinguished acts, "it is necessary to procure a phantom, its spirit, a monstrous abstraction, an all-embracing something which is nothing, a mirage—and that phantom is *the public*."[17] This phantom is Don Draper's audience, the abstract monster Kierkegaard worried about. Draper is thus its puppeteer, creating the curiosity and inauthentic desires of public.

The instrument of the public is the press, whose pronouncements become the opinion of everyone and of no one. Just as a want of character results in *ressentiment*, a lack of real interpersonal relationships results in the public. The public does not consist of real individuals; it is not a numerical sum, which could be eliminated by the realization of inwardness by each of its individuals. Rather, it is an abstraction, an apparition demonstrating the leveling tendency of the age.[18] Draper holds the strings of a new type of "press," the advertising industry, creating opinions and desires of everyone and no one, those of the abstract phantom, the public. The problem is that only individuals can be truly morally responsible for the victims of the public, but, as mere onlookers, individuals can always cast blame on the public,

13. "Guy Walks into an Advertising Agency," S3/E6.
14. Kierkegaard, *Present Age*, 50–51.
15. "The Beautiful Girls," S4/E9.
16. Kierkegaard, *Present Age*, 51.
17. Ibid., 59.
18. Ibid., 60.

much as the Bohemians in the Village can blame the ad men on Madison Avenue for the insincerity of the age. The public turns all decisive action into fodder for its gossip, commentary, and remorseless attacks. The (inauthentic) individual turns into the public by being a third party spectator, whose sole desire is for distraction and amusement, for the embrace of the world Don Draper offers. Further, if the individual receives her formation from this "gruesome abstraction," which tends to flatten moral distinctions, the conditions of salvation and self-recovery, as Kierkegaard describes them, are stifled.[19]

Kierkegaard's *public* runs parallel to Heidegger's depiction of *das Man*. Heidegger's phenomenological ontology aims to let Being be seen from itself in the way in which it shows itself; therefore, he must allow the Being of *Dasein*, the actual inquirer who seeks Being to be seen and uncover the structures of its disclosure in its everydayness. The most powerful constitutive element of *Dasein's* everyday existence is *das Man*, the "they." We find ourselves often in response to and in relation to what Others do and the ways they are. And the care we have in the inevitable distance between Others and us is expressed existentially as "distantiality."[20] This state of distance reveals that *Dasein* actually stands in subjection to Others. We are inconspicuously dominated by advertisements because while the ads are there in everyday Being-with-one-another, the ads are not particular others, "not this one, not that one, not oneself, not some people, and not the sum of them all. The 'who' is the neuter, the 'they' [*das Man*]."[21]

In the world of *Mad Men*, *das Man* registers as the phantom public, which is both the creation and audience of Draper's advertising. The audience engages in its environment by concerning itself with that which lies closest to us in a ready-at-hand way. We ride the train into work, read newspapers and magazines clothed in ads, such that every Other seems monotone, in the mode of anonymity. As Kierkegaard puts it, the public whose instrument is Draper's images and messages, levels particular distinctions, and we lose individual character by engaging in an anonymous mode of existence. Heidegger writes, "We take pleasure and enjoy ourselves as they take pleasure; we read, see, and judge about literature and art as they see and judge [. . .]. The 'they,' which is nothing definite, and which all are, though not as the sum, prescribes the kind of Being of everydayness."[22] We can become the public or the "they," not as a part of an aggregate, but in the mode

19. Ibid., 64.

20. Heidegger, *Being and Time*, 164.

21. Ibid.

22. Ibid.

of anonymity. We drink Pepsi, smoke Lucky Strikes, and fly on American Airlines.

Heidegger calls this mode averageness, and it shares with the public the capacity for leveling by reducing particularity to mere averageness. Distantiality, averageness, and leveling down are the modes of *das Man* that constitute publicness, which defines the ways that the world and *Dasein* are interpreted. Publicness is "insensitive to every difference of level and genuineness, and thus never gets to the heart of the matter."[23] When Roger hears of a parade for astronaut, John Glenn, he is reflective and utterly levels the distinctions between the American hero and himself: "I'd like a ticker tape for pulling out of my driveway and going around the block three times."[24]

This mode of anonymity of *das Man*, by its indifference to genuineness and particularity, relieves *Dasein* of its call to be genuine, particular, or even responsible for its own Being. One cannot find oneself or lose oneself if it persists in this mode. Heidegger calls this way of Being, in failing "to stand by one's Self," *inauthenticity*.[25] But inauthentic existence, acting as the "they," is a "primordial phenomenon," which "belongs to *Dasein's* positive constitution."[26] Only through the mode of *das Man* can Don, *qua Dasein*, be given to himself. He can only recover himself through this average, inauthentic relation to the world. That is, authenticity is always a function of inauthenticity, and only by way of "a clearing-away of concealments and obscurities, as a breaking up of the disguises" which bar its own way will Don discover his own authentic Being.[27] Draper cannot achieve authenticity by any other means but recovering himself from the public he helps create. In contrast to Kierkegaard's approach, Heidegger thinks any authentic existence he can manage will not come by way of turning inward, but by modifying the 'they.'[28] The leveling power, the averageness, and the anonymity of the public and *das Man* run parallel in Kierkegaard and Heidegger, but each describes and prescribes a divergent turn toward authenticity. Kierkegaard thinks Don needs to turn inward, while Heidegger thinks Don can only modify his anonymous modes of existence to achieve authenticity.

How does Don Draper seek himself in and through the modality of the anonymous crowd? As we have seen, he both creates it and lives within it. Early in the show, Draper deals with his discomfort in the world he creates

23. Ibid., 165.
24. "Flight 1," S2/E2.
25. Heidegger, *Being and Time*, 166.
26. Ibid., 167.
27. Ibid.
28. Ibid., 168.

through escapism, which does not yet constitute a modification of the they; rather, it is a turning away from the they. Draper retreats to the Village and shacks up with pot-smoking Bohemians, as if the scales of his existence need to be balanced by non-conformism and artistic pretense. His alcohol abuse is an escape, not only from the mundane stress of Madison Avenue, but from the inanity of the entire public which he authors. He is both master and slave of its surperfices.

But such escapism is no self-recovery. Not only can the counter-culture in which he revels be co-opted and sold back to the masses, but his retreats and affairs only create a greater distance between himself and his relationships, especially with his wife and family, which could disclose the inwardness of his character. But in the early seasons of *Mad Men*, Don's two children seem like foreign little creatures to him. And Betty represents the wishes of the masses that Don's ads make desirable. He helps create the world she embraces, as is illustrated when she is lured to purchase and serve Heineken beer at a dinner party. His guests know that he was pushing that the company market to women, and they laugh when the beer becomes a part of her "trip around the world menu." He creates what she embraces, but he recoils in disgust at what he has given life to and often flees from it.[29] He confesses to Anna in Palm Springs: "I've ruined everything. My family, my wife, my kids." He describes himself as merely "watching" his life. "I keep scratching at it, trying to get into it. I can't."[30] Late in Season 3, Roger tells Don abruptly that his problem is that he doesn't value relationships—the very relationships that could disclose the inwardness of his character.[31]

Acting as the public in a mode of anonymity shows the scope of reflection's leveling net. "Reflective understanding" transforms the individual at one with himself into everything possible, which is really nothing at all.[32] As Don claimed in the inaugural episode, "Advertising is based on one thing, happiness. And you know what happiness is? Happiness is the smell of a new car. It's freedom from fear. Happiness is a billboard on the side of the road that screams with reassurance that 'whatever you're doing it's OK. You are OK.'"[33] Draper's voice then levels all distinctions, hinders authentic individuality, and transforms the consuming public into a reflective one which thinks itself everything possible, which Kierkegaard tells us is really nothing

29. "A Night to Remember," S2/E8.
30. "The Mountain King," S2/E12.
31. "Shut the Door. Have a Seat," S3/E13.
32. Kierkegaard, *Present Age*, 68.
33. Smoke Gets In Your Eyes," S1/E1.

at all. As Draper tells the owners of Lucky Strike, "whatever you are, you're OK."

The path toward inwardness comes by way of decisive action that overcomes the leveling effects if the age of advertising. Kierkegaard claims that *talkativeness* levels the distinction between talking and keeping silent. Silence, "as the essence of inwardness," provides the condition for the possibility of genuine discourse. Talkativeness, as mere gossip, extends its scope so broadly as to cover nothing at all, but brush superficially across every topic. Because mere talk prematurely expresses what is "still in thought," it "weakens action by forestalling it."[34] Talkativeness emerges when people fail to turn inward in their religious lives; instead they turn toward others in talkativeness to find that which they lack. Silence expresses religious inwardness and is the prerequisite for educated discourse. On the other hand, talkativeness creates an external caricature of *a vulgar* inwardness.[35] While Don is the strong and silent type, he lives in a world of petty gossip. Does his resistance to office gossip reveal a glimpse of inwardness? When Paul, Harry, Sal, and Ken make fun of Freddy Rumsen's mishap (he peed his pants in the office), Don tells them to stop gossiping like teenage girls.[36] When Sterling Cooper loses the flagship American Tobacco account, Don decisively tells the creative team to "skip the gossip out there."[37] But his road to recovery is slow, indirect, and (perhaps inevitably) incomplete.

Kierkegaard's discussion of talkativeness runs parallel to Heidegger's analysis of "Idle Talk." Don's understanding and interpretation of the world manifest themselves in the everyday discourse of *das Man*. There he interprets the crowd by directing his understanding to what is "said-in-the-talk."[38] Everyday talk about something does not express the primary relationship-of-Being and the entity talked about. Instead, communication takes the form of idle talk, manifested in gossiping or thoughtlessly passing along what is said-in-the-talk.[39] What is said-in-the-talk, through its diffusion via gossip, takes on the cloak of authoritative consensus, but it is utterly groundless. Advertising traffics in the discourse of idle talk. It excuses the failure of genuine understanding as it becomes the possibility of understanding everything; it understands everything only insofar as it

34. Kierkegaard, *Present Age*, 69.

35. Ibid., 71.

36. "Six-Month Leave," S2/E9.

37. "The Gyspy and the Hobo," S4/E11.

38. Heidegger, *Being and Time*, 221.

39. Ibid., 212.

develops an "undifferentiated kind of intelligibility."[40] It fails to differentiate the primary relationship of Being to the entity understood from the mere talk about it. It does not disclose, but closes off, conceals, and limits the possibility of new inquiries. Idle talk becomes the form of *das Man* which limits the disclosure in a primordial manner with the imperious dominance of a groundless consensus. The modality in which we find ourselves, in the anonymity of the public and the inauthenticity of *das Man*, expresses itself in "idle talk" and "talkativeness," both of which cover everything superficially and nothing genuinely. Here, Kierkegaard and Heidegger paint a dauntingly similar picture of the world that Don Draper helps create, in which he lives, and from which he must recover himself.

The "undifferentiated kind of intelligibility," which idle talk creates, echoes in Kierkegaard's other descriptions of the leveling of distinctions. Of special relevance to *Mad Men* are *superficiality* and *flirtation*. The former denies the distinction between concealment and manifestation.[41] The latter does away with "the distinction between real love and debauchery."[42] Almost every pitch is an act of superficiality and flirtation. What could be characterized or criticized as the immoral use of others as means and not as ends, is better understood as the denial of fundamental distinctions. When Don uses his wife as a pawn to smooth over problems with a client, Utz chips, the entire dinner is a denial of the distinction between concealment and manifestation. Utz's pitchman does not conceal his desire for Betty, nor is the real reason for her presence made manifest. Rather, flirtation ensues so as to deflect the inwardness of anyone's character, including Don's, who hates the affection his wife receives and that he invited by bringing her there. In Don's constant flirtations, he conceals the distinction between true love and debauchery. The audience holds out hope that each new affair is true love only to see many of them as examples of debauchery. This is because Don is an archetype of flirtation in the age of advertising.

In the anonymity of *das Man*, idle talk controls the objects of our curiosity. What is said and passed along points to the ways and the objects of our searches. As curious, Don is caught up in a continuous search for novelty after novelty. Much like the rest of us, he restlessly seeks excitement and distraction. His curiosity reveals a fundamental hesitation and lack of certainty in his existence. Don turns from one woman to the next as a curious and anonymous flirt. In the world of advertising, the public interest is directed to novelty after novelty, to "new and improved," moving at the fast

40. Ibid., 213.

41. Kierkegaard, *Present Age*, 75.

42. Ibid.

drum-beat of idle talk. It constantly limits the possibility of our primordial Being-with-one-another by circumscribing our understanding of each other with advertising's idle talk. However, despite the negatively charged descriptions, Heidegger's explanations of the features anonymity and inauthenticity are not disapproving. Rather, they are inevitable structures of the world into which Don is thrown.[43]

In idle talk, Don loses himself in the anonymity of *das Man* and *falling* into groundlessness.[44] This points to a telling feature of Draper's existence. As the visual introduction to *Mad Men* illustrates, Don represents an extreme archetype of Fallenness. The paintings fall off the walls, the furniture collapses, and an anonymous silhouette falls among dizzying skyscrapers draped in images of female sexuality, post-War suburban life, cocktails, and diamonds. The groundlessness registers as a temptation, and the allusion to Original Sin is not accidental.

The phenomenological kernel in the notion of Original Sin is that finitude disables us from realizing our own Being unaided. Don and the rest of us must inevitably engage in the inauthentic temptations of the world. Our curiosity structures our desires such that we want diamonds, women in bikinis, even the pleasantness of post-war suburban life—the life Don leads. Don needs to recover himself, but is always already tempted by a world constituted by the anonymous modality of the "they." The idle talk Don produces "tranquilizes" him and the rest of us into thinking he is leading a genuine life. When asked, "'We got it all, right?' Draper says, "Yep. This is it."[45] Don is deluded into asking, "Who could not be happy with all this?"[46]

Don's delusion manifests itself in that he must constantly lose himself in his curious and restless attachments to things and flings. But in his avoidance of a genuine self-understanding, Don becomes alienated. Don's fallen state is his separation from his own potentiality-for-Being and from a primordial Being-with-one-another in the world.[47] However, Don's alienation and fall towards groundlessness "remains hidden from [himself] by the way things have been publicly interpreted," so as to invert his understanding of his existence as "living concretely."[48] In the state of fallenness, Don must search for his authentic self by way of retrieval and recovery from inauthenticity.

43. Heidegger, *Being and Time*, 219.
44. Ibid.
45. "Marriage of Figaro," S1/E3.
46. "Ladies Room," S1/E2.
47. Heidegger, *Being and Time*, 222.
48. Ibid., 223.

DESPAIR AND DREAD

The features of the age of advertising illustrated by Kierkegaard and the parallel account of the structures of Don Draper as *Dasein* explained by Heidegger depict the state from which Don must recover himself. Now we must show and define the affective experience that signals the need for such a recovery. Kierkegaard calls this despair, Heidegger dread, or anxiety. Where Heidegger investigates the being of that entity for whom its being is a question, Kierkegaard investigates the existence of the human self, which he says is constituted in a relationship, not of one's own making, and thus Don must find himself by recourse to a standard (God) beyond himself. Heidegger illustrates Don as thrown into a qualitative and relational matrix of worldhood, and thus must inquire into his Being by way of those relations. Both situations signal the need for Don's self-recovery.

But in order to depict Don Draper's road to self-recovery, we have to investigate the question: What is a self? Kierkegaard helps us do that. He tells us that a self is a synthesis, more specifically, a relation which relates itself to itself. Kierkegaard explains that the self is a synthesis or relation between the temporal and eternal, finite and infinite, actual and possible parts of oneself. Don's self is not only a synthesis of who he has been but of who he is right at a given moment and of whom is capable of becoming. In *The Sickness unto Death*, Kierkegaard defines the self as a relation in that it is "a contradiction between the external and the inward, the temporal and the eternal."[49] The negative unity of finitude and infinitude constitutes the relation which is the self. However, in this relation to opposites, it is free, because freedom is the dialectical element in these categorical opposites. Thus, the self is the freedom of movement toward these poles.[50] Both kinds of movement toward the eternal and the temporal, however, are forms of despair because they both attempt to evade the true self, which is a combination of these opposites.[51] The first of these movements toward infinitude results from a lack of finitude. The subject tries to escape his temporality and live (inauthentically) in abstract "categories" such as "ad man," "father," "husband," or "universally sexually desirable."[52]

The second kind of movement toward finitude results from a lack of infinitude. Don lives under the guise that there is nothing more to life than

49. Wyschogrod, *Kierkegaard and Heidegger*, 83.
50. Kierkegaard, *Sickness Unto Death*, 59.
51. Wyschogrod, *Kierkegaard and Heidegger*, 83.
52. Ibid.

the quotidian and fleeting. As he said in Season 1, "This is it."[53] His modality of being is disguised as pragmatic, but underneath this veil is the horror of losing oneself "in the world of non-self," the anonymous world of the public that he creates.[54] In this characterization, Don's despair is both the result of the ontological make-up of the human person and the modality in which he takes up a relationship to himself. The subject devoid of "infinitude" would live un-self-reflectively in the temporal flow. He would have no standpoint from which to posit his self-identity. However, he cannot destroy the eternal in him that founds his despair; instead, he is left as the synthesis and negative unity of the eternal and temporal.[55] Of utmost importance to this understanding of despair is that Don cannot, by his own power, extricate himself from his despair without recourse to the power that constituted the entire relation (of eternal and temporal).

Kierkegaard highlights two variations of despair. The first is passive, or feminine, despair.[56] In this state, Don lives in mere immediacy, as a function of his environment.[57] Things happen to him, and if they are misfortunes as when he is fired, he despairs.[58] He wishes not to be himself, or to be another self, because living in immediacy defines himself as the accidental intersection of the events which confront him.[59] But in fact he despairs because he does not want to be a self and, living in immediacy, has no self. Don has no relationship to the infinitude of his consciousness by which he might genuinely self-reflect. He imagines that he can change selves as easily as he can change his clothes because his conception of himself within immediacy is just this superficial.[60]

In a sense, Don's character is itself one change of clothing. He embodies the mode of anonymity as he has taken on the persona of a fellow fallen soldier. After Betty has discovered his past, he takes his children trick-or-treating, and the question of his self surfaces when his neighbor, Carlton asks Don: "And who are you supposed to be?"[61] Similarly, a reporter from Advertising Age quizzes Don over lunch by asking "Who is Don Draper?"

53. "The Marriage of Figaro," S1/E3.
54. Wyschogrod, *Kierkegaard and Heidegger*, 84.
55. Ibid., 85.
56. Kierkegaard, *Sickness unto Death*, 80.
57. Ibid., 81.
58. Ibid., 82.
59. Ibid., 83.
60. Ibid., 84.
61. "The Gypsy and the Hobo," S3/E11.

Don demurs (despairs) in response.[62] This form of despair is a weakness that leads him to despair over the fact that he is in this state of despair, such that his despair becomes dialectically productive.[63] He becomes more self-aware with the depth of his despair. For instance, in the same moment that Betty reveals to him that she knows his secret past, Don despairs over the suicide of his half-brother, whom he turned away so as not to risk his inauthentic self-creation with Betty and his kids.[64] In this step forward, despair holds open the possibility of salvation.[65] Put this way, Betty's discovery of his persona is an opening, a possibility for self-recovery.

The second variant of despair that Kierkegaard illustrates is the active, or masculine, form of despair. Here, Draper, embodying the masculine form, is in despair to be himself.[66] He defiantly makes use of the eternal, the sense of our internal infinitude within consciousness, in order actively to fashion the self he wants.[67] He realizes the power of the will and of self-control, but by virtue of his finitude (the relation to the external), he can only constantly experiment with himself, escaping to the Village, smoking hashish, or finding a new extra-marital affair. He recognizes no power over himself, and therefore he lacks what Kierkegaard calls "seriousness,"[68] (which lacks humility). Kierkegaard refers to the Stoic as living in this masculine despair.[69] He can accomplish great things. He closes deal after deal, wins CLIOs, and fashions his look with purpose. But "beneath it all there is nothing."[70] His desire to take credit for his constructions is equal only to the speed at which they can all dissolve.[71] Because when some misfortune reveals his true finitude, his strength and self-control turn against him, and he directs his passion at his torment. For example, Don hires a call girl in Season 4, telling her, "Do it," during sex. She slaps his face. "Harder," he demands.[72] His despair in wanting to be himself, because of his lack of humility, turns to demonic rage, and the desire to be himself becomes the

62. "Public Relations," S4/E1.
63. Kierkegaard, *Sickness unto Death*, 92.
64. "The Gypsy and the Hobo," S3/E11.
65. Kierkegaard, *Sickness unto Death*, 93.
66. Ibid., 98.
67. Ibid., 99.
68. Ibid., 100.
69. Ibid., 99.
70. Ibid., 101.
71. Ibid.
72. "Public Relations," S4/E1.

desire to be miserable and to hate existence itself.[73] This turning away from the possibility of salvation and self-recovery announces what Kierkegaard calls "the offense" of the possibility of salvation which Christianity conveys. But, as fallen, Don despairs because he is present to himself and aware of himself in a distorted way.

According to Heidegger's analysis of the structures of Draper's existence, Don's attempts at understanding are "mooded." Don brings an affective attunement to each of his interpretive endeavors. Heidegger claims these moods are founded on the primordial state of mind of "care." When *Dasein* expresses anything about itself, it has "pre-ontologically" interpreted itself as "care."[74] Heidegger's concept of care as the being of *Dasein* mapped onto the character of Don Draper reflects a heavy sense of *worry* about his relationship to his world. However, care is different than other "ontical" states of mind, such as fear, in that fear has an intentional structure. When we fear, we always fear something. But in anxiety, no entity within the world is intended; rather, we are anxious in light of the world as such. Anxiety is a function of the pure possibility of being in a horizonal world.[75] When Megan quits the agency to pursue an acting career, Don chases after her to the elevator, which opens only to reveal the dark chasm of an empty shaft. Here darkness encroaches on the objects of his consciousness—his affect is dread and anxiety. While fear reflects a definite threat, anxiety reflects an utterly indefinite one, which is part and parcel of Don's being. Don's anxiety is captured in Season 1 through the title of Episode 13, "Meditations in an Emergency." Set at the time of the Cuban Missile Crisis and taking its title from Frank O'Hara's book, Don reads:

> Now I am quietly waiting for
> the catastrophe of my personality
> to seem beautiful again,
> and interesting, and modern.

Much like the doctor warned him during a check-up, Don is thirty-six years old, but seems to be waiting for the catastrophe of his personality to seem beautiful again. He is pervaded by the anxiety Heidegger describes.

According to Heidegger, anxiety founds every other affective modality.[76] But in the same way that Don must achieve authenticity via inauthenticity, he realizes his anxiety only in other affective modes. He might

73. Kierkegaard, *Sickness Unto Death*, 105.
74. Heidegger, *Being and Time*, 227.
75. Ibid., 231.
76. Ibid., 232.

rejoice at his success, (and surely success reveals an ontic relationship), but the condition for the possibility of that feeling, of that affect, is his general, pervasive angst. Anxiety reveals affective possibility, which is at the heart of Don's being. Anxiety is an important individuating factor, which announces our "uncanniness."[77] Being thrown into the world not of his own making and as fallen, Don's default mode is the comfort of *das Man*, which brings tranquillized self-assurance—"Being-at-Home" into his average everydayness.[78] However, anxiety is that affect which awakens Don from his slumber in *das Man* because anxiety causes him to feel uncanny and not-at-home. And through his uncanniness, Don loses his familiarity with the world and its anonymity. Draper turns from his absorption in *das Man* and its constant ontical relationships, which generate "meanings," in order to realize in anxiety that his meaning lies in his pure possibilities.

When do we see anxiety and Don's individuation, and when does he realize that his anxiety is an opening to his possibilities in the world? The most obvious opening is Betty's realization of his past. He no longer has to fear that direct object. Instead, the dread of its occurrence can serve as an opening to new possibilities. But at the end of Season 3, the prospect of another "death" of Sterling Cooper in a buyout reminds Don of the death of his father, Archibald Whitman when he was kicked in the face by a horse after resolutely determining to sell wheat by himself (not through the collective). These anxious and mooded prospects of death signal openings. Whitman went it alone. Similarly, Don, Roger, Burt, and Lane hatch a scheme to buy the company and embrace the risk come what may.[79] Similarly, the death of his marriage is contemporaneous with the birth of his new company and all of the possibilities that await. When he insists that Lane retire for his financial dishonesty, Don assures Lane that he himself has started over more than once, and that the new start is always the worst part.[80] We suspect he is giving a mere ad man's pitch, but it rings true in Don's life—that our anxious deaths open possibilities.

As *Dasein*, Don must choose his ownmost potentiality-for-Being. When he does, he projects himself into the future and is always ahead of himself. This projection is not from a position of neutral possibility; rather, his possibilities are realized in a mode of anxiety, which announces itself in and through the ontical engagements in which he is always already. Don

77. Ibid., 233.

78. Ibid.

79. "Shut the Door. Have a Seat," S3/E13.

80. "Commissions and Fees," S5/E12.

is "ahead-of-[himself]-in-already-being-in-the-world."[81] The unity of his projective engagements in the world registers affectively. As Don tells his acquaintance, Dennis, in the delivery waiting room as both of their wives endure birth pains: "Our worst fears lie in anticipation."[82]

Don's constantly mooded and anxious engagement and his regularly projected mode of understanding intersect at his experience of having been thrown into the world. In this fallen state, he feels the need to recover himself into a central unity. But this unity is not factical. Rather, he desires recovery from that which is ontical and factual, but this, for Don, is not self-explanatory. The actuality of these ontical engagements is their need for explanation. But he cannot direct himself to their explanation because he is lost in them and the rules of his engagement have been decided upon by *das Man*.[83] He does not choose these engagements because they are not genuine possibilities. The call to recover himself hurls him back to his possibilities, and only then does he realize his *authentic* potentialities.

RECOVERING OURSELVES

Don's despair, as illustrated by Kierkegaard, and his anxiety, as illustrated by Heidegger, signal the need for recovery such that we can now differentiate the paths of self-recovery Kierkegaard and Heidegger claim he might take. According to Kierkegaard, Don's despair results from taking himself as his own standard. In despair, Don either lacks a self (by being lost in his external immediacy) or fashions a concrete self (by his defiant self-reliance). Kierkegaard states that these despairs are *sins before God*. God is that true standard before which Don may individuate himself. In despair, Don sins spiritually and *universally* by not wanting to be himself or by wanting to be himself in a self-controlled and self-reliant way. But he sins *individually* before God. The possibility of salvation is the offense to reason, which collapses his being in despair. Don cannot in weakness or in defiance want to be or not want to be himself in despair and admit the possibility of forgiveness. That which maintains his despair, his free movement toward either pole of his relational self, is that which is offended by the possibility of forgiveness.

The condition for the possibility of recovering ourselves, according to Kierkegaard, is the realization that despair is sin, and that its opposite is not virtue, but faith. To make this claim clearer, Kierkegaard outlines three stages of individualization: the aesthetic, the ethical, and that of faith. In

81. Heidegger, *Being and Time*, 236.

82. "The Fog," S3/E5.

83. Heidegger, *Being and Time*, 312.

the aesthetic stage, Don instantly gratifies his desires in the present. Don can grow into the ethical stage by forgoing immediate desires and being guided by moral norms, but he must do so through a leap of faith, not by a deliberative decision. The highest stage of an individual, the stage of faith, involves a deep commitment to a life of meaning. The transition into this stage involves a suspension of the ethical.[84] In other words, the motives of the faith stage are not subject or cannot be defined under any ethical requirement. Kierkegaard does not denounce the ethical stage, but he does imply that Don can become an individual only in confrontation with God. Worshipping in faith is not being offended by the paradox of Christianity.[85] That God became man, suffered, and died, puts Don to the test: he needs to make up his mind over the possibility of forgiveness. That the possibility of forgiveness offends reason is the essence of the paradox of Christianity.[86] If Don denies the possibility, he shows that he is offended.[87] His sin intensifies because he fails to see sin as the opposite of faith. Living in faith transparently before God is Kierkegaard's prescription of Don's recovery from despair. He begins in anonymity, consequently enters into despair, and can only choose its opposite, faith, in order to recover himself. As we will see below, Don's attempts at virtuous living, his self-renewal projects, each end in backtracking and succumbing to the despair Kierkegaard diagnoses. Does he ever recover himself as an individual through faith?

In Heidegger's account, the need for Don to recover himself away from factical actualities and toward authentic possibilities emerges from the call of his conscience—the one Roger and Ivan had at the opening of this chapter. The voice of conscience attests to potentiality. Heidegger writes that conscience is disclosive because it calls us to our "ownmost potentiality-for-Being-its-Self."[88] Don is both the caller and the recipient of the call, which emerges from himself and calls himself to stand in attention. In bringing forth his self, the "they-self" dissolves.[89] Don's individuation is a recovery

84. Kierkegaard uses the story of Abraham and his son Isaac to illustrate this stage of faith. Through a message sent from God, Abraham was told to sacrifice his only son. At this moment, he was faced with a moral dilemma. How does one choose between adhering to divine authority, and killing one's son? However, when faced with the choice, Abraham was willing to sacrifice his son. At this stage, the individual becomes higher than the universal (ethics). Kierkegaard, *Fear and Trembling and The Sickness Unto Death*, 116.

85. Kierkegaard, *Sickness Unto Death*, 162.

86. Ibid., 163.

87. Ibid., 165.

88. Heidegger, *Being and Time*, 314.

89. Ibid., 317.

from the anonymity and inauthenticity of *das Man* to his individual authentic potentialities.

Don must understand the call insofar as it registers as his Being-guilty.[90] However, this experience of guilt is not the ontical, which only registers in relation to others. In the everyday experience of guilt, Don feels indebted to others in the same way he might feel a certain failure to follow a moral law, or he might feel that he has broken an obligation to others, such as his wife, Betty or Megan.[91] In Season 4, after hooking up with two different women at the CLIOs, he fails to show up at Betty's to retrieve the children. He forgets much of the night before, and we are tempted to see Don near the bottom where guilt might render him capable of individuation. If he is guilty in this failure, it is still an ontical "not" in the everyday experiences of guilt. But Heidegger describes the ontological "not," which is the condition for the possibility of these ontical examples. He describes the existential idea of guilt as "Being-the-basis of a nullity."[92] Although the call of conscience gives rise to the primordial "Being guilty" which is an ontological nullity, it also calls Don positively, albeit highly formally, to his own possibilities. The call is negative in relation to *das Man* and positive in relation to "nullity of Being-a-basis."[93] With this in mind, we realize that Don was fettered by being lost and fallen in *das Man*. But in the call, the they-self dissolves, and in Being-guilty, he discovers our "ownmost potentialities for Being" himself. If and when he heeds the call, he realizes his existential freedom.

Does Don ever experience the call of conscience and the ontological sense of Being-guilty? Our access to his guilt is indirect, but his most resolute moments in the show provide a window into his conscience. The first comes in his attempt at self-renewal and recovery, including his cutting back on smoking and drinking, his journaling, and his taking up of swimming.[94] The image of Draper in a pool certainly evokes images of his baptism and the cleansing of previous sins, but his and our projects of self-recovery are not so easy, and, according to Kierkegaard, cannot be achieved by this sort of boot-strapping, where he still takes himself as his own standard.[95] The second moment comes when, without consulting his partners, Don writes a letter, which he places as a full-page ad in the *New York Times*, to announce

90. Ibid., 326.

91. Ibid., 326–28.

92. Ibid., 329.

93. Ibid., 331.

94. "The Summer Man," S4/E8.

95. For further exploration of Don Draper and the metaphor of baptism, see Jackson Lashier's Chapter 8.

publicly that he is happy to no longer be advertising a product that kills people and that SCDP will no longer accept tobacco accounts.[96]

These examples illustrate the resoluteness Heidegger claims ensues when the being-toward-death allows the call of conscience to signal the need for decisive action. The call of conscience essentially individualizes Don into his ownmost possibilities as it discloses the anticipation of death as the possibility that is non-relational.[97] If Don resolutely anticipates his own death and takes up its possibility, he can realize his potentiality-for-Being; in this way Don's "authentic existence can no longer be outstripped."[98] Don becomes himself "essentially" when he "constitutes [himself] as anticipatory resoluteness."[99] Heidegger thinks Don's self-recovery is found in these recuperative projects and resolute actions in anticipation of the future. The projects are authentic only if Don resolutely anticipates death and consequently realizes his own finitude.[100] Many of Don's realizations of possibility come in moments when death looms in the foreground. Often death and possibility emerge in the same scene, as when Don folds up his father-in-law's bed with his forthcoming child's crib behind him. Later in Season 3, Don reassures Sally, who is frightened both by Gene's death and by her newborn baby brother sharing Gene's name: "He's only a baby," Don tells her. "We don't know who he is yet or who he's going to be. And that is a wonderful thing."[101] Don values pure possibility, and the insight he gives to Sally emerges contemporaneously with the death of his father-in-law.

Both Kierkegaard and Heidegger articulate a similar state from which Don Draper needs to recover himself. He both creates and falls victim to the anonymity of the public or the "they." Despair and anxiety signal the need for his recovery. But each philosopher describes a different road to that recovery. Kierkegaard's road to recovery temporally privileges the present and ontologically privileges the will in one's ability to choose faith. Heidegger's road to recovery seems more complex, as he temporally privileges the future, and ontologically prioritizes possibility. Don's Being is "essentially futural so that [he] is free for [his] death, and can [. . .] take over [his] thrownness and be in the moment of vision."[102] This moment of vision is meant to allow Don

96. "Blowing Smoke," S4/E12.
97. Heidegger, *Being and Time*, 354.
98. Ibid., 355.
99. Ibid., 370.
100. Ibid., 380.
101. "Guy Walks into an Advertising Agency," S3/E6.
102. Heidegger, *Being and Time*, 437.

to reflect upon himself as temporality, whose essence is the futural nature of his own possibilities.

Despite their differences, however, Heidegger's "call of conscience" serves a similar function as Kierkegaard's "sin before God." In each of Don's *ontic* affective states, encountered in his everyday engagements, he has an access to his *ontological* mood. In anxiety, Don is individualized by facing that which delimits its conception of primordial time—his own death. Heidegger states, "Only by the anticipation of death is every accidental and 'provisional' possibility driven out."[103] In despair, Don sins spiritually (and universally) by not wanting to be himself or by wanting to be himself self-reliantly. But he sins individually before God. Anxiety and despair are the calls to self-recovery by virtue of their individualizing function.

Despite the enduring similarities in each philosopher's characterization of anonymity and dread, Heidegger would still deem inauthentic the attempt to recover oneself by recourse to ready-made theology, even a theology as individualized and existential as Kierkegaard's. Authenticity, for Heidegger, comes in the mode and temporal dimension of anxious resoluteness, which projects Draper ecstatically into the future toward his inevitable death. In this authentic mode, Don can recover himself in his ownmost possibilities. But Kierkegaard would tell us that Don's resolute attempts to recover himself ultimately fail because he is stuck at the ethical stage and he still takes himself as his own measure. Kierkegaard would respond to Heidegger, with Draper as his example, by showing that all of Don's guilt is still ontical and that an ontological realization of being guilty can only result from the realization that faith, not mere resolute action or morally guided action, is the opposite of the sin of despair.

So we must return to the question posed above: *does Don ever recover himself as an individual through faith?* Three moments stand out. The first comes when Don breaks down in tears in front of Peggy after learning of the death of Anna Draper. The second comes at the peak of the series, the final episode of Season 6.[104] He is transparent for the first time with his children—showing them his childhood home. This decision, this passionate action, demonstrates what his other attempts all lack, humility, lack of control, confession, and the possibility of forgiveness. His humble transparency is also relational in an authentic way: it does not traffic in idle talk, curiosity, flirtation, or the modality of the public. It makes a leap of faith. The last moment comes in the series finale when Don listens to Leonard share his fears with other members in the group-therapy session. Don appears genu-

103. Ibid., 435.

104. "In Care Of," S6/E13.

inely moved by Leonard's fear of not understanding love and of not being loved. Don embraces Leonard in a hug and breaks down in tears. In these moments, Don sheds the persona of strong and silent and embraces the paradox wrought by the virtues Kierkegaard highlights, passionate action, humility, and possibly forgiveness.

Using *Mad Men* and Don Draper's road to recovery as a lens to examine our own roads to recovery is helpful but limited. The writers must perpetuate Don's opaque and mysterious self in order to extend the show. Maintaining Don's masks means writing into the show a series of cyclical recoveries, ontical notions of guilt, glimpses of his own death, and merely ethical projects involving resolute actions. One cannot argue that Don Draper achieves the authenticity Kierkegaard and Heidegger call for—his seemingly genuine search for himself in the final episode, we learn, becomes fodder for a Coca-Cola ad. The ad manipulates the search for "perfect harmony" into a call to buy a sugary soda. Don might have been looking for the real thing, but Coke is *not* "the real thing."[105]

But how far from our own lives has that limitation taken us? Each of us must discover ourselves piecemeal and question Being through the process of existence itself. Like Roger in the hospital bed, we all wish we were going somewhere. Don's fallenness into the public, his momentary calls of conscience and passionate actions allow us to ask whether our own age is more representative of Kierkegaard's present age than Don Draper's. If so, it calls us to recover ourselves from the inauthenticity and anonymity of our public. Our social media have their own levelling effects, temptations at flirtation, and envious strivings. The public of social media asks for clever witticisms and reflective repose, much as it can forestall the decisive and passionate action Kierkegaard and Heidegger demanded. We watch revolutions occur online and comment on them in reflective repose, but such discourse is only the idle talk of the public. Few of us reject the world in a revolutionary way. Our present age is still with us, and viewing Don grapple with his, helps us do the same. Whether we turn inward to achieve character in religious humility, or we live in the possibility of our own deaths, we all wish we were going somewhere and must recover ourselves to do so.

105. "Person to Person," S7/E14.

BIBLIOGRAPHY

Heidegger, Martin. *Being and Time*. San Francisco: Harper and Row, 1962.

Kierkegaard, Søren. *Fear and Trembling and The Sickness Unto Death*. Princeton: Princeton University Press, 2013.

———. *The Present Age*. New York: Harper and Row, 1962.

———. *The Sickness Unto Death*. London: Penguin, 2004.

Tolstoy, Leo. *The Death of Ivan Ilyich*. Translated by Richard Pevear and Larissa Volokhonsky. New York: Vintage, 2009.

Wyschogrod, Michael. *Kierkegaard and Heidegger: The Ontology of Existence*. London: Routledge and Kegan Paul, 1954.

Chapter 10

DON DRAPER, DOUBLE CONSCIOUSNESS, AND THE INVISIBILITY OF BLACKNESS

—NSENGA K. BURTON

AMC's *Mad Men* is a media darling. It is the cable television equivalent of the little engine that could. The celebrated show ushered the *American Movie Classics* channel and its aging audience into the new millennium, with an initiative that began in 2003 to re-brand the channel once synonymous with classic black and white cinema as a colorful place for quality scripted television serving a younger, yet equally sophisticated demographic.[1] Premiering in 2007 as the channel's first scripted original show, *Mad Men* follows the complicated personal and professional lives of Madison Avenue advertising executives. AMC tapped *The Sopranos* show runner Matthew Weiner to accomplish this network goal of bringing the channel into the new millennium with this new series. Offering compelling narrative and stylistic elements, a virtually unknown lead actor who became a star and strong support from AMC executives, *Mad Men* overcame weak ratings and an uncertain future to become one of the most celebrated television shows in cable history.

Despite *Mad Men's* commercial and industry success, the show faced scrutiny from audiences and critics for portraying a lily-white New York City, in spite of the racial and ethnic diversity of the city. In 2010, *The Root,*

1. Idov, "Zombies at AMC's Doorstep."

the premiere online destination for upwardly mobile black audiences, introduced a "Mad Men Black-People Counter" to keep track of the limited representations of black characters on the show.[2] Some critics argued that its lack of black characters was unrealistic and disturbing, while others believe that shows like *Mad Men* and, most recently, HBO's *Girls* serve as a true reflection of how life often operates.[3] When you are privileged, you create and live in a world where disenfranchised people do not exist, even when they are physically present.[4] The show that is beloved on many fronts was assailed for failing to pay the same type of attention to race as was given to unpacking gender issues.

While many argued that *Mad Men* avoided complex representations of black people and thus blackness, historian, cultural theorist and social justice activist W.E.B. Du Bois's theory of double consciousness is at the very nexus of the show. The fact that people overlook whiteness as a category of race in critically analyzing the show, speaks volumes about the extent to which dominant ideology permeates all facets of society. The inability or unwillingness of critics to see race as something only related to characters of color, reflects the power, normalcy, and invisibility of whiteness as a category of race. Ironically, the willingness to define race in such narrow terms, to equate race with blackness and woeful inability to see whiteness as a racial category, speaks to the white hegemonic structures that the critics critique. Don is not an outsider based on the color of his skin or national origin; he is an outsider because of his economic identity—one marked by a childhood lived in abject poverty. Don Draper's ability to hide in plain sight, based on a double-consciousness around race, economics, and power remains on full display throughout the show. Race is central to the character development of Don in particular and *Mad Men* as a whole.

AMC'S *MAD MEN*

Mad Men's success was slow but sure. The "engine that could" seemed as if it might run out of steam due to low-ratings during Season 1. *Mad Men* averaged less than one million viewers for its first season, but this number grew with each of its next three seasons. The show, which won four *Emmys* for best drama, peaked with its 2012 season, when an average same-day

2. The Buzz, "Root's Mad Men Black People Counter," *TheRoot.com*. Nsenga Burton served as the Buzz editor and writer for *The Root* from 2009 to 2012. She currently serves as editor-at-large for *The Root*.

3. Colby, "Mad Men and Black America."

4. Burton, "Mad Men and Race."

audience of 2.7 million watched each week's Sunday episode. The May 2015 series finale reached a series high in total viewers (4.6 million) between the ages of eighteen and fifty-four.[5]

Mad Men morphed into a bonafide hit because of the seamless flow of exceptional narrative and stylistic elements such as realistic dialogue, complex storylines and exquisite lighting and art direction. Jon Hamm and the writers' ability to create a lead character that could wound and uplift a fellow character with an expression or a turn of phrase was poetic. Audiences were let in on secrets that the characters didn't know, helping to build suspense and maintain viewers over several seasons. The smart wardrobe, cinematic lighting and camera techniques communicated to audiences, television executives and critics that *Mad Men* was here not only to compete, but also to serve as a benchmark for high-quality programming among cable outlets. Coupled with Jon Hamm's 2015 *Emmy* win for Outstanding Lead Actor in a drama series, an honor that eluded him for seven previous nominations, it is safe to say that the academy, *AMC* executives and audiences agreed that *Mad Men* accomplished what it set out to do—serve as a game changer for AMC and cable programming.

MAD MEN AND DOUBLE CONSCIOUSNESS

While many argued that complex representations of black people and thus blackness were missing from *Mad Men*, DuBois's theory of double consciousness is central to the development of the show. Race played an integral role in the show, despite the absence of complex black characters and omnipresence of blacks in stereotypical roles. The absence of complex storylines regarding black characters is instead represented by the character of Don Draper, who is coded as a black man through narrative and stylistic elements. Cultural critic Ta-Nehisi Coates refers to Draper as "The Negro Donald Draper" because of his double life and thus double consciousness.[6] W.E.B. DuBois's theory of double consciousness complicates the assertion that *Mad Men* neglected the subject of race and the ways in which the character of Don Draper reproduces the black experience.

5. Kissel, "Mad Men Finale Ratings."
6. Coates, "Negro Donald Draper."

Double Consciousness

In his 1903 book *The Souls of Black Folk*, W.E.B. DuBois introduces the concept of double consciousness. Dubois states:

> After the Egyptian and Indian, the Greek and Roman, the Teuton and Mongolian, the Negro is a sort of seventh son, born with a veil, and gifted with second-sight in this American world- a world which yields him no true self-consciousness, but only lets him see himself through the revelation of the other world. It is a particular sensation, this double consciousness, this sense of always looking at one's self through the eyes of others, of measuring one's soul by the tape of a world that looks on in amused contempt and pity. One ever feels his twoness—an American, a Negro; two souls, two thoughts, two unreconciled strivings; two warring ideals in one dark body, whose dogged strength alone keeps it from being torn asunder.[7]

DuBois's theory of double consciousness speaks to the condition of people of African descent who are made to inhabit two spaces, that are literally at war with each other, in order to survive in an America that loathes black people, particularly at the turn of the twentieth century. The veil he refers to symbolizes the divide that is constantly engaged when a member of a disenfranchised group has to interact with his oppressor in order to exist and advance personally and professionally. The veil also speaks to the actual skin color of the Negro; it is simultaneously a type of armor that serves as a visual reminder of the separation between blacks and whites and an indicator of one's status as an outcast. DuBois writes:

> The history of the American Negro is his longing to attain self-conscious manhood, to merge his double self into a better and truer self. In this merging he wishes neither of the older selves to be lost.[8]

DuBois understands that the desire to peacefully co-exist is an impossibility for blacks based on the racial reality of the time and the commitment of those in power to maintain the status quo, which is a separate, unequal and unjust world for blacks in a country dominated by whites, who also have created a power structure supported by dominant institutions. Not only do blacks have to view themselves through the lens of whiteness, they must hide whole parts of themselves from what scholar Frantz Fanon would call

7. DuBois, *Souls of Black* Folks, 215.

8. Ibid.

the "colonial gaze" in his 1952 seminal work *Black Skin, White Masks*, which examines the impact of racial oppression on the black psyche.[9] Fanon, who was born in French Martinique and eventually migrated to Europe, understands DuBois's concept of double consciousness. Fanon's concept of the mask reflects DuBois's theory of double consciousness, layering the psychological impact and damage of having to literally become someone else in order to please those who would otherwise not have you.

THE NEGRO DONALD DRAPER

The character of Donald "Don" Draper represents DuBois's theory of double consciousness mainly through narrative elements including characterization (physical and behavioral), dialogue, storylines, locations, and settings. Don Draper literally reflects dominant beauty ideals. He is tall, lean with broad shoulders, dark hair, alabaster skin, and piercing blue eyes and is always dressed in a tailored business suit. He struts through the office, walking with confidence, head up, shoulders down, winning attention from onlookers. The genius creative director navigates the narrow hallways of the advertising agency Sterling-Cooper, weaving his way through the maze of desks, secretaries, typewriters and fluorescent lights to his office, which is spacious and coveted. Don's second home is furnished with leather and oak trappings, a liquor cabinet, modern sofa, and drawer full of crisp white shirts so the creative genius can work uninterrupted if necessary.

Draper speaks in a direct, confident tone through lips that drag on cigarettes and kiss beautiful women with passion and comfort. Draper is the embodiment of Norman Mailer's "White Negro"[10] without the subversive or hipster leanings. In 1957, Mailer wrote an essay for *Dissent Magazine* discussing the white avant-garde's rejection of whiteness as the country had known it (the atomic bomb, corporate culture, materialism, Eisenhower's politics) in lieu of appropriating black America's version of cool in the form of culture, oppositional politics and style. Mailer had taken Anatole Broyard's 1948 definition of a hipster as a black subcultural figure and applied it to white Americans,[11] connecting whiteness with coolness, ironically subverting Broyard's definition articulated in *The Partisan Review*. Mailer

9. Fanon, *Black Skin, White Masks*, 93.

10. In 1957 Norman Mailer wrote an essay entitled, "The White Negro: Superficial Reflections on the Hipster," for *Dissent Magazine*. Mailer discusses the phenomena of whites that oppose conformist culture to such an extent that they have appropriated black culture.

11. Broyard, "Portrait of the Hipster."

appropriated Broyard's definition, applying it to white men in America, creating a white subcultural figure inextricably linked to coolness. It is this problematic treatment of hipsterism by Broyard and Mailer that is on display in the character of Don Draper, who interestingly embodies that duality as well. Like Mailer's "White Negro," Draper is a man who resides on the margins of society until assuming someone else's identity, thus benefitting from the application of Dick Whitman's identity and status. Draper meets the physical and behavioral style of Mailer's "White Negro" through his dress, cadence, and willingness to reject the status quo like sexually harassing secretaries and participating in fratboy antics and misogynistic behaviors of the affluent white men at work like Peter "Pete" Campbell (Vincent Kartheiser). Don has the look and feel of a hipster, but is quite the opposite in some regards.

Don is a military man—a "square"—who defended his country and then worked his way up from a furniture salesman to the head of one of Madison Avenue's most competitive advertising agencies. He defends the honor of women (Peggy Olsen), keeps little boys masquerading as men in check (Pete Campbell), has a complex relationship with his bohemian lover who lives in the Village, and makes it home to his house in the suburbs to kiss his unquestioning wife and perfect son and daughter. He is a Clark Kent like superhero who keeps the world as we know it safe, helping to preserve the definition of success, what it means to be American, and what it looks like according to those who make the rules. Advertising is to Donald Draper what newspapers are to Superman—a place to help perpetuate the fantasy that makes us believe in the world we wished we lived in as opposed to the world where we actually live. While hiding in plain sight, Clark Kent and Don Draper perpetuate this ideology and representation of American life, in which neither of them fit. In a world defined by a certain type of whiteness that equates to success, Don Draper and his life are the physical manifestations of the American Dream, while his psychological profile reflects that of a black man.

Draper's "perfection" belies the psychological and social conflict that exists within him because, like many (white) Americans, Donald Draper is not the man the world sees or who he pretends to be. Don Draper has not only assumed someone else's identity, he has also assumed someone else's race. He is a character that blends in seamlessly until some peculiarity reveals his true identity in a world where he wants to live by his rules. Like many immigrant groups, it could be language, religion, accent, eye color, hair color or texture—some quality or characteristic that lets the ruling class know that even if you look the part, you are not of the part.

Omi and Winant's Theory of Racial Formation

Psychologists Omi and Winant's theory of racial formation defines race as a social construct that creates a hierarchy of categories of specific types of human beings based on their physical features. They write:

> Race is indeed a pre-eminently socio-historical concept. Racial categories and the meaning of race are given concrete expression by the specific social relations and historical context in which they are embedded. Racial meanings have varied tremendously over time and between different societies.[12] The social construction of race is made even more complicated by the idea that whites are categorized as "pure," while others are considered non-white.[13] As Roediger discusses in his book *Working Towards Whiteness: How America's Immigrants Became White: The Strange Journey From Ellis Island to the Suburbs,* those who are often considered white, were not regarded as such when they first immigrated to the United States in the 20th Century.[14] He discusses how many different ethnic groups including Italians, Irish, Polish and religious minorities like Jews, were discriminated against in jobs, housing and access to resources because they were considered "dark whites" or un-American because of language, religion, cultural practices or a combination of factors.

This discrimination is referenced in *Mad Men* through the characters of Sal, who is Italian and gay, Peggy, who is Irish Catholic, and Rachel Menken, who is Jewish. Like Draper, all of these characters live dual lives in order to participate in the dominant world—the desired world that determines racial categories based on social and power structures in society and where you fit in if at all.[15] While some critics assailed *Mad Men* for making the subject of race invisible, the issue was in fact omnipresent. Don fits squarely into Roediger's discussion of how undesirable whites, known as "dark whites," had to navigate their way to dominant definitions of whiteness and Americanness. Don's illustrious climb from abject poverty and literally living on the margins of society, and all of the psychological and experiential conflict that comes with it, is evident in Roediger's work, DuBois's theory of double consciousness and Frantz Fanon's discussion of the mask.

12. Omi and Winant, *Racial Formation*, 4.
13. Ibid.
14. Roediger, *Working Towards Whiteness,* 5.
15. Ibid., 151.

Don's pride in his success and profession is a source of pain and shame based on how he obtained the prestigious position and trappings that go along with being a classically handsome, white man at the top of an advertising industry predicated on conceptualizing and perpetuating the American Dream that only exists for a small group of people, which excludes Richard "Dick" Whitman. Don Draper is not who he appears to be to his colleagues or would be oppressors which is why he must wear a symbolic mask or veil to conceal his identity. Roediger discusses how "dark whites"[16] were kept out of the mainstream American society because of a number of factors including economics. Dick is the son of a prostitute, who grew up on a farm. He escaped abject poverty by enlisting in the service and assuming the identity of a financially well-off officer. Don's story and experience is not the story of most of the men at the agency.

Roger Sterling (John Slattery) plays the role of a WASP with generational wealth whose physical characteristics make him "whiter" than Don Draper. Roger has a slight frame, white hair, blue eyes and his name is on the proverbial door of the agency. He is a partner—not an employee—someone who can determine the fate of everyone in the agency. Bertram Cooper (Robert Morse) represents the old guard even with his eccentricities and obsession with Japanese culture, which he ostensibly gained while serving in World War II. His appropriation of Asian culture does not undermine his position of authority, as he is always the decision-maker even when removing his shoes before entering a room. The bow-tie clad Cooper is the man in charge whose financial, gender and racial privilege allow him to move about the world as he pleases and to reassume his powerful position as the man whose name is also on the proverbial door of Sterling-Cooper. Don Draper is not the character of Pete Campbell (Vincent Kartheiser), who also wears suits and aspires to greatness at the agency. A socially awkward, misanthrope that lacks empathy, Campbell is tolerated because of his family name and connections, especially after he marries a socialite, the daughter of an executive at Boeing. As much power as Draper wields at Sterling-Cooper, he remains an employee without anything other than his creative genius to keep him there. Donald Draper looks the part but is not of the part.

In addition to physical and behavioral characteristics, the use of dialogue in *Mad Men* buoys Draper's status as an outsider. Unlike the young, male advertising executives who hang on every word that Don Draper utters, the black and Jewish characters seem to notice Draper's outsider status. For example in "Smoke Gets in Your Eyes," S1/E1, an African-American elevator operator says, "going up" as Draper steps onto the elevator. In "Six

16. Ibid., 4.

Month Leave" (S2/E9), Don, his assistant Peggy (Elisabeth Moss), and an older African-American elevator operator in uniform take the elevator up to the agency. Don and Peggy are discussing the shocking death of iconic actress Marilyn Monroe, when the elevator operator says, "Some people just hide in plain sight." The black elevator operator who has to employ double consciousness in order to survive working in this white world that would otherwise not have him in it, recognizes himself in the character of Don Draper. As Coates notes, the black elevator operator functions as a Greek chorus, "unseen, but offering short poetic takes on the themes at work."[17] The largely invisible black character that resides on the margins of the narrative and society becomes hypervisible as he offers insight into the character of Don Draper.

Jewish American Princess Rachel Menken

The character of Rachel Menken (Maggie Siff), a Jewish woman who runs her father's department store, is introduced to viewers in the first episode of the series, "Smoke Gets in Your Eyes." Menken wants to hire Sterling-Cooper to re-brand her aging department store into a contemporary, high-end luxury department store that rivals *Tiffany & Co.* with whom they share a wall. In preparation for the meeting with Menken, Roger Sterling asks the advertising team if they had hired any Jews. Don Draper responds, "Not on my watch." Roger and Don debate on where to "find a Jew" for the meeting, to which Draper says, "Do you want me to run down to the deli shop and grab one?" They finally locate a Jewish man in the mailroom to bring in as a "member of the team" for the meeting.

During the meeting, Draper and Menken get into a heated argument over using a coupon to draw customers to the department store. Roger states, "Coupons work. Housewives love coupons," which angers Rachel who is not interested in housewives, but women who want to shop in the store because they think it's expensive. Pete suggests that she work with a Jewish advertising agency, which would "serve her better." Menken retorts, "I was at a Jewish firm and their research favors coupons too." Menken lets them know that she doesn't want to stay in her position as "the Jewish Department store," which undermines their ability to make more money. Angered by her dismissal of his ideas and oppositional tone, Draper storms out of the meeting. The dialogue in these two scenes demonstrates the marginalized status of Jews in society. Despite being a beautiful, white woman, Menken's identity as a Jew codes her as "other" or a "dark white." Her father's financial

17. Coates, "Negro Don Draper," 5.

success gives her a seat at the table, but Menken is not supposed to be an equal participant in the discussion. Menken's "ethnoracial identity," reflects the desire of Jewish Americans to achieve the status of White Americans through upward mobility, especially post-WWII. Whiteness meant safety, protection and power in a world that was unsafe and rendered many Jews powerless by allowing their systematic and state-sanctioned murder and discrimination throughout the world.

Menken's desire to brand her department store in the image of "Tiffany's and Chanel," reflects her family's desire to move beyond what Karen Brodkin would call her "ethnoracial identity" to one of a national identity.[18] In spite of the large amount of money Menken is prepared to spend in order to change the identity of the department store, and ostensibly her family, she is rebuffed because, at the end of the day, she is still Jewish. Sharing a wall with *Tiffany & Co.* is not the same as being *Tiffany & Co.* and having lots of money is not the same as being a white, Anglo-Saxon, Protestant having lots of money, particularly in 1960. The wall between *Menkens* and *Tiffany & Co.* symbolizes the veil that DuBois discusses which separates the oppressed from the oppressor even in close proximity. Menken is trying to build a business under the colonial gaze of the ruling class. The wall is representative of other walls like the Berlin Wall, the Great Wall of China, and the border walls between Mexico, Arizona and California, that are meant to keep those who may desire a "better" way of life confined to the life the ruling class believes they deserve.

Unlike blacks, who are marked by skin color, Menken, who is marked by religion, can still get a seat at the table based on her white identity. It is also her white identity that affords her the privilege to stand up for herself and her company that escapes the black characters in the same spaces. The Jewish employee from the mailroom can "pass" as a member of the team because of his whiteness and maleness. Unlike characters that are disenfranchised in some way (religion, gender, sexual identity), blacks do not have that luxury because black skin cannot be hidden or changed.

The four walls of the elevator operator are there to not only keep certain groups *out*, but also certain groups *in* their confined spaces. The elevators are designed to keep the black male workers in their place of servitude as they help whites transgress social, gender, and religious boundaries. The black elevator operators literally take white employers on their rise to the top of society, while always returning to their predetermined position of ground floor. Despite being able to witness how the other half lives and the trappings available to them if they "make the right moves" as articulated

18. Brodkin, *How Jews Became White Folks*, 4.

by Joan Harris (Christina Hendricks) and thus reinforcing the status quo, ultimately blacks know that this white lifestyle is not available to them.

Menken's white skin and wealth also allow her to speak candidly to Draper, with whom she meets so he can apologize for his "unacceptable" actions in the meeting. Draper stormed out of the meeting, but all members of the team demoralized and demeaned Menken. Yet and still, Draper, the employee without the trust fund, is sent to "play nice" with Menken. Like Draper and the black elevator operator, Menken also hides in plain sight, which is why she sees herself in Draper. Like him, she has alabaster skin, dark hair, piercing eyes, and is dressed impeccably in tailored dresses, even donning the iconic pillbox hat, made famous by another raven haired beauty, Jacqueline Bouvier Kennedy Onassis, the former first lady of fashion and the United States. Menken's pillbox hat symbolizes her aspirations to be something more than a Jewish woman who inherits a Jewish department store and who is always identified as Jewish first before anything else.

During the meeting where Don asks for Menken's forgiveness, he also asks the proverbial question posed to every single woman of a certain age, "Why aren't you married?" Rachel tells him that she wants to be in love, to which he responds:

> What you call love was invented by guys like me, to sell nylons. You're born alone and you die alone and this world just drops a bunch of rules on top of you to make you forget those facts. But I never forget. I'm living like there's no tomorrow, because there isn't one.

Rachel responds:

> I don't know what it is you really believe in, but I do know what it feels like to be out of place, to be disconnected, to see the whole world laid out in front of you the way other people live it. There is something about you that tells me you know it too.

Rachel sees Don Draper in a way that the black elevator operator sees him—as a man who is at war with himself. Even though Don is trying to blend the two spaces he inhabits in society, he cannot have them both without losing a part of himself. As much as Don tries to pass for a wealthy, WASP who has the perfect American life, he is hounded every second of the day by knowing that if he is exposed, he will be imprisoned, literally and figuratively, to a world in which he is despised. He will be regarded as a fraud, returned to his lower socio-economic position in society and have to hear a chorus of the WASP equivalent of "I knew it" in the colonial gaze of those to which he aspired to be, but would not accept him as Dick

Whitman. He will also have to suffer the knowing looks from those who would claim to know all along that he didn't belong, despite his ability to deceive other members of the desired group.

THE INVISIBLE BLACK WOMAN: DAWN'S TRIPLE CONSCIOUSNESS

Don is clearly coded as black, which is buoyed by a parallel black character named Dawn (Teyonah Parris) who serves as a referential point for Don's struggle, despite the perceived lack of complex black characters on the show. The character of Dawn is the exception to the rule because her character is anything but stereotypical and flat. Introduced in "Tea Leaves," S5/E3, Dawn is a black woman who becomes Sterling-Cooper's first black employee after a civil rights joke gone awry. Perhaps critics are unable to see the centrality of Dawn's character in the topic of race in *Mad Men* because the character was under utilized in Season 6 and only appears once in Season 7, but her characterization and relationship to Don is quite complicated unlike the other stereotypical representations of black characters on the show.

Dawn comes to Sterling-Cooper as a consummate outsider. She is a black woman, from Harlem, attempting to have a career as a secretary at a white company where only black men are allowed to run the elevators and clean the premises as long as they are silent. Prior to Dawn's employment, black women are never present in any capacity at the advertising firm. The other black women are maids, caretakers or the exotic "girlfriends" of white advertising executives experimenting with hippie culture. Motivated by the civil rights movement, Dawn wants a better shot at life. She wants to move about in parts of Manhattan that are the strict domain of whites and only open to people of color that are in service to whites. Everyone in the office is told to treat Dawn respectfully, who attempts to blend in but unlike Don, her gender and race complicate her ability to be a part of the team. It is what Womanist scholar Patricia Hill Collins calls intersectionality.[19]

Hill Collins believes that black women are uniquely situated because of their experience with two exceptionally powerful and prevalent systems of oppression come together: race and gender. The ability to understand this position and the connection between race and gender is something she calls "intersectionality" which opens up the possibility of seeing and understanding many more spaces of overlapping interests. Like Don, Dawn's potential to use all of her life experiences, particularly those not available to whites, make her a great candidate for Sterling-Cooper. Unlike Don, Dawn who is

19. Collins, *Black Feminist Thought*, 228.

eventually joined by Phyllis (Yanni King) in "The Flood," S6/E5 and Shirley (Sola Bamis) in "Lost Horizon," S7/E12, is never able to become a part of the team, even more so than Peggy, who transgresses gender boundaries to become the head of the copywriting department. Dawn's promotion is much slower than Peggy and Joan Holloway's promotions (Christina Hendricks), redheads who slept with powerful men in the firm on their climb to positions as department head (Peggy) and partner (Joan). The intersectionality that gives Dawn the insight needed to succeed as a woman and black person is also what undermines her ability to connect with her colleagues and move up in the world.

Dawn cannot sleep with her supervisors and remain credible in the space because of the troubled history between black women and their white male employers, particularly during segregation. One only has to look at segregationist politician Strom Thurmond, who bore a child named Essie Mae Washington-Williams with his fifteen-year-old African-American maid in 1926. Washington-Williams publicly spoke about Thurmond, who requested that his identity as her father remain secret from the public until after his death. Washington-Williams had also maintained the secret to ensure a peaceful life for the South Carolina senator and herself. Despite the power that Strom Thurmond had at age twenty-six when he impregnated his maid, dominant myths circulated about the promiscuousness of black women.[20]

Black women were characterized as seductresses and "loose" in order to explain away the biracial children born to "the help" of those who worked for white families. A holdover from slavery, instead of seeing the role that power plays in the forced or unforced coupling of white master and black slave woman, the black woman was held in contempt, often being physically punished by white men and women and losing her job, when becoming pregnant, often from rape. Dawn did not have the luxury of sleeping with the partners for advancement because it had already been determined who and what she was in relation to white men. She could not afford to jeopardize her career trajectory by satisfying the troublesome stereotype.

An example of this racial and power dynamic from *Mad Men* is the character of Lane Pryce (Jared Harris), a partner who commits suicide after he is discovered to have been stealing money from the company. Pryce had also been having an affair with a black Playboy Bunny Toni Charles (Naturi Naughton). Once his father found out about the affair, Pryce was beaten and forced to give up the relationship, reflecting the real and potential violence interracial couples faced during this time period. Black women are

20. Curtis, "Strom Thurmond's Black Daughter."

rendered invisible and deemed undesirable personally and professionally in this particular space, so Dawn could not assume that risk, nor would she be supported if discovered to have been breaking society's "rules."

Even though Don is disconnected from his team, the team actually wants to connect with him personally and professionally because of what he represents—whiteness and maleness, both of which are desirable characteristics. Dawn represents blackness and woman-ness, both of which are undesirable characteristics at Sterling-Cooper. Although hesitant, Dawn is willing to try and connect with her peers. The invisibility of black women in the office makes her hypervisible, whereas Don blends in with his colleagues.

In "Mystery Date," S5/E4, Dawn is unable to go home due to riots and protests on a hot, summer night. The subway train she normally takes home will put her in the line of fire, so she opts to sleep at the office instead. Realizing that Dawn plans to sleep at the office, Peggy invites Dawn to stay with her at her apartment. The women enjoy each other's company, drink beer and have conversation, during which Dawn explains to Peggy why she cannot go home. Dawn occupies a different social space now that she works at Sterling-Cooper and while many black people are proud of her, some meet her with suspicion. The story reminds Peggy that she and Dawn come from two vastly different worlds. As she goes to bed, Peggy who has left money on her coffee table, scans the money and then eyes Dawn suspiciously before going into the bedroom. Dawn's glimmer of hope for connecting with someone at work is shattered as evidenced by her shocked and saddened face when witnessing Peggy's behavior.

Peggy, who offers herself up to Don Draper in "Smoke Gets in Your Eyes," S1/E1, does not quite know how to connect with Dawn, who wants nothing from her other than opportunity. Don rejected Peggy's advances and Peggy rejected Dawn because she is disinterested in Dawn's world. The women connect on the level of gender and class (Peggy is an Irish Catholic from Brooklyn who started out in the secretaries pool), but race trumps the other categories causing Dawn to have a truly disconnected experience. Don Draper chooses how and when to connect with people. Dawn does not have that privilege or opportunity. Dawn learns a hard lesson: no matter how well dressed, articulate, educated, or well meaning she is as black woman, those who define the social categories and relationships can also define you. Whereas, all of the men want to *be* Don Draper and all of the women want to be *with* Draper, the same cannot be said for Dawn who remains isolated from her colleagues while being isolated from her community. Thus, there is no benefit to being a black woman at Sterling-Cooper at this time, because the only experience is rejection and suspicion. Men do not necessarily desire Dawn as a romantic partner and few would want to be the first black woman integrating an iconic advertising firm at the height of the civil rights movement. Don and Dawn's connections are strong yet vulnerable based on the power and meaning of whiteness in society and the show.

The Don/Dawn Connection

In order to survive her precarious social and professional position, Dawn must employ a type of triple consciousness to maintain her existence in this white world that has pre-determined her worth. She must work across race, class, and gender lines in order to maintain her limited access to this world of power and privilege. The shattered look on her face when Peggy looks at her in an accusatory way is the dissolution of the mask she employs in order to interact with Peggy. Dawn removes the mask and shows the pain of being judged, rejected, and dismissed because of factors she cannot control—race and gender. The psychological effect that Fanon discusses translates in the devastated look on Dawn's face, when Peggy reminds her that she is white and Dawn is black. Don rarely shows a face of devastation, even when he learns that Rachel Menken has passed away, Betty is dying from cancer and his young wife has abandoned him. He only shows emotion when hooking up with the prostitute in "The Runaways" (S7/E5), who reminds him of his deceased stepmother.

Like Don, Dawn looks the part of the person she wants to be; she is impeccably dressed in business attire, conducts herself as a professional and adheres to the written and unwritten rules of the office. Unlike Don, Dawn does not have the luxury of blowing off work, being unprepared for meetings, coming undone in front of superiors or finding romantic partners at work. Don only has that luxury because he can invoke his whiteness and maleness when necessary and hide his impoverished beginnings. Dawn's blackness cannot be hidden, her gender cannot be changed, and her socioeconomic status is fixed until enough members of society come together to determine that she should be allowed to earn a living in the same place as others, even if it means being paid less and social isolation. Don sleeps in his office because he doesn't *want* to go to his house in the suburbs for a variety of reasons. Dawn sleeps at the office because she *cannot* go home because of the social unrest in her neighborhood, which makes it unsafe for her to travel alone at night.

Dawn's character is the female manifestation of Don's character. Despite limited screen time, narrative and character development, Dawn's presence serves as the referential point for Don's double consciousness. The two seemingly unrelated characters have a shared history, present, and future. Don's ability to wear a mask at all times, and self-medicate his pain through liquor, is rooted in his identity as a white man. He cannot show pain or weakness, as is the case with the men of the ad agency. When the white men show pain, they are discarded like Ken Cosgrove (Aaron Staton) who loses an eye during a client outing gone awry. Lane Pryce (Jared Harris) hangs himself after he is discovered as a thief and fired by Don as does Don's biological brother, who is rejected by Don. People are hanging themselves when Don, who has assumed someone else's identity, meets them with disdain but could not care less if they have any type of relationship with Dawn.

It is Dawn to whom Don turns when he is cast out of the firm after he suffers what appears to be a mental breakdown, later revealing his impoverished background to a client. Dawn partners with Don to keep him in the loop when he is first asked to leave, and continues to help him despite having no one offer her the same level of support or protection at the advertising agency. Dawn's triple consciousness allows her to see and empathize with Don, whose mask is dissolving. It is also her triple consciousness that gives her the ability to determine that Don is her greatest ally in a company that would otherwise not have her, had it not been for his ability to "see."

Despite appearances, Dawn does not necessarily satisfy the tired literary and media trope of the magical negro who serves as the conscious of the conflicted white character; in fact, Don is as much a part of her consciousness as she is a part of his. They both realize that they need each other in the struggle against white supremacy that is undergirded by class. Don is the black man that Dawn would have if the world were equal and just. Dawn is the black woman Don should have if society was not defined by constructed categories that organize knowledge, access and experiences. In a world that is clear that poor people and blacks do not belong, Don and Dawn have each other and it is their ability to "see" each other as DuBois and Coates state, that inextricably link their characters, storylines and outcomes. They wear the masks at Sterling Cooper and in society because of arbitrary factors they cannot control—skin color and economic status.

The power afforded Don because of his race and gender is extinguished for those very same reasons when it comes to Dawn. Don hires Dawn, recognizing that she deserves a chance at a better life despite her skin color in the same way his character deserves a better life despite his impoverished upbringing. Had Don not assumed someone else's identity, which requires an ability to code-switch that is representative of the experiences of marginalized groups, Don might not have gotten his advertising executive position, thus propelling him to an economic status that translated into real power—the ability to change someone else's life.

Draper's connections to whites are always risky because he cannot let them know who he really is or aspires to be, possibly because he does not know anymore. The same can be said of Dawn, who knows who she is, but is not necessarily sure where she wants to be in terms of her personal life or profession. Dick Whitman, an impoverished white man who never finished high school, lives within Don, who appears to be someone else but cannot shake his true identity, no matter how many masks he tries to wear. Dick had no real power over his life because of his economic status, which is why in spite of his success, Don always feels powerless. It is Whitman's economic status that most clearly likens him to blacks in this country, and codes him as black in relation to the whites in the company, many of whom have inherited wealth (Roger and Pete). Dick's white skin, good looks and ability to blend in with dominant society's concept of what they want white America to be gives him the perception of power that allows him to become Don Draper, something Dawn can never have. It is the privilege of whiteness, which allows him to literally reimagine and reinvent his life, which binds him to Menken, yet separates him from the black elevator operator and Dawn. Draper's double consciousness is reflected not only in his characterization, but also those around him.

THE ELEMENTS OF MAD MEN'S TELEVISUAL STYLE

In addition to character development and storylines, the settings and locations of the series, particularly in Season 1, also highlight Draper's double consciousness. The lily-white world that is created by male advertising executives like Draper exists for some members of society but not those like Whitman. In flashbacks, we see Whitman as a boy, who loses his mother who dies in labor, and then his abusive father to a horse injury. He is then taken in by relatives and raised in a brothel, a farm boy with soiled hair and tattered clothes. He walks everywhere by foot, spends time alone or in the company of his younger, redheaded brother, and suffers abuse, at the hands of his step-mother, who allows him to be sexually abused by a prostitute.

Whitman's childhood is marked by locations that are wide, expansive spaces, with small, dilapidated wooden flats, in untouched rural areas that could be anywhere in America. It is clear that this impoverished group of people have been discarded by society. The chiaroscuro lighting of the childhood scenes highlight the contrast in the darkness and light that coexist between the land, the distressed home, and Whitman's childhood experiences. The ground touches the sky, the grass blows in the wind, and the sprawling space symbolizes freedom and space. Outside, Whitman is free to be himself while inside he's trapped in squalor, dankness, and tight spaces, which are mired in violence and death. Living on the margins of society, Whitman and his family are invisible, when inside the shack or brothel, but outside Whitman has the possibility of escape, which he finally makes as a young man. The settings represent the duality of his existence even as a powerless young boy, who knew there had to be more to the world beyond the confines of this house and the sky that seemed to go on forever. Wide, sprawling spaces translated into freedom while tight spaces meant restriction and conformity.

The countryside, which is wild and unkempt, is in direct opposition to Draper's home in the suburbs. The lawns are perfectly manicured, white fences are meticulously painted, and the homes are large with multiple rooms, lots of space and colorful linens, decorations, and finishes. The Draper home is painted in powder blue with white details that offset the blue sky with a white picket fence. Their family home is pristine and marked by abundance from food to furniture to technology. There is even a maid, a black woman named Carla, who cleans the house and takes care of the children, despite the fact that Betty, a former model, no longer works. Like Don, Betty has achieved what appears to be the American Dream with a handsome husband who provides a generous lifestyle, beautiful children, and a lovely home that looks like it could be on a magazine cover. Like Whitman's childhood, the lovely home hides the fact that Betty has a duplicitous husband who rarely comes home, two precocious children with whom she is unable to bond and a debilitating case of anxiety. Betty, a beautiful blond from a wealthy family is the archetypal trophy wife mounted on the mantle of the picturesque home in the suburbs even though the setting contradicts the reality.

Although Don's home in the suburbs represents some level of freedom and openness for him as a man, his job at Sterling-Cooper signifies confinement. The long, narrow hallways, low ceilings with fluorescent lights that highlight every nook and cranny in the building, the tight elevator located in a tall skyscraper that obscures the sky denotes confinement. Space is at a premium in the building, Manhattan and on Madison Avenue, which has scores of people moving about in multiple directions at all hours of the day. Like the brothel, strangers come and go, move in and out of small spaces, in an effort to get what they need—money, sex, power, or all three. A tightly wound Don boozes his way through his charmed and assumed life, sleeping on the modern sofa in his office and remaining disconnected from those around him.

The settings in *Mad Men* physically represent the dual existence that Don Draper has lived and continues to live. They connect his past to his present and future and replicate the same meanings in his life. The confines of his childhood homes are reflected in the confines of his adult workplace, which is fashioned like a second home, as discussed earlier in the chapter. Although Sterling-Cooper is a more expensive version of his childhood home, the experience is largely the same. Because of his assumed identity, Draper experiences a certain level of powerlessness. The fluorescent lights literally expose everything happening in the office, including his double-life. The male executives harass women in the office; they sleep with them and the women are compensated with career advancement or social mobility. The ad agency culture is presented as very similar to the culture of the brothel in which Draper was raised.

The suburbs reflect the freedom he experienced in the expansive spaces of the countryside. While the homes are carbon copies of each other, he has the space to move about freely in his home and his suburban town. He is in control of his life there because he is isolated and disconnected from those who could expose his real identity. Don is relaxed because he is the man of the house and Betty and the children are now confined to the domestic space, which allows him to move about freely. In "The Gypsy and the Hobo," S3/E11, Betty learns of Don's real identity and his freedom is finally compromised. After Betty's discovery of his lies, Don seems to be confined by his suburban home and a wife, who no longer respects him. Don's social and professional status is in jeopardy, returning him to a place of powerlessness and vulnerability as his fate rests in the hands of his betrayed, economically empowered wife. Betty gains power over Don by re-inscribing dominant social categories onto the relationship. She comes from generational wealth, which makes her social position greater than Don's social position, thus giving her the ability to either destroy or help him to maintain his façade.

DON DRAPER AND THE MYTH OF WHITENESS

Don represents what Roediger refers to as the myth of whiteness, the false belief that white Americans are "pure" and one-dimensional.[21] It is this myth of whiteness that defines what it means to be white and who gets to be white in America. Due to his socio-economic status, Dick Whitman is not white in the way that Roger Sterling or Pete Campbell are white. Don Draper is only white because he looks the part, creates the story and setting through his financial gains and is pretending to be of a different socio-economic background. Don's double-consciousness is his ability to see the world in a way that others he works and lives with cannot, based on their

21. Roediger, *Working Toward Whiteness*, 33.

limited perceptions due to limited definitions of whiteness. His brilliance in advertising is not because he is a creative genius; it is because he is able to see the world through different lenses based on his true identity. He is able to pierce the veil because he knows what is on the other side of it and suffers because of that knowledge. Ta-nehisi Coates vividly summarizes Don Draper's double consciousness. He states:

> Don believes his progress is tied to no one ever knowing who he truly is, to no one discovering his true history—his secret identity, if you will. Don Draper is, in the parlance of old black folks, passing. His origins are not proper and gentile—he is the child of a prostitute, who has reinvented himself for the Manhattan jet set. He is Gatsby and Anatole Broyard, no? And yet, the irony that animates Mad Men is the fact that, without that past, Draper would likely be the sort of pampered hack he despises. He'd be Pete Campbell. His double consciousness makes him, indeed, doubly conscious, doubly aware. Don Draper sees more.[22]

Many critics believe that *Mad Men* failed to adequately address race. I suppose that is correct if you only consider non-whites as races. *Mad Men* more than adequately addresses the complexity of whiteness and the audacity of those like Donald Draper who try to transgress its constructs and confines. Don is coded as a black man mainly through dialogue, settings, locations, and characterization. Weiner and company certainly could have done a better job at writing more complex characters for black actors. However, their failure at including black characters in a meaningful way, particularly in the failure to fully develop the character of Dawn (or any black characters for that matter), reflects the power structure in Hollywood and society, where racial minority characters appear in marginal roles that serve white hegemony. Such is also the case with white, female characters.[23]

It is the double-consciousness on full-display through the character of Don Draper that is the greatest indictment of the Hollywood elite because they clearly know better, which means they can do better. The character of Don Draper symbolizes DuBois's theory of double consciousness, and his struggle with whiteness creates meaning typically conveyed through characters of color. Draper is this millenium's "White Negro," whose style, physical and psychological make-up uses blackness as a referential point, further buoying DuBois and Fanon's philosophy on surviving race and racism in a world that needs the existence of black people and black culture

22. Coates, "Negro Donald Draper."
23. Burton, "Mad Men and Race," 2.

to validate whiteness and white culture. Considering the work of DuBois, Fanon, Mailer, and Hill Collins complicates the idea that *Mad Men* failed to adequately race, when in fact, the show is an exercise in the historical and social construction of race as embodied by "The Negro Donald Draper." Race is profoundly present in the show *Mad Men*, and to suggest otherwise is short-sighted at best, and lets whiteness and privilege off the hook at worst. The inability to see how race and whiteness work is as much of an indictment of the limitations of audiences and critics as it is of the creators, writers, and producers of *Mad Men*.

BIBLIOGRAPHY

Brodkin, Karen. *How Jews Became White Folks*. Rutgers: Rutgers University Press, 1998.

Broyard, Anatole. "A Portrait of the Hipster." *Karakorak*. November 5, 2010. Online: http://karakorak.blogspot.com/2010/11/portrait-of-hipster-by-anatole-broyard.html.

Burton, Nsenga. "Mad Men and Race: Lets Talk Gender." *The Root*, June 1, 2012. Online: http://www.theroot.com/articles/culture/2012/06/mad_men_gender_trumping_race_issues.html.

The Buzz. "The Root's Mad Men Black People Counter." *TheRoot.com*, July 23, 2010. Online: http://www.theroot.com/articles/culture/2010/07/the_roots_mad_men_blackpeople_counter.1.html.

Coates, Ta-Nehisi. "The Negro Donald Draper." *TheAtlantic.com*, October 27, 2008. Online: http://www.theatlantic.com/entertainment/archive/2008/10/the-negro-donald-draper/6129/.

Colby, Tanner. "Mad Men and Black America: The Critics Are Wrong," *Slate.com*, March 14, 2012. Online: http://www.slate.com/articles/arts/culturebox/features/2012/mad_men_and_race_the_series_handling_of_race_has_been_painfully_accurate_/mad_men_and_race_the_series_handling_of_race_has_been_painfully_accurate_.html.

Collins, Patricia Hill. *Black Feminist Thought: Knowledge, Consciousness, and the Politics of Empowerment, 2nd edition*. Routledge: New York, 2000.

Curtis, Mary C. "Strom Thurmond's Black Daughter: A Symbol of America's Complicated Racial History." *The Washington Post*, February 5, 2013. Online: https://www.washingtonpost.com/blogs/she-the-people/wp/2013/02/05/strom-thurmonds-black-daughter-a-flesh-and-blood-symbol-of-americas-complicated-racial-history/.

DuBois, W.E.B. *The Souls of Black Folks in Three Negro Classics*. New York: Avon, 1965.

Fanon, Franz. *Black Skin, White Masks*. New York: Grove Weidenfeld, 1952.

Idov, Michael. "The Zombies at AMC's Doorstep: Can the Mad Men Network Survive Its Own Success," *NewYorkMag.com*, May 15, 2011. Online: http://nymag.com/arts/tv/upfronts/2011/amc-2011-5/.

Kissel, Rick. "Mad Men Finale Ratings: DVR Playback Boosts Series Finale to Record High 4.6 Million," *Variety.com*, May 19, 2015. Online: http://variety.com/2015/tv/ratings/mad-men-finale-ratings-amc-drama-draws-3-3-million-sets-demo-high-1201499468/.

Omi, Michael, and Howard Winant, eds. *Racial Formation in the United States: From the 1960s to the 1990s*, 2nd ed. Routledge: New York, 1994.

Roediger, David R. *Working Towards Whiteness: How America's Immigrants Became White: The Strange Journey From Ellis Island to the Suburbs*. Cambridge: Basic, 2005.

Chapter 11

THE EROTIC REDUCTION OF DON DRAPER

Iconicity, Idolatry, and Madness

—CAROLE L. BAKER

> What you call "love" was invented by guys like me—to sell nylons.[1]
>
> I am the least difficult of men. All I want is boundless love.[2]
>
> The only thing keeping you from being happy is the belief that you are alone.[3]

After seven seasons, the opening sequence for *Mad Men* has become so familiar that its imagery has little to no lasting impact on the regular viewer. Yet, it may provide the best conceptual frame with which to consider the question driving the series: Who is Donald Draper?[4] The silhouette of

1. "Smoke Gets in Your Eyes," S1/E1.
2. Frank O'Hara, *Meditations in an Emergency*.
3. "The Mountain King," S2/E12.
4. "Public Relations," S4/E1 opens with a reporter asking Don, "Who is Don Draper?" Don eventually responds, "Well, as I said before, I'm from the Midwest. We were taught it's not polite to talk about yourself."

a man stands facing a disintegrating office. Cut to man slowly descending among skyscrapers displaying an array of images—purportedly advertising images: beautiful women, idyllic families, the accouterments of those who would be well-off. The music alerts us to a tension between the descending man and the images beckoning the consumer's eye to the higher echelons of a wealthy society. Relatively up-tempo, the syncopation gradually becomes more dense. The man falls without struggle, limbs open and relaxed. There is no grasping of air—or anything. Suit intact, he cascades downward. We do not witness the landing. Instead, the sequence breaks and the rhythm changes: the pattern becomes simplified, steady, more predictable when the bass enters lending a grounding to the soundscape. Visually, the sequence ends with a view of the man from behind, sitting relaxed on a couch, smoking a cigarette. Who is this man that has beaten the odds? Who has defied the logical inevitability of the dramatic descent? The musical outro ends with a horn? siren? alarm?—whatever the source or allusion, the dissonance alerts the viewer—or perhaps the man himself: images are not always what they seem.

Mad Men centers on the life of a man, Donald Draper, who is as much a mystery to himself as to others: his wives, his colleagues, his clients, and his lovers. This mystery is constituted by both a lack of information about his past and an elusive ambiguity concerning what drives this man forward. Already in the first episode, "Smoke Gets in Your Eyes," both aspects of the mystery are on full display. We learn something of his past when he accidentally pulls a Purple Heart medal out of his desk that reads "Lt. Donald Francis Draper." We learn something of his penchant for infidelity and professional insecurities when he confides in his then-lover, "I have nothing. I am over and they're finally going to know it. The next time you see me there will be a bunch of young executives picking meat off of my ribs." This confession of insecurity is rarely repeated in the remainder of the series. Even here, it is quickly juxtaposed with Draper's demeanor in the office where he exhibits a cool confidence displayed by his quick-witted repartee with his boss, Roger Sterling; his condescension toward Pete Campbell, his junior colleague who is clearly vying for his approval; and his outburst toward Rachel Menken, a female client who challenges his campaign idea for her department store. Later, in the highly anticipated meeting with Lucky Strike, the meeting that has him straining for a new creative concept, we learn of Draper's uncanny ability to conjure brilliance under pressure and wax eloquently about the one thing driving consumer culture: pursuit of happiness.

> Advertising is based on one thing: happiness. And you know what happiness is? Happiness is the smell of a new car. It's

> freedom from fear. It's a billboard on the side of the road that screams reassurance that whatever you're doing it's OK. You are OK.

Yet, his own definition of happiness reveals an incongruity that marks the lives of his high-earning peers and makes Draper a mystery unto himself. Despite all of his achievement and material acquisition Donald Draper remains unsure if he is okay.

As the character's storyline unfolds, the origins of this lack of assurance emerge. We learn of his troubled childhood, having been raised by adoptive parents after his biological mother dies giving birth to him. We learn of his early years on the struggling farm where tangible signs of love and acceptance were scarce. We learn of his father's death and the subsequent move to a brothel owned and run by his adoptive aunt and uncle. It is little wonder that someone whose formative years were so deeply marked by desperation and lack of love and security would define happiness in terms of material possession and mistake love for momentary sexual gratification. Moreover, it is not too surprising that when the desperation of the war trenches coalesced with the desperation of years of deprivation and shame, transforming himself from Dick Whitman into Donald Draper obtained easy rational justification—even if it meant stealing another man's identity.

Yet, what makes Donald Draper so compelling as a character is not his reduction to these circumstantial facts, which make his troubling decision-making psychologically logical if not morally justifiable. No, what makes Donald Draper so compelling as a character is that in spite of our being privy to the historical circumstances that have conditioned his choices, we are nonetheless faced with our inclination to accept or deny a narrative of victimization and hence our proclivity to want to convict or absolve him of moral misconduct. Even if we refuse any explicit judgment of this fictional man, we find ourselves nonetheless morally confronted by certain aspects of his character because those aspects are not fictional. We struggle because we recognize the reality depicted on screen. We might even recognize ourselves. The boundary between fiction and reality dissolves the moment we begin to reason with or against the protagonist because the substance of our thought presumes the reality of the moral quandaries enacted for us. Therefore, as viewers, we join the cast of characters who similarly encounter the paradox that is Donald Draper.

In this chapter, I do not aspire to dissolve the paradox of Donald Draper. Rather, I want to explore the paradox from the perspective of contemporary philosopher Jean-Luc Marion's phenomenology of givenness and, in particular, what he names the "erotic reduction."[5] My interest in this exercise is to see how pairing Marion's highly sophisticated philosophy with the sophisticated development of a television character might illumine contemporary television's potential for providing philosophical substance while conversely illuminating the significance of Marion's aim to reintroduce love to philosophical discourse which has suffered by its absence. Indeed it may well be that modern philosophy's rejection of its erotic origin is partly why the arts so often display the effects of and seek a remedy for the widespread avoidance of serious reflection on the subject. In other words, while Marion has suggested that philosophy's rejection of its erotic origin as "love of wisdom" (*philo*-sophia) allowed for metaphysics's imperial objectification of all knowledge and consequently paved the way for "technology's hold upon the world," I want to submit the possibility that some forms of technology may be used to undermine or at least problematize the metaphysical imperialism Marion similarly seeks to overturn insofar as those technologies themselves may play a role in the re-eroticizing of contemporary thought. In the very least, it is undeniable that the mediums of film and television, and before these literature and theater, perform a *techne* in contemporary society by (in)forming our erotic discourse. Although not the subject of this essay, it may be worth exploring further the possibility of technology's role in the erotic reduction.[6]

By submitting the character of Donald Draper to Marion's erotic reduction I hope to show that rather than reflecting on the question driving the series, "Who is Donald Draper?" a more fitting question for philosophical reflection transposes this question and asks instead, "What is Donald Draper?" Although it may seem counterintuitive to shift from "who" to "what" as an attempt to shift from the metaphysical to the phenomenological, it is fitting because the "what" in question is provided by Marion's lexicon. In other words, the question pursued here is, "Donald Draper—icon or idol?" This chapter should be read as an exercise in applying some of the main features of the erotic reduction in order to better appreciate the paradoxical

5. For a detailed description of this project see Marion's *Being Given.*

6. I have in mind something along the lines of "Marion meets Marshall McLuhan." Some scholars have already explored McLuhan's phenomenological bent. My interest would be to read McLuhan specifically according to Marion's phenomenology of givenness. McLuhan seems to share Marion's critique of metaphysics' reductionist hold on the world, "In a culture like ours, long accustomed to splitting and dividing all things as a means of control, it is sometimes a bit of a shock to be reminded that, in operational and practical fact, the medium is the message." *The Medium is the Message.* It would be interesting to explore where Marion and McLuhan might overlap and diverge specifically from the perspective of the erotic reduction.

nature of the phenomenology of love by attending to the paradoxical nature of *Mad Men*'s lead character.

METAPHYSICS AND MADNESS

In order to understand the logic of the erotic reduction, it is necessary to begin with a brief introduction to Marion's phenomenological critique of metaphysics. Both metaphysics and phenomenology belong to philosophical discourse. *Philosophy* means "love of wisdom" and has, since Plato, according to Marion, been unapologetic about its "erotic origin." However, a shift occurred in the Middle Ages and reached its apex in the philosophy of Rene Descartes whose privileging of the *ego cogito* introduced a new totalizing paradigm. The primacy of love was replaced by "the primacy of beings as the universal object of knowledge."[7] To know something now meant to know it as object—separate, observable, dissectable, manipulable. This metaphysical reduction requires an epistemology that reduces the things of the world to objects and knowers to the masters (or potential masters) of those objects. Consequently, to know something now meant to exert mastery over it. The eroticism inherent to ancient philosophy was displaced by a methodological hubris that transforms the pursuit of wisdom through the love of wisdom to a pursuit of knowledge through the conquest for certainty.[8]

By contrast, according to Marion, phenomenology allows for an altogether different reduction. To reason according to metaphysics is to reason toward certainty and mastery. To reason according to a phenomenology of givenness is to allow the world to appear on its own terms while not privileging being and objectness. The phenomenological attitude is one of radical receptivity. This means phenomena must be free from metaphysical certainty, which systematically submits all phenomena to predetermined criteria/concepts. When phenomena cannot be controlled or reduced by applying these predetermined concepts the metaphysical attitude insists they must, therefore, be unessential to the quest for knowledge.

But Marion points to a deep problem with this kind of reasoning, this epistemic reduction, namely, what to do with love? How can we account for love and its knowledge according to the metaphysical reduction? We can't.

7. Marion, *Erotic Phenomenon*, 3.

8. Marion is well aware of those philosophers who, since the Middle Ages, have treated love as a worthy subject. His critique is more a methodological one concerned with the shifting telos of the discipline and the underlying methodological presumptions this shift betrays.

There is no room for love within the metaphysical reduction because rather than submitting to metaphysics' reductionist epistemology, love "falls under an erotic rationality."[9] Love follows its own rationality, which is precisely why Cartesian rationalism shifts away from love. The radical receptivity precluded by metaphysics becomes, therefore, the condition of possibility for a phenomenology of givenness and the gift of erotic reduction.

Marion's phenomenology of givenness may best be understood as an intervention. A world may be easily reduced to beings and objects, certainty may be achieved, but who would want to live in this world? Even more, who would want to live with others in this world? Marion's intervention reasserts the legitimacy of such questions. Yet, in order to pursue these questions Marion must show that even more integral to the self's desire for knowledge than certainty is the desire to love and to know oneself as loved. Without this knowledge—without the assurance that I am capable of love and worthy of another's love—even certainty fails. Descartes's *cogito* fails because epistemological certainty cannot lead me beyond my self, which is precisely where I must go to avoid the trap of vanity, "To produce my certainty myself does not reassure me at all, but rather maddens me in front of vanity in person. What is the good of my certainty, if it still depends on me, if I am only through myself?"[10]

LOVE OR VANITY?

I begin with these brief remarks concerning the conceptual background of Marion's phenomenology in order to indicate that the existential crisis illustrated in Donald Draper is indeed a form of madness. Despite the glamour, civility, and apparent stability of the 1960s executive culture, madness reigns. In a world driven by metaphysical certainty, that is, a world run by the ego and its objects, capitalism becomes synonymous with freedom and acts as the sole guarantor of one's pursuit of happiness. When Don defines happiness as "the smell of a new car" he is correct: in a mad world happiness results from being certain that new car is never too far beyond reach. In a mad world—that is, a world wherein I can produce my own certainty—vanity's quest plays out in the myth of the self-made man. But, as Marion shows, the myth of the self-made man requires a correlative mythical autonomy

9. Marion, *Erotic Phenomenon*, 5.

10. Here Marion is exposing the vanity of the ego, which cannot avoid solipsism. "Vanity thus disqualifies every certainty, whether it bears upon the world or upon myself" (*Erotic Phenomenon*, 19).

that results in every man for himself. The chain of events set-off by every man for himself leads to the saddest conclusion: hatred of self and the other.

Love requires an exteriority that cannot be accommodated by the solipsism of every man for himself. Even self-love cannot reach the assurance sought after in the question "Am I okay?" for the question assumes an exteriority; that is, the question already implies an elsewhere—beyond the self—and the response must come from an elsewhere which, by responding, assures me I have been heard. In the absence of this elsewhere, that is, in a world wherein he's responsible for making himself and can only survive according to every man for himself, the mad man responds to the lover's quest most pragmatically, "What's the use?"[11]

Returning to episode one, we find Draper in the unseemly position of having to apologize to his client, Rachel Menken. In an attempt to assuage Roger's fear of losing the Menken account, Don meets her for drinks and a conversation takes place that reveals his cynical outlook on love. When the conversation turns to the subject of Rachel's marital status, she states she is not married because she has never been in love. Draper replies cynically,

> Draper: When you say "love" you mean a big lightning bolt to the heart. You can't eat. You can't work. You just run off and get married and make babies. The reason you haven't felt it is because it doesn't exist. What you call "love" was invented by guys like me—to sell nylons.
>
> Menken: Is that right?
>
> Draper: Sure. You're born alone and you die alone. And this world just drops a bunch of rules on top of you to forget those facts, but I never forget it. I'm living like there's no tomorrow because there isn't one.

11. "The self goes along of its own accord—even to the point of going along as far as itself; the return to the self would bring together the point of departure and the end of the course; the solipsism of consciousness would assure autonomy without transition, right up to the love of self . . . I must take care of myself ('Take care!'), because, we admit implicitly, no one else will do it for me, or be concerned about myself ('Who cares?'). Henceforth, this care of the self by the self becomes strictly a moral duty ('Charity begins at home'), failing which one will compromise every other duty toward anyone else. . ." (Marion, *Erotic Phenomenon*, 43). Marion's discussion of self-love is much more complex and nuanced than I am able to show here. The rejection of self-love runs so counter to popular wisdom I commend this section, "Concerning Every Man for Himself, and His Self-Hatred," to be read slowly while keeping one's rapid objections in check. To relinquish the possibility of self-love does not, in the end, mean the self will not be loved—even in the absence of a lover. Rejecting self-love, for Marion, means freeing oneself for love. Rejecting self-love and self-hatred amounts to rejecting vanity, which must be done in order to advance in the erotic reduction.

> Menken: I never realized until right now, but it must be hard to be a man too.
>
> Draper: Excuse me?
>
> Menken: Mr. Draper.
>
> Draper: Don.
>
> Menken: Mr. Draper, I don't know what it is you really believe in but I do know what it feels like to be out of place, to be disconnected, to see the whole world laid out in front of you the way other people live it. There's something about you that tells me you know it too.
>
> Draper: I don't know if that's true. You want another drink?
>
> Menken: No. But you can tell your boss that you charmed me.

The irony (and profundity) of this conversation unfolds slowly throughout the remainder of the show's seven seasons. Don's snarky dismissal of love as a fabrication and hence the desire for love as a fabricated desire is betrayed by his serial infidelity through which he seeks the very security promised in that so-called fabrication. Moreover, his rejection of love as an unattainable ideal reveals he is not rejecting love as much as he is rejecting the belief that anyone can be certain in love. The irony, in other words, is that the very ideal he rejects becomes the premise for his rejection. He has bought the nylons—if only to discard them.[12]

Don is correct to resist what he takes to be a false security in the certainty of an idealized love. But, again, the rejection reveals a confusion: love does not seek certainty; certainty is vanity's quest. Rather, love seeks assurance. It is love—not vanity—that leads me to ask, "Am I okay?" which, as Marion shows, is but another way of asking "Does anyone love me?" The cynicism produced by vanity's quest for certainty not only denies the question as irrelevant to the self's acquisition of knowledge, it denies the self:

12. Marion describes this stance as an "amorous hatred." One instance of such hatred is found in self-hatred which itself reveals the self's primal desire for love even when denying it to one's self: "Self-hatred cannot be absorbed so easily in the indifference to an absence of love. . . . The hatred of self still refers back, at least as much as the impossible love of self, to the question, 'Does anyone out there love me?' and for an obvious reason: the *self* that one hates is hated precisely because one succeeds neither in making it loved from out there, nor in loving it oneself. The impossibility of love within hatred appears as more inward to hatred than hatred itself" (*Erotic Phenomenon*, 58). This unfortunate situation, that is, self-hatred, therefore is the effect of certainty's 'every man for himself' and ultimately leads to hatred of the other. Cf. §13 Marion, "Passage to Vengeance," 58–64.

> For to give up on even the possibility that someone loves me would be like operating transcendental castration upon myself, and would bring me down to the rank of an artificial intelligence, a mechanical calculator or a demon, in short, very likely lower than an animal, who can still mimic love, at least to our eyes. And in fact, those of my likeness who have given up—in part and only in a certain respect, it is true—on the possibility that someone loves them have, in proportion, lost their humanity. To give up on asking (oneself) the question "Does anybody love me?" or above all to give up on the possibility of a positive response implies nothing less than giving up on the human itself.[13]

Vanity, not hatred, is the enemy of love because vanity leads me to question my quest for assurance. Vanity requires, ultimately, that I abandon myself. By contrast, Marion shows that to ask the question "Does anybody love me?" is the only way of avoiding vanity's objection: "What's the use?" and advance towards assurance. Therefore, asking the question, "Does anyone love me?" marks the beginning of the lover's advance which, finally, permits a further necessary interrogative transposition. The advance outside oneself and toward an other transposes the question from "Does anybody love me?" to "Can I love first?" The lover is the one who risks loving first—and last.

LOVER OR SEDUCER?

Once the ego is de-centered by the lover's advance, that is, by denying vanity's objection ("What's the use?) and asking "Can I love first?" the erotic reduction continues and yields the difference between the lover and the seducer. Both, Marion contends, are rightly viewed as erotic as both advance in the direction of another. They both operate in the same advance and share the same anticipation: "I love first, without any other reason than this one whom I risk loving, without awaiting her response, without presuming reciprocity, without even knowing her."[14] However, they soon diverge according to their opposing modes.

> In seduction, the advance remains provisional and ends by canceling itself out, because once the other is seduced (led to give consent), I no longer love in advance or to the point of loss, but rather with a return, in full reciprocity. . . . Just as the other will

13. Marion, *Erotic Phenomenon*, 20–21.

14. Ibid., 82–83.

> inevitably wind up returning to me the love I initially credited to her, my possession will catch up with it and will assimilate it to my ego, which will once again become the center of the circle.[15]

The mode of seduction short circuits the erotic reduction, Marion explains, because "it mimics the lover's love in order to finally invert it . . . Seduction wants to make itself loved without, in the end, loving . . ."[16] The seducer appears to seek love but only desires a love that will not require anything of him in return. This is the frightful characteristic of the seducer: that he is not manifest as seducer until he refuses to love in return.

Already by Season 2, Don's habitual infidelity has begun to crack the veneer of his public persona as a loving husband and father. Interestingly, he is confronted by the reality of his waning pretense by his then-lover, Bobbie, when she playfully remarks, "I want the full Don Draper treatment. . . . I want it and I got it and it's better than they said." When he responds with genuine surprise ("They?") she continues, "Have no fear. You're known as a connoisseur. You have lots of fans." Don is clearly disturbed and asks who has been talking about him. When he continues to look surprised and says he doesn't recognize the name she gives, Bobbie replies, "Oh stop. This is no one's maiden voyage here. . . . You have a reputation. Enjoy it." But he cannot enjoy it because enjoying it would require acknowledging the lie, which is that he is—or ever wanted to be—a lover: "lying consists here in claiming to be the lovers that we above all do not want to become."[17]

According to Marion, the seducer advances in sincerity under the guise of the lover. Indifference enables his sincerity and is the veil under which he advances seen but unseen, that is, never fully manifest before the object of his advance. Shrouded with indifference, he no longer faces her, and she no longer him. His indifference deprives her of the individuation that is necessary for the erotic reduction.[18] He must now sustain the lie, that is, that he advances as lover in order to perform his advance as seducer lest she disappear completely.

Don Draper is nothing if not sincere in his advance. This is partly what makes him likable despite any reservations we might have about his conduct when viewed solely in the mode of seduction. Yet in the case of the exceptional lie, that is, from the perspective of his infidelity, his sincerity does little to soften regard for his duplicity. Or does it? Are we as

15. Ibid., 83.

16. Ibid.

17. Ibid., 159.

18. See §22 on "Individuality" which concerns how the lover is individualized. It is important to note the role of the face in Marion's erotic reduction.

viewers ultimately led to abandon this character because of his duplicitous nature? We will return to this question momentarily. But first we must see how Don's infidelity affects his erotic reduction. Don's sincerity ultimately becomes a precondition for his madness. In Marion's words, "By insisting upon my sincerity in my eroticized flesh, I not only rave madly, but I lie all the more, because I lie to myself."[19]

Don's infidelity means he must perform the double lie: he must lie publically to his wife(s) according to *the oath*, and privately to the object [I don't say person] of his advance according to *indifference*. The first form of the lie is obvious enough: the oath, according to which there would be no other than *this* beloved, has been broken. The second form we've just discussed above: the seducer advances by masking his indifference and appearing as lover. Each of these lies performs a contradiction of the lover's advance: in infidelity Don can neither advance as lover toward his wife or his then-lover because in the double lie the seducer is kept from appearing fully to either one. But there is a third form of the lie—what Marion calls "the most dangerous lie," namely, the lie to himself. The eventual consequence of his infidelity is Don's tragic loss of self.

> For, in order to elude the two that I deceive, I must not only conceal from them my person, but also construct for myself a double personality for a double life; but by acquiring two personalities, I lose myself as such, precisely because I no longer have anyone to lose or save. The most dangerous lie requires that I truly lose myself in person. Contrary to the two preceding cases, the lie does not conceal the true person, it kills him. Under the mask of the lie, I am no longer anyone. . . . I am no longer anything but a mask, which no longer hides anyone [*qui ne cache plus personne*].[20]

To return to Bobbie's proposal: Don cannot enjoy the reputation he has earned because this would require that he acknowledge his indifference toward all of those affected by his earning it—including himself. To find pleasure in his reputation would mean to find pleasure in his "transcendental castration," to use Marion's phrase, as he cannot enjoy his reputation without admitting his refusal to love—something Don Draper, despite his dismissals of love, is unwilling to admit. Therefore, rather than admit defeat,

19. In this section Marion's "I" is the "I" of the one who attempts to restore truthfulness after the lie by insisting he has been successful in the erotic reduction; ". . . *for me* at least, I accomplish perfectly 'Here I am!' in my flesh; and, what is more, that my flesh, by being eroticized, puts my person in evidence; in short, that I have personal, transparent, and sincere eroticization." Marion, *Erotic Phenomenon*, 163.

20. Marion, *Erotic Phenomenon*, 161

Don rejects the proposal—and Bobbie. The scene ends with him tying her to the bed and departing abruptly, "I told you to stop talking."

The scene brilliantly brings closure to the discussion from the previous episode, which initiated the beginning of their affair. Over dinner, Bobbie asks Don what he likes, "It doesn't have to be business. It can be anything." After an awkward pause she observes, "You can't even answer." The conversation continues,

> Draper: The answer is huge.
>
> Bobbie: I don't think so. Do you like the ocean?
>
> Draper: Yes. Yes, I do.

In the next scene they are driving out to her oceanfront property. The conversation continues as she queries him for more information about what he likes. They agree on bridges and movies. Then Bobbie's mood turns, creating a stark contrast to Don's manifest indifference.

> Bobbie: Why is it so hard to just enjoy things? God, I feel so good.
>
> Draper: I don't feel a thing.[21]

ICON OR IDOL?

Don Draper's advance in the erotic reduction has been completely undermined by his vanity ("What's the use?"), his indifference, and his duplicity (infidelity). In each case, Don is denied (denies himself) full eroticization and, therefore, closer proximity to assurance. He has threatened his own existence, which seems increasingly precarious as he struggles toward some kind of resolution/absolution. Don's brilliance in business meetings and sincerity at home is no longer enough to presume security. Our self-made man is self-unmaking. As the show progresses, the madness informed by his vanity and indifference becomes more visible and begins taking a toll on him personally and professionally. Is the erotic reduction still within the realm of possibility for Donald Draper? Will Donald Draper ever find the assurance that he seeks? Will he ever be assured that he is okay?

In order for the erotic reduction to be accomplished, he must advance towards a specific other[22] such that he 1) does not take possession;

21. "The New Girl," S2/E5.
22. Cf. §22 "Individuality."

2) makes love with "an eternity of intention"[23]; and 3) is, thereby, freed to receive himself from one who gives him himself (his flesh) without ever possessing him in return.[24] It appears, momentarily, that Don will be able to advance successfully as lover in his expressed attempt to fulfill his oath to his second wife, Megan.[25] In an episode entitled "Signal 30" Don proffers advice to Pete Campbell who has just enjoyed a night of "fun" with a high profile client.[26] Pete is put-off by Don's refusal to take part and accuses Don of judging him.

> Pete: You of all people.
>
> Draper: Well, have a good night.
>
> Pete: Boy, this is rich. I can't believe I have to explain I was doing my job to a man who just pulled his pants up on the world.
>
> Draper: Pete, just go home and take a shower and forget this.
>
> Pete: I suppose there are no stone looks for Roger.
>
> Draper: Roger is miserable. I didn't think you were.
>
> Pete (with sarcasm): I have it all. Wait until your honeymoon is over.
>
> Draper: Look, I'm just trying to tell you—because I am who I am and I've been where I've been—that you don't get another chance at what you have.
>
> Pete: Brave words from a man on his second time around.
>
> Draper: Yeah. And if I'd met her first I would have known not to throw it away.

23. The role of time is integral to Marion's account of the erotic reduction and pertains specifically to his account of *the oath.*

24. "The flesh" for Marion is precisely that which distinguishes me from an ontological reduction to "being." In order to know myself as more than a being, I must experience myself as flesh. Moreover, the only way to do this is to encounter the flesh of another—whether through reception or resistance. The radicalization of the flesh which takes place in the erotic reduction occurs through a mutual eroticization with an other who simultaneously receives their flesh in its encounter with mine. "I can only free myself and become myself by touching another flesh, as one touches land at a port, because another flesh can make room for me, welcome me, and not turn me away and resist me—that is, comply with my flesh and reveal it to me by providing it a place. . . . The other gives me to myself for the first time, because she takes the initiative to give me my own flesh for the first time. She awakens me, because she eroticizes me." Marion, *Erotic Phenomenon,* 118–19.

25. See the next chapter for reflections on the erotic aspects of Don's marriage to Megan.

26. "Signal 30," S5/E4.

Don's optimism is short-lived. Season 6 opens with Don reading Dante's *Inferno* while sitting next to Megan on the beach. We learn at the end of the episode the book was on loan from his downstairs neighbor with whom he's having an affair. It is New Year's Eve and they're lying in bed together. She asks, "What do you want for this year?" He replies, "I want to stop doing this." But he doesn't stop. In the following episode she asks him whether it bothers him to have dinner with their two spouses and he replies, "I don't think about it. They're both good company."[27] Don's attempt to suppress the urge to seduce seems to have brought it back with a vengeance. When his then-lover tells him she prays he'll find peace, he simply slides the cross on her necklace out of sight.

As previously indicated, there are two terms in Marion's grammar of givenness to which we should now turn in order to assess the likelihood of Don Draper's advance in the erotic reduction: the icon and the idol. In order to understand how these two terms are operative in the erotic reduction I will provide a brief summary of Marion's prior use of the terms. The primary claim regarding his use of these terms is crucial for understanding how they operate within the context of the erotic reduction, namely, the distinction between the idol and icon for Marion is not ontological. That is, these terms do not describe two different kinds of objects. According to the phenomenology of givenness, anything/anyone can be either an icon or an idol. Stated differently, according to the phenomenological attitude a given phenomenon may easily be both: "the icon and the idol are not at all determined as beings against other beings, since the same beings (statues, names, etc.) can pass from one rank to the other."[28] How then are they distinguished? In their mode of appearance as determined by the gaze. The idol, for Marion, acts as a mirror. When my gaze reduces the phenomenon purely to its objectness, I have transformed it into a mirror whereby it can only reflect my intentionality toward it. Therefore, "instead of presupposing an interobjectivity and an at least teleological communication . . . the idol provokes an ineluctable solipsism. . . . The sequence of gazes that I continually pose on the idol establish so many invisible mirrors of myself. . ."[29] By contrast, the icon resists the idolatrous gaze and instead imposes an encounter; it confronts me and I see myself not according to my own intentionality but according to the intentionality of the other (towards me): "the icon . . . no longer offers any spectacle to the gaze and tolerates no gaze from any

27. "Collaborators," S6/E2.

28. Marion, *God Without Being*, 8.

29. Marion, *Being Given*, 229–30.

spectator, but rather exerts its own gaze over that which meets it. The gazer takes the place of the gazed upon . . ."[30]

In *The Erotic Phenomenon,* these two terms come into play when describing the lover's advance toward an other. As we have seen, both the seducer and the lover advance but the seducer's advance is blocked by his indifference toward the other. Another way of describing this unfortunate miscarriage of love is by naming it an act of idolatry: the seducer reduces the other to an idol by reducing her to a spectacle to be gazed upon and eventually possessed. We have already discussed the unfortunate consequences of this idolatrous act. The seducer's idolatrous gaze eventually renders himself an idol through an unremitting solipsism where he must determine his own certainty. The idol as mirror reflects the seducer's vanity according to which he can never advance beyond "What's the use?" and therefore remains alone in his lack of love's assurance. Thus Marion concludes, "it is not the least accomplishment of the hatred of every man for himself to give rise to his own idol. As if he needed to mime the glory of a god."[31]

By contrast, the lover's advance towards an other is successful precisely because the other functions as an icon: "the other appears as an iconic face (without any façade to see), who envisages me at a distance."[32] As icon, the other is eroticized because she is not possessed (by the lover) and is free to give of herself freely (as beloved). Moreover, as icon the other facilitates the lover's radical eroticization because "the icon . . . opens a teleology [and] therefore exercises (but in a more radical mode) an individuation over the gaze that confronts it."[33] Therefore, "I receive from the other what I do not possess of my own accord (nor she of her own accord), and that I nevertheless give to her (just as she gives it to me)—the rank of lover."[34]

A mutual eroticization is facilitated when the other assumes the status of icon over the idol. Vanity's idol is broken. The lover is now able to recognize himself as both lover and beloved simultaneously. In short, the lover receives assurance and thus himself as a gift from the iconic face of the beloved.

> Loveable as lover, I thus do not love myself, but rather receive my assurance from somewhere else. I do not immediately love myself every man for himself and me for myself, but through the detour of the one who tells me that I love her, that she loves that

30. Ibid., 232.

31. Marion, *Erotic Phenomenon*, 61.

32. Ibid., 25.

33. Marion, *Being Given*, 233.

34. Marion, *Erotic Phenomenon*, 212.

> I love her, and thus that she loves me. I discover myself loveable through the other's grace; and if I finally risk loving myself, or at least no longer hating myself (in short, if I risk forgiving myself), I dare it on the word of the other, through my confidence in her and not in me; I will surmount my hatred for myself as I would walk on water or step forward into the void—because from out there the voice of the other convinced me (or nearly) that I can do it, and that I am worth it.[35]

Returning to our query concerning Don Draper's advance or seeming inability to advance, we must ask whether there isn't more to him than vanity, indifference, and duplicity. The answer to this question relies precisely in our experience of him as paradox. Don Draper is a paradox—to himself as much as to others—precisely because there is more to him than his vanity and indifference. He experiences himself as paradox—and others with him—because he is both idol and icon.

Thus far we have only traced the idolatrous side of Don Draper; those aspects of his person that lead to a failure in his erotic advance. But there is reason to believe and hold out hope for his radical eroticization in those relationships wherein he does manifest as icon in refusing to reduce the other to idol. Moreover, it is not insignificant that the two relationships which display Don's iconicity best are with women. It is in his friendships with Anna Draper and Peggy Olsen that we are assured Don is more than the indifferent seducer. In his friendship with these two women, Don is shown capable of love—even if not erotic love.[36] And it is through these loves that his status as idol is transfigured: when he sees himself through their eyes as icon he recognizes himself and comes the closest to receiving his self.[37]

There are two key episodes that illustrate this transfiguration. "The Mountain King" from Season 2 consists of several flashbacks that tell the backstory of Don's relationship with Anna Draper, the widow of the soldier whose identity Don has stolen. In the first flashback we are shown Don's first encounter with Anna, when she's come to discover the truth about her husband and why Don has stolen his identity. Don is nervous and evasive until Anna demands he tell the truth. He relents and tells her the truth about her husband being killed in combat. She asks what his real name is and

35. Marion, *Erotic Phenomenon*, 213

36. For further discussion of the uniqueness of his relationship to these two women, see Jackson Lashier, Chapter 8.

37. His relationship with his daughter, Sally, exhibits another instance of the possibility of love and would be a worthwhile study. However, the familial context of their relationship requires an altogether different frame for assessment and thus cannot be further explored here.

he tells her, "Dick Whitman." The flashback ends with Anna asking, "Well, Dick, what do I do with you?" He is afraid as he realizes his fate lies in her answer to this question.

The scene transitions into a real-time surprise visit Don pays to Anna. She greets him warmly, invites him in and introduces him to her student, "This is my friend, Dick." Later the two are shown sitting on her front porch. We learn that Don has been providing for Anna. Anna says, "You like the porch? You paid for it." She asks about his children. Then she says, "If you don't want to talk about what happened that's fine. But you can, you know. I always felt that we met so that both of our lives could be better. That's just how it is between us." He tells her, "I've ruined everything. Family. My wife. The kids. My brother came and I told him to go away." He tells her of how he'd kept things from Betty and now he's losing her. Anna asks, "So what are you going to do?" Don replies, "I don't know. I have been watching my life. It's right there. I keep scratching at it and trying to get into it. I can't." Anna says, "I'm sorry."

Later, Don is still staying with Anna. She brings him some clothes and he has another flashback to a scene from a Christmas-past. He tells Anna he's met someone special, referring to Betty. "She's so beautiful and happy." He wants Anna to meet her. Anna is very happy for him, "Look at you. You're in the lavender haze." "I just like the way she laughs. And the way she looks at me . . . I want to ask her to marry me."

Anna: Oh, Dick. That's wonderful.

Draper: Well, I would need you to give me a divorce.

Anna: Oh, well, of course. I never thought of that.

Draper: You should hire a lawyer out here. I'll pay for it. And I want you to know I'm going to take care of you—forever.

Anna: You don't have to do that.

Draper: I do.

Anna: You don't owe me anything.

Draper: This whole life. If it wasn't for Don—if it wasn't. . .

Anna: Stop that. (pause) So, there will be another Mrs. Draper.

Draper: I guess so.

Anna: You'll have a family. That will be good for you.

Draper: I want to.

Anna: This will probably be our last Christmas together.

Draper: Why?

Anna: Dick.

Draper: You can be my cousin.

[Laughter]

Anna: No. This is a chance at a whole new life.

Don: [kisses her on the cheek] Thank you, Anna.

At the end of the episode, we return to real time. Don is sitting with Anna who invites him to let her read his tarot cards. Don, always the cynic, plays along reluctantly. While he waits for her to deal the cards he picks up a copy of Frank O'Hara's *Meditations in an Emergency*.

Draper: Did you read it?

Anna: I did. It reminded me of New York. And it made me worry about you.

Draper: What about the cards? Should I be worried?

Anna: It's all here. You're definitely in a strange place. . . . But here's the sun. [points to a card reading "Sun"]

Draper: [pointing to a card that reads "Judgment"] That can't be good.

Anna: It is.

Draper: It's the end of the world.

Anna: It's the resurrection. Do you want to know what this means or not? [invites him to sit]

Draper: [sitting] No, I don't. [gazing at the open window] I can smell the ocean.

Anna: [pointing to a card that reads "World"] This is the one.

Draper: Who is she?

Anna: She's the soul of the world. She's in a very important spot here. This is you—what you are bringing to the reading. She says, "You are part of the world. Air. Water. Every living thing is connected to you."

Draper: That's a nice thought.

Anna: It is.

Draper: What does it mean?

Anna: It means the only thing keeping you from being happy is the belief that you are alone.

> Draper: What if that's true?
>
> Anna: Then you can change.
>
> Draper: People don't change.
>
> Anna: [picking up the "World" card] I think she stands for wisdom. As you live you learn things.

The episode ends with a suggestive image of an immanent resurrection—a baptism.[38] Don is alone at the beach. He wades deeper into the water. Eyes closed he presses forward against the waves. The musical outro "Cup of Loneliness" begins and concludes, "To suffer with the Savior and when the way is dark and dim, to drink of the bitter cup of loneliness within."[39] Don returns to New York—and his family—renewed. But it is not long before he returns to his old ways.

Don's friendship with Anna comes to the fore again in Season 4. In a pivotal episode entitled, "The Suitcase," Don's friendship with Peggy Olsen, his junior colleague and mentee, is solidified as she is present with him when he learns of Anna's passing. Peggy's presence with Don in this episode mirrors his presence when he visits her in the hospital after she has given birth. He is the only one who visits—an observation which leads her mother to assume it is Don who is responsible for Peggy's "situation."

Don's relationship with Peggy is tumultuous, largely due to the office politic, which they both simultaneously embrace and decry. It is the tension of this simultaneity that largely accounts for their unspoken understanding of one another. Don sees in Peggy something of himself—but more. Her prevalent sense of principle is a reversed image of Don's lack thereof. Yet as a woman in a mad man's world, her principle won't benefit her as much as Don's lack of principle has benefit him in their ascent of the career ladder.

Don is responsible for Peggy's ascent in that hierarchy and as such he holds her to a higher standard than the others—a standard that is expressed in harsh critiques and often sarcasm. In "The Suitcase," the tension between the two begins early as Don quips, "Peggy, I'm glad this is an environment where you feel free to fail." Don's foul mood is partly the result of his drunkenness. He is supposed to be returning a call from California, but he cannot bring himself to do it as he knows it is bad news about Anna. Instead of returning the call, he throws himself into work, the bottle, and his begrudging protégé. It is Peggy's birthday, but she hasn't told Don who presses her to stay late and work until they've settled on an idea for the Samsonite account.

38. See Jackson Lashier, Chapter 8.

39. "The Cup of Loneliness" by George Jones and Burl Stephens (Glad Music Company, Copyright 1959).

She keeps throwing him bad ideas and he keeps passing on them. Don dismisses her after learning it is her birthday—but only after he tells her she's too old to care about birthdays. She stays, eventually at the expense of her steady relationship. She is alone again. They are both miserable.

Their individual miseries build to an explosive conversation where Peggy's resentment of Don and his success is finally brought out into the open. When she accuses him of taking credit for her award-winning idea, Don is stunned by her ingratitude and remarks, "It's your job. I give you money. You give me ideas."

> Peggy: But you never say "thank you"!
>
> Draper: That's what the money is for! You're young. You will get your recognition. And, honestly, it is ridiculous to be two years into your career and be counting your ideas. Everything to you is an opportunity. And you should be thanking me every morning when you wake up—along with Jesus—for giving you another day.
>
> [Peggy begins to cry and leaves.]
>
> Draper: Aw, come on. I'm sorry about your boyfriend, OK?

When she returns, Don is almost giddy having just discovered a taped recording of Roger's book notes wherein he divulges intimate details that Don finds exceedingly funny. Peggy refuses to indulge as she's still upset. Undeterred by her sour disposition, Don invites her to "stay and visit." She initially declines saying she has nothing to talk about because "it's personal." Draper responds "We have personal discussions," to which Peggy retorts, "No we don't. And I think you like it that way. I know I do." She wavers and begins telling him of the fallout with her boyfriend. Don suggests he take her to dinner as it is still her birthday. Over dinner they learn things about one another they'd not discussed before. He saw men die in Korea, and his father die from being kicked by a horse. She saw her father die of a heart attack. She asks about his mother. "I never knew her."

They move the conversation to a bar. There Don assures Peggy she's "cute as hell" and will find somebody. Their banter is candid and endearing.[40]

> Peggy: Men don't exactly stop in the streets.
>
> Draper: Do you want that?
>
> Peggy: That's not what you were supposed to say.

40. A very important moment of this conversation includes their discussion of Peggy's inability to forget about her child.

Draper: What do you care what I think?

Peggy: Everybody thinks I slept with you to get the job. They joke about it—like it's so funny because the possibility is so remote.

Draper: It's not because you aren't attractive. I have to keep rules about work. I have to. You're an attractive girl, Peggy.

Peggy: Not as attractive as some of your other secretaries, I guess?

Draper: You don't want to start giving me morality lessons do you? People do things. Alright?

Later, after they've returned to the office, Don asks Peggy to make him a drink although he has just been sick.

Peggy: How long are you going to go on like this?

Draper: I have to make a phone call and I know it's going to be bad.

Peggy: Oh. Do you want to be alone?

Draper: Just make me a drink.

He never drinks it. Instead he lays his head on her lap and they both fall asleep. While sleeping, Don dreams of Anna holding a suitcase. She smiles at him and exits (without limping).

When he awakens he finally calls California and Stephanie, Anna's niece, tells him, "She's gone." Don hangs up. Peggy has been listening. When their eyes meet he breaks down. She moves closer to him.

Peggy: What happened?

Draper: Somebody very important to me died.

Peggy: Who?

Draper: The only person in the world who really knew me.

Peggy offers him comfort. He tells her she can go home. She goes to her office instead. Later, when she returns to check in on Don he appears refreshed, "Come over here." He shows her his winning idea for the Samsonite account. When she places her hand on the drawing to express her approval, he takes it up in his. They share a prolonged look, indicating an unspoken understanding. She departs his office and pauses at the door, "Open or closed?" Don replies, significantly, "Open."

Don's ability to maintain intimacy with each of these women without reducing either to an idol indicates, according to the erotic reduction, that the possibility of his erotic reduction remains open. He is capable of love. However, it does not follow from his capacity to be friend that Don will necessarily experience the radical eroticization that results from the free giving of one's self (eternally) to a unique other. According to Marion, friendship proceeds along the lover's advance in every way—only it does not advance far enough to reach climax. This is no failure. Friendship is the name given to a form of love that does not aspire to climax; it fulfills its aim precisely by maintaining its distinction from *eros*. Friendship must preserve "its distinctive privilege—the ability to exercise itself at the same moment toward *several* friends . . ."[41] Therefore, friendship at once fulfills its moment in the advance along the univocal path of love while distinguishing itself from the full eroticization of the lover. Marion contends, "The difficulty does not exist in introducing exceptions to the erotic reduction and equivocations into univocal love, but in measuring just how far love's one way extends."[42]

We can at once find hope in Don's capacity for friendship while still asking whether friendship will be enough to provide the assurance Don seeks. The show's final episode provides no definitive answer to the question. Don attends a retreat with Stephanie, Anna's niece and the only other person besides Anna in his life who refers to him as "Dick." Stephanie attends the retreat in an attempt to come to terms with the choices she's made in life thus far. She brings the cynical Don along with the hope it will also be of some help to him. During one of the group sessions Stephanie is told by another participant that her (Stephanie's) child, whom she has left, will spend the rest of his life staring at the door, waiting for her to return. Stephanie runs out. It is too much. Don goes after her.

> Draper: Don't listen to them.
>
> Stephanie: You think this is a big laugh, don't you?
>
> Draper: You weren't raised with Jesus. You don't know what happens to people when they believe in things.
>
> Stephanie: It's true. You think that I don't want to hear the truth?
>
> Draper: I could help you. I could move to LA.
>
> Stephanie: Look, I don't know what you're doing. You show up with some family heirloom. You're not my family. What's the matter with you?

41. Marion, *Erotic Phenomenon*, 219.

42. Ibid., 221.

Draper: I just know how people work. You could put this behind you. It will get easier as you move forward.

Stephanie: Oh, Dick. I don't think you're right about that.[43]

In her contradiction Stephanie reveals and rejects Don's *modus operandi*: to believe the lie that what goes unacknowledged remains unknown.[44] When Don later discovers Stephanie has left without saying goodbye he is very upset, "People just come and go and no one says goodbye." In the next scene, Peggy receives a phone call from Don.

Peggy: Where the hell are you?

Draper: Somewhere in California.

Peggy: Do you know how angry everyone is?

Draper: Did everything fall apart without me?

Peggy: It's not about that. You just took off. People were worried. What have you been doing?

Draper: I don't know. I have no idea.

Peggy: I know you get sick of things and you run, but you can come home.

Draper: Where?

Peggy: McCann will take you back in a second. Apparently it's happened before. Don't you want to work on Coke?

Draper: I can't. I can't get out of here.

Peggy: Don, come home.

Draper: I messed everything up. I'm not the man you think I am.

Peggy: Don, listen to me. What did you ever do that was so bad?

Draper: I broke all my vows. I scandalized my child. I took another man's name and made nothing of it.

Peggy: That's not true.

Draper: I only called because I realized I never said goodbye to you.

Peggy: I don't think you should be alone right now.

43. "Person to Person," S7/E14.

44. When Don visits Peggy in the hospital after she's given birth he offers her similar counsel. He tells her to get out and move forward, "This never happened. It will shock you how much it never happened" ("The New Girl," S2/E5).

> Draper: I'm in a crowd. I just wanted to hear your voice. See you soon.

When he hangs up, Don experiences a break down. He's unable to move. Later, he is persuaded by a seminar leader to attend her seminar. There he listens to another participant, Leonard, describe his unhappiness:

> I've never been interesting to anybody. I work in an office. People walk right by me. I know they don't see me. And I go home and I watch my wife and my kids, and they don't look up when I sit down. . . . I don't know, it's like no one cares that I'm gone. They should love me. Maybe they do. But I don't even know what it is. You spend your whole life thinking you're not getting it—people aren't giving it to you. Then you realize they're trying. And you don't even know what "it" is. I had a dream that I was on a shelf in the refrigerator. Someone closes the door and the light goes off. And I know everybody's out there eating. And then they open the door and you see them smiling. And they are happy to see you, but maybe they don't look right at you. And maybe they don't pick you. And then the door closes again. The light goes off.

Leonard breaks off and begins to cry uncontrollably. Don, who has been listening intently, rises to his feet and embraces Leonard. They cry in each other's arms. It is a pivotal scene that appears to occasion an epiphany for Don, although the content of the epiphany remains undisclosed. The show ends with Don returning to the ocean.[45] The voice of a guided meditation is heard, "Mother Sun, we greet you and are thankful for the sweetness of the earth. The new day brings new hope. Lives we've led to lives we've yet to lead. A new day. New ideas. A new you."

The closing Coke commercial suggests that Don does return to New York, having now learned that he too has spent his whole life in search of love's assurance—but hasn't had the eyes to see it. Leonard's confessed unhappiness occasions Don's epiphany. Perhaps Don finally realizes that happiness is not in the smell of a new car, but rather is in being seen as the lover you aspire to be. Thus we are led to hope that Don is able to return to New York without returning to the madness that forestalled his erotic reduction. But the fulfillment of this hope depends upon Don's ability to reject madness in favor of his own iconicity. Friendship may provide him with the kind of training he will need to resist the idols that will continue to threaten his iconicity. But we cannot be certain it will be enough to secure his radical eroticization. There is no certainty with love; it neither seeks nor secures it.

45. The entire episode recalls significant aspects of "The Mountain King."

The paradox remains. Man's iconicity does not mean a world free of idols; it means only that man is never abandoned to his idols—even the idol of his self. For the universe may be indifferent; love is not.

BIBLIOGRAPHY

Marion, Jean-Luc. *Being Given: Toward a Phenomenology of Givenness.* Translated by Jeffrey L. Kosky. Stanford, CA: Stanford University Press, 2002.

———. *The Erotic Phenomenon.* Translated by Stephen E. Lewis. Chicago: University of Chicago Press, 2006.

———. *God Without Being.* Translated by Thomas A. Carlson. 2nd ed. Chicago: University of Chicago Press, 2012.

O'Hara, Frank. *Meditations in an Emergency.* New York, NY: Grove Press, 1996.

Part 3

POLITICS AND SOCIAL THEORY

Chapter 12

ZOU BISOU BISOU

Feminist Philosophy and Sexual Ethics in *Mad Men*

—JACOB L. GOODSON

INTRODUCTION

Loosely following J. M. E. McTaggart's logical structure concerning the construction of time—A-concept and B-concept—we can suggest a similar logical structure concerning the experience of sexual intimacy. For McTaggart, an A-concept of time involves continuity—the past flows into present, and the present flows into the future. Although time remains a human construction under the A-concept, humans cannot divide up their experience of time into abstract periods or distinct moments. In contrast, the B-concept of time involves an atomistic approach to time where time gets defined by integrals or discrete moments.[1]

How does this theory of time apply to the experience of sexual intimacy between lovers, partners, or spouses of the opposite sex?[2] Sex can be

1. See McTaggart, "Unreality of Time."

2. In this chapter, I focus exclusively on heterosexual sexual activity within marriage in *Mad Men*; for the sake of clarity and brevity, I do not consider other types of sexual relationships.

defined as occurring at the moment of vaginal penetration and coming to an end after either ejaculation or orgasm (B-concept).[3] Or, sex can be described in terms of continuity—between partners—where being together in ordinary life contributes to the experience of sexual intimacy (A-concept). While some of the vagueness of the A-concept description of sex cannot be clarified, we tend to use the following phrases and words to help identify other aspects of sexual intimacy in addition to intercourse: compliments about appearance, cuddling, dancing, dirty talk, exercising together (especially yoga and other meditative routines), foreplay, frontal hugging, heavy petting, holding hands, oral sex, passionate kissing, playfulness and sexual teasing, provocative gesturing, and sexting.

In this chapter, I implement this distinction as a helpful tool for making sense of how sex and sexuality are represented on the AMC television series, *Mad Men*—a series that, while not explicitly pornographic, presents the storylines of several of its characters through their sexual behavior.[4] Oddly enough, scholarly treatments of *Mad Men* fail to attend to and reflect upon the presentation of sex and sexuality within *Mad Men*—an omission indicative of the tendency, identified by feminist philosophers, for philosophy to

3. Some examples of what I consider a B-concept understanding of sex can be summarized very briefly. Thomas Aquinas argues that the uniqueness of sex concerns its momentary nature where the extreme intensity of bodily pleasure—experienced through sexual intercourse—can be neither commanded by reason nor explained in terms of an act of the mind. For Thomas Aquinas, sexual intercourse can be considered "good" when and only when every sexual act comes with the intention of procreation. I interpret Lisa Sowle Cahill's surprising defense of Thomas Aquinas's sexual ethics as a shift from B-concept to A-concept in the sense that she tweaks his argument to say that the marriage must have a "public" and "social" intent for procreation, not each and every act of sexual intercourse (see Cahill, *Sex, Gender, and Christian Ethics*, 199–201). Rollo May claims that sex and sexual activity require a "key moment," and May defends the entrance of the penis into the vagina as the most important "key moment" for a proper psychological understanding of sex and sexuality (see May, *Love and Will*, 75). Gilbert Meilaender writes that penile penetration of the vagina supplies us with "an act in which human beings are present most fully and give themselves most completely to another" (Meilaender, *Limits of Love*, 47).

4. Brief examples include: the first season depicts Peggy struggling with her decision to have sex with Pete Campbell in his office; Roger Sterling has a heart attack while having sex with one of his mistresses; Betty Draper responds to her husband's promiscuity by going to a bar to enjoy random (B-concept) sex herself; Pete Campbell and his father-in-law find themselves at the same brothel, which leads to distrust on both sides; Joan engaged in coitus with a client and uses sex in order to become a partner in the advertising agency; and even Glen Bishop—the young boy played by Matthew Weiner's own son—develops and maintains a strong sexual attraction toward Betty Draper/Francis throughout the series.

refuse to consider the ways that sex and sexual pleasure remain an integral part of life.[5]

In the collection of essays, *Mad Men and Philosophy*, for instance, not a single chapter explores questions of sex and sexuality in *Mad Men*; beyond an entry for "sexism," neither sex nor sexuality appear in the index.[6] Given all of the ways in which sex defines and forms the identity, livelihood, and mistakes of the characters in *Mad Men*, accounting for the problems of sexuality—i.e., "sexism"—falls far short of the obligation scholars maintain in attending to the complexities, positivities, and surprises of sex and sexual relationships within *Mad Men*.

This chapter offers three philosophical lenses for interpreting and understanding the sexual aspects of Don Draper's marriages to Betty and Megan. Two of these lenses come from Immanuel Kant's sexual ethics and Catherine MacKinnon's theory of radical feminism; I demonstrate how both of their theories become helpful for interpreting and appreciating the details of Don's marital relationships. The third interpretive lens comes from

5. See Millett, *Sexual Politics*; Millett distinguishes between thinking philosophically about "sex" and thinking philosophically about "sexual politics," and she claims that Western philosophers tend to do the former but not the latter. Alan Soble makes a charge worse than neglect and accuses philosophers (and theologians) of ridiculousness and silliness when writing about sex: "I have over the years collected a number of apparently absurd or ridiculous claims made by intelligent people [but critiquing these] silly assertions may say more about my own biases and prejudices than about the thoughtfulness of their authors"; additionally, "reader[s] should take much of what is written about sexuality with a grain of salt . . .[because approaching] the philosophy of sex [requires] a light heart and a willingness to poke holes in bubbles" (Soble, "Introduction," xvii and xx). While I am quite aware of the risks of writing on sex and sexuality identified by Soble, I worry much more about the problems that come with Millett's urging for philosophers to write about sex and "sexual politics."

6. See Carveth and South, *Mad Men and Philosophy*. While this collection of essays is extremely accessible and proves helpful for reflecting philosophically on the television series *Mad Men*, its early publication date makes it extremely limited in terms of where the television show ended. *Mad Men and Philosophy* divides its chapters into four sections: section one offers reflections on *Mad Men* through the lenses of epistemology (theories of knowledge) and the metaphysics of morality (theories of freedom); section two offers existentialist reflections on *Mad Men*, and the title of the section uses the line from *Mad Men* also used as the title for this book "The universe is indifferent"; section three offers ethical reflections on *Mad Men*, which includes both business ethics and personal ethics; and section four offers reflections on *Mad Men* through the lenses of political philosophy and social philosophy. In my judgment, the two best chapters in the whole collection are found in section four: Abigail Myers's "'And Nobody Understands That, But You': The Aristotelian Ideal of Friendship among the *Mad Men* (and Women)" and Ashley Jihee Barkman's "Mad Women: Aristotle, Second-Wave Feminism, and the Women of *Mad Men*." While Barkman's chapter implements de Beauvoir's work for interpreting *Mad Men*, she never discusses the sexual experiences of the characters of *Mad Men* (see Chapter 14).

Simone de Beauvoir's existentialist-feminist theory of sex and sexuality—which provides the most useful approach for the particular task of thinking through how both Betty and Megan become *empowered* through the sexual components of being married to Don. I intend for all three theories to remain live options for interpreting *Mad Men*, and I leave it to the practical reasoning of viewers to work with the theory (or theories) that they find most compelling.

I construct a selective chronological account of Draper's sexual experiences within his two marriages.[7] I attend to the scenes depicting the sexual intercourse that led to the conception of his third child with Betty, Megan's display of her sexuality through dancing for Don during his fortieth birthday party, and Don and Megan's final sexual encounter—which also includes Megan's friend, Amy. I interpret each scene in terms of the A-concept/B-concept distinction. I also demonstrate what it means to watch these episodes through de Beauvoir's, Kant's, and MacKinnon's theoretical frameworks. Before attending to the scenes, however, I outline what the basic arguments involved with these three theoretical frameworks.

THREE THEORIES OF SEX AND SEXUALITY

This chapter utilizes three theories of sex and sexuality. The modern philosopher Immanuel Kant (1704–1824) constructs a theory of sexuality that pays close attention to the role of actions, desires, and volition within sexuality.[8] Kant reasons that sex necessarily objectifies the other and, thus, violates the version of the categorical imperative often called the dignity test—never treat persons merely as a means but always as an end in themselves.[9] Kant's solution to the problem of the immorality of sex and the objectification that occurs during sex is quite simple: marriage. Marriage requires partners to treat the other in terms of their *whole* self, transforming sex into an act respectful of their personhood. Sex outside of marriage remains an action that necessarily reduces the other to one of their parts—specifically, sexual

7. Approaching marriage through the trope of ordinary life proves extremely illuminating—as Chapter 13, in this volume, illustrates; the A-concept of sex and sexuality ought to be tied to the significance of ordinary life and some of its particular features.

8. Kant's most thorough treatment of sex and sexuality can be found in *Lecture on Ethics*, 155–62; he also writes about sex within the context of marriage in *The Metaphysics of Morals*, 61–64.

9. The precise wording of this version of the categorical imperative reads: "For, all rational beings stand under the *law* that of each of them is to treat himself and all others *never merely as means* but always *at the same time as ends in themselves*" (Kant, *Groundwork of the Metaphysics of Morals*, 41).

organs. Kant thinks that sex always objectifies the other: the male partner objectifies the female partner, and the female partner objectifies the male partner. The *only* solution to this problem of objectification can be marriage. Marriage provides an institutional context where the parts, the sexual organs, of one's partner are not prioritized over the whole person. The fact that marriage demands the constant fulfillment of obligations toward the other—and those obligations remain directed toward the other as a whole person—allows for sexual activity to become moral because it includes the fulfillment of a duty toward the other person. Marriage provides a context that transforms sexual activity from the objectifying fulfillment of your own desire to the fulfillment of a duty toward your sexual partner.

In her radical feminism, the legal philosopher Catharine MacKinnon (1946–) makes a similar distinction to my structure of A-concept vs. B-concept for understanding sex and sexuality. MacKinnon claims that men tend toward a B-concept understanding of sexual intimacy that emphasizes penetration and ejaculation as the definitive bookends of sex; on the contrary, women's A-concept experience of sexual intimacy includes both "private" and "public" ways of women and men relating to one another. MacKinnon offers this gendered distinction as a way to critique the emphasis on "privacy" in the wording of *Roe vs. Wade*—demonstrating that the Supreme Court justified abortion through the logic of a male-centered understanding of sex and sexuality (B-concept).[10]

Notably, MacKinnon arrives at the same conclusions about sex as Kant does—albeit taking a different path to get there. For MacKinnon, sexual activity remains immoral within a patriarchal society because the male sexual partner brings his patriarchal power with him into the bedroom. In Western society, MacKinnon thinks that men *always* and *necessarily* objectify their female sexual partners.[11] Kant and MacKinnon agree that sex invites immorality. MacKinnon reasons that heterosexual sex always involves the man exercising power over the woman and, thus, renders impossible a genuinely consensual sexual relationship.[12] Other than her exhaustive and invigorating work on how law and policy ought to be written to overcome patriarchy, MacKinnon offers neither concrete nor on-the-ground solutions for individuals within their own lives. The disagreement that arises between Kant's and MacKinnon's diagnoses of sexual activity concerns the possibility for reciprocal objectification: Kant believes that both partners objectify one

10. See MacKinnon, "Privacy v. Equality," in *Feminism Unmodified*, 93–102.

11. For MacKinnon's understanding of objectification, see "Desire and Power," 46–62.

12. This connection between Kant's and MacKinnon's work receives full consideration in Barbara Herman's "Could It be Worth Thinking," 53–72.

another while MacKinnon locates objectification only from male to female. Another disagreement that arises between these two thinkers concerns the solution to the problem of objectification brought about through sex: MacKinnon thinks that marriage simply institutionalizes patriarchal power, within the bedroom, while Kant believes that marriage provides the only solution to the problem of sexual objectification.

The French existentialist philosopher Simone de Beauvoir (1908–1986) provides a phenomenological account of sex—which means that she does not claim to know *a priori* if sex is moral or immoral, consensual or rape,[13] objectifying fulfillment of pleasure for one partner or moral fulfillment of a duty toward the other.[14] For de Beauvoir, patriarchy brings its problems to the bedroom; however, the bedroom also holds the promise for genuine gender equality that can overcome the problems of patriarchy. Toward the end of her important and massive work, *The Second Sex*, de Beauvoir details two kinds of sexual intercourse. Both of these, she claims, lead to equality between individual men and women—and demonstrate how equality in the bedroom leads to equality in society. The first kind of sexual intercourse that promotes gender equality, somewhat counter-intuitively, involves shared domination: the male partner becomes passive so as to be dominated by his female partner, and the female partner becomes passive so as to be dominated by her male partner. The key ingredient here seems to be that one of them *voluntarily* becomes passive for the sake of allowing the other to fight "their own self."[15] The second kind of sexual intercourse that promotes gender equality includes a shared sense of intimacy, pleasure, and vulnerability.[16] Sex ought to serve as an opportunity for the intense experience of being "stalked by death" *together*,[17] having a lustful and uncontrollable "*need of the other*,"[18] and *mutually* savoring orgasmic moments—moments that feel

13. This part of the list refers to MacKinnon's thinking about sex and sexuality; for MacKinnon, most heterosexual sex is a form of rape within Western society: "Men define women as sexual beings; feminism comprehends that femininity 'is' sexual. Men see rape as intercourse; feminists say much intercourse 'is' rape. Men say women desire degradation; feminists see female masochism as the ultimate success of male supremacy and marvel at its failures" (MacKinnon, "Desire and Power," 59).

14. "The truth is that physical love can be treated neither as an absolute end in itself nor as a simple means; it cannot justify an existence: but it can receive no outside justification. It means it must play an episodic and autonomous role in all human life. This means it must above all be free" (de Beauvoir, *Second Sex*, 468).

15. Ibid., 763.

16. Ibid.

17. Ibid.

18. Ibid.; emphasis added.

like they represent the climax and purpose of our whole existence as human beings.[19]

De Beauvoir concludes with her radical claim: if these experiences can be had together in the bedroom, then there would be no temptation "to contend for false" power and privileges—"and fraternity could then be born between them [men and women]."[20] For de Beauvoir, equality between men and women comes about through healthy sexual encounters. Good sex leads to "the good life" between men and women. De Beauvoir emphasizes (in my terms) how the A-concept of sexual intimacy requires the B-concept—the moment involving vaginal penetration, ejaculation, and orgasm—to be dynamic, free, healthy, playful, and well balanced.[21]

Significantly, de Beauvoir thinks that marriage does not guarantee a proper and secure context for sex to achieve gender equality. While not anti-marriage, de Beauvoir's phenomenological method leads her to conclude that marriage tends to serve as protector and sustainer of patriarchy: "the principle of marriage is obscene because it transforms an exchange that should be founded on a spontaneous impulse into rights and duties; it gives bodies an instrumental, thus degrading, side by dooming them to grasp themselves in their generality."[22] Whereas Kant champions the fact that spouses treat each other as whole persons and "grasp themselves in their generality," de Beauvoir finds this same aspect of marriage problematic because viewing one's spouse in "their generality" means limiting the spontaneity required for a healthy relationship.[23] She specifies problems with the dual roles of "husband" and "wife": "the husband is often frozen by the idea that he is accomplishing [his masculine] duty, and the wife is ashamed to feel delivered to someone who exercises a right over her."[24] Marriage sustains patriarchy in the sense that husbands view the livelihood of their wives through the so-called manly or masculine duties that husbands allegedly have, and wives tend to view themselves as passively waiting for

19. Ibid.

20. Ibid.

21. Interestingly, the difference between de Beauvoir's and Kant's theories of sex and sexuality relates to and resembles the differences between Kant's formulaic moral theory and G. W. F. Hegel's creative and freedom-centered moral theory. The best and most thorough analysis of how de Beauvoir borrows from, yet tweaks, Hegel's moral philosophy can be found in Nancy Bauer's *Simone de Beauvoir, Philosophy, and Feminism*, especially chapters 3–6.

22. De Beauvoir, *Second Sex*, 465.

23. For de Beauvoir's arguments on the problems of viewing other persons in terms of generality, static notions of identity, and universal categories, see *Ethics of Ambiguity*.

24. De Beauvoir, *Second Sex*, 465.

their husbands to exercise his rights over her. In troubling ways, marriage makes it a *duty* for a wife to allow her husband to exercise his rights over her.

Given this line of reasoning, how can I defend the judgment that de Beauvoir does not come out as anti-marriage? Because the very next sentence, found in the same paragraph of *The Second Sex*, reads: "Of course, relations can become individualized [in] married life; sexual apprenticeship is sometimes accomplished in slow gradations; as of the first night, a happy physical attraction can be discovered between the spouses."[25] In the honeymoon stage of marriage, and if and only if they both find "a happy physical attraction" with one another, husband and wife become sexual *for each other*—which encourages and leads to gender equality within their life together. Even after the honeymoon stage of a marriage, de Beauvoir reasons that marriage can remain "good" because it "facilitates the wife's abandon by suppressing the notion of sin still so often attached to the flesh; regular and frequent cohabitation engenders carnal intimacy that is favorable to sexual maturity."[26] De Beauvoir regrets the lasting influence of Christianity on how we understand "the notion of sin still so often attached to the flesh,"[27] but for phenomenological reasons this "notion of sin" remains part of her observations—and her observations guide her judgments.

Sex tends to be better and healthier outside of marriage because the role of otherness contributes to the eroticism required for sexuality: "Eroticism is a movement toward the *Other*, and this is its essential character; but, within the couple, spouses become, for each other, the *Same*; no exchange is possible, between them any more, no giving, no conquest."[28] Because married couples tend to become too similar *to* one another,[29] and hence lose their sense of otherness *with* one another, sex within the context of marriage tends to be non-erotic sex—which, for de Beauvoir, lacks the ingredient sex needs for sex to be good, healthy, and well balanced. De Beauvoir continues her train of thought: "If they remain lovers [within marriage], it is often in embarrassment: they feel the sexual act is no longer an intersubjective experience where each one goes beyond himself, but rather a kind of mutual masturbation."[30] On the one hand, the risk taken by entering into marriage

25. Ibid., 466.

26. Ibid.

27. Ibid., 185–86, 765–66 (de Beauvoir commits some of her final thoughts, in *The Second Sex*, to thinking through how the category of sex negatively impacts women's sexuality).

28. Ibid., 467.

29. De Beauvoir's reasoning here problematizes any notion of "soul-mates" within marriage.

30. De Beauvoir, *Second Sex*, 467.

concerns how sex within marriage becomes more like "mutual masturbation" rather than dynamic, free, healthy, playful, and well balanced. On the other hand, while marriage does not make sex good—or moral—sex can become good, in the sense of achieving gender equality, within the institution of marriage as long as otherness refuses its reduction to sameness.[31]

Catherine MacKinnon's theory of sex and sexuality remains much less theoretical than either de Beauvoir's or Kant's positions are: MacKinnon's brilliance appears in her detailed analysis of law and policy—what we might call a feminist hermeneutic of legal texts, neither a phenomenological study of actual sexual encounters (as de Beauvoir offers) nor a deontological account of practical reasoning in regards to one's own sex life (as Kant provides).[32] MacKinnon shies away from constructing moral positions, and she refuses to commit her reasoning to either an ethical theory (deontology) or a philosophical movement (existentialism). While MacKinnon's and Kant's philosophical arguments about sex and sexuality offer helpful and interesting lenses for making sense of Draper's sex life, the primary claim of this chapter concerns how Simone de Beauvoir's feminist *defense* of sex provides the best philosophical lens for properly understanding Draper's sexuality and the two primary women in his life: Betty and Megan.

DON DRAPER'S SEX LIFE

In what follows, I consider three episodes presented in terms of the sexual components of Don Draper's two marriages. The episodes—and the particular scenes in those episodes—under consideration are the following: (1) Don and Betty have sex to "save" their marriage,[33] and their third child gets conceived as a result of this sexual act;[34] (2) Don's second wife, Megan,

31. For more of my thoughts on the problem of reducing otherness to sameness, within feminist philosophy, see my "'The Woman Question': William James's Negotiations with Natural Law Theory and Utilitarianism," in *Feminist Interpretations of William James*, ed. Shannon Sullivan & Erin Tarver, (University Park, PA: Penn St. University Press, 2015), chapter 2.

32. Even her essay, "Not by Law Alone," concludes with the argument that freedom established *through the law* ought to be the goal of feminism (see MacKinnon, "Not by Law Alone," 21–31).

33. See "The Inheritance," S2/E10.

34. I recognize the difference between conception taking place *within* the sexual act vs. *resulting* from the sexual act. In agreement with Kant's theory of sexuality, I think it more defensible to say that conception *results* from the sexual act than to say that it occurs within the sexual act: sex is for pleasure, and it happens to make babies. If the Kantian terms disappoint, then in Aristotelian terms: pleasure remains an internal good of sex; procreation ought to be considered an external good of coitus/sexual intercourse.

performs a dance for Don at his fortieth birthday;[35] and (3) Megan brings a friend to the bedroom for a ménage à trois,[36] which becomes the last time that Megan and Don have sexual intercourse together.

Betty Tries to Save Her Marriage ("The Inheritance," S2/E10)

Betty and Don have separated because Betty has gained knowledge of Don's infidelities. Betty learns that her father has become ill. Without their two children, Betty and Don drive to her father's house. After several awkward familial moments throughout the day, Betty seems to genuinely appreciate Don's presence with her and her family. As night falls, Betty lays down in the guest bed while Don sleeps on the floor beside it. Betty has a difficult time falling asleep. Betty comes down to him, puts her hand on his chest, and waits for his attention. He wakes up, stares into her eyes for a few seconds, and shifts his body in order to be receptive to Betty's advance. Betty lies on top of him and starts to kiss him. Don caresses her hair. They are very receptive toward one another, and this scene depicts deep intimacy and reciprocal vulnerability between the two of them. Surprisingly, though, Don wakes up alone.

Although hurt by Don's promiscuous behavior, Betty reaches out to Don—through sexual intimacy—as a way to show her appreciation for him being present with her through a difficult time in her family and to see if their marriage is salvageable. When they return to their home, Betty makes the decision that they ought to remain separated. This decision confuses Don, but he obliges her request. Don's past infidelities remain too much for her to overcome.[37]

Later in Season 2, however, we learn that they conceived their third child as a result of the sexual intercourse they enjoyed on the floor together at Betty's father's house. They name the baby "Gene," also Betty's father's name, to honor Betty's father and to memorialize his place of conception.

35. "A Little Kiss," S5/E1.

36. "The Runaways," S7/E5.

37. Kirsten Guidero disagrees with my interpretation and offers a counter-argument: "I think Betty realizes that, as Don has treated her poorly, she can genuinely care for him and express her care for him sexually without her sexual expression of care implying any sense of a permanent relationship" (Guidero, correspondence with the author [Feb. 22, 2016]).

Betty and Don from a Kantian Perspective

What happens if we watch this episode with Kant's sexual ethics in mind? Betty seems to be completely justified in both having sex with Don while at her father's house, and making her decision to divorce him—even with the expectation of their third child.

Don's infidelities remain inexcusable from Kant's deontological perspective. Kant writes,

> The second *crimen carnis secundum naturum* is *adulterium*, which occurs only in marriage, when the marriage-vow is broken. . . . all betrayals and breaches of faith, *adulterium*, is the greatest [immoral act], since there is no promise more important than this. Hence *adulterium* is . . . a cause for divorce.[38]

Given Kant's words here, there ought to be no negative judgment on Betty that Don's infidelities lead to their divorce.[39]

However, Betty initiating sex with Don should surprise us. She has no obligations to Don because he has broken the most important promise that a rational person makes to another rational person in the marriage vows. We should deem this sexual act between Betty and Don absolutely *moral* for three reasons. First, they are married; marriage remains the only proper context for sex. Second, Betty initiates sex and refuses to reduce Don to the single part of his sexual organ; they care for each other, as whole persons, in their sexual intimacy. Third, *consequences do not matter* within deontological reasoning—which means that neither the later (positive) consequence of Gene's birth nor the (negative) consequence of their divorce bears on this sexual act.

Betty and Don within MacKinnon's Framework

According to MacKinnon, Betty ought not try to save her marriage through sex but should seek a divorce from Don. Patriarchal society, however, makes it hard for women to seek divorce and be divorced. MacKinnon laments that both "aging" and divorce "devalue a woman economically."[40] While we should not place moral blame on Betty for initiating and having sex with her husband, we ought to recognize that Betty betters herself by divorcing Don

38. Kant, *Lectures on Ethics*, 160.

39. Kant lists two justifications for divorce: adultery committed by either party, or the impotence of the husband.

40. MacKinnon, *Toward a Feminist Theory*, 35.

and escaping the particulars of his patriarchal power—yet simultaneously recognizing that Betty will not escape general patriarchal power. Betty's character does not meet the standards of radical feminism.

Betty and Don within de Beauvoir's Framework

The sexual intimacy between Betty and Don seems to be healthy and well-balanced: Betty initiates, and Don responds; Betty maintains control over how their erotic experience will go; Don does not manipulate the situation to favor himself in any way, and Betty possesses the power and volition to enjoy sex with her adulterous husband. Of course, the adultery matters; at this point in the television series, Betty also has had an affair in the form of a random sexual encounter at a local bar. What would Simone de Beauvoir say if she watched this episode with us?

De Beauvoir reflects upon the differences between adultery committed by a husband and adultery committed by a wife. When a wife commits adultery, she has brought shame on both her marriage and their community; when a husband commits adultery, he demonstrates the freedom given to him by nature.[41] What is the solution to these imbalanced and unfair judgments about "the second sex"? For de Beauvoir, fidelity within marriage is not the answer. De Beauvoir uses the "fact" of adultery as one of her reasons for why marriage remains problematic: "Marriage, by frustrating women's erotic satisfaction, denies them the freedom and individuality of their feelings, drives them to adultery by way of a necessary and ironic dialectic."[42] The solution to adultery, for the wife, becomes either more adulterous affairs or the dissolution of the marriage—Betty tries the first solution and settles for the second solution, especially after she falls in love with Henry.[43] Surprisingly, however, de Beauvoir concludes that—for women—adultery, at its best, provides "only [an] artificial escape . . . that in no way authentically allows the woman to take her destiny into her own hands."[44] Other artificial escapes from marriage include superficial friendships and an active social life for the sake of appearances. De Beauvoir's conclusion about adultery

41. See de Beauvoir, *Second Sex*, 206–8.

42. Ibid., 592.

43. De Beauvoir's case for more adulterous affairs as a solution to adultery can be found here: "If she has no singular attachment to her husband, but he has succeeded in awakening her sexuality, she will want to taste the pleasures she has discovered through him with others" (593).

44. De Beauvoir, *Second Sex*, 598.

strikes me as a reigning endorsement of multiple affairs, even if those sexual encounters tend toward a B-concept understanding of sex.

Within de Beauvoir's framework, does having sex with one's adulterous husband provide any type of redemption within the marriage? Yes, it does. For de Beauvoir, one of the (few) benefits of marriage concerns how marriage makes available erotic experiences and sexual pleasure for women. De Beauvoir seemingly offers three reasons for why a wife should continue to have sex with her adulterous husband.[45] First, given that the husband's mistresses represent a type of "power" and "prestige" for him, de Beauvoir argues that marital sex focused upon the experience and pleasure of the wife removes the vicious types of "power" and "prestige" within sex for her husband. He might seek to manifest his vicious understanding of the role of "power" and "prestige" when he enjoys sexual intercourse with his mistresses, but marriage provides a context where this vicious type of "power" and "prestige" gets challenged and controlled. In this case, sex between a husband and wife brings down the "prestige" of the husband—not because his wife makes him less prestigious but because the "power" and "prestige" can be reconciled and shared on the wife's terms.[46]

Second, when the wife initiates sex—even, perhaps especially, with the knowledge that her husband has wandered from her—means that they have sex on her terms. Post-adultery sexual encounters between husband and wife keep the marriage focused on *her* erotic experiences and sexual pleasure, and it means that *she* refuses to let his infidelity and other vices (intemperance might be as much as a vice than infidelity is) control the terms of their marriage.[47] Third, sex within marriage remains the best way to achieve equality within marriage; moreover, wives should feel the same amount of freedom—as husbands feel—to enjoy sex outside of marriage too.[48] Betty having sex with Don at her father's house achieves the point of all three of these reasons for why having sex with one's adulterous husband provides a type of redemption within marriage. This episode shows Betty in control: she has sex with Don on her terms, and she chooses to divorce him as well. Betty might not be the ideal woman on the standards

45. I say "seemingly" because de Beauvoir does not address this question directly. Rather, I use arguments and reflections found from three different chapters—"The Married Woman," "Social Life," and "The Independent Woman"—to offer the best possible answer to the stated question. These reasons are based upon my interpretations of de Beauvoir's arguments about and reflections upon the role of sex within marriage.

46. See de Beauvoir, *Second Sex*, 439–523.

47. Ibid., 571–98.

48. Ibid., 721–52.

of "second-wave feminism,"[49] but she demonstrates that she manages their marriage on her terms in this episode.[50]

Megan Dances for Don ("A Little Kiss," S5/E1)

The fifth season opens with a ninety-minute episode that feels more like a film about the 1960s than a single episode of an ongoing television series. The first episode of the fifth season begins with a display of Don's and Megan's sexuality. Don's daughter, Sally, observes Megan's nakedness; her facial expressions display that she knows her father recently had intercourse with her beautiful and naked step-mother. Sally sees Megan lying naked in bed, and Don comes to talk to Sally wearing only his towel.

Megan plans a surprise party for Don's fortieth birthday and invites colleagues from their office. Already intoxicated before the party even begins, Don and Megan make out in the hallway of their apartment building. Roger Sterling ruins the "surprise," because he is also in the hallway, but once inside Don claims to the anxious crowd: "I am surprised!" While she sits on the couch talking with a gay man (the dialogue in the show labels him a "homosexual"), Don stares at Megan's legs as he converses with Harry and Roger—publicly displaying his desire for her. Megan interrupts the party to announce, "Okay everyone, my friends, first of all I wanted to thank you all for coming and second of all. . . . I think I've had just enough to drink that I am ready to give my own present to the birthday boy." Don looks deeply worried, and says, "Thank you, thank you all for coming. . . . I think we should call it a night." Megan quickly responds, "Absolutely not!" She continues but directs her speech only toward Don while she sits him down, "Don, you stay there."

"Did you buy him a pony?" asks Cooper. Megan walks up to the stage to join the band, grabs a microphone, counts off in French, and sings the French version of "*Zou Bisou Bisou*"[51]—a song intended as a public proclamation of one's love and translates into English as either "Oh! Kiss, Kiss" or "Oh, you ought to kiss me!" During Megan's provocative performance, Harry cheers Megan on by shouting inappropriate phrases. After the song

49. For a feminist critique of Betty Draper's character, see Ashley Jihee Barkman's "Mad Women: Aristotle, Second-Wave Feminism, and the Women of *Mad Men*," in *Mad Men and Philosophy*, Chapter 14.

50. For a contrarian, but not contradictory, interpretation of Betty's character through de Beauvoir's existentialist-feminist theory, see Chapter 4 in this volume.

51. "*Zou Bisou Bisou*," written by Bill Shepherd, Alan Tew, & Michel Rivgauche; originally recorded by Gillian Hills (Summer 1960).

ends: she sits on Don's lap, gives him a kiss, and tells him "happy birthday." After Megan's performance, Roger offers a toast: "If I may raise a glass. . . . To Megan, for letting us see the Don Draper smile usually reserved for clients. And to Don, you lucky so-and-so. As a wise man once said, 'The only thing worse than not getting what you want, is someone else getting it.'" Roger's words demonstrate his lust for Megan and envy toward Don for being the object of Megan's desire.

Megan's performance is not a striptease but, more accurately, a provocative performance of an intentionally sexy song. She lifts her skirt up and shows off her legs, but neither her undergarments nor even her clothing comes off during her performance. Toward the end of her performance, Megan takes her hand and slides it down from her breasts to her abdomen. She happily celebrates their marriage and their healthy sex life, and she remains comfortable and confident in utilizing her beauty and sexuality to give Don what (she thinks) only she can give to him.

After the party, Don falls into bed and tries to convince Megan not to clean the house. Megan agrees and tells Don that she wants to talk and have sexual intercourse. Don remarks that he has interest in neither talking nor having sexual intercourse because he needs to sleep. But then he does talk; he condemns her, "Don't waste money like that." Megan responds, "It was *my* money, and you don't get to decide what I do with it." Rolling his eyes, Don makes a request driven toward bringing guilt and humiliation onto Megan: "Well, could you please not use it to embarrass me again?" She deflects, "I know why you're upset . . . you're forty," and she puts her lower lip out. Megan tells him that she threw him a party because she loves him, and he informs her that he will go to sleep while she does whatever she wants—which, he knows, sounds contradictory because "she wants" to have sexual intercourse with him yet he continues to refuse her advances. She goes to the balcony, displaying confusion and defeat. The next morning, Don wakes up alone but with the news playing on the radio. He looks at himself in the mirror, while shaving, and this scene hints at the possibility of his feelings of regret.

Towards the end of this ninety-minute episode, Megan cleans the apartment wearing only her undergarments. Don asks her to put clothes on, and Megan tells him not to look at her. "You don't deserve to look at this," she says and then adds: "You're too old. I don't need an old person. You probably couldn't do it anyway." After an aggressive—almost violent—argument, where Don tells her that she wants him, and she tells him to get off her and to sit down, Megan shouts: "You can watch me from over there." He grabs her and kisses her, and they have aggressive coitus on the living room floor.

Afterwards, Megan says that no one at the office likes her. "I didn't want them in our home," Don tells Megan—*finally* offering a reason for why he did not want his own birthday party. This is the first scene where Megan begins to hint at leaving her job with the advertising agency; she works there only because he does, and she wants to be around him all day. Don tells her that he simply wants her to have what she wants, and she again initiates sexual intercourse. The camera allows viewers to see above them: viewers are offered a glance at a portion of her breast and Don's hairy stomach.

Megan's Dance from a Kantian Perspective

If a rational person wants to perform a provocative dance, then performing that dance *for their spouse* becomes the only way to render it a moral—and, therefore, rational—action. In her essay, "Kant and Kinky Sex," Jordan Pascoe rightly claims that Kant's theory of sex and sexuality surprisingly allows for the kinkiest of sexual actions and relationships—*if and only if* these actions take place within the bounds of the marital relationship.[52] For Kant, marriage offers the space and time for rational persons to explore "kinky" urges.[53] Megan's marriage, therefore, creates this space and gives her the

52. Jordan Pascoe writes: "[In] one way . . . Kant was rather radical about sex. Unlike many other philosophers and religious figures of his day, he rejected the idea that sex was about procreation and that sex was permissible only if it was procreative. Kant understood that sex was about pleasure and pleasure alone. . . . Kant thought that sex was about the heedless pursuit of pleasure at the expense of one's humanity and the humanity of one's lover. He thought it was the desire to objectify and be objectified, to debase and be debased, and an appetite so consuming as to be cannibalistic. Kant thought sex was unimaginably kinky. And, given that he knew very little about sex [empirically], this is not surprising. He's wrong to think that sex is inherently kinky. . . . Lots of sex is loving and respectful and even (imagine!) motivated by an appreciation for the humanity of yourself and your lover. . . . But we can think about this in another way: some sex *is* unimaginably kinky. Some sex is about hunger and debasement and objectification. So Kant is right, in a sense: sex is about pleasure, and some pleasure is kinky. And the trouble with kinky sex is that . . . it's awesome and consuming and highly pleasurable—and totally morally dangerous. Often, it means seeking out scenarios in which we are debased and dehumanized just because this is pleasurable. Sometimes, what we want is precisely to be used, to be dominated and devoured, and to take a break from all that [comes with the demands of] respect and dignity. So . . . we *can* take Kant's thoughts on sex seriously, as long as we understand that he's taking on the moral perils of unimaginably kinky sex. And if we read Kant's concerns about sex in this way, his claim that kinky sex can't be *transformed*, but only *quarantined*, seems more reasonable. . . . A relationship that's consistent with kinky sex isn't one that transforms our kinky urges, but one that creates a space in which we can explore them" (Pascoe, "Kant and Kinky Sex," 32–33).

53. See Pascoe, "Kant and Kinky Sex," 33.

time to publicize her sexuality and her sexual desire for Don. The context of marriage allows Megan to get her kink on!

Although Megan chooses a public setting to display her sexuality and sexual desire for Don, that choice does not authorize characters like Harry and Roger to make known their envious (Roger) and lustful (Harry) feelings. Harry's behavior, the next Monday at the office, only worsens in terms of the inappropriateness of his comments about Megan. Much later in the television series, after Megan and Don divorce in the seventh season, Harry meets with Megan over lunch—allegedly for professional reasons—and asks her to have sex with him. Harry lacks moral permission to make his lustful desires known to Megan, both during her provocative performance at Don's birthday party and when they dine together over a professional lunch meeting. Harry's character ought to be judged with harsh criticism on deontological standards.

The problem with this episode does not concern Megan's provocative dancing. Rather, the problem concerns how Don fails to fulfill his obligation toward his wife. Don ought to display gratitude for his wife's provocative performance,[54] and he ought to respond in kind to the sexual advances made by Megan. This is what marriage is for, and Don's failure to have sex with her should be considered a moral failure on his part.

Megan's Dance within MacKinnon's Framework

Interpreting Megan's dance as an emphasis on the "public" nature of Megan's sexuality allows us to connect this scene with MacKinnon's argument that the public nature of sex remains a necessary aspect of female sexuality. The public display of sexuality, in a variety of forms, serves as the pivotal problem of *Roe v. Wade*—which defines sexuality in terms of "privacy,"[55] as well as what I have developed as a B-concept understanding of sexuality. Megan does not define her sexuality strictly in terms of when Don's penis finds its way into her vagina (B-concept), and MacKinnon argues that law and policy ought to cease defining sex in the masculine-centered and reductive terms of vaginal penetration.[56]

54. Instead, he displays disappointment and embarrassment; he attempts to guilt and shame her about it as well. For a review that centers on Don's character in this episode, especially his response to Megan's dance, see Perpetua, "'Mad Men' Version."

55. See MacKinnon, "Privacy v. Equality," in *Feminism Unmodified*, 93–102.

56. I firmly concur with MacKinnon on this particular point: law and policy ought to shift from a B-concept to an A-concept understanding of sexuality. For a consciousness-raising example of the cruelty of the law concerning how a B-concept understanding of sex determines American policy about rape, see the meme

Megan's Dance within de Beauvoir's Framework

David Haglund, of *Slate* magazine, offers an interpretation of this scene closest to de Beauvoir's interests and purposes when he asks, "Could [Matthew] Weiner have found a more perfect and surprising song with which to convey the sexual liberation of Megan and her generational cohort? I doubt it."[57] Within de Beauvoir's framework, Megan's dance exhibits her *freedom* and promotes sexuality on her terms. In my judgment, this whole episode depicts Don's discomfort with the equality occurring within his marriage to Megan; simultaneously, Megan begins to come out of her shell and find her comfort and confidence through her sexual relationship with Don. After they have aggressive sexual intercourse on the living room floor, toward the end of the episode, Megan finally feels the *freedom* to inform Don that she wants to leave her job—the sequence here, how sex leads to freedom in other aspects of life, represents exactly what de Beauvoir has in mind when she discusses (in the "Conclusion" of *The Second Sex*) the equality achieved and freedom gained through heterosexual sexual relationships. Season 5 opens with Megan beginning to experience both this equality and freedom, and she refuses to back down from Don because she possesses the comfort and confidence to enjoy their sexual relationship on her terms.

Megan, Don, and Megan's Friend ("The Runaways," S7/E5)

Although I will not go as far as one commentator by calling it "rape,"[58] the final sexual encounter with Megan involves a ménage à trois where Draper's body language—despite his eventual complicity—can be interpreted as either a lack of desire or an unwillingness (a lack of volition) to participate. Draper's body language in that particular scene begs for a philosophical examination about the power dynamics at play in Megan's bedroom. I describe the pertinent details of this episode and then address the question: how does it help us to make sense of this *ménage à trios* if we view it with de Beauvoir's, Kant's, and MacKinnon's understanding of power in their theories of sex and sexuality?

Toward the end of "The Runaways," Don comes home from the bar and sees Amy and Megan acting very flirtatiously. Amy and Megan became

---

"If Mugging Were Treated the Same as Rape" (http://www.buzzfeed.com/derekj/if-mugging-were-treated-the-same-way-as-rape-r76).

57. Haglund, "What Was That French Song on *Mad Men*?"

58. See Evonne, "Don Draper Is a Rape Survivor."

friends in Los Angeles, and Don asked Megan to not have Amy around when he arrives to see Megan in Los Angeles. Megan throws a Hollywood party, and Amy attends this party. Don leaves the party to go to a bar with Harry and returns to find only Amy and Megan in their home. Amy offers Don marijuana, but Don refuses. He tells them that he feels tired, and he wants to go to sleep. Amy says to Megan: "There goes the fun." Don goes into the bedroom to get undressed. After knocking on the bedroom door, Amy walks into the bedroom to find Don without his shirt. She says, very sexy-like, "I'm supposed to tuck you in." Don looks confused and asks, "And what does your friend say?" Megan enters and says to Don, "I don't want you to be in a bad mood." Megan lies on the bed while Don remains in an upright position, still with no shirt. Don makes a request of Megan, "Stop playing around. . . . you're stoned." Megan smiles at him, tries to kiss him, and asks, "Don't you like Amy?" Amy giggles and becomes shy, "Leave him alone." Megan pats the bed, a gesture suggesting to Amy to sit next to Don. Megan rubs Don's chest and shoulder, and Don watches Amy sit down next to him. Now that Don finds himself between two younger women, Megan tries to persuade him: "This is the best place to be right now." Amy confirms, "Right here." Megan instructs Don to kiss Amy and tells him, "You know you want to." Don clearly states, "I don't want anything right now." Still wearing his pants, Megan takes her hand and grabs Don's groin area. Don's hands remain on the bed and on his own leg, and Megan says to him: "Don't lie." Megan kisses Don, her hand remaining on his covered genitals. Megan nods to Amy, seemingly granting her permission, and Amy leans over to kiss Don. Don's eyes are closed while Amy kisses him, and he seems to reciprocate this kiss. Megan continues to rub Don's chest, and Amy and Megan lean over to kiss one another in front of Don. Don looks even more confused now, and Megan pulls up from Amy to kiss Don again. Megan gently pushes Don onto the bed. Amy and Megan undress one another, and Don gazes upon them—mostly staring at Megan's body. Don tries to stand up, but Megan puts her legs on top of him—placing her own genitals, although still wearing her undergarments, over his groin area. Megan takes Don's hand and puts his fingers around the area of Amy's vagina, and viewers assume that Don touches Amy's clitoris. Amy leans down to kiss Don while he touches her, and Megan remains on top of Don although their clothing prevents penetration at this point. The scene stops here.

Viewers see Megan and Don, the next morning, asleep together—with Megan's hand on Don's shoulder. Don slowly wakes up and still seems confused. The camera broadens out, and Don's confusion intensifies as both Don and the viewers simultaneously see Amy also in the bed asleep. Don gets out of bed, gets dressed, and makes himself breakfast. Megan greets

Don, and Don complains that he cannot locate the coffee. Megan tells him that she will get it, and she looks at him seductively and says "good morning." He replies with a masculine-sounding "good morning," and they embrace to kiss. Amy comes out of the bedroom and says that she needs to go. None of them discuss the night before. While Don and Megan have no way to know this at the time, this ménage à trios turns out to be the final time that Megan and Don have sexual intercourse together.[59]

Amy, Megan, and Don from a Kantian Perspective

What would Kant say if he witnessed this scene with Amy, Megan, and Don? Although Kant fails to consider the moral status of a ménage à trios, his argument against polygamy actually applies to this particular scenario. Kant writes,

> So the sexual impulse creates a union among persons, and only within this union is the use of it possible. This condition upon utilizing the sexual impulse, which is possible only in marriage, is a moral one. . . . [It also follows] that nobody, even in *matrimonium*, can have two wives; for otherwise each wife would have half a husband, since she has given herself totally to him, and thus has a total right to this person as well. There are therefore moral grounds that tell against *vagae libidines*; grounds that tell against concubinage; and grounds that tell polygamy in *matrimoniu*; so in the latter we only have monogamy.[60]

In the sexual encounter depicted in this scene, Don gets only "half a wife" (Megan) because he gives himself totally to her but—by bringing a friend into their marital bed—she splits herself in two from the perspective of Don's sexual obligations. Don must attend to the sexual organs of Amy and his wife, Megan. Therefore, Megan acts immorally toward Don in this scene.

59. Viewers are led to assume that Megan and Don had sexual intercourse because Megan placed herself on top of Don; another reasonable assumption concerns Don and Amy *not* having penile-vaginal intercourse but only the vaginal penetration of Don's fingers into Amy.

60. Kant, *Lectures on Ethics*, 159.

De Beauvoir, MacKinnon, and the Ménage à Trios as a Philosophical Problem

In order to understand how de Beuavoir and MacKinnon might view this scene, I begin with Sarah Evonne's interesting claim that Don Draper was raped in this scene. Evonne begins her review of "The Runaways" with the proclamation: "We need to talk about the rape of Don Draper."[61] Evonne makes two claims to defend her argument that Megan raped Don. First, her definitional claim:

> When we are confronted with a narrative such as Don's, the question of rape is simply not present. We don't understand, and write it off as a steamy threesome, without mentioning that there was no consent given. The definition of consent varies, and in this case, it is best to turn to the definitions that some colleges employ. The definition used by Antioch College sets the foundation for consent as "the act of willingly and verbally agreeing to engage in specific sexual conduct." Reed College goes more in depth, demanding the standard of effective consent as "informed; freely and actively given; mutually understandable words or actions; which indicate a willingness to do the same thing, at the same time, in the same way, with each other." The second definition illustrates a standard sometimes described as "enthusiastic consent." If at the very least we use consent as prescribed by Antioch College, the encounter between Don, Megan, and Amy does not pass.

By these definitions of consent, provided by Antioch College and Reed College, Don did not consent to participate in the *ménage à trios.*

Evonne's second claim involves what she calls the "gender reversal test": "What if Megan was in Don's position, and Don and Harry . . . were in the positions of Megan and Amy? Would it . . . be seen as a sex scene? Would Megan's assertion, 'I don't want anything right now', be brushed over?" Evonne's second claim remains simple: if Megan had been the one wanting to go to sleep and refusing the advances of two men, this scene would be considered a rape—without delay, doubt, or hesitation from the television audience. *Questioning it,* as legitimate sexual intercourse, would be controversial.

Although quite interesting as a claim, there are two logical problems with Evonne's argument. First, the definitional claim remains inadmissible because the intent of these two definitions of consent directs itself at

61. Evonne, "Don Draper Is a Rape Survivor."

college-aged citizens in the twenty-first century. Neither definition takes into account what is required of consent in the context of marriage—or the 1960s/1970s. Although not quite a logical fallacy, this anachronistic argument comes close to fallaciousness.[62] Even putting this logical judgment aside, the second definition emphasizes both "words and actions"; we determined that Don's words refused the ménage à trios, but his actions participated in it—returning Amy's kiss, touching Amy's clitoris, and staying under Megan's straddling position.

Secondly, the narrative of *Mad Men* does not lend itself to the "gender reversal test." We cannot make a proper moral judgment on a hypothetical but only on what actually happened—or, better stated, what occurs within the narrative context of the world of *Mad Men*. The "gender reversal test" involves too much of an unrealistic hypothetical. It also ignores the nuanced character of Don Draper, who would never *share* his lover with another man. Perhaps he does not wish to share his lover with another women—*that* is the question here. Introducing an alternative hypothetical does not allow us to address *that* question but deflects us away from the characters to an abstract hypothetical scenario that ought to be considered nonsensical in relation to the narrative world of *Mad Men*.

For Catharine MacKinnon, men cannot be raped by women in the current context of Western society because power dynamics always favor the male partner.[63] We cannot and should not consider Don to be raped in this scenario because men cannot be raped by women. Rape requires power, and Don goes into the bedroom with patriarchal power. Yes, we can talk about "the rape of Don Draper"; we will conclude, however, that no rape occurred here: only a powerful man put in an uncomfortable situation.

MacKinnon offers no *feminist* definition of consent. She deconstructs several legal definitions of consent.[64] She concludes her chapter, "Rape: On Coercion and Consent," with these words:

> [W]hen an accused wrongly but sincerely believes that a woman he sexually forced consented, he may have a defense of mistaken belief in consent or fail to satisfy the mental requirement of knowingly proceeding against her will. Sometimes his knowing disregard is measured by what a reasonable man would disregard. This is considered an objective test. Sometimes the disregard need not be reasonable so long as it is sincere. This is

62. Some historians, but usually not logicians, call it a fallacy of *nunc pro tunc* (now-for-then or the-now-applies-to-then).

63. See MacKinnon, "Desire and Power," in *Feminism Unmodified*, 46–62.

64. See MacKinnon, *Toward a Feminist Theory*, 172–83.

> considered a subjective test. A feminist inquiry into the distinction between rape and intercourse, by contrast, would inquire into the meaning of the act from a women's point of view. . . . What is wrong with rape in this view is that it is an act of subordination of women to men. It expresses and reinforces women's inequality to men. Rape with legal impunity makes women second-class citizens.[65]

A "feminist inquiry into the distinction between rape and intercourse," in the context of a consideration of Don Draper with two women, would lead us away from the B-concept emphasis on this particular ménage à trios and toward A-concept considerations of Don's and Megan's marriage. How many times did Don's actions and words make Megan feel "subordinate" to him? In what ways does Don's sheer presence, in Megan's life, reinforce her "inequality" in relation to him? Does he make decisions, in his life, that pave the way for her to become a first-class citizen; or, do most of his decisions sustain her status as a "second-class citizen"?

To analyze this *ménage à trois* properly, we ought to consider de Beauvoir's words on the necessity of alterity and asymmetry within a healthy and well-balanced sexual experience:

> [I]n love, tenderness, and sensuality woman succeeds in overcoming her passivity and establishing a relationship of reciprocity with her partner. The asymmetry of male and female eroticism creates insoluble problems as long as there is a battle of the sexes; they can easily be settled when a women feels both desire and respect in a man; if he covets her in the flesh while recognizing her freedom, she recovers her essentialness at the moment she becomes an object, she remains free in the submission to which she consents. Thus, the lovers can experience shared pleasure in their own way; each partner feels pleasure as being his own while at the same time having its source in the other. The words "give" and "receive" exchange meanings, joy [becomes] gratitude, pleasure [turns into] tenderness. In a concrete and sexual form the reciprocal recognition of the self and the other is accomplished in the keenest consciousness of the other and the self. Some women say they feel the masculine sex organ in themselves as part of their own body; some men think they *are* the woman they penetrate; these expressions are obviously inaccurate [because] the dimension of the *other* [needs to] remain . . .; but the fact is that alterity no longer has a hostile character; this consciousness of the union of the bodies

65. Ibid., 181–82.

> in their separation is what makes the sexual act moving; it is all the more overwhelming that the two beings who together passionately negate and affirm their limits are fellow creatures and yet are different.[66]

This passage beautifully and wonderfully captures the dynamics of a healthy sexual relationship, and (in my terms) it comes across as an A-concept understanding of sexuality. There are three important points, from this passage, for considering the philosophical problems of the ménage à trios between Amy, Don, and Megan.

First, while Don may not have verbally consented to the *ménage à trios*, we could interpret Megan's persistence as her attempt to "negate and affirm [Don's] limits"—to take him out of his comfort zone of control and power and to show him another illuminating sexual experience where "This is the best place to be right now." Philosophically, this makes the *ménage à trios* an *epistemological* problem—not a *moral* one. How does Megan know where Don's limits are, and does Don communicate well enough to his wife about his boundaries? Boundaries can be honored only when they are articulated.

Second, the *moral* problem does not seem to involve his consent; rather, the *moral* problem concerns the role of asymmetry within a *ménage à trios*. How does shared *otherness* occur between three people? Is it possible to achieve "a relationship of reciprocity" when three people are involved? Can a "union of bodies" be accomplished between three people? Granted, de Beauvoir does not defend a traditional notion of complementarianism in her theory of sexuality. However, she consistently emphasizes the significance of the male partner giving himself fully to the female partner—while maintaining his masculinity—and the female partner finding true freedom with her male partner. In the context of a ménage à trios, this reciprocity seems to get lost because of the addition of another sexual partner.

Third, while Don may not want this to happen, a *ménage à trios* might be what Megan needs. As de Beauvoir says, "the lovers can experience shared pleasure in their own way; each partner feels pleasure as being his own while at the same time having its source in the other."[67] Megan's attempt at achieving pleasure, in her own way, might require another partner—especially if Megan wants *both* to be with a woman *and* to remain faithful to Don.

66. De Beauvoir, *Second Sex*, 415.

67. Ibid.

CONCLUSION

In this chapter, I have utilized three different theories of sex and sexuality to make sense of Don Draper's sexuality and sexual relationships. For my conclusion, I enter into the record my own judgments about the edification and usefulness of these three theories in relation to watching *Mad Men* and reflecting upon the questions that *Mad Men* poses to its viewers.

MacKinnon leads us, not necessarily to a radical feminist hermeneutic for interpreting the characters and plots of *Mad Men*, but to a despairing and pessimistic stance toward the characters and plots of *Mad Men*. If MacKinnon's arguments and theories are accurate, the male characters remain vicious human beings with no way out of their viciousness except for absolute and disciplined solitude. If her arguments and theories are accurate, the female characters remain trapped in a patriarchal world with little-to-no agency and no way to maneuver based upon their ambition, hopes, and dreams. Perhaps this describes the way that we ought to interpret the characters and plots of *Mad Men*, but we need to identify it as despairing and pessimistic.

Simone de Beauvoir's theory of sex and sexuality gives us a very hopeful and optimistic interpretive lens for making sense of the characters and plots of *Mad Men*. For the female characters, de Beauvoir provides terms sufficient for showing how some of their sexual encounters ought to be judged as edifying and freeing. When attending to her sex life, Betty's character comes out better than other feminist treatments of her character allow. Don may not be completely power mongering in the bedroom, and de Beauvoir offers us the tools to clearly distinguish between and identify Don as a good lover and a terrible one. Through their sexual relationships, some of the female characters achieve equality and freedom—at least on the terms de Beauvoir sets out in the conclusion of *The Second Sex*. Of the three theories of sex and sexuality, I find de Beauvoir's the most helpful and useful for interpreting the characters and plots of *Mad Men*.

While de Beauvoir's existentialist-feminist theory of sex and sexuality proves to be the most helpful and useful, Kant's deontological theory of sex and sexuality provides surprising nuggets for interpreting the sexual aspects of Don's marital relationships. Perhaps the most surprising Kantian interpretation of *Mad Men* involves how Megan's provocative performance becomes justified through a deontological lens while Don's refusal to engage in coitus with her—after she dances for him—violates the marital obligation Don maintains to Megan. Also, Kant's defense of divorce (in *Lectures on Ethics*) helps us see clearly Betty's moral justification for divorcing Don.

Finally, Kant's problem with polygamy surprisingly applies to the *ménage à trios* between Don, Megan, and Amy.

Mad Men certainly teaches us the severe limitations of a B-concept understanding of sex and sexuality. Sex and sexuality cannot and should not be defined in terms of the bookends of vaginal penetration and ejaculation/orgasm. Instead, sex and sexuality ought to be understood in more A-concept ways—which includes the small gestures that lovers give to one another, raising children together, and sharing in one's hopes and dreams. The characters on *Mad Men* present themselves at their best when they perform these everyday tasks with the people they love and when they act intentional about being attentive lovers, playful parents, and sensitive spouses.[68] Through these intentional actions,[69] the characters ensure that their vulnerability during sexual intercourse will not be used to bring them harm but, rather, will contribute to their true end or *telos*—happiness.[70]

BIBLIOGRAPHY

Anscombe, Elizabeth. *Intentions*. Cambridge, MA: Harvard University Press, 2000.

Barkman, Ashley Jihee. "Mad Women: Aristotle, Second-Wave Feminism, and the Women of *Mad Men*." In *Mad Men and Philosophy: Nothing Is As It Seems*, edited by Rod Carveth and James B. South, 203–16. Hoboken, NJ: John Wiley & Sons, 2010.

Bauer, Nancy. *Simone de Beauvoir, Philosophy, and Feminism*. New York: Columbia University Press, 2001.

Cahill, Lisa Sowle. *Sex, Gender, and Christian Ethics*. New York: Cambridge University Press, 1996.

Carveth, Rod, and James B. South, ed. *Mad Men and Philosophy: Nothing Is As It Seems*. Hoboken, NJ: John Wiley & Sons, 2010.

68. Several chapters in this volume hit on these themes: see Chapters 4 and 15 on questions concerning parenthood; see Chapters 13 and 14 on the question of sensitive vs. insensitive spouses.

69. I use this phrase in line with Elizabeth Anscombe's development of "intentional actions" in her *Intention*, 1, 9–11, 30–33, 37–41, 84–94.

70. Thank you to Ann Duncan for her patient and prudent editorial work on this chapter; she encouraged me to write on the subject matter and helped shape the chapter every step of the way. Lindsey Graber played a pivotal role as the research assistant for this chapter, and she helped me achieve a significant amount of clarity on arguments about Simone de Beauvoir's existentialist-feminist theory of sex and sexuality; she also provided helpful editorial insights once the chapter was written. Jessica Boylan, Morgan Elbot, Kirsten Guidero, Kari Nilsen, and Phil Kuehnert greatly improved the chapter with their editorial comments and corrections. Angela McWilliams Goodson watched every episode of *Mad Men* with me, and she gave her approval of the content of this chapter; I remain deeply grateful for our life together.

De Beauvoir, Simone. *The Ethics of Ambiguity*. Translated by Bernard Frechtman. New York: Citadel, 2000.

———. *The Second Sex*. Translated by Constance Borde and Sheila Malovany-Chevallier. New York: Vintage, 2011.

Evonne, Sarah. "Don Draper Is a Rape Survivor." *Medium*, May 12, 2014. Online: https://medium.com/@SarahEvonne/don-draper-is-a-rape-survivor-c238657d2bea#.sh6qfhhuj

Goodson, Jacob L. "'The Woman Question': William James's Negotiation with Natural Law Theory and Utilitarianism." In *Feminist Interpretations of William James*, edited by Shannon Sullivan and Erin Tarver, 57–78. University Park, PA: Pennsylvania State University Press, 2015.

Haglund, David. "What Was That French Song on *Mad Men*?" *Slate*, November 3, 2012. Online: http://www.slate.com/blogs/browbeat/2012/03/25/mad_men_season_5_premiere_why_did_don_s_wife_megan_sing_that_french_song_.html.

Herman, Barbara. "Could It be Worth Thinking About Kant on Sex and Marriage?" In *A Mind of One's Own: Feminist Essays on Reason and Objectivity*, edited by Louise M. Antony and Charlotte E. Witt, 53–72. Boulder, CO: Westview, 2002.

Kant, Immanuel. *Groundwork of the Metaphysics of Morals*. Translated by Mary Gregor. New York: Cambridge University Press, 1997

———. *Lectures on Ethics*. Translated by Peter Heath. New York: Cambridge University Press, 1997.

———. *The Metaphysics of Morals*. Translated by Mary Gregor. New York: Cambridge University Press, 1996.

MacKinnon, Catharine. *Feminism Unmodified: Discourses on Life and Law*. Cambridge, MA: Harvard University Press, 1987.

———. *Toward a Feminist Theory of the State*. Cambridge, MA: Harvard University Press, 1989.

May, Rollo. *Love and Will*. New York: W. W. Norton, 1969.

McTaggart, J. M. E. "The Unreality of Time." In *Time: Readings in Philosophy*, edited by Jonathan Westphal and Carl Levenson, 94–111. Indianapolis, IN: Hackett, 1908.

Meilaender, Gilbert. *The Limits of Love*. University Park, PA: Pennsylvania State University Press, 1987.

Millett, Kate. *Sexual Politics*. New York: Columbia University Press, 2016.

Pascoe, Jordan. "Kant and Kinky Sex." In *What Philosophy Can Tell You About Your Lover*, edited by Sharon M. Kaye, 25–36. Chicago, IL: Open Court, 2012.

Perpetua, Matthew. "'Mad Men' Version of 'Zou Bisou Bisou' in iTunes Store." *Rolling Stone*, March 26, 2012. Online: http://www.rollingstone.com/movies/news/mad-men-version-of-zou-bisou-bisou-in-itunes-store-20120326.

Soble, Alan. "Introduction: The Fundamentals of the Philosophy of Sex." In *The Philosophy of Sex: Contemporary Readings*, edited by Alan Soble, 3–23. Lanham, MD: Rowman & Littlefield, 2002.

Chapter 13

EXITUS ET REDITUS IN MARRIAGE

Mad Men vs. Hollywood Remarriage Comedies

—BRANDON L. MORGAN and JONATHAN TRAN

We want to begin by contrasting two portraits of marriage, one gathered from the investigations of ordinary language by Stanley Cavell and one inferred from the words of *Mad Men*'s Don Draper. The first can be found initially through Cavell's important study of 1930s and 1940s era Hollywood remarriage comedies. These early films bear some common features to their Shakespearean comedy forebears, most noticeably their attention to the domesticity of everyday life as that sphere in which happiness is pursued and hopefully discovered. One of their added features is the contemplation of divorce and remarriage and, thus, of the space of marriage as one of conversation, deliberation, reconciliation and acknowledgement. Marriage in the form of remarriage "recapture[s] the full weight of the concept of conversation, demonstrating why *our* word conversation means what it does, what talk means . . . learning to speak the same language."[1]

While many—for Cavell's interest, Shakespearean—comedies require overcoming certain objective obstacles on the way to marriage, remarriage comedies attempt to overcome marriage itself (or ourselves in marriage) that is, the temptation to avoid the constraints that marriage places on our speech, desires and futures. In Cavell's words, the hope for such domestic

1. Cavell, *Pursuits of Happiness*, 87–88.

conversation, "invokes the fantasy of the perfected human community, proposes marriage as our best emblem of this eventual community—not marriage as it is but as it may be—while at the same time it grants . . . that we cannot know that we are humanly capable of achieving that eventuality, or of so much as achieving a marriage that emblematizes it, since that may itself be achievable only as part of the eventual community."[2] For Cavell and the film genres he describes, marriage names one possibility for the mutuality of voice and the space for self-transformation. It emblematizes the possibility of equality and moral education working in tandem to sustain, or more often to recoup, a trust in our common speech and moral growth. These possibilities (and here an overlap with certain metaphysical claims of Christian theology) remain possible by a fact of language, a fact of how language works out, itself a function of the kinds of creatures we are (hence those metaphysical claims), just as do possibilities of denial and departure.

The second portrait of marriage comes initially in Season 6 of *Mad Men,* as Don analyzes an ad campaign for air fresheners with an image of a happily newlywed couple in the background. The tag line reads, "Love is in the air." Don responds to the concept of "love" this way.

> Don: It's a big word . . . what is this?
>
> Staffer: They're newlyweds.
>
> Don: This couple doesn't exist . . . anything matrimonial feels Paleolithic. . . . let's try to trade on the word love as something substantial . . . [W]hy are we contributing to the trivialization of the word? It doesn't belong in the kitchen. "I love this, I love my oven. You know what I'd love? A hamburger." We're wearing it out . . . we want that electric jolt to the body. We want *eros*. It's like a drug; it's not domestic.[3]

Don allies a "substantial" concept of love with the sparks and spontaneity of the erotic, though willingly unhinging its association with the matrimonial and the domestic, claiming that appealing to such modes of life is radically outdated—prehistoric.

This remark from Don speaks volumes about how he and other characters in *Mad Men* picture marriage for themselves. It is a concept seemingly homeless. Or at least its home has been lost at some point and what we have left is a shell of the concept—no context for its use. This appears to be what *Mad Men* often and perhaps most often says about marriage and domestic life, something Don's claims here summarize. To any dedicated

2. Ibid., 152.

3. "The Doorway," S6/E1.

viewer of the show, *Mad Men* can often subject its viewers to some of the bleakest portraits of marriage in all of television.

Marriage in *Mad Men* can easily be read as a swan song to the institution of marriage as such, of its possibilities to establish the conditions for trust, dependence, self-knowledge—in short, happiness. For Don and others, marriage as an institution is up for grabs, foreshadowing, and ostensibly explaining, social realities commonplace in our time, *Mad Men* as genealogy dramatized. Marriage depicted in *Mad Men* is susceptible to its usefulness or lack thereof in inciting sentimentality or nostalgia among potential buyers of a product. Don's own cynicism here speaks truthfully in that such romantic appeals to nostalgia are empty. Marriage as an institution no longer has authorization or ratification but rests on the maintenance of the participants themselves—a maintenance that *Mad Men* portrays as a lost, ancient art.

Yet, *Mad Men* also raises—periodically and evanescently—notions of recovery, self-knowledge and apology, risking scenes that appear ripe for, if not redemption, at least acknowledgement of what has been forgone in the loss of mutual voice. Read straight, these scenes might indicate hesitation about the show's otherwise steady march into cynicism. We remain intent in this essay to show them as more. For Cavell, marriage preserves the possibility of shared life amid a world that avoids or rejects our everyday commonness as human beings. It names a space in which the temptation toward various skepticisms might be faced and lived through with the help of another. For *Mad Men*, the question of marriage's legitimacy appears to name its failure and outdatedness and, thus, a form of the ordinary and the domestic to which persons no longer find reason consenting.

However, reading *Mad Men* alongside Cavell's claim that marriage resides between the acknowledgement and avoidance of shared speech in conversation reveals its bouts of cynicism about marriage and mutuality as more recoverable, or at least less inevitable, as they might otherwise be. Hence we read *Mad Men*'s cynicism as threatening rather than fated; in this vein, hopefulness—of the kind piqued by Cavell on marriage as comedy—is seen as lying in the same soil as that which saps hope, both emerging from the fructifying conditions of human words and worlds, to the conclusion that the same conditions that allow words and the human worlds constituted by them to atrophy, threatening cynicism, make for other possibilities, provoking hope. We call these conditions, biological as they are, "ordinary." Where *Mad Men* and Cavell's philosophy of the ordinary agree concerns the belief that marriage is struggling to rediscover its legitimacy and, in so doing, adapting itself to a struggle often won (or thwarted) by no more than one's own willingness to go on (or not). The stakes of going on here include not only the possibility of hope (rather than despair), but of talking, and so of companionship. Where philosophies of the ordinary and Christian theology agree, as dramatized in this essay by *Mad Men*'s scenes from a marriage, is in an unrequited claim about the semblance of conversation, life, and hope, according to which Christianity says something about God.

MARRIAGE AS CONVERSATION

Stanley Cavell is one of our most important interpreters of what has gone by the name of ordinary language philosophy, a mode of thinking that follows the path of J. L. Austin and the later Wittgenstein in returning words from metaphysical to everyday uses—or, to put it differently, from the world of fantasy to the world of domestic conversation. Cavell's developed account of marriage, worked out through his interpretation of early Hollywood talkies, is in response to his complicated picture of language and its relationship to skepticism. In philosophical parlance, this is the view that human knowledge is universally insufficient to secure the existence of the world and the presence of others. Most importantly for us, Cavell claims that the meanings of our words do not lay in the establishment of philosophical rules or requirements or in correspondence with universal ideas, mental states, or external world objects—medium-sized dry goods (paraphrasing Austin) or otherwise—but in the ordinary agreements in uses that speakers share. In Cavell's words, "There is nothing deeper than agreement itself."[4] Yet those agreements are not everywhere available for inspection, nor are they arbitrarily established by fiat but reside in an attunement created through learning concepts with other speakers who live in the world they do together. Agreement (or its absence) is not so much established as discovered.

One goal of Cavell's philosophy is to explore the surprising depth to which our agreements in language extend. Such lengths are surprising because agreements do not simply reside at the level of social convention, but reach to those everyday human concepts of eating, walking, thinking, desiring etc. that both constitute our very sense of humanness and grow out of the kinds of creatures humans are—the bodies we are, the environs we inhabit, so on and so forth. Only by appreciating the eventuality and degree of these agreements can one also appreciate what is at stake when agreement comes to an end. A second goal involves exploring these stakes when agreement gives out. Those stakes are artificially raised, and hence flattened, under philosophical skepticism—which views given failures in linguistic agreement as connoting the notion that words as such are useless

4. See Cavell, *Claim of Reason*, 32

in making us intelligible to one another, that that which matters—that beyond language (universals, mental states, or medium-sized dry goods)—is the very thing language cannot convey, the very thing language shrouds or blocks. According to the skeptic, we are trapped in unknowability by the words we use because they can neither grant the knowledge of others or of the world we crave nor the knowledge we crave to grant them of ourselves.

Part of the hope of Cavell's work on language is to investigate this craving and the kinds of responses it elicits. Part of the challenge for any theological investigation into what happens when, as Rowan Williams puts it, words come under pressure (marriage being that eventuality where agreement lives under continuous close quarters pressure) requires remaining attentive to, rather than avoiding (i.e., "I won't understand you because I cannot"), the fragile conditions that make for agreement, and its ending. Such attention comes with some difficulty and often no small amount of pain (e.g., "I don't understand you because I refuse to")—which helps us place the presence of cynicism as the avoidance of these conditions and skepticism as their denial.[5]

The practical outcome of the first goal of Cavell's picture of language is that we speaking animals can appeal to nothing beyond everyday human words and concepts in order to make ourselves intelligible. We have nothing beyond "our sense that we make sense" to others in the language we speak. However, this also means that we *need* nothing beyond our shared concepts for obtaining mutual intelligibility. The sense that we need more than what language can give us is, on Cavell's reading of philosophical skepticism, the tendency for humans to deny their own humanity, deny the very conditions that make them human and so therefore to relinquish the possibilities language grants including a whole range of human intimacy. Here lies the significance of Cavell's famous adage of the crucified body as the ultimate picture of the unacknowledged person.[6]

If we can resist the fantasy of a mutuality beyond or before or after (in other words, better than) the words we have, we can arrive at a practical outcome of the second goal of Cavell's work—namely, the discovery that the end of language entails the end of our world. To acknowledge the end of ourselves in those moments is to recognize our separateness from others, a separateness as much a part of our humanness as is our mutuality—each constituting the possibility of the other. Going on with another beyond this point will require a willingness to see oneself as incomplete, as needing another to recognize one's limits and possibly to deepen or expand those

5. Williams, *Edge of Words*, viii, 3.

6. Ibid., 430.

limits by way of mutual conversation. In the world of human speech, we are "bound to become lost and to need the friendly and credible words of others in order to find [our] way. . ."[7]

Common pursuit of conversation does not negate our incompleteness vis-à-vis others; it does not dispel what Cavell calls the "threat of skepticism"—which attends all human speech. The temptation to repudiate our shared words is not so much overcome as repeatedly faced and lived through, which means refusing again and again taking oneself as exempt from language, affirming again and again our aliveness to speech. To see meaning as residing in our agreements, which is to say, residing with *us*, suggests that the possibility of recouping and maintaining conversation amid disagreement lies in our own patience and generosity, our own humility and hope for moral growth. Conversation's future requires perpetually attending to our future together and to our own habits of recuperation. A remarkable claim of Christian theology is that these same facts of language determine speech about God—since God in Christ has spoken ordinarily when extraordinarily speaking about himself—making theology (that is, our speech about God) as ordinary and hence as dependent, dispossessive, tempting, and threatening as all of our other ordinary occurrences of speaking of and to one another.

Cavell's interpretation of remarriage comedies allegorizes this picture of language by seeing remarriage as exemplifying the importance of ordinary speech through the diurnal and domestic practice of matrimony. For Cavell, remarriage comedies find their inspiration from Shakespearean comedy but diverge from them in (at least) two key ways. They both narrate the coming together of a couple or couples in matrimony. While Shakespearean comedies present an obstacle external to the central pair, remarriage comedies present the couple's relationship itself as the obstacle, an aspect unavailable to the audience until the conclusion. The second difference, following from the first, is that while the principle pair in Shakespearean comedies seeks to achieve marriage, the pair in remarriage comedies seeks marriage *again*—that is, coming back together after separation or divorce. Cavell sees such comedic recouping as expressing the pursuit of shared marital happiness in terms of the search for a common language.

Thus, marriage is always remarriage: always the search for a next step together in the wake of missteps and avoidances—realities that humans inevitably face given their life in words. Cavell often describes such a pursuit as the return to innocence after trespass—which, in turn, depends upon seeing innocence as a something capable of being rewon. In addition,

7. Cavell, *Cities of Words*, 16.

Cavellian marriage is akin to Aristotelian friendship in that what is sought and maintained (in its most noble forms, at least) is the friendship itself and not a form of relation for the sake of external gains.[8]

To still speak about and pursue marriage for Cavell (or friendship, for that matter) is to refuse skepticism its satisfaction in seeing communicative barriers as deserving the repudiation of our shared words, which is to say, of a future for us. It involves reviving our interest in talking and conversing, to overcome narcissism by way of trusting speech to do what we need it to do. Cavell finds conversation in these early Hollywood talkies as central to their plots of matrimonial recuperation, not simply because the capacity to represent human life as simultaneously visual and aural was a technological achievement for film, but that such technology registers a vital cultural achievement (even a philosophical achievement) in (re)turning attention, trust and responsibility to ordinary language as what constitutes marriage. In Cavell's words, "Talking together is for us the pair's essential way of being together, a pair for whom, to repeat, being together is more important than whatever it is they do together."[9]

An outcome of seeing marriage as largely constituted by the repeated human acceptance of ordinary language and the pursuit of common speech amid our separateness raises the question of what constitutes and counts as a marriage—which turns out to be a central concern for Cavell's interpretation of remarriage comedy.

> It is part of our understanding of our world, and of what constitutes an historical event for this world, that Luther redefined the world in getting married, and Henry the Eighth . . . in getting divorced. It has since then been a more or less open secret in our world that we do not know what legitimizes either divorce or marriage. Our genre emphasizes the mystery of marriage by finding that neither law nor sexuality (nor, by implication, progeny) is sufficient to ensure true marriage and suggesting that what provides legitimacy is the mutual willingness for remarriage, for a sort of continual reaffirmation . . .[10]

According to Cavell, the conversation of marriage denudes the attempts to secure marriage institutionally (e.g. ecclesially, politically)—which in

8. Quoting Cavell, "What this pair does together is less important than the fact that they do whatever it is together, that they know how to spend time together, even that they would rather waste time together than do anything else—except that no time they are together could be wasted. Here is a reason that these relationships strike us as having the quality of friendship. . ." (Cavell, *Pursuits of Happiness*, 88).

9. Ibid., 146.

10. Ibid., 141–42.

turn brings into question how society as such is to be legitimized. Cavell has his own complicated answers to that question, but for our concerns suffice it to note that marriage as continual conversation involves seeing the future of marriage as residing in the willingness of the participants to trust that their words have a common future. Part of the challenge of Cavell's philosophy is in seeing that the future of marriage is not granted by something more than this. The other part of the challenge comes in seeing that it is also not denied by something more than this. Its granting and denial will depend largely on the moral possibilities of the persons involved. "The virtues most in request here are those of listening, the responsiveness to difference, the willingness to change."[11]

This issue about marriage as institution leads into questions about the church's proper comportment toward marriage. It is often thought that the church is the home of this institutionalization and, therefore, responsible for many of our cultural attitudes about marriage—including, rightly or wrongly, the overinflated ones. An account of human life that aspires to the ordinary conditions of language deflates the standard romanticisms, including those that follow on what we have characterized through Cavell as a banal Shakespearean narrative arc that presumes marriage insofar as it is marriage (two people as now married) as complete. It will find confounding any conception of marriage that does not view marriage a species of friendship, even if one with very unique features (i.e., duration, sex, begetting children, etc.).

Approaching conversation as the practice and enactment of friendship and marriage as internal to friendship, it prepares a vision of marriage as the kind of conversation (i.e., unfolding over time, as one body, hospitable, etc.) fit to bear the weight of the humanness we have discussed. Inasmuch as it resists marriage as fantasy—or, fantasy as marriage—it admits of a picture of marriage attentive to the fragile conditions that make for attunement and heralds the event of agreement. Unlike the banal romantic sense, this theological sense of marriage does not presume agreement a possession of marriage, but rather a moral achievement along the way. By placing marriage within the life worlds of the kinds of bodies humans are, a theology attentive to ordinary language institutionalizes marriage as natural convention—though, undoubtedly, not the only convention conducive to human life in words. By narrating ordinary language conditions as natural to human life and emplotting sin as that denial which denigrates creaturely life, marriage as church institution initializes a locus of meaning that goes before any marrying couple as the sacramental life—which directs desire towards certain

11. Ibid., 174.

ends, recognizing the all-too-human tendency to deny our humanity. We continuously call ourselves back to humanity, where repentance indicates something like remarriage.

ANOTHER COUNTRY

Our description of Cavell's picture of marriage as constituted by a hope in mutual conversation is meant to cast light on *Mad Men*'s own picture of marriage and its suggested commentary on the trustworthiness of speech. Cavell himself admits that his picture might look utopian. "So many terrible charges can be brought against the institution of modern, or say, bourgeois, marriage, that it can sometimes seem a wonder that sensible people who have choice in the matter continue to seek its blessings and accept its costs."[12] He acknowledges hope in the institution may be but wager on marriage and might be understood best as his faith that words and those animals that speak them have a possible future worthy of pursuit. One way of interpreting *Mad Men*, especially in light of this Cavellian hope, is as an account of why that wager is no longer worth making and, thus, why hope in this case is misplaced. Akin to Cavell, *Mad Men* raises the question of marriage's legitimacy and finds in domestic life a sphere of investigating how marriage might be viewed in terms of pursuing a shared world of words. The difference being that, in *Mad Men*, conversation gives way so often to irony, cynicism, and doublespeak that the reinvention of marriage as a pursuit of mutual voice might be insufficient to save it.

The show raises these questions by exposing its audience, regularly and relentlessly, to a stark division of worlds—suburbia and Madison Avenue.[13] The presence of the competing worlds or spheres of existence presses the audience toward judgments regarding the relevance of concepts within ordinary domestic life, forcing us to decide about marriage and its possibility, directing us to either despair or hope. From the perspective of Don and the advertising industry he embodies, the world of suburbia naïvely theatricalized love and marriage, falsely presumed to carry substance, while the world of Madison Avenue fully understands theatricality as bespeaking an exhaustion of concepts such as "love" and "marriage," opening them up to commodification. In Don's words, "What you call 'love' was invented by

12. Cavell, *Cities of Words*,16.

13. This is revealed most starkly in "Smoke in Your Eyes," S1/E1, where Don's spends his day sneaking away from work to sleep with a woman in town only to return home to Betty and his family.

guys like me to sell nylons."[14] Compare Betty's words upon visiting Sterling Cooper: "It's like walking into another country where I don't speak the language."[15]

Mad Men's competing worlds portray the reality of an impasse in the mutuality of concepts and their dependability for making speakers intelligible to one another. Don's claim that love's association with matrimony is woefully outdated and brazenly overused indicates a waning faith, if not yet full-blooded skepticism, in the ability for marriage to create and sustain the rapid progress in human modes of relation. The show's questioning of everyday concepts like "love" and everyday forms of life like marriage furnishes central insights about language and life, and *Mad Men* is intent on exploiting these insights for the sake of exploring contemporary life. After all, the show takes place at a point in American history where the nation is truly becoming "another country," at point from which we might aptly see the past as nostalgically frozen, no longer apt for cultivating self-knowledge but only for marketing fantasy.[16] Moreover, those, like Don Draper, set to benefit from a certain American future—called it neoliberalism—have little compunction betraying a vested interest in the subliming or ironizing of speech in order to relieve themselves of their responsibility to it. This vested interest in turning language into so many objects of will helps to explain Don's cynical remarks about the concept of "love." Cynicism risks displaying itself viciously as skepticism about the possibility of marital life, as well as the future of conversation as such.

In Cavell's work, one finds a mode of exploration that portrays *Mad Men*'s hopelessness as a version of the philosophical skepticisms that sees our shared words, our mutual voice, as initially or eventually inadequate. Since the legitimacy of marriage rests on the ability of speech to give us one another, the demise of the former, then, is seen as indicating the inadequacy of the latter, eventuating in fantasy and a viciously attending skepticism. Watching the show, one gets the feeling that by trading in mutuality for the theatricalization of mutuality characters like Don Draper and Peter Campbell are denying or refusing to see something particularly human in their wives, something that if seen might also save their own humanity, or conversely relinquish it. They think themselves clever, but the show's brilliant

14. "Smoke in Your Eyes," S1/E1.

15. "5G," S1/E5.

16. Recall at one point that Betty's friend, Francine, calls her "old-fashioned," a term Betty agrees to, which is not surprising given that her character most often embodies a version of matrimonial happiness that Don also sees as "old-fashioned" to a fault.

reflexivity reveals them to be fools or worse. Their claim of impossibility is in all honesty a refusal of possibility.

Mad Men initially assesses the sustainability of marriage as a concept of relation, whether it can survive on the seemingly fragile ground of agreements upon which it rests. But as it goes along, the show arrives, finally, at its last question: should we go further? Should we see *Mad Men* as a representation of agreement having already failed, resulting in skepticism not only about marriage but the capacity of mutual intelligibility itself? In this case, the presumed failure of the latter entails the failure of the former and much else besides. Whether or not we *should* or whether or not *Mad Men* in fact *does* go further here is what we might call *Mad Men*'s voice of temptation, a voice that tempts its audience to see, as Don Draper and Peter Campbell do, particular instances of mutuality's denial as the universal (metaphysical?) rejection of its possibility. That temptation, involving a universal maligning of speech as fit to our whims, will play out as a rejection of the possibility of responsiveness to the separateness of another, blinding one to one's own incomplete moral education. Accordingly *Mad Men* stages what happens when one no longer recognizes the possibility of moral growth in another or in oneself, where exempting myself from language entails exempting myself from a moral future. *Mad Men*'s picture of marriage raises the stakes of those questions while leaving the role of answering them to the viewer.

There are plenty of reasons—supported by scenes of confrontation, manipulation, and avoidance within the show—that would support reading *Mad Men* as sinking in its cynicism about marriage. We will focus on those scenes that highlight the impasse between Betty and Don Draper already mentioned, an impasse that from Don's perspective bespeaks the divergence between America come of age and the naïve childishness of a supposedly original America. In "Three Sundays," Don deals with the firm's American Airlines account after the airline suffers a market-altering plane crash (which so happened to have Peter Campbell's father on board): "American Airlines is not about the past any more than America is. Ask not about Cuba, ask not about the bomb, we're going to the moon. Throw everything out. There is no such thing as American history . . . only a frontier."[17] Don's America as forward facing no doubt contributes to his contempt toward his own suburban life, not to mention his own secreted past. Don is wedded to a vision of American enterprise where maturation entails leaving behind childhood ways as associated with Betty in Season 1. The information he receives from her psychiatrist only reinforces his sense that he is taking care of a child: "Mostly, she is consumed with petty jealousies and everyday ac-

17. "Three Sundays," S2/E4.

tivities. Basically, we're dealing with the emotions of a child here." This is not only a case of men sexually infantilizing women, since Betty herself enables a crush that the young boy Glen develops for her. Limiting the charge of childishness to sexism risks assuming that her desires, if granted a hearing, would follow the path of Don's America, and that does not appear to be the case. What is being denied by Don's condescension toward Betty is her right to social adulthood *and* her ability to voice what that adulthood would cost her, namely a shared pursuit of innocent happiness and a common task of learning what love and marriage could become. Her connection with Glen, as uncomfortable as it is, provides an alternative to Don's vision of America. It is one that recognizes something amiss in the confidence that Don conveys about his own knowledge of others and the license he grants himself for trespassing a common future with another in, say, marriage. Don's repeated affairs are underwritten by his repudiation of the concept of love as tied to matrimony, something Betty takes (understandably) as presumed by the conversation of married life.

This tension between childhood and maturation, along with mutuality and its slow demise, come to a head in Season 1's final and memorable episode, "The Wheel," which includes Betty's confessing to Glen "I can't talk to anyone." The arrival of Don's America comes at, and can only come at, Betty's expense. What Don calls her childishness is, in reality, her unwillingness to surrender all that her life has given her—a home for her concepts of love and marriage. It is a home, no matter his denials, shared with Don. We see this tension poignantly displayed in Don's fraught pitch for a projector ad during which he memorably uses pictures of his and Betty's family life, turning the projector into a "time-machine" that "takes us to a place where we ache to go again. It's not called the wheel. It's called the Carousel. It lets us travel the way a child travels, around and around and back home again, to a place where we know we are loved."[18] Are we to see Don's cavalier use of their family as telling the truth, that their common life, and common life as such, is but advertisement? In this reading, Don appeals to a life of marriage, a concept of love, and a past happiness as only manipulation—where the concept of love has been lost, sold, by way of so much manipulation.

Throughout *Mad Men*, Don appears to place the risk of that loss on Betty and her failure to notice all the times he tells her "I love you," when, in fact, that loss is on the emptiness those words must have in Betty's (and our) ears. We are brought again to the repeated question in *Mad Men* of the significance of love as a concept related to marriage by way of the perpetuity of love as viable concept in Don world. How could Don mean those

18. "The Wheel," S1/E13.

words given his avoidance of the life required to provide a home for them, given his turning "love" to unfamiliar uses. What accounts for the briefest moments of sincerity, or at least, longing, in Don for things to remain how they are? Can we (or he) differentiate in such moments between his twisting of words and his (admittedly failed attempt) to recover them and use them again? How do we (does he) pick up concepts that have been twisted almost irrecoverably and claim to know what is being said? These are some of the stakes that *Mad Men* means to put before us by displaying Don's liminal space between using words and misusing them. And to the extent that the fate of marriage is bound to the fate of conversation, his often-purposeful misuses betray a desire to exist outside language and its demands, and more specifically, outside marriage.

Interestingly, their marriage does not come to an end because of his affairs but because his secret life and his former "Mrs. Draper" is made known to Betty—which suggests that what she despises is not infidelity, in general, but his secretly making her name and their marriage a lie to her. What appeared to be a common pursuit was, from the beginning, constituted on the ability for love and marriage to function ironically—as other than what they appear to be. She cannot remain married to him. In her words, she no longer loves him but he can't hear her confession because he does not know—or has forgotten—what love means and, therefore, what it might look like for it to end. As he claims in Season 1, "Every day I make pictures where people appear to be in love . . . I know what it looks like."[19] What surfaces here is the common need to differentiate appearance from reality, true love from feigned love, a real marriage from a pretend one. And what we may go searching for, perhaps, is a way to secure the presence of marriage in those cases where it has turned out that agreement was absent all along. We may want to secure ourselves from how words can be twisted, secure ourselves from Don's own twisting of them. A plethora of philosophical inventions have surfaced in the search for that kind of security to our shared lives. But it is a telling feature of *Mad Men*, one it shares with Cavell, that there is no security of twisting outside Don's initial willingness to found his marriage with Betty on the confession of his past. There is nothing beyond his own hope in her willingness to accept him in spite of it. Agreement in marriage, like agreement in language, is not willed or enforced but discovered and rediscovered. So the forgoing of marriage by both Betty and Don is expressed by seeing that future as hopeless, as love itself among those concepts where agreement can no longer be found or trusted.

19. "The Hobo Code," S1/E8.

Does *Mad Men* grant us only this conclusion? Does it offer more, require more? We come face to face with *our own* judgment of Don's exemption from moral growth. Are we willing to go on speaking in a world of Don Drapers and are we willing to grant him an eventual future with us? What might be entailed if we buy into his own sense of himself as too beautiful or too ugly for ordinary love and companionship, for an ordinary future? What are the implications of admitting with Don that ordinary concepts are insufficient to the task of mutual intelligibility and that therefore to go on speaking as if they were sufficient is simply to buy into nostalgia? In short, if we cannot save marriage from Don's America, can we at least save language from it, and thus the possibility of, as Cavell says, an eventual marriage? In Season 1 Rachel Mencken remarks that utopia is both "the good place" and "the place that cannot be." Does *Mad Men* claim as much for marriage, if not shared human speech itself?

This question surfaces most clearly in *Mad Men*'s final episode, when he speaks morosely to Betty who is now dying of lung cancer in a subtle moment that brings us face us to face with our judgments of Don.

> Betty: I want to keep things as normal as possible and you not being here is part of that.
>
> Don: Birdie . . . [crying]
>
> Betty: I know.[20]

What does she know? That Don is sorry for his exemptions from their marriage, that he *does* love her and therefore admits submission to love as belonging to married life? Maybe that is hoping too much. It might not be too hopeful, however, to say that Don's silence is his confession that he has reached the end of his words (his world) and her response ("I know") is her acknowledging his silence as his confession of his own inadequacies and moral incompleteness. Taking the latter option denotes at least one moment where silence betokens a submission to language's future, namely, that what cannot now be said may come to be spoken *eventually*. And an underlying implication of the futurity of language is Don's future in speaking, that while words are not yet forthcoming, his potential moral change might allow him to consent to conversation again. Here, *Mad Men* seems to want to lay claim to a place for love still present in domestic life, to a possibility of acknowledging Betty's voice even now as companionable to his? If so, what, we need to ask, licenses his appeal to love at just this moment, when the end seems to have come, when all seems lost? How could he begin to find the words again, and if not for himself, at least for others?

20. "Person to Person," S7/E4.

These questions return us again to the ordinariness of language, for the fact of its convertibility conditions egress of concepts from the homes in which lives are made *and* any return to homes in which lives might be built anew. Indeed it is the self-same words that we believe forgotten that might return to us and return us to the forgotten: words like "love" and "marriage." Our concepts come to us amidst our life in the world, the bodies we are, and the environs we inhabit and so *naturally*; while words are conventional, these conventions arrive naturally suited to and arriving out of life in the world. "Love" and "marriage" may not be eternal, but neither are they arbitrary. One might be straightforward and say that concepts stick because they work, but that would beg the question as to what is meant by "work" since what remains under consideration is where our concepts now find their homes. If putting it this way sounds evasive then are we better off claiming Christian theology as able to view the ground of language as the natural history from which it springs as natural convention? Here ground may speak to literal places of growth rather than the foundation philosophical skepticism craves. Ground speaks to dirt and soil, the local contexts and sources that feed into conceptual development, the mutuality of organic life that reproduces itself domestically again and anew. Conversational life as natural life fructifies further possibilities for mutuality, since the same aliveness that gathered concepts from the raw material of life in the world toward what is called "love" continues stubbornly as it grows allied to new contexts and sources, bearing up new concepts of love and their varied uses.

Mad Men repeatedly—if also minimally and barely—gives voice to this vis-à-vis Betty's use of an outmoded "love" and her recollection of another country almost forgotten. Don senses it too, mostly making use of it, sometimes mourning it, every once in a while giving into it. If, as we have claimed in one way or another, language goes all the way down, then there is no pure *nihil*, no Christian justification for suspending hope, however bleak. One might invoke *creation ex nihilo* upending an eternal return to nothing, a belief in a place where the promise of future words could come to an end. There is no such end and, thus, no ultimate bleakness—only the ongoing structure of creature life: *exitus et reditus*.

It has been tempting to take *Mad Men* as a justification for an intellectual or institutional overhaul of language as such, seeing in Don the epitome of modernity's nihilism about shared meaning. However, that temptation looks too much like admitting with skepticism that meaning has already been lost, that too many concepts are already stained and forgone. To see in *Mad Men* one further example that justifies remaking Christian discourse from the ground up is to succumb to *Mad Men*'s voice of temptation. And that risks missing the fact that there are no established limits to language's ability for being recouped, nor are their definable points in which our attempts to reconcile or redeem conversation are necessarily in vain, necessarily requiring us to start from ground up. What appears more companionable to Christian commitments is the claim that whatever ways in which our current words and worlds have been twisted, they have not yet lost their capacities for an eventual future any more than the murdering of the Son of God forecloses resurrection. The Christian commitment to the latter implies a greater willingness in the pursuit of learning a common language in a context in which intelligibility itself appears difficult to establish. This is not to say that the difficulties of shared words in *Mad Men* are overcome; indeed, the Christian view of marriage supplied earlier lays bear the stakes. Rather, this is to say that *Mad Men* presents a truly cynical world where our notions of things as common as marriage and love and conversation are brought under remarkable pressure in ways that place our willingness to suffer the companionship of characters like Don under severe scrutiny. It has been our purpose to show that such willingness has its source in our capacity to trust in the future of our words and that Cavell's view of language and remarriage helps to strengthen the scope of that trust without succumbing to false securities. This trust gets strengthened by those who recognize their future as fundamentally bound up with the future of a life identified by the ability to continue on. If our America is anything like *Mad Men*'s America, we are going to need all the sources of trust we can get.

BIBLIOGRAPHY

Cavell, Stanley. *Cities of Words: Pedagogical Letters on a Register of the Moral Life*. Cambridge, MA: Harvard University Press, 2004.

———. *The Claim of Reason: Wittgenstein, Skepticism, Morality, and Tragedy*. Cambridge, MA: Harvard University Press, 1979.

———. *Pursuits of Happiness: The Hollywood Comedy of Remarriage*. Cambridge, MA: Harvard University Press, 1981.

Poggi, Jeanine. "As AMC Says Goodbye to 'Mad Men,' It Says Hello to Millennials." *Advertising Age*, March 26, 2015. Online: http://adage.com/article/special-report-mad-men/amc-goodbye-mad-men-seeks-millennials/297772/.

Williams, Rowan. *The Edge of Words: God and the Habits of Language*. London: Bloomsbury, 2014.

Chapter 14

UNEASY BEDFELLOWS

On Pete and Trudy's Marriage

—MATTHEW EMILE VAUGHAN
and CHRISTOPHER J. ASHLEY

In this chapter, we will reflect on the institution of marriage through an investigation of *Mad Men*. We will look at one particular marriage, that of Trudy Vogel to Pete Campbell, as a lens for our reflections on marriage. Grounded in this example, we will pursue an argument—a hopeful one—about the sustainability and virtue of the institution of marriage in the twenty-first century. In limiting our reflections to Pete and Trudy, we wish to provide a common text for concerned viewers of *Mad Men* to reflect on the institution of marriage, and also the impact of marriage on family structures, professional life, and broader discussions of morality. This chapter has a particular eye to how younger audiences, of the so-called Millennial generation, might respond to this marriage in particular. While Millennials did not make up the majority of the viewers of *Mad Men*,[1] they have made up a large portion of the community engaging in discussion of the show—particularly in online settings. In the concluding section of this chapter we will describe why we feel that *Mad Men* generally, and Pete and Trudy specifically, resonates with Millennials.

1. Poggi, "As AMC Says Goodbye."

Beginning from the opening scenes of the pilot, marriage serves *Mad Men* as a social institution and a mode of relationship that both reveals and changes the characters. Every central adult character participates in one or more marriages (save Peggy Olson, whom we discuss below). Through their own marriages, and in responding to the marriages of others, the core characters of *Mad Men* display both their attitudes and opinions toward one another and toward the institution of marriage in general—as it existed in 1960s upper-class New York City. Though the show's main themes primarily emerge through the workplace, the characters' moods, beliefs, and choices around marriage also illuminate themes such as gender and self-creation. Certain characters such as Trudy, Betty Draper (Francis), and Sally Draper, for instance, appear only seldom at the office but respond powerfully to the marriages around them.

Examples abound of the prominence of marriage in the culture and minds of *Mad Men*'s central characters. Take, for example, the conversation that Joan Holloway, Peggy, and the women from the operators' office have regarding D. H. Lawrence's *Lady Chatterley's Lover* in "Marriage of Figaro," S1/E3. Joan takes that infamous novel to be a commentary on marriage: "It's another testimony to how most people think marriage is a joke." It also foreshadows the way she interacts with the institution throughout the show, in her marriage to Greg Harris and her liaisons with Roger Sterling and Richard Burghoff.

The show does not advance a tight, singular argument about marriage. Rather, like any good work of fiction, *Mad Men* poses as many questions to society and its institutions as it does concrete proposals. This chapter, therefore, will begin by teasing out some of the underlying assumptions and challenges regarding marriage that *Mad Men* presents. Pete and Trudy's marriage may well teach viewers something of the moral nuances embedded (pardon the pun) within marriage.

Before we begin, we offer a brief word on our methodology. This chapter treats creator and showrunner Matthew Weiner as *Mad Men*'s auteur: the architect of its plot and setting, but also of its problems, questions, and presuppositions. Auteur criticism is, of course, nothing new in media studies. Its engagement can serve to legitimate a show, elevating it to a level of art worthy of investigation.[2] Television critics Michael Newman and Elana Levine argue that Weiner's influence pervades nearly every aspect of *Mad*

2. Newman and Levine, *Legitimating Television*, 38: "The showrunner is potentially an *auteur*: an artist of unique vision whose experiences and personality are expressed through storytelling craft, and whose presence in cultural discourses functions to produce authority for the forms with which he is identified."

Men's ethos and plot, and they label him as a prime example of an auteur.[3] We likewise acknowledge rival—perhaps *supplemental*—claims about *Mad Men's* auteur. For example, the popular blog "Mad Style: Tom + Lorenzo, Fabulous & Opinionated"[4] implicitly claims that *Mad Men's* costume designers are its true auteurs. Nevertheless, we attempt here a responsible critique of *Mad Men*—not solely to the intention of its auteur, but also to the subtler, more diverse intentions legible in its "text." We will make claims about the show—and about marriage—that we do not necessarily attribute to Weiner—namely, that Trudy serves as an audience surrogate and that the institution of marriage can be redemptive, even as seen in *Mad Men's* narrative.

During the *Mad Men* era, the institution of marriage was at a major crossroads. Indeed, for the United States as a whole, the sexual revolution of the 1960s marks one of the central identity crises of modern history.[5] *Mad Men* had the opportunity—or perhaps, the *responsibility*—to reflect on the larger social, economic, and sexual setting in which those changes took place. As reflective viewers of *Mad Men*, therefore, we ask not only what the characters have to teach us (particularly with regard to marriage) but also what that tumultuous decade means for marriage in the twenty-first century.

CHARACTER PORTRAITS: WHO ARE PETE AND TRUDY?

Mad Men develops its plot in such a way that each of its characters pushes beyond caricature. This achievement suggests much about the show's guiding aesthetic. The characters, as period types, could have been written and perceived as stereotypes. Instead, the show portrays them with psychological and moral complexity—never simply as the objects of intergenerational

3. Ibid., 40: "It's hard to imagine a show like *Mad Men* coming into existence without a central intelligence overseeing production and storytelling, and indeed its *auteur* Matthew Weiner is reported to assert his agency in myriad aspects of production, from writing and casting to set design and costuming." To supplement this argument, they appeal to Witchel, "Mad Men Has Its Moment."

4. Fitzgerald and Marquez, "Tom + Lorenzo." We will use Tom and Lorenzo's keen eye when discussing a particular scene below, without embracing their argument regarding Mad Men's auteur to its fullest extent.

5. For a helpful introduction to the myriad issues at play in shaping marriage in the 1960s and 1970s, see Coontz, *Marriage, A History*, Chapter 15, "Winds of Change: Marriage in the 1960s and 1970s," 247–262. Interestingly, on 252, Coontz argues that it was the *media*—in this case, women's magazines—that shaped women's approaches to marriage during that time. Perhaps, then, Pete's time in the office is just as important as his time at home for understanding what happened to marriage in the 1960s!

irony. Pete and Trudy are no exception to this rule. They completely conform neither to the norms viewers might have expected to prevail in a 1960s marriage nor to the hopes that viewers might have for their own marriages in the early twenty-first century. Nevertheless, Trudy and Pete do represent types, especially through the earliest seasons of the show—which shape initial viewer response to them and make their development and emergence as distinct individuals all the more affecting.

Pete represents old American money, even as his family's fortune dissipates over the course of the show. Not surprisingly, throughout the show, Pete exhibits the greatest sense of entitlement of any main character of *Mad Men*. He constantly evokes his status, Ivy League credentialing, and his mother's maiden name, Dyckman. His Campbell ancestry serves likewise to legitimate him as a white Protestant of old standing—even as a late scene nods to the Scots tribal violence that colors white American imaginations of their Anglo-Saxon ancestors. Pete stands in for the psychic and ethical inheritance of pre-1960s American capitalism and for the prestige and comfort it had once given to its practitioners and beneficiaries. Importantly, that brand of capitalism decays throughout the show. Pete watches his family fortune dwindle to nothing as first his father and then his mother squanders their estate. He is also the most prominent witness on the show to the rise of the West Coast creative economy. Even after moving back to New York City in Season 7, Pete encourages the partners at SC&P to take California money seriously. (His foresight provides an example of the show's characteristic generation-scale dramatic irony: contrary to the predictions of many in the advertising industry at the time, California is now a fulcrum of the global economy as the United States' most populous state and the home to the wealthy creative centers of Hollywood and Silicon Valley.)

Pete's entitlement mirrors and complicates his profound and consistent lack of self-confidence. His inner life, as formed by old-money capitalism, assumes zero-sum conflict over scarce material and emotional resources. He never fully believes that he will come out of that conflict with all that he needs to thrive or deserves to enjoy. As a result, Pete frequently shows himself unable—or, at best, unwilling—to share in the successes of others. Take, as an example, the frequent and bitter jealousy Pete feels toward Ken Cosgrove both for Ken's artistic ability and his business successes. In "Shoot," S1/E9, Paul Kinsey says to Pete—upon hearing of Ken's success at Sterling Cooper: "Suddenly, upon news of success, a young man is filled with ambition." Paul's aphorism describes Pete's response only partially. Ambition does indeed fill Pete at that moment but that ambition is grounded neither in a hearty work ethic nor a desire for communal advancement. Rather, Pete responds with jealousy at anyone else's success.[6] He cannot celebrate with those who do well. All glory must be his and his alone. That the world is actually not his oyster becomes his biggest crisis in life.

Pete treats himself, in turn, in the same way he treats others. He ranks his own achievement by that of others, consistently jealous of their attainments, never satisfied, even provisionally, with his own. Even when things go well for him, Pete displays an affect of melancholy, disappointed in his inability to conquer the world. When he feels his worst, he treats others around him at his worst. He has no regard for others—much less Trudy—and he cannot figure out why his peers do not like him. (They do not, in part, because of this lack of self-awareness which they can see clearly but he cannot.)

A poignant insight into Pete's character comes from a surprising source: the complex character Herman "Duck" Phillips. Duck's addiction and professional desperation had formerly led him to behavior that made Pete look stable and successful by comparison. In "The Better Half," S6/E9, Duck and Pete have an informal conversation in Pete's apartment about Pete's career options outside of advertising. Instead of advising certain business practices, Duck exhorts him to be a better family man. "You got to spend less time in this place, and more at home," he says. Duck calls family the "wellspring of my confidence." Pete says, "My family is a constant distraction." Duck retorts, "You better manage that, or you're not going to manage anything." Importantly, Duck names the family—Pete's marriage included—the wellspring of *confidence*, not the wellspring of success or happiness. Duck sees what the audience sees: Pete hates himself, has no confidence in himself or the system he serves, and as a result cannot seem to progress.

In contrast to Pete's "old money," Trudy Vogel represents the epitome of "new money." Her family's money comes not from land, as did the Dyckmans', but from the twentieth-century chemical industry—namely Vick. Clearasil, the most adolescent of necessities, is the flagship product in the business of Tom Vogel, Trudy's father. Clearasil represents the Vogel family's relative dynastic youth. This *nouveau riche* family pursued the stability and prestige of marrying Trudy into a wealthy family such as the Dyckmans, thus raising their social status without (they believe) imperiling their material security.

6. The pastoral psychologist Ann Ulanov and her husband Barry distinguish between jealousy, the desire for another's possessions, qualities, or achievements, and envy, the desire that others' goods be destroyed. Cf. *Cinderella and Her Sisters*. Pete does not appear to want Ken's story retracted from the *Atlantic*, but he does want to place his own in the *New Yorker*.

For the prosperous-Millennial segment of the *Mad Men* audience, Trudy makes for something of a surrogate. They share her standpoint of material comfort, but aspire to appreciate, and enter aesthetically, an old-fashioned world of class and prestige. She reflects the audience's position *vis à vis* the main action of the story—as a bystander who sees the more active characters more clearly than they can see themselves. As a result, Trudy functions, most crucially, as an *ethical* surrogate for the audience, a figure who shares the audience's conscience and expresses its judgments attractively. Whereas a protagonist surrogate allows the audience to enjoy a charismatic figure (like Don Draper or Peggy) whose actions they do not necessarily wish to emulate, an ethical surrogate like Trudy reflects the audience's reasoned response.

In a story as morally complex as *Mad Men*, an ethical surrogate must be relatively inactive—both in the plot and in the dynamics of their own character. Trudy benefits from the lifestyle of her father and husband, but she does not take part in the work that undergirds and enables their material comfort. She never appears alone in a shot or undertakes an individual plot arc, even though she appears in every season. As a result, Trudy is a relatively flat character. In a show where even peripheral characters often express richly-developed inner lives, Trudy's is notably underdeveloped, especially by comparison to Pete's.

Nevertheless, Trudy's judgment of Pete does change throughout the show. She goes from a posture of compliance and acquiescence to one of disgust and, finally, to one of redemption. In this respect, she reflects the larger ethical arc of the show and the audience's experience with these characters. She moves from a lusty, greedy optimism in the show's early seasons into a disillusionment about not only the lifestyles of those around her but also of the institutions intertwined with those lifestyles. Paying attention to Trudy means paying attention to ourselves as viewers. Does she see good in the people in her life where nobody else does? Or does she see them as a means to an end?

SCENE ANALYSIS: PETE, TRUDY, AND PEGGY IN THE OFFICE

In "For Those Who Think Young," S2/E1, Weiner, in collaboration with director Tim Hunter, presents a scene that demonstrates Trudy's role as surrogate, one that also betrays subtle hints about the nature of Pete and Trudy's marriage. The plot of the scene primarily illuminates Pete and Peggy's relationship, as well as Peggy's relationship with others in the office. Our analysis of this scene below, however, informs our approach to Trudy, particularly regarding how she relates to both Pete and the audience. The scene's intentionality speaks to the role of costume, setting, and dialogue in Weiner's storytelling.

The scene unfolds in Pete's office, where he sits facing Peggy, Sal Romano, and Harry Crane. The scene begins with a close-up shot of Trudy's picture on Pete's desk. She stares at the viewer for a second or two. She knows what the audience knows, and she wants the audience to know it. She sees what the audience has seen and will see. But Trudy sits between the audience and Pete in the scene—signaling her intermediary role. Then the camera blurs her as it moves toward Peggy. In the picture, Trudy wears black and white, with her lips slightly open. She looks almost amused, as if about to offer some witty remark. She wears a coquettish dress, with exposed shoulders and a prominent pearl necklace. Her hair is up. This picture of Trudy remains in view through the entire scene. Pete keeps Trudy's image on his desk facing him, not facing out. He is not showing Trudy to visitors in his office. The picture reminds him that he has a wife.

In this scene, Pete, Peggy, Sal, and Harry are discussing Clearasil, a Tom Vogel product. The theme of the scene is one of youth: how to sell Clearasil to teens, and how to deal with young talent at Sterling Cooper. At Peggy's suggestion, the meeting has to do with whether they can use testimonials from girls who are too young to use Clearasil. They all agree that this ethical dilemma, though, "doesn't matter." Paul then barges in, bemoaning the presence of younger talent in the agency. Peggy claims not to understand this thinking by saying "I am only twenty-two." Paul retorts: "You don't count." Ken also barges in, ready to take Harry out on the town to celebrate his then-wife Jennifer's pregnancy. Ken says, in front of Peggy and Pete—with Trudy in the room, of course—"We've got to get him out of here. Come on, Daddy, how often do you get to celebrate getting some girl pregnant?" Then the finishing touch: "Campbell, you're buying!" Pete has, of course, never celebrated such a thing for himself. He is in the midst of battles with Trudy about this very issue due to their battle with infertility. He does not yet know that he has previously impregnated Peggy. Then, when the meeting concludes, Pete looks at Peggy and says, "Kids, what's the big deal? Do you want to have kids?" With a look of surprise, followed by relief, Peggy says, "Eventually." Pete, validated, replies "Exactly."

The critic Todd Van Der Werff has observed that *Mad Men* prioritizes mood and character over plot.[7] The scene under discussion here illustrates the point. Though short on overtly dramatic incident, it is poignant and revealing. Themes of youth and procreation in the explicit dialogue, combined with the scene's heavy dramatic irony, create a simultaneously hopeful and uneasy mood. Peggy serves both as the focus of that mood and as the dramatic center of the scene. Trudy is present, through her picture, to Pete and to the audience. Peggy is present to the workplace, where her embodiment and speech contrast her both with Trudy and with the men in the office. For Paul, Peggy does not "count," but the viewer knows better. In the moments that follow, Peggy moves directly from Pete's office to Don's, where in a major professional breakthrough, she coins the gorgeous tagline, "What did you bring me, Daddy?"

Consider the wardrobe in this scene from Pete's office. Peggy wears a blouse in what Tom and Lorenzo call her "power color": golden yellow, or mustard.[8] The men's clothes suggest her unspoken, even unconscious influence over them: Sal's tie and pocket square blatantly mimic her blouse, and Paul's tie has hints of mustard. The room itself even echoes her blouse: there are hints of golden yellow in the lamp behind Peggy. In turn, Peggy dictates much of the conversation, making marketing/advertising suggestions that all of the men follow. She even maps out the "next steps" on the Mohawk account with Sal—and he obeys—after the others leave the room. Her status and influence are on the rise, even if the men cannot see or acknowledge the fact.

So what is this scene telling us about Trudy, that silent face in the lower left, situated between Peggy and Pete? Trudy's silence does not render her unimportant in the scene. Take, for example, the way Pete is dressed. Unlike the others in the room, Pete is dressed in black, gray, and white, mirroring Trudy's portrait, not Peggy's blouse. Since Trudy is in black and white—the

7. For instance, from his recap of "New Amsterdam," S1/E4: "Story seems to just happen to *Mad Men* sometimes. A calamity arises from outside the characters, and then they respond to it. That's a very typical method of storytelling in a short-story show, where mood, atmosphere, and slowly coming to understand a character better and better are more important than having the characters endlessly push the plot forward." Van Der Werff, "Mad Men: '*New Amsterdam*.'"

8. Fitzgerald and Marquez, "Mad Style: Signal 30."

very image of old-fashioned domesticity—Trudy and Peggy contrast against one another. But Pete and Trudy call back to one another. The monotonous tone in Pete's attire, however, signals boredom and complacency, not an attempt to honor Trudy. Trudy's choker chain of pearls perhaps speaks to her feeling in that office: one of suffocation and indentured elegance. She has no direct connection to the office or to its goings on. She is, like the audience, an onlooker and an implicit judge of the proceedings, not a participant.

Trudy's portrait sits between Pete and Peggy in this scene. Visually, the camera's perspective contrasts Trudy with Peggy and points to the differences in Pete's relationships with both women. Pete's relationship with Peggy illuminates, largely by contrast, his marriage to Trudy. While a full discussion of Pete and Peggy's relationship is beyond this chapter's scope, a few points are important. At first, and in the larger shape of their character arcs, the show contrasts Peggy and Pete. Peggy's striving and self-creation answers Pete's static response to his privilege at many turns. Nowhere is the contrast greater than in Season 1, which begins with a tryst between them and ends with Peggy's resulting pregnancy. Pete approaches Peggy from a sense of sexual entitlement yet does not learn for years that anything more came of their one-night stand. Peggy's decision to give the baby up for adoption, by contrast, represents an overt affirmation of her professional and personal independence, and at the interpersonal level, a pivotal moment in her mentoring relationship with Don. Where Pete's choices flow from his social location, more or less unobstructed by virtue or deliberation, the show portrays Peggy in terms of her growing sense of agency.

The contrast is heightened and shaded, however, by the choices and priorities Pete and Peggy share in common. Take, for example, Pete's complaint, mentioned above, that his family distracts him. Peggy, like Pete, judges her success in professional rather than familial terms. The show makes her priorities especially obvious in "The Suitcase," S4/E7, in which Peggy bemoans then-suitor Mark's inattention to the seemingly obvious fact that she dislikes her family. Likewise, in the present scene, Peggy and Pete both agree that the time for children may one day come, but has not yet. Peggy's actions with regard to her own pregnancy thus retrospectively affirm not only her own professional self-understanding, but Pete's as well.

Trudy's image in this scene, at this moment of the narrative, thus stands as a silent reproach to Pete. The show's relatively straightforward ethical judgment here, as embodied in Trudy's portrait, is somewhat unusual but aesthetically consistent. Pete's open neediness had rendered his approaches to Peggy unflattering, unlike the visually glamorous affairs of characters like

Don or Roger.[9] Trudy, by contrast, becomes the non-ironic picture of elegance and propriety. Her image signals his transgressions and becomes the occasion and vehicle of the audience's judgment. She is the comeuppance he does not yet know he has received.[10] Extrapolating a general point, then, we would argue here that Pete's *marriage* displays the fruits of the virtues and vices predominantly formed in the workplace.

PETE, TRUDY, AND MARRIAGE

In his very first scene in *Mad Men*'s pilot, Pete reveals his initial attitude toward marriage through a phone conversation with Trudy. We know Pete, not yet as an adman, but first as Trudy's fiancé. As she (correctly) fears, he will spend at least some of the evening of his bachelor party with other women. Knowing this, Pete attempts to prove his virtue: "Of course I love you. I'm giving up my life to be with you, aren't I?" he says.

This claim, though dramatic, does not have any single, obvious meaning. It could be that Pete is promising to give up something in a general sense, or that something unique about Trudy makes his promise so emphatic. Pete's forceful affect could come either from "giving up his life," or from doing so "for you." Indeed, he may not mean what he says. If he means that marriage represents a professional sacrifice, he is misguided even by his own standards. He knows that at many of his business dinners he will need a beautiful, doting wife by his side. He likely does not mean that he will end his promiscuous habits, either. As the viewer learns shortly, he has no intentions of curtailing his philandering. It may be monogamy he bemoans specifically. He should not be judged as a simple philanderer, let alone a sex addict, in spite of his liaisons and his time in brothels. For Pete the problem lies much deeper within himself. His first words to Trudy are meant to empower him by belittling her. He is telling his wife-to-be: "I have to come down in the world to be with you, and I am willing to do it because I am just *so good*."

9. Beyond Pete and Trudy, whether *Mad Men* glamorizes (heterosexual, male) adultery is debatable. It does not lack clarity about the personal and intergenerational costs of Don and Roger's infidelities. Other characters, like Harry, are rendered pitiable buffoons by their attempts at womanizing. A counter-argument would suggest that the show's luxurious visual sensibility cannot help but flatter the sexual choices of its conventionally-attractive men.

10. This role, we should add, is very difficult to fill, especially for women, whom some audiences are predisposed to receive as shrewish or scolding (cf. many responses to Betty). That Trudy can play this role effectively, even in a still photograph, is a testament to the subtlety and precision of Alison Brie's performance.

Pete has a particular understanding of the status quo of the institution of marriage, one that he wants to defend and from which he wants to benefit. To him, the wife serves the interests of the husband, and in return the husband does lip service to love for the wife. Pete is not interested in a relationship between equals. He is interested in being in charge. He displays this desire throughout the show—not just in this introductory scene. Perhaps the most poignant example is his experience in a brothel in "Signal 30," S5/E5. A prostitute suggests possible "fantasy" liaisons for Pete, several of which he rejects. He seems uninterested in the fantasy of a supporting wife or that of an inexperienced partner. The fantasy he accepts, however, crystalizes everything else he wants. She says to him, "You're my king" as she crouches before him. He desires power, not to seduce women, but to command the world. He wants power over *everything*—to be king, even if just within a brothel. Extramarital sex becomes his way of reclaiming the power and prestige that is (as he sees it) lost to him professionally and personally.

Trudy's ideal for marriage could not, on the surface, be more different than Pete's. Trudy does not want power over Pete—or, really, power over *anything*. She wants a co-conspirator, a partner, in Pete. She wants that power that he feels so strongly about to be shared among the two of them. It is her quest for camaraderie—indeed, her expectation of it—that leads her to some of her strongest problems with Pete. Take, for example, the vignette in "Red in the Face" (S1/E7) in which we see Trudy excoriating Pete for his decision to exchange the chip 'n' dip for a rifle (a rifle, we might add, that remains in his possession throughout the show). We do not see Trudy in this scene; we only hear her voice. Pete sits there like a small child with his rifle sitting on his lap, embarrassed, impotent, crushed. She screams, "That was for *us*. You are always telling me to grow up! I can't believe you!" Is she angry because he should have bought something they could both enjoy or because he made a decision without her (or both)?

For Trudy, there remains an unspoken social contract that they are in this together—not only in marriage but also with regard to his professional life, at least before their separation. She sees her presence in his life as having a great deal to do with his success. Take, as a cursory example, another scene in "Signal 30" (S5/E5), one in which Trudy hosts a dinner party for the Drapers and Cosgroves. For Trudy, this dinner is a watershed moment. For the first time, she can actively participate in Pete's professional success (a success marked by Don finally deeming them worthy of a personal visit). Trudy's unspoken social contract, however, carries with it other implications about what she wants. She wants to assist in Pete's advancement but also wants to present the illusion of happiness to the outside world.

The most poignant example of the marital dynamic and social contract between Pete and Trudy comes in "Collaborators" (S6/E2) after Trudy takes Pete's battered lover to safety while he sits at home brooding. As he leaves for work the next day, he tries to pretend that nothing is wrong. Trudy, having ensured that Tammy is away from the house, awaits him.

> Trudy: Couldn't you just pretend? I let you have that apartment. Somehow I thought that there was some dignity in granting permission. All I wanted was for you to be discreet.
>
> *He denies the liaison.*
>
> Trudy: There's no way for me to escape, to not be an object of pity while you get to do whatever you feel like. I have never said no to you.
>
> . . .
>
> Trudy: We're done, Peter. This is over.
>
> Pete: You want a divorce?
>
> Trudy: *I refuse to be a failure.*[11] I don't care what you want anymore. . . . This is how it's going to work: you will be here only when I tell you to be here. I'm drawing a fifty-mile radius around this house. And if you so much as open your fly to urinate I will *destroy* you. Do you understand?
>
> Pete: Do you know what? You're going to go to bed tonight and realize you don't know anything for sure.
>
> Trudy: I'll live with that.

In the course of this breakup fight, the show offers viewers a rare entry into Trudy's mindset. What provokes her to anger is not his cheating, but her embarrassment. She punishes him: he may not return home. She will not, however, seek a divorce, which would constitute defeat. Trudy makes explicit, for the first time, her private hope for their marriage. She wants, of course, for them to be happy. More than that, she wants outside appearances to be settled. While she has displayed her strength of character before, her forceful expression and righteous indignation are new, surprising, and gratifying—expressing as they do the show's long-withheld judgment and, implicitly, the audience's too.

Many factors shape their respective approaches to marriage, particularly their surrounding culture and the respective family systems from which they came. As we noted above, at no time in American history have the forces of social change altered marriage more than in the period of Pete

11. Emphasis added, not only for how it reveals Trudy's character, but for how Brie sells this potentially-melodramatic line.

and Trudy's marriage. The 1960s and 1970s saw the rise of birth control, divorce reform, shifting paradigms of marriage, and massive changes in female labor patterns.[12] The marriages that surround Pete and Trudy, as microcosms of that culture, are as diverse as the people in them. Returning to the dinner that Trudy hosts for the Cosgroves and Drapers in S5/E5, we see two varying poles of marital stability. At one pole is the marriage of Cynthia and Ken Cosgrove, whose fidelity and happiness are unexpected, profound, and consistent throughout the show; at the other, that of Megan and Don Draper, an intergenerational marriage fraught throughout with conflict, adultery, and misaligned expectations.

Trudy lives, in many ways, as a traditional woman—even in spite of these evolving standards all around her. She does not ever seem to resent her role as a young housewife and mother through the entire show. Her problems with Pete have more to do with his trustworthiness than with the life he and she lead. Yet, Trudy remains in control of her marriage—perhaps due to the financial security enabled by her father. Never dependent upon Pete for her livelihood, she never fears the personal implications of a life without him. Instead, she fears that others will see her as a failure at marriage.

Pete and Trudy are the only couple in *Mad Men* that Weiner has portrayed in the context of their larger families. Throughout the show, the audience meets both Pete's and Trudy's parents. Indeed, Pete and Trudy have been reacting to the marriages of their respective parents throughout the show. In this respect, they display a much broader human tendency to be deeply influenced in our own approach to the institution of marriage by the approach of our parents. Both Pete and Trudy go into their marriage essentially expecting what their respective parents have. Pete wants distance and independence; Trudy expects everyday involvement. Crucially, however, both Pete and Trudy eventually go against the norms set by their parents' marriages. Pete deviates by coming back to Trudy after their separation and, ostensibly, corralling his actions in the move to Wichita. Trudy, for her part, goes against Tom Vogel's wishes by taking Pete back.

It would be unfair to *Mad Men*, and to Pete and Trudy, to suggest that their marriage is always unpleasant. In fact, when Pete and Trudy's marriage is going well, it is highly attractive. A few examples will illustrate. In "A Little Kiss" (S5/E1), Trudy says, "This becomes a home the minute you walk through that door," after seeing Pete come in after a long day. Alison Brie's performance suggests a woman wholly sincere, at least in her desire to please her husband. Vincent Kartheiser's suggests a man glad to accept

12. Again, see Coontz, *Marriage, A History*.

her offering at face value, even if it does not address his actual desires. As we noted above, Pete will later spurn a similar interaction when mimicked by a prostitute, but that suggests the distinct contribution of his marriage to his actual happiness. Perhaps there is a compromise at the heart of their marital happiness here, but successful compromise is precisely at the heart of many long-term relationships.

As a second example, Pete and Trudy are never happier than during their elaborate dance at Roger Sterling's Kentucky Derby party in "My Old Kentucky Home" (S3/E3). They have obviously choreographed their dance in advance, and enjoy one another in their successful, mutual execution of their plan. Clearly, they have done this before. Even Roger, Harry, and Ken appear jealous of their happiness together. That the dance is obviously rehearsed, a conscious performance of marital happiness, does not make it any less a *successful* performance, all the more striking on a show where the marriages are so often wholly unhappy.

Finally, their parenting seems essentially cooperative. Both Pete and Trudy love Tammy, and she loves them. Their shared care for her even after their separation provides for rich interactions and lays the foundation for their eventual reconciliation. Take, as a cursory example of their cooperative parenting, the vignette in "The Milk and Honey Route" (S7/E13)—in which Trudy validates and supplements Pete's care for Tammy in light of her bee sting (note that this scene occurs *before* their reconciliation later in the episode).

REFLECTIONS ON *MAD MEN* AND MARRIAGE IN THE EARLY TWENTY-FIRST CENTURY

The so-called Millennial generation did not make up the primary audience for *Mad Men*, but the strong response by some Millennials to the show helped shape its immediate critical and cultural reception. Marketers, perhaps naturally, were among the first to pick up on this thread. Since 2007, photo spreads in blogs or magazines have shown stylish, often affluent, young people throwing *Mad Men* parties, where they might encourage vintage dress and serve classic cocktails. Banana Republic even offered a clothing line in its spring 2013 collection modeled after the costumes of *Mad Men*.

The Millennial audience, in turn, provides a window into the show's generationally-relevant themes. Without underplaying the show's sensual appeal, so critical to its storytelling, we suggest that its portrayal of an earlier period's marriages directly contributes to the show's popularity. Think, for instance, of the generational placement of *Mad Men*'s characters. Don, Peggy, Pete, and Trudy were born between the late 1920s and early 1940s, so that their ages during the show range from their early twenties to their early forties. Don and Betty's children, in turn, were born in the 1950s and 1960s. Over the show's run, many Millennial cohorts would have begun to age beyond their college years and thus naturally would have begun to understand earlier generations as fellow adults. At just the right moment, *Mad Men* offered a highly entertaining glimpse into the environments that shaped their grandparents and parents, including taboo hints about the kinds of marriages their Greatest and Silent Generation grandparents might have had.[13]

Mad Men appeared, too, in a cultural moment in which Millennials were increasingly changing the face of marriage in the United States. Most directly, Millennials have changed American marriage by marrying—and just as importantly, to a degree unprecedented among previous generations, by delaying marriage. Millennials, across lines of race and class, have married later than previous cohorts. More educated Millennials have accordingly delayed or avoided childbearing. Among other causes at play, sociological studies have suggested that these choices reflect their lessened material prospects, the worst since those that faced their grandparents during the Great Depression.[14]

We argue that delayed marriage among Millennials seems not, however, to reflect a cultural turn against marriage as an institution. Consider, notably, the very strong Millennial support for the provision of civil marriage to same-sex couples. Without mooting the question of the proper definition of marriage, polling data clearly demonstrates where majorities of Millennials have taken their stand: around 70% of American Millennials favor the legalization of gay marriage.[15] They see the married estate, as they understand it, as desireable. The right to marry is something to aspire to, even fight for if needed, and to celebrate when attained. Intriguingly, though no doubt coincidentally, the spread of civil marriage for same-sex couples in the United States occurred largely during *Mad Men*'s run. In 2007, when

13. An auto-biographical note: We, the authors, were both born in the early 1980s, making us both demographically Millennial. In the case of our own families, the parallel is particularly exact. Our grandparents on all sides are roughly Don and Betty's ages, and our parents are exactly the same age as Sally.

14. One accessible summary of the large literature on this topic is in Friedman, *Baby Bust*.

15. Pew Research, "Gay Marriage": "Currently, 70% of Millennials favor allowing gays and lesbians to marry legally, compared with 59% of Gen Xers, 45% of Boomers and 39% of the Silent generation."

the show premiered, Massachusetts was the sole U.S. state with marriage equality for same-sex couples. By *Mad Men*'s finale in May 2015, thirty-four other states had joined Massachusetts in granting marriage equality; the rest would follow a few weeks later with the Supreme Court's ruling in *Obergefell v. Hodges*.

If the Millennial audience of *Mad Men* were using the show to reflect on marriage, what lessons might they glean? What is *Mad Men* saying about the institution of marriage itself? In the case of Pete and Trudy, based on their reconciliation in "The Milk and Honey Route," S7/E13, the show offers grounds for optimism. Pete and Trudy, perhaps surprisingly, succeed as marriage partners. Marriage is not necessarily a "joke" (per Joan's remark in Season 1) or a guaranteed failure. Over the show's arc, marriage serves as a site for the characters' growth and development. In spite of Pete's often-indefensible behavior, Trudy's choice to take him back (and the show's implicit request to viewers to accept him too) does not immediately lessen or dishonor her character or to render her the "failure" she has refused to be.

Trudy's role as the audience's ethical surrogate, wise before her times and years, provides the clue for understanding how she has construed her own integrity for this moment. We suggest that Trudy displays a certain "second naiveté" toward not only her own marriage to Pete but also toward the institution of marriage as a whole.[16] What Ricoeur holds for biblical hermeneutics, we attribute to Trudy's approach to her husband and the institution that unites them. Ricoeur claims:

> What we need is an interpretation that respects the original enigma of the symbols, that lets itself be taught by them, but that, beginning from there, promotes a meaning, forms the meaning in the full responsibility of autonomous thought . . . we can, we modern men (sic), aim at a second naiveté in and through criticism. In short, it is by *interpreting* that we can *hear* again.[17]

16. By using the term "second naiveté," we are co-opting the language and thought of Paul Ricouer and his interpreters. Of particular focus is Ricoeur's oft-quoted progression on biblical hermeneutics, of which here we offer the barest possible sketch. Ricoeur's concept was initially intended to map out a trajectory for interpreting the Bible in light of criticism. Broadly stated, the progression moves from the reader holding a primitive naiveté (acceptance of text at face value), experiencing a critical distance from the text, and then embracing a second naiveté (acceptance of text, but only in light of interpretation). For his most succinct articulation of the concept of second naiveté, see the conclusion of Ricoeur's *Symbolism of Evil*, particularly 348–51.

17. Ricoeur, *Symbolism of Evil*, 349–50, 351. Emphasis in original.

From this dense philosophical passage, we would extrapolate a simple thesis—spoken as if Trudy were saying it: The institution of marriage provides a context in which people change and develop. Where Ricoeur calls for a critical approach to symbols, Trudy calls, congruently, for an honest approach to the institution of marriage. Marriage, like the reading of Scripture, offers an opportunity to be taught, to be shaped and transformed—not because of the perfection of either the institution or the symbol—but because the reader and the couple find meaning by responsibly reinterpreting them. Trudy's example opens to audiences discerning the meaning of marriage an opportunity to *hear* from the institution through her own experience of critical transformation.

Trudy's posture of post-critical interpretation of Pete and of marriage makes her approach appealing to the audience generally—but, particularly, to Millennials. Keenly aware of the challenges that face her marriage, she knows what she can and cannot expect—both from marriage and from the men in her life. She has experienced, by virtue of her surrounding culture, a certain "critical distance" from marriage—in the Ricoeurian sense of the term. Alison Brie's performance conveys an astute and knowing ethical candor to accompany its WASPy, aspirational sensual appeal. Throughout the show, Trudy knows the secret lives of both her husband *and her father.* Thus, she is able to approach the institution in general knowing what it really is—which allows her to engage it attractively and well.

This critical distance does not necessarily render virtuous all of her choices, including her choice to recouple with Pete. It becomes clear, even at the end, that she outmatches Pete in her understanding of the shape of their marriage. Pete, in his naïve entitlement, has not achieved that same critical distance—which means he cannot fully understand either his own life or his wife's. Trudy knows how Pete lives apart from her, but Pete knows nothing about Trudy's own separate life—nor does he demonstrate the desire to do so throughout the show. To the extent that Trudy recognizes this fundamental imbalance in her marriage, she willingly turns a blind eye. There are understandable reasons why she might make this choice, not the least of which is to protect the quality of life for herself and her child.

Nevertheless, Trudy's knowing posture—her critical distance—earns her the benefit of the doubt, particularly with Millennial viewers who bring that same second naiveté both to *Mad Men* and to marriage. The audience knows, throughout, that Trudy was right about Pete all along. He will philander; he will lie; but he will still come home. The Millennial audience, in particular, also knows a good deal about the world and its institutions. They have experienced the failed marriages of previous generations, not least their parents', and have seen the institution at its most divisive and destructive.

Due to the myriad cultural, sociological, and technological advances of the past thirty years, young people today seem to know more about the world than even Trudy does about marriage.

Still, nobody is completely over their naiveté—even Trudy. But Trudy's is a *second* naiveté, one that remains hopeful about a complex institution. This second naiveté renders her appealing—both to Pete and to the audience. It is this second naiveté we advocate for marriage. Indeed, this second naiveté allows people to see good in one another—despite so many reasons to the contrary. Trudy never stops loving Pete, in a similar way that Millennials have never stopped completely believing in marriage.

To be fair to Millennials, the challenges that invite our critical distance are different from those that Trudy faced. Without downplaying the very real problems faced by young people in the 1960s, we pose, as an example of the difference between the generations, the precarious financial state of most Millennials. The privileged scions of the postwar boom did not have to deal with student debt, poor job prospects, or stagnant (indeed, deflating) wages in the way that Millennials do. The main characters of *Mad Men*—regardless of economic background—share materially in the fruits of the postwar US economy and its attendant opportunity to pursue professional and personal satisfaction.[18] Perhaps for Millennials eager to find success and satisfaction in both their professional lives and the shaping of their households, the fantasy of *Mad Men* serves as an attractive space in which they can reflect on what success in those realms might mean to them if they had better access to good jobs and affordable housing. Younger characters—like Peggy, Pete, and Trudy—approximate Millennials' ages and life stages, making them congenial foils for viewers' reflection and discernment.

Our reading of Pete and Trudy's marriage, however, presents practical implication. Pete and Trudy ultimately succeed in their marriage in part because both characters change throughout the show—shifting through critical distance and naiveté. Trudy, having merely observed Pete's infidelities for years, finally confronts him over the course of Season 6 and reshapes their relationship to suit her needs. As a character, she changes from a powerless observer, a relatively passive influence on Pete like the picture on his desk in the scene we discussed above, to a powerful agent. In her function as audience ethical surrogate, she suggests a generation coming of age and taking a newly active role in the governance of their institutions. For Millennial viewers—who value marriage in society while often finding it unattainable

18. They also share the privileges attendant to being White and (presumptively) heterosexual. Racial and LGBT themes are less prominent in *Mad Men* than those around social class and gender roles, an artistic choice beyond this chapter's scope but still worth noting.

in their own lives—Trudy's role in Pete's marriage suggests a possible model for discovering and claiming their place in the world.

Pete, for his part, changes most crucially through the realization of his professional potential. The show characterizes him throughout as an ambitious and effective account manager. His effectiveness lets him rise through company ranks, even as his excessive ambitions, grounded in his upbringing, are routinely and obscurely frustrated. Over the course of the show, he earns a partnership in SC&P, makes a tremendous windfall on the firm's sale, and finally becomes a top-level corporate executive with Learjet. Only then is his desire to be "king," as revealed in "Signal 30," finally satisfied: who but "royals" could have jet planes, particularly in the early 1970s?[19] He could never hope to find such royal treatment in his home life, certainly not from a wife so much his equal (if not superior). He comes to realize that his seemingly impossible entitlement can only find fulfillment through his professional achievement.

Nevertheless, we would not attempt to argue that Pete's marriage to Trudy changes his fundamental character. He remains a largely disagreeable figure through the entire show. As with most of the show's central characters, the virtues of the capitalist workplace shape him more than those of the household. Trudy's development, by contrast, does happen in their marriage—a particularly intriguing fact, in light of her audience surrogacy.[20] In the marriages of *Mad Men*, sympathetic but externally powerless people can discover their agency and shape their own destiny.

Not all change is for the better. Pete and Trudy succeed at marriage in part because each has provided the other—*unchangingly*—something they both need. In other words, Pete and Trudy did some things well in marriage; those moments clearly serve to bolster the credibility of their eventual reunion. Pete consistently provided Trudy with what she wanted most throughout the show: a home. She can take his financial support and her financial stability for granted and build a life on it. Trudy, for her part, never removes Pete from Tammy's life—though she had plenty of opportunity.

19. Cf. Lorde's 2014 pop song "Royals," whose chorus names airplane ownership as among the paradigmatic forms of conspicuous consumption.

20. In this respect, a contrast between Trudy and characters like Peggy, Joan, and Betty is obvious, both at the level of their development on the show and in terms of their engagement with the feminisms of their period. The critic Alyssa Rosenberg has suggested Peggy and Joan's development through their work experiences is signaled by changes in their respective feminist consciousnesses between seasons four and seven. See her "'Mad Men' and the Radicalization of Joan Holloway." Betty ultimately pursues graduate education in psychology, and is actually seen reading Betty Friedan's *The Feminine Mystique*. If Trudy has any life outside her household, let alone an explicit, feminist awakening, the audience is not privy to it.

Their reunion, then, opens the possibility of a more comprehensive return to the kind of full partnership that had already characterized their marriage at its best. Pete proves willing to join, to choose, the home he has been providing. Their new life together in Wichita may come off as brilliantly as their dancing.

CONCLUSION

The marriage of Trudy and Pete represents a piercing dynamic that is both generationally remote, yet surprisingly contemporary. Its aesthetic and thematic power comes through, however, only when viewers attend to it carefully and take the characters themselves as seriously as Weiner intended. As critics and theologians, we have tried to "read" Pete and Trudy with both sympathy and critique—as the writers portrayed them and Alison Brie and Vincent Kartheiser brilliantly performed them. In a testament to the richness of *Mad Men's* artistic achievement under the direction of Matthew Weiner—even peripheral characters like Trudy repay close reading both aesthetically and ethically—and so contribute to the larger themes and narratives of the show.

We have suggested that Pete and Trudy's *marriage* formed the core of their shared storyline. In light of their marriage—as depicted in a workplace scene in the first episode of Season 2—we have suggested that Trudy serves *Mad Men* as an ethical surrogate for the audience, a moral compass that represents what the show's creators assumed the audience would feel. Also, we argued that Pete and Trudy's larger story offers a broader implication about the institution of marriage: namely, that a successful marriage is possible between flawed, even naïve, partners. This message matters particularly to Millennial viewers, many of whom are now engaging in marriages of their own. Perhaps Millennials will look to marriages like the Campbells' and see, not that marriage is a "joke," but rather that it is indeed life-giving. Marriage, in *Mad Men*, serves as a site where both privileged and powerless people can discover their agency and shape their own lives. For Millennial viewers, who value marriage in society while often finding it unattainable in their own lives, Trudy and Pete's marriage suggests a possible model for discovering and claiming their place in the world.

BIBLIOGRAPHY

Coontz, Stephanie. *Marriage, A History: How Love Conquered Marriage.* New York: Penguin, 2005.

Fitzgerald, Tom, and Lorenzo Marquez. "Mad StyleSignal 30." *Tom and Lorenzo.* Online: http//tomandlorenzo.com/2012/04/mad-style-signal-30/.

Friedan, Betty. *The Feminine Mystique.* New York: Norton, 1963.

Friedman, Stewart D. *Baby Bust: New Choices for Men and Women in Work and Family.* Philadelphia: Wharton Digital, 2013.

Newman, Michael Z., and Nancy Levine. *Legitimating Television: Media Convergence and Cultural Status.* New York: Routledge, 2012.

Pew Research. "Gay Marriage." Online: http://www.pewresearch.org/data-trend/domestic-issues/attitudes-on-gay-marriage/.

Poggi, Jeanine. "As AMC Says Goodbye to 'Mad Men,' It Says Hello to Millennials." *Advertising Age,* March 26, 2015. Online: http://adage.com/article/special-report-mad-men/amc-goodbye-mad-men-seeks-millennials/297772/.

Ricouer, Paul. *The Symbolism of Evil.* Translated by Emerson Buchanan. New York: Harper & Row, 1967.

Rosenberg, Alyssa. "'Mad Men' and the Radicalization of Joan Holloway." *Washington Post,* April 6, 2015. Online: https://www.washingtonpost.com/news/act-four/wp/2015/04/06/mad-men-and-the-radicalization-of-joan-holloway/.

Ulanov, Ann Belford, and Barry Ulanov. *Cinderella and her Sisters: The Envied and the Envying.* Philadelphia: Westminster, 1983.

Van Der Werff, Todd. "Mad Men: '*New Amsterdam*.'" *AV Club,* November 27, 2013. Online: http://www.avclub.com/tvclub/new-amsterdam-106060.

Witchel, Alex. "Mad Men Has Its Moment." *New York Times,* June 22, 2008. Online: http://www.nytimes.com/2008/06/22/magazine/22madmen-t.html?pagewanted=all&_r=0.

Chapter 15

MAD MEN, BAD PARENTS

Representations of Parenting in *Mad Men*

—SUSAN E. FREKKO

At first blush, parenting may seem like an odd topic for a chapter about *Mad Men*, given that children are marginal figures on the show. Yet, I argue that to the extent that children do appear we can discern a clear pattern: they serve as foils for viewers to experience the bad parenting of the main characters. That adults' interactions with children are meant to be viewed as condemnable becomes clear when we examine examples whose shock value is particularly obvious. To list only a few examples from recent seasons: Sally's grandmother giving her a sleeping pill, Betty insisting that Bobby eat all of the candy for which he has traded her lunch, and Megan leaving Sally home alone where she becomes prey to a burglar.

This chapter argues that these brief moments of bad parenting permit an air of self-congratulation and superiority among viewers. We would never do *that*! *Mad Men* gives us permission to revel in its dark social world because it offers us the illusion that today's society has advanced far beyond it. As a professional in her late 30s recently posted on Facebook, "Although I would never want to take a step back to Mad Men era non-equality . . . drinking lunches would be nice some days." The poster seems to take for granted that the changing times have swept away both 1960s inequality and workplace drinking. While she depicts the former as unattractive and the former as attractive, both appear as quaint and outmoded things of the past.

Critical and popular responses to the series have focused on inequality between men and women and between whites and African Americans on the show. Several authors have argued that *Mad Men* allows us to revel in the dramas of past inequalities because it offers the illusion that we have overcome them (see below). I argue that the social world that the show depicts (and caricatures) is dominated not only by males and whites but also by adults. Many of today's parents were themselves raised largely on the "permissive" approach of Dr. Spock and are now thoroughly entrenched in the trend of "intensive mothering." They see in *Mad Men* examples of shockingly bad parenting. The show allows us to pat ourselves on the back for our own progress/progressiveness, while at the same time taking for granted a set of neoliberal values—including responsibility and self-regulation—that form the basis of contemporary "good" parenting and the ideological basis of global capitalism. In the neoliberal era, parents face pressure to raise the next generation of children as responsible citizen-workers. While the personal traits associated with neoliberal values seem unquestionably attractive, I argue that taking them for granted may in fact harm children in two ways. Firstly, some of the techniques designed to instill these values in children have been shown to be detrimental; secondly, to the extent that parents hold sole responsibility for their children's development, social programs for families seem less and less necessary. This chapter uses *Mad Men* to think about what twenty-first-century American television viewers take for granted about parents and children and its social implications.

USING *MAD MEN* TO THINK ABOUT SOCIAL ISSUES

Diana Davidson's analysis of motherhood in *Mad Men* places the show in its "post-war/pre-feminist era."[1] Davidson notes that in "Shoot" (S1/E9), Don effusively praises Betty's mothering abilities: "You're a mother to those two little people and you are better at it than anyone else in the world." He goes on to say, "I would've given anything to have had a mother like you. Beautiful, and kind, and filled with love like an angel." But Davidson points out that, in fact, Betty chafes in her role as a mother. What begins as a "detached and obligatory"[2] approach to parenting evolves into neglect and occasional cruelty. As Davidson puts it, "Betty is not the kind of loving mother Don idealizes her to be. And her children notice. For instance, when she catches

1. Davidson, "A Mother Like You," 138.
2. Davidson, "A Mother Like You," 141.

Sally smoking her cigarettes, Betty locks her eight-year-old daughter in a closet. The little girl tells her mother that Don has left because she is 'mean and stupid.'"[3]

Davidson lists the many improvements for women since that time, but she goes on to note the continuing problems of gender, racial and economic injustice that plague the country. Davidson writes, "We watch *Mad Men*'s representation of motherhood knowing the show is portraying an America on the cusp of second-wave feminism: we watch it with a sense of nostalgia, 'the pain from an old wound.'[4] As an audience, we can sit back and feel triumphant that things are much better for women now."[5] Deborah Tudor pushes the argument about nostalgia further, saying that it hides the need for further social change:

> The sexism depicted in this series is overt, but its setting consigns it safely to the past. This allows the series to perform a postmodern sleight of hand—to package the past spectacle of men behaving badly as a salable commodity for contemporary audiences who visit past oppression as postmodern tourists but can return safely to the present, where our postmodern, postneoliberal culture reminds us that feminism is no longer needed.[6]

In a similar vein, Benjamin Schwarz argues in an *Atlantic Monthly* article that *Mad Men* sets up a straw man version of the 1960s—which it then smugly depicts as reprehensible. The show's stance "amounts to a defiant indictment of sexism and racism, sins about which a rough moral consensus would now seem to have formed." He goes on to say, "We all like to congratulate ourselves, and as a group, *Mad Men*'s audience is probably particularly prone to the temptation."[7]

The Drapers' bad parenting gets just the kind of self-congratulatory uptake that Schwarz describes. For example, Erin Hill's post on their parenting lists "9 Parenting Lessons from Mad Men's Betty and Don Draper," which includes "1. Dry-cleaning bags are not a toy" (in reference to a scene in which a young Sally is seen playing with a dry-cleaning bag over her head and "2. Secondhand smoke is bad for your kids; firsthand smoke is worse"

3. Ibid.
4. Here Davidson is citing Don's definition of "nostalgia."
5. Davidson, "A Mother Like You," 152.
6. Tudor, "Selling Nostalgia," 337.
7. Schwarz, "Mad about *Mad Men*."

(in reference to a scene in which Betty lights up while driving pre-teen Sally and then offers Sally a cigarette).[8]

While scholarly and mass media discussions of *Mad Men* have tended to focus on representations of gender and race, it is worth taking a closer look at representations of parenting through a similar lens. Additionally, I argue that there is another element to viewers' condemnation of the Drapers' parenting. It relates not to increasing equality between adults and children but rather to the neoliberal economy, and in particular the Drapers' failure to raise their children according to its ideals.

TWENTIETH-CENTURY PARENTING: CHILDREN'S RIGHTS, PARENTS' ANXIETY, AND THE NEOLIBERAL FAMILY

In part, our horror and self-congratulation when we watch *Mad Men* relates to increased equality in American society. The show pays most attention (implicitly) to improvements between men and women and whites and African Americans, but improvements in equality between adults and children also emerged over the course of the twentieth century. Examples include definitions of child abuse, restrictions on how parents may treat their children, and the idea that children have a right to dignity and respect. A key document in this trend was The League of Nation's 1924 Declaration of the Rights of the Child, updated and expanded in 1959 by the United Nations. The increasing belief in children's rights is one reason why we inwardly condemn Betty for locking Sally in the closet.

The series makes us think of Betty and Don's parenting as "typical" of the 1960s,[9] but Peter Stearns's (2004) history of twentieth-century American parenting makes it clear that even by 1960 when the show begins, Betty's mothering would have been condemnable.[10] He shows that by the 1920s, parenting manuals turned away from shaming, guilt and physical punishment (even scolding, in some cases) as studies about children's psychologi-

8. Ibid.

9. Schwarz argues that the inclusion of iconic historical footage and the show's careful attention to accuracy in material details such as wardrobe and furniture encourages the audience to think of the show as a historically accurate representation of the 1960s; it is not.

10. Schwarz also points out the sway that Dr. Spock had over families like Betty and Don's and the unlikelihood that real-life couples of their social standing would have acted as they do; he mentions for example a scene in which a neighbor parent hits a Draper child at a birthday party and the other adults do not react. The scene shocks today's viewer, but Schwarz argues it would have been equally shocking in 1960: "Nice people—the educated and affluent—didn't hit other people's kids."

cal development revealed the damage these practices could inflict.[11] This was part of a shift from the Victorian view of children as "sturdy" beings possessed of inner goodness, who merely needed guidance and protection from bad example[12] (this belief was preceded by the pre-nineteenth-century belief that children possess inner evil as a result of original sin that must be actively overcome through discipline, a view that persists in some religious communities). In the twentieth century, children instead came to be viewed as "vulnerable," in part in reaction to studies that revealed their actual susceptibility to accidents and abuse, and in part in reaction to lower birth rates and parental uncertainty about children's roles, as those children shifted from being economic contributors to economic burdens.

It was an ironic turn, considering that by all objective measures children's wellbeing improved dramatically over the course of the twentieth century. However, these shifts made children both precious and needy of protection. This belief in children's vulnerability results in what Stearns identifies as the main affective mode of parents in the twentieth century: anxiety, both about keeping their children safe and about their own performance as parents.[13] Beginning in the 1920s, parents' anxiety about how to raise their children without traditional methods of harsh discipline led them to turn to expert advice. Early, simplistic applications of strict behaviorism in the 1920s and 1930s involved rewards combined with mechanistic manipulations of the environment to avoid conflicts. The rigidity and coldness of these methods showed their own drawbacks and gave way to the "permissiveness"[14] of Dr. Spock, which combined manipulations of the environment with a relaxation of certain standards to avoid conflicts, along with a careful consideration of how to uphold important standards.[15] Popular authors in this vein encouraged parents, when they came up against standards they weren't willing to relax, to use reason and if all else failed,

11. We see this trend in the US's government's best-selling pamphlet *Infant Care*, first issued in 1914 (Children's Bureau, 1914) and in the 1926 founding of *Parents Magazine*.

12. See, for example, Alder, *Moral Instruction of Children*.

13. Stearns, *Anxious Parents*. Greenstone's (2005) analysis of the Curious George book series shows how the early books from the 40s maintain a Victorian view of children. George is sturdy and plucky, a monkey version of the Victorian child. The later books (50s and 60s) transform George and he acts more like a twentieth-century child—timid, fearful, and prone to anxiety—a projection of shifting adult views of children.

14. I place "permissive" in quotes because this style of parenting is not really permissive in the everyday sense of the word. It required a lot of work and careful consideration (see Stearns, *Anxious Parents*).

15. See Spock, *Common Sense Book*.

grounding and fines, but without inflicting anger or emotional withdrawal on their children.[16]

The anxiety identified by Stearns is key to the current mainstream trend of "intensive mothering," a term coined by Sharon Hays to describe "the idea that correct child rearing requires not only large quantities of money but also professional-level skills and copious amounts of physical, moral, mental, and emotional energy on the part of the individual mother."[17] Hays points out that this kind of parenting has been solely women's responsibility and that they perform it often to the detriment of their careers (sacrificing potential gains in the present in favor of an "investment" in their children). She asks why women would make this choice in a society in which the rational pursuit of self-interest is so highly valued. Her answer is that intensive mothering both expresses our societal ambivalence about the dogged pursuit of self-interest and supports continued gender, racial and class inequality (because these axes of difference influence who provides and benefits from intensive mothering).

The rational pursuit of self-interest is a key tenet of neoliberalism, which Pierre Bourdieu describes as a political order that aims to advance free-market global capitalism by removing state and international regulations and by encouraging a view of workers as "individuals" who succeed or fail on the basis of their own merit and who are kept in line by precarious work conditions that constantly threaten unemployment.[18]

Several authors have explicitly linked current parenting trends to the neoliberal ethos. Through an analysis of a Flemish online parenting support discussion board, Geinger et al. discuss the ways that current parenting trends express a neoliberal emphasis on individual autonomy and responsibility, the role of a the "good parent" in producing children as future value-producing citizens.[19] Drawing on Foucault's notions of self-regulation, they identify an ideology of parenting in which the "good parent" confesses his or her (but especially her) insecurities and mistakes. In fact, the act of confession alone can protect parents from the accusation of being "bad." These traits dovetail with what the authors call the "parenting trend" in social policy in Europe which looks to parents to solve social problems, as funds are withdrawn from the welfare state. Pressure is put on individual families

16. Stearns, *Anxious Parents*, 76–77.
17. Hays, *Cultural Contraditions*, 4.
18. See Bourdieu, "L'essence du néoliberéralisme."
19. Geinger et al., "Parenting as a Performance."

to be self-regulating. In a similar vein, children become precious "investments" in the future who must be protected from risk.[20]

While Geinger et al. write about Flanders in particular and Europe in general, these trends are clearly identifiable in the North American context as well. Writing about Canada, Wall explicitly links intensive parenting with neoliberalism, noting the close fit between the high demands intensive parenting places on parents (particularly mothers) and neoliberalism's focus on "individual self-management, self-enhancement, and personal responsibility."[21] In addition to regulating themselves and their children, parents are also responsible for turning them into citizen-producers—"children are cast as the potential self-reliant entrepreneurs of the future."[22] In a similar vein, Aihwa Ong has argued that in today's America, "citizenship is defined as the civic duty of the individual to reduce his or her burden on society, and instead to build up his or her own human capital."[23]

Wall has pointed out that the failure to parent one's children according to these ideals is a social failure, "Like breastfeeding, engagement in intensive parenting has become an important standard by which mothers are judged, and mothers' social accountability (and potentially their own identity as good mothers) is tied up with their performance in this regard."[24] In this sense, intensive mothering concerns the mother's presentation of self as much as the wellbeing of her child. The fact that mothers are more susceptible to critique than fathers reveals a continuing double standard between expectations for mothering and fathering.

Writers have criticized the show's postfeminist self-congratulation for its facile claim that gender inequality is all behind us now. Similarly, there are significant ways in which today's intensive parenting is adult-centric, disrespectful and even harmful. For example, Carol Dweck's research has famously shown that praising children for their academic results or intelligence can result in performance anxiety and the loss of motivation.[25] Alfie Kohn depicts this sort of praise as manipulative and controlling, giving children a sense that their worth is conditional on doing what adults want them

20. We see this in child welfare policies such as the British New Labour "social investment state," as described by Lister, "Children (But Not Women)." It is also reflected in the name of Canada's Invest in Kids Foundation (see Wall, "Is Your Child's Brain.").

21. Wall, "Is Your Child's Brain," 41.

22. Ibid., 46.

23. Ong, *Buddha is Hiding*, 12.

24. Wall, "Mothers' Experiences," 258.

25. Dweck, *Mindset.*

to do.[26] We must therefore also question the self-congratulatory attitude about parents and children that the show promotes.

I examine examples that suggest that we condemn the Drapers not only because they don't live up to today's (supposedly) egalitarian ideals but also because simultaneously, and ironically, they fail to live up to neoliberal ideals of responsibility and self-regulation. We take these behavior ideals for granted, yet they emerge from and support a particular political and economic status quo and instilling them in our children can have its own drawbacks despite the apparent advantages that these traits offer. While there is nothing wrong with being responsible or self-regulating, of course, my interest concerns pointing out how these traits belong to a broader constellation of values that link up with a neoliberal political regime—which may, in fact, undermine children's wellbeing.

THE DRAPERS AS BAD PARENTS: FAILURES OF RESPONSIBILITY AND SELF-REGULATION

Dr. Spock's Baby & Child Care appeared in 1946 (originally as *The Common Sense Book of Baby and Child Care*) and became a best seller. As Schwarz points out in his *Atlantic Monthly* article cited above, someone of Betty's social class would most likely have been familiar with its advice ("Dr. Spock's permissive parenting notions exercised a near-hegemonic sway over childrearing practices in the bedroom communities of the Northeast's professional class"). However, Betty does not treat her children as needing protection and, in some instances, her treatment of them corresponds most closely to a belief in their inherent inner corruption (see above discussion of Stearns). On the other hand, she conforms to the Victorian model when she asks Don to hit the children. This approach is Victorian both in its positive opinion of physical punishment and in making the father the final authority figure, trends that began to subside in the 1920s.[27] We do see Betty restricting Bobby's TV viewing for a week, a common form of grounding by the late 1950s.[28]

Clearly, Betty is not an intensive mother, although of course her being one in the fictional timeframe of the show would be an anachronism. The ideal of the intensive mother can only emerge, paradoxically, when mothers are no longer expected to quit work and other obligations. Betty instead is a woman who does give up paid employment in order to be a mother, a role

26. Kohn, *Unconditional Parenting*.

27. See Stearns, *Anxious Parents*, 58.

28. Stearns, *Anxious Parents*, 78.

in which she displays evident dissatisfaction. Moreover, according to 1960s norms described above, she performs poorly as a mother.

If Betty fails according to 1960s ideals, then she has no chance against the neoliberal ideals that are taken for granted in good parenting practices of the late twentieth and early twenty-first centuries. A particularly interesting example that runs throughout the series is Betty's relationship with smoking, both as an individual and as a mother. Smoking is set up as a theme early in the series when in Season 1 the Drapers' pregnant neighbor Francine appears smoking a cigarette (the gasp of the twenty-first-century post-smoking audience is nearly audible in the scene). Betty smokes throughout the series, often in front of her children. On the one hand, this is a failure of her responsibility to protect her children from risks—both health risks and bad influence. It is also a clear violation of her obligation to self-regulate; Wall shows how self-regulation in the area of health is a key responsibility of the neoliberal citizen.

She is not indifferent, however, to her children's relationship with cigarettes. She punishes young Sally for lighting a cigarette by locking her in a closet (she accompanies this with more typical twentieth-century punishments of restricting TV, toys and time with friends). The scene unfolds as follows, as Betty seethes with rage:

> Betty: What do you think you're doing?
>
> Sally: Nothing.
>
> Betty: Get over here. You could burn the house down.
>
> Sally (in Betty's grip): You're hurting me.
>
> Betty: Good. I can't believe you were smoking.
>
> Sally: I wasn't.
>
> Betty: Go upstairs. You're not going to be watching television. You're not going to be playing with your friends. I'm taking away Barbie.
>
> Sally: You're mean.
>
> Betty: You betcha. Get in there (gesturing to the closet).
>
> Sally: I'm telling Daddy when he gets home.
>
> Betty: Go ahead.
>
> Sally: He left because you're stupid and mean.
>
> Betty: You want to sleep in there?[29]

29. "The Mountain King," S2/E12

Betty protects Sally from direct contact with cigarettes, but she will not win any parenting points here because her handling of the situation inflicts on Sally what in today's America would be considered child abuse. Moreover, her concerns about cigarettes seem to relate not to a concern for the health consequences of smoking but rather with Sally claiming an adult privilege for herself. Betty's concern about Sally burning down the house seems to point to this; Sally is not qualified for the adult privilege of lighting matches. At this point in the series, the research on the health risks of smoking is recent (and, in fact, a major plotline concerns the cigarette companies that Don's firm represents and how to handle this unwelcome news in its campaigns).

Several years later, when Sally is a pre-teen, Betty offers her a cigarette in the car scene mentioned above. Seemingly aware that she should protect her child from the risk of smoking, Betty justifies herself saying, "I'm sure your father's given you a beer." The statement suggests that Sally's father has similarly failed to protect her from risk. Sally responds, "My father's never given me anything" (a comment expressing her anger over his own failures of self-regulation, described below).

Apparent instances in which Betty tries to protect Sally seem to respond to a different logic than the neoliberal one of protecting the future of a potential value producer. Rather, she seems most interested in protecting her value as a desirable future wife. When Sally is caught kissing a male friend in a game of "grownup," Betty warns her, "You don't kiss boys. Boys kiss you."[30] When Sally sustains a nose injury, Betty chastises her for putting her appearance at risk; Sally, evidently picking up on the fact that her mother's concern is for her social capital rather than her wellbeing, responds sardonically, "It's a nose job, not an abortion."[31]

We do see an adult figure that treats Sally as an "investment" in the future—her grandfather and Betty's father, Gene. He explicitly contrasts his own opinion of Sally with that of his daughter's. He compares Sally to her grandmother, who performed professional work outside of the home in the 1920s. "You can really do something," he tells Sally. "Don't let your mother tell you otherwise." Since he is comparing Sally to her grandmother and contrasting her to her mother, "doing something" implicitly refers to participation in the labor force, which, according to Gene, Sally can prepare for if she doesn't let her mother hold her back.

Betty's failure to regulate her own smoking generates another risk for her children: at the end of Season 7, she is diagnosed with lung cancer. Even

30. "Souvenir," S3/E8.

31. "The Runaways," S7/E5

worse, from the neoliberal point of view that makes individuals responsible for regulating their health, she refuses medical treatment. Giving the excuse that she doesn't want to be a burden on her children, she opts to leave her three children motherless rather than attempt to save her life through treatment. This is a big neoliberal no-no, and Sally accuses her of "loving the tragedy." In apparent confirmation of this accusation, Betty leaves narcissistic instructions for her burial. Sally now steps into the role of the responsible adult. She comforts a sobbing Henry. She intervenes on behalf of her brothers to arrange for them to stay with Henry after their mother's death. One of the final scenes of the series depicts Sally doing the dishes while Betty sits smoking.

Don is a much better developed and more complex character than Betty and his strivings and failures make him a sympathetic character, while Betty comes off as shallow and infantile. Betty's moments of tenderness with her children are few and generally double-edged: She cares for Sally when she gets her first period and then uses this moment of connection to compete with her ex-husband's current wife, Megan ("I think she just needed her mother").[32] She enjoys the start of a field trip with Bobby, before throwing a grown-up tantrum that ruins the rest of the day.[33] While Betty's parenting is rather consistently reprehensible from both 1960s and current-day viewpoints, Don's oscillates. He is frequently unavailable and irresponsible—neglecting the children and exposing them to risk. At other times, he comes closer to meeting twentieth-century expectations in that he refuses to punish his children, expresses concern for risk and acts as their "pal," which, as Stearns points out, increasingly defines fathers' role in the twentieth century).[34] Some examples: he brings Sally a puppy after missing her birthday party; he takes Sally and Bobby to Disneyland; he takes Sally to a Beatles concert.

In one episode, Betty complains about her children unsympathetically, "I'm here all day alone with them. Outnumbered."[35] Later in the episode, she encourages Don to spank Bobby. "Do you think you'd be the man you are today if your father didn't hit you?" she asks. Don does not comply with her wishes, allowing the audience to identify with him as a "good parent" (or at least a better one than Betty!). He also expresses anxiety similar to the one identified by Stearns; for example, he disapproves of Betty allowing his chil-

32. "Comissions and Fees," S5/E12.

33. "Field Trip," S7/E3. For more discussion of this scene and Betty as mother, see Ann Duncan's Chapter 4.

34. Stearns, *Anxious Parents*, 57.

35. "Three Sundays," S2/E4

dren to watch television footage of the Kennedy assassination and he does not like Sally attending funerals. We even see him confessing his concerns about his own parenting abilities to his colleague Faye.[36] These are all signs of the twentieth century's good parent, and these are parent positions with which today's parent can identify.

There are other times in which Don's lack of responsibility comes across as shocking to the viewer as Betty's. For example, Don fails to protect Bobby and Sally from risk in a set of scenes in which he first leaves Bobby unsupervised—resulting in Bobby burning himself on the stove; while Betty takes Bobby to the emergency room, Don takes Sally (age eight) to work. His only instructions to employee Joan, with whom he leaves Sally, are "Keep her away from the paper cutter." She asks Joan about breasts and Stan about sex before, unsupervised, she gets drunk on whiskey and falls asleep.[37] Considering that sex and alcohol are two major concerns of twenty-first century American parents, this choice of risks has particular shock value.

If Don sometimes manages to meet our expectations of a responsible father that protects his children from risks, he fails miserably at meeting the other standard of the neoliberal parent addressed in this chapter: self-regulation. In "Favors,"[38] Sally walks in on Don having sex with his neighbor Sylvia Rosen. This is an egregious failure of self-regulation; it's not only an affair, but also a high-risk one that takes place in close proximity to his own home and family. Stunned, Sally runs out of the apartment. Later that evening, she screams at Don in front of her family and Sylvia's husband and son, "You make me sick!" Here we see Sally begin to emerge as the person who will regulate her parents, people who fail to regulate themselves.

Don tries to repair the relationship with Sally, but makes it worse by lying (rather than confessing as the "good" twentieth-century parent would): "I know you think you saw something," he says. "I was comforting Mrs. Rosen. She was upset. It's very complicated." Soon after this series of events, Sally says she will no longer visit her father and her stepmother Megan and requests to attend boarding school. Don agrees and offers to pay her tuition.

By Season 7, Sally has emerged as the person responsible for regulating Don. Sally drops in on Don at the office and inadvertently discovers that he is no longer working there. Confused, she goes to look for him at his apartment, which she finds empty. When he arrives home, he says that he has been at the office and has come home early because he does not feel well. She lets his obvious lie pass, not revealing what she knows until her father

36. "The Chrysanthemum and the Sword," S4/E5.

37. "Three Sundays," S2/E4.

38. "Favors," S6/E11.

presses her later in the day. As he is driving her back to boarding school, Don questions Sally about the funeral she has attended in the city, and she responds with annoyance:

> Sally: Why are you interrogating me?
>
> Don (raising his voice): Because I know you went to my office. So I find your story a little suspicious.
>
> Sally: My story? There's some man in your office.
>
> Don: We're not talking about me.
>
> Sally: Did you lose your job?
>
> Don: Why were you in the city? I want to know this minute.
>
> Sally: I don't have to tell you anything. Just stop the car.
>
> Don: I'm talking to you.
>
> Sally: You're yelling at me.
>
> Don: Why would you just let me lie to you like that?
>
> Sally: Because it's more embarrassing for me to catch you in a lie than it is for you to be lying.
>
> Don: So you just lay in wait like your mother?[39]

Here Don chides Sally for "letting him lie" to her rather than immediately pointing out his lie about his whereabouts. Here he has transferred his own task of self-regulation onto his daughter; she should regulate him by not "letting him" lie. He moreover compares her to her mother, the implication being that Betty knew of his marital transgressions and waited to ambush him rather than confront him directly. First, his wife was supposed to regulate Don; now that tasks falls to his daughter. Later, at a restaurant, Don confesses his actions at work to Sally—an appropriate sign of neoliberal self-reflection—and the two make up. They talk comfortably and Don jokes that they should leave the restaurant without paying (pals again). Don drops Sally off at school and when she exists the car, she turns to Don with nonchalance and resignation, saying, "Happy Valentine's Day. I love you." Don has failed at self-regulation, and his regulation is placed in the hands of his weary teenage daughter.

In the final episode of the series, Sally calls Don to report Betty's cancer (against Betty's wishes) and make arrangements for her brothers after their mother's death.

39. "A Day's Work," S7/E2.

> Sally: I'm telling you because she wants Gene and Bobby to live with Uncle William.
>
> Don: Don't worry. You're all going to live with me. I promise.
>
> Sally: Let me finish. I've thought about this more than you have. You have to tell her that you think it's best that they stay with Henry.
>
> Don: No.
>
> Sally: Daddy, it's going to be so hard for them already. They should at least be in the same bed and at the same school.
>
> Don: Sally, grown-ups make these decisions.
>
> Sally: Do you understand I'm betraying her confidence? I'm not being dramatic. Now, please, take me seriously.[40]

In the absence of parents who are responsible and self-regulating, Sally fulfills this role herself—thinking through the various possibilities and arguing in favor of the boys' current stable home with their self-regulating stepfather, Henry. Despite Betty's and Don's parenting failures, Sally seems well poised to grow into her role as a neoliberal citizen.

CONCLUSION

As Norman Fairclough points out, the neoliberal ethos so permeates American social, political, and economic life that it goes undetected.[41] My goal in this chapter has been to articulate our view of the past through *Mad Men* and our ideas about good parenting in the present with the political-economic regime in which we live and may take for granted. The particular form that "ideal" parenting takes in the present day fits the neoliberal project like a glove and the glee we express at seeing "bad" parenting from the past, takes a series of neoliberal ideals for granted.

On one level our *schadenfreude* at watching these bad parents probably helps assuage our own anxiety—that central emotion of contemporary parenting.[42] We may not know what we're doing, but we know we're doing a better job than the Drapers! This of course places us squarely within a neoliberal framework in which everyone competes—even parents (See, for example, Wall's "Mothers' Experiences with Intensive Parenting and Brain

40. "Person to Person," S7/E14.
41. Fairclough, "Language and Neo-liberalism."
42. See Stearns, *Anxious Parents*.

Development Discourse" on parents' competition with respect to their ability to stimulate their children's brain development).

Taking neoliberal values for granted also has bigger consequences for the wellbeing of children and families. As Wall points out,

> A neo-liberal understanding of parental responsibility for, and control over, child outcomes at the very least supports the shift of blame for the problems of encountered by poor children and their families from the social to the individual level, and legitimates cutbacks to social supports."[43]

As laudable as goals such as responsibility and self-regulation may be, the more that individuals take on these roles and the more that parents become responsible for instilling them in their children, the less that public policy or funding for family wellbeing seems necessary.

As Fairclough points out, neoliberalism pervades all mainstream political stances, even most leftist ones.[44] This fact makes the sleight of hand all the more effective—*Mad Men* looks progressive as it discredits gender and racial inequality and the mistreatment of children. It depicts these problems as part of the past, while simultaneously lauding neoliberal values and condemning fictional parents for failing to embody them. In the midst of so much self-congratulation, it becomes easy to overlook that the neoliberal order is rife with entrenched inequalities and threats to the wellbeing of children and families.[45]

BIBLIOGRAPHY

Alder, Felix. *The Moral Instruction of Children*. New York: D. Appleton and Co., 1892.

Bourdieu, Pierre. "L'essence du néolibéralisme." *Le Monde Dipomatique* (1998).

Children's Bureau. *Infant Care*. Washington, DC: United States Government, 2014.

Davidson, Diana. "'A Mother Like You': Pregnancy, the Maternal, and Nostalgia." In *Analyzing Mad Men: Critical Essays on the Television Series*, edited by Scott F. Stoddart, 136–54. Jefferson, NC: McFarland, 2011.

Dweck, Carol. *Mindset: The New Psychology of Success*. New York: Random House, 2007.

Fairclough, Norman. "Language and Neo-liberalism." *Discourse and Society* 11.2 (2000) 147–48.

43. Wall, "Mothers' Experiences," 261.

44. Fairclough, "Language and Neo-liberalism," 147.

45. I thank editors Ann Duncan and Jacob Goodson for inviting me to participate in this book project and for their comments on my chapter. Thanks also to Erika Hoffmann-Dilloway and Chantal Tetreault for their suggestions.

Geinger, Freya et al. "Parenting as a Performance: Parents as Consumers and (De) constructors of Mythic Parenting and Childhood Ideals." *Childhood* 21.4 (2014) 488–501.

Greenstone, Daniel. "Frightened George: How the Pediatric Educational Complex Ruined the Curious George Series." *Journal of Social History* 39.1 (2005) 221–28.

Hays, Sharon. *The Cultural Contradictions of Motherhood*. New Haven, CT: Yale University Press, 1998.

Hill, Erin. "9 Parenting Lessons from Mad Men's Betty and Don Draper." *The Mid*, April 2, 2015. Online: http://www.scarymommy.com/parenting-lessons-mad-men-don-betty-draper/.

Kohn, Alfie. *Unconditional Parenting: Moving from Rewards and Punishments to Love and Reason*. New York: Atria, 2005.

Lister, Ruth. "Children (But Not Women) First: New Labour, Child Welfare and Gender." *Critical Social Policy* 26.2 (2006) 315–35.

Ong, Aihwa. *Buddha is Hiding: Refugees, Citizenship, the New America*. Berkeley, CA: University of California Press, 2003.

Schwarz, Benjamin. "Mad About *Mad Men*." *The Atlantic Monthly*, November 2009. Online: http://www.theatlantic.com/magazine/archive/2009/11/mad-about-mad-men/307709/.

Spock, Benjamin. *The Common Sense Book of Baby and Child Care*. Oxford: Duell, Sloan & Pearce, 1946.

Stearns, Peter N. *Anxious Parents: A History of Modern Childrearing in America*. New York: New York University Press, 2004.

Tudor, Deborah. "Selling Nostalgia: *Mad Men*, Postmodernism and Neoliberalism." *Society* 49.4 (2012) 333–38.

Wall, Glenda. "Is Your Child's Brain Potential Maximized?: Mothering in an Age of New Brain Research." *Atlantis* 28.2 (2004) 41–50.

———. "Mothers' Experiences with Intensive Parenting and Brain Development Discourse." *Women's Studies International Forum* 33 (2010) 253–63.

Chapter 16

"I CAN'T BELIEVE THAT'S THE WAY GOD IS"

Sexism, Sin, and Clericalism in Peggy's Pre-Vatican II Catholicism

—HEIDI SCHLUMPF

It's October 1962, and the characters of *Mad Men* are dealing with the possibility that the world might come to an end. Though the Russians have not yet attacked, bombs of other kinds are dropping during the episode that concludes Season 2. One character escapes the threat of the Cuban Missile Crisis with anonymous sex in a bar restroom; others distract themselves by worrying about the impending corporate takeover at Sterling Cooper. Junior copywriter Peggy Olson, a Catholic, goes to church.

There she hears an old-fashioned, fire-and-brimstone sermon from the young associate pastor, Father John Gill, who reminds the congregation to "take charge of [their] own souls and prepare [themselves] for the most important summit meeting of all."[1] But Peggy, who has become the priest's personal project, will get more than the general reminder for repentance in the sermon. After Mass, while Peggy delivers a boxful of baked goods from her mother ("in case we get stuck here"), Gill tries one last time to get her to confess her sins—and one sin, in particular: the fornication, or sex outside

1. "Meditations in an Emergency," S2/E13.

of marriage, that resulted in her giving birth to a baby she then chose not to raise.[2]

"Hell is serious and very real," Gill says to Peggy. "And unless you unburden yourself, you cannot know peace." In an era when lay people, especially women, did not talk back to priests, Peggy nonetheless challenges Gill, politely telling him that his words are upsetting her. "That's your guilt, Peggy," he shoots back. "All God wants is for you to reconcile with him. Don't you understand that this could be the end of the world, and you could go to hell?"[3]

Peggy's response, the last thing she will say to Father Gill as the three-episode story arc ends, is bold. "I can't believe that's the way God is," she says, challenging his—and the church's—theology of a vengeful, punishing God.[4] This statement of belief represents a shift in Peggy's relationship to the church, from the more traditional "pray, pay, and obey" to one in which she can come to her own conclusions about God and the afterlife. Earlier in the season, when Peggy's mother was nagging her about not going to church, Peggy's reply sounded like that of a petulant teenager: "It doesn't mean the same thing to me that it means to you."[5] But by the end of 1962, Peggy's image of a more forgiving, understanding God is stronger, more internalized, and likely the result of spiritual reflection on her own experiences.

Perhaps no other character experiences the level of transformation that Peggy Olson does over the course of *Mad Men*'s seven seasons. Just shy of twenty-one when the show opens,[6] she grows up from 1960 to 1970: from a freshly-scrubbed, naïve girl from Brooklyn to a confident Manhattan career woman. Her faith and connection to the Catholic Church also transform. Although later seasons do not delve into Peggy's spirituality and religious practices, the topics are explored in depth in Season 2, one in which creator Matthew Weiner admitted he "definitely wanted to talk about faith."[7] In earlier episodes, Peggy is a young woman influenced by the pre-Vatican II church and trying to navigate the often tumultuous changes in society, especially those related to gender and sexuality. But as Peggy trans-

2. Ibid.

3. Ibid.

4. Ibid.

5. "Flight 1," S2/E2.

6. "The Suitcase," S4/E7. The day of the second Clay-Liston fight (May 25, 1965) is depicted as Peggy's twenty-sixth birthday, which would mean she was born in 1939. The first episode of *Mad* Men is set in March 1960, which would mean she was twenty, just two months shy of her twenty-first birthday.

7. Sepinwall, "Matthew Weiner Q&A." In this interview, Weiner also noted that he consulted with Catholic clergy to get the details right in the show.

forms into a more liberated woman, her connection to institutional religion is left behind. This storyline is realistic, since many Catholics left the church in the aftermath of the Second Vatican Council.[8] With its focus on women's limited roles, on sin and sexuality, and on hierarchy and clericalism, the depiction of Catholicism in *Mad Men* accurately portrays the devotional faith of the pre-Vatican II church. The series fails, however, to take into account the flourishing of liberal Catholicism in the late 1960s and 1970s. Examining the depiction of Peggy's Catholicism through a historical, theological, and feminist lens, it becomes clear that this is not as complexly portrayed as other aspects of her life.

Curiously, the inclusion of Peggy's religious background on the show prompted considerable debate on fan sites as to why a character with the obviously Norwegian (and most likely Lutheran) last name of "Olson" would be a practicing Catholic. Weiner reveals the simple answer to this question in the commentary track for Season 4, by explaining that Peggy's father (who died of heart attack in front of Peggy when she was twelve years old) was a Norwegian-American Lutheran who married Katherine, an Irish-American Catholic, and they raised their children in the Catholic Church.[9] The Olsons attend the fictitious Holy Innocents Church[10] in Brooklyn's Bay Ridge, an Irish, Italian, and Norwegian neighborhood that also was the setting for the movie *Saturday Night Fever*. The mid-twentieth century church of *Mad Men* is one of second-and third-generation Catholic immigrants who proudly celebrated their religion and their Americanism. Yet, it was also a church still heavily influenced by the immigrant mindset of the nineteenth- and early twentieth-century. This "devotional Catholicism," as church historian Dolan calls it, is characterized by a Catholic ethos of authority, sin, ritual, and the miraculous.[11]

Mad Men is set during a decade of massive social change not only in America, but in the Catholic Church as well. The Second Vatican Council (1962–65) ushered in significant changes: the language of the Mass, the clothing of religious women, and how Catholics interacted with other religions. According to McBrien, "the council is regarded by many as the

8. Regular Mass attendance declined significantly after Vatican II, as did the number of clergy, women religious, parochial schools, and Catholic colleges. For more, see the research of Davidson et al.

9. Adamson, "Is the Pope Catholic? Is Peggy?"

10. The Holy Innocents are the children killed in Bethlehem by Herod the Great in an attempt to destroy the newborn Jesus, as told in the biblical book of Matthew, Chapter 2, verses 16–18. The "Holy Innocents" are often referenced by Catholic prolife/anti-abortion activists. The actual parish in Bay Ridge is called "St. Anselm."

11. Dolan, *American Catholic Experience.*

most significant religious event since the sixteenth-century Reformation and certainly as the most important of the twentieth century."[12] Although Pope John Paul XXIII announced his intention to call a council in 1959 and officially convoked it with a document called an "apostolic constitution" in 1961, the first of four, months-long sessions did not begin until October of 1962. Held in St. Peter's Basilica in Rome, the meetings were attended by 2,600 bishops from all over the world, as well as their private advisers and other guests, including twenty-three women.[13] The documents produced and approved during Vatican II addressed doctrinal and moral issues, the position of other Christian churches, and liturgy. In short, the Second Vatican Council brought the Catholic Church into the modern era. Peggy comes of age in the midst of this historic change in the church.

Vatican II was also the first church council to be held in the time of modern communication technology. Journalists were not allowed in the sessions, but daily press handouts and inside sources resulted in substantial media coverage of the council in newspapers, magazines, and on radio and television.[14] Still, most everyday Catholics were unaware of the details of the council until more visible changes began to happen in their parishes. *Mad Men* begins in March 1960, and the episodes that deal most overtly with Peggy's Catholicism, in the second season, are set in 1962, just as the council is opening and at a time when many American Catholics were unaware of the momentous changes in the church to come. Scenes set inside of a Catholic Church[15] accurately portray a pre-Vatican II liturgy. The celebrant faces the altar rather than the congregation and speaks in Latin, while the parishioners pray with rosary beads during the Eucharistic prayer. They receive Communion, on the tongue with a paten under their chins, while kneeling at an altar rail. In addition to getting many of the liturgical details correct (as is essential in a visual medium such as television), Weiner and the writers paint a fairly accurate portrait of the pre-Vatican II church's traditional views of women, its emphasis on sexual sin, and its hierarchical structure throughout the first two seasons.

SEXISM IN THE CHURCH

One of the most significant societal shifts portrayed in *Mad Men* is that of the changing roles and status of women as part of second-wave

12. McBrien, "Vatican Council II," *Encyclopedia of Catholicism*, 1299.
13. McBrien, *Catholicism*, 646.
14. McBrien, "Vatican Council II," *Encyclopedia of Catholicism*, 1301.
15. "Flight 1," S2/E2.

feminism—which took on issues of work, sexuality, reproductive rights, and other gender inequalities in the 1960s. Peggy Olson is emblematic of the changes achieved for women in this generation. As a Catholic, however, Peggy did not experience any feminist transformation in her church during those years. Although a strong Catholic feminist theology would emerge in the decades after the Second Vatican Council, as the show opens in 1960, it is still to come. Mary Daly—who is considered the founder of Catholic feminist theology—is still eight years from publishing her first book, *The Church and the Second Sex*, and her more influential *Beyond God the Father* does not come out until three years after the show's timeframe ends.

Daly strongly critiques sexism, misogyny, and women's repression in the Christian and Jewish traditions, and she calls for an overthrow of the patriarchy that has informed and infused those traditions for three millennia. Although Rosemary Radford Ruether and other feminist theologians have pointed out that Jesus subverted the gender hierarchy of his time,[16] the church nonetheless adopted the patriarchal attitudes of the broader cultures with which it intermingled—where women's experience and power were subordinated to that of men.

This sexism is evident not only in a structure that emphasizes an ordained, male-only clergy but also in many dogmatic and moral teachings in Catholicism—all undergirded by the predominantly masculine imagery of the divine.[17] From Thomas Aquinas's acceptance of women as "defective males"[18] to the contemporary "Theology of the Body"[19] that holds women's procreative abilities above all others, the Catholic Church is deeply and systemically patriarchal. Women can participate in only six of the seven sacraments in the church, since they are not eligible for ordination to the priesthood or the diaconate. Since most leadership positions in the church are commonly filled by the ordained, women's access to power in the institutional church is limited. Throughout history, males also filled liturgical roles, from presiders to readers to altar boys to ushers. Translations of liturgical prayers and scriptures have used non-inclusive language, both horizontally (referring to all persons as "men") and vertically (with predominantly male language and metaphors—i.e., "Father" —for God).

The church contributes to women's oppression outside the church, in families and in society, by encouraging women to joyfully accept their lesser

16. See, for example, Ruether, *Women and Redemption.*

17. As Daly famously wrote, "If God is male, then the male is god" (Daly, *Beyond God the Father*).

18. Aquinas, *Summa Theologiae.*

19. For more on Theology of the Body, see West, *Theology of the Body Explained.*

status on earth in exchange for an eternal reward in the afterlife. Schneiders has noted that the church was "a prime legitimator of patriarchal marriage and its attendant abuses."[20] Throughout history, the Catholic Church has discouraged women from seeking divorce from abusive husbands, forbade the use of contraception to control their fertility, and even counseled them to accept marital rape and abuse as God's will. "By its romantic reduction of women's identity and role to motherhood and its definition of the family in patriarchal terms of male headship, church authorities constricted women's self-image, loaded women's emergence into the public sphere with guilt, and legitimated patriarchal structures of economic discrimination designed to keep women out of the work force and dependent on the male head of the household."[21] Despite fledgling movements toward reform in the church, including feminist reform, the Catholic Church in the late 1950s and early 1960s retained the more traditional ethos of devotional Catholicism, including an ideal of womanhood as submissive and second-class. This is the image Peggy would have internalized from her childhood in the church—and it is a stereotype she ultimately sheds.

Yet, in early depictions of her character, Peggy appears as that ideal Catholic woman—submissive, passive, and deferential to authority—though she begins to test the waters of breaking out of those roles. In her first scene, Peggy averts her eyes and ignores the flirtatious, yet sexist comments from her future coworkers while riding the elevator to work on her first day. Dressed in a frumpy skirt and conservative sweater, hair pulled back in a ponytail with little-girl bangs, Peggy contrasts with the overtly sexual Joan, who struts confidently in a fitted green dress that accentuates her curves. Joan's advice to Peggy after giving her a tour of the office: "Go home, take a paper bag, and cut some eye holes out of it. Put it over your head, get undressed, and look at yourself in the mirror. Really evaluate where your strengths and weaknesses are, and be honest."[22] Joan also reminds Peggy to "always be a supplicant" when introducing her to the telephone operators, who also comment on her unrevealing skirt length. The men in the office are more blunt about Peggy's lack of sex appeal. When Pete first meets her, he asks, "Are you Amish or something?" adding, "It wouldn't be a sin for us to see your legs."[23] In a later episode, when Ken and Pete see Peggy with Father Gill, Ken jokes that, "It makes a lot of sense. She's an undercover

20. Schneiders, *Beyond Patching*, 33.

21. Ibid.

22. "Smoke Gets in Your Eyes," S1/E1.

23. Ibid.

nun."[24] Peggy is not a nun, but she was mostly taught by them. Catholic girls in the 1940s and 1950s learned it was indeed a sin to tempt men with their beauty and sexuality—and their responsibility to dress modestly to make it easier for men to avoid the sin of lust.[25]

Early on, there are hints of Peggy's desire to break free of the good, Catholic girl mold, as she makes attempts to try on the more worldly values of Madison Avenue. On Joan's recommendation, Peggy uses her first lunch hour to visit a doctor and get a prescription for birth control pills—prohibited by the Catholic Church to this day.[26] She also presumably believes part of her job is to be sexually available to her boss, so she makes a pass at Don Draper at the end of the workday. But in both instances, her attempts at sexual promiscuity are tempered by her words—as she reassures the moralizing doctor that she is "a very responsible person" and responds to Don's rejection of her advances with, "I hope you don't think I'm that kind of girl."[27] Throughout the series and in other aspects of her character, Peggy will be torn between the moral values of her youth and those of the changing world around her, as perhaps many Catholics of her day were. Yet, later that first evening, when Pete arrives drunk and dejected at her apartment, Peggy invites him inside her bedroom. After a day of being mocked for her uptight clothes and persona, an insecure Peggy breaks perhaps the most important of Catholic moral rules—to reserve sexual intercourse for marriage.

Marriage and motherhood were the two vocations held up by the pre-Vatican II church for women who did not choose the "higher" vocation of religious life as nuns. Yet, Peggy—viewers later learn—"wanted other things" besides early marriage and children.[28] Though she occasionally complains about her lack of marriage-material dates, Peggy is determined to pursue her career—a choice highlighted when Joan shows off her engagement ring while Peggy shows off her new office.[29] Peggy's choices are also contrasted with those of her older sister, Anita Olson Respola, who is trapped in a loveless marriage with an unemployed husband. When Anita goes to confession about her jealousy and anger over her younger sister's freedom, she complains, "What about me, Father? My troubles? What about

24. "A Night to Remember," S2/E8.

25. The Baltimore Catechism, Lesson 19.

26. For an explanation of Catholic teaching against artificial contraception, see Pope Paul VI's encyclical "Humanae Vitae" ("On Human Life," 1968).

27. "Smoke Gets in Your Eyes," S1/E1.

28. "Meditations in an Emergency," S2/E13.

29. "The Mountain King," S2/E12.

me being good—for what?" Typically, the priest counsels her, "You will get your reward in heaven."[30] This frustration with "good girls finishing last" echoes Peggy's words in Season 1, when she breaks down in front of Don, crying, "I don't understand. I try to do my job. I follow the rules. And people hate me. Innocent people get hurt, and other people—people who are not good—get to walk around doing whatever they want. It's not fair."[31] Peggy is beginning to believe that the church has sold her a bill of goods, yet her actions (and her dress) in early seasons indicate that she has not completely appropriated secular culture's values either.

Peggy eventually loses the ponytail (literally, at the shears of her gay colleague), moves to Manhattan, and tries to become "one of the boys" at the office. While climbing the career ladder, she dates inappropriate men (journalist Abe Drexler, who refuses to settle down but does not respect her work either, and her married boss Ted Chaough), yet she starts to realize that work alone won't satisfy her. "You know, I just saved this company . . . But it's not as important as getting married,"[32] she says in Season 4. She also realizes her own maternal feelings through her relationship with the neighbor boy, Julio, in Season 7.

In later episodes, the Catholic Church will play a diminishing role in dictating to her what the ideal woman should be, but Peggy has few role models to replace the church as she navigates her new identity. Conflicting advice comes from her mother, sister, and parish priest on the one hand, and Don, Joan, and Don's mistress, Bobbie Barrett, on the other. The latter tells her to "start living the life of the person you want to be. . . . You won't get that corner office until Don sees you as an equal. And no one will tell you this. But you can't be a man. Don't even try. Be a woman. It's powerful business when done correctly."[33] But what does it mean to be a woman? Both the church and Madison Avenue communicate the message that womanhood and sexuality are inextricably linked.

SIN AND SEXUALITY

A white-gloved hand holds a bulletin for "Passion Sunday," while the monsignor's voice bellows from the pulpit: "But when the flesh lusts against the spirit, and the spirit against the flesh, the rational soul is supported by the cross of Christ. Nor does it, when seduced by evil desires, consent to

30. "Three Sundays," S2/E4.
31. "Nixon vs. Kennedy," S1, E12.
32. "Tomorrowland," S4/E13.
33. "The New Girl," S2/E5.

evil doings. For it is pierced by the nails of countenance and the fear of God . . . "[34] Sin was a hot topic in the pre-Vatican II church; and sexual sin was a near obsession. It was the topic of sermons, of religious instruction, and of books and pamphlets for young Catholics during the era of devotional Catholicism in which Peggy was growing up. The Baltimore Catechism, the primary religious education text of the time, emphasized Augustine's tradition of "original sin" as well as "actual sin"—which was categorized into "mortal" or "venial."[35] The world was seen as a dark, dangerous place, full of constant temptation, with guilt as a natural result of failing to avoid that temptation. Catholicism offered hope in that humanity's naturally evil inclinations could be curbed through submission to the church's laws, rules, and practices, including confessing one's sins to a priest. As Dolan has pointed out, "A culture of sin demanded a culture of authority."[36] The church, especially in the person of the priest, mediated between Catholics and God, especially when it came to issues of sin and morality.

Of course, some sins received more attention than others, and the Catholic moral code of the late nineteenth and twentieth centuries mirrored that of Protestants in their focus on the sins of drunkenness and sexual impurity.[37] The latter was deemed especially dangerous and was the focus on the church's message to the young, especially women. The sinless Virgin Mary and other saints were held up as models for young women, and Catholic books for young people preached antiquated views of sex and sexuality. One, written by a Jesuit priest, railed against scantily clad girls as "occasions of sin" and praised saints who covered themselves modestly even as they were being martyred.[38] Another, also written by a Catholic priest, had chapters on "Holy Purity," evil thoughts, and custody of the eyes; two chapters were necessary to cover "Modesty in Dress."[39] Sex, it was taught, was only for procreation and thus limited to married couples whose duty it was to bear and rear children.[40] As taught in a catechism published in 1951, the sixth commandment reserved sex for "only husband and wife

34. "Three Sundays," S2/E4.

35. Baltimore Catechism, Lessons 5 and 6.

36. Dolan, *American Catholic Experience*, 226.

37. Ibid.

38. Plus, *Facing Life*, as cited in Farrell, *Twentieth-Century Global Christianity*, 358.

39. Deshon, *Guide for Catholic Young Women*. For an example of an earlier, though still relevant guide, see also *The Sinners Guide* by Venerable Louis of Granada, OP (1504–1588), especially chapter, "Remedies Against Lust."

40. Although this teaching has been since been expanded to include a "unitive" purpose to sexual intercourse, the procreative aspect of sex must still be retained. For more, see "The Catechism of the Catholic Church" (1992), #2363.

who are validly married to each other, and only in the natural manner, with their proper marriage partner, and only in a manner that leaves open the conception of a child."[41] Sins against this commandment included fornication, adultery, "self-abuse" or masturbation, looking at impure pictures, impure dancing, and keeping company with people who are a temptation.[42] Given this indoctrination, it's likely that an unmarried woman facing an unplanned pregnancy would see her transgression as "a sin that is like murder, and maybe worse for a young girl, probably the worst sin a young girl could make," as Weiner described it.[43]

The storyline around Peggy's pregnancy, though cliché, is not an inaccurate depiction of how a young, unmarried Catholic girl might have reacted to an out-of-wedlock pregnancy in the late 1950s or early 1960s. Although Peggy's coworkers (and *Mad Men* viewers) couldn't miss Peggy's obvious weight gain throughout Season 1, she seemingly was oblivious to her symptoms of pregnancy. In the final episode, upon being escorted to her new office after promotion to junior copywriter, Peggy remarks that she doesn't feel well. "My stomach hurts. I think I had a bad sandwich," she tells Joan.[44] When the doctor tells her she is pregnant, she insists, "That's impossible," and tries to rush out of the office after the doctor forces her to feel the baby kicking in her abdomen.[45] Ashamed and disconnected after the birth, Peggy refuses to hold or feed her newborn son. Although later episodes hint that the child is being raised by her sister Anita, Weiner revealed in an interview that Peggy's baby was instead placed for adoption.[46]

The period in which massive numbers of children were placed for adoption from the end of World War II until the 1970s is referred to by some as "the baby-scoop era" or "The Mother's Project," in which white, usually middle-class, unwed mothers were pressured by their families or by others essentially to serve as "breeders" for adoptive parents.[47] An estimated four million children were placed for adoption from 1945 to 1973, two million during the 1960s alone, not counting those adopted by relatives.[48] Those adoptions represented some 80 percent of illegitimate births, as compared

41. Cogan, *Catechism for Adults*.
42. Ibid.
43. Sepinwall, "Matthew Weiner Q&A."
44. "The Wheel," S1/E13.
45. Ibid.
46. Ryan, "Peggy's Baby on 'Mad Men,'" para. 1–4.
47. Billhartz, "About."
48. The Adoption History Project, "Adoption Statistics."

to only 4 percent today.[49] During that period, young pregnant women were told adoption was their only option, with many of them being sent away to maternity homes to have their babies.[50] The legalization of abortion and option of single motherhood would come in later decades.

Although Peggy's situation was complicated by her denial of her pregnancy, it was not uncommon for unwed mothers at the time to be labeled as mentally ill.[51] When Peggy is in the hospital after the birth of her child, the doctor references Peggy's "psycho-neurotic disorder" and reveals that she has been medicated so she is "more relaxed and feels like talking."[52] His comment that "she's in no condition to be making decisions for herself" hints that the adoption decision may have been forced or at least heavily coerced, perhaps while drugged, as critics of the "baby scoop era" have argued.[53] Her sister's comments during a later argument also hint that perhaps Peggy had little say in her child's future. When Peggy asserts that she can make her own decisions, Anita snaps back, "Really? State of New York didn't think so. The doctors didn't think so."[54]

Despite changing sexual mores and the release of the birth control pill in the 1960s, American culture still shamed young women with unplanned pregnancies. Labeled as promiscuous and scorned by families who thought their pregnant daughters reflected poorly on them as parents, the young women were often pressured to keep their situations secret. (Meanwhile, in the 1960s more than a quarter of births to *married* women under thirty occurred before marriage. A quick wedding absolved women of the guilt and shame.[55]) This judgmental attitude about sex outside of marriage is evident shortly after Peggy returns to work after giving birth, when Harry Crane tries to sell Belle Jolie on the idea of sponsoring an episode of the television show "The Defenders," whose storyline involves an abortion doctor. During a meeting, Peggy is forced to watch the scene in which the father who discovers his daughter was one of the doctor's patients shouts, "Eighteen years old and you have to have an illegitimate child taken away from you by a filthy doctor" before he slaps her.[56] The episode of "The Defenders" was

49. Brodzinsky, "Surrendering an Infant," 297.

50. This was mostly for white, middle-class women. For more see, Fessler, *Girls Who Went Away*, and Solinger, *Wake Up Little Susie*.

51. Ellison, "Authoritative Knowledge," 326.

52. "The New Girl," S2/E5.

53. "The New Girl," S2/E5.

54. "Flight 1," S2/E2.

55. Billhartz, "The Mothers Project."

56. "The Benefactor," S2/E3.

an actual one in 1962 (and one in which sponsors really did pull out), and while its existence hints at growing social acceptance for a public discussion of abortion, the plot and sponsor reaction prove that traditional cultural values about sexuality still held.[57]

For a Catholic girl, an out-of-wedlock pregnancy was even more guilt-and-shame-producing, given devotional Catholicism's emphasis on sin—especially sexual sin. Catholic charities around the world ran maternity homes where girls could go to hide the evidence of their sexual immorality, lest their growing bellies brand them a public sinner. "The beliefs the Catholic Church had about premarital sex and the judgmental approach the church had, made it particularly aggressive in pressuring women into putting their children up for adoption," according to an attorney who sued the church for its treatment of unwed mothers in Quebec.[58] While Peggy does not go to a maternity home, her family keeps her pregnancy secret by telling Don that she is sick with tuberculosis, and he advises her to move past this mistake by forgetting about it: "This never happened. It will shock you how much it never happened," he says after tracking her down at the hospital.[59] While Don does not shame Peggy about her mistake, he encourages her to move forward as if it never happened—reinforcing the secrecy that surrounded unplanned pregnancies at the time.

That all this judgment and shame about sex outside of marriage fell disproportionately on women (witness that Pete is not shamed for getting Peggy pregnant) resulted not merely because of the biological reality of pregnancy. The church's association of women with sexuality in general, and with sexual sin in particular, goes back to interpretations of Genesis that blame Eve for original sin and hold up Mary, a virgin mother, as the unattainable ideal woman.[60] Spong has argued that Augustine, influenced by the dualism of neoplatonic thought and guilt for his own sexual sins before his conversion to Christianity, is responsible for the belief that original sin was transmitted through sex. Redemption thus requires a savior not touched by this sin, necessitating Jesus's birth through a virginal conception, with the Holy Spirit as the male progenitor. It was not enough that Mary be virginal before and during the conception of Jesus; church fathers also argued that she remained a perpetual virgin after the birth of Jesus and, perhaps most unbelievably, that she remained a "postpartum virgin," in that she retained

57. "The Benefactor," S1/E30 of *The Defenders*.

58. Merchant, "Adopted or Abducted?"

59. "The New Girl," S2/E5.

60. Simply put: "Eve was sexual and evil. Mary was sexless and good." Spong, *Born of a Woman*, 210.

her unbroken hymen during childbirth.[61] Declaring Mary's Immaculate Conception (in 1854) and her bodily Assumption (in 1950) complete the myth of the ideal woman uncorrupted by sexual sin. Previously, Mary Magdalene, a powerful woman, had been seen as the central woman in the Christian story. Replaced by the submissive Virgin Mary, Magdalene is re-cast as a prostitute and sexual sinner.

These historical teachings have resulted in a devaluing of the vocation of marriage and of sexual intercourse itself, as well leaving a legacy of sexual guilt for men and women. But their most toxic effects involve the oppression of women. Defining women in biological categories "has served to legitimize as God-given the second-class status of women,"[62] according to Spong, who cites research by Warner that found that cultures where Mary was popular had particularly low status for women.[63] Since actual women could not be "virgin mothers," they could try to attain holiness by being either virgins *or* mothers. "The cultural assumption was made that the only proper way for a moral woman to conduct herself was to remain safely inside the sexually protective barriers provided first by her father and second by her husband."[64] (In *Mad Men*, the dualistic, sexualized categories for women are also evident in the sexist Playtex pitch that forced every woman to be either a sultry "Marilyn" or a virtuous "Jackie." To Peggy's objections, Paul Kinsey re-emphasized women's lack of agency in these labels: "Women want to see themselves the way men see them."[65])

In the church, these teachings also would be used to justify a male-only priesthood and to disapprove of divorce, artificial contraception, abortion, sex outside of marriage, and women working outside the home. Later Catholic feminists will retrieve a more positive image of Mary as a powerful, feminist woman, while others will rescue Mary Magdalene's reputation from disrepute.[66] Still, although neither Peggy or even her mother or sister display devotion to Mary, they inherit the legacy of these teachings. Even today, in the post-Vatican II church, the hierarchy's obsession with the so-called "pelvic issues" of sex, contraception, and abortion are evidence of the persistence of the church's patriarchal history. Instead, Peggy could have benefitted from a later feminist theologian's insight that sin needs to be re-

61. Ibid.

62. Ibid., 204.

63. Warner, *Alone of All Her Sex*, 183, as cited in Spong, *Born of a Woman*, 219.

64. Spong, *Born of a Woman*, 202.

65. "Maidenform," S2/E6.

66. For more on reclaiming Mary as a strong, feminist role model, see Johnson, *Truly Our Sister*. For more on restoring the reputation of Mary Magdalene, see Schlumpf, "Who Framed Mary Magdalene."

defined, in light of women's experience, to downplay selflessness and instead name women's lack of asserting themselves as sinful.[67] Alas, Peggy will not hear that from the priests in her life.

HIERARCHY AND CLERICALISM

If women are at the bottom of the patriarchal pyramid, then ordained males—priests, bishops, deacons, and popes—represent the pinnacle in the Catholic Church. Although the Second Vatican Council called for increased involvement in the church by the laity, the pre-Vatican II church heavily emphasized the hierarchal nature of the institutional church, in which ordained clergy not only were seen as closer to God but also possessing more earthly power, especially in church matters. Although the Vatican II document, *Lumen Gentium*, would highlight the importance of the "people of God" as an essential part of the church,[68] equality between clergy and laity has yet to be achieved in the church, even today, and certainly not in the years of *Mad Men*. Peggy's frustration about women's lack of power, in general but also in her church, is highlighted in her interactions with the young priest at her parish, Father John Gill. Gill is portrayed as a hip, young associate pastor (and a Jesuit, so presumably bright and forward-thinking). Yet despite his relative open-mindedness, Gill still represents the authority in the church to which lay people—especially women—must submit.

Although the early church was more loosely organized, the roots of the hierarchal structure of Christianity go back to the second century. The clericalism of the twentieth-century church, however, can be traced the church's reaction to the modernist crisis. McBrien has noted that "much of twentieth-century Catholic theology in the years before Vatican II was written while authors were looking back over their shoulders at the Modernist crisis."[69] In the nineteenth century, the rise of industrialization, science, and democracy had threatened the church's beliefs and structure. In response, Pope Pius X condemned modernism as heresy and required every priest to swear an anti-Modernist oath before ordination.[70] Among the ideas condemned as modernist were those that promoted responsibility of laypeople, in contrast to the privileged status of priests and bishops. Dolan also writes that modernist challenges to the church's authority in the late eighteenth

67. Saiving, "Human Situation," 25–42.

68. For more about the reimagined role of the laity, see the Vatican II document *Lumen Gentium*, Chapter 2 (Pope Paul VI).

69. McBrien, *Catholicism*, 646.

70. McBrien, *Encyclopedia of Catholicism*, 877–78.

and nineteenth centuries resulted in the church reasserting its authority with "excessive vigor," culminating in the declaration of papal infallibility in 1870 at the first Vatican Council.[71] The American spirit of independence was seen as particularly threatening to the church's authority.

Especially for Catholics of the pre-Vatican II era, the local priest was the chief authority, in that priests received their authority from the bishop, who in turn received it from the pope. Lay Catholics needed the clergy to access God, especially through the Eucharist, which could take place only through the action of the priest at Mass. Pope Pius X made this two-tiered system clear: "It follows that the Church is essentially an *unequal* society, that is, a society comprising two categories of persons, the Pastors and the flock . . . the one duty of the multitude is to allow themselves to be led, and, like a docile flock, to follow the Pastors."[72] Lay Catholics were taught to be docile and submissive. Dolan notes, "In such a culture, the rights of the individual conscience were deemphasized, as each person was conditioned to submit to the external authority of the church."[73] Although the early and mid-twentieth century saw the rise of some lay reform groups in the United States, the Catholic Church before Vatican II (and to some extent still today) was highly clerical in that it emphasized its institutional, patriarchal structures to the exclusion of lay people, especially women.

In Gill's first appearance, he is portrayed in a position of authority, chastising two young boys for misbehaving in church. A hungover Peggy is sneaking out of church when she runs into the new associate pastor at Holy Innocents. "If you see me again and I'm not smiling, I don't think you're going to like it," Gill tells the two boys standing with their noses to the wall.[74] Although he is less stern, even flirtatious, toward Peggy, she is nonetheless deferential toward him. She calls Gill by his title, "Father," while he says she is "Katherine Olson's girl," terminology that sets her up as childlike, although the two are likely close in age.[75] The portrayal of clericalism continues when Gill comes for dinner at Peggy's sister's home. Anita and Katherine gush over Gill, asking him to say grace, praising his singing voice, and even requesting a commemorative photo with him. He eats up the attention and adoration, practically flirting with the older women while regaling them with stories from Rome that exaggerate his closeness to the pope.

71. Dolan, *American Catholic Experience*, 219.

72. Pope Pius X, "Vehementer Nos," No. 8.

73. Dolan, *American Catholic Experience*, 224.

74. "Three Sundays," S2/E4.

75. Ibid.

However, the power differential between priest and laywoman is somewhat reversed when Gill requests a favor from Peggy that seems to recognize and respect her professional experience. In the car while driving her to the subway after dinner, he asks for her advice on public speaking for his next sermon. She is reluctant, yet he persists, saying that despite his training, her experience giving presentations makes her the expert. Her tips are helpful—be prepared, have confidence in what you're selling, and make eye contact with one audience member—but she doesn't mince words. "The sermon is the only part of mass that's in English, but I can hardly tell sometimes," she says. "Maybe be simpler? Give us a chance of understanding."[76] A later scene indicates that Gill has taken her advice, when Katherine praises his sermon: "I felt like I was the only one you were talking to."[77]

Gill again calls upon Peggy's professional expertise in helping to promote the parish Catholic Youth Organization (CYO) dance, but his ulterior motive of bringing her back to the church complicates the power dynamics between them. Peggy initially tries to politely decline, but Gill doesn't take no for an answer. "Think of it as *pro bono*," he says. "Talk to your boss."[78] The assumption is that she can't authorize *pro bono* work on her own. When he calls to give feedback on the "A night to remember" flyer, she pretends to be her own secretary and is assertive with her opinions, yet treats Gill like a client to be appeased. Gill seems to both admire and respect Peggy's professional achievements (being impressed during her tour of the office, for example), but he also sees her as a lost lamb who needs rescuing to return to the flock, the latter job being more urgent. She may work on Madison Avenue, but he has a more prestigious Boss.

Peggy has grown up in a church that equates priests with God (One argument against admitting women to the priesthood is that they can't image Jesus, as male priests do[79]), yet she is surprisingly assertive in challenging him, given Catholics' tendency to be deferential to clergy even today. On the phone, when he mentions the CYO committee's concerns about the flyer, she responds that he should "tell them this is the way it works. And I know better than them."[80] Later, during a meeting with the committee, Peggy begins with a confident pitch of the campaign, which Gill initially supports,

76. Ibid.

77. Ibid.

78. "A Night to Remember," S2/E8.

79. See *Inter Insigniores* (Declaration on the question of admission of women to the ministerial priesthood).

80. "A Night to Remember," S2/E8.

saying, "It's just what we talked about."[81] But when the committee members express concerns about sexual innuendo and the campaign's effectiveness, Gill throws Peggy under the bus, concluding, "Maybe we should try something else."[82] After the committee members leave, Peggy boldly expresses her disappointment to Gill: "You asked me to do this based on my expertise. . . . You're supposed to tell them that they should trust me. That's *your* job."[83] Neither Peggy's mother or sister would ever talk to a priest in such an assertive way; it's likely Peggy wouldn't have either before meeting Gill, who has blurred the hierarchal lines in his interactions with her. However, in the end, he has the final say, telling her, "I'm sure you can turn it around."[84] Peggy resumes the deferential posture of lay person to priest by apologizing, "I'm sorry, Father. I don't mean to be disrespectful."[85] Peggy faces the same dance at Sterling Cooper, knowing she needs to be assertive to earn respect as a copywriter, yet recognizing that as a woman she clearly has less power than her male colleagues. Patriarchy, in the form of clericalism, is even more entrenched in the Roman Catholic Church—where patriarchy becomes sanctioned by God.

If Gill can trump Peggy on the CYO advertising campaign, as a priest he reigns as the clear expert in spiritual matters. In the inappropriate setting of her office, Gill makes a clumsy attempt at pastoral counseling, by noting that he sees that Peggy does not take Communion. Catholics in a pre-Vatican II church, and even today, may refrain from Communion for a variety of reasons, such as not completing the required fast. Gill's implication, however, is that Peggy must be skipping Communion because she has not confessed any mortal sin, as required by church rules.[86] The young priest attempts to offer compassion, yet his repeated attempts to get her to "spill all" are counterproductive and result only in Peggy's defensiveness and recognition of the advantages of clericalism. "Father, you don't have to live life like the rest of us. Maybe you're lucky," Peggy says.[87] Gill is so busy playing the spiritual expert that he fails to understand Peggy's real challenges, including the sexism she faces in the office and excessive guilt resulting from the church's association of women with sexual sin. Although Gill's character does serve to represent the culture of clericalism in the pre-Vatican II

81. Ibid.
82. Ibid.
83. Ibid.
84. Ibid.
85. Ibid.
86. Code of Canon Law, #916 and 919.
87. "A Night to Remember," S2/E8.

church, this "add priest and stir" formula for the Catholic storyline in *Mad Men* ironically also perpetuates it.[88] The addition of a priest to the cast in Season 2 legitimized the Catholic storyline and captured the attention of Catholic commentators. Popular writer Father James Martin uncovered many Catholic themes in *Mad Men* while cohosting a podcast about the show, yet the initial impressions from this fellow Jesuit were focused, not surprisingly, on Father Gill, not Peggy. For the most part he likes Gill's character, though, like many Catholic commentators, is concerned about Gill breaking the seal of the confessional.[89] After Peggy's sister confesses to Gill her anger over Peggy's apparently lack of consequences for her sexual sin, Gill repeatedly hints that he knows something about her past—most notably when he hands Peggy a blue egg during the Easter egg hunt, saying it's "for the little one."[90] Even such an indirect reference to information from a confession would be prohibited by the church. Other commentators were more concerned about a possible romantic storyline between Gill and Peggy, especially since a priest-parishioner affair was part of Weiner's previous series, *The Sopranos*.[91] The *Mad Men* writers do not pursue that clichéd plot, but the use of Gill does portray the clericalism of the pre-Vatican II church.

CONCLUSION

Despite the unequal nature of their relationship, the portrayal of Gill and Peggy contains hints of a more post-Vatican II theology that downplays the clericalism of Catholicism in the earlier twentieth century. Gill's desire for Peggy to return to the church seems to go beyond merely getting her obey the rules and saving her soul from damnation. "You know, when you distance yourself from the church, you are distancing yourself from everything. That's why it's called Communion. It's not just being with God. It's being with people," he says in his impromptu counseling session in her office.[92] When she nods, he asks, "Well, then why are you pushing everyone away?"[93] His final exhortation ends with a poignant question that seems to linger with her: "There is no sin too great to bring to God. You can reconcile yourself with him and have a whole new start. . . . Do you feel you don't deserve

88. Schlumpf, "'Mad Men': A Truly Catholic Storyline."
89. Martin, "Mad Men, Priests, and the Catholic Worldview."
90. "Three Sundays," S2/E4.
91. Murray, "*Mad Men*, 'A Night to Remember.'"
92. "A Night to Remember," S2/E8.
93. Ibid.

his love?"[94] That episode ends with Peggy deep in thought in the bathtub, while Gill's guitar-accompanied version of "Early in the Morning" exhorts God to "help me find the way."[95] As usual, the writers don't make Peggy's spiritual struggle explicit, but it's clear she has been affected by Gill's words.

Gill, representing the institutional church, is offering redemption, but does Peggy need saving? Ultimately, Peggy rejects clericalism and the judgmentalism of the institutional church, confessing her pregnancy not as a sin, but as a fact—and not to Gill, but to Pete. After he professes his love to her in the midst of the Cuban Missile Crisis scare, she tells him about the baby, noting that she could have shamed him into marrying her but chose not to. "I wanted other things," she tells him, then emotionally describes the shift that has happened to her: "Well, one day you're there, and then all of a sudden, there is less of you. And you wonder where that part went. If it's living somewhere, outside of you, and you keep thinking, maybe you'll get it back. And then you realize, it's just gone."[96] Peggy is not talking about her baby,[97] nor is she talking about her faith, as the closing shot of the episode features her making the sign of the cross before falling asleep. What is "just gone" is her naïve, innocent self of Season 1—and her commitment to the institutional church, since she has found devotional Catholicism lacking. The church's judgmental attitude, especially about sexual sin, and its inability to see women as full, equal members renders it irrelevant to a young, single working woman.

In fact, the final episode of Season 2 is the last time viewers will see Gill, Holy Innocents, or any direct references to Peggy's Catholicism. While she never explicitly states that she has left the church, it's safe to assume that organized religion is no longer a relevant part of her life. Yet it's apparent that Peggy has not lost her faith. Her clearly Eucharistic language for the Popsicle campaign, "Take it, break it, share it, love it,"[98] indicate that such deeply held beliefs are still there; they simply are not be mediated by the institutional church anymore. Her assertion to her colleagues that ritual is "Christian, as in behavior, not religion" implies she is able to separate the two. "Let me tell you something, the Catholic Church knows how to sell things," she adds.[99] But Peggy is not buying it anymore. She will continue to grow, change, and transform, and while some of her future decisions will be informed by the

94. Ibid.

95. Ibid.

96. "Meditations in an Emergency," S2/E13.

97. Sepinwall, "Mad Men: Matthew Weiner Q&A."

98. "The Mountain King," S2/E12.

99. Ibid.

morals she learned from the church, religion is now one more product that can be chosen or not. Ultimately, the Catholic Church has not persuaded Peggy on the benefits of pre-Vatican II, devotional Catholicism.

What the Catholic storyline in Season 2 of *Mad Men* misses, however, is a more complex depiction of American Catholicism just before, during, and after the Second Vatican Council. While the portrayal of the church's sexism, sexual sin, and clericalism comprise an accurate depiction of much of American Catholicism in the 1960s, it missed the growth in the church that follows, including the burgeoning reform movements, especially in major cities. How about the priest turning around at a Mass at Holy Innocents? Or Anita becoming active in the Christian Family Movement? Or Gill joining other priests and religious in marching on Selma? Even more egregiously, *Mad Men* never even mentions the momentous Second Vatican Council, which many fans and Catholic commentators had assumed would be part of a future storyline.[100] For Weiner and the writers, the Catholic Church was a one-season vehicle to move the character of Peggy forward. By not mentioning it again, it exists only in the past.

By the final season, Peggy has "come a long way, baby," although her spiritual life has been largely unexplored since Season 2. She clearly has succeeded in her career, with her Burger Chef pitch about the changing nature of American families comparable to her mentor's iconic "Carousel" one. "You're starving," she says of the typical late sixties family, "and not just for dinner."[101] Once again, as with the Popsicle pitch, she hints at Eucharistic language, suggesting that families find "the connection that [they're] hungry for" at Burger Chef.[102] Yet, the underlying ideologies in the pitch are consumerism and individualism, not faith. Burger Chef is "where everybody gets what they want, when they want it."[103] Peggy has reached her professional potential, and her hair, clothing, and demeanor indicate a one hundred eighty-degree turn from the mousy secretary of a decade ago.

On many levels, Peggy gets the "happily ever after" treatment in the series' final episodes. Her slow-motion entrance into McCann Erikson—clad in dark sunglasses, with an erotic octopus painting under her arm, and strutting like she owns the place—portrays her as the epitome of a confident, sexy woman. In addition to professional success, she also finally finds love—right at the office where it's been for years, in art director Stan Rizzo, her "work husband" and soul mate. Although the two have kissed and seen

100. Rasmussen, "'Mad Men's Catholic Confessions."

101. "Waterloo," S7/E7.

102. Ibid.

103. Ibid.

each other naked, Stan and Peggy kept their relationship a friendship, one connected primarily through phone conversations— though their offices are just feet away. So it's not surprising that Stan finally confesses that he loves her on the phone, and that she realizes she loves him, too, because, "You make everything OK. You always do. No matter what."[104] Of course, Stan is the only person outside Peggy's family (besides Don) who knows about her child. But Peggy's revealing of her secret to Stan was not a confession; in fact, it was accompanied by a feminist sermon: "No one should have to make a mistake just like a man does and not be able to move on. . . . Maybe you do what you thought was the best thing," she says before sharing her own experience: "I'm here . . . and he's with a family, somewhere. I don't know, but it's not because I don't care. It's because you're not supposed to know, or you couldn't go on with your life."[105]

Peggy's pregnancy and decision not to parent her child are part of her life, but she refuses to live as a perpetually repentant sinner—as the Catholic Church seemed to require. As she takes a final skate around SC&P, it remains unclear if she has found a spiritual path to replace the devotional Catholicism of her youth. Will she explore New Age spirituality with Stan, become an evangelical Christian at a megachurch, or, perhaps, return to a reformed, post-Vatican II Catholic Church? More spiritual transformation awaits.

BIBLIOGRAPHY

Adamson, Rondi. "Is the Pope Catholic? Is Peggy?" *Minneapolis Star-Tribune*, March 27, 2012. Online: http://www.startribune.com/is-the-pope-catholic-is-peggy/144467135/.

The Adoption History Project. "Adoption Statistics." University of Oregon. Online: http://pages.uoregon.edu/adoption/.

Baltimore Catechism, Revised Edition (1941). Lesson 19: "The Fourth, Fifth, and Sixth Commandments of God."

Bednowrowk, Mary Farrell. *Twentieth Century Global Christianity: Volume 7 of A People's History of Christianity*. Minneapolis: Fortress, 2010.

Billhartz, Celeste. "About." *The Mothers Project*. Online: www.themothersproject.com.

Brodzinsky, A. "Surrendering an Infant for Adoption: The Birthmother Experience." *The Psychology of Adoption*, edited by D. Brodzinsky and M. Schechter, 295–315. New York: Oxford University Press, 1994.

Catechism of the Catholic Church. *Vatican Archive*. Online: http://www.vatican.va/archive/ENG0015/_INDEX.HTM, 1992.

Code of Canon Law. Washington, DC: Canon Law Society of America, 1983.

104. "Person to Person," S7/E14.

105. "Time & Life," S7/E11.

Cogan, William J. *A Catechism for Adults*. Chicago: ACTA Foundation, 1958.

Daly, Mary. *Beyond God the Father*. Boston: Beacon, 1973.

———. *The Church and the Second Sex*. New York: Harper & Row, 1968.

"Dan Rather Reports." Season 7, Episode 12. Yahoo News, May 1, 2012. Online: http://news/yahoo.com/forced-adoptions-for-unwed-mothers-around-the-globe.html.

Davidson, James D., et al. *The Search for Common Ground: What Unites and Divides Catholic Americans*. Huntington, IN: Our Sunday Visitor, 1997.

Defenders. "The Benefactor." Season 1, Episode 30, April 28, 1962, *CBS*.

Deshon, George. *Guide for Catholic Young Women: Especially for Those Who Earn Their Own Living*. New York: Columbus, 1910.

Dolan, Jay P. *The American Catholic Experience: A History from Colonial Times to the Present*. Garden City, NY: Doubleday, 1992.

Ellison, M. "Authoritative Knowledge and Single Women's Unintentional Pregnancies, Abortions, Adoption and Single Motherhood: Social Stigma and Structural Violence." *Medical Anthropology Quarterly* 17.3 (2003) 322–47.

Fessler, Ann. *The Girls Who Went Away: The Hidden History of Women Who Surrendered Children for Adoption in the Decades Before Roe v. Wade*. New York: Penguin, 2006.

Johnson, Elizabeth E. *Truly Our Sister: A Theology of Mary in the Communion of Saints*. New York: Continuum, 2003.

Martin, James A. "Mad Men, Priests, and the Catholic Worldview." *America Magazine*, October 27, 2008. Online: http://americamagazine.org/content/all-things/mad-men-priests-and-catholic-worldview.

McBrien, Richard P. *Catholicism: New Study Edition*. New York: HarperCollins, 1994.

———, ed. *The HarperCollins Encyclopedia of Catholicism*. New York: HarperCollins, 1995.

Murray, Noel. "Mad Men, 'A Night to Remember.'" *A.V. Club*, September 14, 2008. Online: http://www.avclub.com/tvclub/mad-men-a-night-to-remember-13174.

Plus, Raoul. *Facing Life: Meditations for Young Women*, London: Burns & Oates, 1928,

Pope Paul VI. *Humanae Vitae* ("On Human Life"). Online: http://w2.vatican.va/content/paul-vi/en/encyclicals/documents/hf_p-vi_enc_25071968_humanae-vitae.html.

———. *Lumen Gentium* ("The Dogmatic Constitution on the Church"). Online: http://www.vatican.va/archive/hist_councils/ii_vatican_council/documents/vat-ii_const_19641121_lumen-gentium_en.html.

Pope Pius X. "Vehementer Nos." Online: http://w2.vatican.va/content/pius-x/en/encyclicals/documents/hf_p-x_enc_11021906_vehementer-nos.html.

Ruether, Rosemary Radford. *Women and Redemption: A Theological History*. Minneapolis: Augsburg Fortress, 1998.

Ryan, Maureen. "Peggy's Baby on 'Mad Men': Clearing Up that Confusion." *The Chicago Tribune*, October 28. 2008. Online: http://featuresblogs.chicagotribune.com/entertainment_tv/2008/10/mad-men-peggy-b.html.

Saiving, Valerie. "The Human Situation: A Feminine View." In *WomanSpirit Rising: A Feminist Reader in Religion*, edited by Carol Christ and Judith Plaskow, 25–42. New York: HarperCollins, 1979.

Schlumpf, Heidi. "Mad Men": A Truly Catholic Storyline," *NCR Today* (blog), *National Catholic Reporter*, August 18, 2009. Online: http://ncronline.org/news/mad-men-truly-catholic-storyline.

———."Who Framed Mary Magdalene," *U.S. Catholic* 65.4 (2000) 12–16.

Schneiders, Sandra M. *Beyond Patching: Faith and feminism in the Catholic Church.* New York: Paulist, 1991.

Seper, Cardinal Franjo. *Inter Insigniores* (Declaration on the question of admission of women to the ministerial priesthood). Online: http://www.vatican.va/roman_curia/congregations/cfaith/documents/rc_con_cfaith_doc_19761015_inter-insigniores_en.html.

Sepinwall, Alan. "Mad Men: Matthew Weiner Q&A for Season Two." *What's Alan Watching,* October 26, 2008. Online: http://sepinwall.blogspot.com/2008/10/mad-men-matthew-weiner-q-for-season-two.html.

Solinger, Rickie. *Wake up Little Susie: Single Pregnancy and Race Before Roe v. Wade,* New York: Routledge, 2000.

Spong, John Shelby. *Born of a Woman: A Bishop Rethinks the Birth of Jesus.* San Francisco: HarperSanFrancisco, 1992.

Venerable Louis of Granada. *The Sinners Guide.* Online: https://www.ewtn.com/library/SPIRIT/GRANADA.HTM.

West, Christopher. *Theology of the Body Explained: A Commentary on John Paul II's Man and Woman He Created Them.* Boston: Pauline, 2007.

Chapter 17

"WE DON'T KNOW WHAT'S REALLY GOING ON"

Mad Men as a Bellwether of the Politics to Come

—JARED D. LARSON

Matthew Wiener, along with the other writers and producers of *Mad Men*, have masterfully tricked us into thinking that the series is about advertising in the 1960s. It seemingly is: the advertising revolution of the Post-WWII era, which laid the foundation for our branded society today, remains in the foreground with the politics and societal transformations of the time playing out in the background.[1] In reality, these subjects are reversed. The genesis of modern advertising provides the backdrop upon which the collective bad conscience of American culture is on full display through its socio-political realities at home and abroad. As the showrunners brilliantly capture the style and *zeitgeist* of the period through convincing sets, props, and costumes, they concurrently remind us that this was one of the darkest times in American history.

The transitional and turbulent times of the 1960s are well documented: the Cold War, racial tension and out-right conflict, overt and unapologetic sexism and anti-Semitism, the rise of Richard Nixon, the assassinations of Medgar Evers, John F. Kennedy, Martin Luther King, and Bobby Kennedy,

1. Klein, *No Logo*, 3–13.

class privilege in the land of supposedly equal opportunity, the war in Vietnam, counterculture movements, etc. The showrunners do more than indirectly document, as *Mad Men* is clearly not a documentary.[2] They proffer a major indictment of American society and politics, one that continues to be instructive today. In many ways, the 1960s ushered in a change in political climate and tone, for the better and for the worse; over the course of the series we see some evidence of the positive changes, yet if we are honest with ourselves as we watch the show from our twenty-first century couches, the case can be made that in many ways less progress has been made then we would like to have seen.[3]

Without a doubt, from the pilot to the finale, as television serial dramas go, the series continually stuns and surprises us both visually and emotionally. It tells the story of the 1960s with an elegance and grace in the great tradition of earlier forms of Western drama. However, the sets, props, costumes, and storylines of advertising, interpersonal relationships and conflict (both of the vertical and the horizontal varieties), and the daily machinations of 1960s New Yorkers (albeit the predominantly white and wealthy), easily mask the fact that this is much more than a soap opera for the casual viewer.[4] While there are clearly "soap operatic" elements throughout the entirety of the series, soap operas are atemporal and thus can be set in any age; this is a series of dramatic, and dramatically intertwined, stories told within a very specific socio-political context.[5] *Mad Men* effortlessly bridges the gap between pseudo-docudrama and soap opera in this way, leading at least one commentator to note the atemporality of the show through the lessons it has to teach us, in spite of the aesthetics and design of its setting.[6]

But what are the lessons to learn from seven seasons of *Mad Men*? Previous examinations of the program, as well as the other chapters of this

2. Beail and Goren, "*Mad Men* and Politics," 4. Although, it should be noted, in his years as a film student at the University of Southern California, creator Matt Weiner had some success with documentary film making (Bigsby, *Viewing America*, 358).

3. See, for example, Colton Josephson, "Tomorrowland," 261–65. She discusses how *Man Men* deromanticizes our nostalgic view of the past while also forcing to us to confront the politics of the present.

4. Perlman notes the contrast between the "gorgeousness of its mise en scène" and the "ugliness of its characters' actions" ("Strange Career of *Mad Men*," 209).

5. In fact, Guilfoy argues that the affluence of the U.S. during the golden age of advertising portrayed in *Mad Men* provides a fertile context for an interesting discussion of the differing views on capitalism and freedom between the economic philosophies of John Kenneth Galbraith and Milton Friedman. While I certainly do not disagree, I will argue throughout that taking a broader view of the show, beyond economics, we can see its applicability to every decade from the 1960s to today.

6. Vasquez, cited in Marcos, "El final de 'Mad Men,'" 38.

book have all dealt in different ways with many of the problems and issues it presents, from questions of race[7] and gender,[8] to the business and theory of capitalism,[9] even through the lens of Mills's *Power Elite*,[10] and to American nationalism[11] and identity.[12] Here in the twenty-first century, we know that as a society we have yet to effectively deal with what unfolds on our screen as we watch the show, we realize that we are witness to the acting out of a "return of the repressed" for many of the main characters at that specific time, but also within American society.[13] From the outset of the series, beginning in 1960, right at the end of what was a run of fifteen post-war years of prosperous and stable economic growth and social development, Americans find themselves under a cloud of unresolved guilt as their collective bad conscience has come back to haunt them, a state which, again, *Mad Men* captures beautifully.

We find a greater depth among the intertwined stories presented, both fictional and historical, which I will argue herein is further evidence of the brilliance of Weiner and the other writers of *Mad Men*. While I do maintain that the series may be viewed as a period piece and, thus, also a document of history in its own way, the period in which it is set marked much more than a revolution in advertising. What we are shown is a multi-level socio-political revolution as well, as many of the political problems that were indeed on display in the foreground in *Mad Men* are also, unfortunately, many of the same political problems that have remained willfully unresolved and with which we have continued to deal since the 1960s. In terms of structure, they are the same political issues mentioned above. In terms of process, through the benefit of hindsight and the cover provided by four decades of chronological distance, *Mad Men* provides some insights into how American politics itself changed shortly after World War II. If we take a minute to honestly compare now and then, and indeed if we consider every intervening decade,

7. See Chapter 10 in this volume as well as Carveth, "'We've Got Bigger Problems,'" and Beail, "Invisible Men."

8. See Chapters 4, 5, 6 in this volume as well as Fuehrer Taylor, "You Can't Be a Man," and Vázquez, "Me llamo Peggy Olson."

9. See Chapter 1 in this volume as well as Guilfoy, "Capitalism and Freedom."

10. Goldman, "*The Power Elite*."

11. Heyman, "Appearances."

12. Beail and Goren, "*Mad Men* and Politics," and Colton Josephson, "Tomorrowland."

13. In an interview with Terry Gross on National Public Radio's *Fresh Air*, in a discussion on death and US history, portrayed at the outset of Season 6 (which takes place in the violent year of 1968), Weiner mentions Freud's *Civilization and Its Discontents* as an influence on his writing and the development of series (Gross, "Matthew Weiner on 'Mad Men' and Meaning").

while styles and technology have consistently changed and evolved, how politics works has changed so very little since. In the United States, we find ourselves repeating many of the same mistakes at present that are on display in the series, never learning our lessons: the politics of fear of the other both at home and abroad,[14] the politics of racial and gender privilege (an omnipresent debate at present), the close connection between the moneyed and the political classes,[15] the commodification of nearly everything,[16] and the commercialization of the political process itself (which will only get worse after the infamous *Citizens United* U.S. Supreme Court decision of 2010).

This combination of docudrama and soap opera makes *Mad Men* so riveting, which is enhanced if we zoom out in order to see its bigger picture and better understand our recent history. In this sense, more than merely a hybrid between a "docudramatic" period piece and soap opera, the program is much more profound and its applicability all the more wide-reaching, much like a Greek tragedy, a work of Shakespeare, or an eighteenth-century opera, so many of which are ultimately atemporal. All of these contain interpersonal drama, politics of some kind, and moral lessons to be learned both for the individual and for society writ large, and the stories they tell are almost always applicable to other times and in other contexts. Indeed, the great lessons to be learned in Hamlet, Macbeth, or Richard III might not have been written were it not for the Greek tragedies, and in turn the operas of the seventeenth and eighteenth centuries and beyond without Shakespeare. Again, in the case of the Post-WWII United States, the many events and issues portrayed in *Mad Men* marked the beginning of a new and very consequential economic and socio-political era in U.S. history, out of which we have not yet evolved. Whether or not the writers intended this, if we pay close attention, the program is much more than an indictment of the 1960s. Following the great works of the Western literary and theatrical canons, it is an indictment of our contemporary society for failing to learn the obvious lessons of our relatively recent past, and its stories can be told and retold again and again.[17]

14. See Robin, *Fear*, Part 2.

15. See Phillips, *Wealth and Democracy.*

16. See Klein, *No Logo.*

17. Bigsby documents several references to all the literature, poetry, art, and criticism in the show, from or indicative of the late 1950s and through the 1960s: Yates, Cheever, Rand, John Kenneth Galbraith, Rachel Carson, Mills, Mamet's Glengarry Glen Ross, *The Great Gatsby*, *Meditations on an Emergency*, Hemingway, Miller, Bob Dylan, Edward Hopper, etc. (*Viewing America*, 365–85). Let us not forget that Season 6 opens with Don Draper on the beach in Hawaii reading the first lines of Dante's *Inferno* ("The Doorway, Part 1," S6/E1).

The following sections will examine the rise of the "politics of fear" that took root with the Cold War, the nascent politicization of foreign policy and the political polarization of the time, the socio-political and practical manifestations of male, class, and white privilege. The next will turn more specifically to the morally corrupt nexus of big business and politics, and in particular to the Republican Party, that is a subtle ongoing subtext of the series. The final section of this chapter will then show that what began in the 1960s has only intensified throughout the decades: marketing, money, manipulation and polarization of interests in American society and its political system. More than simply a critique of the present day through the fictitious portrayal of the lives of those in the advertising industry of forty-plus years ago, as some observers of and commentators on the show have argued,[18] I believe that *Mad Men* could and should be seen through a much longer lens. The bigger-picture warnings to be heeded and lessons to be learned both by the individual and by society in general, much in the way that we read the great parables of all types of fiction in our collective Western tradition, separate *Mad Men* from its television peers in terms of content, depth, and delivery.

THE POLITICS OF FEAR

There is no doubt that from the outset, both the time and place of *Mad Men* are consequential to the stories that are to evolve. Internationally, the Post-WWII economic and military strength of the United States are paramount. Domestically, equally crucial are the socio-political changes that will occur throughout the decade of the 1960s, with New York City—and to a lesser extent the state of California—as the vanguard. Several commentators on the show, and the characters (or, the writers) themselves, have stressed this. It was a time of great economic development, as Bert Cooper notes upon the death of Ida Blankenship, his former secretary: "She was born in 1898 in a barn. She died on the thirty-seventh floor of a skyscraper."[19] Upon the conclusion of war and the effective end of fascism and perhaps because of the perceived threat of Communism,[20] the US at this time did what it often does

18. See Beail and Goren, "*Mad Men* and Politics"; Heyman, "Appearances, Social Norms, and Life in Modern America"; and Colton Josephson, "Tomorrowland."

19. "The Beautiful Girls," S4/E9.

20. Two Fascist regimes were of course left in place in Europe, in Portugal and in Spain. As the Cold War heated up, the Salazar and Franco regimes, respectively, were accepted as US allies in exchange for access to military bases on strategically located islands in the North Atlantic. Paralleling the experience of the various iterations of the Sterling Cooper agency and many of the individual characters throughout the series,

best: it sought to reinvent itself, this time as a wealthy global superpower, in order to lead both the expansion of capitalism against Soviet influence and in the exploration of space.[21] With hindsight, the viewers know the magnitude of the many changes underway throughout the program. Many observers have praised how the showrunners captured such a pivotal time. Bigsby points out that "change would be the keynote of *Mad Men*,"[22] which is seconded by Beail and Goren: "*Mad Men*'s conscious 1960s setting allows it to chart many of the socio-political changes that transpired within the United States during that period."[23] In this pinnacle moment of change, Colton Josephson asserts that "*Mad Men* is really about us at the beginning of America's third century."[24]

For many reasons, it seems that this period of massive economic growth and expansion of American power also fostered great fear, even chronic paranoia, within American society. The first two seasons of the program demonstrate the international roots of this fear: the looming shadow of the Cold War. In "Ladies Room," set in 1960, a proposal to include an astronaut and a rocket ship in an ad for Right Guard deodorant—to be sold in a "space-aged" aerosol can—gets rejected by the protagonist, ad executive Don Draper.[25] His concern that such a celestial approach would make consumers think of a bomb shelter, clearly a reference to the existential fear posed by nuclear war with the Soviet Union. By the end of the second season, everything in the concluding episode, which takes place in late October of 1962, is completely overshadowed by the Cuban Missile Crisis, when, according to first-hand accounts from those present in the White House, the danger of nuclear war with the Soviets was indeed at its most imminent.[26]

POLITICAL POLARIZATION

One of the most important changes in the history of U.S. politics was portrayed right under our noses in these two early seasons. Distinguishing ours from the two previous centuries (following Colton Josephson's chronology),

---

for Washington, convenience and self-preservation proved more important than values and ideals.

21. Recall "Wee Small Hours," S3/E9—in which Conrad Hilton foresees Hilton hotels on the moon.

22. Bigsby, *Viewing America*, 362.

23. Beail and Goren, "*Mad Men* and Politics," 23.

24. Colton Josephson, "Tomorrowland," 260.

25. "Ladies Room," S1/E2.

26. "Meditations in an Emergency," S2/E13 and *The Fog of War*.

the politicization of American foreign policy has become an undeniable, if not the defining, political reality of this "Third American Century." The aftermath of the 1962 crisis ushered in, and in practice legitimized, the now typical political posturing and Monday-morning quarterbacking over any given president's foreign policy decisions. For nearly two hundred years prior, this had been politically out of bounds. This shift is a spillover of the hyper-politicized national election of the Nixon-Kennedy election of 1960, in which Sterling Cooper had sought to participate as media advisors to the Nixon campaign (throughout Season 1, starting with the second episode of the series, which will be discussed in further detail below).[27] While perhaps not so obvious to the audience of *Mad Men*, this has proven to represent a paradigmatic change in the tenor and practice of American politics. Prior to this early milestone of the Cold War, still in the afterglow of victories in the World Wars, which had for the most part united patriotic sentiment among most sectors of American society, the political elite continued to respect Republican Senator Arthur Vandenberg's adage that "politics ends at the water's edge." From that point on, as policy makers in Washington have ignored Washington's warning against "foreign entanglements," conflict abroad and the politicization of foreign policy has become the new normal, from the Vietnam War, to Iran (the Hostage Crisis, Iran-Contra, the recently brokered nuclear deal), to how to handle Russia (from the USSR to today), to the Invasions of Iraq, etc. Beyond the reaction to the attacks of September 11, 2001, which, for the time being, continues to foster a semblance of national cohesion, squabbling over foreign policy in Congress, on cable news networks, and over dinner tables is now commonplace.

Further accentuating Colton Josephson's demarcation of the 1960s as the beginning of the third American century, and as alluded to above, this change in *how* we discussed foreign policy neatly coincided with the rise of television advertising in electoral politics and thus, ultimately, it marked the beginning of the hyper-polarization of domestic politics in the United States. Evidenced throughout *Mad Men*, signs of today's all-too-typical polarization are at times overt yet usually very subtle and we see this through criticism of policy measures to down-right snide remarks on the part of the characters regarding public figures. Among the more understated, ambient observations, often times the journalistic commentary on the Civil Rights Movement, the Vietnam War, and domestic politics "merely play in the background on the television or radio," yet occasionally these provoke a response from the characters who are paying attention.[28] In the case of the

27. See Larson, "Un declive auto-imposto."

28. Colton Josephson, "Tomorrowland," 263.

former, in "The Quality of Mercy" (S6/E12), taking place in the run-up to the 1968 presidential election, Don watches a Nixon ad that plays on the fear of rising crime rates at home, indirectly blaming his perspective opponent, incumbent president Lyndon B. Johnson, for this dangerous state of affairs. Similar subtle treatment of Vietnam is also proffered throughout,[29] during which the evening news describes how the war is intensifying as Don's second wife Megan leisurely prepares dinner in the kitchen of their comfortable high-rise apartment.

In terms of more overt commentary on the changing politics of the time—or how we "do" domestic politics in the US—one of the running themes throughout Season 1 is the 1960 presidential election between Richard M. Nixon and John F. Kennedy. In "Ladies Room," there was debate, leading to outright reluctance on the part of some in the office (including Don Draper), regarding Sterling Cooper's working with the Nixon campaign.[30] Advocating in favor of this revolutionary merger of politics and advertising and foreshadowing today's poll- and focus group-driven class of professional political advisors, Senior Partner Bert Cooper, who Goldman, in her chapter on Mad Men seen through Mills's *Power Elite*, rightly points out is never seen "doing anything productive"[31] proclaims that "we know better what Dick Nixon needs, better than Dick Nixon."[32] By the seventh episode of that season, the consensus is that the firm is officially on board with the Republican, preparing attacks on Kennedy before he has even secured the nomination of the Democratic Party.[33] Marking an evolution in electoral politics in the US, two episodes later, we observe the Sterling Cooper crew derisively dissecting a political ad featuring Jacquie Kennedy speaking in Spanish, complete with a comment from art director Sal Romano that "women will hate" her.[34] In that same episode, also foreshadowing the politics of the present day, Pete Campbell figures out how to benefit both the Nixon campaign and the agency by encouraging a client to buy peak commercial time in swing states in order to drive up the cost of television advertisement for Kennedy. By the tenth episode, we watch as Don and Pete catch a Kennedy attack ad on Nixon which includes footage of then President Dwight D. Eisenhower himself telling the viewers that as his Vice

29. See "Lady Lazarus," S5/ E8.
30. "Ladies Room," S1/E2.
31. Goldman, "*Power Elite*," 71.
32. Ibid.
33. "Ladies Room," S1/E2.
34. "Shoot," S1/E9.

President, Richard Nixon did not make any decisions of consequence.[35] In "Nixon vs. Kennedy," the office holds a party to watch the election returns and we learn through an offhand conversation among Pete, Ken Cosgrove, Harry Crane, and Paul Kinsey that the Nixon Campaign ended up not hiring Sterling Cooper but that, nonetheless, a Nixon loss would be a "damper on things" (Harry) both for the agency and "for the nation" (Paul).[36] While Bert Cooper alleges electoral fraud on the part of the Kennedy operatives in Chicago, he knows what we twenty-first-century viewers know now all too well: that the commercialization of the democratic process was a win for his sector of the power elite, because "for even if elite lose, the elite always wins."[37]

Set in the late summer of 1963, perhaps the most important single episode both in terms of its commentary on the times in which it takes place and also through its forewarning of the politics to come, comes to us in "Wee Small Hours."[38] The highlight of this episode for Don, professionally, sheds light on the international politics of the time while also foreshadowing another important movement for the Republican Party. Through his personal proximity with Conrad Hilton—whom he had met by chance[39]—Don officially lands Hilton Hotels as a blue chip account at Sterling Cooper. In an awkward late night meeting to which Mr. Hilton calls Don, the former shares with the latter his vehemently anti-communist views which are rooted in nationalistic and religious sentiment, none of which we have reason to believe that Don shares. Hilton pontificates, "It's my purpose in life to bring America to the world, whether they like it or not. You know, we are a force of good, Don, because we have God. The communists don't. It's their most important belief. Did you know that?" Don replies, "I'm not an expert." Citing the Marshall Plan, Hilton goes on to extol the "generosity" of American foreign policy, adding that "Everyone who saw our ways wanted to be us." What we see here is the sewing of the seeds of the Reagan Revolution. These were the ideas upon which organizations like Jerry Falwell's Moral Majority and James Dobson's Focus on the Family were founded in the late 1970s. The Cold War fear of the atheist communists was of course a motivating factor for such groups, as was their disappointment with President Jimmy Carter, elected in 1976 as the first Evangelical Christian to hold the office, who, despite also being a Southerner, was not seen to be true to the faith.

35. "Long Weekend," S1/E10.
36. "Nixon vs. Kennedy," S1/E12.
37. Goldman, "*Power Elite,*" 70–71.
38. See "Wee Small Hours," S3/E9.
39. See "My Old Kentucky Home," S3/E3 .

As the age of political advertising and marketing continued to consolidate, the stage was set for former Hollywood actor Ronald Reagan to end Carter's bid for reelection in 1980, relying on his own anti-communist, religiously-motivated saber rattling.[40]

In another scene from the same episode, regarding domestic politics, Don and Suzanne Farrell, his daughter's former third grade teacher with whom he is having an affair, listen to Martin Luther King's "I Have a Dream" speech.[41] Moved, Suzanne declares that she will play it for students on first day of school; Don asks if they will understand. To his surprise, Suzanne informs him that they already do. The day after the speech, we observe Don's soon-to-be-former wife Betty Draper at home reading a *New York Times* front page article, albeit not one about King's speech in Washington. Betty reads about the impossibility of Nelson Rockefeller, a liberal Republican, sitting governor of New York, and a divorcé of the "Eastern Establishment," earning the GOP nomination for president in the election of 1964.[42] Nonetheless, this bleak forecast for Rockefeller does not prevent Betty from organizing a fundraiser for the Rockefeller primary campaign.[43] Later in the episode we listen in to attendees of the fundraiser discuss the politics of the day: one declares that she voted for Kennedy and will again; another wonders how Rockefeller will "deal with the South;" a third, in reference to the March on Washington, asks the question "Do you know how bad it must

40. In this episode, there is also something completely inconsequential to the plot yet worth noting for its relevance to international politics. Don's proposed international campaign for Hilton Hotels, "How do you say X in language Y? 'Hilton'" includes "How do you say towels in Farsi?" in an ad for the Hilton Hotel in Teheran. We must not forget that at this time the Iranian Revolution was still some sixteen years away. Ten years prior, according to its aim to contain communism, the United States (along with the British) had overthrown the democratically elected Mohammed Mosaddegh. Mosaddegh's crime had been to nationalize US and British oil interests in Iran. With him out of the way, Washington invested the political power of the country in the secular Shah Reza Pahlavi, which in turn fostered the anti-Western sentiment that led to the 1979 Islamic Revolution, which continues to simmer today in the country.

41. "Wee Small Hours," S3/E9.

42. Although never portrayed in the show, we viewers know that this eventually opens the path to the nomination for the radically conservative Barry Goldwater, Senator from Arizona, who eventually lost to Lyndon Johnson in one of the largest landslide reelections in US history.

43. At this point, we know that Betty's interest in the Rockefeller campaign is more than political. Her future second husband, Henry Francis, for whom she will leave Don at the end of Season 3 ("Shut the Door. Have a Seat," S3/E13), is a close advisor to the governor. Already planning to leave Don, who we know to be rather apolitical, is not happy about his wife's insistence on hosting the fundraiser at their home, Betty sees this as a way to get closer to Henry. However, to the great disappointment of the still Mrs. Draper, Henry sends someone else to address the attendees.

be for the negroes to descend on Washington like that just to be heard?" A fourth proclaims, "Segregation is uncivilized. It's that simple," to which the third adds, "You know what my father says about the South? It's not 1963, it's 1863." What we as viewers are witness to is the reenactment of the setting of the scene for Nixon's "Southern Strategy" in the subsequent presidential election of 1968 and the remaking of the GOP as the party of the South, which it clearly remains today.[44]

We should not interpret this as an endorsement of the North either. This defining episode of the series, politically speaking, ends with Carla, the Draper's African-American maid and nanny, listening to a broadcast of the funeral for three of the victims of the 16th Street Baptist Church Bombing in Birmingham, Alabama, on what Betty refers to as "your [Carla's black] station." Betty concludes that, while she hates to say it, the attack really made her wonder about Civil Rights. "Maybe it's not supposed to happen right now," she says to Carla just as Don enters the kitchen, having returned home from work.[45] The conversation conveniently ends there.

One defining moment that ties the essence of the politics of fear and polarization, both regarding communism and race, comes in a brief scene in "Christmas Comes but Once a Year."[46] Taking place in December of 1964, with President Johnson having pushed through and signed the Civil Rights Act into law in July of that year and then having been reelected the month prior, the newly formed Sterling Cooper Draper Pryce holds its annual office Christmas party. As the agency pursues new avenues of marketing, it has hired two consumer researchers as consultants; they exchange some fanciful and paranoid reflections over cocktails with Bert Cooper (a disciple of Ayn Rand) and the newest addition as junior partner, the Englishman Layne Price:

> Pryce: So, we are not part of this herd you're talking about.
>
> Geoffrey Atherton: You're from Great Britain. I'd think you'd be familiar with the perils of socialism.
>
> Cooper: Civil Rights is the beginning of a slippery slope.
>
> Atherton: If they pass Medicare, they won't stop until they ban personal property.
>
> Dr. Faye Miller: Storm our houses and rape our wives.

44. Lieven, *America Right or Wrong*.
45. "Wee Small Hours," S3/E9.
46. "Christmas Comes but Once a Year," S4/E2.

PRIVILEGE

Along with a growing politics of fear and polarization, the decade of the 1960s saw the lowest level of economic inequality ever in the United States. Compound this with an expanding role for women in the workplace and the Civil Rights movement and perhaps it is then not at all surprising that many of the characters, most of them wealthy, white, elite, Republican males, fear that their advantaged positions in society could further slip away. It is not that racism, sexism, and classism did not exist prior to the 1960s—which is to say that this moment in history is not necessarily all that different from any other, despite the changes experienced throughout the decade—but the open discussion of and overt confrontation to them at the societal level did take root around this time. Again, although many studies of and commentaries on the program, both published previously and within this volume, deal with questions of race and racism, and gender and sexism, religion and bigotry, sexual orientation and homophobia, it behooves us to mention a few instances of privilege depicted in *Mad Men*, especially regarding gender, race, and class and including abuse of power by the police and the dependence on them on the part of the privileged, as we continue to deal with these issues today.

Citing a new "equal pay" law—the Equal Pay Act, signed into law by President Kennedy on June 10, 1963—after having proved herself as an equal or better among her male peers on copy, Peggy Olson asks her immediate boss, Don Draper, for a raise.[47] Don, of course, does not care and dismisses the request because "it's not a good time" for him or for the company as he is "fighting for paperclips" around the office. What is crystal clear is that the Equal Pay Act of 1963 did not have its intended effect and the right to access the information to prove discrimination in pay without time restriction or statute of limitation was not upheld through legislation until after the Supreme Court decided the case of Ledbetter vs. Goodyear Tire and Rubber Company in 2007, which led to the controversial Lilly Ledbetter Fair Pay Act, passed and signed into law in early 2009 by the 111th Congress after the inauguration of President Barack Obama.

Because of all of the sexism and racism, to which we will also turn momentarily, perhaps what is not easily noticed throughout the series is the blatant class privilege exhibited by many of the protagonists. In a conversation with a neighbor woman in the suburbs named Beth, who Pete then seduces, they share their thoughts on "all the hobos and the panhandlers" in the city who accost them. When she was little, Beth's father told her "We

47. See "The Fog," S3/E5.

couldn't take care of everybody," to which Pete replies, "I guess we're supposed to get used to not seeing them."[48] And in many ways, as a society, we certainly have.

The police, on the other hand, from the beginning to the end of the show, do not miss seeing anyone. In the first season, on a visit to his then mistress's apartment while she has a group of counterculture friends over, Don has a flashback to his rather humble and poor upbringing. An unrelated disturbance in the building then draws a strong police presence. Because they have been doing drugs and do not want to draw the attention of the policemen, one of the friends tells Don that he cannot leave the apartment. Don, well aware of his appearance and having successfully climbed the social ladder, retorts that while he indeed can walk right past the police, *they* cannot.[49] Another incident occurs between Betty (now Francis) and the police in the first episode of Season 6, only in this case, Betty assumes that the police should leave her alone. After having been pulled over for speeding in poor road conditions, Betty and her mother-in-law, Pauline, are indignant with the state trooper because he insists on issuing Betty a ticket despite Pauline having informed him that Betty's husband (at the time) works for the Mayor Lindsay of New York City. As the officer steps away from the car, Pauline histrionically declares that "That ruined it. That ruined everything. I can't imagine it getting any darker than this." Upon arriving home, ticket for reckless driving in hand and oozing an air of entitlement, Betty tells her husband (who eventually decides not to call in favors to fix the ticket) that this cop was "verbally abusive."[50]

As another example of how so little has changed, such treatment by police, in the form of deference to Don or courtesy with Betty, could only be hoped for by the less wealthy and the less white, both as depicted in *Mad Men* and on today's nightly news. Indeed, the wealthier and the whiter seem to rely on the police to control the poorer and the darker while also giving themselves carte blanche to abuse others, which we rarely so overtly see in *Mad Men* but rather hear through conversation between and among the characters, as described above regarding politics on the television and radio.[51] Taking place in May of 1966, Part One of Season 5's first episode (in two parts), opens with African-American protesters on Madison Avenue, marching with some white clergy in support of equal opportunity

48. "Lady Lazarus," S5/E8.
49. "The Hobo Code," S1/E9.
50. "The Doorway, Part 1," S6/E1.
51. See also Colton Josephson, "Tomorrowland," 263.

in employment.[52] A low-level employee at a rival ad agency yells from his office window that the marchers "shut up," another that they should "get a job," while a third asks that someone call the cops. The second responds, in exasperation, that that is all there is outside, "cops and negroes and priests." Given that cops do not seem to be handling the situation to the liking of the young mad men, they proceed to throw water bombs out of the window and onto the protesters. While they are caught red-handed in the office, there is no indication that the police took any action, protecting the white office workers over the rights of peaceful protesters.

However inert they were with the junior ad men, the police were not always so when, two years later, in both New York and Chicago, they openly employ violence. In "The Better Half," upon being stabbed in the hand, Peggy's then soon-to-be ex-boyfriend, the countercultural journalist Abe, claims that we are living in a "fucking police state" as the "fascist pig" of a cop who took the report of the incident asks him if the assailant was "colored or Puerto Rican."[53] The policeman in question suggested that Peggy get a nightstick to protect herself in their lower class Brooklyn neighborhood. The very next episode, "A Tale of Two Cities," in what is either confusion or silent horror, from a hotel room in Los Angeles, Don watches the riots and clashes with police taking place outside of the 1968 Democratic National Convention in Chicago.[54] While on the phone with his wife Megan who is back in New York, he asks, "Can you imagine a policeman cracking your skull? It would change your whole life." While the DNC riots were not exclusively racially motivated, they did take place in a very turbulent year which included riots in hundreds of cities, both in the U.S. and abroad. That the DNC riots provided the backdrop for this episode served to show the viewers that from their privileged perch in society, few of the mad men themselves had any idea as to what middle and lower class actually felt about what was going on at the time.

Perhaps the clearest single example of privileged aloofness is brought to us earlier on in the series, from Bert Cooper, cited above as the most "power elitest" of them all.[55] In "The Chrysanthemum and the Sword," already replete with Orientalist language and imagery, during a conversation before a partner's luncheon, with newspaper in hand, Roger observes and asks, "This Selma thing is not going away. You still don't think they need a civil rights law?" Cooper responds, "They got what they wanted. Why aren't

52. "A Little Kiss," S5/E1.

53. "The Better Half," S6/E9.

54. "A Tale of Two Cities," S6/E10.

55. See Goldman, "*The Power Elite.*"

they happy?" Pete follows, "Because Lassie stays at the Waldorf and they can't."[56] While true, Pete's statement also quite incorrectly assumes that just any African American, or anyone else outside of the very rich for that matter, could afford to stay at such a hotel, as though just changing a law would solve the problem of inequality within American society.

Evidence of such willful ignorance of the plight of others has arguably spilled over to the white middle class as well, as we have seen in the reactions to the recent spate of police violence against unarmed black men caught on video. Consider the media and conservative outrage over the vandalism of a CVS Pharmacy in Baltimore in the wake of the death of Freddie Gray, an African American, while in police custody there in April of 2015. Add to this the Cooper-esque incredulity on the part of some who would ask how people could be so upset at the police when at the same time Baltimore had a black mayor, a black police chief, and three of the six police officers accused of abusing Mr. Gray were black as well, completely misunderstanding the complexities of the inherent structural racism of the system itself. Months later, as debate over the removal of the Confederate Battle Flag brewed after the cold-blooded murder of nine people at bible study by a white supremacist in Charleston, South Carolina, a slew of arson attacks on black churches took place throughout the South. The media coverage of the church burnings was next to absent compared to the coverage of the looting of the one CVS. As if we needed another example, consider the fact that the motive for Freddie Gray's arrest was that he was supposedly in possession of an illegal switchblade knife. Compare this to police indifference in dozens if not hundreds of protests of heavily armed whites since the inauguration of President Obama in January of 2009. Among the haves within U.S. society, which in practical terms includes many more than just the power elites themselves,[57] there is a horrible disconnect between perception and reality for the majority of the minority have-nots. Although this disjuncture is not new, at times, *Mad Men* reminds us of how openly such comments used to be made while, at the same time, it should also remind us that we have not come as far as we would like to think.

PROXIMITY OF BIG BUSINESS AND POLITICS

As described above, close ties between capital and the political class is nothing new, in the United States or anywhere else in the world, so what we see

56. "The Chrysanthemum and the Sword," S4/E5

57. Now is not the time or place for a discussion of income inequality in *Mad Men*, but this would be a fertile area of future inquiry.

in *Mad Men* is not exactly novel either.[58] However, the time period marks a change in how politics was, and continues to be, "done" in the U.S. In a bizarre reversal of the racism discussed above, which before and during the time of *Mad Men* was blatantly out in the open and now, while still present, is more subtle, money's role in the political system used to be more behind the scenes yet now is the order of the day, the new normal in U.S. democracy, both in terms of electoral politics and the legislative process. While we have described the birth of the role of advertising in electoral politics above, which was especially present throughout the *Mad Men*'s first season, by the final episodes of Season 2, which picks up in February of 1962 and, again, tellingly ends with the Cuban Missile Crisis in October of that year, we begin to see how corporations will take advantage of pork barrel spending in Washington. This continues to occur through the infrastructure of the budding Post-WWII Military-Industrial Complex, about which President Eisenhower had warned against in his farewell address just over a year earlier. Several interventions by the industrious Pete Campbell demonstrate exactly how advertising, big business, and congressional spending will soon find themselves in bed together from that point on to our present day.

While the team prepares for a trip to California to attend an aerospace conference, a seemingly innocuous conversation among those involved with marketing to the budding industry with regard to business strategy sums up the future of budgetary spending and electoral politics in the United States:

> Don: You're there to sell, but you're also there to listen. Every scientist, engineer, in general, is trying to figure out a way to put a man on the moon or blow up Moscow, whichever one costs more. We have to explain to them how we can help them spend that money.
>
> Paul Kinsey: Well, it certainly would behoove us to find out . . .
>
> Don: Campbell, you do the talking. Kinsey, you do the listening.
>
> Peggy: Don, you wanted me to remind them about the congressmen.
>
> Pete: We know. They control the purse strings.
>
> Don: No, they are the customer. They want aerospace in their districts. Let them know that we can help them bring these

58. Think, for example, of the Gilded Age of the late nineteenth century, or the protection of the interests of the white, propertied elite enshrined in the US Constitution itself.

> contracts home. Did you read anything that she prepared? Maybe I should send her.[59]

In the following episode, while at the convention, Don and Pete attend a presentation about MIRV missiles and nuclear warheads, after which, having connecting the dots regarding the profit potential, Pete gleefully points out that one defense company spends more annually on media than "three Lucky Strikes."[60]

Although not continuously explored throughout the duration of the series, we see other flashes of the strengthening of this bond between business and politics, catalyzed by the interpersonal connections among the power elite themselves, such as this exchange between Don and Pete:

> Don: What happened with North American Aviation?
>
> Pete: It's going well. My friend Russ who deals with McNamara's office said the orders are through the roof for helicopters, carbines, and especially jets for Vietnam. I think I've convinced North American they're going to need to spend more if they want to get out of NASA and into the Pentagon.[61]

We know that Pete's family has old money as one of the earliest settlers in what became New York City, which saved him from being fired by Roger Sterling early in the series.[62] We also know that he is a graduate of Deerfield Academy and Dartmouth, both prestigious institutions, and we can assume that he has his own connections throughout the business and political worlds. It is moments like this one in the program when everything observed herein truly comes together: the politics of fear, polarization, privilege, and profit. Unfortunately, this is a recipe with which we continue to be all too familiar today.

CONCLUSION: THE PAST, PRESENT, AND FUTURE

Many other observers and commentators, many of whom have been cited herein, have noted the parallels between the socio-political realities portrayed in *Mad Men* and those of today. Often stressing that the series was set during a very crucial period in U.S. history, consequently, we must conclude that we too are presently living in an important age. According to Colton

59. "The Inheritance," S2/E10.
60. "The Jet Set," S2/E11.
61. "Seven Twenty Three," S3/E7.
62. "New Amsterdam," S1/E4.

Josephson, "optimistically speaking," the postwar economic growth and social change of the 1960s, "changed the country and gave us a fresh start"; at the same time, she correctly warns us that the window into the sixties that *Mad Men* provides us is of an era that is clearly "not as glorious or as free as we remember or would like to believe."[63] In a chapter on *Mad Men* and American nationalism, citing the paradoxical strength and political failures experienced by the U.S. from the end of WWII to the twenty-first century, Heyman observes that we feel the echoes of the same dilemmas from the 1960s resonating today.[64] Colton Josephson claims that the show asks "us to reflect on the consequences and possibilities of a reconfigured America in the 1960s and today"[65] because, as Beail and Goren point out, "there are so many dimensions of the current cultural landscape that are reflected within the series itself."[66]

Once we start seriously drawing comparisons, we see that these scholars are indeed correct. Just as the Cold War overshadowed the political socialization of more than one generation, I fear that the so-called War on Terror will do the same. More specifically, within the context of fear, polarization, and elite decision-making, the parallels between the specific wars in Vietnam and the so-called "Second Gulf War" are instructive. At the conclusion of "Meditations in an Emergency," as everyone is sweating the Cuban Missile Crisis, a brief exchange between Roger and Don foreshadows what is our present reality regarding foreign policy:

> Roger: Kennedy is daring them to bomb us.
>
> Don: We don't know what's really going on. You know that.[67]

While successive presidential administrations since then have assured as much public transparency as possible, by the time of the invasion of Iraq in 2003, we should have learnt our lesson. Our government lies to us. In the War on Terror, we were told that the government did not torture captives or spy on its citizens. This is not an indictment or criticism of any kind but rather a now timeless fact which, no matter the setting, we do not seem to want to understand and which Don clearly did back in 1962. As with the other lingering and unresolved social issues discussed throughout the present chapter, collectively, we continue to repress that which we do not want to face, willfully ignoring the fact that until we decide to act differently,

63. Colton Josephson, "Tomorrowland," 261.
64. Heyman, "Appearances, Social Norms and Life in America," 145.
65. Colton Josephson, "Tomorrowland," 263.
66. Beail and Goren, "*Mad Men* and Politics," 11.
67. "Meditations in an Emergency," S2/E13.

the same problems will continue to resurface and/or manifest themselves in different ways, further chipping away at our collective conscience.

Ultimately, what I am arguing herein is that despite the seeming perfection of the staging of *Mad Men* in terms of time and place, much of what it portrays is not new. As the same stories portrayed in the classical tragedies, the works of Shakespeare, and many European operas may be set in times before and after what their authors originally intended, so too can the social and interpersonal experiences of the protagonists of this series. This is yet another distinction between *Mad Men* and most if not all of its contemporary television dramas which are often set in the present day yet tend to avoid dealing head on with the socio-political problems of our time. Take, for example, another groundbreaking anti-hero, "prestige television" hit, also from the AMC cable network: *Breaking Bad*. This program premiered on January 20, 2008, one year to the day before the inauguration of Barack Obama as the 44th President of the United States. Its premise began with a cancer-afflicted high school chemistry teacher, who also has a disabled son, becoming a millionaire meth cooker and drug kingpin, motivated by the fact that he could not afford health care. It revolved entirely around the drug trade and included a great amount of gun violence. Despite the obvious connections to the ongoing political debates throughout its run, until September 2013, not once did any of the characters say a word about, nor did the show contribute to any meaningful discussion of, drug control and/or legalization, gun control, or the Affordable Care Act.

On the contrary, perhaps *Mad Men* can provide the blueprint for the television dramas of the near future, set in different decades of the past in order to play on the television-viewing public's appreciation for nostalgia,[68] historical context, and the evolving socio-political climate(s) of the time,[69] while depicting over time similar interpersonal stories among the protagonists that hook us as viewers within the "microcosm" of a relevant business sector or industry.[70] In doing so, we would inevitably see evidence of a continued "return of the repressed" in the American psyche, in which the interpersonal drama facilitates the willful ignorance of the socially obvious on the part of the protagonists as it would on the viewer, in a way that often cannot be done without the help of temporal distance. As Bigsby points out, "For of the specificity of time and place in *Mad Men*, the existential dramas played out in the lives of those unsure of their identities and searching outside of themselves for the meaning of their lives are scarcely restricted

68. Heyman "Appearances, Social Norms and Life in America," 119–22.

69. Beail and Goren, "*Mad Men* and Politics," 22.

70. Heyman, "Appearances, Social Norms and Life in America," 121.

to that time and place."[71] And after all, for all of the power that the Bert Cooper's of the ad world seem to wield among the elites, "*Mad Men* reminds us that advertising firms are ultimately the speculative middlemen of mass consumption, at the beck and call of a business order they only sometimes understand and over which they exert control at the pleasure of the real motors."[72]

If this is the case, let us entertain a few possibilities for future television storytelling which would both aid in our understanding of the past and our future present. For example, consider a fictitious Detroit-based car company in the 1970s, which was so important throughout *Mad Men*, especially in the later seasons. This would be a pivotal decade for the auto industry in the U.S. for all the wrong reasons: the oil crises of 1973 and 1979, conflict in the Middle East, and the rise of Japanese imports. Politically, the country was continuously shaken by scandal as mistrust in government grew under Nixon, anti-war sentiment festered as the war ended, and physical abuse by security forces remained unchecked on the campuses of Kent State and Jackson State universities.

Subsequent blocks of time offer equally fertile ground for historical dramas. The 1980s offer us the banking industry, complete with the savings and loan scandals and Wall Street greed. The optimism of the Reagan Revolution brought us the New Right/Televangelists and the end of Equal Rights Amendment (after ten years of push). The Iranian Revolution and the Soviet invasion of Afghanistan led to covert aid to Afghan rebels (which became Al-Qaeda), the Iran-Contra affair, massive defense spending under the Star Wars program, and the end of the Cold War. A program set in the mid-1990s to mid-2000s might revolve around the machinations of a cable news company. Economic growth and the recurrent sexual scandals surrounding President Clinton at home overshadowed genocide in Rwanda and the Balkans while also blinding most investors, from the most wealthy to anyone with a retirement plan, to the looming dot com bust of the late 1990s. Quite obviously, a revisionist view of the 2000 Presidential Election, the attacks of September 11, 2001, and the run-up to the invasion of Iraq in 2003 would presumably hold the attention of those born in the late 1970s and early 1980s. And finally, the mid-2000s to the present might present to us a twitter-like social media startup in the era of torture and government surveillance as support for the war in Iraq dissipates, along with that for George W. Bush, all of which facilitates the rise of Obama and the then-hoped for "post-racial society." While we would relive some milestone changes, such

71. Bigsby, *Viewing America*, 397.

72. Goldman, 75.

as expanded health care and the recognition of equal rights for the LBGT community, we would also be reminded that this society is clearly not post-anything. Supreme Court decisions will ensure that money will continue to shape politics and not the other way around (*Citizens United*, 2010) while ending the protections of the Voting Rights Act for minority voters (*Shelby County v. Holder*, 2013), several states will debate so-called religious freedom legislation to effectively legalize anti-gay bigotry (the gay wedding cake debate), and police abuse will continuously be caught on mobile phones. All of these scenarios would provide the backdrop for myriad interpersonal stories of love, loss, betrayal, advancement, perseverance, and privilege, as well as for the socio-political change of the period in question.

Mad Men premiered nearly fifty years after its story began. The chronological distance has allowed the writers more overtly to both depict and criticize the truths and realities of the time, in a way that would not have been acceptable for the exact same series to have been made in their own time. Such a history lesson is interspersed with the interpersonal drama we have come to expect from this seemingly new medium of prestige television. For today's viewers who were young when the program took place, they can see the time now as adults;[73] but the existential dramas played out are "scarcely restricted to that time and place."[74] This reenactment of the quotidian functions as a form of dramatic process tracing that shows us how and why we are where we are today. As has been maintained throughout this chapter, while the contexts might shift, the issues, problems, and mistakes are often the same, both for society and the individual, and for some reason, since we started telling stories in the West, we have refused to learn from the past. Instead of the indifference of the universe, *Mad Men* re-reveals the indifference of humanity.

BIBLIOGRAPHY

Beail, Linda. "Invisible Men: The Politics and Presence of Racial and Ethnic 'Others' in *Mad Men*." In *Mad Men and Politics: Nostalgia and the Remaking of Modern America*, edited by Linda Beail and Lilly J. Goren, 231–55. New York: Bloomsbury Academic, 2015.

Beail, Linda, and Lilly J. Goren. "*Mad Men* and Politics: Nostalgia and the Remaking of America." In *Mad Men and Politics: Nostalgia and the Remaking of Modern America*, edited by Linda Beail and Lilly J. Goren, 3–33. New York: Bloomsbury Academic, 2015.

Bigsby, Christopher. *Viewing America*. Cambridge: Cambridge University Press, 2013.

73. Heyman, "Appearances, Social Norms and Life in America," 143.
74. Goldman, "*Power Elite*," 397.

Carveth, Rod. "'We've Got Bigger Problems to Worry about Than TV, Okay?' Mad Men and Race." In *Mad Men and Philosophy: Nothing is as it Seems*, edited by James B. South and Rod Carveth, 217–27. Hoboken, NJ: Wiley, 2010.

Colton Jospehson, Rebecca. "Tomorrowland: Contemporary Visions, Past Indiscretions." In *Mad Men and Politics: Nostalgia and the Remaking of Modern America*, edited by Linda Beail and Lilly J. Goren, 259–75. New York: Bloomsbury Academic, 2015.

The Fog of War. DVD. Directed by Errol Morris. 2003. Culver City, CA: Sony Pictures Home Entertainment, 2004.

Fuehrer Taylor, Natalie. "'You Can't Be a Man. So Don't Even Try': Femininity and Feminism in *Mad Men*." In *Mad Men and Politics: Nostalgia and the Remaking of Modern America*, edited by Linda Beail and Lilly J. Goren, 207–30. New York: Bloomsbury Academic, 2015.

Goldman, Loren. "*The Power Elite* and Semi-Sovereign Selfhood in Post-War America." In *Mad Men and Politics: Nostalgia and the Remaking of Modern America*, edited by Linda Beail and Lilly J. Goren, 63–94. New York: Bloomsbury Academic, 2015.

Gross, Terry. "Matthew Weiner on 'Mad Men' and Meaning" *NPR*, April 20, 2015. Online: http://www.npr.org/2013/04/25/178832854/matthew-weiner-on-mad-men-and-meaning.

Guilfoy, Kevin. "Capitalism and Freedom in the Affluent Society." In *Mad Men and Philosophy: Nothing is as it Seems*, edited by James B. South and Rod Carveth, 34–50. Hoboken, NJ: Wiley, 2010.

Heyman, Lawrence. "Appearances, Social Norms, and Life in Modern America: Nationalism and Patriotism in *Mad Men*." In *Mad Men and Politics: Nostalgia and the Remaking of Modern America*, edited by Linda Beail and Lilly J. Goren, 119–46. New York: Bloomsbury Academic, 2015.

Klein, Naomi. *No Logo*. New York: Picador, 2000.

Larson, Jared D. "Un declive auto-imposto: a averia institucional na política estadounidense." *Tempo exterior: Revista de análise e estudios internacionais* 29 (2014) 7–19.

Lieven, Anatol. *America Right or Wrong: An Anatomy of American Nationalism (2nd Editon)*. New York: Oxford University Press, 2012.

Marcos, Natalia. "El final de 'Mad Men', ¿el adiós de una era?" *El País* (Spain), May 11, 2015.

Perlman, Allison. "The Strange Career of *Mad Men*: Race, Paratexts and Civil Rights Memory." In *Mad Men*, edited by Gary R. Edgerton, 209–25. New York: I.B. Tauris, 2011.

Phillips, Kevin. *Wealth and Democracy: A Political History of the American Rich*. New York: Broadway, 2003.

Robin, Corey. *Fear: The History of a Political Idea*. New York: Oxford University Press, 2004.

Vázquez, Isabel. *Me llamo Peggy Olson*. Barcelona: Ediciones B., 2015.

CONTRIBUTORS

CHRISTOPHER J. ASHLEY, NewYork-Presbyterian Hospital, Department of Pastoral Care & Education

CAROLE L. BAKER, Duke Divinity School

NSENGA K. BURTON, Associate Professor of Media Studies, Goucher College

ANN W. DUNCAN, Associate Professor of Religion, Goucher College

SUSAN FREKKO, Independent scholar, translator and writing coach, Barcelona, Spain

JACOB L. GOODSON, Assistant Professor of Philosophy, Southwestern College

GABRIEL HALEY, Assistant Professor of English, Concordia University, Nebraska.

KRISTEN DEEDE JOHNSON, Associate Professor of Theology and Christian Formation, Western Theological Seminary

JARED D. LARSON, Lecturer, Department of Politics, Humbolt State University, and Research Associate, Centre for Geographic Studies, Universidade de Lisboa, Portugal

JACKSON LASHIER, Assistant Professor of Religion, Southwestern College

DAVID MATZKO MCCARTHY, Fr. James M. Forker Professor of Catholic Social Teaching, Mount St. Mary's University

BRANDON MORGAN, Baylor University

JENNIFER PHILLIPS, Principal, JLP Strategy

HOWARD PICKETT, Assistant Professor of Ethics and Poverty Studies and Director of the Shepherd Program for the Interdisciplinary Study of Poverty and Human Capability, Washington and Lee University

HEIDI SCHLUMPF, Associate Professor of Communication, Aurora University

SARAH CONRAD SOURS, Assistant Professor of Religion, Huntingdon College

JONATHAN TRAN, Associate Professor of Religion and Faculty Master of the Honors Residential College, Baylor University

SETH VANNATTA, Associate Professor of Philosophy and Religious Studies, Morgan State University

MATTHEW EMILE VAUGHAN, Instructional Designer & Project Manager, Columbia University School of Professional Studies

SUBJECT INDEX

abortion, 84, 94, 269, 338, 347, 355–57
acceptance, 3, 80, 95, 123, 239, 298, 323, 349, 356
action, 17, 43–48, 63, 132–40, 170, 193–202, 208–13, 268, 280, 313, 359, 381
adoption, 81, 165, 316, 352–58, 365
adulthood, 114, 233, 309, 332–42
adultery, 73, 275, 277, 317, 320, 354
advertising, 4, 6, 8, 14, 16, 18, 21, 23, 26, 29–34, 38, 40, 43, 48, 50, 73, 77, 90, 101, 107, 123, 125, 129, 133, 135, 139, 146, 151, 173, 176, 190, 194, 196, 199, 201, 203, 211, 215, 219, 222, 226, 229, 231, 234, 238, 266, 280, 300, 307, 311, 315, 328, 361, 368, 370, 372, 374, 377, 383, 387
aesthetics, 32, 45, 48, 155, 208, 310, 327
anonymity, 57, 192–213
anxiety, 31, 100, 192–213, 232, 332, 335, 339, 342
art, 5, 15, 26, 30, 34, 40, 45, 82, 127, 131, 133, 142, 155, 158, 197, 217, 294, 309, 364, 371, 375
attraction, 105, 160, 266, 272
auteur, 309
authenticity, 10, 14, 21, 27, 129, 159, 162, 166, 179, 183, 187, 192–213
authoritarian, 35
autonomy, 4, 6, 17, 26, 37, 91, 154, 157, 242, 334
baptism, 163, 171, 180, 182, 183, 185, 188, 210, 255
birth Control, 97, 107, 320, 345–65
blackness, 215–35
business ethics, 3–25, 28–48, 52–76, 78–99, 101–20, 123–41

caregiving, 102, 106, 110, 112, 114, 116, 119
character, 4, 7, 9, 13, 21, 25, 32, 40, 49, 57, 62, 67, 71, 75, 79, 81, 82, 87, 89, 96, 107, 110, 126, 131, 138, 140, 149, 153, 158, 164, 170, 173, 181, 189, 193, 195, 197, 199, 201, 204, 206, 213, 216, 219, 222, 226, 229, 230, 231, 234, 239, 247, 272, 276, 278, 281, 286, 289, 301, 309, 311, 313, 315, 319, 323, 325, 339, 345, 347, 350, 361, 364
Christianity, 145–66, 169–90, 206, 209, 272, 295, 345–65
clericalism, 345–65
Cold War, 368–88
comedy, 40, 147, 150, 154, 160, 162, 166, 292–307
communication, 55, 200, 250, 348, 392
community, 40, 49, 62, 83, 103, 115, 117, 119, 125, 127, 131, 133, 134, 136, 140, 153, 179, 183, 189, 228, 276, 293, 308, 388
confession, 19, 165, 171, 179, 181, 182, 184, 187, 212, 238, 304, 334, 351, 362, 365

confidence, 10, 18, 37, 57, 69, 165, 219, 238, 252, 282, 303, 311, 342, 360
confusion, 73, 171, 244, 279, 283, 366, 381
consciousness, 24, 31, 61, 204, 206, 215–35, 243, 281, 287
consent, 39, 245, 285, 288, 305, 352
consequences, 5, 8, 150, 183, 251, 275, 338, 343, 362, 385
consumerism, 46, 156, 364
contraception, 97, 107, 320, 345–65
convention, 68, 70, 73, 75, 295, 299, 306, 381, 384
conversation, 6, 43, 57, 61, 65, 67, 76, 96, 102, 113, 129, 136, 228, 243, 248, 256, 292, 294, 297, 299, 301, 303, 305, 307, 309, 312, 315, 317, 376, 378, 379, 381, 383
corruption, 159, 336
counterculture, 145–66, 369, 380
courtesy, 52–76, 380
creativity, 28–48, 75, 125, 127, 129, 134, 138
creator, 20, 38, 40, 49, 85, 100, 111, 121, 123, 142, 146, 176, 309, 346, 369
Cuban Missile Crisis, 177, 206, 345, 368–88
culture, 5, 26, 37, 41, 46, 50, 54, 81, 96, 103, 110, 114, 118, 120, 123, 125, 128, 137, 145–66, 199, 219, 222, 226, 233, 235, 238, 240, 242, 309, 319, 324, 331, 353, 355, 359, 361, 368

Dasein, 192–213
deception, 5, 7, 8, 11, 63, 67
desire, 4, 6, 8, 15, 18, 21, 24, 26, 47, 63, 70, 73, 78, 80, 83, 85, 87, 90, 93, 95, 98, 111, 131, 138, 146, 149, 152, 155, 159, 161, 162, 165, 167, 195, 201, 205, 218, 224, 228, 242, 244, 269, 278, 280, 282, 286, 299, 304, 311, 318, 320, 324, 326, 351, 362
despair, 192–213
diversity, 20, 29, 35, 38, 40, 41, 46, 48, 50, 215
divorce, 108–10, 183, 185, 253, 275, 277, 281, 289, 292–307, 319, 350, 357, 377
domestic life, 293, 300, 305
duplicity, 12, 246, 248, 252

economy, 24, 29, 38, 46, 48, 103, 105, 108, 110, 117, 120, 125, 142, 311, 325, 332
ego, 152, 154, 241, 245
emotion, 62, 83, 229, 342
employment, 39, 71, 101, 104, 106, 108, 110, 112, 226, 336, 381
epistemology, 10, 149, 241, 267
equality, 80, 103, 109, 121, 125, 164, 269, 271, 273, 277, 281, 289, 293, 323, 329, 332, 358
erotic, 16, 237–61, 265–90, 293, 364
ethics, 3, 5, 6, 8, 10, 14, 22, 24, 26, 30, 33, 42, 44, 56, 65, 77, 139, 142, 174, 209, 265, 267, 268, 271, 275, 284, 289, 291, 392 *(Also see "business ethics")*
ethos, 30, 310, 334, 342, 347, 350
excellence, 132
existentialism, 163, 192–213, 273

false self, 169–90
family, 19, 23, 37, 44, 55, 58, 62, 72, 78, 80, 83, 86, 88, 90, 92, 95, 97, 99, 109, 115, 118, 120, 125, 129, 137, 139, 170, 174, 177, 179, 181, 187, 195, 199, 222, 224, 232, 253, 255, 258, 274, 300, 303, 308, 311, 316, 319, 328, 340, 343, 350, 356, 364, 376, 384
fear, 15, 89, 131, 173, 182, 199, 206, 209, 213, 239, 243, 246, 353, 371, 373, 375, 378, 384, 389
feminism, 78–99, 101–20, 267, 269, 270, 273, 276, 278, 281, 286, 290, 331, 349, 367, 389
feminist, 39, 79, 87, 91, 95, 97, 99, 226, 235, 265–90, 326, 330, 347, 349, 357, 365
freedom, 15, 32, 48, 75, 95, 117, 121, 131, 179, 199, 203, 210, 231, 233,

239, 242, 267, 271, 273, 276, 282, 287, 289, 351, 369, 388

God, 89, 113, 116, 120, 147, 150, 153, 161, 166, 171, 176, 180, 189, 192, 203, 208, 212, 248, 250, 261, 295, 297, 307, 345, 349, 353, 357, 360, 362, 365, 376
guilt, 69, 78–99, 193, 210, 212, 279, 281, 332, 346, 350, 353, 355, 357, 361, 370

happiness, 15, 18, 60, 81, 86, 90, 124, 131, 138, 152, 159, 175, 195, 199, 238, 242, 260, 290, 292, 294, 297, 301, 303, 307, 312, 318, 321
harassment, 39, 41, 109, 112
hierarchy, 154, 221, 255, 347, 349, 358
hero/heroic, 124, 127, 140–41, 173, 176, 198, 386
hippie, 91, 156, 226
history, 4, 11, 21, 26, 34, 46, 50, 82, 96, 100, 105, 121, 133, 142, 146, 170, 176, 179, 181, 191, 215, 218, 227, 229, 234, 301, 306, 310, 319, 327, 332, 344, 349, 354, 357, 365, 368, 370, 373, 377, 379, 384, 388
honor, 12, 126, 139, 217, 220, 274, 316
hope, 20, 37, 44, 48, 56, 86, 89, 91, 115, 125, 135, 141, 146, 166, 169, 177, 180, 183, 189, 194, 201, 228, 240, 252, 258, 260, 292, 294, 296, 300, 304, 306, 319, 326, 351, 353
hopelessness, 166, 301
household, 72, 103, 106, 110, 114, 117, 119, 120, 126, 326, 350
housekeeping, 58, 104, 114, 116, 120
hypocrisy, 12–13

icon, 11, 237–61
idealism, 150
identity, 3, 8, 10, 12, 15, 17, 21, 24, 37, 49, 58, 64, 80, 83, 88, 92, 96, 123, 126, 130, 132, 133, 145, 170, 173, 179, 182, 186, 189, 204, 216, 220, 222, 223, 227, 230, 233, 239, 252, 267, 271, 310, 335, 350, 352, 370
idol, 240, 248, 250, 251, 258, 261
infallibility, 359
infidelity, 8, 24, 131, 141, 173, 179, 181, 183, 189, 238, 244, 246, 248, 277, 304
institution, 117, 126, 273, 294, 299, 308, 310, 318, 320, 322, 324, 327
intelligence, 36, 83, 93, 245, 310, 335
intention, 8, 171, 176, 249, 266, 290, 310, 248
intimacy, 70, 74, 108, 183, 258, 265, 269, 272, 274, 276, 296

jealousy, 60, 74, 195, 311, 351
Judaism, 165
judgment, 9, 114, 133, 164, 239, 254, 267, 272, 275, 282, 286, 305, 313, 316, 319, 356

knowledge, 5, 10, 27, 34, 58, 98, 130, 132, 230, 234, 237–61, 267, 274, 277, 294, 301, 303, 355, 366

leadership, 34, 41, 102, 349
love, 6, 14, 18, 20, 61, 73, 82, 90, 93, 95, 107, 112, 115, 117, 119, 121, 127, 132, 137, 153, 159, 161, 165, 167, 180, 195, 201, 213, 223, 225, 237, 239, 241–47, 251, 258, 260, 266, 270, 276, 278, 287, 290, 293, 300–307, 317, 321, 327, 330, 341, 344, 363, 388
lying, 6, 9, 11, 16, 79, 82, 94, 131, 136, 246, 250, 278, 294, 340

mainstream, 147, 155, 157, 159, 163, 166, 222, 334, 343
manipulation, 6, 10, 67, 302, 372
manners, 52–76
market, 7, 15, 17, 23, 31, 40, 103, 105, 108, 110, 117, 120, 126, 134, 162, 199, 302, 334
marketing, 31, 34, 43, 128, 301, 315, 372, 377, 383

marriage, 39, 63, 72, 82, 84, 86, 88, 93, 95, 104, 106, 108, 110, 112, 119, 137, 145, 148, 152, 153, 155, 157, 159, 161, 183, 202, 204, 207, 249, 265, 268, 271, 273, 275, 277, 279, 280, 281, 284, 287, 292–307, 308–27, 346, 350, 354, 356
masculinity, 175, 190, 288
maternity, 91, 355
media, 6, 17, 45, 48, 79, 213, 215, 230, 309, 332, 348, 374, 382, 387
metaphysical, 151, 157, 240, 242, 293, 295, 302
millennial, 47, 308, 313, 321, 324, 327
minorities, 221
monastic, 169, 171, 172, 175, 179, 181, 183, 189
money, 21, 32, 35, 86, 125, 139, 183, 223, 227, 232, 256, 279, 311, 334, 372, 383, 388
morality, 56, 75, 163, 257, 267, 307, 353
motherhood, 78–99, 137, 330, 344, 350, 355

naiveté, 323, 325
narrative, 4, 8, 14, 24, 71, 90, 96, 113, 126, 133, 136, 151, 161, 170, 172, 175, 177, 189, 215, 217, 219, 223, 229, 239, 285, 299, 310, 316
negotiate, 54, 79, 93, 102, 109
nihilism, 306

objectification, 128, 240, 268, 280
objectivism, 163
ontological, 204, 210, 212, 249
oppressed, 97, 224
oppressor, 218, 224
optimism, 153, 155, 250, 313, 323, 387
order, 3, 11, 17, 31, 34, 36, 46, 52, 56, 58, 60, 66, 71, 93, 107, 116, 123, 125, 126, 136, 138, 154, 161, 169, 173, 175, 180, 187, 195, 203, 205, 207, 209, 213, 218, 221, 224, 227, 229, 240, 242, 243, 246, 248, 250, 266, 274, 285, 296, 301, 334, 336, 343, 371, 375, 383, 386

parenting, 84, 86, 321, 329–43
peace, 20, 48, 62, 81, 90, 97, 124, 128, 139, 172, 179, 185, 250, 346
pessimism, 15
philosophy, 31, 164, 170, 178, 190, 234, 240, 265, 267, 271, 273, 278, 290, 294, 299
pleasure, 63, 70, 92, 139, 197, 247, 266, 270, 273, 277, 280, 287, 387
politics, 24, 71, 91, 105, 118, 190, 219, 235, 266, 270, 274, 278, 282, 286, 290, 294, 298, 302, 306, 310, 314, 318, 322, 326, 330, 334, 338, 342, 348, 352, 356, 360, 364, 368–88
poverty, 20, 22, 24, 48, 171, 188, 216, 221
power, 5, 11, 13, 19, 34, 39, 43, 45, 54, 65, 69, 71, 75, 102, 119, 147, 153, 156, 160, 198, 204, 216, 218, 221, 224, 227, 229, 230, 232, 234, 269, 276, 282, 286, 288, 289, 315, 318, 327, 349, 358, 360, 370, 373, 375, 376, 379, 381, 384, 387
pregnancy, 62, 82, 88, 94, 108, 314, 316, 343, 354, 356, 363, 365
prestige, 186, 218, 277, 311, 388
privacy, 169, 281
privilege, 29, 41, 81, 97, 222, 224, 228, 231, 235, 258, 316, 338, 369, 372, 379, 384, 388
professional, 4, 22, 32, 39, 57, 69, 80, 91, 94, 114, 124, 126, 137, 140, 151, 162, 215, 229, 233, 238, 281, 308, 312, 315, 317, 318, 325, 329, 334, 336, 338, 360, 364, 375
progress, 38, 41, 136, 150, 155, 162, 165, 171, 234, 301, 312, 330, 369
psychological, 8, 78, 92, 170, 219, 221, 229, 234,266, 310
public, 5, 7, 43, 53, 59, 64, 76, 103, 107, 110, 113, 115, 117, 119, 120, 135, 196, 198, 199, 201, 204, 211, 213, 227, 237, 246, 266, 269, 278, 281, 343, 350, 356, 360, 370, 374, 385
purity, 353
purpose, 18, 21, 38, 43, 53, 81, 124, 126, 180, 186, 205, 271, 307, 353, 376

race, 11, 24, 48, 58, 81, 96, 215–35, 322, 332, 378
racism, 234, 331, 379, 382
rape, 53, 84, 86, 227, 270, 281, 285, 350, 378
reason, 18, 99, 108, 124, 150, 178, 201, 208, 239, 241, 243, 245, 252, 266, 280, 294, 298, 307, 332, 376, 388
redemption, 85, 131, 138, 151, 170, 172, 182, 277, 294, 313, 349, 356, 363
rejection, 56, 93, 114, 187, 219, 228, 240, 243, 302, 351
relationships, 4, 55, 57, 60, 79, 90, 93, 96, 115, 119, 125, 127, 137, 145, 179, 182, 185, 187, 189, 195, 199, 207, 228, 252, 265, 267, 280, 282, 289, 298, 316, 321, 369
religion, 18, 24, 54, 88, 193, 220, 224, 345–65, 379
remarriage, 292, 297, 300, 307
representation, 5, 24, 127, 147, 220, 302, 331
resentment, 73, 256
respect, 16, 62, 74, 88, 186, 245, 280, 287, 313, 320, 326, 332, 343, 352, 360, 374
responsibility, 7, 29, 36, 42, 44, 46, 48, 54, 125, 130, 298, 301, 310, 323, 330, 334, 336, 340, 343, 351, 358
revolution, 48, 95, 102, 104, 106, 108, 110, 119, 194, 310, 368, 370, 376, 387
romantic, 29, 37, 39, 76, 91, 94, 96, 118, 127, 138, 140, 149, 153, 228, 294, 299, 350, 362
Roman Catholicism, 114, 345–65

second-wave feminism, 80, 87, 267, 278, 290, 331
security, 39, 65, 75, 95, 98, 239, 244, 248, 287, 304, 312, 320, 287
seduction, 16, 21, 32, 146, 245
self, 4, 7, 10, 12, 14, 18, 22, 24, 32, 38, 54, 58, 60, 61, 62, 64–67, 75, 79, 81, 88, 95, 114, 117, 120, 129, 134, 136, 138, 140, 146, 151, 153, 154, 157, 162, 164, 169, 171, 172, 175, 177, 179, 182, 184, 185, 188, 190, 193, 197, 202, 204, 206, 208, 209, 210, 212, 218, 230, 242, 244, 247, 252, 258, 268, 270, 287, 293, 301, 306, 309, 311, 316, 329, 331, 334, 336, 340, 341, 343, 350, 354, 363, 373
self control, 60, 63, 67, 205
selfishness, 179, 195
self-regulation, 330, 334, 337, 340, 343
sensual, 321, 324
sex, 15, 18, 24, 73, 75, 80, 86, 95, 107, 109, 117, 128, 179, 190, 205, 232, 265–90, 299, 317, 322, 340, 345, 349, 349, 353, 355, 357
sexism, 4, 39, 42, 43, 79, 88, 137, 267, 303, 331, 345, 348, 350, 357, 361, 364, 368, 379
sexual sins, 348, 353, 356, 361, 363
sexuality, 39, 73, 79, 84, 96, 154, 202, 266, 268, 269, 271, 273, 276, 278, 280, 282, 288, 290, 298, 346, 349, 352, 356
sin, 19, 64, 72, 160, 171, 175, 180, 182, 183, 193, 202, 208, 212, 272, 299, 333, 345, 347, 350, 352, 354, 356, 361, 363
sincerity, 3, 10, 13, 18, 21, 25, 27, 64, 246, 248, 304
skepticism, 10, 295, 298, 301, 306
sociology, 131
social construct, 221
society, 16, 18, 28, 42, 44, 45, 53, 55, 57, 60, 71, 74, 80, 87, 90, 95, 98, 102, 105, 116, 119, 125, 164, 171, 180, 184, 216, 220, 223, 225, 229, 231, 234, 238, 240, 269,275, 286, 299, 309, 325, 327, 329, 332, 334, 343, 346, 349, 359, 365, 368, 371, 373, 379, 381, 387
spiritual/spirituality, 19, 150, 182, 186, 190, 346, 361, 364
status quo, 41, 48, 88, 218, 225, 318, 336
stereotype, 227, 350

success, 12, 24, 29, 36, 39, 47, 70, 73, 75, 81, 90, 93, 99, 107, 116, 126, 207, 215, 220, 224, 231, 256, 270, 311, 316, 318, 325, 343, 364, 369
surrogate, 310, 313, 323, 325, 327
suspension, 209

technology, 30, 37, 41, 43, 115, 232, 240, 298, 348, 371
telos, 241, 290
temptation, 129, 160, 202, 271, 292, 294, 297, 302, 306, 331, 353
theory of time, 265
theology, 104, 112, 120, 147, 150, 169, 171, 172, 179, 181, 183, 189, 212, 293, 297, 299, 306, 346, 349, 358, 362
tradition, 113, 124, 153, 170, 172, 179, 353, 369, 372
truth, 4, 6, 8, 11, 13, 15, 21, 89, 94, 114, 135, 159, 173, 179, 182, 186, 252, 258, 270, 303

violation, 55, 70, 73, 337
virgin, 353, 356
virtue, 5, 8, 12, 17, 31, 63, 157, 162, 164, 205, 208, 212, 308, 316, 324
vocation, 81, 93, 119, 351, 357
volition, 268, 276, 282
vulnerable, 30, 38, 128, 179, 229, 333

weakness, 70, 205, 208, 230
whiteness, 215–35
woman/women, 5, 10, 16, 25, 29, 39, 41, 58, 68, 71, 79, 81, 85, 87, 92, 95, 97, 98, 101–20, 124, 127, 136, 145, 150, 152, 154, 158, 161, 177, 199, 210, 219, 223, 226, 228, 233, 238, 252, 258, 267, 269, 270, 273, 275, 277, 283, 286, 290, 303, 309, 316, 318, 331, 334, 347, 349, 351, 353, 355, 357, 358, 359, 361, 363, 375, 379

youth, 60, 312, 314, 351, 360, 365

AUTHOR INDEX

Adamson, Glenn, 132
Alder, Felix, 333
Alger, Horatio, 9
Alighieri, Dante, 145–66, 250, 371
Allyn, David, 107
Andjelic, Ana, 41
Anscombe, Elizabeth, 290
Aquinas, Thomas, 63, 266, 349
Aristotle, 65, 139
Augustine, 147, 166, 353, 356

Bahr, Linsey, 91
Baier, Kurt., 4
Baker, Samm Sinclair, 5
Barkman, Ashley Jihee, 267, 278
Bauer, Nancy, 271
Beail, Linda, 369–73, 385–86
Berry, Wendell, 117
Bigsby, Christopher, 170, 177, 369, 371, 373, 386–87
Billhartz, Celeste, 354–55
Booth, Wayne C., 12
Bourdieu, Pierre, 334
Bowie, Norman E., 4
Brodkin, Karen, 224
Brodzinsky, A., 355
Butler, Joseph, 11

Cahill, Lisa Sowle, 266
Carveth, Rod, 370
Cavell, Stanley, 292–307
Chodorow, Nancy, 80
Coates, Ta-nehisi, 217, 223, 230, 234
Cogan, William J., 354
Colby, Tanner, 40, 216
Collins, Patricia Hill, 226, 235
Colton Jospehson, Rebecca, 369–74, 380, 384–85
Coontz, Stephanie, 79, 310, 320
Crisp, Roger, 17
Curtis, Mary C., 227

Dahle, Cheryl, 35
Daly, Mary, 349
Davidson, Diana, 330–331
de Beauvoir, Simone, 80–81, 86–87, 92, 265–90
Della Femina, Jerry, 73
Deshon, George, 353
Diller, Vivian, 101
Dolan, Jay P., 347, 353, 358–59
Douglas, Dean, 37
Drucker, Peter, 28, 34, 26
Drumming, Neil, 41
Dubner, Stephen, 148
duBois, W. E. B., 215–35
Dweck, Carol, 335–36

Elder, Sean, 164
Ellison, M., 355
Emerson, Ralph Waldo, 22
Evonne, Sarah, 282, 285

Fairclough, Norman, 342–43
Ferreira, M. Jamie, 16
Fessler, Ann, 355
Firestone, Shulamith, 95
Fitzgerald, F. Scott, 9
Florida, Richard, 28–29, 36–38, 41–42, 45–46

Frank, Thomas, 45
Freeman, R. Edward, 23
Friedan, Betty, 87–88, 91, 97–99, 104–6, 326
Friedman, Stewart, 37, 322

Galbraith, John Kenneth, 17–18, 44–45, 369, 371
Godin, Seth, 4
Goldman, Loren, 370, 375, 376, 381, 387
Greenblatt, Stephen, 10
Greenstone, Daniel, 333
Guilfoy, Kevin, 369–70

Hansen, Jim, 170
Harman, Gilbert, 8
Hays, Sharon, 334
Hegel, G. W. F., 271
Heidegger, Martin, 8, 192–213
Henry, Astrid, 96
Herman, Barbara, 269
Heyman, Lawrence, 370, 372, 385, 386, 388
Hochschild, Arlie, 133
hooks, bell, 97

Idov, Michael, 215

Jones, Caroline A., 138
Joshi, Aditya, 31

Kaganovsky, Lilya, 170
Kant, Immanuel, 6, 11–13, 265–90
Kieran, Matthew, 31–32
Kierkegaard, Søren, 8, 192–213
King, Martin Luther, 48, 377
Kissel, Rick, 217
Klein, Naomi, 368, 371
Kohn, Al, 335–36
Krouse, Tonya, 146

Lednicer, Lisa, 38, 40
Levinas, Emmanuel, 16
Levitt, Eodore, 18
Lewis, C. S., 147
Lieven, Anatol, 378
Lincoln, Abraham, 48
Lipp, Roberta, 83
Lister, Ruth, 335
Lombardi, Elena, 161

MacIntyre, Alasdair, 5, 163
MacKinnon, Catherine, 265–90
Madison, Alex, 40
Mahon, James, 6
Marion, Jean-Luc, 237–60
Martin, James A., 362
May, Rollo, 266
McBrien, Richard P., 347–48, 358
McCarthy, David Matzko, 117–20
McDermott, John, 40
McLuhan, Marshall, 75
McTaggart, J. M. E., 265–68
Meek, James, 152
Meilaender, Gilbert, 266
Merton, Thomas, 169–90
Miller, Arthur, 21
Millett, Kate, 267
Murray, Noel, 362

Newton, Adam Zachary, 4
Norris, Kathleen, 114–15
Nussbaum, Martha, 5

O'Hara, Frank, 148, 237, 254
Ong, Aihwa, 335
Ovans, Andrea, 30

Pascoe, Jordan, 280
Perlman, Allison, 369
Peterson, Margaret Kim, 113–14
Phillips, Kevin, 371
Poggi, Jeanine, 308
Poniewozik, James, 99
Pope John Paul II, 124–26
Pope John Paul XXIII, 348
Pope Leo XIII, 124
Pope Paul VI, 351
Pope Pius X, 359
Pseudo-Macarius, 181–82
Pye, David, 133
Ricouer, Paul, 323–24
Robin, Corey, 371
Rosenberg, Alyssa, 326
Rosin, Hannah, 160

Ruether, Rosemary Radford, 349
Ryan, Maureen, 354

Saiving, Valerie, 358
Sandberg, Sheryl, 102, 115–16
Scarry, Elaine, 154
Schawbel, Dan, 37
Schleiermacher, Friedrich, 18
Schulte, Bridget, 47
Schwarz, Benjamin, 331–32, 336
Sennett, Richard, 131
Shakespeare, William, 10, 292, 298–99, 371, 386
Shiner, Larry, 127
Snyder, Benjamin, 123
Soble, Alan, 267
Solomon, Robert C., 8
Spock, Benjamin, 333
Stearns, Peter N., 332–36, 339, 342
Steinbeck, John, 10
Stillman, Jessica, 35

Thomas, Maura, 35
Tolstoy, Leo, 193
Trilling, Lionel, 10
Tudor, Deborah, 331
Turner, Julia, 164

Upton, Candance L., 9

van der Wer , Todd, 315
Vanderkam, Laura, 35, 37
Varon, Jeremy, 170, 181
Vázquez, Isabel, 370
Velasquez, Manuel, 5, 17

Warner, Judith, 98
Wartzman, Rick, 28
Wells, Samuel, 58
Whitman, Walt, 22
Williams, Rowan, 296
Witchel, Alex, 310
Wyschogrod, Michael, 203–4

EPISODE INDEX

SEASON 1

Episode 1, "Smoke Gets In Your Eyes," 5, 18, 31, 68, 82, 128, 131, 158, 164, 169, 175, 178, 194, 199, 222, 228, 237, 243, 300, 301, 350, 351
Episode 2, "Ladies Room," 64, 73, 74, 158, 195, 202, 373
Episode 3, "The Marriage of Figaro," 155, 202, 204, 309
Episode 4, "New Amsterdam," 315, 384
Episode 5, "5.G.," 173, 301
Episode 6, "Babylon," 32, 134, 155, 158, 173, 176
Episode 7, "Red in the Face," 82, 318
Episode 8, "The Hobo Code," 7, 17, 24, 148, 156, 157, 304
Episode, 9, "Shoot," 82, 83, 105, 133, 375, 380
Episode 10, "Long Weekend," 145, 192, 376
Episode 11, "Indian Summer," 173
Episode 12, "Nixon v. Kennedy," 13, 52, 69, 130, 173, 178, 352, 376
Episode 13, "The Wheel," 10, 15, 17, 21, 88, 139, 174, 179, 303, 354

SEASON 2

Episode 1, "For Those Who Think Young," 148, 313
Episode 2, "Flight 1," 12, 74, 176, 198, 346, 348, 355
Episode 3, " The Benefactor," 355
Episode 4, "Three Sundays," 73, 174, 302, 339, 340, 352, 354
Episode 5, "The New Girl," 12, 65, 72, 74, 81, 89, 164, 183, 248, 352, 355, 356, 359, 363
Episode 6, "Maidenform," 68, 174, 357
Episode 7, "The Gold Violin," 45, 174, 175, 176, 178
Episode 8, "A Night to Remember," 89, 177, 199, 351, 360, 361, 362
Episode 9, "Six Month Leave," 123, 132, 200, 223
Episode 10, "Inheritance," 83, 274–78, 384
Episode 11, "The Jet Set," 178, 384
Episode 12, "The Mountain King," 19, 74, 84, 178, 179, 180 184, 199, 237, 252–55, 337, 351, 363
Episode 13, "Meditations in an Emergency," 84, 89, 181, 345, 351, 363, 373

SEASON 3

Episode 1, "Out of Town," 10, 16, 61, 65, 66, 69, 89, 158, 162, 181, 195
Episode 2, "Love Among the Ruins," 6, 61
Episode 3, "My Old Kentucky Home," 57, 60, 70, 72, 195, 376
Episode 5, "The Fog," 68, 84, 208, 379
Episode 6, "Guy Walks into an Advertising Agency," 107, 129, 196, 211

Episode 7, "Seven Twenty Three," 7, 14, 75, 384
Episode 8, "Souvenir," 69, 85, 182, 338
Episode 9, "Wee Small Hours," 69, 376, 377, 378
Episode 10, "The Color Blue," 148
Episode 11, " The Gypsy and the Hobo," 11, 182, 200, 204, 205
Episode 12, "The Grown-ups," 61, 73, 134
Episode 13, "Shut the Door. Have a Seat," 130, 134, 199, 207, 377

SEASON 4

Episode 1, "Public Relations," 64, 65, 135, 139, 205, 237
Episode 2, "Christmas Comes But Once a Year," 53, 61, 70, 74, 378
Episode 3, "The Good News," 94, 186
Episode 4, "The Rejected," 62, 75, 90
Episode 5, "The Chrysanthemum and the Sword," 60, 71, 340, 382
Episode 6, "Waldorf Stories," 67, 75, 139, 183
Episode 7, "The Suitcase," 139, 174, 184, 346
Episode 8, "The Summer Man," 66, 68, 210
Episode 9, "The Beautiful Girls," 43, 44, 196, 372
Episode 10, "Hands and Knees," 64, 66
Episode 11, "Chinese Wall," 64, 74
Episode 12, "Blowing Smoke," 23, 129, 130, 211
Episode 13, "Tomorrowland," 53, 60, 72, 135, 152, 183, 352

SEASON 5

Episode 1, "A Little Kiss," 60, 67, 90, 94, 155, 275, 278–82, 320, 381
Episode 3, "Tea Leaves," 94, 226, 321
Episode 4, "Mystery Date," 133, 228
Episode 5, " Signal 30," 56, 65, 134, 249, 318
Episode 6, "Far Away Places," 13, 73, 74, 133, 161
Episode 7, "At the Codfish Ball," 63, 72
Episode 8, "Lady Lazarus," 111, 161, 375, 381
Episode 9, "Dark Shadows," 134, 162
Episode 10, "Christmas Waltz," 22, 67
Episode 11, "The Other Woman," 3, 16, 24, 109, 134, 146
Episode 12, "Commissions and Fees," 85, 162, 183, 207, 340
Episode 13, "The Phantom," 101, 112

SEASON 6

Episode 1, "The Doorway," 19, 56, 61, 73, 84, 148, 293, 380
Episode 2, "Collaborators," 73, 74, 250, 319
Episode 3, "To Have and to Hold," 6
Episode 5, "The Flood," 227
Episode 6, "For Immediate Release," 132, 135
Episode 7, "Man with a Plan," 61, 69, 187
Episode 8, "The Crash," 148
Episode 9, "The Better Half," 75, 131
Episode 10, "A Tale of Two Cities," 381
Episode 11, "Favors," 75, 151, 186, 340
Episode 12, "The Quality of Mercy," 14, 16, 65
Episode 13, "In Care Of," 136, 151, 157, 212

SEASON 7

Episode 2, "A Day's Work," 58, 69, 91, 151, 341
Episode 3, "Field Trip," 85, 105, 138, 339
Episode 4, "The Monolith," 136, 156
Episode 5, "The Runaways," 71, 85, 105, 138, 186, 229, 274, 282–88, 338
Episode 6, "The Strategy" 78, 95, 136, 137, 187

Episode 7, "Waterloo," 22, 90, 95, 99, 132, 138, 364
Episode 8, "Severance," 58, 59, 68, 69
Episode 9, "New Business," 55, 65
Episode 10, "The Forecast," 33, 111, 151
Episode 11, "Time and Life," 91, 151, 165, 365
Episode 12, "Lost Horizon," 16, 39, 41, 86, 132, 152, 157, 185, 186, 227
Episode 13, "The Milk and Honey Route," 20, 86, 186, 323
Episode 14, "Person to Person," 20, 92, 94, 95, 165, 186, 213, 259, 305, 342, 365

Printed and bound by CPI Group (UK) Ltd, Croydon, CR0 4YY

16/09/2026

14963445-0004